IIIA	IVA	VA	VIA	VIIA	2 He 4.003
			5 B 10.81	6 C 12.011	7 N 14.007	8 O 15.9994	9 F 18.998	10 Ne 20.18
	IB	IIB	13 Al 26.98	14 Si 28.09	15 P 30.97	16 S 32.06	17 Cl 35.453	18 Ar 39.95
28 Ni 58.71	29 Cu 63.55	30 Zn 65.37	31 Ga 69.72	32 Ge 72.59	33 As 74.92	34 Se 78.96	35 Br 79.90	36 Kr 83.80
46 Pd 106.4	47 Ag 107.87	48 Cd 112.40	49 In 114.82	50 Sn 118.69	51 Sb 121.75	52 Te 127.60	53 I 126.90	54 Xe 131.30
78 Pt 195.09	79 Au 196.97	80 Hg 200.59	81 Tl 204.37	82 Pb 207.2	83 Bi 208.98	84 Po (210)	85 At (210)	86 Rn (222)

63 Eu 151.96	64 Gd 157.25	65 Tb 158.93	66 Dy 162.50	67 Ho 164.93	68 Er 167.26	69 Tm 168.93	70 Yb 173.04	71 Lu 174.97
95 Am (243)	96 Cm (245)	97 Bk (247)	98 Cf (249)	99 Es (249)	100 Fm (255)	101 Md (256)	102 No (254)	103 Lr (257)

Chemistry of Our World

CHEMISTRY OF

John K. Garland
Washington State University
Pullman, Washington

OUR WORLD

MACMILLAN PUBLISHING CO., INC.
New York
COLLIER MACMILLAN PUBLISHERS
London

Copyright © 1975, John K. Garland

Printed in the United States of America

All rights reserved. No part of this book may be reproduced or transmitted in any form or by any means, electronic or mechanical, including photocopying, recording, or any information storage and retrieval system, without permission in writing from the Publisher.

Macmillan Publishing Co., Inc.
866 Third Avenue, New York, New York 10022

Collier-Macmillan Canada, Ltd.

Library of Congress Cataloging in Publication Data

Garland, John K
   Chemistry of our world.

  1. Chemistry. I. Title.
QD31.2.G33   540   73-19492
ISBN 0-02-340520-1

Printing: 12345678   Year: 567890

To my family and their skill at disappearing whenever I began work on this manuscript.

Preface

Students taking first-year college chemistry courses are extremely diverse in background, interests, and career objectives. Because of that diversity, course offerings and textbooks cannot be tailored exactly to the needs of each individual. We can, however, respond to the general needs of a broad category of students, and instructors can modify subject offerings to best fit the particular group of students in their classes. This book is a conscious effort to speak to the needs of the largest category of first-year chemistry students. It has been intentionally developed with coverage beyond the needs of a single course sequence, to allow the instructor flexibility in selecting coverage most appropriate to his or her objectives and class. The instructor's manual has further comments about available alternatives.

My teaching of about 10,000 first-year chemistry students and coordinating courses for about 10,000 more has led me to conclude that most students fit into one of three main categories. They enroll (1) for an exposure to science (often because it is required), (2) to prepare for later work outside of chemistry, or (3) to prepare for further work in chemistry.

There are many good textbooks for students in the third category.

Needs of the first category may be satisfied by recent writing efforts by a number of concerned chemists. Several very short "chemistry-for-citizens" books have been produced to show a broad group of people how simple chemistry aids in understanding many events affecting their lives. Unfortunately, in such books the chemistry content must be very abbreviated. The somewhat more complete coverage in this book is necessary to show a more realistic picture of chemistry.

The second group, those preparing for nonchemistry subjects where chemical knowledge is needed, have been least adequately served by the available textbooks. The weakness of texts in this area is surprising because a majority of the students taking first-year college chemistry courses are in that category. This book is addressed primarily to such students. The suitability of the same material for many students in the first group (general science requirements) is a pleasant side effect.

Common needs of the nonchemists who take a limited amount of chemistry are (1) some help in seeing why chemical knowledge is important to them, (2) some assurance that chemistry is a subject they can manage, (3) some moderation in subject matter (particularly in arithmetical subjects), and (4) some human touch to keep science from seeming cold and set apart.

These common needs are hard to meet, particularly during the first part of a chemistry course. There are basic subjects that simply must be learned because without them later important information cannot be properly described. These basic subject areas require some arithmetic, but we have not avoided them. Many students *must* know the basics, and every student needs them for a realistic perspective of chemistry. But we can and do try to ease the burden. Arithmetical examples are limited to the instances where they are necessary, and simple numbers are used. In most cases we offer examples of how the calculations can be used with more complex problems (similar to real cases), but these are merely survey examples without the level of repetition needed for mastery beyond the level of the simpler examples.

Definite statements of the skills students must learn and sample test questions to help check progress are provided. These skills are within the capability of any average student and are sufficient for successful completion of a course. Students who want to go beyond the level of merely adequate completion will find some additional suggestions in the accompanying study guide.

You will also find some forward references to use of the basic concepts later in the book. The later portions of the book take subjects of wide interest—the common things in our world and things of importance to life—and discuss their chemical nature. Because many of the students needing this course have interests in the biological sciences or fields related to biology, our slant is toward the chemical basis of life and environment.

If writing this book had involved my efforts alone, I would have been quite incapable of generating the necessary mixture of completeness with moderation and basic ideas with applied examples. I acknowledge the help of review work and suggestions from Harry Batey (Washington State), O. T. Benfey (Guilford College), Norman Duffy (Kent State), William Erbelding (Purdue, Fort Wayne), Michael Farona (University of Akron), Verl Garrard (University of Idaho), Harry Hopkins (Georgia State, Atlanta), Malcolm Renfrew (University of Idaho), Frank Ruddy (Washington State), and Ralph Yount (Washington State). The enthusiastic response from 2,000 students who used portions of the manuscript on a trial basis provided stimulating encouragement. The accurate typing of Mrs. Joanna C. Farnsworth lessened my labors and left more time for working on content appropriate to students needing an interesting, moderate-level course.

J. K. G.

Contents

Part I
Introduction to Science and Its Tools
1

1. Science and Human Technology   3

- **1-1** Technology   *3*
- **1-2** Science and the Scientific Method   *5*
- **1-3** Tool Making and Productive Technology   *9*
- **1-4** Abuses of Our World   *11*
- **1-5** Explanations of Natural Science   *11*
- **1-6** Our World Is Chemical   *13*
- **1-7** Establishing Our Knowledge   *13*
    Skills Expected at This Point   *14*
    Test Yourself   *15*

2. Atoms and Chemical Combination   16

- **2-1** Physical and Chemical Changes in Matter   *16*
- **2-2** Mixtures   *18*
- **2-3** Pure Substances   *19*
- **2-4** Definite Composition and Multiple Proportions   *19*
- **2-5** Atomic Theory   *20*
- **2-6** Symbols, Formulas, and Nomenclature   *22*
- **2-7** Chemical Equations   *28*
    Skills Expected at This Point   *30*
    Exercises   *30*
    Test Yourself   *31*

3. Measurements and Units   32

- **3-1** Measuring Quantities   *32*
- **3-2** Metric System and the System International (SI)   *32*
- **3-3** Unit Conversions   *35*

3-4 Exponential Numbers and Significant Figures *37*
3-5 Temperature Scales *38*
3-6 Properties of Matter *40*
Skills Expected at This Point *41*
Exercises *42*
Test Yourself *43*

4. Arithmetic of Chemical Combination 47

4-1 Atomic Weights *47*
4-2 Moles, Molecular Weights, and Formula Weights *48*
4-3 Weight Relations in Formulas *51*
4-4 Formulas from Weights *54*
4-5 Weight Relations in Equations *56*
4-6 Molar Volume of Gases *58*
4-7 Energy Relations in Equations *61*
4-8 Quantities in Solution *64*
Skills Expected at This Point *66*
Exercises *66*
Test Yourself *68*

**Part II
Principles of Chemistry
45**

5. Into the Heart of Matter 71

5-1 Using Experiments as Our Guide *71*
5-2 Electricity *72*
5-3 Cathode Ray Tubes *72*
5-4 Properties of Electrons *76*
5-5 The Nuclear Atom *78*
5-6 Atomic Number and Mass Number *80*
5-7 Protons, Neutrons, and Electrons *81*
Skills Expected at This Point *82*
Exercises *83*
Test Yourself *83*

6. Electronic Configurations and Periodic Behavior 85

6-1 The Periodic Table I *85*
6-2 Spectral Evidence of Regularities *87*
6-3 Bohr Atom *87*
6-4 Quantum Numbers and the Exclusion Principle *89*
6-5 Electronic Configurations *90*
6-6 Electronic Orbitals *92*

6-7	Ionization Potentials	*95*
6-8	Paramagnetism	*99*
6-9	Electron Affinity	*99*
6-10	Atomic and Ionic Sizes	*100*
6-11	The Periodic Table II	*101*
	Skills Expected at This Point	*103*
	Exercises	*103*
	Test Yourself	*103*

7. Bonding   106

7-1	Conductors and Nonconductors	*106*
7-2	Metals	*107*
7-3	Ionic Substances	*109*
7-4	Arrhenius and the Theory of Ionization	*111*
7-5	Electrical Evidence for Ions	*113*
7-6	Chemical Evidence for Ions	*114*
7-7	Oxidation States in Simple Ions	*117*
7-8	Covalent Bonding	*119*
7-9	Electronegativity and Predicting Ionic or Covalent Character	*121*
7-10	Lewis Dot Diagrams	*123*
7-11	Complex (More Than One Atom) Ions	*124*
7-12	Resonance	*125*
7-13	Molecular Geometry	*128*
	Skills Expected at This Point	*136*
	Exercises	*136*
	Test Yourself	*137*

8. General Trends in Chemical Behavior   140

8-1	Oxygen	*140*
8-2	Oxidation and Reduction	*147*
8-3	Hydrogen	*148*
8-4	Representative Elements: Chemical Families	*149*
8-5	Irregularities in Family Patterns	*151*
8-6	The Noble Gases	*154*
8-7	Transition Elements	*155*
8-8	Effect of Oxidation State on Acid-Base Behavior	*158*
8-9	Inner Transition Elements	*159*
	Skills Expected at This Point	*160*
	Exercises	*160*
	Test Yourself	*161*

9. Phases of Matter 163

- **9-1** Gases *163*
- **9-2** Gas Laws *164*
- **9-3** Ideal Gases *166*
- **9-4** The Kinetic Molecular Theory *167*
- **9-5** Arithmetic of Gas Laws *168*
- **9-6** Real Gases *170*
- **9-7** Liquids *171*
- **9-8** Evaporation and Heat of Vaporization *174*
- **9-9** Vapor Pressure: An Introduction to Le Chatelier's Principle *174*
- **9-10** Partial Pressures *177*
- **9-11** Boiling *178*
- **9-12** The Critical Point *179*
- **9-13** Transitions Involving Solids *180*
- **9-14** Crystals and Supercooling *182*
- **9-15** Phase Diagrams *184*
  Skills Expected at This Point *186*
  Exercises *186*
  Test Yourself *187*

10. Solutions 192

- **10-1** Nature of Solutions *192*
- **10-2** Gases in Liquids *194*
- **10-3** Solids in Liquids *195*
- **10-4** Concentrations *196*
- **10-5** Colligative Properties *202*
- **10-6** Transfer of Water and Nutrients in Plants and Digestion *211*
- **10-7** Some Problems Illustrating Solution Phenomena *213*
  Skills Expected at This Point *218*
  Exercises *218*
  Test Yourself *220*

11. Mixing and Pollution by Disorder 222

- **11-1** Spontaneous Changes and Irreversible Changes *222*
- **11-2** Entropy: the Universal Tendency to Disorder *224*
- **11-3** Natural Degradation of Energy *226*
- **11-4** Disorder Produced by Human Activity *229*
- **11-5** Wasting Entropy Increases *231*
  Reference Reading *233*

Skills Expected at This Point *233*
Exercises *233*
Test Yourself *234*

12. Kinetics and Equilibrium (Qualitative Basis and Principles)   236

- **12-1** Coverage in This Book *236*
- **12-2** Factors Needed for Reaction *237*
- **12-3** Changing Reaction Rates *243*
- **12-4** Back Reactions *248*
- **12-5** Equilibrium *250*
- **12-6** Law of Mass Action *251*
  Skills Expected at This Point *254*
  Exercises *254*
  Test Yourself *255*

Part III Common Things in the World Around Us *257*

13. Water   259

- **13-1** Abundance and Transport Cycle *259*
- **13-2** Unusual Properties of Water *259*
- **13-3** Hydrogen Bonding and the Structure of Water *262*
- **13-4** Acids and Bases *267*
- **13-5** Natural and Man-Made Uses of Water *270*

14. Nutrient Materials   284

- **14-1** Limiting Nutrients *284*
- **14-2** Nitrogen *284*
- **14-3** Phosphorus *288*
- **14-4** Fertilizer and Food Production *292*
- **14-5** Fertilization of Waters *293*
- **14-6** Secondary Effects of Fertilizer *294*
- **14-7** Human Effluents and Biological Oxygen Demand *297*
- **14-8** Oxygen Demand of Industrial Effluents *299*
  Skills Expected at This Point *300*
  Exercises *300*
  Test Yourself *301*

15. Nuclear Stability and Chemical Abundance   304

- **15-1** Radioactivity *304*
- **15-2** Nature's Preferences *310*

15-3 Conversion of "Mass" to Energy *314*
15-4 Artificial Radioactivity *315*
15-5 Fission and Nuclear Power *318*
15-6 Binding Energy and Fusion *321*
15-7 Operation of Stars *323*
15-8 Abundance of the Elements *324*
Skills Expected at This Point *326*
Exercises *326*
Test Yourself *327*

16. Use of Metals   331

16-1 Occurrence of Metals *331*
16-2 Transition Metal Chemistry *337*
16-3 Complex Ion Formation *339*
16-4 Methods of Ore Enrichment *344*
16-5 Production of Aluminum *346*
16-6 Production of Iron *351*
16-7 Steel *356*
16-8 Tempering Steel *357*
16-9 Steel Alloys *359*
16-10 Production of Copper *359*
16-11 Side Effects of Metal Production *361*
16-12 Poisons Associated with Metal Use *362*
16-13 Metal Disposal and Recycling *366*
Skills Expected at This Point *367*
Exercises *368*
Test Yourself *369*

Part IV
Chemistry of Life
371

17. Carbon and Its Compounds   373

17-1 Bonding with Carbon *373*
17-2 Hydrocarbons *377*
17-3 Naming Hydrocarbons and Isomers *380*
17-4 Hydrocarbon Derivatives *390*
Skills Expected at This Point *396*
Exercises *397*
Test Yourself *398*

18. Organic Reactions   401

18-1 Reactions of Alkanes *401*
18-2 Reactions of Alkenes *403*

18-3	Mechanisms of Reactions *405*	
18-4	Reactions Involving Hydrocarbon Derivatives *412*	
18-5	Polymers *419*	
	Skills Expected at This Point *424*	
	Exercises *424*	
	Test Yourself *425*	

19. Biological Building Blocks   427

19-1	Biopolymers *427*
19-2	Reactivity of Organic Acids and Amines *427*
19-3	Amino Acids *428*
19-4	Structure and Reactivity of Naturally Important Amino Acids *431*
19-5	Polypeptides *434*
19-6	Structure of Proteins *435*
19-7	Nucleic Acids *439*
19-8	Heterocyclic Compounds *439*
19-9	The Pyrimidine and Purine Bases *440*
19-10	DNA and RNA *442*
19-11	Replication and Heredity *447*
19-12	Synthesis of Proteins *449*
19-13	Functions of Proteins *451*
	Skills Expected at This Point *453*
	Exercises *453*
	Test Yourself *454*

20. Energy Reservoirs   456

20-1	Sugars *456*
20-2	Starches and Cellulose *463*
20-3	Fats *465*
20-4	Fossil Fuels *468*
20-5	Recovery of Energy Sources *469*
20-6	Problems Associated with Fossil Fuels *471*
	Skills Expected at This Point *473*
	Exercises *474*
	Test Yourself *474*

21. Using Energy Reservoirs   477

21-1	Types of High Energy Compounds in Cells *477*
21-2	Formation of ATP in Oxidation *480*
21-3	Digestion *481*

21-4	Carbohydrate Metabolism Producing ATP *483*	
21-5	Storage of ATP Energy in Creatine Phosphate *494*	
21-6	Use of ATP Energy in Muscle Contraction *495*	
21-7	Oxygen Availability Via Complexes *502*	
21-8	Use of Fossil Fuels *503*	
21-9	Greenhouse and Dust Effects *505*	
	Skills Expected at This Point *507*	
	Exercises *508*	
	Test Yourself *509*	

22. Biologically Active Molecules   511

22-1	Insulin: A Hormone Balance *511*
22-2	Steroids *513*
22-3	Sex Hormones *514*
22-4	Artificial Substitutes *516*
22-5	Cortisone and Function of Antibodies and Inflammation *519*
22-6	Vitamins *520*
22-7	Essential Amino Acids *528*
22-8	Other Essential Substances *529*
	Skills Expected at This Point *529*
	Exercises *530*
	Test Yourself *530*

23. Changes from Natural Chemistry   532

23-1	Stimulating Natural Growth *532*
23-2	Substitutes for Natural Balances *533*
23-3	Disrupting Natural Problems *535*
23-4	Identifying and Treating Some Hereditary Problems *540*
23-5	Selective Poisons as Medicines *545*
23-6	Selective Poisons as Pesticides *549*
23-7	Losses of Selectivity *534*
23-8	Side Effects of Control Measures *555*
23-9	Compensating for Artificial Changes *556*
	Skills Expected at This Point *557*
	Exercises *557*
	Test Yourself *558*

A. SI Units   560   **Appendixes**

A-1	SI Units *560*
A-2	Notes on the Use of SI Units *562*
A-3	Some Conversion Factors *562*

B. Form in Which Equilibrium Constants Are Used   564

- **B-1** Writing and Using K *564*
- **B-2** Special Forms of K: K_a, K_b, K_w *575*
- **B-3** Hydrolysis   *576*
- **B-4** Solubility Products   *580*
  Skills Expected at This Point   *580*
  Exercises   *581*
  Test Yourself   *581*

C. Quantitative Basis and Applications of Equilibrium   583

- **C-1** Free Energy and Equilibrium   *583*
- **C-2** Using K in More Complicated Problems: Simplifying Approximations   *592*
  Skills Expected at This Point   *608*
  Exercises   *608*
  Test Yourself   *608*

D. Nomenclature Variations: Acids and Compounds Containing Only Nonmetals   610

- **D-1** Nomenclature in This Book   *610*
- **D-2** Acid Nomenclature   *610*
- **D-3** Nomenclature of Nonmetal Compounds   *611*

Answers to Exercises   612
Index   621

Part I

Introduction to Science and Its Tools

1 Science and Human Technology

1–1 Technology

Technology is the knowledge and method of systematic productive labor. (*Note:* These marginal statements are intended as guides to assist you in locating the sections you wish to reread for review. Suggestions on their use appear later in this chapter—Section 1–7. The separate student study guide also uses these statements of the principles covered in the text.)

Systematic labor can be effective with an antlike mentality without innovations or achievement of optimum conditions.

Technology is defined as the branch of knowledge that deals with the industrial arts or the sciences of the industrial arts. The industrial arts are the arts of systematic productive labor. But productive labor is not always systematic and technology is not always scientific.

Human experience with things like cultivation of crops and mass production has clearly shown the advantages of being systematic, but there is nothing uniquely human about a systematic technology. Ants can do as well. They are extremely systematic in their technology. For instance, a worker ant searching for food leaves a trail that can be followed. When he finds food, as in Figure 1–1, he can follow the trail back home and other ants can follow it back to the food. Systematic following of this trail is successful, even though we can see it is not always direct and efficient. We will refer to this sort of slavish following of the path to past success as the technology of ants. We know people sometimes also limit themselves to the old paths, but they should be capable of something better.

An animal with greater reasoning power, such as a chimpanzee, might figure out the real location of the food and develop a new and better path to it. Such an animal must have a strong tendency to seek out new approaches. Desire for a new approach is often developed through emotional responses such as boredom, curiosity, anger, and joy, which encourage deviations from the established path to success. Emotional responses may cause waste effort, which could have been used systematically following the old path, but they sometimes save a lot of effort, as illustrated in Figure 1–2. The chimpanzee who has already learned how to reach bananas via a maze discovers a shortcut path. He then chooses to use the better path on later trips. Emotions can be helpful in stimulating exploration, self-defense, and other desirable behavior, but emotions can also cause problems.

In Figure 1–3 we see an experiment in which the chimpanzee, who now knows about the shortcut, is blocked off by increasing numbers of boxes that must be removed. This irritates the chimp. Finally the shortcut is covered by a very large number of boxes with bananas inside. By this time the chimp is so angry at the interfering boxes that he throws them aside violently without

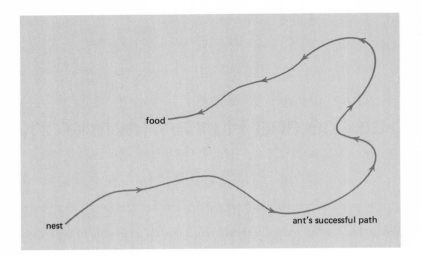

Figure 1-1 The ant's approach: repetition of the first successful path.

bothering to investigate them. That takes a lot more energy than going around through the maze. These reactions (an avoidance of the "old" path even when it is the best, a determination to go the way originally planned without considering any other new possibilities, and a loss of efficiency under emotional stress, particularly anger) are common among people and sometimes even glorified as the wave of the future. We will refer to such emotional responses as the technology of chimpanzees and claim that people are capable of something better.

Man has some unique abilities. He can reason using abstract concepts. He can set up plans and organize his activities around them. He may choose to follow set patterns repetitiously—the technology of ants—or he may choose the emotional responses seeking new paths—the technology of chimpanzees. In either case, his present numbers and powers allow him to do things on such a large scale that the consequences of these technologies may get out of control. But man's reasoning ability opens up a third possibility. He can

Innovation can be achieved by emotional reactions like a chimpanzee's without good rational control.

Figure 1-2 The chimpanzee's approach: discovery of better routes.

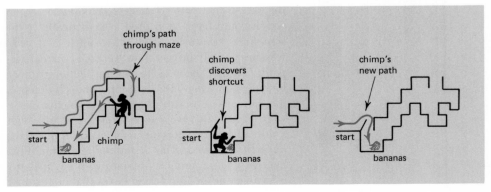

4   Science and Human Technology   [1]

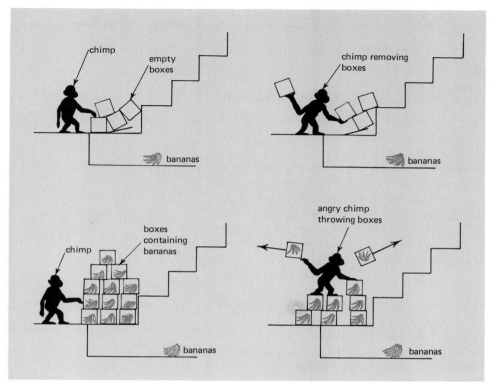

Figure 1-3 How emotions can interfere with efficiency.

Science uses human skills to accomplish both innovative and systematic technology.

use both his experience with old patterns and continuing investigation of new ones to build up a base of knowledge for use in making decisions and adjusting to needs. This is the path of science, and it is the only one that leads to a technology that can be both controlled and improved. Technology will always be with us (even ants need it) and will increase as people and their products increase. Therefore, it becomes increasingly important for us to abandon the technologies of ants and chimpanzees and develop a controllable scientific technology. We must also control it wisely. It is vital for all those who can influence human affairs to understand science and the processes by which our knowledge grows and can be used. In a democracy that means everyone needs an understanding about science.

1-2 Science and the Scientific Method

A science is a branch of knowledge made up of a body of facts systematically arranged and showing the operation of general laws. It is also the study by which that knowledge and those explanations are revised and improved. Man's successes at gathering and improving his knowledge have been greatly aided by using his reasoning and organizational abilities in a particularly desirable pattern. This pattern is called the **scientific method.** In it the following steps are carried out:

1. Observations are made.
2. Facts are examined and organized into ideas.
3. The ideas are tested by conducting experiments.
4. The observations from the experiments are used to start the whole cycle again.

The scientific method organizes the gathering of knowledge.

We can show the effectiveness of the scientific method for gaining knowledge and developing technology by a hypothetical example. In our example we will consider the fortunes of three cavemen who must set out each day from the shelter of their cave to seek food. Their cave is located in the side of a canyon wall near a small side canyon, as shown in Figure 1–4. The three cavemen have different personalities. We will call them Ant-man, Chimp-man, and Science-man to fit their different behaviors. We are going to take them through a series of events where each time period corresponds to one of the steps in the scientific method.

1–2.1   Observations Are Made.   On the first day all three cavemen start up the main canyon in search of food. They eventually struggle through the rugged canyon, as shown in Figure 1–5, reach the plateau, and find the area with abundant food. They then spend the rest of the day struggling back down the same route. They arrive back at the cave exhausted, but fed, and spend the night sheltered from wild animals. All three have made the same

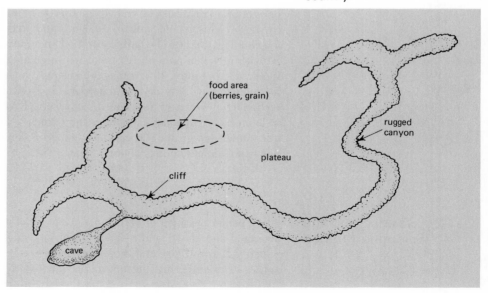

Figure 1–4   Cavemen's country.

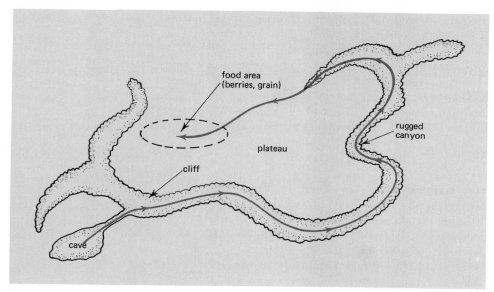

Figure 1-5 Original route to food.

observations: food is available at the end of their route up the canyon (taking the left branch) and it is an all day trip to get it and return.

1-2.2 Facts Are Examined and Organized into Ideas. That night, all three cavemen examine the experience of the first day and reach the same set of facts: there is food on the plateau, which can be reached by going up the canyon, which is a very difficult trip. Ant-man uses the facts to organize the idea "Going back the same way will lead me to food again." Chimp-man and Science-man each form two ideas, "Food is on the plateau," and "The way up the canyon is a long, hard way to reach the plateau."

1-2.3 Ideas Are Tested by Conducting Experiments. On the second day, Ant-man tests his ideas by following the original route. He succeeds, returns exhausted, and continues to do the same thing each day until one day on the way home a sabre toothed tiger catches him when he is too exhausted to get away.

On the second day both Chimp-man and Science-man set out to test their ideas by looking for a new way to reach the plateau. They can see it is right on top of the cliff opposite their cave. They try to climb the cliff, but it is too steep so they slip back down. The attempt to climb the cliff is an **experiment.** It has a purpose, to see if the plateau can be reached by a direct route, and a result, the cliff is too steep to climb.

Chimp-man's reaction is anger. He has decided that the food can be reached by the shorter route up the cliff, and he is determined to prove his idea correct.

All ideas, including established accepted ones and proposed new ones, must be subjected to testing by experiments and evaluation of the results.

He attacks the cliff, tearing rocks loose and repeatedly scrambling part way up and falling back. We will leave it to your imagination to decide whether the rocks he knocks loose help the cavemen by damming up a nice pool of water by their cave or ruin them by flooding out their only shelter. In our story he wears himself out completely, without success, and the noise he made attracts the neighborhood sabre toothed tiger to the scene.

That leaves us with Science-man who sees the result of the experiment, the cliff is too steep to climb, and goes on to the next step in the scientific method.

1-2.4 Observations from Experiments Are Used to Start the Whole Cycle Again. When Science-man makes the observation that the cliff is too steep to climb, he adds that to all his earlier observations and goes through step 2 again, examining the facts and organizing ideas. He still wants to reach the plateau for food; he still knows the way up the canyon is long and hard; he now also knows that he cannot go directly up the cliff because it is too steep. The idea he reaches from these observations is that a route to the plateau must be less steep than the cliff and should be easier and shorter than the route Ant-man is using. He may follow Ant-man up the original route one or two more times to keep fed while he is forming new ideas and testing them, but he will always continue to use some effort to apply the scientific method cycle to unproven ideas, such as trying to climb other points in the cliff. Eventually he will discover the small side canyon where he can test by experiment his idea that some place less steep than

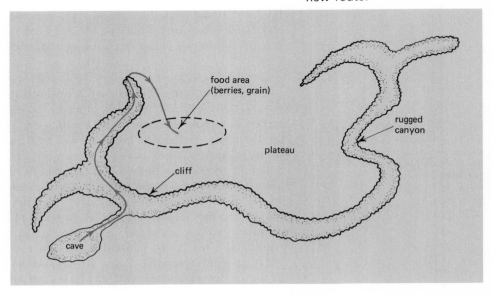

Figure 1–6 Science—man's new route.

the cliff might be climbable and shorter and easier than Ant-man's route. He then establishes the new route shown in Figure 1-6.

Science-man's adventures cannot stop with finding the new route. To be truly scientific he must continue to experiment, observe, and form new ideas. He might develop a way to cut steps in the cliff so it could be climbed or find or build a shelter nearer to the food supply. He has time and energy to use working on new ideas because of his new, easier route to the food supply.

We will stop our story before Science-man makes any spectacular breakthroughs like house building. But we will note that he was able to use some of his free time to experiment with making sharp points on sticks. The fur of the sabre toothed tiger he killed with the sharp sticks was a big help in keeping warm on cold nights. Especially since he couldn't stay warm by huddling together with Ant-man and Chimp-man. Science-man **observed** that they didn't come to the cave any more, formed the **idea** that something had happened to them, **tested** the idea by the experiment of looking for them, **found** two groups of bones from which he formed some **new ideas,** and so on, using the scientific method.

1-3 Tool Making and Productive Technology

Tools increase productivity but may discourage innovations or make dangerous mistakes possible.

It would be wrong to suppose that all of the differences between human technology and animal technologies are due to reasoning and use of the scientific method. Man's skill and ingenuity at making and using tools plays a large role in the state of his technology. Many of these tools have been developed or improved by applying the scientific method. They contribute greatly to productivity, but sophisticated tools and highly productive technology can occur without being backed up by science and a scientific attitude. In fact, they sometimes encourage us to settle for an antlike technology. We become so adept with tools designed for the old routes to success that all new ideas seem pointless.

For instance, in our caveman story, Ant-man might have devised a tool to help him get more food. If he built a device out of sticks, which he could load with food and drag down the canyon behind him, he might become more successful than Science-man with his new route. If Science-man's new route had a tight place such a device could not go through, he might give up and go back to the old route. We can see it would be better for him to develop new tools suitable to the new, better route, but that might not be apparent to Science-man when he saw Ant-man's greater success.

Although tool making is usually a good and desirable activity, it can lead to serious problems when it is not controlled by the continued critical questioning and testing of science. Some of these problems occur because of the great power tools give us. Sometimes a technology prospers because it disrupts natural patterns. An example is the building of dams by beavers. Beavers are somewhat limited in their dam building abilities, so their lack of knowledge about the damage they do to forests and some other life forms in the streams and meadows that are flooded is usually not a serious problem. But men can use great machines to build dams 700 feet (ft) high; their power makes lack of knowledge very dangerous. The collapse of such a dam during an Italian earthquake and the disruptions of some important fish migrations elsewhere prove that our power has sometimes exceeded our knowledge.

A second problem caused by intensive tool making without new scientific inquiry is that the conditions for which the tools were designed may change. The tools themselves may even cause the changes that make them obsolete. In our caveman example, a tool dragged down the canyon by Ant-man might wear a trail leading to the cave. Wild animals might then be attracted to watch the trail, making the trip very hazardous for someone burdened by a load. Even worse, the animals might be attracted to the cave. That could have disastrous effects on the whole pattern of gathering food on the plateau while spending nights in the safe cave.

Tools may cause changes which make the tools obsolete.

A problem of changing conditions recently occurred in corn production in the United States. For many years a tool—highly productive hybrid corn varieties—had been developed and refined. The best and most successful of these hybrid corn varieties all had one characteristic in common because breeding this superior corn had proven easier when it had what is called T cytoplasm. Over 80% of the corn planted in the United States in 1970 had T cytoplasm.

In 1970 a disease, southern corn leaf blight, developed a strain that thrived on any corn with T cytoplasm. Because there was so much susceptible corn around, the disease spread rapidly. Over 50% of the corn crop was lost in the Mississippi area where the disease first appeared. In July it spread to the corn belt, where about 10% of the corn crop was lost even though it was diseased only for a short part of the year. By September the disease had spread to all the important corn growing areas of the United States and part of Canada. Because the soils were now disease infested, it was clear that any crops of T cytoplasm-type hybrid corn in 1971 would be seriously threatened. If 80% of the 1971 crop had been planted with T cytoplasm corn, there would have been serious danger of famine in the United States.

The factors preventing disaster in 1971 were scientific knowledge, the existence of other corn varieties, and a change of conditions (weather) making the blight less serious. Enough was known about corn to identify the problem. Scientists were still studying ways to make other types of corn, without T cytoplasm, more productive. These scientists were aware that natural disasters, like the disease, can be caused by large scale technological success. But their principle goals were simply sound science, trying to expand our knowledge of everything about corn.

Scientific knowledge can protect us from some technological mistakes.

By working on kinds other than the proven successful T cytoplasm corn, they were laying themselves open to criticism from those who call for "relevance" in all research. It would be more economical to support only the research giving the best return for the effort, the T cytoplasm corn. The spirit of Ant-man lives on in those who call for that kind of "stick to the successful path" relevance.

1-4 Abuses of Our World

Abuses of our world occur when the power of technology is used without the rational control of science.

Man has a powerful technology, based on effective tools and the ability to produce new ones, and a frequent tendency to stop short of a truly scientific approach. The combination often leads to abuses of our world. Examples of these abuses will be brought up throughout this book. The analogies to ants and chimps used in this chapter may help us recognize some of the most common and most dangerous errors.

We can recognize the technology of ants whenever someone says, "We've always done it that way." People who say that may even think their position is scientific because it is based on conclusions from scientific investigations at some other time or place. An example of this sort of abuse is waste disposal practices. Our grandfathers found dumping wastes into rivers was a simple and effective solution. We continue to do it on a huge scale. Previous generations found it particularly important to gather sewage which might otherwise spread disease, and they found that the streams purified themselves fairly quickly (except for a few bacterial problems). In Chapter 14 we will have some fairly long discussions of the ways in which a water system that could handle a number of moderate-sized waste inputs without noticeable problems can be wrecked by the huge concentrated inputs from large city sewage systems. The common modern technology of waste disposal has invalidated the facts that originally caused us to choose that technology.

The technology of chimps is also very much with us whenever people cry out for *immediate* solutions to technical problems. The impatience and anger of such cries encourage hasty, poor solutions, and the proposals tend to be snapped up as the ultimate, unchangeable answers. In Chapters 14 and 16 we will discuss how concern over water pollution almost forced a rapid conversion from phosphates in detergents to incompletely tested substitutes. Discontent, even anger, about technological problems is all right, but we cannot afford to bypass the reasoning, testing, rethinking, and retesting stages of the scientific method, particularly on large scale changes.

1-5 Explanations of Natural Science

Those sciences concerned with nature are classified as natural sciences. Although there is much we have not yet learned, we have observed that much about nature is regular and reproducible. If we conduct an experiment and observe the result, we can often conduct the experiment in the same way again and observe the same result. If we get a different result, experience has taught us to look for some small difference between the experiments and use that difference to reason out an explanation.

The ideas suggested to explain some part of nature may be based on just one observation (or even none) or on many observations. To judge how good and important an idea is, we need to know something about how it was reached and how it has been tested. We also need to direct our efforts toward new experiments, observations, and reasoning to advance our knowledge. We keep track of current status of science by communication, principally by written records. In this way each scientist has the observations and explanations of others to use in forming new ideas and planning new experiments.

Exchange of information does not always lead to agreement among scientists. Disagreements (sometimes quite heated) often stimulate progress. Scientists who are aware of that make no attempt to avoid arguments, but they try to keep their arguments within boundaries consistent with the scientific method. All stages of the scientific method are argued about vigorously, but the arguments must be supported by facts and reasoning. Arguments for the existing idea because "everyone knows that is right" or for a new idea because "it would be a much nicer explanation" are not adequate.

Disagreements contribute to scientific progress.

We find definitions of some terms describing the status of scientific ideas helpful. The term we use for the least supported ideas is **hypothesis.** Any idea can be a hypothesis. Some hypotheses are good; others are bad. All hypotheses need to be tested by experiments. If a hypothesis is tested and the observed result agrees with what that hypothesis would predict, the hypothesis is strengthened. If the experiment leads to a different result, the hypothesis must be discarded or modified to fit. The strongest support comes when a hypothesis agrees with accurately made observations showing facts not previously known that disagree with some or all of the competing hypotheses.

Hypotheses are ideas suggested to explain observed facts.

The arguments of science rage about whether the observations were really accurate, whether the reasoning about what the hypothesis predicts is sound, whether the observation is showing something new or just repeating the observations the hypothesis was based on, and whether the other possible hypotheses have been considered. To defend their arguments, scientists are forced to make their observations very carefully, prove that they see what really happens, check the reasoning on their hypotheses, plan their experiments carefully to be sure they will see something really different, and look carefully for other reasonable hypotheses that could explain the same results.

Eventually some hypotheses emerge with very strong support. They usually have had extra details and modifications added since they were first proposed. These hypotheses, extensively tested and never found wrong, are called **theories.** Theories are also questioned and tested by new experiments. Usually they are supported by the new results or slight modifications adjust them to fit the new result and all previous results. Occasionally we find an experiment that disproves a theory we thought was well established. When a theory is disproved (or even modified somewhat) the reasoning behind that theory and many hypotheses dependent on the truth of that theory must be examined and revised.

Theories are hypotheses established by extensive testing over a long period without being found wrong.

Some principles are even more certain than theories. Mathematical relationships, like $2 + 2 = 4$, can often be proven to be true. Theories are ideas that have never been proven false, but they cannot be positively proven true. Therefore, we use a third term, **law,** to describe those mathematical relations that cannot be false. These are not the sort of laws passed by legislatures and enforced by courts. Scientific laws are simply statements of fact.

Scientific laws are statements of fact, usually mathematical relations, which have positive proof that they are true.

1–6 Our World Is Chemical

Chemistry determines the nature of the matter on earth.

The natural sciences have been divided into several main areas. We will use ideas from several of these sciences, but our main topic area will be chemistry. Chemistry is concerned with some particular properties of matter and how it interacts with other matter. Some properties and interactions of matter are classified as physics, but under the conditions of our earth chemistry determines the nature of the materials around us. Chemical forces determine the nature of substances we find as the solids, liquids, and gases from which our world is made. Chemical interactions bring about the changes by which plants and animals live and grow. Natural processes like cave formation and human activities like production of metals, plastics, fertilizers, and energy from fires involve chemistry. Study of theories and laws of chemistry will show us both how the orderly approach of science is used and a great deal about the world in which we live.

1–7 Establishing Our Knowledge

Simplified models are useful for understanding science.

Lists of principles and skill objectives can be used to plan study and review.

Our understanding of our world will be better if we benefit from the observations, hypotheses, experiments, and theories accumulated over a long period of time. Therefore, we will first study the science of chemistry and then apply what we have learned to the nature of the world.

We will study a limited number of points serving our main purposes. Often we will have to shorten and simplify our discussion of these points by omitting details. Actually, shortcuts and simplifications are common in science. They are used to get enough ideas about what is expected to plan good experiments and progress to more detailed understanding. The simplified patterns are called **models.** As our best understanding of a fact is improved, we can construct new and better models to help us picture what is happening. These models may also allow us to derive statements of laws that must be true if the model holds. Because the laws cannot fail, deviations show us truth about where the models are failing. There will be times when our best and simplest description of reality will be presentation of an ideal model (with related laws) and statements about how it fails.

The general approach of science and our simplifications will lead us to some basic principles that describe our best present understandings. Even though we know these principles may be modified by later experiments and discoveries, we need to learn them and use them to understand other points. A separate study guide entitled *Learning the Chemistry of Our World* gives emphasis to main points of this book by providing listings of the principles covered in each chapter, the skills students are expected to master, and some study approaches. Brief lists of important skills are given at the end of each chapter in this book. They are intended to encourage conversations with your instructor and review of text and lecture notes in the areas where the most serious confusions may arise. They are *not* a complete summary of the course.

The statements of principles that appear in the text margins and in the study guide are also only a starting point for further study and thought. Their placement in the margins may help you find sections you need to review, but your instructor will certainly require much more than knowing these brief statements. The study guide will show some examples of how knowledge of the basic principles can, with thought, lead to broader understanding. Your instructor will use tests and discussions to make you progress to the stage of thought and broader understanding. Anyone who thinks rote memory is enough because "it always worked in high school" is using Ant-man mentality and will probably meet the neighborhood sabre toothed tiger—an F grade.

Exercises are provided at the end of each chapter for review and practice on problem solving. (There are no exercises for this chapter because the narrative material does not require that kind of review.) The Exercises are followed by a Test Yourself section offering samples of reasonable exam questions on the material. That section will be most useful if you complete your study of the chapter and exercises first (as if preparing for a test) and then use this review aid to reveal any remaining weaknesses. The Answers to Selected Exercises at the back of this book lists answers to many of the exercises (or to selected parts when there are several parts in the exercise). If you have doubts about your exercise answers you may wish to check them against Answers to Selected Exercises. You should not come to depend upon answers you can check, so we will offer answers to only part of the exercises and no answers to any Test Yourself questions. You should *know* your answers to the Test Yourself section are correct or else review the subject again. Additional Exercises and Test Yourself questions are available in the study guide, *Learning the Chemistry of Our World*.

When you find you are unsure about a question, you can use the marginal notes listing principles as a guide to the areas of the text you should review. You will find the listed principles may not directly answer questions, but you will be able to find related sections. For example, question 5 in the Test Yourself section at the end of this chapter requires some thinking beyond the basic principles. The marginal statements on pages 9 and 10 do not answer the question but they show a part of the text that is related to Test Yourself question 5.

Skills Expected at This Point

1. You may be required to describe the steps of the scientific method and the meanings of hypothesis, theory, and law.
2. You should understand the difference between science and technology.
3. You may be required to describe the differences between technology being applied scientifically and technology that is not scientific, and you may be asked to give examples illustrating the differences.

Exercises

Narrative material such as this chapter is better reviewed by rereading than by working problems.

Test Yourself

1. After observed facts have been examined and a reasonable explanation has been proposed, what is the next stage in a scientific inquiry?
2. **(a)** What is the word used to describe ideas proposed with only moderate supporting evidence? **(b)** When ideas have been extensively tested and always found correct, although there is always some possibility that some later test will prove them false, what are they called? **(c)** Those mathematically stated relationships that have been proven positively are given what name?
3. Describe, in 25 words or less, those characteristics of a useful but not particularly scientific technology that make it useful and that keep it from being scientific.
4. A scientific theory is
   (a) an hypothesis not yet subjected to experimental test.
   (b) an idea that correctly predicts the results of many experiments.
   (c) a guess.
   (d) a figment of the imagination.
   (e) a statement of religious belief.
5. A proper use of tools at this point in history is to
   (a) permit rapid expansion of current methods of production and thus increase goods available and living standards.
   (b) allow rapid, complete conversion to new technologies that avoid specific known problems.
   (c) provide enough goods efficiently to permit spending substantial time and effort seeking new methods and testing their consequences before widespread adoption.
   (d) determine the conditions that existed in nature prior to human civilization and restore them.
   (e) convert the entire earth to raising only the most efficient food-producing variety of each plant and animal.
6. What do you think is the most important point in this chapter that is not covered by any of the above questions? Write a good test question covering that point.

Atoms and Chemical Combination

Most observations and measurements about our surroundings are observations about matter. In these observations we notice both what the matter is and the conditions it is under. For instance, a sample of gasoline could be examined. We might find that it had a temperature of 20°C, was a clear liquid, had a particular odor, and had other characteristics that helped us identify it as gasoline. If we heated the gasoline, we could change it into gaseous vapors, but it would still be the same kind of matter, gasoline. Simply cooling it down would put it back into the clear liquid form. This kind of change is called a **physical change.** Physical changes affect the form and energy state of matter but do not change the type of matter.

If we took our sample of gasoline and burned it in air, we would get another kind of change. When the sample was cooled off to the original temperature, it would not return to the clear liquid we know as gasoline. Instead we would get a different clear liquid, water. We would also find the air had been changed, losing the matter we know as oxygen and gaining matter we know as carbon dioxide. This change in the types of matter present is called a **chemical change.**

Physical changes (and their important related energy effects) are the general topic of the science of physics, and chemical changes (and their related changes in energy and identification of matter) are the topic of the science of chemistry. These very closely related fields are called the physical sciences. We will not really separate them from each other, but we will indicate by our emphasis on chemistry that we are particularly concerned with the identifications and characteristics of particular kinds of matter.

Although physical changes can alter the form of matter and chemical changes can alter both the form and identity of matter, there are two important, observable properties of matter that do not change at all. These properties are **mass** and **energy.** Mass is the quantity-related property of matter we usually measure by weighing. (Mass and weight are discussed in Section 3–2 of this book.) The **law of conservation of mass** states that mass can neither be created nor destroyed in any process. The law of conservation of mass tells us that the quantity of matter is not changed by either chemical or

2–1 Physical and Chemical Changes in Matter

Physical changes affect the form and energy state of matter, but do not change the type of matter.

Chemical changes affect the type of matter.

Total mass and energy remain constant during chemical and physical changes.

physical changes. Only the arrangement of the matter is altered. Therefore if 100 grams (g) of a solution of salt in water is evaporated to form solid salt and water vapor there must still be 100 g. If 2 g of solid salt are formed, the other 98 g will be water vapor, giving the same total mass we started with—100 g. If there are 2 g of salt, there cannot be either more or less than 98 g of water vapor. Similarly, if 6 g of coal requires 16 g of oxygen to burn completely, there must be 22 g of product formed. (The product is a gas, carbon dioxide.) These examples are illustrated in Figure 2–1.

A similar conservation law occurs for energy. Energy occurs in many forms: heat, energy of motion (kinetic energy), and potential energy in various forms, such as mass which could be allowed to fall from a height, coiled springs, or chemical energy. The **law of conservation of energy** states that energy can be converted from one form to another but the total amount of energy remains constant. Therefore, when heat is released by a chemical reaction, the chemical potential energy goes down by the same amount that the heat energy goes up. Figure 2–2 shows how energy is conserved in the same two examples with which we illustrated conservation of mass. The water evaporation example is expanded somewhat to further emphasize the conversions possible from one energy form to another. In Section 15–3 we will describe the "conversion" of mass to energy and show how the laws of conservation of mass and conservation of energy can be replaced by a single law. However, for most everyday cases there is no substantial conversion of mass to energy. Under these normal conditions the two laws each hold true separately as stated above.

Figure 2–1 Conservation of mass.

A. Physical change: the solution of 2 g of salt homogeneously mixed with 98 g of water has a total mass of 100 g. When separated the mixture forms 2 g of solid salt plus 98 g of water vapor and still has a total mass of 100 g.

B. Chemical change: as 6 g of coal is consumed by burning in air, the air loses 16 g of oxygen and gains 22 g of carbon dioxide. The total mass stays constant.

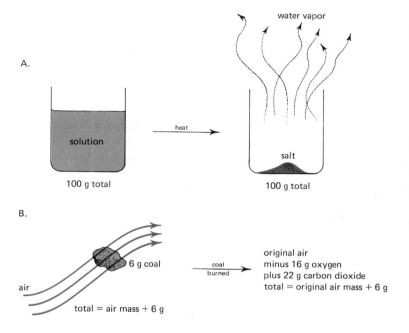

[2–1] Physical and Chemical Changes in Matter

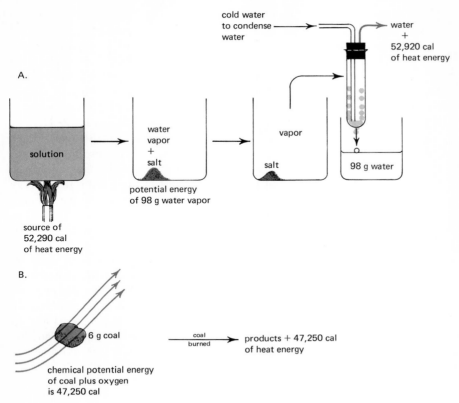

Figure 2-2 Conservation of energy.

A. A physical change: 52,920 calories (cal) of heat energy must be supplied to evaporate 98 g of water, but that energy can be recovered as heat if the 98 g of water vapor is condensed. While the 98 g of water is vapor the energy is in a potential energy form, but the total energy content stays constant throughout the whole process.

B. A chemical change: 6 g of coal plus 16 g of oxygen have a potential energy content (chemical energy) 47,250 cal larger than the 22 g of carbon dioxide that can form when the coal burns. As the coal burns, the potential energy is converted into heat energy, but the total energy content stays constant.

Most of the matter we observe is in the form of mixtures. Two or more substances can be mixed (a physical change) without changing their identities. A chemical change would be necessary to change their identities. A chemical change might convert everything into a single substance (one identity) or to a new, different mixture of two or more substances.

Some mixtures have clearly different regions which can be physically separated from each other. These are called **heterogeneous mixtures.** An example is a mixture of white sand and iron powder. We can identify the matter in individual pieces as either only sand or only iron. In this case the mixture can be separated by using a magnet to draw out the iron, leaving the sand behind. Not all heterogeneous mixtures can be separated so easily. If we mixed powdered carbon with white sand, we would also get a heterogeneous mixture of clearly different black (carbon) and white (sand) particles. A direct physical separation would be possible, perhaps by use of a magnifying glass and delicate tweezers, but it would be difficult. We would probably just leave it as "dirty sand" or "sandy carbon black."

One characteristic of heterogeneous mixtures is the complete freedom in the amounts that can be mixed together. In the case of the carbon-sand mixtures, we could have almost pure sand with a tiny amount of carbon,

2-2 Mixtures

Heterogeneous mixtures have different regions and complete freedom in the proportions mixed.

almost pure carbon with a few grains of sand, or any other combination. Some parts of the mixture may also differ from other parts of the mixture. There is no way to know accurately the quantities in the mixture unless it is taken apart in some way.

Some other mixtures, such as air or salt dissolved in water, can be mixed to the point where all regions of the mixture are alike. These are called **homogeneous mixtures** or solutions. We will discuss some features of these mixtures in Chapter 10. Although freedom to vary the amounts usually exists (similar to the freedom of heterogeneous mixtures), there are sometimes limits on the relative amounts mixed homogeneously. The nature of the matter present may also be significantly changed in solution (as noted in Chapter 7) even though physical changes (such as evaporating water away to leave salt behind) can separate the mixture.

> Homogeneous mixtures are uniform throughout and can vary in proportions subject to certain limits.

2-3 Pure Substances

If we separate mixtures into their parts, eventually we obtain pure substances. If we began with a heterogeneous mixture, we might first separate it into two or more homogeneous mixtures. That would move us toward matter that is more uniform throughout. The next step in making the matter more uniform throughout is separation into pure substances—those for which no further separation is possible by physical changes. Most of the pure substances we know are **compounds.** Compounds can be broken down into different parts only by chemical changes. They differ sharply from mixtures because they require a different type of change to separate them into parts and because they have no variation at all in the relative amounts of products that can be obtained.

> Compounds are uniform throughout, have a single fixed composition, and their components cannot be separated by physical changes alone.

If separations by chemical changes are continued, eventually products are reached that cannot be separated further by either chemical or physical changes. These basic building blocks of matter are called **elements.** They are few in number compared to the known compounds, but their characteristics and abilities to form compounds determine the forms and behavior of all matter.

> Elements cannot be separated into constituent substances by chemical or physical changes.

2-4 Definite Composition and Multiple Proportions

Each compound has a definite ratio of elements. We know this must be so because each compound has exactly fixed ratios of products to which it could be broken down. Since everything can eventually be broken down to the elements, the ratios of elements must be exactly fixed. This fact of nature is called the **law of definite composition.** It is most commonly used in the form of weight relations—that the proportions by mass of the elements in any one compound are always the same.

The mass relationship is not a very good way to state the law of definite composition. The ratios of atoms, not mass, are the true constants, but the mass relationship works for everything in our normal experience and is quite useful. The mass relationship is also historically the way in which the law

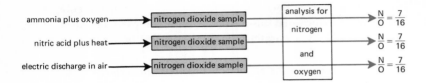

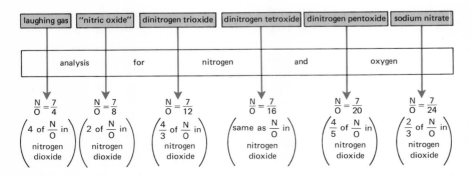

Figure 2-3 Definite composition and multiple proportions.

was discovered. For example, samples of water obtained from different locations and from different kinds of sources (sea water, lakes, streams, plants, animals, rainwater, and so on) always gave the same mass relationship. When nonwater impurities have been removed, water from any natural source on Earth has 11.2% of its mass as the element hydrogen and 88.8% of its mass as the element oxygen. Every other compound was also found to have a definite proportion by mass of the elements from which it was formed.

When the definite proportions by mass of many compounds were determined, a second important fact of nature was noticed. All of the proportions between any two elements are simple multiples of each other. For instance, carbon dioxide, the common gas formed from the elements carbon and oxygen, has a proportion (by mass) of three parts carbon to eight parts oxygen. But another gas exists with six parts carbon to eight parts oxygen, exactly two times the carbon to oxygen ratio of carbon dioxide. All compounds are found to give similar relationships between any two elements. The ratio in any compound can be converted to the ratio in any other compound by multiplying by some ratio of whole numbers, usually a ratio of small whole numbers. This principle is called the **law of multiple proportions.** Figure 2-3 illustrates the laws of definite composition and multiple proportions.

Law of definite composition: The proportions of elements in any given compound are always the same.

Law of multiple proportions: The ratio between two elements in one compound can be converted to the ratio in any other compound by multiplying by some fraction of small whole numbers.

2-5 Atomic Theory

The laws of definite composition and multiple proportions are now explained by a theory of the atomic nature of matter. The earliest records of a proposal that matter consisted of small pieces date from Greek philosophers. We call

The atomic theory provides a reasonable explanation for the laws of definite composition and multiple proportions.

the pieces **atoms,** from the Greek *atomos,* meaning uncut. Leucippus and Democritos had some beautiful ideas (beginning about 450 B.C.) about different kinds of small pieces leading to the different kinds of matter, but their ideas were more often ignored than accepted during the next 2000 years. Their ideas had little influence because they were not tested and refined by the scientific method. The Greeks did not prove the truth of their basic idea through experiments. Experiments were also needed to show the errors that needed to be corrected in their proposed details. With neither proof nor progress, their idea of atoms was eventually displaced by other ideas which seemed beautiful but left out the really important truth.

In the early part of the nineteenth century, the atomic theory was reintroduced by John Dalton. He used the information about matter that was available to state the theory in a form that could be tested by the scientific method. The errors in details of his theory could be found and corrected. The correctness and importance of the idea of atoms has since been shown by testing the theory through experiments. The main points in Dalton's atomic theory, plus comments about how some details have been revised to fit more recent experiments, are as follows:

Basic points of the atomic theory are stated as given by Dalton and later modified.

1. All matter is made up of tiny, indestructible units called atoms. In other words, they are *atomos,* cannot be cut.
   We now know that atoms are not indestructible, but the conditions needed to change atoms (Chapter 15) are not present in most of our experiences. Therefore, Dalton's original idea is a correct and useful statement about matter as we usually observe it.
2. All atoms of a given element are alike.
   We now know that atoms of a given element may differ in mass but not in their ability to form compounds. We might restate this point as "all atoms of a given element are chemically alike."
3. Compounds are formed by the union of atoms of two or more elements.
4. During chemical reactions atoms may combine or combinations may break down, but the atoms themselves remain unchanged.
5. Atoms combine in small whole numbered ratios such as $1:1$, $1:2$, $2:3$, and so on.

Dalton's prediction of ratios of small whole numbers is particularly interesting because it came *before* the law of multiple proportions had been observed and stated as a scientific law. The law of definite composition was known in Dalton's time, and the atomic theory explains that law. Since Dalton needed specific ratios of atoms (plus having atoms of the same element alike) to get the observed definite compositions, he reasoned that the ratios should be simple ones and other simple ratios should also be possible. Later measurements showed that this was usually true and led to the law of multiple proportions. We now know that some of the combinations get complex enough to involve fairly large numbers, but simple ones with small whole numbers are quite common. We also know that only certain kinds of combinations are possible, and much of chemistry concerns the determination of what combinations can occur and why (Chapter 7).

In our hindsight the atomic theory (supported by the results of numerous experiments since Dalton's time) seems so obviously correct that it is hard to imagine anyone overlooking it. When we do recognize the difficulties in thinking of something completely new and different, we can err in the other direction by thinking that only a few mental giants can make basic discoveries. Both of these conclusions are dangerous because they can lull us into thinking of our present science as infallible truth. A review of the history of Dalton's atomic theory may help us resist the pitfalls of overdependence on existing theory.

Although Dalton was very clever, particularly in anticipating the law of multiple proportions, he had no crystal ball to show him all the right answers. He actually overlooked one of the most important and useful concepts resulting from his basic assumptions. He failed to differentiate between the basic units of elements (atoms) and the basic units of compounds, which we call **molecules**. The ratios of small whole numbers in the definite compositions of various compounds of the same elements often occur because they have whole-numbered groups of each kind of atom. The specific combination of a group of atoms forms one type of molecule, and identical molecules are all of the same compound just as identical atoms are all of the same element.

Atoms are the smallest units of elements, and compounds may exist in discrete combinations of atoms making up the basic units called molecules.

The atomic theory was not obvious enough to gain general acceptance until supporting evidence was gathered. It was the scientific method, not the idea alone, that established the theory in modern times. The scientific method involves both thinking of ideas that fit the available information and testing the ideas by continued planned experimentation. We should learn to respond to the results of science in the same way. We must *think* about what is known and the proposed explanations. Then we must *question* and test the ideas, preferably by planned experiments.

Testing and revision using the scientific method was necessary to get acceptance of the idea of atoms.

2-6 Symbols, Formulas, and Nomenclature

Identification of elements and compounds was followed by assignment of names. The number of elements to be named was fairly small. Their names and abbreviations for their names have been established and agreed upon. The names were not chosen in a very systematic fashion, partly because people did not understand what the underlying system was when many elements were identified. Sometimes the same element has been given different names in different languages, but an agreement has been reached to use the same abbreviations everywhere. For some elements, particularly the most common ones, the abbreviation is simply the first letter of the element's name. That letter (written as a capital) is called a chemical **symbol**. It represents the element and can be defined to represent any given quantity of the element. Since we know that atoms are the fundamental parts of elements, one of the best quantities to choose is one atom of the element. We will use that concept in this section and frequently throughout the course. Sometimes it

Symbols are used as abbreviations, usually representing one atom of the elements.

**Table 2–1
Some Common Elements with Single Letter Symbols**

Element	Symbol	Element	Symbol	Element	Symbol
oxygen	O	sulfur	S	boron	B
hydrogen	H	phosphorus	P	vanadium	V
carbon	C	fluorine	F	uranium	U
nitrogen	N	iodine	I		

will be more convenient to use each symbol to represent a larger (but fixed) number of atoms, as described in Chapter 4.

Some examples of names and symbols of elements with single letter symbols are given in Table 2–1.

Because there are more than 26 elements, some must have symbols using two letters instead of only one. In these symbols, the first letter is written as a capital and the second as a small letter. In the groupings of symbols we will use to represent compounds, each capital letter is therefore the start of a new symbol. Many elements have the first two letters of their names as symbols. Examples are given in Table 2–2.

In other cases the first letter and some other letter of the name are used. Usually this is because more common elements already have the one letter and first two letter symbols [example: cadmium Cd because C (carbon) and Ca (calcium) are already in use] or because two elements of similar importance have the same first two letters (example: chlorine Cl and chromium Cr—neither is given the Ch symbol). Common elements in this category are listed in Table 2–3.

The usual pattern for selecting symbols is helpful in learning the most common symbols, but some symbols do not fit the elements' names as we know them. These symbols are based on names in other languages. The names, symbols, and sources of these symbols are listed in Table 2–4. All are Latin except wolfram (German).

When each symbol represents one atom, we can easily use a group of symbols to represent the combination of atoms in a particular compound. These symbol combinations are called **formulas.** If the symbols are written down with no spaces between them (example: HCl) they represent elements chemically combined to form a compound. Since each symbol represents one atom, HCl represents one hydrogen atom combined with one chlorine atom. The two symbols form the smallest possible unit of the compound, one

**Table 2–2
Some Common Elements with Symbols Using Their First Two Letters**

Element	Symbol	Element	Symbol	Element	Symbol
calcium	Ca	aluminum	Al	silicon	Si
argon	Ar	cobalt	Co	neon	Ne
barium	Ba	helium	He	nickel	Ni
beryllium	Be	lithium	Li	titanium	Ti
bromine	Br				

[2–6] Symbols, Formulas, and Nomenclature

Table 2-3
Some Common Elements with Symbols Other Than Their First Two Letters

Element	Symbol	Element	Symbol
arsenic	As	magnesium	Mg
cadmium	Cd	manganese	Mn
chlorine	Cl	strontium	Sr
chromium	Cr	zinc	Zn

molecule of HCl. If a greater number of atoms than one is needed of a particular element, it is shown by a subscript after the symbol: H_2O represents two atoms of hydrogen combined with one atom of oxygen. If more than one molecule of the compound is needed, the number is written in front of the entire formula: 2 HCl represents 2 hydrogens and 2 chlorines combined for form 2 HCl molecules.

Formulas that represent the combination of atoms actually held together as molecules are called **molecular formulas.** Molecular formulas often give useful information. For example, many elements (particularly those existing as gases) have diatomic molecules—combinations of two atoms of the same element. These are shown by the molecular formulas H_2, O_2, N_2, Cl_2, F_2, Br_2, and I_2.

Molecular formulas show the smallest basic unit of a molecular compound.

Some formulas that give us useful information are not really molecular formulas. The simplest ratio of atoms that fits the observed definite composition is called an **empirical formula.** Empirical formulas can be determined and used even if the true molecular combination is not known. An example is the formula P_2O_5 (which has been used for many years to report the phosphorus content of fertilizers). The molecular formula of the compound is now thought to be P_4O_{10}. That does not affect the weight ratio, so the empirical formula P_2O_5 is still used.

Empirical formulas show the ratio of atoms in a compound in the simplest ratio, even if that is not the molecular arrangement or if the compound is not molecular.

The concept of molecules is so useful to our understanding of nature that it is easy to forget that many formulas are only empirical. Chemists (who should know better) often look at formulas like MnO_2 and think of MnO_2 molecules instead of the billions of atoms in a complex solid arrangement that MnO_2 really represents. The MnO_2 case is a very special one which shows the danger of carelessness in science.

Very careful measurements show the atomic ratio to be closer to $MnO_{1.98}$.

Table 2-4
Some Common Elements with Symbols Based on Foreign Names

Element	Foreign Name	Symbol	Element	Foreign Name	Symbol
antimony	stibium	Sb	potassium	kalium	K
copper	cuprum	Cu	silver	argentum	Ag
gold	aurum	Au	sodium	natrium	Na
iron	ferrum	Fe	tin	stannum	Sn
lead	plumbum	Pb	tungsten	wolfram	W
mercury	hydrargyrum	Hg			

The person assuming MnO_2 molecules would probably decide the measurements are wrong (they are correct) and ridicule the person who made them. Human history, in science and in other areas, is full of examples of such blindness to the truth. If the evidence became irresistible, the person assuming MnO_2 molecules might give up on the whole atomic theory. The person recognizing MnO_2 as an empirical formula would simply recognize that the measurements used to get such formulas are usually slightly inaccurate. The closeness to MnO_2 would still tell him something about the combination, and the deviation would tell him something about nature's complexities—in this case, crowding and structure problems in a solid. In this course we will use the formula MnO_2 and many other simplifications. These will help us see important general trends without being confused by excessive details. You will not be required to know the omitted details, but you must recognize the possibility of oversimplification or error in any statement.

Another kind of formula that is not truly molecular is represented by NaCl (common salt). NaCl is not an empirical formula for a large molecule. Instead it shows the ratio of smaller electrically charged pieces (ions) of chlorine and sodium that are present in a definite ratio but are not linked into specific molecules. Ionic bonding is discussed in Chapter 7.

The separate ions have names, and these can be combined to name the compound the ions form. All compounds are electrically neutral (no excess of positive or negative charge). Therefore the relative numbers of positive and negative ions must be in a ratio that makes the total of positive charges equal the total of negative charges. The formula is the result of that balancing of charges. The positively charged ion is always placed first in both formula and name. This naming system is also used to name nonionic compounds by treating the part that has the greater tendency to be positive as the positive part.

Compounds are named by systematically naming their positive and negative parts.

The positive part is named using the name of the element. Therefore, Na is sodium. If there are several ways for one element to combine, a Roman numeral in parentheses is added after the element's name to show the oxidation state of that element (see Chapter 7). In CuCl copper has an oxidation state of $+1$, so the name is copper(I) chloride, whereas in $CuCl_2$ copper has an oxidation state of $+2$ and the name is copper (II) chloride. Unfortunately, there is another, old-fashioned method for naming some of the more common compounds where elements have more than one positive oxidation state. The most common names are listed in Table 2–5, but you may also see the endings -ous (the lower of two common oxidation states) or -ic (the higher of two) used in some other cases. It is useful to be able to recognize that type of name, but you will find it easier to use the simpler modern names when given a formula and asked to name the compound.

There is only one complex (more than one element) positive ion, NH_4^+, ammonium ion, whose name you need to learn at this point. (Other, very complex positive ions can be assembled. Some of them are described in Section 16–3)

Negative parts are named using a root portion (taken from the element's

Ion	Systematic Name	Old Name	Ion	Systematic Name	Old Name
Fe^{2+}	iron(II)	ferrous	Sn^{2+}	tin(II)	stannous
Fe^{3+}	iron(III)	ferric	Sn^{4+}	tin(IV)	stannic
Cu^{1+}	copper(I)	cuprous	Co^{2+}	cobalt(II)	cobaltous
Cu^{2+}	copper(II)	cupric	Co^{3+}	cobalt(III)	cobaltic

Table 2–5 Modern and Old-Fashioned Names of Common Metal Ions

name) plus endings with particular meanings. The ending -ide normally means a single atom as a negative ion. Complex ions with many oxygens attached to another element usually end in -ate, whereas those with less oxygens end in -ite. If more than two complex ions with oxygen form, the prefix per- may be added to mean even more oxygens than the -ate ending alone, and the prefix hypo- may be added to mean even less oxygens than the -ite ending alone. Table 2–6 uses the case of chlorine to illustrate all of these endings and prefixes.

The negative ion endings -ide, -ate, and -ite are used in a systematic arrangement with the same root names.

root name = chlor	
Cl^-	chloride
ClO^-	hypochlorite
ClO_2^-	chlorite
ClO_3^-	chlorate
ClO_4^-	perchlorate

Table 2–6 Naming Chlorine Containing Ions

There are some negative ions whose names do not fit in this system. An example is CN^-, a complex ion not involving oxygen. It cannot fit the -ate or -ite scheme, so it simply acquired another name, cyanide, which does not fit any element root name. Positive ions involving more than one atom of the same element can be named using the oxidation state method [example: Hg_2^{2+}, mercury (I)], but there is no such method for negative ions. Therefore, N_3^- also fits no element root name and is called azide. Some other names come from substitution of one element for another in a well-known ion

Ion	Name	Ion	Name
CO_3^{2-}	carbonate	PO_4^{3-}	phosphate
O^{2-}	oxide	NO_3^-	nitrate
OH^-	hydroxide	S^{2-}	sulfide
Cl^-	chloride	SO_3^{2-}	sulfite
Br^-	bromide	SO_4^{2-}	sulfate
I^-	iodide	CN^-	cyanide
F^-	fluoride	$C_2H_3O_2^-$	acetate
		ClO_4^-	perchlorate

Table 2–7 Common Negative Ions

It is necessary to memorize some names of common negative ions.

(example: $S_2O_3^{2-}$, thiosulfate, from SO_4^{2-}, sulfate). Although most common negative ion names are fairly systematic, the need to learn root names makes memorization of a few of the common ion names the simplest starting approach for most people. Table 2-7 lists a group of common negative ions. You should gradually expand the list you know as other names appear in the text or laboratory work.

Even the brief list of negative ions in Table 2-7 is enough to let you name many compounds (given their formulas) or write the formulas for many compounds (given their names). You will have to pay close attention to the **charges** of the ions as well as the names. Remember that the total of positive charges in a compound must always equal the total of negative charges. Here are some examples of naming exercises.

Example 2-1. Name the compounds $CuSO_4$ and $Mg(NO_3)_2$.

(a) $CuSO_4$:

We know SO_4^{2-} is a common ion called sulfate. Therefore, this is copper sulfate.

There is *one* negative ion (one sulfate) with a charge of 2−, so the total of negative charges is 2. The total of positive charges must therefore also be 2.

There is one positive ion (a copper ion). To give a total of positive charges as 2, the positive ion must be Cu^{2+}.

Therefore, this is copper(II) sulfate.

(b) $Mg(NO_3)_2$:

We know NO_3^- is a common ion called nitrate. Therefore this is magnesium nitrate.

There are *two* negative ions (nitrates), each with a charge of 1−, so the total of negative charges is 2. To balance charges, the *one* positive ion must be Mg^{2+}.

At this point you might call $Mg(NO_3)^2$ magnesium(II) nitrate. However, by the time you finish Chapter 7 you will be expected to know that magnesium is *always* Mg^{2+} in compounds. You will then be expected to stop putting down the unnecessary (II) and call it simply magnesium nitrate.

Some elements forming positive states form only one oxidation state.

In order to provide a good selection of sample exercises to use before you reach Chapter 7, Table 2-8 gives a short list of very common positive ions of elements that form only one common positive oxidation state.

Table 2-8
Some Common Positive Ions Not Requiring Oxidation States in Their Names

Na^+	Mg^{2+}	Zn^{2+}	Ag^+
K^+	Ca^{2+}	Al^{3+}	Li^+

Here are some examples of formula writing exercises.

Example 2-2. Write the formulas of iron(III) chloride and zinc perchlorate.

(a) Iron(III) chloride:
Iron(III) tells us we have Fe^{3+}.
Chloride is Cl^-.
We need three Cl^- to balance out one Fe^{3+}, so the formula is $FeCl_3$.

(b) Zinc perchlorate
Zinc is Zn^{2+} and perchlorate is ClO_4^-.
We need two ClO_4^- to balance out one Zn^{2+}, so the formula is $Zn(ClO_4)_2$.

2-7 Chemical Equations

Since atoms are "indestructible" units, there is no change in the number and kinds of atoms during chemical changes. That fact can be used to determine the relative amounts of compounds used up in a reaction (**reactants**) or formed (**products**). When the correct numbers are put in front of each formula, the expression is called a **balanced equation.** It shows the amounts of elements, compounds, or molecules that are equivalent to each other in that chemical reaction. An arrow is used to show the direction of the reaction, pointing from reactants toward products. Sometimes the formulas of reactants and products are put down without balancing the amounts. These are also called equations and used to represent reactions. An arrow is used to show the direction of change from reactants to products. The most accurate representation of a reaction is given when the equation is both **complete** (has all reactant and product formulas written) and **balanced** (has the equivalent amounts of each).

Completing and balancing an equation can be illustrated by the example of the reaction of H_2 with Cl_2. Since H_2 and Cl_2 are the reactants, they appear on the left side of the arrow.

$$H_2 + Cl_2 \longrightarrow$$

To complete the equation, we must know any other reactants or products. H_2 and Cl_2 are the only reactants in this case, and there is only one product whose molecular formula is known to be HCl. Therefore, the completed (but unbalanced) equation is

$$H_2 + Cl_2 \longrightarrow HCl$$

To balance the equation, we take each element (one at a time) and make the number of atoms the same on each side of the arrow. There are two hydrogen atoms in H_2, so we need two HCl's to have an equal number on the right side of the equation.

$$H_2 + Cl_2 \longrightarrow 2\,HCl$$

Next we check the chlorine atoms and find we already have two on each side. No further changes are needed, so the equation is balanced.

Chemical equations show the reactants and products of chemical reactions, and they can be balanced to show the relative amounts.

Sometimes it is necessary to back up and change previous numbers to balance atoms on later elements. If we started by balancing hydrogens in the equation

$$HCl + O_2 \longrightarrow H_2O + Cl_2$$

we would go through the following steps:

1. To get two hydrogens on the left to match the two on the right

$$2\,HCl + O_2 \longrightarrow H_2O + Cl_2$$

2. The chlorines already balance at two, thus no change.
3. To get two oxygens on the right to match the two on the left:

$$2\,HCl + O_2 \longrightarrow 2\,H_2O + Cl_2$$

4. But that changes the hydrogens on the right, so they must be balanced again (four on each side):

$$4\,HCl + O_2 \longrightarrow 2\,H_2O + Cl_2$$

5. Now that changed the chlorines on the left, so they must be balanced again (four on each side):

$$4\,HCl + O_2 \longrightarrow 2\,H_2O + 2\,Cl_2$$

The equation is now balanced.

In some cases one element may appear in two or more forms on the same side of the equation. If so, you will find it easier to balance the elements that appear only once on each side first. In such cases you may also need to settle for a fractional number of molecules such as $\frac{3}{2}\,O_2$, and then multiply everything by the denominator to make it a whole number. Here is an example:

$$NH_3 + O_2 \longrightarrow NO + H_2O$$

1. Try nitrogen and hydrogen before oxygen.
2. Nitrogens originally balanced
3. To balance hydrogens

$$2\,NH_3 + O_2 \longrightarrow NO + 3\,H_2O$$

4. That upset the nitrogen balance, so we need

$$2\,NH_3 + O_2 \longrightarrow 2\,NO + 3\,H_2O$$

5. That does not let the oxygens balance, but it is too complicated to try to straighten out in one step, so settle for a fractional value to balance oxygens.

$$2\,NH_3 + \tfrac{5}{2}\,O_2 \longrightarrow 2\,NO + 3\,H_2O$$

6. Multiply *everything* by 2 to get rid of the fraction

$$4\,NH_3 + 5\,O_2 \longrightarrow 4\,NO + 6\,N_2O$$

The equation is now balanced.

There are more complex methods for balancing some equations involving changes in oxidation state (see Chapter 7), but the above technique is adequate for most equations you will need.

Skills Expected at This Point

1. You may be required to describe the forms of matter and the effects on form of chemical and physical changes.
2. You may be required to list points of the atomic theory, state the laws of conservation and composition, describe the development and significance of atomic theory and the laws of conservation and composition, and recognize situations that illustrate each of the laws.
3. You are expected to use correctly the symbols, formulas, equations, and systematic nomenclature specifically covered in this chapter and steadily to expand your vocabulary of names to include other positive and negative units as they appear in your later study.
4. You are expected to know and use the charges as well as the names of the common ions listed so you can write correct formulas (when given the systematic name) as well as correct names (when given the formulas).
5. You are expected to balance equations of similar complexity to the examples in the chapter.

Exercises

1. (a) List two examples of physical changes. (b) List two examples of chemical changes.
2. What is a compound?
3. Write the names of the following elements: C, Al, Si, Cl, Na, K.
4. Write the names of the following compounds: $KClO_3$; $LiCN$; $Ca(OH)_2$; Cu_2O (*note:* Cu has more than one possible oxidation state).
5. Write the formulas of lead(II) acetate and iron(III) oxide.
6. NCl_3 reacts with H_2O to form NH_3 and $HOCl$. Write and balance the equation for that reaction.
7. Balance the following equations:
    (a) $CO + O_2 \longrightarrow CO_2$
    (b) $PCl_3 + Cl_2 \longrightarrow PCl_5$
    (c) $CH_4 + O_2 \longrightarrow CO_2 + H_2O$
    (d) $C_2H_6 + O_2 \longrightarrow CO_2 + H_2O$

Test Yourself

1. List the significant differences between a homogeneous mixture and a compound.
2. Ice and water are both pure substances made from the same combination of the elements hydrogen and oxygen. **(a)** What is the class of pure substance comprising both ice and water? **(b)** What type of change is involved when water freezes to ice? **(c)** Because a glass of ice and water has clearly nonidentical regions (some ice, some water), how must it be classed? **(d)** Salt in water is a homogeneous mixture; how can the various regions be described? **(e)** Salt and water can be separated by what kind of change?
3. **(a)** Name the compounds with the formulas $NaNO_3$ and KI. **(b)** Give the formula for hydrogen sulfide. **(c)** Balance
$$Al + FeO \longrightarrow Al_2O_3 + Fe$$
4. Decide which of the following statements are true.
   (a) All matter is made up of atoms.
   (b) All atoms of an element are identical with respect to chemical behavior.
   (c) The atom is the smallest unit of an element that can be said to have the chemical properties of that element.
   (d) All of the above are correct.
   (e) None of the above are correct.
5. The formula of potassium chlorate is
   (a) KCl **(c)** $KClO_3$ **(e)** KC
   (b) $PClO_2$ **(d)** PCl
6. Which of the following is heterogeneous?
   (a) A solution of sodium chloride in water.
   (b) A mixture of 80% nitrogen gas and 20% oxygen gas.
   (c) A highly purified sample of gold.
   (d) A mixture of ice and water.
   (e) None of the above.
7. To which of the following does the law of definite composition apply?
   (a) A heterogeneous sample.   **(d)** An element.
   (b) A mixture.   **(e)** A compound.
   (c) A solution.
8. What do you consider the most important subject in this chapter which was not covered in any of the above questions? Write a test question covering that subject.

Measurements and Units

Science requires accurate measurements. Measurements are not necessary for antlike repetition of a successful path. A general searching for new ways can be done without precise measurements. But the reasoning and planning stages of the scientific method are effective only if observations are measured accurately and results are recorded in a way that can be understood when they are needed again.

Any measurements of quantity require systems of units. Measurements of the natural world require units to describe length, volume, mass, time, temperature, and a number of properties that can be described by combinations of these units. We also need methods to tell how many we have of the units and how accurately we have made the measurements.

3–1 Measuring Quantities

Accurate and properly recorded measurements are necessary in science.

Systems of units are needed to describe measurements.

We state quantities in numbers, using a decimal number system (ones, tens, hundreds, and so on). This decimal system makes some numbers much easier to work with than others. For instance, the quantity 0.2 mile is easier to write and use than 0.22727272727 mile. When we measure distances in units of miles using the speedometer of a car 0.2 mile is convenient, but when we use smaller sized units, which are not decimal fractions of a mile, the quantity is less convenient (376 yd, 1128 ft, or 13,536 in.). If we chose the quantity 400 yd to fit the small units better, we would get the inconvenient number 0.2272727 . . . (with the numbers 27 repeating forever) for the same quantity in miles. As a result, miles, yards, feet, inches, and most of the other units that have been used traditionally in the United States and Canada are not the best possible choices.

Most of the world now uses a system of measurement units called the **metric system.** It avoids many of the problems in changing units by using decimal relationships between units. The metric system has been the generally accepted system for scientific measurements with a basic unit of length, the **meter,** a unit of volume, the **liter,** and a unit of mass, the **gram.** It also

3–2 Metric System and the System International (SI)

The metric system is the most generally accepted system of units for mass, volume, and length.

gives us a naming system for larger or smaller units decimally related to the basic units.

In 1960, the recognized international authority on units (Conference Generale de Poids et Measures) adopted a variation of the metric system intended to simplify conversions even more. This official system is called System International d'Unities (in French) or the International System of Units (in English). By agreement, it is abbreviated SI in all languages. SI units are a convenient overall system and the product of an international agreement. Most countries have already adopted these units or are in the process of officially adopting them. They are becoming the official units used in science. Therefore, they will be described here and used extensively throughout this book.

SI attempts to relate each unit to other kinds of units or some important characteristic in nature. Some of the original metric relationships were imperfect because of errors in the measurements used to work them out. A meter was intended to be one ten millionth of the distance from the earth's equator to pole, measured along a meridian. The fact that the measurement was a little off is of no practical importance now, since that is a very inconvenient distance to measure. One meter (1 m) is now defined as 1,650,763.73 times the wavelength of the orange-red light emitted by an isotope of the element with atomic number 36, krypton-86 ($^{86}_{36}$Kr). In its effort to be simple, SI omits commas in numbers, leaving only empty spaces when needed to make large numbers clearer. The reasoning is that writing the commas is an unnecessary extra effort. Therefore, 1 m is 1 650 763.73 times the wavelength of the orange-red light emitted by $^{86}_{36}$Kr. Many people, including this author, will find the comma habit hard to break. We now use the SI defined meter as the standard against which all other lengths are compared. There are 39.37 in. in 1 m.

When we want length units larger or smaller than meters, we put a prefix in front of the word meter. These prefixes tell us how many places we must move the decimal point to get the new unit. For instance, moving the decimal point one place in the direction making the unit smaller is represented by the prefix **deci.** Therefore, a decimeter is one tenth of a meter. We will use the same prefixes for making all metric or SI units (not just meters) larger or smaller.

Three of the many possible prefixes are used often enough to justify being memorized. They are **centi** (meaning 1/100), **milli** (meaning 1/1000), and **kilo** (meaning 1000 times). Therefore, centimeters and millimeters are small units of length and kilometers are useful long units. Since these three most common prefixes start with different letters and the basic metric units of length, volume, and mass also begin with different letters, we can abbreviate them by using their first letters (as small letters). Therefore 5.5 m is 5.5 meters, 4.0 l is 4.0 liters, and 10 g is 10 grams. When there are two letters in the abbreviation, the first letter refers to the prefix and the second letter refers to the unit. Therefore, 55 ml is 55 milliliters, 18 cm is 18 centimeters, and 12 kg is 12 kilograms. We can easily convert from one kind of unit to

There are standard prefixes for decimally related metric system units.

another, and we will do so when one of the units we are familiar with is more convenient than the others. If we were told that the distance from New York to San Francisco was 5,000,000 m, we might write down:

New York-San Francisco, 5000 km

If we were told that the distance from one end of an eyeglass lens to the other was 0.05 m, we might write down:

Eyeglass width, 5 cm

If we were told that the eyeglass lens was 0.002 m thick in the center, we might write down:

Eyeglass thickness, 2 mm

Length cubed equals volume, so we do not really need another basic unit for volume. Therefore SI does not use the metric system volume unit, the liter. However, the volume that is 1 m long, 1 m wide, and 1 m high is often larger than desired for volume measurements. The volume of 1 cubic decimeter (0.1 m on each side) was chosen in the older metric system and given the name liter. When smaller units are desirable, milliliters (ml) are used: 1 ml equals 1 centimeter cubed (cm^3). The cubic centimeter is a volume unit in SI whereas milliliters and liters are not officially recognized. Liters and milliliters are still used on much laboratory equipment and in textbooks, lab books, and reference books. You can always substitute milliliters for cubic centimeters or centimeters cubed for milliliters in any problem.

The metric mass unit, gram, was chosen to fit the mass of a definite volume of a common substance, water. Mass is a property of matter related to what we observe as weight. Weight is a force exerted by gravity on an object. Mass is a measure of the extent to which the object can interact with gravity. Objects have weight only when they are being acted on by gravity. They always have mass, the potential to be acted upon if brought into a gravitational field. Figure 3–1 illustrates the constant mass and varying weight of a spacemen. The mass is a property of the spaceman. If he does not change, it does not change.

The mass of 1 cm^3 of water is set equal to 1 g. Water expands or contracts as it is heated or cooled, so a particular temperature had to be specified. The temperature chosen was the temperature of maximum density of water, about 4°C. (Density is defined in Section 3–6.) At every temperature above or below that point water expands somewhat from the volume it occupies at 4°C. Therefore, the largest mass of water we ever get into 1 ml becomes 1 g. At every other temperature 1 g of water occupies more than 1 cm^3, but the difference is small enough to ignore in many practical problems.

A particularly important unit of mass is 1 kg. It is the mass of 1000 cm^3 (or 1 liter) of water at maximum density, but, more important, it is the mass unit that can be used along with meters and seconds to derive the series of units for energy, force, and so on, used in SI. Therefore, the kilogram (not the gram) has been declared the official standard unit of mass in SI.

There are interrelations between meters, liters, and grams.

Figure 3–1 Weight depends upon gravity but mass is constant.

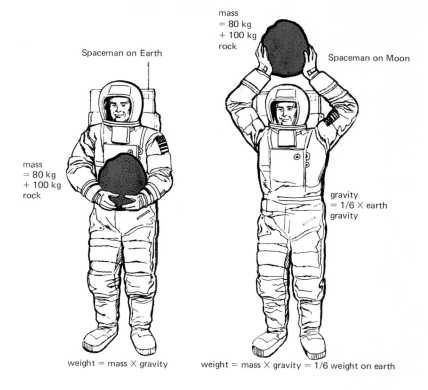

Although this fact has no effect on the gram or the kilogram, it does help emphasize the relationship between SI units more complex than the ones we discuss here.

3–3 Unit Conversions

The unit factor method provides an effective system for converting units and solving many problems.

One effective method of converting from one kind of unit to another is the unit factor method. This method depends on two simple facts. First, any quantity can be multiplied by the number 1 without changing the amount. Second, any number divided by itself is 1. Any time we can write down an equality we can set up unit factors. We can write down the equality

$$2.54 \text{ cm} = 1 \text{ in.}$$

From it we can write either

$$\frac{2.54 \text{ cm}}{1 \text{ in}} = 1 \quad \text{or} \quad \frac{1 \text{ in}}{2.54 \text{ cm}} = 1$$

Both of them are a quantity being divided by itself.

Table 3–1 lists some of the equalities we find very useful in relating English and metric system units. You will notice that it is not necessary to know

Table 3-1
Equalities Useful in Converting Units

Length	Volume	Mass
12 in. = 1 ft	2 pt = 1 qt	16 oz = 1 lb
3 ft = 1 yd	4 qt = 1 gal (U.S.)	2000 lb = 1 ton
5280 ft = 1 mile		

Conversions between systems

2.54 cm = 1 in.	1.06 qt = 1 liter	454 g = 1 lb
1000 mm = 1 m	1000 ml = 1 liter	1000 mg = 1 g
100 cm = 1 m	1 cm^3 = 1 ml	1000 g = 1 kg
1000 m = 1 km	1 ml H$_2$O (4°C) = 1 g	

every possible equality. As long as we know one conversion from English to metric for each type of unit (length, volume, and mass) we can multiply our original amount by a series of unit factors until we reach the desired answer. Because each equality gives us two unit factors, we pick the ones that cancel out the units we started with and lead us toward the answer.

You will want to practice some simple one-step conversions, but the real advantages of the method lie in the more complex problems. Some thought in planning the approach on each individual problem can keep the amount of necessary memory work quite low. We will describe planning the steps as "road mapping" the problem. It is somewhat like using a map to choose the route for a car trip. The metric system simplifies many problems, particularly those needing a length-to-volume relationship (use 1 cm^3 = 1 ml) or a mass-to-volume relationship (use 1 ml H$_2$O at 4°C = 1 g).

Example 3–1. An example of planning and carrying out a unit conversion is given in the following case. Congressman Jones is visiting an Army base. His guides are from an artillery company that uses 105 mm howitzers. They are approaching a display of various sized artillery pieces, and it might be worth 20 or 30 votes in the next election if the Congressman can pick out and admire the 105 mm gun. How can he estimate its size in inches?

Step 1: What is the given information?
Answer: The size (inside diameter) is 105 mm.
Step 2: What units are needed?
Answer: Convert to inches.
Step 3: Map the route from given information to answer.
  (a) He does not know the relationship between millimeters and inches
  (b) He does know there are 2.54 cm in 1 in.
  (c) He can convert millimeters to centimeters because he knows their relationships to the same unit, meters; that is, 1000 mm = 1 m and 100 cm = 1 m. Therefore, starting from millimeters

$$\text{mm} \longrightarrow \text{m} \qquad \text{m} \longrightarrow \text{cm} \qquad \text{cm} \longrightarrow \text{in.}$$

Step 4: Start from the beginning and put in the unit factors so units cancel out.
Answer:

$$105 \text{ mm} \times \frac{1 \text{ m}}{1000 \text{ mm}} \times \frac{100 \text{ cm}}{1 \text{ m}} \times \frac{1 \text{ in}}{2.54 \text{ cm}}$$

$$= \frac{105 \cdot 100}{1000 \cdot 2.54} \text{ in.} = \frac{105}{25.4} \text{ in.} = \text{about 4 in.}$$

Example 3-2. More complex problems can be attacked in the same general way. Let us calculate how long a sewage disposal plant would be put out of commission by a cyanide release that killed off all the useful bacteria which do the sewage disposal work. If the cyanide was carried through with the water, we would calculate the time from the volume input and volume in the sewage plant. We will assume the working bacteria are incapacitated for the same time as the time the cyanide is in the plant.

Given: Input rate = 3000 gallons per hour (gal/hr).
Water tank dimensions: length, 85.0 ft; width, 40.0 ft; depth, 12.0 ft.
Plan: 1. Calculate tank size.
2. Convert tank size to time for water replacement.
3. Key step—go to metric system to get from lengths to volume. Our "road map" plan:

$$\text{ft}^3 \longrightarrow \text{in.}^3 \longrightarrow \text{cm}^3 \longrightarrow \text{ml} \longrightarrow \text{liter} \longrightarrow \text{qt} \longrightarrow \text{gal} \longrightarrow \text{hr}$$

Set-up:

$$(85.0 \text{ ft})(40.0 \text{ ft})(12.0 \text{ ft}) \left(\frac{12 \text{ in.}}{1 \text{ ft}}\right)^3 \left(\frac{2.54 \text{ cm}}{1 \text{ in.}}\right)^3 \left(\frac{1 \text{ ml}}{1 \text{ cm}^3}\right)$$

$$\left(\frac{1 \text{ liter}}{1000 \text{ ml}}\right)\left(\frac{1.06 \text{ qt}}{1 \text{ liter}}\right)\left(\frac{1 \text{ gal}}{4 \text{ qt}}\right)\left(\frac{1 \text{ hr}}{3000 \text{ gal}}\right) = 102 \text{ hr}$$

3-4 Exponential Numbers and Significant Figures

Often we will write numbers in an exponential form like 5.4×10^3 to represent 5400. This form has two big advantages. First, it saves space in writing very large or very small numbers. For example, 2.0×10^8 takes less space than 200 000 000 and 1.0×10^{-14} takes less space than 0.000000000000010. Second, it lets us show clearly how accurately we know the number.

There are two kinds of numbers used in science, pure numbers and measurement numbers. If we multiply something by 2 we mean exactly 2. The number 12 in Example 3-2 is an example of an exact number, because there are exactly 12 in. in 1 ft. But a measurement of 2 cm may mean only that the distance is closer to 2 cm than it is to 1 cm or 3 cm. A number like 5400 is confusing as a measurement number because the last two zeros must be there to show us the decimal point. The measurement may have been good enough for all four of the figures to be **significant figures.** That would mean the value was closer to 5400 than to 5401 or 5399. If only the first three figures are significant, the value would be closer to 5400 than to 5410 or 5390. If only the first two figures are significant, the value would be closer to 5400 than 5500 or 5300. When the number is written in exponential form, we can show how many figures are significant by only writing down the ones that are significant. The above three examples would be written 5.400×10^3, 5.40×10^3, and 5.4×10^3, respectively.

<small>Numbers are written in exponential form to show how many figures are significant figures.</small>

It is normal scientific procedure to limit the significant figures in each answer to the *least number* of significant figures in any of the numbers used to get the answer in multiplication or division. In addition or subtraction we use the place of the last significant figure of the least accurate number used to get the answer.

<small>The significant figures in an answer are limited by the significant figures in the given data and a set of rules.</small>

Example 3-2.   Multiplication

$$\frac{(3.5 \times 10^1)(3.762 \times 10^2)}{7.00 \times 10^5}$$

The smallest number of significant figures given was two (in 3.5×10^1), so the answer must be rounded off from 1.881×10^{-2} to show only two significant figures, 1.9×10^{-2}.

Example 3-3.   Addition

$$3.55 \times 10^3 + 3 \times 10^{-2} + 2.8 + 3.44 \times 10^2$$

$$\begin{array}{r} 3550 \\ .03 \\ 2.8 \\ \underline{344} \\ 3896.83 \end{array}$$

The answer is only significant to the ten's place because 3550 is only significant to the ten's place. Therefore, the answer must be rounded off to 3900 or (showing the significant figures better) 3.90×10^3. Note that 3×10^{-2} has less significant figures than 3.55×10^3, but it is a more accurate number for addition; 3×10^{-2} is known within 0.01 whereas 3550 is only known to within 10.

3-5 Temperature Scales

The temperature unit in common use in the United States and Canada is Fahrenheit degrees. Just as English system units have been replaced in most of the world by the more orderly metric system, Fahrenheit temperatures have been replaced by the Celsius temperature scale. In converting Fahrenheit to Celsius we are faced with a new kind of conversion problem. These temperature scales not only have units (degrees) of particular sizes, they also

Temperature scales have both unit sizes and reference points.

have **reference points.** To change from one temperature scale to another we must change both the unit size (multiplying by the correct unit factor) and the reference point.

Figure 3–2 shows the reference points used to define Fahrenheit and Celsius temperatures. Both are now defined using the temperature at which ice melts and the temperature at which water boils at the standard (average at sea level) pressure of 1 atmosphere (atm). On the Celsius scale the point where ice melts is called zero and there are 100 degree units between that and the normal boiling point. An obsolete standard put zero where it is on the Fahrenheit scale. The original standards were the lowest temperature Gabriel Fahrenheit could reach using salt, water, and ice as zero and normal body temperature as 100°, but the values were later adjusted to a definition based on the melting point of ice in pure water as 32°F and the normal boiling point as 212°F. That left exactly 180 degree units between the two points.

Those definitions make 180 Fahrenheit-sized degrees = 100 Celsius-sized degrees. That leads to the unit factors

$$\frac{180°F}{100°C} \quad \left(\text{or } \frac{9}{5} \frac{°F}{°C}\right) \quad \text{and} \quad \frac{100°C}{180°F} \quad \left(\text{or } \frac{5}{9} \frac{°C}{°F}\right).$$

To complete the conversion, we need to adjust for the fact that there are 32°F even when there are 0°C. This is done by adding or subtracting the 32 *while we are in Fahrenheit-sized degree units.* The following equations can be used.

$$°C = \frac{5}{9}(°F - 32) \quad °F = \frac{9}{5}°C + 32$$

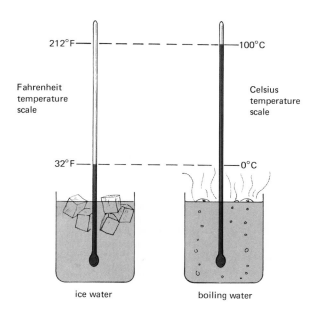

Figure 3–2 Temperature scales.

In Chapter 9 we will see that temperature is really a measure of the average energy of motion of the parts of a substance. It is therefore both convenient and necessary for some purposes to define another scale of temperature in which zero is the point where there is zero energy of motion. (To be precise, zero translational motion—the movement of units such as molecules from one place to another. Energy of vibrational motion—movement of parts of a molecule back and forth within the molecule—does not ever reduce to zero.)

Any temperature scale with a zero point at zero energy of motion is called an absolute temperature scale. You can have such a scale using Celsius-sized degrees, Fahrenheit-sized degrees, or some other sized degrees. The only absolute temperature scale we will use is the Kelvin temperature scale. It uses Celsius-sized degrees. The melting point of pure ice is 273.15 degrees on the Kelvin scale, so Kelvin temperatures can be easily found from or converted to Celsius temperatures by the formulas

$$°K = °C + 273.15 \quad \text{or} \quad °C = °K - 273.15$$

The Kelvin scale is the one that has been adopted in SI. SI also drops use of the word "degrees." The melting point of ice (273.15°K) is stated simply as "273.15 K."

Fahrenheit, Celsius, and Kelvin temperature scales are described and related to each other.

3–6 Properties of Matter

We are going to be discussing the matter of our universe, particularly as we find it on earth, and the changes it undergoes. We will be concerned with a number of properties of matter and properties related to changes in the form of matter. Among the most fundamental of these properties are the mass associated with matter, the space occupied by the matter, and the energy released or used when the forms of the matter are changed.

3–6.1 Mass. Mass is a fundamental property of all matter. As noted in Figure 3–1, it remains constant for any portion of matter that remains the same. The mass of matter also remains the same during any change in the form of matter. If the cars in a city burn 1 ton of gasoline per hour, that ton of material is changed in form but neither increased nor decreased in mass. It actually combines with about $3\frac{1}{2}$ tons of oxygen to form about $4\frac{1}{2}$ tons of water, carbon dioxide, carbon monoxide, and some other products. We can represent the process by the equation

1.0 ton gasoline + 3.5 tons oxygen ⟶

CO_2 + H_2O + CO + other products (total mass = 4.5 tons)

The mass of the products is exactly equal to the mass of the reactants, the matter that came together to form them. This constant value for mass is referred to as the law of conservation of mass, which was mentioned earlier in Section 2–1. It is a fact of nature without exceptions, although we can sometimes deceive ourselves into thinking otherwise. In Chapter 15, we will show that the law of conservation of mass holds even for cases like atomic bomb explosions where some measurements seem to say it does not hold. The law of conservation of mass has helped in organizing our understanding of many of the other facts of nature.

3-6.2 Density. All matter has mass and occupies space. The relationship between the amount of mass and the amount of space is often an important property of matter. We usually state this property as the ratio mass/volume which is called **density**. Because the unit of mass in the metric system, the gram, was defined from 1 ml of water at the temperature of maximum density, grams per milliliter (g/ml) is a particularly common and convenient unit for density. The density of water is 1 g/ml at 4°C and slightly less than 1 g/ml at any other temperature. Mercury (a liquid) has a density of about 13.6 g/ml. Material of low density will float on a liquid of higher density. The special nature of water, with its maximum density at 4°C, has a tremendous effect on our world. Because this density phenomenon in water is so important, it is discussed in detail in Chapter 13 and in chapters following Chapter 13.

> Density is the ratio mass/volume.

3-6.3 Energy Units. Energy is associated with all matter just as definitely as the facts that matter has mass and occupies space. The energy can be present as kinetic energy (energy of motion) or potential energy (stored energy). One of the forms of potential energy is chemical. Changes in the form of energy will occur during chemical reactions and other changes in the form of matter. Often energy will be released from potential energy to appear as heat (which is energy of translational motion). That heat will raise the temperature. We often measure the energy by the unit of heat called the **calorie**; 1 calorie (cal) is the amount of heat needed to raise the temperature of 1 g of water by 1°C. This definition is made more precisely accurate by choosing a particular 1°C, the degree between 14.5°C and 15.5°C, but almost the same value would have been found at any temperature where water is a liquid. We will use the unit calorie to measure energy even though it is not the official SI unit. SI uses the unit joule, which is directly derived from the SI units meter, kilogram, and second; 1 cal is 4.1855 joules.

> Energy can be measured in calories; 1 cal raises the temperature of 1 g of liquid water by 1°C.

Skills Expected at This Point

1. You should be able to describe the movement to SI units and the reasons for needing SI.
2. You should know the relationships between metric units and between English and metric units and use those relationships in problems.
3. You should use unit factor conversions (with proper significant figures) and "road map" planning of multiple step conversions.
4. You may be asked to define and use the three common temperature scales, density, and calories.

Exercises

1. Name the basic units for length, volume, and mass in the metric system and describe how they are related to each other.
2. Name the three most common metric system prefixes and state the meaning of each of them.
3. Write the name and abbreviation for each basic metric system unit and for each unit you could make from the basic units and the three common prefixes.
4. Some of the units you wrote down in exercise 3 are almost never used. Pick one that is not very useful and explain why it is less desirable than the others.
5. Make the following conversions:
    (a) 3.55 m to cm
    (b) 0.56 liter to ml
    (c) 452 kg to mg
    (d) 2,156 mm to cm
    (e) 4.0×10^3 km to m
    (f) 2.5×10^{-2} liter to cm^3
    (g) 454 g to kg
    (h) 2.45 mg to g

    Check your answers and put a special double circle around those that must be written in exponential form to avoid confusion about the number of significant figures.
6. If you had a problem about a swimming pool in which you were given length, width, and depth and asked questions about the water it could hold: (a) What relationship in the metric system would let you get from length units to volume units? (b) What relationship in the metric system would let you get from volume units to the mass of water the pool could hold?
7. Write down the "road map" plan for solving each of the following using the equalities given in Table 3–1.
    (a) 100 yd to m
    (b) 247.5 miles to km
    (c) 1 kg to lb
    (d) volume in ml of 12.5 lb of water
    (e) 1 yd^3 to liters
    (f) 20 gal to liters
8. Calculate the answers to the problems given in exercise 7.
9. How many significant figures are there in each of the following?
    (a) 3.55
    (b) 2.4×10^4 (which is 24,000)
    (c) 1.02×10^{-6} (which is 0.00000102)
    (d) 1.000×10^{10} (which is 10,000,000,000)
    (e) the answer from $(1.02 \times 10^{-6})(2.4 \times 10^4)$
    (f) the answer from $3.55 \div 1.000 \times 10^{10}$
    (g) the answer from $3.55 + 2.4 \times 10^4$
    (h) the answer from $3.55 + 8.65$

10. Write the formulas you would use for (a) converting °F to °C; (b) converting °C to °F; (c) converting °C to K.
11. Convert
    (a) 68°F to °C
    (b) 35°C to °F
    (c) 98.6°F to °C
    (d) 27°C to K
    (e) 50°F to K
    (f) 273°C to K
    (g) 400 K to °C

12. The density of mercury is 13.6 g/ml. Calculate the volume occupied by 500 g of mercury.
13. If 500 g of water at 25°C is heated to the boiling point of 100°C, how many calories of heat had to be supplied?

Test Yourself

1. What do each of the following prefixes mean:

    milli   centi   kilo

2. What unit is represented by each of the following abbreviations?

    m   kg   °C
    ml   cm   K

3. Convert 77°F to °C.
4. A road map lists the distance for a trip as 356 miles. How far is it in kilometers?
5. Show a reasonable plan for finding volume, in quarts, of 2.75 kg of water.
6. An automobile race is limited to cars with an engine displacement of 7.00 liters or less. Calculate the maximum allowed displacement in cubic inches.
7. Which of the following "set ups" should be used to convert 8.00 miles to kilometers? (*Given:* 1 in. = 2.54 cm, 5280 ft = 1 mile, 12 in. = 1 ft.)

    (a) $\dfrac{8.00 \times 5280 \times 12 \times 2.54}{100 \times 1000}$ (c) $\dfrac{5280 \times 2.54 \times 12}{8.00 \times 100 \times 10}$

    (b) $\dfrac{8.00 \times 5280 \times 12}{2.54 \times 100 \times 100}$ (d) $\dfrac{100 \times 1000 \times 2.54}{5280 \times 12 \times 8.00}$

    (e) None of these

8. If 2.0 g of NaCl (weighed to the nearest 0.1 g on a triple beam balance) is added to 0.379 g of KBr (weighed to the nearest 0.001 g on an analytical balance), which of the following is the best expression of the total sample weight?

    (a) 2.379 g   (c) 2.38 g   (e) 1.621 g
    (b) 2.4 g   (d) 2.5 g

9. Which of the following is the highest temperature?

    (a) 100°F   (d) 200°F
    (b) 250 K   (e) Two of the above are
    (c) 100°C   tied for the highest

10. Choose the best estimate of the mass of the water in a swimming pool, which is 25.0 yd long, 15.0 yd wide, and 5.00 ft deep at all points. (*Given:* 1 yd = 3 ft, 1 ft = 12 in., 1 in. = 2.54 cm.)

    (a) Less than 10^2 kg   (d) Between 10^4 and 10^5 kg
    (b) Between 10^2 and 10^3 kg   (e) More than 10^5 kg
    (c) Between 10^3 and 10^4 kg

^{12}C is now used instead of oxygen because individual oxygen atoms have masses significantly different from 16 atomic mass units. The value of 16 for oxygen used in early measurements was the *average* value for all oxygen atoms. There are three isotopes (types of atoms differing in mass) of oxygen in nature, so the average mass was not the mass of any one isotope. When a large number of atoms from any natural sample on earth are weighed together, the mixture of isotopes is always in the same proportion as their total abundances on earth (or very, very close to the same). The average mass of any large sample of oxygen atoms is therefore the average of all oxygen atoms on earth. When ^{12}C is set equal to exactly 12.000 atomic mass units, the average of all oxygen atoms is 15.9994 atomic mass units, a value close enough to 16 to permit us to use 16 in all problems of only moderate accuracy, such as all problems in this course.

> The mass of a single atom is its atomic weight in atomic mass units.

We will use the term **atomic weight** (abbreviated at. wt.) to indicate that these values are usually obtained by weighing and are, in fact, averages of all the atoms present. It would be more accurate to call these values atomic masses instead of atomic weights. They would still be the same in a gravitational field different from Earth, which would change weight. Because we are really concerned with comparison of relative values, the difference between mass and weight is of no practical importance to us here. The atomic weights of all the elements are listed in the table on the inside back cover of this book and again (above the symbols) in the periodic table on the inside front cover of this book.

> The atomic weight of an element is the average atomic weight of all atoms of that element.

4–2 Moles, Molecular Weights, and Formula Weights

Atomic mass units are not convenient for most real problems. Masses of single atoms are much smaller than the smallest masses measurable on balances or even the masses of the smallest objects that can be seen. Therefore, most real measurements are of very large numbers of atoms or molecules as a group. Comparisons between these groups can be made more easily if the atoms or molecules are counted in batches of about the size handled in most real problems. To do that, we must define a standard quantity to be used for comparisons. A particular number has been selected and put into widespread use. That particular number has been given the name **mole** and has been recommended for adoption as one of the fundamental SI units.

The mole was selected because (1) it (being a specific number) allows comparison of numbers of atoms, molecules, or other units, and (2) it allows use of convenient mass units (grams) related to SI instead of atomic mass units without giving up the atomic weight scale based on ^{12}C as 12.000. The mole is simply the number, 6.023×10^{23}, of ^{12}C atoms needed to have a mass of 12.00 g instead of 12.00 atomic mass units. (The exact number is known to more significant figures than the 6.023×10^{23} shown here, but we choose to stop at four significant figures to leave an easily memorized number that is accurate enough for all problems in this course.)

> The mole unit (6.022×10^{23}) is the unit usually used to count and compare quantities of atoms and molecules.

One mole of atoms is sometimes referred to as a **gram atomic weight.**

48 Arithmetic of Chemical Combination [4]

4 Arithmetic of Chemical Combination

Study Suggestions

Many students find the arithmetic of chemistry the most difficult part of the course. If you are in that group, this chapter may seem too concentrated at the first reading. Don't let that frighten you. If you attack the material by stages, you will find it is not as difficult as you first thought it was. First read through the chapter looking only for the types of things which can be accomplished but without trying to follow all the math. Then go back and try to pick out and master the easiest examples you find. Feel free to ask your instructor for help, even on simple points. Any experienced chemistry teacher expects such questions. As you develop skill with the simpler parts (such as molecular weight calculations) you will be able to handle the other parts your instructor assigned. Your instructor can also direct you to other sources of problems and guidance on those subjects he wishes to emphasize.

4–1 Atomic Weights

A scale of atomic weights, defined relative to ^{12}C as 12, is used in solving problems.

The laws of definite composition and multiple proportions (Chapter 2) were first formulated and stated in terms of weight relations. The observed weight relationships in compounds are related to differences in mass between atoms of different elements. A listing of the relative masses of the atoms of each element is one of our most useful tools in chemistry.

The scale of atomic masses went through some adjustments to make it as accurate and useful as possible. When the mass of the lightest element, hydrogen, is made about 1, many other atomic masses are also close to whole numbers. The most common element on earth, oxygen, was set as an exact whole number (16), so the atomic masses of the many other elements that combine with oxygen in compounds could be determined by direct measurements. That scale has now been replaced by an almost identical one that can be used even more accurately with modern instruments. Atomic masses are now defined relative to the mass of the isotope of carbon, ^{12}C, as exactly 12. A single atom of ^{12}C has a mass of 12.000 **atomic mass units.** Hydrogen atoms have a mass of about 1.008 atomic mass units, and oxygen atoms have a mass of about 16.0 atomic mass units.

Principles of Chemistry

The mass of 1 mole of atoms is the atomic weight in grams.

It is the quantity with a mass in grams equal to the numerical value of the atomic weight of that particular element. Therefore, 1 mole of hydrogen atoms has a mass of 1.008 g, 1 mole of oxygen atoms has a mass of 16.0 g, and 1 mole of each other element has a mass equal to its atomic weight value of grams.

However, most substances occur in units that are not single atoms. For instance, hydrogen and oxygen usually exist as diatomic molecules, H_2 and O_2. If we really wish to compare the number of units of hydrogen or oxygen, we should count the moles of molecules instead of atoms. That quantity, 1 mole of molecules, is sometimes referred to as a **gram molecular weight.** For H_2 molecules, 1 mole of molecules equals 2 moles of atoms and has a mass of 2 times the gram atomic weight. Therefore, 1 mole of H_2 has a mass of 2 times 1.008 g or 2.016 g. Similarly, 1 mole of O_2 has a mass of 2 times 16.0 g or 32.0 g.

Even when more than one element is involved, the gram molecular weight can be found by adding the atomic weights of all the atoms in each molecule. Because 1 mole of H_2O molecules has the mass of 2 moles of hydrogen atoms plus 1 mole of oxygen atoms, we must take 2 times 1.008 g plus 1 times 16.0 g for a total of 18.0 g. (If we used the more accurate 15.9994 g for 1 mole of oxygen atoms, we would have been able to state the answer to more significant figures, 18.015 g. You can see that our use of the value 16.0 g causes little change.)

The molecular weight is the sum of the atomic weights of the elements in a molecular substance with each atomic weight taken as many times as there are atoms of that kind in the molecular unit.

Gram molecular weights are related to the masses of molecular units in the same way that gram atomic weights are related to the masses of atoms. The numerical value by itself is called the **molecular weight.** The molecular weight of H_2 is 2.016, the molecular weight of O_2 is 16.0, and the molecular weight of H_2O is 18.0. These numbers are simply relative values compared to ^{12}C as exactly 12.000. If they are given the unit label "grams," then the groups being compared are each 1 mole; 1 mole of H_2 molecules (2.016 g) is the same number of units (6.023×10^{23}) as 1 mole of ^{12}C atoms (12.000 g). If the numerical value is given the unit label "atomic mass units" (amu), then the groups being compared are the average values for single units. The average H_2 molecule (2.016 amu), is the same number of units (one) as a single ^{12}C atom (12.000 amu). Because these relative numbers do not depend on the units being used, atomic weights can be summed to give molecular weights without using grams or limiting ourselves to mole quantities. For example, given the atomic weights of hydrogen and oxygen we have been using and the atomic weight of sulfur as 32.0, we can calculate the molecular weight of H_2SO_4 as 2×1.008 plus 1×32.0 plus 4×16.0 for a total of 98.0. Similar calculations will be the first step in most of the chemical arithmetic problems described in this chapter.

O_2 and H_2 are formulas for actual molecular units. Problems involving gases and some other problems require use of such molecular formulas. In other problems the answer depends only on the ratio of different kinds of atoms. In those cases we can use **empirical formulas** and their advantages are described in Section 2–6.

The dilemma in which we must sometimes have molecular formulas and at other times molecular formulas are unnecessary and unavailable is resolved by simply using the formula we write down in each case. If we need molecular formulas (or are working with compounds where the molecular formulas are used often enough to be in common use), we will write the molecular formulas. In other cases we will use empirical formulas. In every case we can add the atomic weights of the atoms in the formula to give a **formula weight**. Formula weights are obtained and used just like the molecular weights, and 1 mole of any formula could be called a gram formula weight. The formula weight of SiO_2 is 1×28.1 (from silicon) plus 2×16.0 (from oxygen) for a total of 60.1, and 60.1 g of SiO_2 is 1 mole of SiO_2. This number represents 3 moles of atoms and less than 1 mole of giant molecules, but it is 1 mole of SiO_2 formula units.

> The formula weight is the sum of the atomic weights of the elements in a pure substance with each atomic weight taken as many times as that element is shown to appear in the formula.

The term mole originally referred to a gram molecular weight of molecules. As the word mole was adopted for use with nonmolecular units, such as atoms, confusion was avoided by specifying which nonmolecular unit was being counted. When mole is used without specifying units, the usual kind of molecules are assumed. The statement, "1 mole of oxygen contains 2 moles of oxygen atoms," illustrates this convention. When no unit is specified ("1 mole of oxygen"), the usual molecules, O_2, are assumed to be the subject. If that is not so ("2 moles of oxygen *atoms*"), the unit must be stated. For reasons stated in the previous paragraphs, empirical formulas often must be used. Therefore, we will extend the use of mole (without stating a unit) to refer to the *common* formula used for a substance, with either the molecular formula (if it is known and used) or the empirical formula (if that is the only common formula) acceptable. That allows us to refer to "1 mole of sodium chloride" without needing to specify that we mean NaCl formula units. However, we are always free to put down the formula unit to which we are referring. In cases where both empirical and molecular formulas are used (such as P_2O_5 and P_4O_{10}), it is best to state the formula each time.

The formulas, empirical or molecular, used in solving problems show the relationships needed to solve the problems. Therefore, the formula weights become important quantities in almost every problem. Sometimes these formula weights are also molecular weights. That causes some confusion between formula and molecular weights. Both formula and molecular weights are also calculated by adding atomic weights of the atoms in the formula, so they often appear to be identical and interchangeable quantities. Sometimes they are "interchanged" incorrectly as in the expression "the molecular weight of NaCl is 58.5." NaCl has no molecules, so the 58.5 must be a formula weight. The best way to avoid such errors is to use "formula weight" consistently and simply take care to use the molecular formula whenever the true molecular weight is needed (as in gas problems).

4-3 Weight Relations in Formulas

Weight relationships from formula weight calculations can be used in many problems.

The fixed relationship (law of definite composition) in a compound's formula leads to weight relationships that can be found from the atomic and formula weights. One kind of information that can be found from a formula and atomic weights is the percentage composition. In H_2O, 16.0 out of each 18.02 weight units is oxygen. When that fraction is multiplied by 100%, we get the percentage of H_2O that is oxygen.

Example 4-1. Find the percent oxygen in H_2O.

$$\frac{16.0}{18.02} \times 100\% = 88.8\% \text{ O in } H_2O$$

The same thing can be done with hydrogen, using 2.016 (which is 2 times the atomic weight of hydrogen), out of each 18.02 (the formula weight of H_2O).

Example 4-2. Find the percent hydrogen in H_2O.

$$\frac{2.016}{18.02} \times 100\% = 11.2\% \text{ H in } H_2O$$

In every case we need 100% times the atomic weight of the element multiplied by the number of atoms of that element in the formula and divided by the formula weight. In order to have convenient units to work with, we often choose to use 1 mole of the compound and write the weights in grams. Gram units are not necessary here, but that approach is consistent with the pattern of using mole quantities, which we will find helpful in later problems.

Some examples of percentage composition problems follow.

Example 4-3. Find the per cent of sodium in $NaClO_3$.

Given: atomic weight of Na = 23.0; number of Na in formula = 1.

Then the formula weight of $NaClO_3$ is

$$\begin{aligned}
1 \text{ Na} &= 23.0 \\
1 \text{ Cl} &= 35.5 \\
3 \text{ O} = 3 \times 16.0 &= \underline{48.0} \\
&106.5
\end{aligned}$$

$$\frac{1 \times 23.0 \text{ g Na}}{106.5 \text{ g NaClO}_3} \times 100\% = 21.6\% \text{ Na}$$

Example 4-4. Find the per cent phosphorus in $(NH_4)_3PO_4$.

Given: atomic weight of P = 31.0; number of P in formula = 1.

Then the formula weight of $(NH_4)_3PO_4$ is

$$3\,N = 3 \times 14.0 = 42.0$$
$$3 \times 4\,H = 12 \times 1.008 = 12.1$$
$$1\,P = 1 \times 31.0 = 31.0$$
$$4\,O = 4 \times 16.0 = \underline{64.0}$$
$$149.1$$

$$\frac{31.0 \text{ g P}}{149.1 \text{ g (NH4)}_3\text{PO}_4} \times 100\% = 20.7\% \text{ P}$$

Example 4–5. Find the per cent nitrogen in $(NH_4)_3PO_4$.

There are three N's of atomic weight 14.0 so that $(NH_4)_3PO_4$ contains 42.0 g N per mole. Thus

$$\frac{42.0 \text{ g N}}{149.1 \text{ g (NH}_4)_3\text{PO}_4} \times 100\% = 28.1\% \text{ N}$$

Example 4–6. Sometimes percentage composition problems are complicated by having only part of the material as the compound containing the element whose percentage is being calculated. An example would be calculation of the percentage of copper in an ore that is 5.0% CuS. We then use the actual percentage (5% in this case) instead of 100%.

Problem. Find the percent copper in 5% CuS.

The formula weight of CuS would be

$$1\,Cu = 63.54$$
$$1\,S = \underline{32.06}$$
$$95.60$$

$$\frac{63.54 \text{ g Cu}}{95.60 \text{ g CuS}} \times 5.0\% \text{ CuS} = 3.3\% \text{ Cu}$$

Example 4–7. Sometimes the problems are further complicated by giving the amount of an element as the percentage that would have been there if it was all in some particular compound (which may not actually be there at all). The main nutrients in fertilizers are reported in this way. A typical example would be a material that was 42% $(NH_4)_3PO_4$. We could calculate the percentage phosphorus in the material by using the $P/(NH_4)_3PO_4$ numbers from our earlier example.

Problem. Find the per cent phosphorus in 42% $(NH_4)_3PO_4$.

$$\frac{31.0 \text{ g P}}{149.1 \text{ g (NH}_4)_3\text{PO}_4} \times 42\% \text{ (NH}_4)_3\text{PO}_4 = 8.7\% \text{ P}$$

However, phosphorus is normally given in percentages as P_2O_5, so we have to set up another formula weight relation and multiply by that factor.

Revised problem. Find the percent phosphorus (as P_2O_5) in 42% $(NH_4)_3PO_4$.

The formula weight of P_2O_5 is

$$2\,P = 2 \times 31.0 = 62.0$$
$$5\,O = 5 \times 10.6 = 80.0$$
$$142.0$$

so the factor needed is

$$\frac{142.0 \text{ g } P_2O_5}{62.0 \text{ g P}}$$

The whole problem is then

$$\frac{31.0 \text{ g P}}{149.1 \text{ g } (NH_4)_3PO_4} \times \frac{142.0 \text{ g } P_2O_5}{62.0 \text{ g P}} \times 42\% \ (NH_4)_3PO_4 = 20.0\% \ P_2O_5$$

Notice that our 42% $(NH_4)_3PO_4$ gets reported with a number (20.0% P as P_2O_5) which sounds about the same as the real phosphorus content of *pure* $(NH_4)_3PO_4$. To protect the public from deception, all manufacturers must report their numbers in the same way. Figure 4–1 shows the deception possible if different methods of reporting were used.

Example 4–8. Sometimes one element is present in more than one form. An example would be nitrogen in the case where 42% of the sample was $(NH_4)_3PO_4$ and the other 58% was $(NH_4)_2SO_4$. We can calculate the nitrogen from each source and add them together. In this case, fertilizers are given

Figure 4–1 Labeling standards.

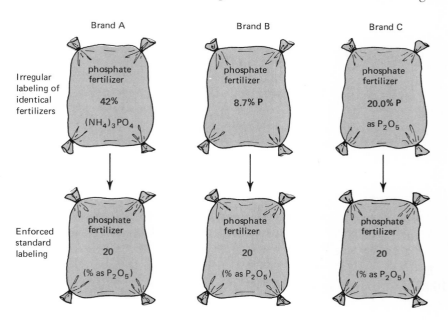

[4-3] Weight Relations in Formulas

in nitrogen as NH_3, so we will calculate it in that form. Check the formula weight-element fractions on your own.

Problem. Find the percent nitrogen (as NH_3) in a mixture of 42% $(NH_4)_3PO_4$ and 58% $(NH_4)_2SO_4$.

$$\frac{42.0 \text{ g N}}{149.1 \text{ g }(NH_4)_3PO_4} \times \frac{17.0 \text{ g }NH_3}{14.0 \text{ g N}} \times 42\% \ (NH_4)_3PO_4 = 14.4\% \text{ as } NH_3$$

$$\frac{28.0 \text{ g N}}{132.1 \text{ g }(NH_4)_2SO_4} \times \frac{17.0 \text{ g }NH_3}{14.0 \text{ g N}} \times 58\% \ (NH_4)_2SO_4 = 14.9\% \text{ as } NH_3$$

adding 14.4% and 14.9%, the total equals 29.3% as NH_3.

This fertilizer might be described as 29-20-0. The numbers refer to nitrogen as NH_3, phosphorus as P_2O_5, and potassium as K_2O. The bag would probably also carry (in small print) the percentage sulfur present. You should be able to calculate the percentage of sulfur (as S) now without any further information by using the methods we have been describing.

Formula weight relations also let us calculate the mass of one element from that of another element or the mass of the compound in which the elements are combined. Typical calculations are given in Examples 4–9 and 4–10.

Example 4–9. Calculate the mass of O_2 that will combine with 15.0 g carbon to form CO_2.

$$1 \text{ C} = 1 \times 12.0 = 12.0$$
$$2 \text{ O} = 2 \times 16.0 = 32.0$$

$$15.0 \text{ g C} \times \frac{32.0 \text{ g O}}{12.0 \text{ g C}} = 40.0 \text{ g O}$$

Example 4–10. Calculate the mass of iron in 85.0 tons of Fe_2O_3.

$$2 \text{ Fe} = 2 \times 55.8 = 111.6$$
$$3 \text{ O} = 3 \times 16.0 = \underline{\ 48.0}$$
$$1 \text{ Fe}_2O_3 = 159.6$$

$$85.0 \text{ tons Fe}_2O_3 \ \frac{111.6 \text{ g Fe}}{159.6 \text{ g Fe}_2O_3} = 59.4 \text{ tons Fe}$$

4–4 Formulas from Weights

Empirical formulas can be calculated from given weight compositions and from atomic weights.

Chemical formulas were originally worked out from data on weight relationships. We are often still faced with the same kind of calculation, and accurate atomic weight values let us get correct empirical formulas quite easily.

Example 4–11. Calculate the formula of a pure compound for which 19.81 g was found to contain 4.28 g sodium, 6.60 g chlorine, and 8.93 g oxygen.

Our first step is to convert each of these quantities into moles of atoms by dividing by the atomic weight in grams.

$$\frac{4.28 \text{ g Na}}{23.0 \text{ g Na/mole Na}} = 0.186 \text{ mole Na atoms}$$

$$\frac{6.60 \text{ g Cl}}{35.5 \text{ g Cl/mole Cl}} = 0.186 \text{ mole Cl atoms}$$

$$\frac{8.93 \text{ g O}}{16.0 \text{ g O/mole O}} = 0.558 \text{ mole O atoms}$$

Next, we make the smallest number 1 by dividing it by itself. We must also divide each other number of moles by that smallest number.

$$0.186 \text{ mole Na}/0.186 \text{ mole} = 1.00 \text{ Na atom}$$
$$0.186 \text{ mole Cl}/0.186 \text{ mole} = 1.00 \text{ Cl atom}$$
$$0.558 \text{ mole O}/0.186 \text{ mole} = 3.00 \text{ O atoms}$$

If all the numbers are now close to whole numbers, they give the formula. In this case it is $NaClO_3$. If one or more are far from whole numbers, we must look for the simple fraction they are closest to ($\frac{1}{2}$, $\frac{1}{3}$, $\frac{2}{3}$, $\frac{1}{4}$, and so on) and multiply everything by the whole number that makes the fraction a whole number. Remembering that many measurements are somewhat inaccurate, we ignore small differences. 1.04 is assumed to be 1.00 with an error in the data. 1.19 is too far off to assume it is 1 and would be assumed to be $\frac{6}{5}$, the nearest simple fraction.

Example 4–12. Often we start from percentages of the elements. One convenient approach in such cases is to change each percentage to an equal number of grams. That gives us a definite quantity (100 g) of the compound from which we can calculate the moles of atoms of each element. Given data on a compound that is (by weight) 43.3% phosphorus and 56.7% oxygen, the solution would be as follows.

Problem. Find the formula of a compound that is (by weight) 43.3% phosphorus and 56.7% oxygen.

$$\frac{43.3 \text{ g P}}{31.0 \text{ g P/mole P}} = 1.40 \text{ mole P}/1.40 \text{ mole} = 1.00 \text{ P}$$

$$\frac{56.7 \text{ g O}}{16.0 \text{ g O/mole O}} = 3.54 \text{ mole O}/1.40 \text{ mole} = 2.46 \text{ O}$$

2.50 is the closest simple fraction to 2.46, so we multiply each number by 2. Rounding the 4.92 off to 5, we conclude the compound is P_2O_5.

This method always gives us empirical formulas. In the last case the real molecular formula happens to be P_4O_{10}, but we simply were not given enough information to find that out.

Atomic and formula weights can also be used to solve problems involving chemical reactions. In this case the balanced equation for the reaction is needed. An example would be: calculate the mass of O_2 produced from 10.0 g $NaClO_3$ in the reaction

$$2 \text{ NaClO}_3 \longrightarrow 2 \text{ NaCl} + 3 \text{ O}_2$$

We will solve this problem twice, as Examples 4–13a and 4–13b.

In problems involving equations it is particularly convenient to convert to mole quantities. Because the number we call the mole can be used with any units, we can use it with events like chemical reactions as well as with atoms or molecules. We will use the term "mole of reaction" to refer to 1 mole of the events shown in an equation.

The equation in this problem shows 2 $NaClO_3$ units forming 2 NaCl units plus 3 O_2 molecules. Therefore 1 mole of the reaction would require 2 moles of $NaClO_3$ and give 2 moles of NaCl and 3 moles of O_2. This relationship can be used to set up **chemical equivalents,** which can be used to get unit factors for the calculations. In this reaction, 2 moles of $NaClO_3$ is chemically equivalent to 3 moles of O_2, so we can set 2 moles $NaClO_3$ = 3 moles O_2. From that equality we get the unit factors

$$\frac{2 \text{ moles NaClO}_3}{3 \text{ moles O}_2} \quad \text{and} \quad \frac{3 \text{ moles O}_2}{2 \text{ moles NaClO}_3}.$$

We could also calculate the relative masses of $NaClO_3$ and O_2 that are chemically equivalent and set up direct mass unit factors, but we will choose to go through the mole relationship, which is easy to see from the balanced equation.

Example 4–13a. To use moles we must break our problem down into three parts—converting $NaClO_3$ to moles, changing from $NaClO_3$ to O_2, and converting moles O_2 to the units needed in the answer. Using the formula weights calculated from atomic weights in the usual way (106.5 for $NaClO_3$ and 32.0 for O_2) our setup becomes

$$10.0 \text{ g NaClO}_3 \times \frac{1 \text{ mole NaClO}_3}{106.5 \text{ g NaClO}_3} \times \frac{3 \text{ moles O}_2}{2 \text{ moles NaClO}_3} \times \frac{32.0 \text{ g O}_2}{1 \text{ mole O}_2}$$
$$= 4.51 \text{ g O}_2$$

Example 4–13b. This particular problem could also be done as a weight relationship in the formula. Since all the oxygen in the $NaClO_3$ appears in the O_2 product, the mass of oxygen atoms in the $NaClO_3$ must equal the mass of O_2 that can be formed in the reaction. We chose to do it by the method for equations to introduce the general methods needed for other

4–5 Weight Relations in Equations

Problems on weight relations in equations use both formula weights and the coefficients of the balanced equation.

problems but still allow a check using the formula methods. The amount of oxygen atoms in 10.0 g NaClO$_3$, as shown by the formula, is

$$10.0 \text{ g NaClO}_3 \times \frac{48.0 \text{ g O}}{106.5 \text{ g NaClO}_3} = 4.51 \text{ g O}$$

Example 4-14. A more typical example, where the formula alone would not allow solution of the problem, is calculating the mass of CaO produced by heating 10.0 g CaCO$_3$.

Problem. Find the mass of CaO formed when 10.0 g CaO are decomposed by the reaction

$$\text{CaCO}_3 \longrightarrow \text{CaO} + \text{CO}_2.$$

The set-up becomes

$$10.0 \text{ g CaCO}_3 \times \frac{1 \text{ mole CaCO}_3}{100 \text{ g CaCO}_3} \times \frac{1 \text{ mole CaO}}{1 \text{ mole CaCO}_3} \times \frac{56.0 \text{ g CaO}}{1 \text{ mole CaO}}$$
$$= 5.60 \text{ g CaO}$$

Notice that two more formula weights had to be calculated and used. We will not write out any more examples of formula weight calculations, although they are used in almost every problem.

Example 4-15. In one of the possible problem variations, the equation weight relation can be used to calculate the extent of a chemical reaction.

Problem. Find what percentage of a pure CaCO$_3$ sample decomposed to CaO and CO$_2$ during heating if 22.50 g of CaCO$_3$ was reduced to 18.30 g of CaCO$_3$ plus CaO.

The weight lost (22.50 g − 18.30 g = 4.20 g) is the weight of CO$_2$ gas produced. We can use the equation to calculate how much CaCO$_3$ was needed to produce that amount of CO$_2$.

$$4.20 \text{ g CO}_2 \times \frac{1 \text{ mole CO}_2}{44.0 \text{ g CO}_2} \times \frac{1 \text{ mole CaCO}_3}{1 \text{ mole CO}_2} \times \frac{100 \text{ g CaCO}_3}{1 \text{ mole CaCO}_3}$$
$$= 9.55 \text{ g CaCO}_3 \text{ decomposed}$$

Since 22.50 g CaCO$_3$ was 100% of the original sample

$$9.55 \text{ g CaCO}_3 \text{ decomposed} \frac{100\% \text{ of sample}}{22.50 \text{ g CaCO}_3}$$

$$= 42.4\% \text{ of sample decomposed}$$

Example 4-16. When all of a compound reacts according to the given equation, the equation weight relation can be used to calculate how much was originally present in an impure mixture. An example (frequently used as a lab experiment) is calculating the percentage of NaClO$_3$ in a mixture of NaCl and NaClO$_3$.

Problem. If 12.51 g of sample weighs only 10.35 g after all the oxygen is driven off, what was the percentage of $NaClO_3$ in the sample?

$$(12.51 \text{ g} - 10.35 \text{ g}) \; O_2$$
$$\times \frac{1 \text{ mole } O_2}{32 \text{ g } O_2} \times \frac{2 \text{ moles } NaClO_3}{3 \text{ moles } O_2} \times \frac{106.5 \text{ g } NaClO_3}{1 \text{ mole } NaClO_3} \times \frac{100\% \text{ sample}}{12.51 \text{ g sample}}$$
$$= 2.16 \times \frac{1}{32} \times \frac{2}{3} \times \frac{106.5}{1} \times \frac{100\%}{12.51} \; NaClO_3 = 37.4\% \; NaClO_3$$

4-6 Molar Volume of Gases

All of the examples in the preceeding section involved equations in which gaseous products (O_2 or CO_2) were produced. In the laboratory it would be easier to measure the volume of these gases than their mass, and the volume of a gas is as effective a measure of quantity as mass. Measurement of gas quantities will be described in more detail in Chapter 9, where we will see the effects of changing pressure and temperature on the volume occupied by a given amount of gas. But at this point we will use the relationship between volume and quantity while pressure and temperature remain unchanged to show the value of the mole concept in gas problems.

In 1811 and Italian scientist, Amadeo Avogadro, made a suggestion which proved vital to later understanding of molecules and atomic and molecular weights. **Avogadro's hypothesis** can be stated: *when measured at the same temperature and pressure, equal volumes of different gases contain equal numbers of molecules*. This principle can be used to compare quantities of different gases. The numbers of molecules can be compared more conveniently by using mole quantities in samples of usual sizes. Relationships to moles also allow conversions between mass in grams and volume. Figure 4-2 shows Avogadro's hypothesis and some of the related logical principles.

Equal volumes of gases at the same conditions contain equal numbers of molecules.

Example 4-17. In the simplest applications of Avogadro's hypothesis, the quantity of one gas can be used to calculate the quantity of another gas. For example, a large new balloon for the Macy's Thanksgiving Parade may have been tested for leaks by filling it with nitrogen gas. If we know that 5.60×10^2 kg N_2 were required to fill the balloon, we can calculate the

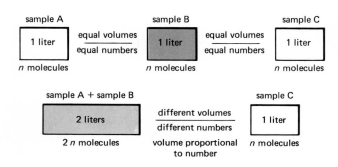

Figure 4-2 Avogadro's hypothesis.

mass of helium needed to fill it to the same pressure at the same temperature. We can convert the mass of N_2 to moles N_2 and later convert moles He to mass of He by the kind of formula weight relationship used in the previous sections. The conversion of moles N_2 to moles He is given by Avogadro's hypothesis—they must be equal to each other.

> **Problem.** Find the mass of helium needed to fill a balloon which holds 5.60×10^2 kg N_2 under the same conditions.

$$5.60 \times 10^2 \text{ kg } N_2$$
$$\times \frac{1000 \text{ g}}{1 \text{ kg}} \times \frac{1 \text{ mole } N_2}{28 \text{ g } N_2} \times \frac{1 \text{ mole He}}{1 \text{ mole } N_2} \times \frac{4.0 \text{ g He}}{1 \text{ mole He}} \times \frac{1 \text{ kg}}{1000 \text{ g}}$$
$$= 5.60 \times 10^2 \times \frac{4.0}{28} \text{ kg He} = 8.0 \times 10^1 \text{ kg He}$$

Example 4–18. Volume calculations need not be limited to cases where the volumes are equal. As noted in Figure 4–2, Avogadro's presumption of equal numbers in equal volumes leads to the conclusion that different numbers must occupy different volumes, with the volumes and numbers proportional. An application of that principle would be calculation of the mass of helium needed to bring one balloon of known volume to the same pressure (and temperature) as another balloon of known volume.

> **Problem.** Given that 80.0 kg of He was needed to fill a 1200-ft³ balloon, calculate the mass needed to fill a 3200-ft³ balloon to the same pressure at the same temperature. The calculation is simply

$$80.0 \text{ kg He in balloon A} \times \frac{3200 \text{ ft}^3 \text{ in balloon B}}{1200 \text{ ft}^3 \text{ in balloon A}} = 213 \text{ kg in balloon B}$$

Changes in volume can also be used to calculate the fraction of the gas of a particular kind. For example, if burning magnesium in air could be done under conditions that removed all the O_2 and nothing else, the percentage volume change would measure the percentage of the air molecules that were O_2. The volume of normal air would decrease by about 20.7% during such a reaction, because about 20.7% of the *molecules* in air are O_2. (*Note:* more than 20.7% of the mass is O_2, but the volume is proportional to moles, not mass.)

Volume relationships are more useful in problems when they can be directly compared to mass without comparison to a known mass-volume relationship. Chapter 9 will describe how this is done at all temperatures and pressures. However, there is one particular condition of temperature and pressure used as a standard. We will use the known relationship between volume and moles at that standard condition for some problems here and defer similar problems at nonstandard conditions until Chapter 9.

The condition of 0°C and 1.00 atmosphere (atm) pressure is defined as **standard temperature and pressure** and represented by the abbreviation

1 mole of any gas occupies 22.4 liters at STP, which is 0°C and 1.00 atm.

STP. (Chapter 9 defines the pressure unit "atmosphere" and other common units of pressure.) At STP, 1 mole of any gas occupies about 22.4 liters. (Chapter 9 describes some minor deviations from that value and their cause, but we can disregard them here.) That volume, 22.4 liters at STP, is described as the **molar volume** of a gas. It can be used to introduce the unit factors

$$\frac{1 \text{ mole}}{22.4 \text{ liters STP}} \quad \text{or} \quad \frac{22.4 \text{ liters STP}}{1 \text{ mole}}$$

for use in calculations.

Example 4–19. The molar volume at STP allows us to calculate the volume at STP for gases produced in reactions. For example, the $NaClO_3$ decomposition problem in Example 4–13 could be changed to ask for volume of the O_2 produced instead of mass.

Problem. What volume at STP of O_2 can be formed from 10.0 g of $NaClO_3$ in the reaction

$$2\ NaClO_3 \longrightarrow 2\ NaCl + 3\ O_2$$

The problem is solved in the same steps as those shown in Section 4–5 except for the last step, where the unit factor 22.4 liters STP/1 mole is used. The solution becomes

10.0 g $NaClO_3$

$$\times \frac{1 \text{ mole } NaClO_3}{106.5 \text{ } NaClO_3} \times \frac{3 \text{ moles } O_2}{2 \text{ moles } NaClO_3} \times \frac{22.4 \text{ liters STP } O_2}{1 \text{ mole } O_2}$$
$$= 3.15 \text{ liters STP of } O_2$$

Example 4–20. Similarly, we can calculate the volume of CO_2 at STP generated from 10.0 g $CaCO_3$ in the reaction $CaCO_3 \longrightarrow CaO + CO_2$ (instead of the calculation of mass of CaO done in Example 4–14).

Problem. Find the volume at STP of CO_2 formed by decomposition of 10.0 g $CaCO_3$.

$$10.0 \text{ g } CaCO_3 \times \frac{1 \text{ mole } CaCO_3}{100 \text{ g } CaCO_3} \times \frac{1 \text{ mole } CO_2}{1 \text{ mole } CaCO_3} \times \frac{22.4 \text{ liters STP } CO_2}{1 \text{ mole } CO_2}$$
$$= 2.24 \text{ liters STP of } CO_2$$

Example 4–21. Sometimes data from gas volumes simplifies problems. The example of partial reaction of $CaCO_3$ in Example 4–15 was complicated by the need to subtract one mass from another and convert to mass of CO_2, which was not directly stated in the problem. The same problem is more straightforward when stated directly in terms of the volume of CO_2 gas produced. It would then read and be solved as follows: Calculate the percentage of a 22.50 g sample of $CaCO_3$ that decomposed to CaO and CO_2 if the CO_2 gas produced has a volume at STP of 2138 ml.

$$\times \frac{\dfrac{1 \text{ liter}}{1000 \text{ ml}} \times \dfrac{1 \text{ mole}}{22.4 \text{ liters STP}} \times \dfrac{1 \text{ mole CaCO}_3 \text{ decomposed}}{1 \text{ mole CO}_2 \text{ formed}}}{22.50 \text{ g CaCO}_3 \dfrac{1 \text{ mole CaCO}_3}{100 \text{ g CaCO}_3}} \times 100\%$$

2138 ml STP of CO_2

Chapter 9 describes methods for converting gas volumes at nonstandard conditions to standard conditions or to the mole quantities needed to solve problems similar to the ones shown here.

4–7 Energy Relations in Equations

Energy released (or used) in chemical reactions is quantitatively related to the amounts of reacting materials.

The masses and volumes of the elements and compounds in a chemical reaction are not the only quantities with fixed relationships to each other. Energy, in the form of heat, is also a reactant or product of each chemical reaction, and the amount of heat is quantitatively related to the amount of chemical reaction. The amount of heat is stated in kilocalories (kcal, thousands of calories) per mole of reaction. For instance, the equation for the reaction of nitrogen with hydrogen could be written

$$N_2 + 3 H_2 \longrightarrow 2 NH_3 + 22.0 \text{ kcal/mole}$$

The heat produced per mole of reaction (as the equation is written) is also the heat produced per mole N_2 reacted; therefore, to eliminate any possible confusion the equation could have been written

$$N_2 + 3 H_2 \longrightarrow 2 NH_3 + 22.0 \text{ kcal/mole } N_2 \text{ reacted}$$

Energy relation problems have three types of complications not usually seen in weight relation problems.

Energy relations in equations are complicated by three main types of problems not usually present in weight relation problems.

First, the equations are sometimes balanced using fractions instead of whole numbers. The N_2-H_2 reaction is often written

$$\tfrac{1}{2} N_2 + \tfrac{3}{2} H_2 \longrightarrow NH_3 + 11.0 \text{ kcal/mole } NH_3$$

Because this is the reaction for formation of NH_3, the heat is called the **heat of formation** of NH_3. We like to state the energy per mole of the NH_3 for which it is the heat of formation. This also lets us save space by writing simply kilocalories per mole (meaning kilocalories per mole of reaction as written) and still have the heat per mole of the product we are talking about, NH_3. Since we always work with mole quantities, $\tfrac{1}{2} N_2$ and $\tfrac{3}{2} H_2$ are acceptable numbers. One-half mole of N_2 is still a very large number of N_2 molecules, so no break-up of N_2 molecules (which would change them to something very different) is necessary.

Second, it is necessary to state whether each substance is a liquid, solid, or gas. Energy would be required or would be released if any substance was changed from one form to another, so we must know which form we have before we can say how much energy is released (or needed) by the reaction.

We use the abbreviations (g) for gas, (l) for liquid, and (s) for solid behind each substance to indicate its form. The N_2-H_2 reaction involves only gases, so it is written

$$\tfrac{1}{2} N_2(g) + \tfrac{3}{2} H_2(g) \longrightarrow NH_3(g) + 11.0 \text{ kcal/mole}$$

Third, the heat is often given a name and symbol from thermodynamics. The heat of reaction under the conditions normally used for chemical reactions is the thermodynamic term called change in enthalpy. It is given the symbol ΔH (where Δ is the Greek letter "delta" which is used as a symbol for "change in"). In Appendix C, Section C–1, we discuss how ΔH interacts with the tendency toward increasing entropy (Chapter 11) to determine which reactions occur. In thermodynamics it is customary to define ΔH as the heat being put in, so ΔH has the opposite sign from the heat we have been writing as a product (coming out) of reaction. The complete N_2-H_2 reaction would therefore usually be written

$$\tfrac{1}{2} N_2(g) + \tfrac{3}{2} H_2(g) \longrightarrow NH_3(g) \qquad \Delta H = -11.0 \text{ kcal/mole}$$

Because this particular heat is the heat of formation of a compound (in its usual form, a gas) from elements (in their usual forms, gases), it is given a special symbol, ΔH_f. This special form, which must be referred to that particular equation and conditions, is handy for making up tables of facts. There are large compilations of such facts, and in them we would find the entry: $\Delta H_f \, NH_3 = -11.0$ kcal/mole. The values of ΔH_f change somewhat with temperature change. Most tables of data and all examples used in this book are for reactions at 25°C.

Sometimes a reaction can be broken down into two or more parts. If so, the energy related to that reaction is also broken down into parts. However, just as the number of atoms cannot be changed by any series of reactions, the total amount of energy cannot be changed by any choice of the way a reaction is completed. The energies for the parts must add up to the energy for the whole regardless of the reaction path chosen. This statement is called **Hess's law.** To make it work we must add (or subtract) reactions together using correct coefficients to make the whole reaction. Here are two examples with diagrams in Figure 4–3 to illustrate the steps.

The energy released (or used) in reaction is not affected by the choice of reaction path (Hess's law).

Example 4–22a. Find ΔH for equation 4–3 from the data given for equations 4–1 and 4–2.

$$H_2(g) + \tfrac{1}{2} O_2(g) \longrightarrow H_2O(g) \qquad \Delta H = -57.8 \text{ kcal/mole} \qquad (4\text{–}1)$$
$$H_2(g) + \tfrac{1}{2} O_2(g) \longrightarrow H_2O(l) \qquad \Delta H = -68.3 \text{ kcal/mole} \qquad (4\text{–}2)$$
$$H_2O(l) \longrightarrow H_2O(g) \qquad (4\text{–}3)$$

Equation (4–3) results when equation (4–2) is *subtracted* from equation (4–1), so ΔH for $H_2O(l) \longrightarrow H_2O(g)$ must be ΔH for equation (4–1) minus ΔH for equation (4–2).

Equations can be combined by addition or subtraction, with amounts adjusted by multiplying through by coefficients. The overall heat of reaction can be obtained by combining the heats of reaction in exactly the same sequence of additions, subtractions, and multiplication by coefficients as the equations.

Figure 4–3 Hess's law cycles.

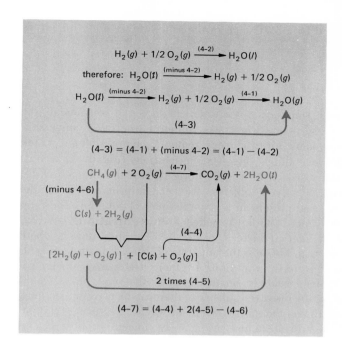

$$\Delta H = -57.8 \text{ kcal/mole} - (-68.3 \text{ kcal/mole}) = +10.5 \text{ kcal/mole}$$

for $H_2O(l) \longrightarrow H_2O(g)$

Example 4–22b. Find ΔH for equation 4–7 from the data given for equations 4–4, 4–5, and 4–6.

$$C(s) + O_2(g) \longrightarrow CO_2(g) \quad \Delta H = -94.1 \text{ kcal/mole} \quad (4\text{–}4)$$
$$H_2(g) + \tfrac{1}{2} O_2(g) \longrightarrow H_2O(l) \quad \Delta H = -68.3 \text{ kcal/mole} \quad (4\text{–}5)$$
$$C(s) + 2 H_2(g) \longrightarrow CH_4(g) \quad \Delta H = -17.9 \text{ kcal/mole} \quad (4\text{–}6)$$
$$CH_4(g) + 2 O_2(g) \longrightarrow CO_2(g) + 2 H_2O(l) \quad \Delta H = ? \quad (4\text{–}7)$$

Then

$$(4\text{–}7) = (4\text{–}4) + 2(4\text{–}5) - (4\text{–}6)$$

(You can check this.)

$$\Delta H \,(4\text{–}7) = \Delta H \,(4\text{–}4) + 2\, \Delta H \,(4\text{–}5) - \Delta H \,(4\text{–}6)$$

$\Delta H \,(4\text{–}7) = -94.1 \text{ kcal/mole} + 2\,(-68.3 \text{ kcal/mole}) - (-17.9 \text{ kcal/mole})$
$\Delta H \,(4\text{–}7) = -94.1 \text{ kcal/mole} - 136.6 \text{ kcal/mole} + 17.9 \text{ kcal/mole}$
$\Delta H = -212.8 \text{ kcal/mole } CH_4 \text{ burned in reaction 4–7}$

Problems like the last one are not as frequent nor as important to us as weight relations, but Hess's law and the concept that heats of reaction can be added, subtracted, or multiplied by a coefficient in the same way as other quantities in the equations are significant.

We have already shown some examples where mass is the convenient way to measure quantities and others where gas volume is the more convenient measure. A third convenient form of measurement is liquid volume. Often we have reactions between things dissolved in solutions (usually in water). The volume of solution used is the easiest thing to measure in such cases, but that is not enough to solve problems. We need information about the amount of reactive material present in a given amount of solution. The ratio of dissolved material to total solution is called a **concentration.**

Although there are as many possible types of concentration units as there are combinations of ways to measure the amounts of dissolved substance and solution, a few types of concentration units are more useful than others. We can see from our problem examples in this chapter that mole units would be particularly useful for the dissolved substance that may be used in reactions. The volume of the solution is the most convenient way of measuring the total quantity. Therefore, we will use almost exclusively concentrations stated in units we call **molarity,** abbreviated *M*, which consist of moles of dissolved substance per liter of solution. In Chapter 10 we will discuss solutions in more detail and use molarity (and two other types of concentration units) in a number of problems.

Examples 4–23 through 4–27 will illustrate how concentrations in molarity can be related to weight or volume in problems.

4–8 Quantities in Solution

Molarity, moles per liter, is a convenient concentration unit for use in calculations.

Example 4–23. Calculate the molarity of NaCl in a solution made by dissolving 2.00 moles of NaCl in water to make 5.00 liters of solution.

$$\text{molarity} = \frac{\text{moles}}{\text{liter}}$$

$$M \text{ NaCl} = \frac{\text{moles NaCl}}{\text{liter solution}}$$

$$\frac{2.00 \text{ moles NaCl}}{5.00 \text{ liters solution}} = \frac{0.400 \text{ moles NaCl}}{1.00 \text{ liter solution}} = 0.400 \ M \text{ NaCl}$$

Example 4–24. Calculate the mass, in grams, of Na_2SO_4 that must be dissolved in water to form 500 ml of 0.250 M Na_2SO_4 solution.

This problem can be set up as a conversion of units. We know the volume of solution, the molarity of Na_2SO_4 in the solution, and the formula for Na_2SO_4. We want the mass in grams of the Na_2SO_4. The formula weight gives us a conversion factor between units of mass and moles. Molarity can be used as a similar conversion factor between moles and volume of solution.

Those two factors plus the relationship of milliliters to liters will allow us to convert the given volume of solution to mass as follows:

$$500 \text{ ml soln.} \times \frac{1 \text{ liter soln.}}{1000 \text{ ml soln.}} \times \frac{0.250 \text{ moles Na}_2\text{SO}_4}{1 \text{ liter soln.}} \times \frac{142 \text{ g Na}_2\text{SO}_4}{1 \text{ mole Na}_2\text{SO}_4}$$
$$= 17.8 \text{ g Na}_2\text{SO}_4$$

Figure 4–4 shows how this solution could actually be prepared in the laboratory.

Example 4–25. Calculate the molarity of KCl in a solution made by dissolving 14.90 g KCl in water and then filling a 250-ml volumetric flask to the mark.

This problem (the most common type in making up laboratory solutions) involves the same factors as Example 4–24 but a slightly different arrangement.

$$\frac{14.90 \text{ g KCl}}{250 \text{ ml soln.}} \times \frac{1 \text{ mole KCl}}{74.5 \text{ g KCl}} \times \frac{1000 \text{ ml soln.}}{1 \text{ liter soln.}} = 0.800 \, M \text{ KCl}$$

Check the cancellation of units (as you should check your own calculations) to confirm that the answer is in molarity, moles KCl/liter solution.

Example 4–26. Calculate the volume of 6.0 M NH$_3$ solution that can be made from 5.60 liters of H$_2$ gas at STP if it is completely used up by

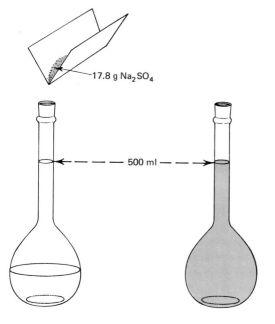

Figure 4–4 Preparation of an 0.250 M Na$_2$SO$_4$ solution. [from R. H. Petrucci: *General Chemistry.* The Macmillan Company, New York, 1972.]

reaction with excess N_2 in the reaction $N_2 + 3 H_2 \longrightarrow 2 NH_3$. The answer is

$$5.60 \text{ liters STP of } H_2 \times \frac{1 \text{ mole}}{22.4 \text{ liters STP}} \times \frac{2 \text{ moles NH}_3}{3 \text{ moles H}_2} = \frac{1}{6} \text{ mole NH}_3$$

$$\frac{\frac{1}{6} \text{ mole NH}_3}{\text{volume solution}} = 6.0 \; M$$

$$\text{volume} = \frac{\frac{1}{6} \text{ mole NH}_3}{\frac{6.0 \text{ moles NH}_3}{\text{liter}}}$$

$$\text{volume} = \tfrac{1}{36} \text{ liter}$$

Example 4-27. Calculate the mass of $H_2C_2O_4$ that would react with 250 ml of 6.0 M NH_3 in the reaction

$$H_2C_2O_4 + 2 NH_3 \longrightarrow (NH_4)_2C_2O_4$$

$$250 \text{ ml} \times \frac{1 \text{ liter}}{1000 \text{ ml}} \times \frac{6.0 \text{ moles NH}_3}{1 \text{ liter}} \times \frac{1 \text{ mole H}_2C_2O_4}{2 \text{ moles NH}_3} \times \frac{90.0 \text{ g H}_2C_2O_4}{1 \text{ mole H}_2C_2O_4}$$

$$= 67.5 \text{ g H}_2C_2O_4$$

If your lab or other exercises require more understanding of concentrations than these brief examples provide, you should refer to Chapter 10.

Skills Expected at This Point

1. You should be able to work the exercises at the end of this chapter.
2. You should be able to work any set of problems of similar complexity and on the same general topics as the problems at the end of this chapter.
3. You should be able to make up good problems of each of the types shown in the chapter and the exercises. You should be able to supply yourself and your friends with good extra drill problems.

Exercises

Sections 4-1, Atomic Weights, and 4-2, Moles, Molecular Weights, and Formula Weights.

1. What is the official standard for atomic weights?
2. The isotope ^{35}Cl has an atomic weight of 34.97. **(a)** What is the mass of one atom of ^{35}Cl? **(b)** What is the mass of 1 mole of ^{35}Cl atoms? **(c)** How many ^{35}Cl atoms are there in 1 mole of ^{35}Cl atoms? **(d)** How many ^{35}Cl atoms are there in 1 g of ^{35}Cl atoms?

3. Natural chlorine is 75.4% ^{35}Cl (at. wt. 34.97) and 24.6% ^{37}Cl (at. wt. 36.97). **(a)** What does the statement "atomic weight of chlorine" mean? **(b)** Calculate the atomic weight of chlorine. Show clearly the method you are using. *Note:* No example of such a calculation was done for you, but you have enough information here to plan a way to do it and carry it out. Think about your answer to part (a) and use it.
4. Calculate the formula weight of each of the following. (You may need values for atomic weights that are listed on the inside back cover of the book.) **(a)** CO_2; **(b)** $KHSO_4$; **(c)** $KAl(SiO_3)_2$; **(d)** NH_4NO_3; **(e)** $Ca_3(PO_4)^2$.

Section 4–3, Weight Relations in Formulas.
5. Calculate the percentage of **(a)** potassium in KOH; **(b)** potassium in $KAl(SiO_3)_2$; **(c)** calcium in 5.00% $CaCO_3$, 95.00% SiO_2; **(d)** potassium as K_2O in 10.0% K_2HPO_4, 90.0% $(NH_4)_2HPO_4$.
6. Calculate the mass of **(a)** sulfur in 25.0 g H_2SO_4; **(b)** aluminum in 56.3 g Na_3AlF_6.

Section 4–4, Formulas from Weights.
7. Calculate the empirical formula of a compound containing 18.56 g magnesium and 7.54 g nitrogen.
8. Calculate the empirical formula of a compound found to be 30.3% calcium, 21.2% nitrogen, and 48.5% oxygen.

Section 4–5, Weight Relations in Equations, and Section 4–6, Molar Volume of Gases.
9. What is the meaning of "chemical equivalence?"
10. How many grams of $NaClO_3$ can be produced from 4 moles of Cl_2? The reaction is

$$6\ NaOH + 3\ Cl_2 \longrightarrow 5\ NaCl + NaClO_3 + 3\ H_2O$$

11. How many grams of aluminum can be made from 85.5 g Al_2O_3? The reaction is

$$2\ Al_2O_3 \longrightarrow 4\ Al + 3\ O_2$$

12. 2.54 tons of iron were produced from the complete reaction of Fe_2O_3 by the equation $Fe_2O_3 + 3\ C \longrightarrow 2\ Fe + 3\ CO$. If the starting material was 5.25 tons of ore in which Fe_2O_3 was the only part containing iron, calculate the percentage of Fe_2O_3 in the ore.
13. **(a)** Calculate the volume at STP of the O_2 gas produced from 85.5 g Al_2O_3 in exercise 11. **(b)** Calculate the volume at STP of the CO gas produced along with 2.54 tons iron in exercise 12.

Section 4–7, Energy Relations in Equations.
14. Briefly state Hess's law.
15. Heat is given off when coal burns in air. What is the sign (+ or −) of ΔH for the reaction? Why?

16. Given:

$$C_2H_2(g) + 2 H_2(g) \longrightarrow C_2H_6(g) \quad \Delta H = -74.4 \text{ kcal/mole}$$
$$C_2H_4(g) + H_2(g) \longrightarrow C_2H_6(g) \quad \Delta H = -32.7 \text{ kcal/mole}$$

Calculate ΔH for

$$C_2H_2(g) + H_2(g) \longrightarrow C_2H_4(g)$$

Section 4–8, Quantities in Solution.

17. How many grams of NaCl must be dissolved in 500 ml of solution to give 3.0 M NaCl solution?

Test Yourself

1. Calculate the percentage (by mass) of sulfur in H_2SO_4.
2. Calculate the mass of H_2O produced when 78 g of C_6H_6 are burned in the reaction

$$2 C_6H_6 + 15 O_2 \longrightarrow 12 CO_2 + 6 H_2O$$

(an excess of O_2 is available).
3. Calculate the volume of O_2 at STP needed to react completely with 78 g of C_6H_6 in the reaction shown in problem 2.
4. The best approximation of molecular weight of H_2CO_3 is
   - (a) 29
   - (b) 50
   - (c) 62
   - (d) 87
   - (e) 90
5. Which of the following statements concerning the mole is *incorrect?*
   - (a) A mole contains 6.022×10^{23} molecules.
   - (b) At 0°C and 1 atm, a mole of a gas occupies 22.4 liters.
   - (c) The mole is a conversion factor between molecular dimensions and laboratory dimensions.
   - (d) A mole of a compound has a weight in grams numerically equivalent to its formula weight.
   - (e) A mole of one element has as much mass as a mole of another element.
6. Chlorine reacts with NaOH solution by the reaction

$$3 Cl_2 + 6 NaOH \longrightarrow 5 NaCl + NaClO_3 + 3 H_2O$$

(At. wt.: Na = 23, Cl = 35.5) The mass of $NaClO_3$ which can be made from 10.0 g Cl_2 is given by

- (a) $10.0 \times \dfrac{1}{35.5} \times \dfrac{5}{3} \times \dfrac{58.5}{1}$
- (b) $10.0 \times \dfrac{1}{71.0} \times \dfrac{1}{3} \times \dfrac{106.5}{1}$
- (c) $10.0 \times \dfrac{1}{35.5} \times \dfrac{106.5}{3}$
- (d) $10.0 \times \dfrac{1}{71.0} \times \dfrac{106.5}{1}$
- (e) $10.0 \times \dfrac{1}{35.5} \times \dfrac{5}{3}$

7. What volume of O_2 gas at STP can be produced by decomposing 12.25 g $KClO_3$ by the reaction $2\ KClO_3 \longrightarrow 2\ KCl + 3\ O_2$? (at. wt. K = 39, Cl = 35.5)
   (a) 3.36 liters
   (b) 11.2 liters
   (c) 7.72 liters
   (d) 2.24 liters
   (e) 0.1 liter

8. According to the equation $N_2 + 3\ H_2 \longrightarrow 2\ NH_3$, how many liters of NH_3 gas could be produced by the reaction of 2 liters of nitrogen and 6 liters of hydrogen?
   (a) 6
   (b) 8
   (c) 4
   (d) 2
   (e) 5

9. What is the mass of 1 liter of ethane (C_2H_6) gas at 0°C and 1 atm?
   (a) $\dfrac{30 \times 1}{22.4} = 1.34$ g
   (b) $\dfrac{22.4}{30 \times 0.082} = 91$ g
   (c) $\dfrac{273.2}{22.4 \times 15} = 0.814$ g
   (d) $\dfrac{1 \times 30}{273.2} = 0.11$ g
   (e) $\dfrac{22.4}{30 \times 1} = 0.75$ g

10. Which of the following statements may *not* be inferred from the chemical equation

    $$2\ CH_3OH + 3\ O_2 \longrightarrow 2\ CO_2 + 4\ H_2O$$

    (a) 2 moles of CH_3OH is exactly consumed by 3 moles of oxygen.
    (b) 3 liters of O_2 gas will yield 2 liters of CO_2 (measured at the same pressure and temperature).
    (c) 2 g of CH_3OH will yield 4 g of water.
    (d) one molecule of CO_2 is produced from one molecule of CH_3OH.
    (e) 1 mole of CH_3OH will require $1\frac{1}{2}$ moles of O_2 for combustion

11. According to the equation given in problem 10, how many liters of CO_2 (at standard temperature and pressure) will result from the combustion of 10 g of CH_3OH?
    (a) $\dfrac{10 \times 2 \times 22.4}{32 \times 2} = 7.0$ liters
    (b) $\dfrac{10 \times 22.4}{44} = 5.1$ liters
    (c) $\dfrac{10 \times 22.4}{2 \times 33 \times 2} = 1.75$ liters
    (d) $\dfrac{10 \times 32 \times 2}{22.4 \times 2} = 14.3$ liters
    (e) $\dfrac{10 \times 32}{22.4 \times 2} = 7.15$ liters

12. Again, from the same equation, if 10 g of CH_3OH were reacted with 11.2 liters of O_2 (0°C, 1 atm) until one of the reagents were exhausted, which of the following would be true?
    (a) $11.2 - \dfrac{10 \times 22.4}{32} = 4.2$ liters O_2 will remain
    (b) $11.2 - \dfrac{10 \times 3 \times 22.4}{32 \times 2} = 0.7$ liters O_2 will remain

(c) $11.2 - \dfrac{10 \times 2 \times 22.4}{32 \times 3} = 5.9$ liters O_2 will remain

(d) $11.2 - \dfrac{10 \times 2 \times 32}{3 \times 22.4} = 1.7$ g CH_3OH will remain

(e) $11.2 - \dfrac{10 \times 22.4}{32} = 4.2$ g CH_3OH will remain

13. When 1.50 g of $H_2C_2O_4 \cdot 2H_2O$ are dissolved in 250 ml, the concentration of the solution is
    (a) $0.5\ M$
    (b) $0.05\ M$
    (c) $0.0666\ M$
    (d) $0.1666\ M$
    (e) $0.0476\ M$

14. If 20.8 g of chromium combines with 9.6 g of oxygen to form 30.4 g of a certain oxide, what is the simplest formula for that oxide? (At. wt. $Cr = 52$)
    (a) CrO
    (b) Cr_2O_3
    (c) CrO_2
    (d) Cr_2O
    (e) none of the above

15. The most common uranium oxide has the formula U_3O_8. What is the percentage (by weight) of uranium in U_3O_8? (At. wt. $U = 238$)
    (a) 27.3%
    (b) 37.5%
    (c) 71.3%
    (d) 84.8%
    (e) 93.7%

16. Form a study partnership with another member of the class. Write a set of problems and have your partner write a separate set of problems. Exchange problem sets. Can you work the problems your partner wrote? Is the set you wrote fair (without being too simple) and reasonably complete in covering the principles in the chapter?

5 Into the Heart of Matter

5–1 Using Experiments as Our Guide

Understanding atoms is the most efficient way to understand matter.

Weight relations and the atomic theory have shown us that matter can be explained by the combinations of a limited number of elements. The properties of those elements are determined by their fundamental parts, atoms. Because there are many less elements than compounds, an understanding of the nature and behavior of atoms provides by far the easiest and most efficient way to come to an understanding of the matter from which we and our surroundings are made. The necessary information about atoms is available to us. A series of experiments in the late nineteenth and early twentieth centuries were particularly crucial to understanding atoms and will be discussed in this chapter.

A systematic analysis of the parts can make complex experiments easier to understand.

Because the concepts shown by these experiments are so important, we are going to try to break them down into their key points. By separating the points and emphasizing what is important we can give you small, easy to learn parts that lead to the important facts. This approach has helped a large number of students master these concepts.

The following study guides should help you approach this material effectively.

Steps for efficient study of experiments are listed.

1. Begin your study of experiment descriptions by asking yourself the following questions.
   (a) How do I pick out the important facts from descriptions?
   (b) What am I supposed to be learning from this material? (In this chapter question (b) can be answered by (1) to learn something about electrons and other matter, (2) to learn to observe experiments, and (3) to learn how knowledge has been accumulated.)
2. Recognize that you cannot reach real understanding by either (a) memorizing results alone or (b) memorizing details alone.
3. Memorize enough details in each experiment to know what happened.
4. Search for the key points and results.
5. Think about how these results are related to your earlier information.

5-2 Electricity

Even before Dalton's atomic theory, electricity was known to exist and to flow. Benjamin Franklin's experiments with electricity, beginning in the 1740s, contributed to the concept of flow of electricity. His famous experiment with a kite in lightning could be broken into details [(1) a kite was flown in a cloud where there was obvious lightning activity and (2) a key on the kite string was used to attract the lightning], a result (the lightning could be guided down the kite string), and eventual conclusions. Franklin proposed a flow of electrical fluid in a particular direction so people could talk about the flow of electricity more precisely. He made it clear that his choice of positive and negative regions between which there was a flow of "electric fluid" was arbitrary. Franklin's choices of positive and negative are still used, and will be used in this chapter to refer to positive or negative electrodes in various experiments. Later workers were therefore able to test and improve the idea without facing arguments that they were violating the known facts. They found that actual flow was of negative charges moving in the opposite direction from Franklin's "electric fluid." The smooth progress that resulted is strikingly different from the long delays caused by some other early scientific arbitrary choices, such as the phlogiston theory, which were accepted and defended as "known facts."

Electricity flows with a direction to its flow.

In the early nineteenth century Michael Faraday passed electricity through solutions, causing chemical reactions. He measured the amounts of electricity and chemical reaction and found they were related to each other. These results laid the foundation for later explanation of chemical reactions by electrical properties.

The amount of chemical reaction in electrolysis is proportional to the amount of electricity that flows.

5-3 Cathode Ray Tubes

As experimentation with electricity continued, various ways were discovered to cause glowing in tubes. The tubes of greatest interest in explaining matter are called cathode ray tubes. These are sealed tubes containing only low pressures of gas that are subjected to high electrical voltages (a large difference between the more positive and the more negative) between two separate pieces of metal called electrodes. In these tubes rays were found to move from the **cathode** (negative electrode) to the **anode** (positive electrode). The rays coming from the cathode were called cathode rays. Such tubes can be made in various shapes and arrangements. We will describe experiments in three different cathode ray tubes.

When high voltage is applied to electrodes in a vacuum tube, cathode rays flow from the cathode to the anode.

Figure 5-1 shows a shadow-casting cathode ray tube. The details necessary for this tube to work are

1. It is a sealed vacuum tube (very little air pressure left).
2. A high voltage is applied to the two metal electrodes in the tube. (Every cathode ray tube has the above two features.)
3. The electrode that is given the positive voltage (positive as defined by Franklin) has an identifiable shape and does not cover the whole end of the tube.

Figure 5–1 Shadow-casting cathode ray tube.

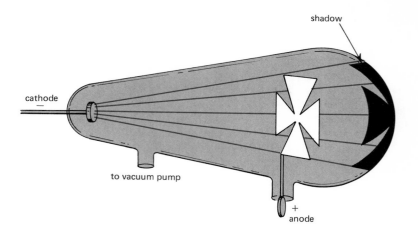

Cathode rays missing the anode cause the tube walls to glow, and the anode casts a sharp shadow.

When the voltage is turned on, the following points are observed.

1. Some cathode rays pass by the positive electrode and hit the wall. They make the glass tube glow where they hit the wall.
2. The electrode cuts off the rays hitting it, thus casting a shadow.
3. The shadow has sharp edges and the same shape as the electrode.

The tube shows us the following points:

1. The flow of rays is from the negative electrode toward the positive electrode.
2. The flow is in straight lines (as shown by the sharp edge on the shadow).
3. The flow past the target electrode to the wall implies fast flow—not enough time to change direction.

The second cathode ray tube, in Figure 5–2, has a solid barrier (blocking the cathode rays) with a slit in it. It also has a slanted piece of material beyond the slitted barrier on which a thin line is made to glow by the stream of cathode rays passing through the slit. Figure 5–3 shows a view from above

Figure 5–2 Cathode ray tube with slit, side view.

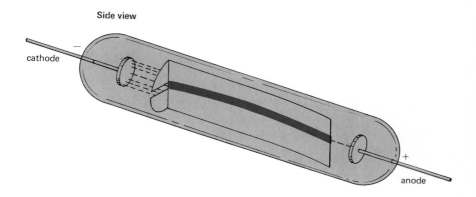

[5–3] Cathode Ray Tubes   73

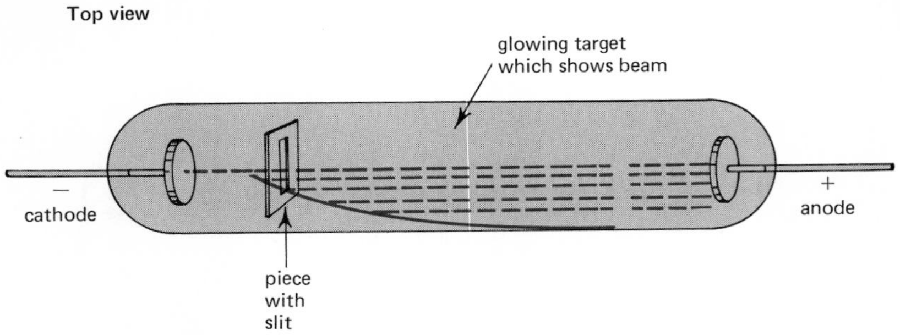

Figure 5-3 Cathode ray tube with slit, top view.

so you can see how the tube works. Voltage is applied to this tube and, while it is running, a magnet is brought up and then taken away.

The important details in this experiment are

1. This tube can form a thin beam of cathode rays.
2. The path of that beam of cathode rays is shown by the glow on the target.

Figure 5-4 shows the beam without a magnet present, and Figure 5-5 shows the beam with a magnet present. The points demonstrated by this tube are

1. It confirms the negative to positive direction and straight line flow shown by the shadow-casting tube.
2. It shows the beam can be bent by a magnetic field.
3. Where the rays hit the glass tube they give the same color and type of glow as the shadow-casting tube (and every other cathode ray tube).

A beam of cathode rays normally moves in a straight line, but the path can be bent by a magnetic field.

The observations made upon these two cathode ray tubes lead to the following conclusions.

1. Cathode rays are moving, electrically charged particles (shown by their being deflected by a magnetic field).
2. The electrical charge is negative (shown by the flow from negative to positive).
3. The negative particles appear to be the same in every case.

Cathode rays are moving, negatively charged particles of the same kind in every case.

A third cathode ray tube is shown in Figure 5-6. The important details about this tube are

1. It has holes or slits in the cathode.

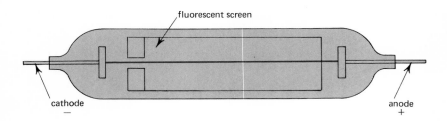

Figure 5-4 Cathode ray tube with slit and no magnet.

Figure 5-5 The effect of a magnet on a Cathode ray tube with slit.

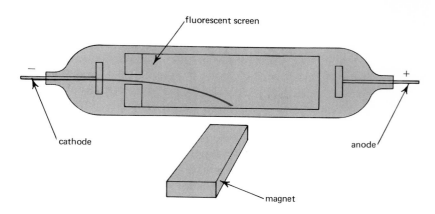

2. It has a large space behind the cathode.
3. It has a low pressure of a particular kind of gas.

The kind of gas can be changed in making up different tubes of this same design.

The points observed when voltage is applied are

1. Something passes through the holes in the cathode and causes a reddish glow in the gas behind the cathode.
2. These rays may not reach all the way to the walls. They are much less penetrating than cathode rays.
3. When the gas is changed, the color of the glow and other characteristics of these rays change.

The conclusions from these observations are

1. These rays are positive particles, flowing toward the negative electrode and passing through holes.
2. The positive particles move more slowly and may have time to stop and turn back toward the cathode before reaching the walls.
3. Although the negative particles are alike in each case, the positive particles vary. The positive particles depend on the gas present.

Positively charged particles can be formed in cathode ray tubes and can pass through holes in the cathode, but they are less penetrating than cathode rays and they are not the same when different gases are used.

Figure 5-6 Canal rays (positive ions).

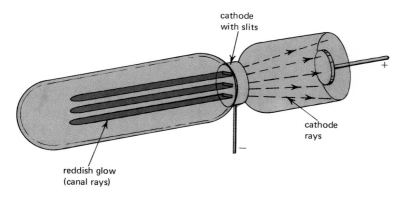

[5-3] Cathode Ray Tubes 75

We will come back to these slower moving charged particles in Section 5–6 and in Chapter 7, Bonding. First we will study the always-identical negative particles, which are now called electrons.

5–4 Properties of Electrons

In 1897, J. J. Thomson carried out an experiment that indicated that electrons were really identical in at least one important property and measured that property. Figure 5–7 shows his experiment. Important details in Thomson's experiment are

1. A beam of electrons was formed (a single dot beam, not a line beam like the one in the slit cathode ray tube).
2. The beam was deflected by a magnetic field.
3. The beam was deflected by an electrical field.
4. The deflections of the beam were measured.

The observations were

1. The deflections fit the pattern expected if the velocities of the particles varied but all the particles had the same electrical charge and mass.
2. The ratio of electrical charge to mass could be (and was) calculated from the deflections.

We will use e as an abbreviation for the charge of an electron and m as an abbreviation for the mass of an electron. The ratio of e/m, which can be determined from Thomson's experiment, was needed by later workers.

The ratio e/m for electrons was determined by simultaneous deflection of a cathode ray beam by electrical and magnetic fields.

Figure 5–7 Thomson's experiment to determine e/m for electrons.

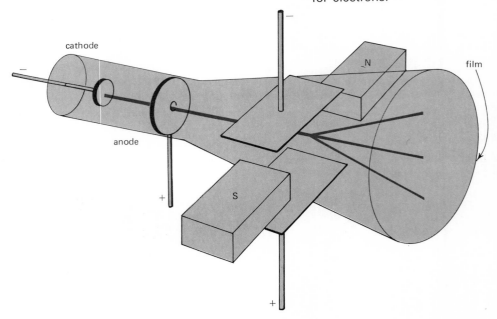

Figure 5-8 Millikan's experiment to determine the charge of an electron.

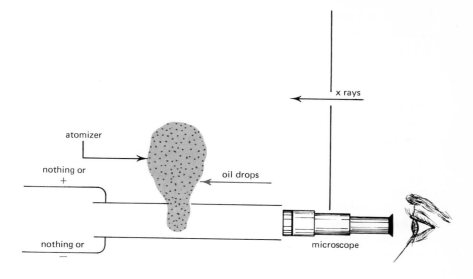

The mass of an electron is much smaller than the mass of the lightest atoms.

At this point we can introduce an example of interaction between two experiments to give additional information not shown by either one alone. Michael Faraday's work on electricity and chemical reaction showed a relationship between charge and mass. When Thomson's e/m was compared to Faraday's data, it was clear that the electron could not be an atom known to man. The lightest known atom (hydrogen) would have a mass almost 2000 times as large, relative to the electricity needed to cause reaction, as the mass of electrons having that same amount of electricity as their charge. Electrons could be obtained from all the different materials used to make cathode ray tubes. That fact made light particles with charges similar to the charge needed to cause reactions seem the most likely explanation. The conclusion reached was that atoms contained light, negatively charged pieces. Thomson proposed a model for atoms that suggested atoms could be taken apart by pulling off electrons.

The availability of Thomson's e/m measurement led to an experiment by Millikan in 1909 to find one of these values (e) so the other could be calculated. This was an agonizingly slow and difficult experiment. The apparatus is shown in Figure 5-8.

The important details were

1. Small droplets of oil were formed.
2. Ionizing radiation induced electrical charges on some of the drops of oil.
3. With a microscope, a marked scale, and a stopwatch, the rate of fall of a drop was measured.
4. An electrical field was then turned on, and the rate of fall (or rise) was measured again.
5. The rates of fall (plus other data such as the resistance of air to the

[5-4] Properties of Electrons 77

falling drop) were used to calculate the mass of the drop, the force applied by the electrical field, and finally the electrical charge on the drop.

The points observed were

1. Other than zero, no charge smaller than 1.6×10^{-19} coulombs (+ or −) was ever observed.
2. All other observed charges were whole number (1, 2, 3, 4, 5, and so on) multiples of the smallest charge found.

The charge of an electron was measured by the oil drop experiment.

From this Millikan concluded that 1.6×10^{-19} coulombs was the charge of a single electron. He was then able to calculate m, the mass of the electron.

Although Millikan's experiment was planned to interact with Thomson's, it also set up a valuable interaction with Faraday's observations on electricity and chemical reaction. It allowed for the first time (over 100 years after Dalton proposed the atomic theory) calculation of the mass of individual atoms and determination of the number we call a mole.

Knowledge of the electrical charge permitted calculation of the mole.

5–5 The Nuclear Atom

The preceding experiments established that atoms have parts, in particular negatively charged particles called electrons. They also showed that flow of electricity is related to chemical reaction. The next major advance was an experiment by Rutherford in 1911 that demonstrated how the electrons and other parts of atoms were arranged.

Rutherford set out to test the model of atoms proposed by Thomson in 1898. Figure 5–9 shows the Thomson or "plum pudding" model of the atom. Positive electrical charge and most of the mass of the atom is spread out

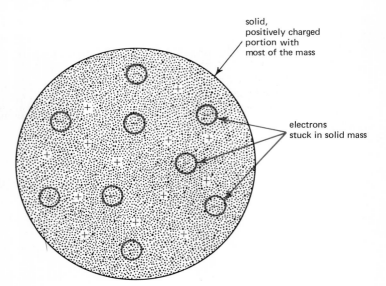

Figure 5–9 Thomson's "plum pudding" model of the atom.

Figure 5-10 Rutherford's experiment leading to the nuclear model of the atom.

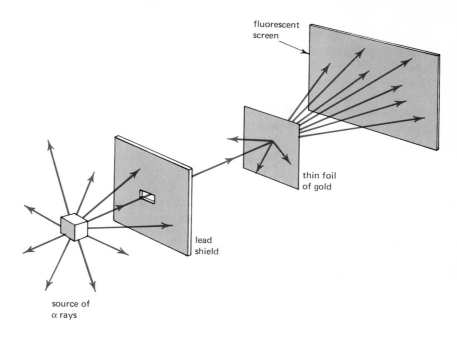

in a jelly like mass filling up the space the atom occupies. Electrons are stuck into this mass like the plums in a plum pudding.

Figure 5–10 shows the apparatus for Rutherford's experiment. Important details are

1. It was known that alpha (α) particles from radioactivity could penetrate matter in thin layers.
2. The α particles ($^4\text{He}^{2+}$ ions) have a known mass and electrical charge.
3. A thick lead shield was used to form a beam of α particles.
4. The beam of α particles penetrated a thin gold foil.
5. Deflections of α particles were measured.

The points observed were

1. As expected, the positively charged α particles were deflected by the massive positively charged part of the gold atoms in the foil.
2. The deflections were less frequent but sharper (some actually being turned back toward the source) than expected if the α particles were going through massive positively charged regions filling the whole atoms.
3. Calculations showed the deflections fit a nuclear atom with
   (a) Most of the atom empty space or filled only with the light electrons.
   (b) All positive charge and most of the mass in a small nucleus.

Deflection of α particles showed the positive charge and almost all mass of atoms were contained in a very small nucleus.

In gold, the nucleus has a radius of about 10^{-12} cm, whereas the atom has a radius of about 10^{-8} cm. In terms of space occupied (which depends on radius cubed) only about one part in 10^{12} was filled by the nucleus. Figure 5–11 shows the relative sizes of a nucleus and the empty space in an atom.

[5–5] The Nuclear Atom

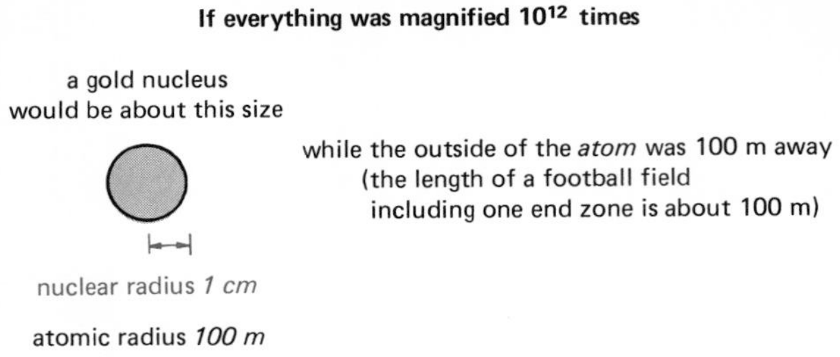

If everything was magnified 10^{12} times

a gold nucleus would be about this size

while the outside of the *atom* was 100 m away
(the length of a football field
including one end zone is about 100 m)

nuclear radius *1 cm*

atomic radius *100 m*

Figure 5–11 Comparative sizes of nucleus and atom.

Since electrons are always the same but atoms of different elements are clearly different, the nuclei of atoms (in which the mass and positive charge are concentrated) must be different. Experiments showed how they differ from each other in both mass and positive charge.

Electrons are so light that the mass of an atomic nucleus and the mass of the whole atom are essentially the same thing. These masses can be measured very accurately by a mass spectrometer, a device very similar to Thomson's device for measuring e/m of the electron. There is more than one way to build a mass spectrometer. Figure 5–12 shows one made of parts similar to those used in the experiments we have already described. The important details of this device are

1. Electrons are knocked off atoms to form ions, many of which have $+1$ charge because they have lost just one electron from the neutral atom.
2. The positive ions are attracted toward a cathode, and some of them pass through a small hole to form a beam. This is like the cathode ray tube with holes in the cathode. It is also like that cathode ray tube because the positive ions are usually formed by bombarding a gas with cathode rays.
3. A second cathode with a hole is usually used to make the beam a very tiny straight beam.

5–6 Atomic Number and Mass Number

Atomic masses are measured by mass spectrometers.

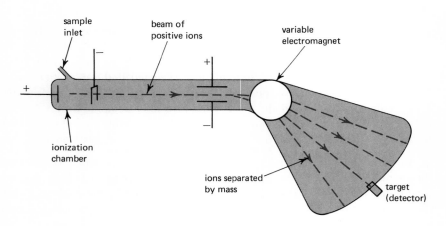

Figure 5–12 Mass spectrometer.

80   Into the Heart of Matter   [5]

4. The beam (in a good vacuum so nothing else deflects any ions) is passed through electrical and magnetic fields like those in Thomson's experiment.
5. The deflection of the beam is measured very precisely. (*Note:* In modern instruments it is easier to have a fixed target and change the electrical or magnetic field to make the beam reach the target.)
6. The cathode voltage that attracted the positive ions, the electrical and magnetic fields, the deflection, and the fact that the electrical charge is known for +1 ions are used to calculate the mass.

Points observed from mass spectrometer measurements are

1. A single element often has several kinds of atoms differing in mass. These are called **isotopes.**
2. Using the atomic weight scale based on ^{12}C isotope as exactly 12, all atoms have masses close to whole numbers.

All isotopes have atomic weights close to whole numbers; these numbers are called the mass numbers.

The whole number close to the observed mass is called the **mass number** of that isotope. Mass numbers are used to identify which isotope we are talking about. The mass number is written as a superscript on the left side of the element symbol, as in ^{12}C.

The differences in positive charge of nuclei can be shown for light elements by taking all the electrons off and measuring the charge on the ions. We find that each element has a different nuclear charge, and one particular element exists for each whole number (1, 2, 3, and so on) times the basic unit of charge determined by Millikan. It is not possible to get all the electrons off atoms with large nuclear charges, but other information confirms that there continues to be one particular element for each whole number value of charge. The number of positive charges determines which element the atom is, and this number is called the **atomic number.**

Each element has a whole number (called the atomic number) of units of positive charge in its nuclei, and each whole number corresponds to an element.

5–7 Protons, Neutrons, and Electrons

Measurements of atomic numbers and mass numbers led to a dilemma in explaining the composition of nuclei. The whole number values imply some definite relationship, and one of the possibilities is that each unit of charge is connected with one unit of mass. The lightest atoms, ^{1}H, do have one unit of mass and one unit of charge, and all isotopes have at least as many units of mass as they do charges. The ^{1}H units are given the name **protons.** But there are usually some extra units of mass, and the existence of isotopes shows that the number of extra mass units varies in atoms with the same number of charges. This problem was solved by the discovery of neutrons in 1932 by Chadwick. His experiment will not be described here because it requires too much information we are not ready for yet. His conclusion was that there were electrically neutral particles, which we call **neutrons,** with about the same mass as protons.

Identification of the basic units protons, neutrons, and electrons allows us to determine the pieces from which each atom or ion is made. The atomic number of the element equals the number of protons in one of its atoms. The mass number of an isotope equals the number of protons plus the number of neutrons. The number of electrons is determined from the atomic number and the overall charge. For an atom, which is neutral in charge, there must be the same number of protons and electrons, so the number of electrons is equal to the atomic number. For ions there must be either more electrons than protons (negative ions) or less electrons than protons (positive ions). The number of electrons can be calculated by *subtracting* the charge of the ion from the atomic number. Here are some examples of calculation of protons, neutrons, and electrons. We will use the abbreviations p (protons), n (neutrons), and e (electrons). The atomic numbers can be found from the list on the inner back cover of the book.

Atoms or ions are composed of protons, neutrons, and electrons, and the numbers of each can be calculated from the atomic number, mass number, and net charge.

^{19}F atom

    Atomic number of F = 9, so there are $9\,p$
    Mass number = $19 = p + n = 9 + n$, so there are $10\,n$
    Neutral atom, so electrons = atomic number = $9\,e^-$

^{16}O^{2-} ion

    Atomic number of O = 8, so there are $8\,p$
    Mass number = $16 = p + n = 8 + n$, so there are $8\,n$
    Electrons = atomic number − charge =

$$8 - (-2) = 8 + 2 = 10\,e^-$$

^{23}Na$^+$ ion

    Atomic number of Na = 11, so there are $11\,p$
    Mass number = $23 = p + n = 11 + n$, so there are $12\,n$
    Electrons = atomic number − charge = $11 - (+1) = 10\,e^-$

You should practice until you are able to do these calculations quickly in your head, particularly calculation of the number of electrons since you will need to do that at later points in this course.

1. You should be able to list (from memory) the basic details of each experiment described in this chapter. Exact reproduction of the wording is not expected.
2. You should be able to list (from memory) the important points observed in each experiment described in this chapter. Exact reproduction of the wording is not expected.
3. You should be able to list (from memory) the conclusions reached from the experiments described in this chapter. Exact reproduction of the wording is not expected.
4. You should be able to calculate the numbers of protons, neutrons, and electrons in an atom or ion.

Skills Expected at This Point

Exercises

1. Pick a news story from a current newspaper and analyze it, using the questions "How do I pick out important facts?" and "What am I supposed to be learning?" (a) Write out an outline summarizing the points you find in your analysis. (b) Do you find analysis of news similar to or very different from analysis of the science experiments in this chapter? (c) Did the "scientific" method of analysis help you understand the news story better or confuse you?
2. List, in outline form, the details you must know to understand the experiments (a) with the cathode ray tube with a beam-forming slit; (b) making the measurement of e/m (by Thomson); (c) showing the nuclear atom (by Rutherford).
3. List the reason or reasons why each of the following was done: (a) Franklin defined a direction of flow for electricity. (b) Millikan carried out his oil drop experiment. (c) Rutherford carried out his α particles into gold foil experiment.
4. List the key points that were observed (a) by Faraday in his electrolysis work; (b) in the shadow-casting cathode ray tube; (c) in the cathode ray tube with holes in the cathode; (d) in mass spectrometers.
5. Calculate the numbers of protons, neutrons, and electrons in each of the following: ^{14}N atom; ^{37}Cl atom; ^{238}U atom; $^{3}H^+$ ion; $^{32}S^{2-}$ ion; $^{9}Be^{2+}$ ion.
6. Describe the mass and charge characteristics of (a) an electron; (b) a proton; (c) a neutron.
7. Define the terms (a) cathode ray; (b) cathode; (c) e; (d) nucleus; (e) mass number; (f) atomic number.
8. Sketch the apparatus and label the key parts in (a) a shadow casting cathode ray tube; (b) Thomson's e/m experiment; (c) Millikan's determination of e.

Test Yourself

1. About how many times the mass of an electron is the mass of a proton? About how many times the mass of a neutron is the mass of a proton?
2. Masses of individual atoms can be measured by ionizing them (removing an electron) and passing them through what? A single element may have atoms of two (or more) different masses. What are these atoms of different masses called?
3. List the characteristics (and parts) found in all operating cathode ray tubes.
4. Sketch the apparatus of Rutherford's gold foil experiment. (a) Label clearly all the important parts and (b) state the conclusion from the experiment and explain what observation led to that conclusion instead of a different "expected" result.
5. Which of the following *cannot* be derived from cathode or canal ray experiments?
   (a) The size of the charge of an electron
   (b) The sign of the charge of an electron

(c) The charge to mass ratio of an electron
(d) The sign of the charge on an ionized atom
(e) The deflection of an electron by a magnet

6. The experiment that established the electrical charge of one electron was conducted by
   (a) Dalton
   (b) Thomson
   (c) Millikan
   (d) Rutherford
   (e) Faraday

7. Which of the following important points could *not* be shown in the demonstration using the cathode ray tube with a slit?
   (a) Cathode rays are charged species.
   (b) Cathode rays can transfer momentum.
   (c) Cathode rays flow from negative to positive electrodes.
   (d) Beams of cathode rays can be bent by a magnetic field.
   (e) Objects in the path of cathode rays cast sharp shadows.

8. Rutherford's α particle experiment, which established the nuclear model of the atom, disproved the earlier "plum cake" (solid) atom model because
   (a) The α particles were not expected to penetrate through matter.
   (b) The α particles showed the existence of radioactivity.
   (c) The particles were deflected less frequently and more sharply than expected.
   (d) The particles were deflected less frequently and less sharply than expected.
   (e) None of the above.

9. Which of the following points was determined by Millikan's oil drop experiment?
   (a) Atoms are made up of protons, neutrons, and electrons.
   (b) All electrons have the same mass.
   (c) Atoms consist of a small nucleus containing the positive charge and almost all the mass.
   (d) The charge of the electron is -1.6×10^{-19} coulombs.
   (e) None of the above.

10. A $^{37}Cl^-$ ion contains
    (a) 18 neutrons, 17 protons, and 17 electrons
    (b) 20 neutrons, 17 protons, and 18 electrons
    (c) 18 protons, 19 neutrons, and 19 electrons
    (d) 17 protons, 20 neutrons, and 17 electrons
    (e) 20 protons, 17 neutrons, and 18 electrons

6 Electronic Configurations and Periodic Behavior

6–1 The Periodic Table I

When elements are arranged by atomic weight, there are periodic recurrences of similar chemical behaviors.

The periodic table arranges similar elements in the same column as the elements are placed in order across the rows.

As the elements were discovered and their reactions to form chemical compounds were examined, it became clear that there were groups of elements that behaved very similarly to each other but very differently from other groups of elements. As the available information increased, chemists tried to group the similar elements and search out the pattern leading to similarities or differences. In 1829 Dobereiner noted that there were several groups of three similar elements, such as chlorine, bromine, and iodine. When the atomic weights of the elements were measured, the middle member (by atomic weight) of each triad had an atomic weight near the average of the other two. This and other efforts led some chemists to believe there was a pattern in the elements related to their atomic weights. Therefore, the known elements were listed in order according to their atomic weights and the appearance of similarly behaved elements at fairly regular intervals was noted.

The pattern was made more apparent in 1862 by Beguyer de Chancourtois, who made a spiral out of the sequence of elements. The spiral could be arranged so the elements were all adjacent (above or below) to elements similar to themselves, even though they were on different turns of the spiral. In 1864 John Newlands made a similar arrangement using a table with similar elements in the same columns. Between 1868 and 1870 Dimitri Mendeleev in Russia and Lothar Meyer in Germany independently developed better tables to show the relationships. In these tables the elements were listed in a row across until an element chemically similar to an earlier one was reached. At that point a new row was started. Eventually a table was completed in which similar elements were in single columns. Meyer and Mendeleev recognized some problems caused by the incomplete data of that time and took steps to correct them. Mendeleev, who is usually given the principal credit for the periodic table, made two particularly significant changes in his table. (Meyer and, to a lesser extent, Newlands also made these changes, but Mendeleev made the greatest use of the changes to predict properties of unknown elements.)

First, Mendeleev noticed that the order of atomic weights in a few pairs of elements was reversed from the pattern with elements of the most similar chemical behavior in the same columns. In those cases he chose to trust the chemical behavior instead of the mass and reversed the atomic weight order in those pairs. He also corrected the atomic weights of some elements that seemed incorrect and out of place. We now know that he was putting the elements into atomic number order instead of atomic weight order. All the earlier patterns related to atomic weight had occurred only because, in all but the few exceptions, atomic weight order happens to be the same as atomic number order.

Second, when Mendeleev reached points where no element in that atomic weight range fit the chemical behavior of a particular column, he left an empty space. He assumed that elements would be discovered to fill these holes. He even predicted the atomic weights and densities of these missing elements and properties of some of their compounds. Mendeleev's suggestions stimulated a flurry of activity to search for the missing elements. The searches were very successful, and the holes were rapidly filled. These discoveries were a classic example of the progress that can be stimulated by a good scientific theory, but there is also a lesson about the dangers of over enthusiasm. One

By reversing atomic weight order where needed to match chemical similarities, Mendeleev produced a table in order by atomic number.

By predicting that holes in the table should be filled, Mendeleev stimulated the discovery of new elements.

Figure 6–1 Periodic table of the elements.

Number above symbol = atomic weight
Number below symbol = atomic number

IA	IIA	IIIB	IVB	VB	VIB	VIIB		VIIIB		IB	IIB	IIIA	IVA	VA	VIA	VIIA	VIIIA
1.008 H 1																	4.003 He 2
6.939 Li 3	9.012 Be 4											10.81 B 5	12.01 C 6	14.01 N 7	16.00 O 8	19.00 F 9	20.18 Ne 10
22.99 Na 11	24.31 Mg 12											26.98 Al 13	28.09 Si 14	30.97 P 15	32.06 S 16	35.45 Cl 17	39.95 Ar 18
39.10 K 19	40.08 Ca 20	44.96 Sc 21	47.90 Ti 22	50.94 V 23	52.00 Cr 24	54.94 Mn 25	55.85 Fe 26	58.93 Co 27	58.71 Ni 28	63.54 Cu 29	65.37 Zn 30	69.72 Ga 31	72.59 Ge 32	74.92 As 33	78.96 Se 34	79.91 Br 35	83.80 Kr 36
85.47 Rb 37	87.62 Sr 38	88.91 Y 39	91.22 Zr 40	92.91 Nb 41	95.94 Mo 42	97 Tc 43	101.1 Ru 44	102.9 Rh 45	106.4 Pd 46	107.9 Ag 47	112.4 Cd 48	114.8 In 49	118.7 Sn 50	121.8 Sb 51	127.6 Te 52	126.9 I 53	131.3 Xe 54
132.9 Cs 55	137.3 Ba 56	138.9 La* 57–71	178.5 Hf 72	180.9 Ta 73	183.9 W 74	186.2 Re 75	190.2 Os 76	192.2 Ir 77	195.1 Pt 78	197.0 Au 79	200.6 Hg 80	204.4 Tl 81	207.2 Pb 82	209.0 Bi 83	210 Po 84	210 At 85	222 Rn 86
223 Fr 87	226 Ra 88	227 Ac** 89–103	261 (Ku) 104	260 (Ha) 105													

*Lanthanide series	140.1 Ce 58	140.9 Pr 59	144.2 Nd 60	147 Pm 61	150.4 Sm 62	152.0 Eu 63	157.3 Gd 64	158.9 Tb 65	162.5 Dy 66	164.9 Ho 67	167.3 Er 68	168.9 Tm 69	173.0 Yb 70	175.0 Lu 71
**Actinide series	232.0 Th 90	231 Pa 91	238.0 U 92	237 Np 93	242 Pu 94	243 Am 95	247 Cm 96	247 Bk 97	249 Cf 98	251 Es 99	254 Fm 100	253 Md 101	256 No 102	254 Lr 103

element (at. no. 61) was "discovered" at least three different times when it was not there. Experiments with radioactivity in the midtwentieth century allowed man to make element 61 artificially and prove that all of its isotopes are radioactive and none could possibly exist in nature.

There are several possible forms of the periodic table. One of them (sometimes called the long form) is shown in Figure 6–1. Some of the columns do not begin until farther down in the table than others. In order to save space, one large group of "late-starting" columns is placed at the bottom of the table. The points where these elements would fit in atomic number order are noted on the main part of the table. The atomic number (below) and atomic weight (above) of each element are included with the symbol at the proper point in the table.

The "long form" of the periodic table is a common modern form.

6–2 Spectral Evidence of Regularities

The cases where atomic weight order is reversed raise a question about whether the elements in the periodic table are all in the proper order. In much of the periodic table, the correctness of the order by atomic number can be shown best by spectral evidence. Light and other electromagnetic waves, such as x rays, can be split into different wavelengths by physical means. The arrangement of different wavelengths obtained from a particular source is called a spectrum. Shortly before the first World War, Moseley discovered that the x ray spectrum of an element was a sensitive measure of the atomic number. He concluded that each change of one in atomic number caused a characteristic shift (both in size and direction) in x ray wavelength. X ray data provide convincing proof that the present arrangement of elements in the periodic table is in direct order by atomic number.

Regularities in x-ray spectra confirm that the periodic table is in order by atomic number.

Spectra of the light emitted from lamps containing different gases also provide some interesting evidence of regular patterns. In lamps where excited individual atoms emit light, the spectra give very sharp patterns of particular wavelengths. Figure 6–2 shows how a prism would break such a spectrum up to give sharp lines in a film. The spectrum obtained from a lamp containing excited hydrogen atoms is particularly simple. Several series of regular lines can be identified from the hydrogen spectrum. These series are shown in

Figure 6–2 Light spectrum.

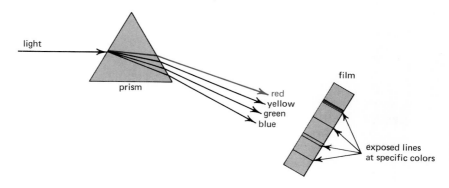

Figure 6–3. The series names (which came from their discoverers) are not important. A relationship exists between the wavelength of light and the energy of the light units (called quanta). This relationship was used to calculate the energy of the observed light. The results fit the transitions between a very small number of energy levels, as shown in Figure 6–4. This energy level pattern was used to develop a model for the structure of atoms in general and the hydrogen atom in particular.

Light spectra show a limited, regular pattern for the energy of light emitted by atoms.

Emission spectra lead to a pattern of specific energy levels in atoms.

6–3 Bohr Atom

The nuclear model of the atom resulted from Rutherford's experiment (see Section 5–5) in 1911. It presented some very serious problems in physics. The electrons in an atom, which is mostly empty space, should fall into the small nucleus because of the attraction of the positive charge there. That would cause the whole atom to collapse to the size of the nucleus. Obviously this does not happen. Atoms continue to occupy space much larger than their nuclei. The collapse is avoided if the electrons are in motion. Centrifugal force would balance out the attraction of the positive nuclear charge. That is the sort of balance that keeps the Earth from falling into the Sun. But an electrically charged particle moving in a circle should generate a magnetic field. That would use up energy and slow down the electron's motion. As it slowed down, the attraction to the nucleus would pull it in closer and eventually the same result should occur as with no motion—collapse to the size of the nucleus.

In 1913, Niels Bohr proposed a model to get around this problem. He suggested that no energy was used up to generate a magnetic field because there was no new energy level to move to with the slightly reduced energy. The hydrogen spectrum showed only a few discrete energy levels, and Bohr reasoned that no change could take place unless a jump was made all the way from one to another of those levels. Bohr was able to calculate a series of orbits that fit both some reasonable assumptions about atoms and the observed energy levels in hydrogen.

We now know that the idea of orbits is oversimplified. Presently the electron

Bohr proposed that changes in electron energies had to be in jumps between a few fixed energy levels represented by certain allowed orbits.

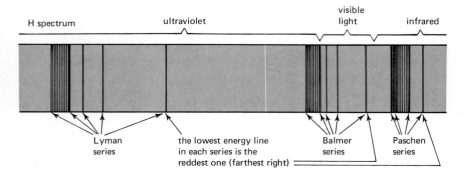

Figure 6–3 Hydrogen atom spectrum.

Figure 6-4 Energy levels in the hydrogen atom.

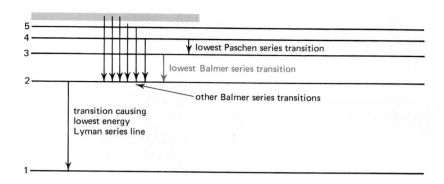

is described by a wave function that provides a mathematical statement of the probability of the electron's being at any given point around the atom's nucleus. However, we still do have certain average distances (and no others) that can exist and the key whole number values in Bohr's calculation (n) that describe the allowed orbits will reappear as a quantum number later in this chapter.

Courses in physics describe Bohr's model, its successors, and their mathematical relationships more properly than we do here. The key fact from the Bohr atom is that electrons in atoms have sharp and very systematic limitations on what they can do, particularly in terms of energy content.

6-4 Quantum Numbers and the Exclusion Principle

Electrons in atoms can be described by a series of four quantum numbers.

Pauli exclusion principle: No two electrons in an atom can have the same four quantum numbers.

The n quantum number is related to distance and can have whole number values 1, 2, 3, and larger.

The l quantum number is related to shape of motion and can have whole number values 0, 1, 2, and larger, up to a maximum of $n - 1$.

The specific places allowed for electrons in atoms are now best described by use of a series of four numbers called **quantum numbers** and a rule called the **Pauli exclusion principle**. The Pauli exclusion principle states that no two electrons in the same atom can have exactly the same set of four quantum numbers.

The quantum numbers are as follows.

n (the radial quantum number). This number is related to the distance of the electron from the nucleus and is related to size. It is the number that appeared in Bohr's calculations about the Bohr atom. n is the most important single factor in determining electron energy levels, but it is not the only factor except in the special case where there is only one electron (hence the simple pattern for hydrogen atoms).

l (the azimuthal quantum number). This number is related to the way the electron moves around the nucleus or, in more modern terms, the shape of the probability distribution. It affects electron energy levels enough to be chemically significant.

m (the magnetic quantum number). This number is related to the magnetic field set up by the electronic motion. It affects the number of electrons possible, but m only affects energy levels through interaction with neighboring groups in directional bonding arrangements such as complexes (see Section 16–3)

s (the spin quantum number). This number is related to something similar to Earth's spinning on its axis. It has some effect on energy levels.

The following rules limit the possible combinations of quantum numbers:

$n =$ whole number values, 1 or larger $\quad$ 1, 2, 3, 4, . . .
$l =$ whole numbers 0 or larger up to $n - 1$ $\quad$ 0, 1, 2, . . . , $n - 1$
$m =$ whole numbers from $-l$ to $+l$ $\quad$ $-l, -l + 1, \ldots, 0, \ldots, l - 1, +l$
$s = +\frac{1}{2}$ or $-\frac{1}{2}$ $\quad$ $+\frac{1}{2}, -\frac{1}{2}$

m and *s* quantum numbers also follow set rules.

When $n = 1$ (the most favorable place for an electron) the only value allowed for l is 0. When l is 0, m must be 0, and s can be $+\frac{1}{2}$ or $-\frac{1}{2}$. Therefore, there are only two possible combinations of quantum numbers with $n = 1$: 1, 0, 0, $+\frac{1}{2}$ and 1, 0, 0, $-\frac{1}{2}$. When $n = 2$, there are eight possible combinations. When $n = 3$, there are 18 possible combinations. The electrons tend to go into the most strongly held energy levels, but as the number of combinations rises, factors other than having the lowest n value begin to affect which combinations are held most strongly.

6–5 Electronic Configurations

We need to keep track of the combinations of quantum numbers that differ enough in energy to affect chemistry, but we do not need to know about quantum numbers that do not affect chemistry. Therefore, we use a shorthand which tells us only about the n quantum number, the l quantum number, and how many electrons we have. We also need to pay some attention to s values, but they are so simple ($+\frac{1}{2}$ or $-\frac{1}{2}$) that we can keep track of all we need about them without including s in our shorthand notation. Our code uses the number for n, a letter for l, and a superscript for the number of electrons of that kind. The letters used for l are an illogical group inherited from old spectral work. You should simply memorize that s electrons have $l = 0$, p electrons have $l = 1$, d electrons have $l = 2$, and f electrons have $l = 3$. Beyond that point an alphabetical order would be used, but we never get beyond f electrons in chemistry of elements occurring naturally on earth.

Electronic configurations are shorthand notations showing the information about electrons which is chemically important.

Using this convention, we say that the one electron in a hydrogen atom is a $1s$ electron and that the **electronic configuration** of the hydrogen atom is $1s^1$. The number (1 in this case) of electrons of that type is always written as a superscript following the name of the electron type ($1s$ in this case). Both of the two electrons in the helium atom are $1s$ electrons, so the electronic configuration of helium is $1s^2$. Lithium has a $1s^2 2s^1$ configuration, beryllium is $1s^2 2s^2$, and boron is $1s^2 2s^2 2p^1$.

The order in which energy levels fill can be determined from the scheme shown in Figure 6–5. By following the arrows in order, starting with the one through $1s$, we pass through the groups in the same order that the levels fill. We simply put as many electrons as possible in each level and then go

The method for determining electron energy level order is described and used.

Figure 6-5 Order of filling of electron energy levels.

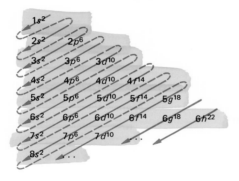

Figure 6-6 Electronic configurations (for atoms in their normal ground state—the lowest energy configuration).

to the next level. Each s type level can hold a maximum of two electrons, each p type level a maximum of six, each d type level a maximum of 10, and each f type level a maximum of 14. When we run out of electrons, we stop.

There are a few minor exceptions to the pattern for electronic configurations. You will not be held responsible for any exceptions from what you would predict from Figure 6-5 but you may want to compare your predictions with the actual electronic configurations, given in Figure 6-6, and see if you can identify nature's exceptions. We will come back to these exceptions in Chapter 8, but the scheme of Figure 6-5 is adequate for our purposes at

Symbol	Atomic number	Electronic configuration	Symbol	Atomic number	Electronic configuration
H	1	$1s^1$	Ni	28	$1s^2\ 2s^2\ 2p^6\ 3s^2\ 3p^6\ 4s^2\ 3d^8$
He	2	$1s^2$	Cu	29	$1s^2\ 2s^2\ 2p^6\ 3s^2\ 3p^6\ 4s^1\ 3d^{10}$
Li	3	$1s^2\ 2s^1$	Zn	30	$1s^2\ 2s^2\ 2p^6\ 3s^2\ 3p^6\ 4s^2\ 3d^{10}$
Be	4	$1s^2\ 2s^2$	Ga	31	$1s^2\ 2s^2\ 2p^6\ 3s^2\ 3p^6\ 4s^2\ 3d^{10}\ 4p^1$
B	5	$1s^2\ 2s^2\ 2p^1$	Ge	32	$1s^2\ 2s^2\ 2p^6\ 3s^2\ 3p^6\ 4s^2\ 3d^{10}\ 4p^2$
C	6	$1s^2\ 2s^2\ 2p^2$	As	33	$1s^2\ 2s^2\ 2p^6\ 3s^2\ 3p^6\ 4s^2\ 3d^{10}\ 4p^3$
N	7	$1s^2\ 2s^2\ 2p^3$	Se	34	$1s^2\ 2s^2\ 2p^6\ 3s^2\ 3p^6\ 4s^2\ 3d^{10}\ 4p^4$
O	8	$1s^2\ 2s^2\ 2p^4$	Br	35	$1s^2\ 2s^2\ 2p^6\ 3s^2\ 3p^6\ 4s^2\ 3d^{10}\ 4p^5$
F	9	$1s^2\ 2s^2\ 2p^5$	Kr	36	$1s^2\ 2s^2\ 2p^6\ 3s^2\ 3p^6\ 4s^2\ 3d^{10}\ 4p^6$
Ne	10	$1s^2\ 2s^2\ 2p^6$	Rb	37	[arrangement of Kr] $+\ 5s^1$
Na	11	$1s^2\ 2s^2\ 2p^6\ 3s^1$	Sr	38	[Kr] $+\ 5s^2$
Mg	12	$1s^2\ 2s^2\ 2p^6\ 3s^2$	Y	39	[Kr] $+\ 5s^2\ 4d^1$
Al	13	$1s^2\ 2s^2\ 2p^6\ 3s^2\ 3p^1$	Zr	40	[Kr] $+\ 5s^2\ 4d^2$
Si	14	$1s^2\ 2s^2\ 2p^6\ 3s^2\ 3p^2$	Nb	41	[Kr] $+\ 5s^1\ 4d^4$
P	15	$1s^2\ 2s^2\ 2p^6\ 3s^2\ 3p^3$	Mo	42	[Kr] $+\ 5s^1\ 4d^5$
S	16	$1s^2\ 2s^2\ 2p^6\ 3s^2\ 3p^4$	Tc	43	[Kr] $+\ 5s^1\ 4d^6$
Cl	17	$1s^2\ 2s^2\ 2p^6\ 3s^2\ 3p^5$	Ru	44	[Kr] $+\ 5s^1\ 4d^7$
Ar	18	$1s^2\ 2s^2\ 2p^6\ 3s^2\ 3p^6$	Rh	45	[Kr] $+\ 5s^1\ 4d^8$
K	19	$1s^2\ 2s^2\ 2p^6\ 3s^2\ 3p^6\ 4s^1$	Pd	46	[Kr] $+\ 4d^{10}$
Ca	20	$1s^2\ 2s^2\ 2p^6\ 3s^2\ 3p^6\ 4s^2$	Ag	47	[Kr] $+\ 5s^1\ 4d^{10}$
Sc	21	$1s^2\ 2s^2\ 2p^6\ 3s^2\ 3p^6\ 4s^2\ 3d^1$	Cd	48	[Kr] $+\ 5s^2\ 4d^{10}$
Ti	22	$1s^2\ 2s^2\ 2p^6\ 3s^2\ 3p^6\ 4s^2\ 3d^2$	In	49	[Kr] $+\ 5s^2\ 4d^{10}\ 5p^1$
V	23	$1s^2\ 2s^2\ 2p^6\ 3s^2\ 3p^6\ 4s^2\ 3d^3$	Sn	50	[Kr] $+\ 5s^2\ 4d^{10}\ 5p^2$
Cr	24	$1s^2\ 2s^2\ 2p^6\ 3s^2\ 3p^6\ 4s^1\ 3d^5$	Sb	51	[Kr] $+\ 5s^2\ 4d^{10}\ 5p^3$
Mn	25	$1s^2\ 2s^2\ 2p^6\ 3s^2\ 3p^6\ 4s^2\ 3d^5$	Te	52	[Kr] $+\ 5s^2\ 4d^{10}\ 5p^4$
Fe	26	$1s^2\ 2s^2\ 2p^6\ 3s^2\ 3p^6\ 4s^2\ 3d^6$	I	53	[Kr] $+\ 5s^2\ 4d^{10}\ 5p^5$
Co	27	$1s^2\ 2s^2\ 2p^6\ 3s^2\ 3p^6\ 4s^2\ 3d^7$	Xe	54	[Kr] $+\ 5s^2\ 4d^{10}\ 5p^6$

Symbol	Atomic number	Xe configuration plus	Symbol	Atomic number	Rn configuration plus
Cs	55	$6s^1$	Fr	87	$7s^1$
Ba	56	$6s^2$	Ra	88	$7s^2$
La	57	$6s^2\ 5f^1$	Ac	89	$7s^2\ 6d^1$
Ce	58	$6s^2\ 4f^2$	Th	90	$7s^2\ 6d^2$
Pr	59	$6s^2\ 4f^3$	Pa	91	$7s^2\ 6d^1\ 5f^2$
Nd	60	$6s^2\ 4f^4$	U	92	$7s^2\ 6d^1\ 5f^3$
Pm	61	$6s^2\ 4f^5$	Np	93	$7s^2\ 6d^1\ 5b^4$
Sm	62	$6s^2\ 4f^6$	Pu	94	$7s^2\ 5f^6$
Eu	63	$6s^2\ 4f^7$	Am	95	$7s^2\ 5f^7$
Gd	64	$6s^2\ 5d^1\ 4f^7$	Cm	96	$7s^2\ 5f^7\ 6d^1$
Tb	65	$6s^2\ 4f^9$	Bk	97	$7s^2\ 5f^8\ 6d^1$
Dy	66	$6s^2\ 4f^{10}$	Cf	98	$7s^2\ 5f^{10}$
Ho	67	$6s^2\ 4f^{11}$	Es	99	$7s^2\ 5f^{11}$
Er	68	$6s^2\ 4f^{12}$	Fm	100	$7s^2\ 5f^{12}$
Tm	69	$6s^2\ 4f^{13}$	Md	101	$7s^2\ 5f^{13}$
Yb	70	$6s^2\ 4f^{14}$	No	102	$7s^2\ 5f^{14}$
Lu	71	$6s^2\ 4f^{14}\ 5d^1$	Lr	103	$7s^2\ 5f^{14}\ 6d^1$
Hf	72	$6s^2\ 4f^{14}\ 5d^2$		104	$7s^2\ 5f^{14}\ 6d^2$
Ta	73	$6s^2\ 4f^{14}\ 5d^3$		105	additional configurations can be estimated (as a few of the above have been estimated) by extrapolating trends seen in the above cases
W	74	$6s^2\ 4f^{14}\ 5d^4$			
Re	75	$6s^2\ 4f^{14}\ 5d^5$			
Os	76	$6s^2\ 4f^{14}\ 5d^6$			
Ir	77	$6s^2\ 4f^{14}\ 5d^7$			
Pt	78	$6s^1\ 4f^{14}\ 5d^9$			
Au	79	$6s^1\ 4f^{14}\ 5d^{10}$			
Hg	80	$6s^2\ 4f^{14}\ 5d^{10}$			
Tl	81	$6s^2\ 4f^{14}\ 5d^{10}\ 6p^1$			
Pb	82	$6s^2\ 4f^{14}\ 5d^{10}\ 6p^2$			
Bi	83	$6s^2\ 4f^{14}\ 5d^{10}\ 6p^3$			
Po	84	$6s^2\ 4f^{14}\ 5d^{10}\ 6p^4$			
At	85	$6s^2\ 4f^{14}\ 5d^{10}\ 6p^5$			
Rn	86	$6s^2\ 4f^{14}\ 5d^{10}\ 6p^6$			

Figure 6-6 (Continued)

this point. Writing the electronic configuration is a frequent starting point for discussion of chemical behavior.

An example of this kind of exercise would be to write the electronic configuration of iron (at. no. 26). A neutral iron atom has 26 electrons, so we must fill levels until we reach a total of 26 electrons. $1s^2$ would be only two electrons so we continue: $1s^2 2s^2$ would be four so we still continue, and so on. When we reach $1s^2 2s^2 2p^6 3s^2 3p^6 4s^2$ we have 20 electrons, and we need places for only six more. Even though the $3d$ level could hold up to 10 electrons, we have only six to put in it. Therefore the electronic configuration is $1s^2 2s^2 2p^6 3s^2 3p^6 4s^2 3d^6$.

6-6 Electronic Orbitals

One reason for using names like $1s$ and $2p$ for electrons is to identify the shape of their position around the nucleus. The probability distributions of electrons (which you can think of as the region they move around in) have shapes determined by the l quantum number. Therefore the shapes are the same for all s type electrons. The shape for a p type electron is different from s type but the same as any other p type electrons.

Figure 6–7 *s* orbital.

Any orbital can hold two and no more than two electrons.

The shapes and directions of *s* and *p* orbitals are described.

Figure 6–7 shows the shape for *s* electrons. This distribution in space is called an orbital. It holds two electrons because each of the two different *s* quantum number values are possible for the same distance (*n* value) and shape (*l* value). The *s* orbital is spherically symmetrical, which means there is an equal probability of an electron being in any direction from the nucleus so long as the distance is the same. There is a most likely distance, but there is also some chance of being closer to the nucleus or farther away.

The most likely single point for electrons in the *s* orbital is at the nucleus, but the most likely distance is farther away. As we move away from the nucleus, the number of possible positions (or, more accurately, the size of the volume element) at any given distance goes up fast enough to more than compensate for the drop in probablilty at each point. The most likely distance is the distance where the product of probability per volume unit times number of volume units reaches its maximum. Figure 6–8 uses an analogy with a two-dimensional figure to show how this most likely distance occurs. The analogy of darts being thrown at a target is similar to the electron probability in an *s* orbital except the "targets" for the electrons are spherical volumes instead of the flat circle areas on the dart board. Notice that there is some chance of being far away from the target. For example, the dart in the dart board analogy could even slip from the thrower's hand and fly in the opposite direction. Electrons in *s* orbitals may also be far from the most likely distance. As points are chosen farther from the most likely distance, the probability of being there becomes smaller, but it never actually becomes zero. An atom on Earth has a small probability for its electrons to be on the moon. The probability can be ignored because it is so ridiculously small, but it does exist.

Figure 6–9 shows a comparison between a 1*s* orbital (which can hold two electrons) and a 2*s* orbital (which can also hold two electrons). The 2*s* electrons are (on the average) farther from the nucleus than the 1*s* electrons, but the shape of the orbital is the same.

A complete set of three *p* type orbitals can hold six electrons, as shown in Figure 6–10. A set of three *p* type orbitals, each concentrated along a particular direction with the three directions perpendicular to each other, exist with identical energy levels. Figure 6–11 shows a single *p* orbital and a set of three *p* orbitals. The dumbbell type shape means that the electrons are likely to be on one side or the other of the nucleus but spend little time halfway in between. Each of the two "ball" shaped halves of a *p* orbital are

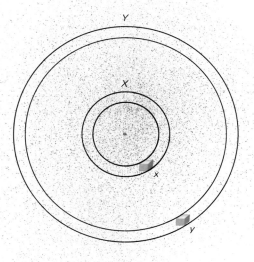

Figure 6-8 (A). Electron Probability in a 1s Orbital.

The electron probability for all the volume elements in shell Y is greater than in shell X. Shell Y represents the most probable location of the electron. Note that the orbital is a three-dimensional spherical region surrounding the nucleus and not just the two-dimensional cross section shown here.

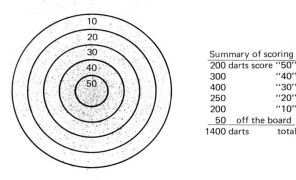

Summary of scoring	
200 darts score	"50"
300	"40"
400	"30"
250	"20"
200	"10"
50	off the board
1400 darts	total

Figure 6-8 (B). Dart Board Analogy to a 1s Orbital

The dart board corresponds to a 1s electron orbital. It includes practically all of the dart holes (1350 out of 1400). The nucleus of the atom lies at the center of the "50" region, where there exists the greatest probability of finding dart holes in a small unit of area. But the most probable score is "30" (400 out of 1400). The "30" region corresponds to the region of maximum electron probability [shell Y in Figure 6-8 (A)]. [From R. H. Petrucci, *General Chemistry*, Macmillan Publishing Company, 1972.]

"solid center" with the exact centers of each half the highest probability points in the electron distribution.

There are five *d* type orbitals of complex shape in each set with identical energy levels, and seven *f* type orbitals per set with identical energy levels. We will not attempt to represent their shapes. The important point is that in every case there are orbitals (each containing up to two electrons) to hold all the electrons in any given energy level. This grouping of electrons in pairs will continue to appear and be important in bonding (Chapter 7) and complex formation (Chapters 16 and 20). The actual shapes of the orbitals are not vitally important to us, although they are related to molecular geometry (Chapter 7). We are asking you to learn the shapes of the simpler orbitals

Figure 6-9 Shell effect.

Figure 6-10 Possible 1s, 2s, and 2p electrons in an atom.

Each "box" represents one possible set of *n*, *l*, and *m* quantum numbers. Each "box" can hold two electrons (one with each of the two possible *s* quantum numbers). Each box corresponds to one of the orbitals in the atom.

(*s* and *p* types) principally to learn the concepts that there are spread out shapes and that there is an orderly pattern to these shapes.

6-7 Ionization Potentials

Ionization potential is a measure of the energy needed to pull an electron away.

There are several measurable properties of atoms that show the effects of the changes in electron energy levels and in nuclear charge. The patterns probably show most clearly in the **ionization potentials.** Ionization potential is a measure of the energy needed to pull an electron away from the attraction of the nucleus. Examples would be the reactions

$$M \longrightarrow M^+ + e^- \quad \text{and} \quad M^+ \longrightarrow M^{2+} + e^-.$$

As such, this quantity should be called ionization energy. But since the charge being pulled away (one electron) is the same in every case it can also be measured as the voltage (or electrical potential) through which that charge has to be moved. Charge times voltage gives units of energy.

Figure 6-11 *p* orbitals.

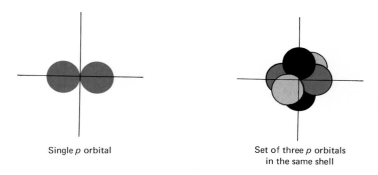

When one electron is pulled off a neutral atom (M $\longrightarrow$ M$^+$ + e^-) leaving an ion with +1 charge, the voltage required is called the **first ionization potential.** If a second electron is pulled off (that is an electron being pulled off a +1 ion, M$^+$ $\longrightarrow$ M^{2+} + e^-), the voltage required is called the **second ionization potential.** The voltage for a third electron would be a third ionization potential, for a fourth electron would be a fourth ionization potential, and so on. A series of ionization potential values (first, second, third, and so on) would be called the **consecutive ionization potentials** of an element.

Figure 6–12 shows a graph of the consecutive ionization potentials for the element carbon. The values are, in order, in V (volts) 11.3, 24.3, 47.4, 64.19, 389.9, and 487. Two kinds of trends show up in Figure 6–12. First, the potential goes up rapidly as more highly charged ions are formed. It is harder to pull an electron away from a +6 charge on the ion left behind than from a +5 charge left behind, and both are much harder than taking an electron away when only a +1 ion is left behind. Second, there is a definite grouping of values. We see larger changes when the type of electron being pulled off changes ($2p$ to $2s$ or $2s$ to $1s$). The change is particularly great when the n quantum number changes from two to one, as noted by the color emphasis on Figure 6–12. There is a tremendous amount of information in plots of consecutive ionization potentials, and we may be confused by the similar effects of increasing charge and changes in electron type.

The differences in the plot caused by the charge on the ion being formed

First ionization potentials, second ionization potentials, and consecutive ionization potentials are described.

Ionization potentials go up as the charge on the produced ion goes up.

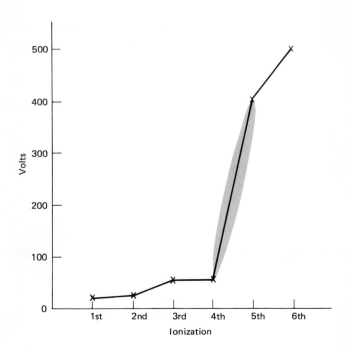

Figure 6–12 Consecutive ionization potentials of carbon.

Figure 6-13 First ionization potentials of the first 20 elements.

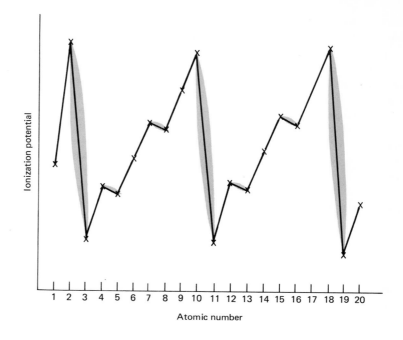

A graph of the first ionization potentials of the elements shows effects of the four main factors affecting electron energy levels.

can be eliminated by plotting a series of different elements each forming ions of the same charge. Because the elements have different numbers of electrons, we can still see the changes in the plot caused by changes in the type of orbital from which the electron is being removed. Figure 6-13 shows the first ionization potentials of the first 20 elements plotted against the atomic number. This graph shows clearly the four most important factors affecting electron energy. You should find the parts of the graph showing each of these four points and be able to reproduce a graph (not to exact scale) that would include these features in the correct places.

The four main factors affecting electron energy are discussed in the following subsections.

6-7.1 Nuclear Charge Effect. For electrons from the same type of orbital bonding, an increase in nuclear charge always increases the strength with which the electrons are held. Notice how the graph of ionization potentials usually slants upward as we move to larger nuclear charge. As long as the type of electron (1s or 2s or 2p, and so on, and with the same s quantum number) stays the same, the upward slant is steady. All of the exceptions to the upward slant are at points where there is a difference in the type of electron being removed. Since the nuclear charge effect is present whenever we compare two elements, we will notice other kinds of effects by looking for any break in the steady upward slant.

6–7.2 Shell Effect. This is often called the n quantum number effect. All electrons with the same n quantum number have the same average distance from the nucleus. The grouping of electrons spending most of their time in the same region forms what is called a **shell.** The $n = 1$ shell is closer to the nucleus than the $n = 2$ shell, $n = 2$ is closer than $n = 3$, and so on. Notice from the graph that there is a very large change in ionization potential when there is a change in the shell the electron comes from. Between helium (which loses a $1s$ electron) and lithium (which loses a $2s$ electron), between neon (which loses a $2p$ electron) and sodium (which loses a $3s$ electron), and between argon (which loses a $3p$ electron) and potassium (which loses a $4s$ electron, there are sharp decreases in the ionization potential. That fits the normal pattern for electricity, that the attraction between positive and negative charges decreases as they become farther apart.

6–7.3 Subshell Effect. This is often called the l quantum number effect. Each electron shell, except the first shell, is divided into subgroups with different l quantum numbers. These subgroups are called **subshells.** The electronic configurations we have been writing are simply the identification of and numbers of electrons in the subshells. Notice that there is a small break in the expected increase from nuclear charge effect whenever there is a change in subshell. The difference from $2s$ to $2p$ (beryllium to boron) or $3s$ to $3p$ (magnesium to aluminum) is tiny compared to shell effects, but there is clearly a difference between s and p electrons.

6–7.4 Half-filled Subshell Effect. This is often called the s quantum number effect. In the p subshells (where there are enough electrons to establish a trend) there is another small break in the pattern at the halfway point. A very careful examination of nature would show that s, d, and f subshells also have a difference between the first half of their possible electrons and the second half. This happens because all of the electrons with one of the two s quantum numbers are held slightly more strongly than those with the other s quantum number. Because of this difference, each of several equivalent orbitals will get one electron before any of them get two. For example, if the three $2p$ orbitals were labeled separately as $2p_x$, $2p_y$, and $2p_z$, the electronic configuration of nitrogen would be $1s^2 2s^2 2p_x^1 2p_y^1 2p_z^1$. Arrangements such as $1s^2 2s^2 2p_x^2 2p_y^1$ would be less favorable. This tendency toward using the maximum number of equivalent orbitals is sometimes called Hund's rule of maximum multiplicity.

The four factors were listed in order of decreasing importance. We will use those factors to explain natural phenomena. Sometimes the two most important factors, nuclear charge effects and shell effects, will be enough to explain things. In other cases subshells and spin (half-filled subshells) will also have to be considered.

Nuclear charge effects and shell effects are the most important factors affecting electron energy levels.

6-8 Paramagnetism

Magnetism is an important property of matter. Some substances interact with magnetic fields, showing that they have a net magnetic field of their own. These substances are said to be **paramagnetic.** Other substances do not interact with magnetic fields, showing they have no net magnetic field. The noninteracting substances are said to be **diamagnetic.**

Atoms have internal magnetic fields associated with the m and s quantum numbers of their electrons. When all of the electrons of an atom are in pairs (two in each orbital that has any electrons), the magnetic fields are arranged so they cancel each other and the atom is diamagnetic. But when one or more electron in an atom is unpaired, the magnetic fields cannot cancel and the atom must be paramagnetic. Hund's rule of maximum multiplicity, as mentioned above in Section 6-7.4, assures us that atoms (and the ions and molecules we will discuss later) will be paramagnetic whenever there are enough equivalent energy levels to permit electrons to remain unpaired. This measureable magnetic property will be an important clue to the true electronic levels in later examples in Section 8-1, Section 8-7, and Sections 16-2 and 16-3.

Unpaired electrons cause paramagnetism, and electrons remain unpaired whenever there are enough equivalent energy levels to permit unpairing (Hund's rule).

6-9 Electron Affinity

Another property that shows the effects of electron energy levels is **electron affinity.** Electron affinity is the opposite of ionization potential. Ionization potential measures the energy *cost* to pull an electron *away* and make a *positive* ion. Electron affinity measures the energy *gain* when an extra electron is brought *in* to form a *negative* ion in a process such as $X + e^- \longrightarrow X^-$.

Electron affinity is affected by the factors described under ionization potential. As nuclear charge increases, more energy can be released when an electron comes into a particular energy level. When the electron has to be put in a shell farther from the nucleus, much less energy can be released. Those are the nuclear charge and shell effects. Subshell and half-filled subshell

Electron affinity is a measure of energy of attraction for an additional electron.

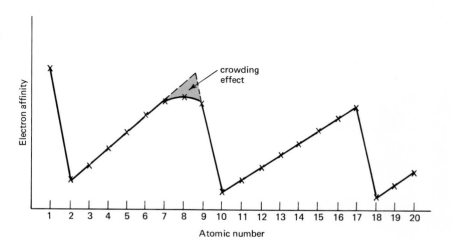

Figure 6-14 Electron affinity.

effects are hard to see in electron affinity, but another kind of effect does show up. Because an electron is going into the atom, there must be room to place it. As shells become filled it becomes difficult to place another negative charge at the same average distance from the nucleus as the many other electrons already there. The result is a **crowding effect** that prevents release of as much energy of attraction as expected from the increasing nuclear charge. The graph of electron affinities in Figure 6–14 shows that this is particularly important toward the end of the second shell, the smallest shell that can hold eight electrons before the next shell begins to fill. This crowding effect could be added as item 5 on the list in Section 6–7 of main factors affecting electron energy.

A graph of electron affinities shows that crowding can also affect electron energy levels.

6–10 Atomic and Ionic Sizes

The sizes of atoms and ions are also affected by nuclear charge and shell effects (Figure 6–15). Within any given shell, the sizes of atoms decrease as the nuclear charge increases (nuclear charge effect). The stronger attraction toward the larger nuclear charge holds the electrons at a closer average distance. Notice that the distance for a given n quantum number is not fixed. In any atom a $2s$ electron averages farther from the nucleus than a $1s$ electron, closer than a $3s$ electron, and at the same average distance as a $2p$ electron, but this distance is not the same as $2s$ or $2p$ electrons in atoms of other elements.

The shell effect causes sharp increases in size whenever a new shell is added. The combination of nuclear charge and shell effects leads to steady decreases in size in each shell followed by increases in size each time we move to a new shell. As we go down the periodic table the atoms become larger because the shell effect causes increases greater than the decreases caused by the nuclear charge effect.

Crowding effects are also important to atomic sizes, as can be seen from the graph in Fig. 6–15. Crowding can be seen at the end of the second shell and in all of the elements just before a new shell is begun. The latter cases

Nuclear charge effects, shell effects, and crowding effects have influences on atomic and ionic sizes.

Figure 6–15 Sizes of atoms.

This figure is a simplified adaptation of Figure 8–5. Used by permission of J. Arthur Campbell, Harvey Mudd College.

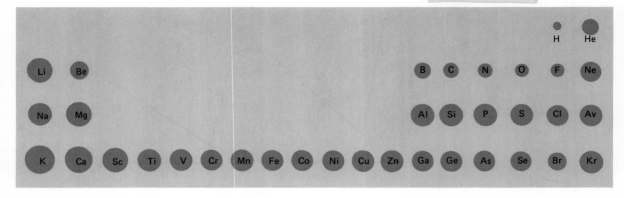

Figure 6–16 Size of negative ions.

Size of atom

Size of negative ion formed from that atom

are not really fair comparisons; different methods of measuring size must be used because of the failure of the noble gases to form chemical bonds as readily as other elements (see Chapter 7).

The sizes of ions are affected by the same factors. The most important points about their differences from atoms are

Three general points about ionic sizes are given.

1. All negative ions are larger than neutral atoms of the same elements. Crowding increases the tendency of the electrons to push apart from each other, and there has been no increase in nuclear charge to pull them in more tightly. Figure 6–16 shows an example.
2. All positive ions are much smaller than the atoms from which they are formed. Less crowding while the nuclear charge stays constant would make them somewhat smaller, but they usually have another major factor making them smaller. The atoms with only a few loosely held electrons in their outermost shell are most likely to form positive ions. If they lose all of the electrons in the outer shell, they drop back to the size of the next shell. There is an important advantage for small ions (see Chapter 7). The ions where the outer shell has been lost become very small and are the only positive ions that form in nature. Figure 6–17 shows an example.
3. A series of ions with the same electronic configurations (called an isoelectronic series) can show the nuclear charge effect on sizes independent of shell or crowding effects. Figure 6–18 shows an example.

6–11 The Periodic Table II

The electronic configurations and electron energy levels we have been describing provide an explanation for the periodic behavior of the elements. The periodic table is a neat way of summarizing the electronic levels of atoms.

Figure 6–17 Size of positive ions.

Size of Mg atom

Size of Mg^{1+} ion (not formed in nature)

Size of Mg^{2+} ion (formed in nature)

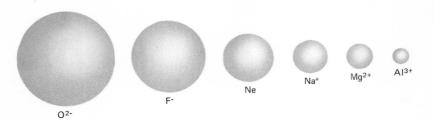

Figure 6-18 An isoelectronic series.

The elements in any one column are called a family of elements because of their similarities.

All of the elements found in the first column of the periodic table (H, Li, Na, K, Rb, Cs, Fr) have a single electron (an s type electron) in their outer shell. As noted in Chapter 8, hydrogen does not fit this group because of the uniqueness of its electron being the only one and being in a shell that can hold only two electrons, but the rest of the group forms a chemical "family." This family of elements are chemically similar because they can do those things that happen with one s electron in the outer shell. Generally speaking, they tend to lose that electron to form $+1$ ions because of their low ionization potentials. They do not lose a second electron to form $+2$ ions because too much energy would be required (shell effect).

All of the elements found in the next to the last column of the periodic table (F, Cl, Br, I, At) have seven electrons in their outer shell (an s^2p^5 configuration in that shell). This family of elements is chemically similar because they can do those things that happen with an s^2p^5 arrangement in the outer shell. Generally speaking, they tend to gain an electron to form -1 ions because of their high electron affinities. They do not tend to gain a second electron because they have little attraction for one in the next shell (shell effect).

All of the elements found in the last column of the periodic table have filled their outer shell to the point where the next electron would go into a new shell. This arrangement is called a **closed shell** because no more electrons can enter the outer shell without also adding a new outer shell. They are not good at losing electrons because all of their electrons are strongly held (nuclear charge effect). They are not good at gaining electrons because the orbitals at the next energy level would not hold electrons as strongly (shell effect). Under most conditions, they do not gain or lose electrons. That makes them a family chemically similar to each other.

Each column in the periodic table is tied together by the similarity in the electronic configurations of its elements. Their energy levels may not be quite as similar as their configurations because shell effects change sizes as we go down each column. But the general family relationship in each column must still be fairly close. With this background on electronic arrangements, we are now ready to consider the actual bonding processes through which elements form compounds.

Periodic behavior of elements occurs because of similarities in electronic configurations.

Skills Expected at This Point

1. You should be able to describe the significance of n and l quantum numbers.
2. You should be able to describe the shapes of s and p orbitals.
3. Given the atomic number, you should be able to write the electronic configuration of any element as predicted by the scheme of Figure 6–5.
4. You should be able to sketch a graph of the first ionization potentials of the first 20 elements and use it to point out and describe the four main factors affecting electron energy levels.
5. You should be able to recognize the effects of crowding of electrons.
6. You should be able to describe the periodic relationships in ionization potential, electron affinity, sizes of atoms and ions, and arrangement in the periodic table.

Exercises

1. List two important features in Mendeleev's periodic table besides arrangement of elements in a table.
2. (a) What is the significance of a row (across) in the periodic table? (b) What is the significance of a column (up and down) in the periodic table?
3. Describe the first two quantum numbers and the rules about their allowed values.
4. Why are there possible places for six $2p$ electrons in an atom? A statement of the Pauli exclusion principle should be part of your answer.
5. Write the electronic configurations of the following elements: (a) Li (at. no. 3); (b) Si (at. no. 14); (c) Ti (at. no. 22); (d) Bi (at. no. 83).
6. What does the name $3d$ mean when used to describe an electron?
7. (a) What is the shape of an s orbital? (b) How are p orbitals arranged relative to each other? (c) How many electrons can a single d orbital hold?
8. Sketch a graph of the first ionization potentials of the elements.
9. Label the graph in exercise 8 to show the effects of the four main factors affecting electron energy levels.
10. Describe the ways in which a graph of the second ionization potentials of the elements would differ from the graph of the first ionization potentials.
11. Why is the crowding effect more important to electron affinities than it is to ionization potentials?
12. Which do you expect to be larger in each of the following pairs? Why?
    (a) Li and C   (c) F and Na   (e) Na and Na^+
    (b) Li and K   (d) F and F^- (f) F^- and Na^+

Test Yourself

1. Sketch a graph of the first ionization potential versus atomic number for the first 20 elements and *circle* each point showing a half-filled subshell effect.

2. For each of these questions provide a very short answer.
   (a) Give the quantum number value always characteristic of p electrons.
   (b) What is the number of electrons possible in the $4f$ subshell?
   (c) What is the shape of an s orbital?
   (d) Give the number of electrons possible in one of the $3d$ orbitals.
   (e) Give the number of $3d$ orbitals in the $3d$ subshell.
   (f) What is the factor that causes silicon to be smaller than sodium?
   (g) What is the factor that causes silicon to be larger than carbon?
   (h) Give the electronic configuration of chlorine.
   (i) Give the electronic configuration of O^{2-}.
   (j) Give the electronic configuration of K^+.
3. The "lines" in line spectra represent discrete energies. Which phrase best completes the following sentence? The energy of a particular line is
   (a) related to the energy required to reverse the spin of the nucleus.
   (b) related to the energy required to unpair a pair of electrons.
   (c) related to the energy difference between an excited state and another state which is less excited.
   (d) greater than the energy required to ionize an atom.
4. The electronic configuration of nickel (at. no. 28) is
   (a) $1s^2 2s^2 2p^6 3s^2 3p^6 4s^2 3d^8$
   (b) $1s^2 2s^2 2p^6 3s^2 3p^6 3d^{10}$
   (c) $1s^2 2s^4 2p^6 3s^8 3p^8$
   (d) $1s^2 2s^4 2p^4 3s^6 3p^6 3d^6$
   (e) none of the above
5. The electronic configuration, $1s^2 2s^2 2p^6 3s^2 3p^6$ describes which species?
   (a) K^+
   (b) S^{2-}
   (c) Ar
   (d) Cl^-
   (e) All of the above
6. The element with electronic configuration $1s^2 2s^2 2p^6 3s^2 3p^6$ is
   (a) Ar (argon)
   (b) P (phosphorus)
   (c) S (sulfur)
   (d) Cl (chlorine)
   (e) Al (aluminum)
7. Which of the following has the lowest ionization potential?
   (a) Be
   (b) F
   (c) Ne
   (d) Ca
   (e) Br (at. no. 35)
8. The steadily increasing first ionization potentials in the series boron, carbon, nitrogen is caused by
   (a) difference in nuclear charge.
   (b) difference in number of shells.
   (c) difference in subshell within the same basic shell.
   (d) difference in electron repulsions (crowding).
   (e) difference in spin (half-filled subshell).
9. The decrease in the first ionization potential of boron compared to beryllium is caused by
   (a) difference in nuclear charge.
   (b) difference in number of shells.
   (c) difference in subshell within the same basic shell.
   (d) difference in electron repulsions (crowding).
   (e) difference in spin (half-filled subshell).

10. Identify the cause of the points indicated on this graph of first ionization potentials versus atomic number.

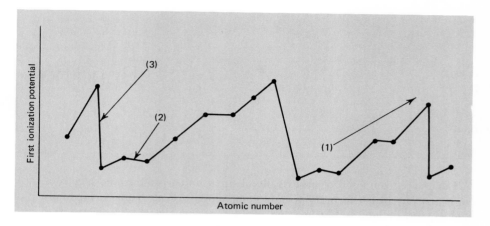

(a) (1) Trend caused by removal of a greater number of electrons, (2) effect of a change in shells, (3) nuclear charge effect.
(b) (1) Trend caused by a change in shells, (2) effect of a change in subshells, (3) effect of half-filled shells.
(c) (1) Trend caused by nuclear charge effect, (2) effect of a change in shells, (3) effect of half-filled shells.
(d) (1) Trend caused by removal of a greater number of electrons, (2) effect of a change in shells, (3) effect of half-filled subshells.
(e) (1) Trend caused by nuclear charge effect, (2) effect of a change in subshells, (3) effect of a change in shells.

11. Which of the following series is in order by size, starting with the largest?
(a) F, Cl, Br, and I
(b) Na^+, Al^{3+}, S^{2-}, Cl^-
(c) Na, Al, Cl, Br
(d) S^{2-}, Cl^-, F^-, F
(e) F^-, F, Na, Na^+

12. From the point of view of electronic configuration, which of the following describes the significant relationship between elements in the *same column* of the periodic table?
(a) Their outermost occupied shells and subshells have the same electronic configuration.
(b) They have the same electronic configurations.
(c) They have to lose one electron to achieve a noble gas configuration.
(d) They obey the octet rule.
(e) The number of electrons in the outermost occupied subshell increases from the top to the bottom of the column.

Bonding

7–1 Conductors and Nonconductors

All matter is composed of electrically charged units, protons and electrons, and neutrons which form atoms and compounds. All elements, compounds, and mixtures contain electrons and could experience a flow of electricity if those electrons or the ions (formed when there are more or less electrons than protons) are transferred to different parts of the matter. But matter differs sharply in its ability to permit a flow of electricity. Flow of electricity through matter is called electrical conductivity. Matter that can permit flow of electricity is called an electrical **conductor.** Matter that prevents the flow of electricity is called a **nonconductor.** Sometimes a given sample of matter will be a conductor under some conditions and a nonconductor under other conditions. The observed differences in conductivity are related to differences in chemical bonding.

Substances that permit flow of electricity are called conductors, whereas those that prevent flow of electricity are called nonconductors.

Metals, such as copper or aluminum, are conductors both as solids and when they are melted to form liquids. They often permit flow of electricity with little resistance, and the electricity does not change the chemical nature of the metal. Copper wire is still copper wire after electricity flows through it.

Some other substances, such as salt, are nonconductors as solids but become conductors when they are melted. These substances also differ from metals because chemical reactions occur while they conduct electricity. When molten salt (formed by melting the solid) carries electricity, separate chemical reactions occur at the points where the electricity flows into and out of the molten salt. These contact points (usually pieces of metal) through which the electricity enters or leaves are called electrodes. The amount of chemical reaction at the electrodes depends upon the amount of electricity that flows. The flow of electricity in molten salts requires a truly molten and salt-like condition, and the resistance depends on the chemical reactions needed at the electrodes.

Metals are conductors under liquid or solid conditions; ionic substances are conductors but only in fluids such as molten salts or solutions; nonionic solutions are nonconductors.

Some other substances are nonconductors as solids and as liquids. They are also nonconductors when mixed to form homogeneous mixtures (solutions) with most other similar nonconductors. But when solutions are formed from solid salts and some nonconducting liquids (particularly water), the solution is a conductor. Conductivity of these solutions depends upon the combination of liquid character and the ionic character of the salt. Ionic character means the salt exists as positive and negative ions. Some substances that are nonionic (and therefore nonconducting) as pure liquid or gases become ionic when they form solutions with water or some other liquids. Those solutions are also conductors, even though none of their parts can be conductors by themselves. All solutions that are conductors will have chemical reactions at the electrodes as electricity flows and a resistance to the flow of electricity which depends both on the electrode reactions and on the concentration of ions available.

7–2 Metals

Elements with low ionization potentials may form metallic crystals.

7–2.1 Metallic Crystals.

Those elements with only a few electrons in their outer shells or with a moderate number of outer electrons but a large number of shells have the lowest ionization potentials, as explained in Chapter 6. Elements with low ionization potential are metals. When present as the elements (not reacted to form compounds) they exist as **metallic crystals.** The ability to conduct electricity and other properties we associate with metals (conduct heat, shiny, able to be formed into new shapes by mechanical working) are all properties of metallic crystals.

Metallic crystals exist as a group of positive ions in a "sea of electrons" which they all share, as shown in Figure 7–1. The metal atoms that form the metallic crystal become positive ions by giving their outer shell electrons to be shared among all the metal atoms in the metallic crystal. The freedom of those electrons to be anywhere in the crystal, instead of on a particular atom all the time, increases the uncertainty about the electrons. That causes an increase in entropy (see Chapter 11) which makes formation of metallic crystals favorable.

On the other hand, taking the electrons away from the individual atoms requires energy (the ionization energy) greater than the heat or other energy provided by letting the electrons be held randomly by the positive ions of the metallic crystal. That requirement of extra energy is unfavorable to formation of metallic crystals. When the ionization energies are low enough, the favorable entropy dominates over the unfavorable energy and metallic crystals form. Elements with the lowest ionization energies (those with the

Figure 7–1 Metallic crystal.

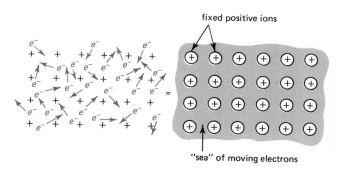

least electrons in the outer shell and the greatest number of shells) should be most metallic. As the electrons in the outer shell become too numerous, the increasing nuclear charge prevents metallic crystal formation, and as the number of shells decreases (for atoms with the same number of outer electrons) a shell effect prevents metallic crystal formation. Figure 7–2 shows the approximate borderline between the metals (toward the left and bottom of the periodic table) and the nonmetals.

7-2.2 Conduction Bands. When electrons are shared among many atoms, a new set of energy levels for electrons is formed. The energy levels the electrons could have had in individual atoms are rearranged into an equal number of energy levels shared among all the atoms in the metallic crystal. This formation of new energy levels is somewhat like the rearrangement into molecular orbital energy levels shown for O_2 in Section 8–1. However, in

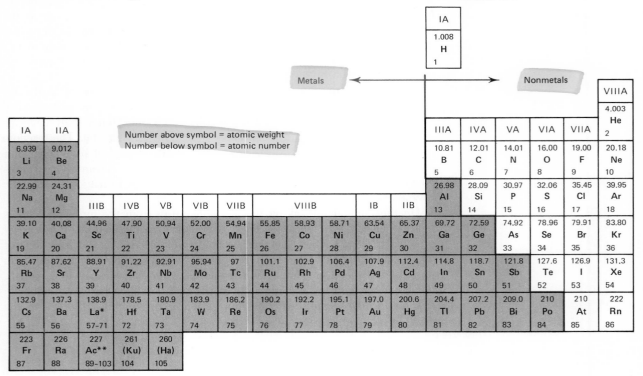

Figure 7–2 Metals and nonmetals.

Figure 7–3 Formation of a conduction band.

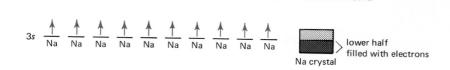

a metallic crystal there are millions of almost identical energy levels. Exclusion rules of nature prevent any two of these energy levels from being *exactly* alike, so they must spread out over a small energy range. That energy range filled with many energy levels is called a **band.** Figure 7–3 shows the formation of a band.

Metallic crystals form conduction bands that allow electrons to move among the atoms.

Metallic crystals have **conduction bands** made from atomic energy levels that were only partly filled with electrons. Therefore, there is room for more electrons in the metallic crystal band than there are electrons to fill the band. The electrons go into the lowest energy parts of the band, but they can easily be "excited" into energy levels in the upper part of the band by addition of very small amounts of energy. The electrons in these bands make exchange or flow of electrons through the metallic crystal very easy. It is that exchange which lets metals conduct electric current. It also lets metals conduct heat by exchanging high kinetic energy electrons from one part of the metal for low kinetic energy electrons from another part. Because of their role in conducting electricity and heat, the partly filled bands in metallic crystals are called conduction bands.

7–2.3 Softness. A true metallic crystal has very little need to take on a particular shape. Unlike covalent bonds, where the electrons being shared must be concentrated in particular directions (see Section 7–12), the metallic crystal has the positive ions surrounded by the electron "sea" in every direction. Therefore, the positive ions can be moved about fairly easily as forces are applied. The electron sea moves with the positive ions. The result is a new shape for the metal whenever force is applied. That sort of behavior is what we call softness in a substance. Metallic crystals tend to be very soft. They are also easily melted at fairly low temperatures because the freedom of motion of liquid phase does not interfere with the attractive forces between the electrons and positive ions of the metallic crystal.

Metals are soft if they are bonded only by metallic crystal forces.

The description we have just given holds quite well for metals like sodium and potassium, which have the most perfectly metallic crystals. But we know it does not fit many other metals, such as iron and nickel. Actually, strength and hardness in metals are signs that their crystals involve bonding forces other than the conduction band electrons of metallic crystals. We will discuss metal strength further in Chapter 16.

7–3 Ionic Substances

Metallic elements, such as sodium, have low ionization potentials. They can give up electrons to form positive ions with only a moderate cost in energy.

As noted in Chapter 6 some elements, such as chlorine, have high electron affinities. They can provide a fair amount of energy when they pick up extra electrons. But even atoms of elements with the largest electron affinities do not give as much energy as is lost in providing the ionization energy for even an atom of the lowest ionization potential elements. If a sodium atom could lose one electron by giving it to a chlorine atom, the Na$^+$ and Cl$^-$ ions formed would be attracted to each other, releasing energy. That energy would both hold the parts together in a definite composition (which is required for any compound) and provide more than enough energy to make up for the difference between the ionization potential of the sodium atom and the electron affinity of the chlorine atom. The elements would then form a compound. The excess energy given off in the process provides a force to make the reaction happen. Figure 7–4 shows this reaction.

The above paragraph is a fair description of bonding between sodium and chlorine. An actual reaction between sodium and chlorine would begin with sodium as a metallic crystal and chlorine as Cl$_2$ molecules, but the additional

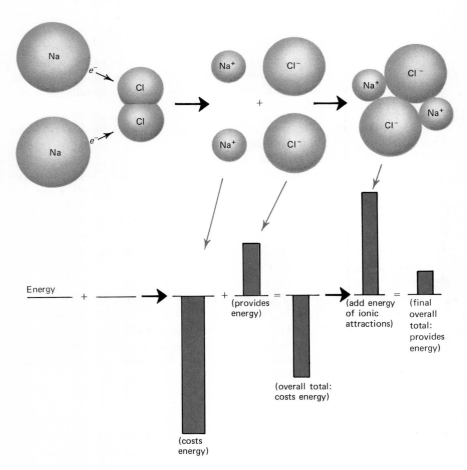

Figure 7–4 Reaction of sodium and chlorine.

110 Bonding [7]

Some compounds exist as ions and are held together by the attractions between positive and negative ions.

need to provide energy equivalent to that needed to convert the sodium and chlorine to atoms does not change the general conclusion. Attraction between positive and negative ions must provide the energy needed for the other steps. The compound formed, NaCl, is common table salt. The bonding force holding it together is the electrical attraction between positive and negative ions. It is the easiest kind of bonding for us to explain from atomic structure. But to be properly scientific in our approach we should insist on more evidence than just a comfortable fit to the theories of atomic structure. If there actually are separate positive and negative ions in compounds (which is true of some but not all compounds), then there should be some definite tests that can show their presence. Human stubbornness and unwillingness to see the facts have left us a neat and quite complete historical record of that evidence.

7–4 Arrhenius and the Theory of Ionization

In Sweden in the 1880s a young graduate student, Svante Arrhenius, was working on his doctorate in physics. He was extending the work of Michael Faraday on electrolysis of solutions.

When a source of electric current is connected to electrodes in a solution, there may or may not be a flow of current. Pure water and solutions of some compounds are nonconductors. Molten salts, solutions of saltlike materials, acids, and bases are conductors, which allow current to flow. While the current flows, chemical reactions occur at the anode (positively charged electrode) and cathode (negatively charged electrode). The amount of chemical reaction is related in a definite way to the amount of electricity that flows. Using values for atomic weights (which had been measured after Faraday's time), Arrhenius was able to confirm a whole-number relationship between the electricity in Faradays (which we now define as units of 1 mole of electrons) and the number of moles of product from chemical reaction at each electrode. Each mole of product seemed to require either 1 Faraday, 2 Faradays, 3 Faradays, or some other whole number of Faradays.

Arrhenius puzzled over how the current was carried and how it was related to reaction at the electrodes. He saw no sign of rays like those which carried current in cathode ray tubes or of contacts like metal wires, and he knew that water alone would not conduct the current. He saw reaction at each electrode but no reaction anywhere else in the solution where the current also had to pass. He concluded that the current was not being carried in the same way as in wires and formulated a theory to explain the behavior.

Arrhenius proposed the existence of ions in solution as an explanation of electrolysis.

The conducting materials (salts, acids, or bases) must be contributing electrically charged particles (ions) to carry the current. Since the compounds are electrically neutral, they must produce equal numbers of positive and negative charges in solution. The positively charged ions would be attracted to the negative electrode (cathode) and the negatively charged ions would be attracted to the positive electrode (anode). When they reached the electrodes, electrical charges had to be transferred so the current could be carried on through the connecting wire in the usual way (which we now know is by a flow of electrons). The transfer of charges caused chemical reaction. The number of Faradays per mole of product was a whole number because some definite number of charges had to be transferred for each ion or molecule that reacted. The electrolysis of NaCl is shown in Figure 7-5.

Let us quote from Arrhenius himself to hear the response his theory received. "I came to my professor, Clive, whom I admired very much, and I said, 'I have a new theory of electrical conductivity as a cause of chemical reactions.' He said, 'This is very interesting,' and then he said, 'Good-bye.' He explained to me later that he knew very well that there are so many different theories formed, and that they are almost all certain to be wrong, for after a short time they disappeared; and therefore by using the statistical manner of forming his ideas, he concluded that my theory would not exist long."

Actually, Arrhenius's theory met with vigorous opposition which led to delaying his doctorate. We will attempt to show the nature of the arguments by a hypothetical conversation between Arrhenius and an opponent.

The theory of ionic solutions was strongly opposed until additional evidence for ions was presented.

Arrhenius: The electric current in electrolysis of salt solution is carried by Na^+ and Cl^- ions being attracted to the oppositely charged electrodes where they react, accounting for the observed chemical reactions.

Opponent: It is generally agreed that electricity is a fluid that flows. There are no such things as ions.

Arrhenius: The salt must be providing particles to carry the flow, some negative and some positive to balance each other. Cl_2 appears at the anode, so the chlorine must be in negative parts. Sodium has to be in positive parts to balance the charges.

Opponent: Salt is composed of molecules. If you separated it into sodium and chlorine, the sodium would be a soft metal that would cause an explosion on contact with water and the chlorine would be

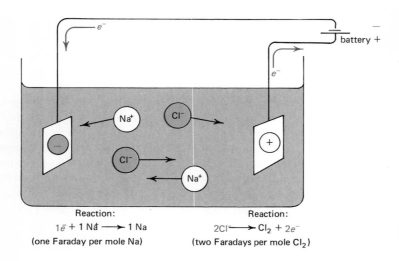

Figure 7-5 Electrolysis of sodium chloride (NaCl).

112 Bonding [7]

Arrhenius: a greenish gas, poisonous in large amounts. I eat salt and water together often and neither explode nor die of poisoning.

Arrhenius: It is not there as sodium and chlorine. It is Na^+ and Cl^-, which act differently, as a salt solution.

Opponent: Ridiculous. Sodium is sodium, chlorine is chlorine, and salt is salt. There is no evidence for these ions you propose.

Arrhenius: How about the conductivity of salt solutions? How about the fixed amounts of chemical reaction from electric current? These provide evidence.

Opponent: The salt solution and electric current work to encourage each other. The salt encourages the electrical fluid to move through the solution. The fluid encourages molecules to react. The more fluid that flows, the more molecules which receive reaction encouragement. It is still the fluid which flows. Ions are not needed for any of this.

A sensible young man concerned about his future might have given up at that point and selected another research problem, but Arrhenius was stubborn. He felt his arguments were correct so he set out to find other evidence for the existence of ions. We will list the main categories he was able to present.

7–5 Electrical Evidence for Ions

The separate movement of positive and negative ions toward the oppositely charged electrodes can be shown in electrolyses where mixing is carefully avoided.

Arrhenius set out to gather evidence that ions were in fact moving separately toward the anode and cathode during electrolysis. Figure 7–6 shows an experiment demonstrating movement of ions. As shown in Figure 7–6A, the center portion of a tube is filled with a blue $CuSO_4$ solution in gelatin. The gelatin is to prevent stirring, convection currents, or other mixing. Each end of the tube is then filled with a colorless conducting solution such as NaCl solution. The tube is hooked up for electrolysis. After several days or weeks the situation in Figure 7–6B is reached. The blue color has clearly moved

Figure 7–6 Movement of ions.

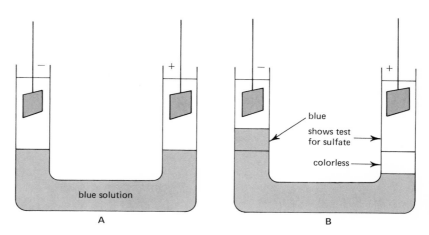

toward the negative electrode, so it is presumed that the positive copper ion is the blue material. At the same time the colorless region around the positive electrode now shows the presence of sulfate through chemical tests such as precipitation of $BaSO_4$.

The response of Arrhenius's opponent might have been something like this:

The blue color is $CuSO_4$ *molecules*. The color moves because they are pushed by the flow of the electric fluid. A reaction must occur which makes a sulfate that is pushed the other direction.

7–6 Chemical Evidence for Ions

Several classes of chemical evidence were obtained.

7–6.1 Color of Ions in Solution. Figure 7–7 shows one of many examples of color similarities in solution for materials that are chemically different as compounds but contain the same ion. In solid form, these compounds are not identical in color, so the color must be associated with the independent ion surrounded by water.

Opponents could note that molecules might also be surrounded by water to give a particular color, and if one copper compound gave blue it seems reasonable that other molecules containing copper are similar and might do the same.

7–6.2 Heat of Neutralization of Acids and Bases. Acids and bases react with each other to form salt solutions. For each equivalent amount of acid (enough to neutralize the same amount of base in each case) neutralized

Individual ions have characteristic colors in solution that are independent of the particular compound dissolved to get them into solution.

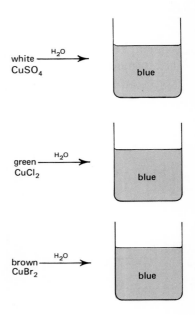

Figure 7–7 Color indicating the same ion from different salts.

by a base, the heat release is identical. For example, the reactions

$$\text{NaOH solution} + \text{HNO}_3 \text{ solution} \longrightarrow \text{NaNO}_3 \text{ solution} + \text{H}_2\text{O}$$

and

$$\text{KOH solution} + \text{HCl solution} \longrightarrow \text{KCl solution} + \text{H}_2\text{O}$$

both give off the same heat, 13.6 kcal/mole (of H_2O formed). In Arrhenius's picture, the strongly conducting acids and bases were present in solution as ions and the salts remained in solutions as ions. Since nothing had happened to the Na^+, NO_3^-, K^+, and Cl^- ions, the only reaction in each case was $H^+ + OH^- \longrightarrow H_2O$. Naturally, the same amount of the same reaction gave the same amount of heat.

The consistency of acid-base neutralizations led Arrhenius to new conclusions about the nature of acids, bases, and neutralization. An acid was simply anything that gave H^+ in solution, a base was anything that gave OH^-, and neutralization was simply the formation of water from ions. This constitutes the Arrhenius theory of acids and bases. Figure 7–8 illustrates the idea.

This made the young, rejected graduate student the author of two major theories, each dependent on the concept of ions. A reaction from his opponents might well have been an unprintable comment about his gall in suggesting so many wild schemes. (Section 13–4 points out some more modern acid-base theories that have become popular because Arrhenius's theory does not explain all acid-base systems perfectly.)

7–6.3 Precipitation Reactions. When salt solutions are put together, precipitates (solids) sometimes form very quickly. Any set of salt solutions containing both parts of a particular precipitate will form it quickly. The rapid pulling together of ions with strong attractions for each other fits the picture, as shown in this example:

$$(Ag^+ + NO_3^-) + (Na^+ + Cl^-) \longrightarrow AgCl\downarrow + Na^+ + NO_3^-$$
$$(2\,Ag^+ + 2\,NO_3^-) + (Cu^{2+} + 2\,Cl^-) \longrightarrow 2\,AgCl\downarrow + Cu^{2+} + 2\,NO_3^-$$

Opponents would note that less favored molecules should form the more favored molecule whenever possible, even if no ions are involved.

7–6.4 Extent of Ionization. Some solutions are less effective conductors of electricity than others. It was suggested that the compounds in these less

The heat of neutralization of acids and bases is independent of the particular acid or base used.

The Arrhenius theory of acids and bases identifies anything giving H^+ in solution as an acid and anything giving OH^- in solution as a base.

Precipitation of solids from solutions is explained in terms of the reaction of ions.

Figure 7–8 Arrhenius's concept of acids and bases.

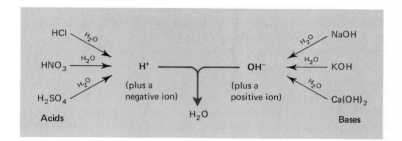

Table 7–1

Condition	Result
AgNO$_3$ solution (conducting) + CCl$_4$ (nonconducting)	no AgCl precipitate
AgNO$_3$ solution (conducting) + HCl solution in H$_2$O (conducting)	AgCl precipitate
AgNO$_3$ + HCl solution in benzene (nonconducting)	no AgCl precipitate
HCl in H$_2$O (strongly conducting)	very acidic behavior
acetic acid (CH$_3$COOH) in H$_2$O (weakly conducting)	only moderately acidic behavior

conductive solutions had only partial ionic character—that they existed partly as neutral molecules. The other properties associated with ions were also less evident under the same conditions. Some examples are shown in Table 7–1. HCl is called a **strong acid**; acetic acid (vinegar) is called a **weak acid**. There are also strong and weak bases; the difference between them is in their extents of ionization.

Those properties caused by ions vary in extent as the extent of ionization of substances vary.

7–6.5 Catalysis. Some chemical reactions are catalyzed (made to go faster) by acid. An example would be the formation of an ester (see Chapter 18).

$$CH_3-\overset{\overset{O}{\|}}{C}-OH + CH_3OH \longrightarrow \text{reaction slow or not seen at all}$$

$$CH_3-\overset{\overset{O}{\|}}{C}-OH + CH_3OH \xrightarrow{\text{in acid}} CH_3-\overset{\overset{O}{\|}}{C}-O-CH_3 + H_2O$$

These reactions were also affected by the extent of ionization, but in this case both the extent of ionization and the catalytic efficiency could be measured numerically. Table 7–2 shows a comparison of extent of ionization (as measured by conductivity) and catalytic efficiency (as measured by rate of reaction) for a series of acids in the liquid ethanol, where some acids are quite weak and acid catalysis of some reactions can be measured.

These data are very impressive to a chemist who is acquainted with measurements of catalysis. The work of Arrhenius on catalysis was brought to the

Reactions catalyzed by acid have rates proportional to the extent of ionization of the acid.

Table 7–2

Acid	Extent of Ionization Compared to HCl	Catalytic Efficiency Compared to HCl
hydrochloric acid (HCl)	100	100
picric acid	10.4	10.3
trichloroacetic acid	1.00	1.04
trichlorobutyric acid	0.35	0.30
dichloroacetic acid	0.22	0.18

attention of Wilhelm Ostwald, a very well known chemist. It convinced Ostwald that the theory of ionization was a correct and important idea. Ostwald was one of the most prominent European scientists of that time, so his opinion was very influential on Arrhenius's advisors. Arrhenius finally received his doctorate quite soon after Ostwald came out in support of the theory of ionization.

Another category of evidence for ions, freezing point lowering and boiling point elevation, is discussed in Chapter 10.

7–7 Oxidation States in Simple Ions

When ions do form, the number of electrons gained or lost is dictated by the electronic configurations discussed in Chapter 5. The charge on a simple (one atom by itself) ion is called the **oxidation state** of that element. Examples are Na^+ ion where the oxidation state of sodium is $+1$, Cl^- ion where the oxidation state of chlorine is -1, and Ca^{2+} ion where the oxidation state of calcium is $+2$. Creation of separate positive and negative charges is always a difficult thing, so the most favorable possible conditions are needed. These favorable conditions are

1. A low ionization potential for formation of the positive ion.
2. A high electron affinity for formation of the negative ion.
3. A very strong attraction between the positive and negative ions.

Favorable conditions are most easily satisfied in substances like NaCl where each element is only one electron removed from a closed shell (the arrangements of the last column in the periodic table). Sodium has just one electron beyond a closed shell, so that electron has a particularly low ionization potential. Chlorine has just one electron less than a closed shell, so an electron can be put in that vacant, low energy level with a particularly high electron affinity. The Na^+ and Cl^- are attracted to each other. Although the Cl^- is quite large, the positive and negative charges get to be closer together (and therefore more strongly attracted to each other) because of the very sharp size reduction when Na^+ was formed from sodium. Small size occurs because the entire outer shell of sodium (one electron) was lost when the Na^+ ion was formed. The energy advantage for small positive ions, with the outer shell completely removed, is one of the most important factors in ionic bonding. The situation that meets the three favorable conditions listed earlier in this section turns out to be

1. One element (a metal) comes from the left side of the periodic table and has a low ionization potential.
2. One element (a nonmetal) comes from the right side of the periodic table (but not the last column) and has a high electron affinity.
3. The outer shell is lost completely when the positive ion is formed, maximizing the attraction between ions.

With the above general rules we can begin to understand some of the cases that are less ideal than NaCl. Magnesium is also an element with low ionization potentials. One electron could be removed from magnesium to form Mg^+. Although it would not be as easy to remove that electron as it was to remove one electron from sodium, it is still easier than for most elements. It is also much easier to remove that first electron from magnesium (first ionization potential) than a second electron (second ionization potential). Therefore, we might expect to find an ionic compound MgCl. We do not find it because Mg^+ has not lost its entire outer shell. To form a really small ion and maximize attractions, magnesium must lose another electron and form Mg^{2+}. That second electron is almost twice as hard to remove (14.96 V ionization potential versus 7.61 V for the first ionization potential), but the extra attractions when it forms are very great. Therefore, we never stop at Mg^+ under natural conditions. We always find the Mg^{2+} which has a closed shell electronic configuration. We cannot stop before reaching the closed shell because we need the size advantage of the small ion. We cannot go beyond the closed shell because there would not be much more advantage in size (no loss of another shell) and the ionization potentials would be prohibitively high. The third ionization potential of Mg is 79.72 volts. Therefore $+2$ is the only oxidation state of magnesium in compounds.

Positive ions must lose their outermost shell to give the small ions maximizing the attractions between positive and negative ions.

The tendency to lose the outer shell dominates most positive ion formations, even in cases where it does not lead to a closed shell configuration. In transition elements like iron (electronic configuration $1s^2 2s^2 2p^6 3s^2 3p^6 4s^2 3d^6$) the $3d$ electrons should be the easiest to remove. However, the preference for losing the outer shell makes the $4s$ electrons come off first. Fe^{2+} ion has an electronic configuration $1s^2 2s^2 2p^6 3s^2 3p^6 3d^6$. Because the $3d$ electrons are not very strongly held, they may or may not also be removed (but that does not change the number of shells). Factors that affect the oxidation states of transition elements will be discussed further in Chapter 8.

Although there is no change as striking as the size change in positive ions, negative ions also tend to reach closed shell configurations. Generally speaking, if the electron affinity for one extra electron is good enough for ion formation, the attraction for a second (or third) electron into the same energy level is also good enough when there is another vacant position. Once again, it is not possible to go beyond the closed shell because the energy factors become very unfavorable. The number of elements that can form simple negative ions is actually fairly limited. Instead of forming highly charged negative ions, they usually fill the low lying energy levels by covalent bonding (see Section 7-8). The covalent bonding forms groups of atoms, which may be neutral molecules or may be ions with the net charge spread out over several atoms. These complex units can tolerate a large negative charge better than most single atoms. We will discuss them further in Sections 7-11 and 7-12.

Ions tend to form by gaining or losing a small number of electrons to reach a closed shell configuration.

7-8 Covalent Bonding

Covalent bonding allows both atoms to fill low lying energy levels by sharing pairs of electrons.

Many combinations of elements cannot satisfy the conditions required for ionic bonding, yet they form compounds. They do so by effectively taking advantage of one favorable part of ionic bonding while avoiding the unfavorable part.

The attraction of electrons into the low lying energy levels of atoms is energetically favorable. It is the removal of electrons to form positive ions that is energetically costly and hard to arrange. If we take two elements that cannot form an ionic compound, they may still be able to bond if they could *share* electrons to fill the low lying levels without either element actually giving up electrons. This bonding would be particularly favorable between two nonmetals where each has very good electron affinity but neither is any good at losing electrons.

The process of sharing electrons is called **covalent bonding.** It occurs with pairs of electrons being shared between atoms. The pair of electrons being shared is in an orbital created by the interaction of a particular orbital on one atom and another orbital on the other atom. Because each atom has an attraction for electrons in the available low lying orbitals, the pair of electrons is held by both atoms and that holds the atoms together. The bonding force can be understood by an analogy to two opposing football players going after a fumbled ball. If they both get a hold on the ball, they will be held together until one of them lets go. If they both have a very good grip on the ball (like atoms with good electron affinity holding the pair of electrons), an extremely large force may be needed to pull them apart. Covalent bonds between atoms can be very, very strong.

In order to bond atoms covalently, each atom must have positions to hold the electrons strongly. Usually that means that each atom could take on extra electrons until it reaches a closed shell. Beyond the closed shell arrangements, the attractions are usually not strong enough for effective bonding. Therefore, covalent bonding is usually the formation of shared pairs until each atom present reaches a closed shell configuration.

One of the simplest examples of covalent bonding is the bonding of an element to itself to form **diatomic molecules** (two atom units). In chlorine each atom needs one more electron to reach a closed shell arrangement. In a Cl_2 molecule each chlorine atom has made one electron available to a pair being shared. This forms a covalent **single bond** holding the unit together. A single bond has one pair of electrons being shared. Figure 7-9 illustrates the sharing of pairs of electrons to form covalent bonds.

More than one pair of electrons may need to be shared to reach closed shells. Oxygen can bond to chlorine to form Cl_2O molecules. Each chlorine atom needs one more electron and can obtain it by sharing one pair of electrons in a single covalent bond very similar to the bond in Cl_2. But the oxygen atom needs two more electrons to complete its closed shell. Since it can only gain one electron by sharing one pair of electrons with a chlorine atom, it forms two such bonds. Figure 7-9 illustrates the formation of two single covalent bonds in Cl_2O.

In Cl_2O, oxygen had to form two separate bonds to gain two electrons.

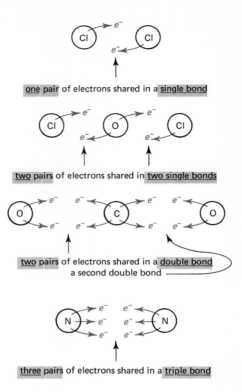

Figure 7-9 Covalent bonds.

In some other cases it can gain two electrons by forming one bond in which four electrons are shared, two from each of the bonded atoms. The bonding of carbon and oxygen in CO_2 has bonds in which two pairs (four electrons) are shared. These bonds are called **double bonds.** They should provide about twice as much bonding as sharing one pair of electrons. Since carbon needs four extra electrons to reach a closed shell (and has four outer shell electrons to use in sharing), it can form two double bonds whereas each oxygen can form only one double bond. The bonding of CO_2 shown in Figure 7-9 is the result.

In cases where at least three electrons are needed to reach a closed shell, a **triple bond** may form. A triple bond has three pairs (six electrons) shared between two atoms. Nitrogen atoms form N_2 molecules by each nitrogen atom contributing three electrons to a triple bond as shown in Figure 7-9.

When two (or more) different elements are bonded covalently, the numbers of each kind of atom are fixed so that each atom can reach a favorable arrangement. When nitrogen bonds to hydrogen, each nitrogen needs three more electrons to reach a closed shell. Each hydrogen has only one electron to share and only needs one more electron to reach a closed shell (the $1s^2$ configuration of helium). Therefore three hydrogen atoms each contribute an electron to the same nitrogen and receive one electron from that nitrogen to complete the pair for a single bond.

Covalent bonds are single bonds when one pair of electrons is shared, double bonds when two pairs are shared, and triple bonds when three pairs are shared.

Sometimes electrons are shared unevenly. In the case of the NH_3 molecule just described, six of the eight electrons in the completed outer shell of nitrogen are being shared in bonds. The other two electrons could also be shared, but the nitrogen could not accept the share of any more electrons without exceeding a closed shell. Additional hydrogen atoms would not bond to NH_3. The electron on each hydrogen is already in the orbital that would have to take the shared pair, and the nitrogen would have no place to put that electron which the hydrogen would have to try to share. However, an H^+ ion could bond to NH_3. The H^+ ion has no electrons to share. It has room for two electrons, so it can share a pair of electrons, both of which came from nitrogen. This kind of one-sided sharing is called a **coordinate covalent bond.** It can occur when one atom has a pair of unshared electrons in its outer shell and the other atom has room for two more electrons in its outer shell. After a coordinate covalent bond forms, we lose track of whether the electrons came from only one atom or both and have simply a covalent bond in which a pair of electrons is shared. Therefore, in NH_4^+ each hydrogen is covalently bonded in the same way (a single bond) with no way to distinguish the one added as H^+ from the other three.

<small>Covalent bonds are coordinate bonds when only one atom provides the pair to be shared.</small>

Another kind of unequal sharing occurs when the pair of electrons in a bond is held more strongly by one element than the other. In such cases there is a concentration of the negative charges of the bonding electrons on the atom that holds them more strongly. That creates positive and negative portions in the molecular unit. This separation of charges is called a dipole, and the bond is called a **polar covalent bond.** The attraction between positive and negative parts in a polar covalent bond provides a bonding force similar to ionic bonding. At the same time, the unequal sharing must mean that the covalent bonding force is less than it would be if the electron pair was shared equally. (The total of covalent plus ionic bonding may, however, be as much or more than the covalent bonding force would be if the sharing were equal.) If the sharing becomes far from equal, the ionic type attraction becomes greater than the covalent attraction. It is customary simply to call any bond where over half the attraction is ionic an ionic bond. There is often some contribution of covalent bonding to the total force of bonding in these ionic bonds. Bonds where over half the attraction is covalent are called covalent bonds, with the special qualification polar covalent reserved for those in which the ionic contribution is noticeably significant.

<small>Polar bonds have the shared electrons held more strongly by one atom than by the other.</small>

7-9 Electronegativity and Predicting Ionic or Covalent Character

It is possible to determine the borderline (the 50% ionic, 50% covalent condition) between ionic and covalent bonding by use of a property called **electronegativity.** Electronegativity is a measure of how strongly an atom attracts electrons in covalent bonds. Because the negative charge is shared with the other bonded atom, electronegativity is less affected by crowding than electron affinity. Figure 7–10 shows a set of measured values for electronegativity. On this scale, a difference of 1.7 in electronegativity between

H 2.1																	He —
Li 1.0	Be 1.5											B 2.0	C 2.5	N 3.0	O 3.5	F 4.0	Ne —
Na 0.9	Mg 1.2											Al 1.5	Si 1.8	P 2.1	S 2.5	Cl 3.0	Ar —
K 0.8	Ca 1.0	Sc 1.3	Ti 1.5	V 1.6	Cr 1.6	Mn 1.5	Fe 1.8	Co 1.8	Ni 1.8	Cu 1.9	Zn 1.6	Ga 1.6	Ge 1.8	As 2.0	Se 2.4	Br 2.8	Kr —
Rb 0.8	Sr 1.0	Y 1.2	Zr 1.4	Nb 1.6	Mo 1.8	Tc 1.9	Ru 2.2	Rh 2.2	Pd 2.2	Ag 1.9	Cd 1.7	In 1.7	Sn 1.8	Sb 1.9	Te 2.1	I 2.5	Xe —
Cs 0.7	Ba 0.9	57-71 La-Lu 1.1-1.2	Hf 1.3	Ta 1.5	W 1.7	Re 1.9	Os 2.2	Ir 2.2	Pt 2.2	Au 2.4	Hg 1.9	Tl 1.8	Pb 1.8	Bi 1.8	Po 2.0	At 2.2	Rn —
Fr 0.7	Ra 0.9	89- 1.1-1.7															

Figure 7–10 Electronegativities of elements.

the two atoms being bonded is considered the approximate point for the borderline between ionic and covalent bonding. It does not always give the right answer.

Figure 7–11 shows a simplified set of electronegativity values. You can memorize this set easily and use it to get reasonable first estimates about whether a bond should be considered ionic or covalent. It is important to remember that Figure 7–11 will give you only crude estimates. You will find some borderline predictions from this method incorrect.

The procedure for use with Figure 7–11 is as follows: (*Note:* Check first to see if your instructor wants you to use this method.)

> Electronegativity differences can be used to estimate covalent, polar covalent, or ionic character.

1. Memorize the values 4.0 for fluorine and 2.5 for hydrogen (The 2.5 value for hydrogen is simply an arbitrary choice selected because it gives correct results more often than the measured 2.1 value does.)
2. In the seven columns of the periodic table omitting group 8 and all the "late starting" columns (transition elements and inner transition elements) the values vary as follows (starting from the 4.0 of fluorine):
   (a) Decrease by 0.5 each time you move one space left or one space down until the value gets down to 1.0.
   (b) Do not reduce the values below 1.0 for any element not in the first two columns.
   (c) Use the value 0.9 for any element in the first or second column for which the 0.5 decrease per move left or down would give a value less than 1.0.
3. Take the elements bonded to each other and subtract the smaller "electronegativity" from the larger. (Do *not* multiply by the number of atoms.) In NH_3, nitrogen is still 3.0 and hydrogen 2.5. Each co-

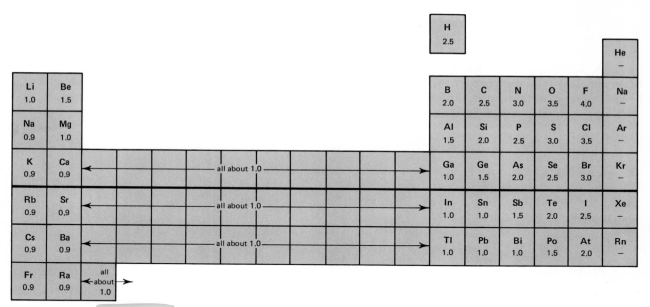

Figure 7–11 Simplified approximate electronegativities.

Elements below the dark line are less abundant than those above the line. This low abundance region includes those elements for which the approximate values are farthest from the real electronegativities.

valent bond always involves only two atoms, in this case one nitrogen and *one* (not three) hydrogen.

4. If your answer is 0 to 0.5 the compound is *covalent*. It is not strongly polar (although any difference at all makes it slightly polar).
5. If your answer is 0.6 to 1.5, the compound is *polar covalent*. (A few of those with 1.5 may actually be ionic in nature. LiI is such a case.)
6. If your answer is 1.6 or higher, the compound is *ionic*. (A portion of those with 2.0 or less may actually be polar covalent in nature. $AlCl_3$ is such a case, actually existing as Al_2Cl_6 molecules.)

Some instructors will have their classes skip the above procedure because of the difficulty in fitting every case. Others will include it to give students a feeling for the ionic-covalent borderline in nature. Those students who use it should recognize its limitations.

7–10 Lewis Dot Diagrams

The usual pattern of covalent bonding, sharing of pairs with each atom reaching a closed shell, lends itself to a convenient shorthand. In this shorthand the symbol of each element is used to represent the nucleus plus all electrons in shells other than the outermost shell. That grouping of nucleus and inner electrons is called the **kernel** of the atom. The electrons in the outer shell are represented by dots around the symbol. These outer electrons are called **valence electrons.** One dot is used for each valence electron. Since there must be two electrons in each covalent bond, the dots are grouped in pairs. Except for hydrogen and helium, where two electrons complete a closed shell, each atom tends to go to eight electrons to complete its outer shell. That can be represented neatly by four pairs of dots, one pair on each of the four sides around the symbol. This kind of diagram, originally proposed by G. N. Lewis at about the time of World War I, is called a **Lewis dot diagram.** Lewis dot diagrams are very useful in describing bonding and will be used throughout this course. An example of the representation of a closed shell configuration is shown here for the neon atom.

Lewis dot diagrams are a useful way of showing electronic arrangements.

$$:\!\ddot{\text{Ne}}\!:$$

The symbol Ne represents the neon nucleus and the two inner electrons ($1s^2$). The outer eight electrons ($2s^2 2p^6$) are represented by the four pairs of dots around the Ne symbol. Here are some more examples of dot diagrams of neutral atoms.

$$\text{Na}\cdot \qquad :\!\ddot{\text{Cl}}\!\cdot \qquad \text{H}\cdot \qquad :\!\dot{\ddot{\text{O}}}\!\cdot$$

When ions are formed, the symbol continues to represent what was the kernel of the neutral atom. Therefore, most positive ions have no dots because they have no electrons left in the outer shell. Most negative ions complete the outer shell and therefore have eight dots. The exception would be H$^-$ ion where only two dots are needed for a closed shell. Here are some examples of dot diagrams of simple ions.

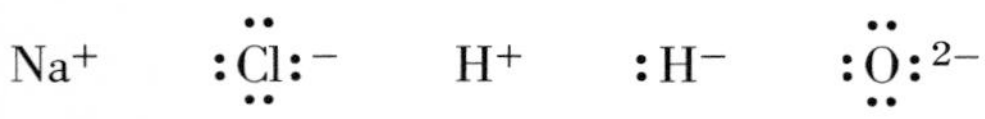

$$\text{Na}^+ \qquad :\!\ddot{\text{Cl}}\!:^- \qquad \text{H}^+ \qquad :\!\text{H}^- \qquad :\!\ddot{\text{O}}\!:^{2-}$$

When molecules are formed by covalent bonding, the shared electrons appear between the symbols of the atoms being bonded and are part of the outer shells of both. If more than one pair of electrons are being shared, more than one pair of dots appears between the atoms. Here are some examples of dot diagrams for molecules.

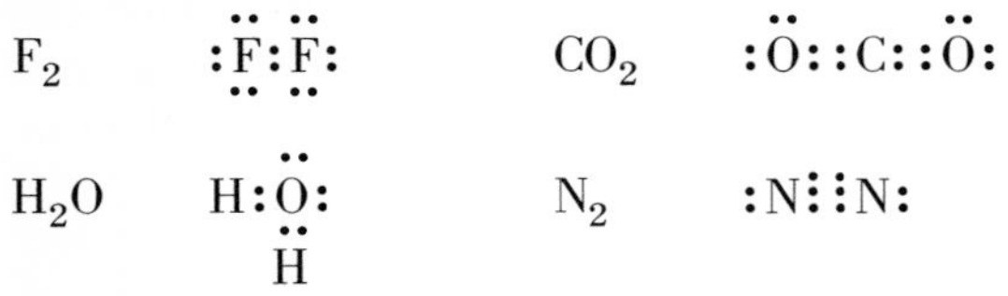

$$\text{F}_2 \qquad :\!\ddot{\text{F}}\!:\!\ddot{\text{F}}\!: \qquad \text{CO}_2 \qquad :\!\ddot{\text{O}}\!:\!:\!\text{C}\!:\!:\!\ddot{\text{O}}\!:$$

$$\text{H}_2\text{O} \qquad \text{H}\!:\!\ddot{\text{O}}\!: \qquad \text{N}_2 \qquad :\!\text{N}\!:\!:\!:\!\text{N}\!:$$
$$\phantom{\text{H}_2\text{O} \qquad }\text{H}$$

Further examples of dot diagrams will be used as other bonding concepts are introduced in the following sections and as important examples are described later in the book.

7–11 Complex (More Than One Atom) Ions

Sometimes combinations of atoms can bond together covalently in an especially good arrangement if they have either a few extra electrons or a few less electrons than the number originally on the atoms. These units will tend to form with the appropriate net charge. They are called **complex ions** since

Complex ions are covalently bonded units with a net electrical charge.

the ion is an entire unit rather than a[...]
see why and how these complex ions f[...]
the dot diagrams of their atom parts al[...]
the need for more or less electrons.

$$:\!\dot{\underset{\cdot}{N}}\!\cdot + 4\,H\cdot - e^- \longrightarrow$$

$$:\!\overset{\cdot\cdot}{\underset{\cdot\cdot}{S}}\! + 4\,:\!\overset{\cdot\cdot}{\underset{\cdot\cdot}{O}}\! + 2\,e^- \longrightarrow$$

$$:\!\dot{\underset{\cdot}{P}}\!\cdot + 4\,:\!\overset{\cdot\cdot}{\underset{\cdot\cdot}{O}}\! + 3\,e^- \longrightarrow$$

7-12 Resonance

Resonance is a variation of covalent sharing superior to those which can be shown by dot diagrams in some cases.

Anytime two equally good "best" dot diagrams can be written for the same unit, resonance exists.

Electrons shared between atoms in σ bonds are concentrated near the bond axis.

The tendency for bonded units to form in the m[...]
usually results in closed shells for every atom. Ho[...]
several possible arrangements from which to cho[...]
nomenon called resonance becomes important. Res[...]
tion that cannot be represented accurately by a Lew[...]
the existence of resonance can be quickly and easily [...]
of a group of dot diagrams. None of these dot diag[...]
but as a group they show the presence of a resona[...]
bonding arrangement that actually exists for that mol[...]
in between them. With the help of our artist, we [...]
the actual situation of resonance. You may later us[...]
the "incorrect" but simpler dot diagram representat[...]
should be careful not to let the convenience of the d[...]
you into thinking resonance is what that method show[...]
best bonding arrangement, and no dot diagram can acc[...]
arrangement if it is resonance bonding.

The common cases of resonance involve interaction [...]
positions where covalent double bonds could occur. To u[...]
we must begin with a more detailed picture of covalen[...]

Figure 7-12 shows formation of a covalent single bond [...]
is concentrated along the axis of the bond (the line betw[...]
In each atom only one pair of electrons in the outer shell [...]
in that direction. The direct overlap of where the electr[...]

Figure 7–12 Covalent single bond (a σ bond).

and where they should be for the other atom allows effective sharing.
of sharing directly down the bond axis is called a **sigma (σ) bond.**
ent single bonds are σ bonds.
a double bond forms, only one electron pair can be shared in the
etween the atoms. The closest other possible positions for sharing
orbitals perpendicular to the direction of the σ bond. If p orbitals
toms are lined up parallel to each other, there can be some sharing
ns between them, as shown in Figure 7–13. This arrangement, called
bond, has some side effects, particularly in holding the atoms rigidly
so the p orbitals remain parallel and close to each other. It does
te to holding the atoms together.

Electrons shared between atoms in a π bond are concentrated in a pair of regions parallel to the bond axis.

rrangement of electrons on atoms does not involve atomic p orbital
s unless the p electrons can be shared to form a π bond. They can
form hybrid orbitals involving combinations of s and p orbitals (and
als when available). These hybrid orbitals are discussed in Chapter
Chapter 17. They permit grouping of the pairs of electrons in the
individual orbitals in positions farther apart from other pairs of
ns. Those hybrid orbitals can be used to form σ bonds. The tendency
their pairs of electrons away from other pairs is a dominant factor
rmining molecular geometry, as discussed in the next section (Section
In Figures 7–14 and 7–15 we have shown the electrons in these hybrid
positions (groups of two electrons with maximum separation of groups)
ver they are not involved in π bonds.
ure 7–14 shows the two "likely" dot diagram structures for SO_2 and
tual bonding these represent. The two dot diagrams for SO_2 show that
uld have resonance. Actually, neither of the dot diagrams represents
rue bonding arrangement. Figure 7–15 shows the actual resonance
ition of SO_2. The p orbitals on the sulfur and both oxygens are lined

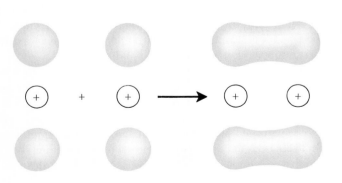

Figure 7–13 Pi (π) bond.

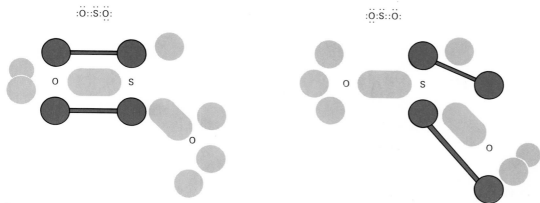

Figure 7–14 Dot diagram structures of sulfur dioxide.

Resonance has π bonds spread over more than two atoms at the same time.

up parallel. If there is an advantage to spreading two electrons out over a space that could hold four (the π bond part of the bond), there is even more advantage to spreading four electrons out over a larger space that could hold six. This arrangement is not just equal to those shown in Figure 7–14, it is better than either of them. Therefore, the bonding pattern chooses the most favorable arrangement (Figure 7–15), and the less favorable patterns shown by the dot diagrams never form. The resonance bond is a high entropy condition (see Chapter 11). The positions of the electrons become less certain when they are shared among three atoms instead of belonging to one atom alone or being shared by two atoms. There is a preference in nature for the less certain (high entropy) conditions.

If spreading electrons in covalent sharing over three atoms is better than two, spreading them over four or six is even better. Figure 7–16 shows examples of these, NO_3^- and C_6H_6. In each case we could recognize that there was resonance (even though we are not picturing it correctly) by writing out the dot diagrams and discovering more than one "best" dot diagram in each

Figure 7–15 Actual resonance bonding in sulfur dioxide.

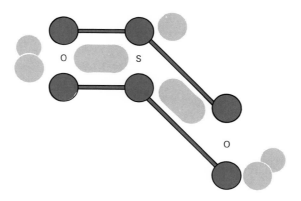

[7-12] Resonance 127

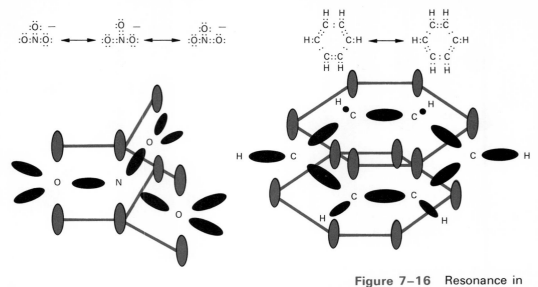

Figure 7–16 Resonance in nitrate ion and benzene.

case. We symbolize resonance by separating the dot diagrams of each of these structures with double-headed arrows.

$$\left[\begin{array}{c} :\ddot{\text{O}}: \\ :\ddot{\text{O}}::\text{N}:\ddot{\text{O}}: \end{array} \right]^{-} \longleftrightarrow \left[\begin{array}{c} :\ddot{\text{O}}: \\ :\ddot{\text{O}}:\text{N}::\ddot{\text{O}}: \end{array} \right]^{-} \longleftrightarrow \left[\begin{array}{c} :\ddot{\text{O}} \\ :\ddot{\text{O}}:\text{N}:\ddot{\text{O}}: \end{array} \right]^{-}$$

In terms of dot diagram structures, the resonance in C_6H_6 (only two structures) looks less impressive than the resonance in NO_3^-, but the artist's drawing in Figure 7–16 shows that C_6H_6 really has considerably more extra sharing of electrons. The extra bonding from resonance is particularly high in C_6H_6. That should help remind us that the dot diagram picture is just a clue to resonance, not a picture or measure of the real thing.

7–13 Molecular Geometry

In our discussion of double bonds and resonance (Section 7–11) we noted some sharp limitations on where electron pairs in an atom could be located. When electron pairs are used to form bonds, the limitations on electron positions force the covalently bonded molecules (or complex ions) to have particular shapes. Since the geometrical shapes of molecules are often very important in determining behavior of matter, we need to understand at least the most common shapes.

The important subject of molecular geometry has received (and continues to receive) a great deal of attention. Both experimental measurements and theoretical explanations have been developed in several ways, some of which would require long and complex explanations. These complex theoretical models are used so extensively that we will begin with very brief comments about the two main approaches so you can recognize references to them. Fortunately, however, a simple model, which gives us the correct answers virtually every time, is also available.

The valence bond approach uses hybrid orbitals concocted from combinations of the basic s, p, and d orbitals of atoms. This is the most widely used model. The pictures in Figures 7–12, 7–13, 7–14, 7–15, and 7–16 are based on the valence bond approach. It has the advantage of telling us where the combining abilities came from, and therefore the language of its hybrid orbitals (sp^2, d^2sp^3, and so on) will be useful in Chapter 16, when we are having trouble keeping track of bonding in complexes, and in Chapter 17, when we discuss the variety of bonding with carbon. The disadvantages of the valence bond system include a complexity that gives us little extra information. Our simple approach will give us essentially the results of the valence bond model without requiring memorization of a set of hybrid orbitals and their shapes, show us the reason for their shapes, and even show us where the shapes vary a bit from the valence bond predictions. Both the valence bond model and our model stop short of several bits of information the molecular orbital model gives.

The molecular orbital model abandons the electronic arrangements of atoms and calculates new orbital distributions for the entire molecule. These calculations can become extremely difficult because of the several centers of positive charge (nuclei). Complexity of some calculations and the fact that molecular orbitals were introduced more recently have held down the use of molecular orbitals. Older approaches, particularly valence bond, were already well established in books and in habitual use and are hard to displace. But the fact remains that molecular orbitals are a much more accurate way to look at matter than the other models. This model gives accurate answers to several cases where the other models fail. Anyone who seriously wants to be an authority on details of chemical bonding must become acquainted with and use the molecular orbital model. The molecular orbital models for H_2 and O_2 molecules are described in Section 8–1.

A third kind of model, which has become increasingly popular in the past few years, emphasizes the repulsions between the regions of negative charge (electrons). Although more sophisticated statements are possible, this model lends itself to the sort of simplification that fits our goals in this course. The electrons are grouped in pairs or groups of two pairs (double bonds) or three pairs (triple bonds) as previously noted in our discussions of orbitals and bonding. These groups follow a very simple rule. They get as far away from

The valence bond and molecular orbital models are used extensively to describe bonding.

Table 7-3
Shapes of Molecules and Complex Ions

Number of Electron Groups Around Central Atom	Geometry
two	linear
three	trigonal
four	tetrahedral
five	trigonal bipyramid
six	octahedral

each other as they can. The results predict shapes that are amazingly accurate.

When a group of atoms is bonded to a single central atom, the arrangement of electrons around the central atom determines the arrangement of the group. There must be at least two other atoms attached for there to be a shape to talk about. (A two-atom unit, such as Cl_2, can have only one shape—two connected to each other.) Therefore, in NH_3 molecules there is no shape around the hydrogen atoms. They are simply attached to nitrogen. It is the shape around the nitrogen that determines the shape of NH_3 molecules. That shape is determined by the number of groups of electrons in the outer shell around the nitrogen. To find that shape we must know both what the shape would be for each different number of electron groups and how many electron groups there are around nitrogen in NH_3.

The basic shapes of molecules are listed in Table 7-3. These shapes of maximum separation are shown in Figure 7-17. Sometimes one or more of these directions is not used to form a bond. The pair of electrons left unbonded is called a **lone pair.** Because a lone pair does not point toward a nucleus with a positive charge, it tends to repel other electron groupings somewhat more strongly than bonded electron groupings are repelled from each other. This leads to some distortion away from the basic shapes. The lone pairs also fall in the positions farthest from each other when there are two lone pairs on the same atom. The basic shapes remain the same, but special names are sometimes given to the common shapes obtained when some lone pairs are present. These are listed in Table 7-4 and shown in Figure 7-18. Notice the maximum separation of lone pairs in those cases where there is a choice of possible arrangements. Pyramidal and square planar shapes are quite common.

Table 7-4
Shapes Involving Lone Pairs

Basic Shape (Electron Groupings)	Name for Arrangement
trigonal with one lone pair	bent
tetrahedral with two lone pairs	bent
tetrahedral with one lone pair	pyramidal
trigonal bipyramid with two lone pairs	trigonal
octahedral with two lone pairs	square planar

Figure 7-17 Molecular shapes.

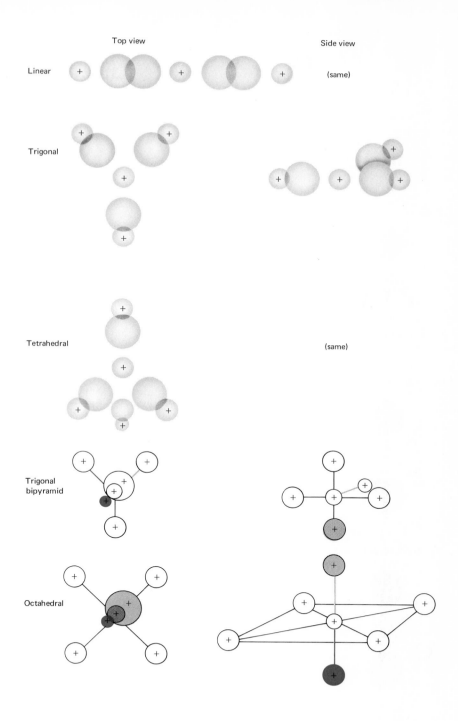

[7-13] Molecular Geometry

Figure 7-18 Shapes with lone pairs.

Trigonal with one lone pair

Tetrahedral with two lone pairs

= bent

= bent

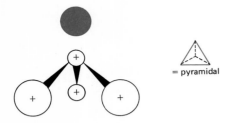

Tetrahedral with one lone pair

= pyramidal

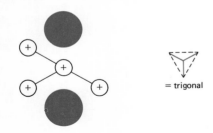

Trigonal bipyramid with two lone pairs

= trigonal

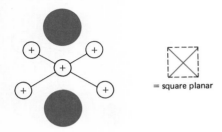

Octahedral with two lone pairs

= square planar

132 Bonding [7]

Lewis dot diagrams and an electron repulsion model can be used to predict geometry of molecules and complex ions.

It should be noted that only unbalanced molecules with either dissimilar bonded atoms or with a lone pair not balanced by another lone pair on the opposite side can be **polar molecules.** Our earlier discussion let us predict when we had polar covalent *bonds*, but if these are distributed evenly about the center they cancel each other's effects. Figure 7–19 shows examples of polar molecules and nonpolar molecules with polar bonds.

The number of electron groups around each atom can be determined from a correct dot diagram. In the case of NH_3, the dot diagram tells us there are *four* groups of electrons around the nitrogen. Therefore, NH_3 should have a tetrahedral shape. Even though there are only three hydrogens attached, the shape is definitely not the trigonal shape for things with three groups. Since one of the four groups around nitrogen is a lone pair, the shape should be described as "tetrahedral with one lone pair" (or if you prefer, "pyramidal"). In resonance cases, any one of the two or more "equally good" dot diagrams will give the correct answer. Table 7–5 lists some examples of molecular geometry determinations using dot diagrams. Note that for purposes of counting groups each single bond, double bond, or triple bond

Figure 7–19 Polar and nonpolar molecules.

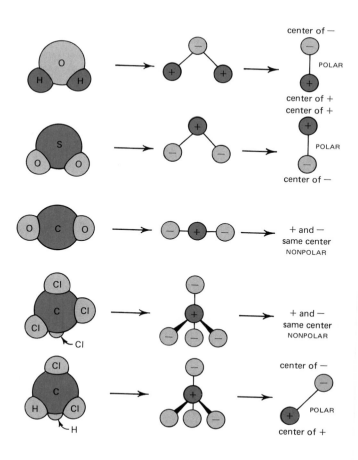

[7–13] Molecular Geometry

Table 7-5 Determinations of Molecular Geometry

Formula	Dot Diagram	Geometry
CCl_4	:Cl:C:Cl: with :Cl: above and :Cl: below	tetrahedral
SO_4^{2-}	[:O:S:O: with :O: above and :O: below]$^{2-}$	tetrahedral
SO_3	:O::S:O: with :O: below	trigonal
SO_3^{2-}	[:O:S:O: with :O: below]$^{2-}$	tetrahedral with one lone pair (pyramidal)
NH_3	H:N:H with H below	tetrahedral with one lone pair (pyramidal)
NH_4^+	[H:N:H with H above and H below]$^+$	tetrahedral
CO_2	:O::C::O:	linear
H_2O	H:O: with H below	tetrahedral with two lone pairs (bent)
NO_3^-	[:O:N::O: with :O: above]$^-$	trigonal
ClO_3^-	[:O::Cl:O: with :O: below]$^-$	tetrahedral with one lone pair (pyramidal)

is *one* group. A group may therefore be two electrons (single bond or lone pair), four electrons (double bond), or six electrons (triple bond).

Sometimes atoms are forced into bonding arrangements with more than eight electrons in the outer shell. The 12-electron arrangement of six pairs in an octahedral geometry is a particularly favored arrangement in nature. Available d electron levels are required to go beyond the eight electrons in s and p levels in the usual closed shell arrangements. These cases force us to change our thinking a little about writing dot diagrams. You can guide yourself to correct answers by using the rule that you always use closed shells unless the formula forces you to go beyond a closed shell. All the extra

Table 7-6
Molecular Geometry in Cases Exceeding Closed Shells

Formula	Dot Diagram	Predicted Geometry
SF_6	:F: :F: :F:—S—:F: :F: :F:	octahedral
XeF_4	:F: :F:Xe:F: :F:	square planar (octahedral with two lone pairs)
IF_3	:F: :F:I:F:	trigonal (trigonal bipyramid with two lone pairs)

electrons beyond closed shells go in pairs on the central atom. Table 7-6 shows some examples.

Geometry can also become complicated in cases where there are several "central" atoms. The basic shape rules still apply in each case. Figure 7-20 shows the case of $Cr_2O_7^{2-}$ ion. The dot diagram for a transition element with partially filled d electron levels requires more information than you have available to you. However, if the dot diagram shown was given, it predicts tetrahedral geometry around each chromium and a bent (tetrahedral with two lone pairs) link between the chromium atoms via the linking oxygen. The oxygen holding the chromiums together is called a bridging oxygen. Most of us find it much easier to describe the geometry of such a unit in words, taking one central atom at a time, rather than attempt drawing a picture of it. It has been said that a picture is worth a thousand words. But in describing geometry, the word tetrahedral is easier to produce than the picture.

Figure 7-20 Shape of dichromate ($Cr_2O_7^{2-}$) ion.

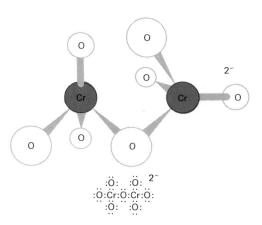

[7-13] Molecular Geometry 135

Skills Expected at This Point

1. You should be able to describe metallic crystals and use conductivity to differentiate between the behavior of metals, ionic compounds, and covalent compounds.
2. You should be able to describe the energy conditions leading to ionic bonding.
3. You should be able to state the electron sharing conditions present in single, double, triple, and coordinate covalent bonds.
4. You should write proper Lewis dot structures for molecules and complex ions and identify the difference between the dot diagrams and actual bonding in resonance.
5. You should use the described approximate methods to predict ionic, polar covalent, or covalent character and the geometry of molecules and complex ions.

Exercises

1. Describe the conditions (if any) under which each of the following are conductors and the conditions (if any) under which each of them are nonconductors.
   (a) metals   (b) ionic compounds   (c) covalent compounds
2. (a) List two important characteristics of a metallic crystal. (b) Using magnesium and sulfur as examples, describe why some elements are metals and others are not.
3. $CaCl_2$ is an ionic compound. Use the formation of $CaCl_2$ to illustrate (a) any and all energetically favorable factors in ionic compound formation; (b) any and all energetically unfavorable factors in ionic compound formation.
4. What electrical evidence is there for the existence of ions?
5. What chemical evidence is there for the existence of ions?
6. State the Arrhenius theory of acids and bases.
7. What oxidation state would you predict for each of the following elements in simple ions?
   (a) calcium   (c) oxygen   (e) sulfur
   (b) lithium   (d) bromine   (f) aluminum
8. The electronic configuration of cobalt is $1s^2 2s^2 2p^6 3s^2 3p^6 4s^2 3d^7$. Would you expect the most likely oxidation state of cobalt in a simple ion to be $+2, +7, -3, -6,$ or -9. Why?
9. $CaSO_4$ precipitates when the salt solutions of $CaCl_2$ and Na_2SO_4 are mixed. (a) What happens when a $Ca(NO_3)_2$ solution (conducting) and an H_2SO_4 solution (conducting) are mixed? Why? (b) What happens when water which was shaken up with $Ca(NO_3)_2$ to give a conducting solution is mixed with water which was shaken up with $BaSO_4$ to give a nonconducting solution? Why?
10. Write dot diagrams to represent each of the following reactions:
    (a) $Na + Cl$ to give ionic $NaCl$ (c) $C + Cl_2$ to give CCl_4
    (b) $H_2 + O_2$ to give H_2O (d) $NH_3 + BF_3$ to give NH_3BF_3

11. Predict whether each of the following should be covalent, polar covalent, or ionic according to the scheme described in this chapter.
    (a) NO
    (b) H_2O
    (c) LiF
    (d) $CaCl_2$
    (e) BN
12. Which of the following should have resonance?
    (a) CCl_4
    (b) SO_3
    (c) SO_3^{2-}
    (d) N_3^-
    (e) NO_2^-
13. Write the dot diagram and predict the geometry for each of the following:
    (a) CH_4
    (b) PO_4^{3-}
    (c) N_2O_4
    (d) ClO_2^-
    (e) SF_2
    (f) PCl_5
    (g) NCl_3
    (h) PCl_4^+
    (i) PCl_6^-
    (j) NO_2^-

    For (c) show N bonded to N, two O's bonded to each N, and no bonds from O to O.

Test Yourself

1. How can you use the electrical conductivity of substances to distinguish between metals, ionic compounds, and covalent compounds?
2. What is meant by the following terms?
   (a) covalent bond
   (b) double bond
   (c) coordinate covalent bond
   (d) polar bond
3. (a) Write the Lewis dot diagram structures for HCl, N_2, H_2O, SO_2, and PO_4^{3-}. (b) Which of these molecules has actual bonding quite different from that shown by the dot diagram structure? What is that kind of bonding called?
4. Arrhenius showed movement of ions while an electric current flowed by putting blue $CuSO_4$ solution between two colorless NaCl solutions and observing the blue color move toward the *negative* electrode. Which sentence best describes what happened to the colorless region between the original blue $CuSO_4$ region and the *positive* electrode?
   (a) It became diluted as the dissolved ions moved away toward the negative electrode.
   (b) the NaCl solution remained unchanged.
   (c) It became negatively charged as Na^+ ions moved away but Cl^- ions remained.
   (d) It began to give a positive chemical test for sulfate.
   (e) None of these.
5. Choose the two phrases that best express how the Arrhenius theory of acids and bases describes acids and bases upon ionization in water.
   (a) H^+ donors; H^+ acceptors
   (b) sources of H^+; sources of OH^-
   (c) electron pair donors; electron pair acceptors
   (d) H^+ acceptors; H^+ donors
   (e) electron pair acceptors; electron pair donors

6. Choose the two phrases that best describe a strong acid and a weak acid.
   (a) is present in high concentration; is present in low concentration
   (b) reacts to give H^+; reacts to give OH^-
   (c) ionizes completely; ionizes to only a small extent at one time
   (d) has strong bonds that cannot easily be broken; has weak bonds that are easily broken
   (e) is covalently bonded; is ionically bonded
7. Choose the phrase that best completes the sentence: When ionic compounds can form it is because
   (a) one element has a low ionization potential.
   (b) one element has a high electron affinity.
   (c) the electron affinity of one element is higher than the ionization potential of the other element.
   (d) strong attractions between ions make up for the difference between electron affinity of one element and the ionization potential of the other.
   (e) all of these occur.
8. A covalent single bond is
   (a) one electron shared between two atoms.
   (b) two electrons shared between two atoms.
   (c) one electron held by one atom.
   (d) two electrons held by one atom.
   (e) eight electrons in the outermost shell.
9. A coordinate covalent bond is
   (a) a bond using a p orbital concentrated along a single coordinate.
   (b) a bond coordinating electrons in two or more p orbitals of the atom.
   (c) a bond where two atoms share a pair of electrons, both of which came from the same atom.
   (d) a bond where two (or more) different dot diagrams are equivalent as the best available dot structures and the real structure is somewhere in between those shown by the dot diagrams.
   (e) none of these.
10. A resonance bonding situation involves
    (a) a bond using a p orbital concentrated along a single coordinate.
    (b) a bond coordinating electrons in two or more p orbitals of the atom.
    (c) a bond where two atoms share a pair of electrons, both of which came from the same atom.
    (d) a bond where two (or more) different dot diagrams are equivalent as the best available dot structures and the real structure is somewhere in between those shown by the dot diagrams.
    (e) none of these.
11. According to the prediction method described in this course, choose the phrases that best describe BP, $MgCl_2$, and H_2O.
    (a) covalent, ionic, polar covalent
    (b) polar covalent, ionic, ionic
    (c) polar covalent, polar covalent, covalent

(d) covalent, ionic, covalent
(e) none of these

12. Given the following dot diagrams, choose the phrases that best describe the shape (around the S) of SO_3 and the shape (around the Xe) of XeF_4.

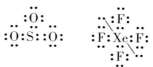

(a) trigonal, square planar (octahedral with two lone pairs)
(b) trigonal, trigonal bipyramidal
(c) tetrahedral, octahedral
(d) pyramidal (tetrahedral with one lone pair), tetrahedral
(e) pyramidal (tetrahedral with one lone pair), trigonal bipyramidal with two lone pairs

13. In which of the following species will resonance be involved in the bonding?
(a) ClO_3^-
(b) NO_3^-
(c) BF_4^-
(d) all of these
(e) none of these

General Trends in Chemical Behavior

8-1 Oxygen

8-1.1 The O_2 Molecule. The patterns of electronic configurations and the principles of bonding allow us to make some broad generalizations about the chemical behavior of elements. However, the chemical reactions of any element depend on the other elements and compounds available for it to react with. On Earth the most abundant material is oxygen. Therefore we will begin our discussion with oxygen.

Oxygen has the electronic configuration $1s^2 2s^2 2p^4$. Formation of a closed shell configuration by gaining two electrons to form O^{2-} ion would fit the bonding trends described in Chapter 7. Oxygen is far toward the upper right hand corner of the periodic table where electron affinities are highest, so ionic bonding as O^{2-} seems likely. However, no energy is released when an O^{2-} ion is formed. The buildup of negative charge makes formation of O^{2-} (and all negative ions with charges greater than -1) require energy. The O^{2-} sometimes does form anyhow, because more than enough energy is supplied by the much larger attractions between the -2 ion and the positive ions. Compounds such as MgO and Na_2O are ionic. However, the same large attractive forces that overcome the energy requirement for O^{2-} formation tend to hold these substances in a rigid solid form. Very high temperatures are needed to melt them to observe the expected electrical conductivity of molten ionic compounds.

Oxide ions have an unfavorable electron affinity but often form anyhow.

The difficulty of building up an energetically unfavorable negative charge could be avoided by having the oxygen covalently bonded instead of ionically bonded. As an element, oxygen exists as covalently bonded diatomic units, O_2. O_2 is a gas at normal conditions on Earth and makes up about 20.8% of the molecules in air (208 of each 1000 molecules are O_2). Most of the remainder are N_2 molecules. The O_2 and N_2 can be separated by cooling the air to liquid form and carefully distilling it. The liquid N_2 boils at 77.35 K ($-195.8°C$) and therefore can be boiled off while most of the O_2, which boils at 90.19 K ($-182.96°C$), remains behind as a liquid. The pure liquid O_2 has a very slight bluish color (liquid N_2 is colorless). O_2 has the very surprising property of **paramagnetism** (interacts with magnetic fields—see Section 6–8) and is very reactive chemically.

O_2 can be separated from air by distillation.

Figure 8–1 Structures for O_2.

$:\ddot{O}::\ddot{O}:$ $:\dot{O}:\dot{O}:$
A   B

A. Predictions for paired electrons: diamagnetic (no magnetic field), closed shells, strong bond, relatively inert.
B. Predictions for two unpaired electrons: net magnetic field, unfilled shells, average strength bond, very reactive.

O_2 acts like a single bonded molecule instead of a double bonded one.

Normally we write dot diagrams by making each atom reach a closed shell configuration. The double bonded O_2 dot diagram structure shown in Figure 8–1A is the result of the usual assumption of closed shells. It would be **diamagnetic** (no magnetic field) and not particularly reactive. Instead O_2 behavior is more like that predicted by the single bonded structure in Figure 8–1B. This structure does not complete the closed shells.

For a less common element we might choose to ignore this strange behavior and go on to other things. For oxygen, the most abundant element on Earth, we must learn and remember the fact that it is clearly different from everything else. You may decide simply to accept this difference on faith and use the single bond dot diagram as a reminder of what seems to be happening. But that is an imperfect solution because the single bonded dot diagram also has shortcomings. The bond in O_2 is stronger than the single bonded structure would predict. For those who need some supporting explanations we will resort to the more accurate molecular orbital picture of bonding.

8–1.2 Molecular Orbital Representation. In the molecular orbital model, the total number of possible orbitals (each capable of holding two electrons) remains the same as it was in the separate atoms. However, groups of orbitals (from each atom in the unit) are rearranged into new combinations. Some of these are in the favorable positions between atoms and are called **bonding orbitals.** In diatomic molecules they may be sigma (σ) bonds along the axis (the straight line passing through the centers of both atoms) or pi (π) bonds parallel to the axis. An equal number of orbitals are in the remaining unfavorable positions away from the point where the atoms bond together. These energetically unfavorable positions tend to pull the atoms apart and are called **antibonding orbitals.** They are set apart by asterisks. In diatomic molecules they may be σ^* or π^*.

Figure 8–2 shows how the electrons in an H_2 molecule (the simplest covalent compound) can be arranged in molecular orbitals. When the two atomic $1s$ orbitals (one from each hydrogen atom) are brought together, they can hold a total of four electrons. But the H_2 molecule can arrange the electron holding capacity into two new groupings, each able to hold two electrons. One of these groupings (the bonding σ_{1s} orbital) is better, energy-wise, than the atomic $1s$ orbitals. The other grouping (the antibonding σ_{1s}^* orbital) is

Figure 8–2 Hydrogen (H_2) Molecular orbital levels.

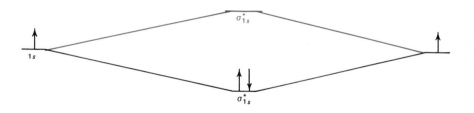

worse than the atomic $1s$ orbitals. The total energies of the σ_{1s} and σ_{1s}^* add up to the same as the total for the two atomic $1s$ orbitals.

Because the H_2 molecule has only two electrons, both of them can be placed in the favorable σ_{1s} orbital. They are held there more strongly than they could be held by the $1s$ orbitals of separate hydrogen atoms. The energy of that extra attraction is the bond energy of the H_2 molecule. Since both of the σ_{1s} positions are used and there are no extra electrons (which would have to go in the unfavorable σ_{1s}^* positions), H_2 molecules have very strong bonds.

Figure 8–3 shows how the atomic orbitals of two oxygen atoms combine to form molecular orbitals. Diatomic molecules such as H_2 and O_2 have the simplest of all possible molecular orbital schemes.

But even the O_2 case begins to show complications in the molecular orbitals formed from the $2p$ atomic orbitals. One of the $2p$ orbitals on each atom is pointed *toward* the other atom. Those $2p$ orbitals are rearranged into a bonding orbital directly between the atoms, called a σ_{2p} orbital, and an

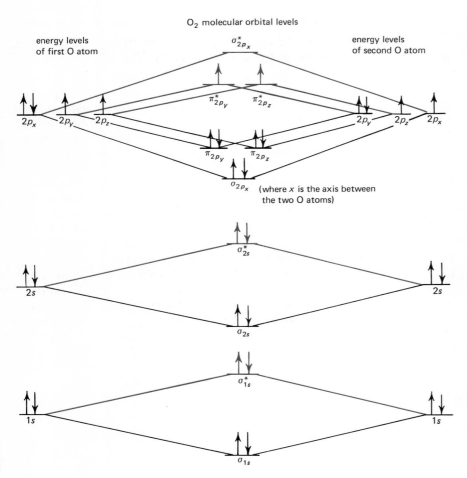

Figure 8–3 Oxygen (O_2) molecular orbital levels.

142 General Trends in Chemical Behavior [8]

antibonding orbital (σ_{2p}^*) directly on the opposite sides (outside) of the atoms from the σ_{2p} position. There is a large difference in the energy of σ_{2p} and σ_{2p}^*.

The other $2p$ atomic orbitals are perpendicular to the direction between the atoms. These four $2p$ orbitals (two from each atom) are rearranged to give two bonding orbitals, called π_{2p} orbitals, with positions somewhat toward the center of the two atoms but above and below the center. There are also two antibonding orbitals, called π_{2p}^* orbitals, with positions above and below and somewhat away from the direction between the atoms. There is a difference in the energy of π_{2p} and π_{2p}^* orbitals, but it is not as great as the difference between σ_{2p} and σ_{2p}^* orbitals.

The molecular orbital model explains the properties of O_2.

Figure 8–3 shows the resulting energy levels for O_2 molecules. When the available electrons are put in the lowest possible levels, we form pairs (shown by the pairs of opposing arrows) of all electrons except the last two. These unpaired electrons account for the magnetic characteristics of O_2 (see Section 6–8). Their unfavorable positions (antibonding) show why the bond is much weaker than in N_2 where there are two less electrons. The empty positions at the same level (which is not too strongly unfavorable) show why O_2 can easily accept other electrons. The fact that there are four more electrons in bonding than antibonding orbitals explains the strength of the O_2 bond, which corresponds to a double bond.

Molecular orbitals have similar successes in explaining details (particularly magnetic) in other cases, but the energy level interactions become quite complex. You can imagine the complications in an octahedral complex where the effects of seven nuclei must be considered at once. The symmetrical pattern helps some but not enough to make it simple enough for our use in this book. (Even the O_2 case is subject to some known complications which we chose to ignore in our description above.)

8–1.3 Oxygen Bonding. One effect of the reactive nature of O_2 is to allow another negative ion besides oxide, O^{2-}. The other (less common) negative ion is peroxide, O_2^{2-}. It could be represented by this dot diagram:

Oxygen can form diatomic peroxide ions, O_2^{2-}.

$$\left[:\overset{..}{\underset{..}{O}}:\overset{..}{\underset{..}{O}}: \right]^{2-}$$

Oxygen also has the ability to react with individual oxygen atoms to form a resonance stabilized O_3 unit, ozone. Dot diagram structures of ozone would be

Ozone, O_3, is stabilized by resonance.

$$:\overset{..}{\underset{..}{O}}:\overset{..}{\underset{..}{O}}::\overset{..}{O}: \longleftrightarrow :\overset{..}{O}::\overset{..}{\underset{..}{O}}:\overset{..}{\underset{..}{O}}:$$

Ozone is formed extensively in the upper atmosphere (where sunlight breaks O_2 molecules up to oxygen atoms) and in electrical discharges. It is very reactive chemically (enough so to be dangerous to humans in large amounts) and plays an important role in photochemical smog. Ozone in the upper atmosphere protects the Earth from some parts of sunlight that would harm living organisms. The controversies over supersonic transports (SST's) center around their effect on that protective ozone layer.

Covalent bonding of oxygen may involve a double bond, a single coordinate covalent bond, in which the strongly electron attracting oxygen shares an electron pair the other atom provides, or two single covalent bonds. Because oxygen is very electronegative, many of its covalent bonds are quite polar. Because its basically tetrahedral arrangement in the single bonded cases makes the shape bent, this polar character cannot be balanced by other polar bonds on the opposite side of the molecule. Therefore H_2O and many other oxygen compounds have polar molecules. The consequences of that polar character make up a large part of Chapter 13.

Oxygen bonds covalently with one double bond, two single bonds, or one coordinate bond where it accepts the electrons.

The electronegativity and geometry of oxygen bonding leads to polar molecules.

8–1.4 Preparation of O_2. Large quantities of pure (or almost pure) O_2 are prepared commercially by separating the O_2 from the other components of air. As we noted in Section 8–1.1, this separation can be done by cooling air to very low temperatures and using the difference in boiling points of N_2 and O_2 (fractional distillation). There is no chemical change in this process, but it presents some interesting problems in engineering design. Cooling the air enough to liquify it is difficult and therefore fairly expensive. And simply warming liquid air would not give a good separation of N_2 and O_2. Large processing plants achieve a much purer product by repeated separations, each producing more highly enriched material for the next separation. The need for cooling is also minimized by making heat flow from incoming air to the product gases being warmed back to room temperature and finally (after the incoming air has already been cooled considerably) to the liquid air being distilled. The result is very cheap purification of O_2, but only when done on a large scale with elaborate equipment.

O_2 can be prepared cheaply by distilling air in large plants or more expensively by simple, small scale laboratory reactions.

Oxygen can also be prepared by chemical reactions. Chemically produced oxygen is more expensive than oxygen from fractional distillation of air, but chemical reactions can be used on a small scale without elaborate equipment.

Chlorate compounds, such as $NaClO_3$ or $KClO_3$, are often used for production of small amounts of oxygen in the laboratory. Chlorates rearrange when heated to form chlorides plus O_2. The decomposition of chlorates (and also of perchlorates) can be very dangerous. The oxygens must be moved into new arrangements, and the chlorate may be heated to a very high temperature before the reaction begins. After O_2 begins to form, the reaction may spread so quickly that an explosion results. This dangerous situation is avoided by adding a **catalyst** to make the reaction occur at a lower temperature where it is less violent. The usual catalyst is MnO_2. The MnO_2 is not changed during the reaction but apparently provides a mechanism where oxygen atoms from ClO_3^- can get together to form O_2 molecules. The mechanism on MnO_2 allows O_2 to form at lower temperatures and provides a steady stream of O_2 as the oxygens reach the MnO_2 instead of the sudden, explosive formation of O_2 from heating the chlorate alone. *Never heat a chlorate without MnO_2 added.* The decomposition of $NaClO_3$ can be shown by the equation

$$2\ NaClO_3 \xrightarrow[MnO_2 \text{ catalyst}]{\text{heat}} 2\ NaCl + 3\ O_2$$

Peroxide compounds are also used to produce small amounts of oxygen. Hydrogen peroxide, H_2O_2, is sold commercially as dilute solutions in water. Hydrogen peroxide can rearrange to form the more stable water molecules and release O_2 by the reaction

$$2\ H_2O_2 \longrightarrow 2\ H_2O + O_2$$

H_2O_2 is quite stable when it is alone and dilute in water, but addition of salt catalyzes the decomposition. The MnO_2 catalyst used with chlorates works even better than salt as a catalyst for decomposition of H_2O_2. The fact that a catalyst is needed to decompose H_2O_2 should be taken as a warning. If H_2O_2 is concentrated (it cannot be prepared pure), its decomposition, once started, can be explosive.

Solid peroxides are also available. Sodium peroxide, Na_2O_2, is easily prepared and provides a simple source of O_2. Addition of water to Na_2O_2 gives O_2 promptly by the reaction

$$2\ Na_2O_2 + 2\ H_2O \longrightarrow 4\ NaOH + O_2$$

The ionic Na_2O_2 serves as the catalyst for its own decomposition, probably with an effect similar to the catalysis of H_2O_2 decomposition by dissolved ionic salts.

Each of the above reactions has the disadvantage of a costly chemical starting material. Cheaper compounds containing oxygen, such as H_2O, are not reactive enough to release O_2. We can get around that problem, however, by providing energy to force the reaction from an outside source, an electric current. Ionic solutions in water conduct electricity, and chemical reactions occur at the electrodes as the current flows. At the negative electrode electrons are transferred to something in solution, either to a positive ion that carried current by moving to the negative electrode or to something that forms a negative ion to carry current by moving away from the electrode. At the positive electrode something must give electrons to the electrode as current flows. The reactions that can most easily take on (at the negative electrode) or give up (at the positive electrode) electrons are the ones which take place. By selecting the ionic solution to avoid anything which gives up electrons more easily, we can make the reaction at the positive electrode be

$$2\ H_2O \longrightarrow O_2 + 4\ H^+ + 4\ e^-$$

Sulfuric acid, H_2SO_4, is one of the readily available materials that form water solutions giving O_2 when they are electrolyzed (electric current passed through, causing chemical reaction). This is a convenient and fairly cheap method of producing O_2.

The reaction at the negative electrode during electrolysis of H_2SO_4 in water is

$$2\ H^+ + 2\ e^- \longrightarrow H_2$$

The electrolysis is more important as a source of H_2 than O_2 because sources of H_2 are limited, whereas O_2 is available cheaply from air.

8-1.5 Reactions of Oxygen.

The most common reaction of oxygen is combustion. Oxygen forms bonds to other elements providing greater bond energies than the O_2 molecules and molecules of the fuel had, and the extra bond energy is released as heat. We use the heat released by combustion to heat our homes, power our cars, generate much of our electricity, and for various other purposes.

Some of the new bonds form more readily than others. When wood burns, the hydrogen content reacts with O_2 to form H_2O whereas the carbon content is left behind as a char. When combustion nears completion, the carbon char can react with O_2 to form CO and eventually form CO_2. In each of those cases the oxygens end up with covalent bonding providing them closed shells, as shown by the dot diagram approach. They also end up with a somewhat greater share of the bonding electrons than they could get from their partners in O_2 molecules.

Oxygen can also achieve a closed shell and a greater share of the available electrons in ionic bonds. Magnesium reacts violently with O_2 in flashbulbs to form MgO, in which ionic bonding predominates.

The general pattern in oxygen reactions is to *take* electrons from other elements to *complete the outer shell* of the oxygen atoms. The bonds may be ionic or polar covalent.

Oxygen usually reacts to take electrons from other elements.

When reacting with very reactive metals an O_2 molecule may accept two electrons and form an ionic bond without breaking the oxygen-oxygen bond. The result is a peroxide. Sodium and barium are two of the reactive metals* that form peroxides by direct reaction with O_2 as shown by these equations.

$$2\,Na + O_2 \longrightarrow Na_2O_2$$
$$Ba + O_2 \longrightarrow BaO_2$$

BaO_2 reacts with sulfuric acid to form solid $BaSO_4$ and hydrogen peroxide solution. These peroxides have already been mentioned as sources for small quantities of pure O_2.

Peroxides and other highly oxygenated compounds such as chlorates and perchlorates can also serve as oxygen atom donors for some reactions where direct combustion with O_2 would not work. The bond in peroxide is easier to break than the bond in O_2 molecule. Ozone, O_3, is also a good source of oxygen atoms that can react without having to break the O_2 bond.

The combustion reactions we observe usually take place at high temperatures and keep themselves going by releasing heat which maintains or even raises the temperature. But oxygen can also react at low temperatures. When iron is left exposed to air (and with some moisture to serve as a catalyst), it gradually reacts with O_2 to form rust. Our atmosphere provides an oxygen-rich environment that slowly converts many of the materials in contact with air to oxygen containing forms. Disruption of this normal oxygen-rich condition is a key factor in water pollution problems described in Chapter 14.

---

* Some metals that have even lower ionization potentials than sodium and barium can form superoxides containing O_3^{2-} units. K_2O_3 is an example. If superoxides form, they are very good as sources of reactive oxygen.

Some of the most important slow oxidations are the energy releasing reactions in living organisms. These reactions are catalyzed so they can occur at normal temperatures, and the energy is stored in forms usable by the organism. In Chapter 21 we will be concerned with these processes and with the availability of the oxygen required for the energy releasing reactions.

8–2 Oxidation and Reduction

8–2.1 Oxidation. In Section 8–1.5 we noted that the general pattern in oxygen reactions was to *take* electrons from other elements. Because of oxygen's high electronegativity, the bonds tend to be either polar covalent, with oxygen getting the greater share of the electrons, or even ionic with oxygen as the negative ion. Oxygen can also receive extra negative charge by accepting electron pairs in coordinate covalent bonds. In every case (except for reaction with F_2) oxygen tends to pick up extra electrons and negative charge. The elements reacting with oxygen therefore tend to lose electrons, either completely to form ions or partially to form covalent bonds where their share of the bonding electrons is less than they had before bonding.

Chemical changes can be described in terms of transfer of electrons, either to other atoms or to covalent bonds between atoms. We can arbitrarily assign the electrons in each covalent bond between different elements as belonging to whichever one of the atoms holds them more strongly (the more electronegative). That arbitrary assignment makes both ionic and covalent bonding fit the same pattern, transfer of electrons with one atom *losing electrons* and the other atom *gaining electrons*. We are going to use this model of electron gain or loss and assign terms to represent the processes of losing electrons and gaining electrons.

Because oxygen is the most abundant element on earth and very reactive to take electrons, the most common way to *lose electrons* is to *react with oxygen*. A substance reacting with oxygen is **oxidized.** The process is called **oxidation.** The word oxidation has now been adopted to describe *any* loss of electrons in chemical reactions, whether oxygen is involved or not.

> Any reaction in which the substance being discussed loses electrons is called an oxidation.

Please note that covalent bonds to a more electronegative element are considered losses of electrons and therefore oxidations. Here are some examples of oxidation.

$$2\,H_2 + O_2 \longrightarrow 2\,H_2O \qquad \text{oxidation of } H_2 \text{ by } O_2$$
$$4\,Na + O_2 \longrightarrow 2\,Na_2O \qquad \text{oxidation of Na by } O_2$$
$$2\,Na + Cl_2 \longrightarrow 2\,NaCl \qquad \text{oxidation of Na by } Cl_2$$
$$Mg + S \longrightarrow MgS \qquad \text{oxidation of Mg by S}$$
$$2\,FeCl_2 + Cl_2 \longrightarrow 2\,FeCl_3 \qquad \text{oxidation of Fe(II) by } Cl_2$$

oxidation = loss of electrons

8–2.2 Reduction. The opposite of oxidation is a gain of electrons. The word **reduction** is now used to describe any *gain of electrons*. Metals are often produced by reduction of metal oxides. Removal of the oxygen reduces the mass remaining, and that is probably where the word reduction originally came from. We use reduction to describe any gain of electrons, whether the mass is decreased or not. The substance gaining electrons is said to be **reduced.** Here are some examples of reduction.

Any reaction in which the substance being discussed gains electrons is called a reduction.

$$FeO + C \longrightarrow Fe + CO \qquad \textit{reduction of Fe(II) by C}$$
$$2\,H_2 + O_2 \longrightarrow 2\,H_2O \qquad \textit{reduction of } O_2 \textit{ by } H_2$$
$$4\,Na + O_2 \longrightarrow 2\,Na_2O \qquad \textit{reduction of } O_2 \textit{ by Na}$$
$$2\,Na + Cl_2 \longrightarrow 2\,NaCl \qquad \textit{reduction of } Cl_2 \textit{ by Na}$$
$$Mg + S \longrightarrow MgS \qquad \textit{reduction of S by Mg}$$
$$2\,FeCl_2 + Cl_2 \longrightarrow 2\,FeCl_3 \qquad \textit{reduction of } Cl_2 \textit{ by Fe(II)}$$

reduction = gain of electrons

As you can see by our selection of examples, every oxidation is also a reduction and every reduction is also an oxidation. The two must come in pairs. The reactions in which they occur are called **oxidation-reduction reactions.** Chapter 16 will describe several important commercial oxidation-reduction reactions and include some further discussion of oxidation-reduction in Section 16–5.3.

8–3 Hydrogen

All elements except the first two, hydrogen and helium, have their outer electrons in a shell that would need eight electrons to be a closed shell. (There is one exception, palladium. See Section 8–7 for an explanation of palladium.) We can use their electronic similarities to arrange them in patterns. Hydrogen and helium will not fit the same patterns. Helium is not much of a problem because it has a closed shell and no significant chemistry. Its $1s^2$ closed shell can simply be grouped with the s^2p^6 closed shells. Hydrogen does have plenty of significant chemistry and will simply have to be treated as a separate case.

Hydrogen is a separate, one-element chemical class.

Hydrogen has only one electron in its outer (and only) shell. Therefore it might be expected to have a low ionization potential and lose that electron to form H^+ in compounds. But that electron is held in the lowest possible energy level with no other electrons to interfere with its attraction to the nucleus. It is fairly difficult to remove the electron from hydrogen, and, as we shall see in Chapter 13, it is almost impossible to get a free H^+ ion in nature.

Hydrogen needs only one electron to reach a closed shell ($1s^2$). Therefore it might be expected to have a high electron affinity and gain an electron to form H^- in compounds. But the nuclear charge of only $+1$ does not provide a very strong attraction for an additional electron, and the electron already there cancels most of that attraction by its repulsion of the extra electron. Hydride compounds form only with very reactive metals, and they react with H_2O to form H_2 gas and OH^- ions. An example of hydride formation is

$$2\,Na + H_2 \longrightarrow 2\,Na^+ + 2\,H^-$$

We will have to consider both the H$^+$ and H$^-$ ions described above, but almost all bonding of hydrogen will actually be covalent instead of ionic. The covalent bonding of hydrogen is very simple. Hydrogen brings one electron, the other atom must contribute one electron, and the two atoms share that pair in a single covalent bond. In most of these compounds the hydrogen is less electronegative than the element with which it bonds. Using the standard described in Section 8–2.1, we can therefore think of the electrons in the bond as belonging to the other atom and say hydrogen loses electrons as it is converted from H$_2$ to the covalently bonded form. The hydrogen is oxidized (loses electrons). Something else must therefore be reduced (gain electrons). Hydrogen is therefore useful as a reducing agent, causing reduction. An example is the reaction

$$H_2 + CuO \longrightarrow Cu + H_2O \qquad \textit{reduction of CuO to Cu by } H_2$$

8–4 Representative Elements: Chemical Families

Representative elements have their last electron in an outer shell needing s^2p^6 to be a closed shell.

Columns in the periodic table are chemical families.

Families of representative elements are numbered as groups.

Group VII are called halogens and group I are called alkali metals.

Most of the common chemical elements have electronic configurations with the last electron an s or p electron in a shell needing eight electrons (s^2p^6) to be a closed shell. The eight columns of these elements in the periodic table are called the **representative elements.** All the elements in any one column have the same number of electrons in the outer shell. A Roman numeral stating the number of electrons in the outer shell is used to name each group. Therefore, those with one outer electron are in group I, those with six outer electrons in group VI, and so on. Each of these groups has chemical similarities caused by their similar outer electron arrangements. Each group is referred to as a **chemical family** of elements.

The group VII family of elements (F, Cl, Br, I, At) is called the **halogens.** They share the ability to reach a closed shell by gaining one electron, so they all tend to form -1 ions. They can all reach closed shells in covalent bonding by forming one regular single covalent bond. They form these single bonds in many compounds, including the diatomic molecules they form with themselves. They all have unshared electron pairs. Sometimes they can use these lone pairs to form coordinate covalent bonds. The kinds of bonding combinations found in one halogen are often possible for the others. For example, the structures ClO_3^- and ClO_4^- are the most common involving some apparent coordinate covalent bonding of chlorine to oxygen. Both bromine and iodine form structures like ClO_3^- (BrO_3^- and IO_3^-) and like ClO_4^- (BrO_4^- and IO_4^-). These complex ions do not allow assignment of oxidation states as clearly as in simple ions, but oxidation states are assigned as if ions had been formed in the covalent bonding. Oxygen is called -2 (as if the electrons in the coordinate covalent bond were simply given to oxygen, since oxygen is the more electronegative atom). The halogens are left with positive oxidation states. Since these positive oxidation states are formed by covalent bonding, the most likely states are $+7$ (the number of electrons in the outer shell of the neutral atom) and numbers 2, 4, or 6 smaller than that ($+5$, $+3$, and $+1$).

Although properties of elements in a family are not completely identical, the differences between them usually follow a regular trend in the order of their positions in the periodic table. For example, the tendency to make electron pairs available in coordinate covalent bonding (which effectively causes positive oxidation states) increases as we go down the list of halogens. Iodine forms IO_3^- more readily than bromine forms BrO_3^-, BrO_3^- forms more readily than ClO_3^-, and fluorine forms no oxygen containing ions at all. Electronegativities vary regularly from fluorine (the largest) through steadily decreasing values for chlorine, bromine, and iodine. (The element astatine would presumably also fit these trends, but it is so rare it is usually ignored.)

There are some exceptions to regular trends. Crowding in F^- cuts down the electron affinity of fluorine by so much that chlorine has a larger electron affinity. However, once the crowding is eased by another shell, the expected regular downward trend is followed. The electron affinity of bromine is less than chlorine and iodine is less than bromine. Differences within families will be discussed further in Section 8–5.

Group I, the **alkali metals,** also shows family relationships clearly. These elements tend to lose one electron to form $+1$ ions. Covalent bonding is essentially impossible because they have many electrons less than a closed shell and so little attraction for electrons in their outer shells.

The tendency to form $+1$ ions is so strong that alkali metals tend to do this even as the pure elements. They form solids (or liquids when heated moderately) in which positive ions are surrounded by a sea of free electrons. The electrons balance out the charge of the positive ions but are not attached to particular atoms. This arrangement forms a **metallic crystal,** as described in Section 7–2. Movement of the electrons can conduct electricity or heat through the metal, the material can be reshaped without breaking, and these elements have the characteristic shiny metal gloss. To be metallic, an element must be able to give away some electrons and form positive ions quite easily.

Alkali metals follow a shell effect trend. The larger atoms lose the outer electron more easily. The inner electrons effectively shield the outermost electron from all but one unit of nuclear charge. Each atom then has one electron held by an effective $+1$ charge. The farther the electron is away, the easier it is to pull off. That does not always mean that the larger atoms are more reactive to form $+1$ ions. Sometimes the attractions between the small Li^+ ions and negative ions is enough more than the larger Na^+ ions, K^+ ions, and so on, to make lithium the most reactive element in the family.

In the above examples we have seen two major types of effects. The main effect was a **configuration effect.** The identical outer electron configurations tend to make elements in the same family alike. The **shell effect** tends to cause regular trends with differences caused by size differences. In the halogens and alkali metals the arrangements are so close to closed shells that the configuration effect dominates. As we move away from closed shells other factors will become more significant.

Configuration effects lead to similarities in families.

Shell effects lead to regular trends in differences.

8–5 Irregularities in Family Patterns

In every chemical family there is an increasing tendency to give electrons away (forming positive oxidation states) as we go down the column. In the halogens this showed up as a decreasing electronegativity (less tendency to hold electrons) and increasing tendency toward positive oxidation states like IO_3^-. In the alkali metals the tendency to lose electrons showed up in decreasing ionization potentials. In each case this tendency is definitely a shell effect. The electrons get farther from the nucleus and are shielded by inner electrons so the increasing nuclear charge does not affect them.

When elements are good at forming positive ions, they are metallic. As we go down any column, elements become better at losing electrons so that they become more metallic. The halogens are so nonmetallic that none of the common halogens act like metals. Alkali metals are so metallic that even lithium at the top of the column is metallic. But somewhere between group I and group VII we pass a transition between metallic and nonmetallic elements. The general trend of shell effects tells us that transition point must move to the right as we go down the table. Several families will begin with nonmetals at the top and end with metals at the bottom. Figure 8–4 shows

Elements become more metallic as we move down any family.

Figure 8–4 Division between metals and nonmetals.

IA	IIA	IIIB	IVB	VB	VIB	VIIB	VIIIB			IB	IIB	IIIA	IVA	VA	VIA	VIIA	VIIIA
1.008 H 1																	4.003 He 2
6.939 Li 3	9.012 Be 4											10.81 B 5	12.01 C 6	14.01 N 7	16.00 O 8	19.00 F 9	20.18 Ne 10
22.99 Na 11	24.31 Mg 12											26.98 Al 13	28.09 Si 14	30.97 P 15	32.06 S 16	35.45 Cl 17	39.95 Ar 18
39.10 K 19	40.08 Ca 20	44.96 Sc 21	47.90 Ti 22	50.94 V 23	52.00 Cr 24	54.94 Mn 25	55.85 Fe 26	58.93 Co 27	58.71 Ni 28	63.54 Cu 29	65.37 Zn 30	69.72 Ga 31	72.59 Ge 32	74.92 As 33	78.96 Se 34	79.91 Br 35	83.80 Kr 36
85.47 Rb 37	87.62 Sr 38	88.91 Y 39	91.22 Zr 40	92.91 Nb 41	95.94 Mo 42	97 Tc 43	101.1 Ru 44	102.9 Rh 45	106.4 Pd 46	107.9 Ag 47	112.4 Cd 48	114.8 In 49	118.7 Sn 50	121.8 Sb 51	127.6 Te 52	126.9 I 53	131.3 Xe 54
132.9 Cs 55	137.3 Ba 56	138.9 La* 57-71	178.5 Hf 72	180.9 Ta 73	183.9 W 74	186.2 Re 75	190.2 Os 76	192.2 Ir 77	195.1 Pt 78	197.0 Au 79	200.6 Hg 80	204.4 Tl 81	207.2 Pb 82	209.0 Bi 83	210 Po 84	210 At 85	222 Rn 86
223 Fr 87	226 Ra 88	227 Ac** 89-103	261 (Ku) 104	260 (Ha) 105													

Number above symbol = atomic weight
Number below symbol = atomic number

*Lanthanide series

140.1 Ce 58	140.9 Pr 59	144.2 Nd 60	147 Pm 61	150.4 Sm 62	152.0 Eu 63	157.3 Gd 64	158.9 Tb 65	162.5 Dy 66	164.9 Ho 67	167.3 Er 68	168.9 Tm 69	173.0 Yb 70	175.0 Lu 71

**Actinide series

232.0 Th 90	231 Pa 91	238.0 U 92	237 Np 93	242 Pu 94	243 Am 95	247 Cm 96	247 Bk 97	249 Cf 98	251 Es 99	254 Fm 100	253 Md 101	256 No 102	254 Lr 103	257

the approximate borderline between metals and nonmetals. An example of the transition is shown by group IV. Carbon is definitely nonmetallic. Silicon is somewhat borderline. It does not form positive ions in compounds, but it does conduct in a somewhat metallic way under certain conditions. Borderline elements such as silicon are sometimes called **metalloids.** Silicon is used as a **semiconductor** in transistors. Germanium is also a useful semiconductor, not fully metallic. Tin and lead are metals.

There is one shell effect that is much greater than others. The difference between the first representative row (the second shell) from lithium to neon and the next row from sodium to argon is particularly sharp for two reasons. The first is simply size. The difference between two shells and three shells is a bigger fraction than the difference between three shells and four or four shells and five.

This emphasis on the size difference from the lithium-neon row to the sodium-argon row is accentuated by other factors. At the left hand side of the table, the elements are reacting by losing their outer shells. Therefore, the really important difference becomes the difference in the ions. One shell of two electrons is very different from two shells with eight electrons in the outer one.

When we get far enough right to get anything but positive ions, another factor appears. Every row after the sodium-argon row has an intervening group of transition elements (see Section 8–6). As these transition elements go in, a nuclear charge effect is making the atoms smaller. Therefore, every row beyond the sodium-argon row has a nuclear charge effect neutralizing most of the size increase expected from the extra shell. Only the transition from the lithium-neon row to the sodium-argon row has the full expected shell effect. Figure 8–5 shows the sizes of atoms. The decrease in size caused by the transition elements can be seen quite clearly.

The sharp shell effect between second and third shells causes diagonal relationships and other sharp size effects.

Figure 8–5 Sizes of atoms and ions (atomic and ionic radii in Angstrom units, which are 10^{-8} cm). [Courtesy of J. A. Campbell, Harvey Mudd College; from E. R. Dillard and D. E. Goldberg: *Chemistry, Reactions, Structure, and Properties.* The Macmillan Company, New York, 1971.]

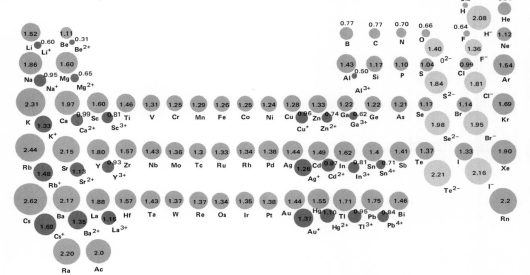

Figure 8-6 Diagonal relationships.

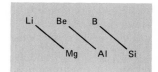

The sharp size differences between the lithium-neon and sodium-argon rows cause behavior differences large enough to break up the expected family resemblances. Elements in the lithium-neon row are sharply less metallic than the rest of their families. They often seem to fit better with the family one column farther to the right. This phenomenon is called the **diagonal relationship.** Figure 8-6 summarizes the diagonals linking similar elements in the first and second rows. This is the only place in the periodic table where such strong diagonal similarities are found.

The differences in number of possible covalent bonds prevent the more nonmetallic columns from having as much diagonal similarity, but there are still some sharp size effects.

Availability of d levels affects reactions and bonding.

The second major cause of differences between the lithium-neon and sodium-argon rows is the availability of d electron levels starting with the sodium-argon row. Even though these levels are not used to reach a closed shell, the availability can have a tremendous influence on chemistry of the sodium-argon row. The difference can be shown by comparing the reactions of NCl_3 and PCl_3 with water. Chlorine is more electronegative than phosphorus and about equal to nitrogen in electronegativity. You may notice from Figure 7-10, page 122, that the measured electronegativities show a sharp shell effect between the lithium-neon and sodium-argon rows. In any case, if the NCl_3 and PCl_3 units were pulled apart in reactions, it seems more likely that the shared electrons would remain with the chlorine (as Cl^-) than leave the chlorine (forming a Cl^+ state). When PCl_3 reacts with water the electrons do remain with the chlorine, as shown in Figure 8-7. In order to maintain its closed shell configuration, the phosphorus must acquire another pair of electrons for each Cl^- lost. These electrons are supplied by the lone pairs on H_2O molecules. Because the phosphorus atom can hold more than eight electrons in its outer shell by using d electron levels, the exchange can begin by a lone pair from H_2O being shared with phosphorus, as shown in step A of Figure 8-7. The Cl^- can then come off, as shown in step B of Figure 8-7. Finally, H^+ can be lost from the water to restore a neutral product, as shown in step C of Figure 8-7. When the process is repeated three times, the products are H_3PO_3, 3 H^+, and 3 Cl^-, or three hydrochloric

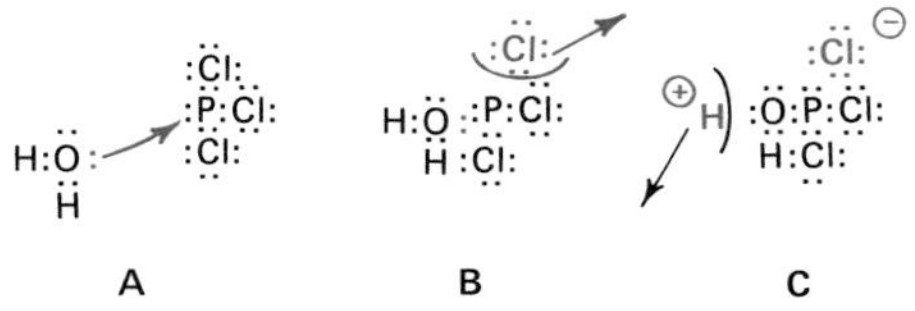

Figure 8-7 Hydrolysis of phosphorus trichloride (PCl_3).

acid units plus H_3PO_3. The phosphorus never has less than a complete closed shell, and the chlorine remains in the favored Cl^- state.

As shown in Figure 8–8, the same process is impossible for NCl_3. Nitrogen has no d levels available in its outer shell, so it cannot go beyond eight electrons in that shell. (A similar reaction is observed for some compounds, but is thought to proceed as a concerted process with steps like 8–7A and 8–7B simultaneous rather than one at a time.) With that path closed, the reaction shown in Figure 8–9 becomes the best one available. In step A of Figure 8–9 the chlorine bonds to a lone pair on oxygen, leaving behind the electrons in its bond to nitrogen. In step B an H^+ is lost from the oxygen and eventually, step C, bonds to the electron pair the chlorine left behind. Three repetitions lead to NH_3 and three $HOCl$ molecules.

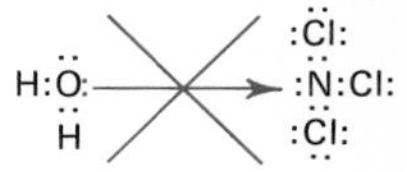

Figure 8–8 Inability of nitrogen trichloride (NCl_3) to accept extra electrons.

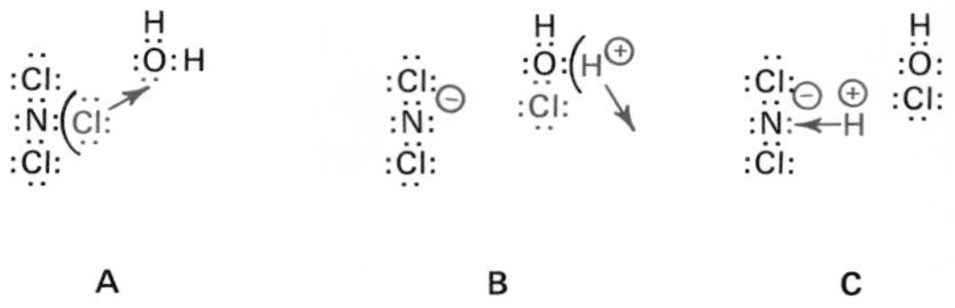

Figure 8–9 Hydrolysis of nitrogen trichloride (NCl_3).

A similar contrast occurs between the reactions of $SiCl_4$ and CCl_4 with water. In CCl_4 the less electronegative carbon cannot manage a reaction like that shown in Figure 8–9, so nothing happens to it at all. The $SiCl_4$ reacts rapidly. Notice that carbon should react like phosphorus if this was just a diagonal relationship due to the less metallic nature of carbon and nitrogen. The total disappearance of reaction in CCl_4 proves the importance of d electron levels.

The availability of d electron levels also makes possible bonding arrangements other than closed shells. The octahedral bonding pattern allows up to six bonds to form instead of four. SF_6 is such a compound. A comparable OF_6 compound is clearly impossible.

8–6 The Noble Gases

The last family of elements is not usually included with the representative elements. These elements already have closed shell configurations, so they have none of the chemical reactions by which representative elements reach closed shells. By setting group VIII apart it is also possible to include helium, which also has an inert closed shell arrangement, although it does not have the same number of outer electrons.

Because of their chemical inertness, group VIII elements were not discovered until 25 years after development of the periodic table. They were so nonreactive that the group was given the name inert gases, and chemical reactions were described in terms of other elements seeking to reach the closed shell arrangements of group VIII. However, we have noted that there are some other electronic arrangements that are also favored in nature, particularly the octahedral pattern. Finally, in 1962, a compound containing xenon was formed. Other compounds of xenon and krypton have since been prepared, and the name of group VIII has been changed from inert gases (never react) to noble gases (rarely react). Because these compounds all involve bonding in which d electron levels take part, helium and neon might still deserve the title inert gases. But the compounds of xenon and krypton dealt such a shocking blow to earlier ideas about bonding that chemists have developed a strong (and very healthy!) respect for the difference between popular theories and absolute truth. Very few would make any absolute statements about what could not be found in future experiments.

The noble gases already have closed shells but may bond in arrangements using d levels.

8–7 Transition Elements

Transition elements are filling a d level one shell below the outer shell.

Those elements where the last electron added went into a d level (in the next to last shell) instead of the outer shell are called transition elements. The first row of these elements (scandium to zinc) are fairly common on Earth. The others are rare enough to be of limited importance. The chemistry of these elements is strongly influenced by two factors.

Transition elements have two nearly equal energy levels and identical outer configurations.

1. They have two energy levels almost equal in energy with their highest energy electrons in one of them.
2. Except for some fine details we will discuss here, they have the same outer electronic configurations regardless of how many electrons they have in the d level.

We have already mentioned that the outer shell must be lost when positive ions are formed (Section 7–7). As each transition element loses two electrons from the outer shell, these electrons are each subjected to an effective charge of $+2$ attracting them. Because they are in the same energy level and subject to the same attracting charge, they ought to be at the same distance (same size) and require the same ionization potential. That is not far from the truth. The increasing nuclear charge as we move to the right is almost completely shielded by the identical increase in inner electrons. But the shielding is not quite completely effective. The attraction for the outer electrons increases slightly as the nuclear charge increases. This is accompanied by increasing ionization potentials and a tendency toward decreasing size as seen in Table 8–1 and Figure 8–10.

Size and ionization potentials vary only slowly across a transition series.

The nuclear charge effect has an even greater influence on the electrons in the d shell being filled, since they are not shielded from the increased charge. Extreme crowding of that shell probably causes the slight increase in atomic sizes at the end of each row seen in Figure 8–10, but the d electrons are clearly held more strongly as we move to the right. Since the outer s electron level and the inner d level being filled were almost equal in energy

The inner d level shifts to lower energy than the outer s level as a transition series is crossed.

Sc	Ti	V	Cr	Mn	Fe	Co	Ni	Cu
19.36	20.40	21.39	23.25		24.08	24.91	25.78	28.01

Table 8–1 Total of the First Two Ionization Potentials (in volts) of Transition Elements (Energy Needed to Form +2 Ion)

at the start, the order of the levels is eventually changed as the d level becomes held more strongly. The outer s level is also being held more strongly with increasing nuclear charge, but the change is slower because of shielding by the added inner electrons. The results of this change in order can be seen in the ground state electronic configurations of Figure 8–11.

Both the $3d$ and $4s$ levels are split into two different energy levels. That half of the electrons with spins lined up favorably to other magnetic fields has slightly lower energy than the other half (half-filled subshell effect, Section 6–7). Partially filled levels are paramagnetic because the magnetic fields are not cancelled out by pairing (see Section 6–8). As the $3d$ level passes the $4s$ level to become the more strongly held, the bottom half of the $3d$ level may get below the top half of the $4s$ level, as shown in Figure 8–12. The $3d$ level must then be half-filled before the second electron goes into the $4s$ level. That situation occurs at chromium. Eventually both halves of the $3d$ level get below the top half of the $4s$ level. That occurs in copper and causes the situation shown in Figure 8–13. In the second transition series the $4d$ level gets below both halves of the $5s$ level at palladium, resulting in no outer s electrons at all.

The changes in order shown by the configurations in Figure 8–11 alone would not be of enough practical importance to deserve our time here.

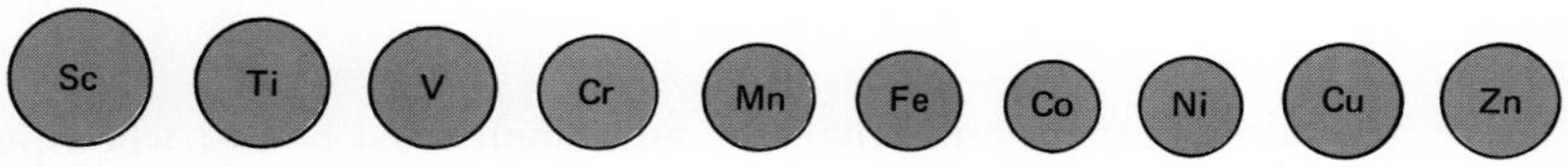

Figure 8–10 Sizes of atoms in the first row transition elements.

First row Ar configuration	Sc	Ti	V	Cr	Mn	Fe	Co	Ni	Cu	Zn
+ this many $4s$	2	2	2	1	2	2	2	2	1	2
+ this many $3d$	1	2	3	5	5	6	7	8	10	10
Second row Kr configuration	Y	Zr	Nb	Mo	Tc	Ru	Rh	Pd	Ag	Cd
+ this many $5s$	2	2	1	1	1	1	1	–	1	2
+ this many $4d$	1	2	4	5	6	7	8	10	10	10
Third row Xe configuration plus $4f$ electrons plus	Lu	Hf	Ta	W	Re	Os	Ir	Pt	Au	Hg
+ this many $6s$	2	2	2	2	2	2	2	1	1	2
+ this many $5d$	1	2	3	4	5	6	7	9	10	10

Figure 8–11 Ground state electronic configurations of transition elements.

Figure 8–12 Comparative energy levels in vanadium and chromium.

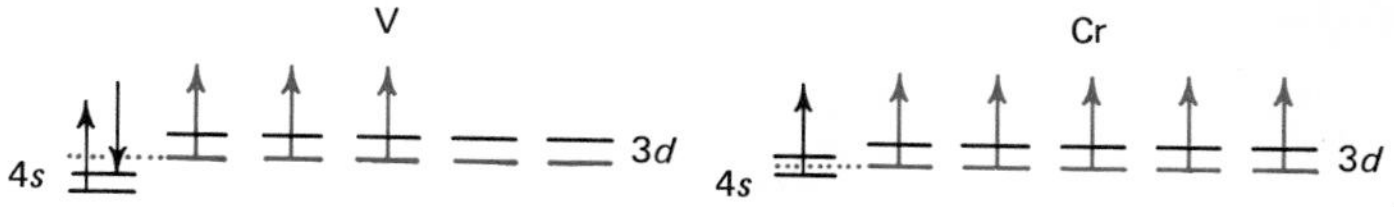

However, the trend they show—that the energy of the 3d level is falling compared to that of the 4s—is important to the very practical question of what oxidation states these elements form. This phenomenon could be called an **energy level shift.** Let us go across the first transition series and look at the effect of the shifting d level.

The first transition element, scandium, has 4s electrons held more strongly than its 3d electron. However, it must lose the 4s electrons first to form the favored small positive ions (Section 7–7). In scandium anything that can remove the 4s electrons also removes the 3d electrons. Sc^{3+} is the only ion formed.

The outer s electrons of transition elements must be lost in reactions, and we may get none, some, or all d electrons lost.

As we move to titanium, the 3d electrons are being held more strongly. It is then possible to remove the 4s electrons alone to form Ti^{2+}. It is also possible to remove the 3d electrons to obtain titanium in a +4 oxidation state. (+4 titanium is not believed to exist as Ti^{4+}. It and many other high oxidation states are found only in lower charged complex ions such as TiO^{2+}.) However, because loss of the 3d electrons does not remove a whole shell, we do not have to take them both at once. We may remove only one 3d electron and get Ti^{3+}. Titanium is the first element to follow a general pattern for transition elements. The outer s electrons must come off, and we may get none, some, or all of the d electrons off.

+2 tends to be the lowest oxidation state of transition elements.

Vanadium also follows that pattern, giving +2, +3, +4, or +5 oxidation states.

The configuration of chromium suggests a +1 ion might be possible, but the only oxidation states we find are +2, +3, and +6. +4 and +5 should also occur, but in water they react with themselves to form the more stable +3 and +6 arrangements. The +3 state is particularly stable because of complex ion formation, which will be discussed in Chapter 16, Section 16-3.

Manganese forms +2, +3, +4, +6, and +7 oxidation states.

The highest oxidation state and number of states rise with increasing d electrons but then fall as the d electrons can no longer be removed.

When we reach iron we find that the 3d electrons are now held so strongly that we cannot get them all off. Fe^{2+} forms easily and Fe^{3+} is quite common. But the other five 3d electrons are held more strongly than the first one (which was the only one in the upper half of the 3d energy level). Oxidation states above +3 for iron are so rare they can be considered nonexistent for practical purposes.

Cobalt and nickel both form +2 and do not easily get beyond that point. Cobalt can get to +3 with the help of favorable complex ion formation and nickel gets to a higher state only in a complex, poorly understood solid.

Figure 8–13 Comparative energy levels in copper.

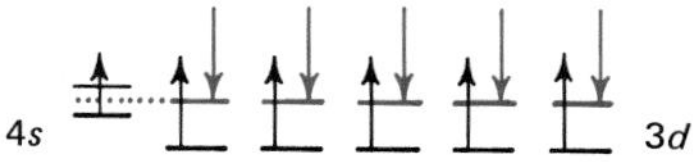

Copper has a configuration suggesting a $+1$ ion, and in this case it does form. Cu^{2+} also forms, but no further d electrons are removed to form higher states.

Zinc forms only Zn^{2+}.

The overall trend can be summarized in the following points.

1. The lowest oxidation state is $+2$. (Exceptions scandium and copper for reasons noted.)
2. As the number of d electrons increases, the highest oxidation state and the number of possible oxidation states rise because of the possible removal of some or all d electrons.
3. Eventually a point is reached where the d electrons can no longer be removed.

Another general trend can be noted between the different transition series. The shell effect makes later series more likely to reach higher oxidation states. Therefore the point where d electrons cannot be removed shifts to the right as we go down. Iron is never found in a $+8$ oxidation state, but osmium in a $+8$ oxidation state is easily prepared.

The point where d electrons cannot be removed shifts right as we go down the periodic table.

8–8 Effect of Oxidation State on Acid-Base Behavior

The variety of oxidation states in some transition elements allows us to see part of the reason for substances being acidic or basic. Low oxidation states give basic behavior; high oxidation states give acidic behavior; oxidation states in the middle sometimes show **amphoterism,** an ability to act as acids sometimes and bases at other times. This is a general pattern among acids and bases.

The oxidation states of manganese serve as an example. The $+2$ and $+3$ states can form hydroxides that are basic. $Mn(OH)_2$ or $Mn(OH)_3$ would ionize in water to give OH^- ions, therefore fitting Arrhenius's conception of what a base is. The $+7$ and $+6$ states are acidic. $HMnO_4$ can ionize to give H^+. The $+6$ state exists only in strong base in the MnO_4^{2-} form corresponding to acidic ionization of the nonexistent acid H_2MnO_4. The $+4$ state forms a compound, MnO_2, which is neither a good acid nor a good base.

Figure 8–14 compares the ionization of $HMnO_4$ ($+7$ state) and $Mn(OH)_2$ ($+2$ state). Each has OH^- units bonded to manganese. In $Mn(OH)_2$ the OH^- units can leave, taking with them the electrons that could have formed bonds to manganese. The Mn^{2+} is an ion with only a moderate attraction for the OH^- ions. In $HMnO_4$, the oxidation state of manganese is so high that it cannot let the electron pairs be pulled away. Therefore, it becomes a covalently bonded unit. As these electrons are attracted strongly toward manganese, the very electronegative oxygen has need for some extra negative charge. Because the oxygen cannot get this negative charge from the manganese, it gets it by taking the bonding electron pair completely away from hydrogen. That leaves an H^+ free to go off and act as an acid.

Low oxidation states tend to be basic and high oxidation states acidic.

Figure 8–14 Ionizations of $HMnO_4$ and $Mn(OH)_2$.

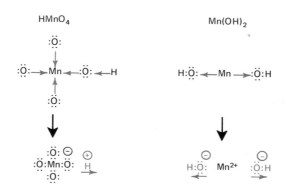

8–9 Inner Transition Elements

Inner transition elements are adding f electrons two shells from the outer shell and have almost identical size and chemistry.

The elements shown in Figure 8–15 have their last electron added in an f level two shells removed from the outer shell. Most of the factors mentioned in transition metal chemistry (Section 8–6) are repeated here except with a much greater constancy throughout. Figure 8–15 shows there is a little rearrangement of levels, but the chemistry is mostly similar to that of scandium. There are two outer electrons and one easily removed d or f electron in each case. The f electrons are buried so deep in the atom that they are not involved in covalent bonding in the first inner transition series, and it appears they cannot be simply lost to form ions of greater than $+3$ charge.

There is a very slight contraction in size across the series that affects the sizes of the transition elements following. Those with a taste for Greek mythology can note that tantalum (Ta) and niobium (Nb) are named for the characters Tantalus and Niobe. Tantalus was sent to a Hell where all sorts of desirable things were kept barely out of his reach; Niobe wept eternally for her slain children. Tantalum and niobium are essentially identical in size as well as outer and next-to-outer electronic configurations, and the frustrations of the chemists who finally separated them found an outlet in their names. Much of the chemistry of inner transition elements and those immediately following them is similarly frustrating and of little practical importance. When one of these elements must be prepared in pure form (principally europium for use in color television set phosphors), it is very expensive.

The second inner transition series has more variety in oxidation states. The usual shell effect toward higher oxidation states increases the tendency to remove some of the varying number of f electrons. This chemistry is also

Figure 8–15 Electronic configurations of inner transition elements.

Xe configuration plus	La	Ce	Pr	Nd	Pm	Sm	Eu	Gd	Tb	Dy	Ho	Er	Tm	Yb
this many 6s	2	2	2	2	2	2	2	2	2	2	2	2	2	2
this many 4f	-	2	3	4	5	6	7	7	9	10	11	12	13	14
this many 5d	1							1						

of greater practical importance because the fissionable nuclear fuels fall in this series.

Skills Expected at This Point

1. You should be able to recognize and to describe the bonding characteristics of O_2, O_2^{2-}, and O_3 and to describe the common methods of preparing O_2 and the general pattern of oxygen reactions.
2. You should be able to define and illustrate the meanings of oxidation and reduction.
3. You should be able to describe the unique position of hydrogen and the general patterns of hydrogen reactions.
4. You should be able to identify the regions of the periodic table corresponding to the representative elements, describe the basis for family relationships, describe trends within families using examples from the alkali metals and halogens, and identify and explain diagonal relationships and other sharp differences between the lithium to fluorine and sodium to chlorine rows.
5. You should be able to identify the regions of the periodic table corresponding to the transition elements and the regions corresponding to inner transition elements, explain why they differ less from column to column than neighboring representative element columns, and describe the trend in first transition series oxidation states and its relationship to shifting energy levels.
6. You should be able to state, give examples of, and explain the relationship between oxidation state and acid-base behavior.

Exercises

1. What are the conditions needed for efficient commercial preparation of purified O_2?
2. What is unusual (not expected from the Lewis dot diagram) about the bonding in O_2?
3. Write examples of **(a)** a typical combustion reaction; **(b)** reaction of O_2 with a very reactive metal; **(c)** reduction of a metal oxide by H_2.
4. What oxidation states might reasonably be expected for each of the following elements?
   - **(a)** magnesium
   - **(b)** phosphorus
   - **(c)** sulfur
   - **(d)** iodine
   - **(e)** aluminum
   - **(f)** vanadium
   - **(g)** osmium
5. List a property of halogens that illustrates the configuration effect.
6. List a property of alkali metals that illustrates the shell effect.
7. Explain why a piece of copper conducts electricity but a piece of sulfur does not.

8. What major chemical difference do you expect between nitrogen and bismuth? Why?
9. Why are boron and silicon so much alike chemically?
10. Phosphorus can form PCl_3 or PCl_5. Nitrogen can form NCl_3 but not NCl_5. What causes this difference?
11. Why is neon less likely to form compounds than argon?
12. List some evidence for a shift in the $3d$ energy level compared to the $4s$ energy level as we cross the first transition series.
13. If the $3d$ electron in scandium is really easier to remove than the $4s$ electrons, why doesn't Sc^+ form?
14. List the three principles that summarize the oxidation state trends of transition elements.
15. Why does $HMnO_4$ not ionize to form OH^- ions?

Test Yourself

1. List the factors involved in the general pattern of oxygen reactions.
2. Identify which of the following are representative elements, which are transition elements, which are inner transition elements, and which are halogens. Each could fit zero, one, or more than one category. (You may look for the symbols in a periodic table.)

O	Fe	Pb
H	Si	U
Cl	Ca	K

3. What oxidation states could be predicted as likely for each of the following, based on position in the periodic table? (They may not all actually occur.)
    (a) strontium **(b)** sulfur **(c)** zirconium
4. Which is the stronger acid, H_3PO_2 or H_3PO_3? Why?
5. The formula of ozone is
    (a) Os_2O_3 **(c)** H_2O_2 **(e)** O_3
    (b) H_3O^+ **(d)** OxO_2
6. Given the electronic configurations of two atomic species:

    $1s^2 2s^2 2p^6 3s^2 3p^6 3d^1 4s^2$ and $1s^2 2s^2 2p^6 3s^2 3p^6 3d^2 4s^2$

    These are
    (a) nonmetals **(d)** inner transition elements
    (b) transition elements **(e)** representative elements
    (c) isotopes
7. A *transition* or *inner transition* element is one whose last electron is added to
    (a) An s orbital **(d)** an f orbital only
    (b) a p orbital **(e)** either a d or an f orbital
    (c) a d orbital only

8. The usual dot diagram structure for O$_2$ is somewhat unsatisfactory because
   (a) The actual bonding is less than a double bond.
   (b) the dot diagram fails to reach closed shells for each atom.
   (c) the two oxygen atoms do not contribute equally to the bond.
   (d) the actual bonding has two unpaired electrons.
   (e) O$_2$ is not a stable molecular structure.
9. Which of the following *has* an s^2 outer shell electronic configuration but *does not* fit in group II of the periodic table?
   (a) magnesium
   (b) carbon
   (c) helium
   (d) nitrogen
   (e) neon
10. O$_2$ can be prepared by
    (a) heating KClO$_3$ in the presence of MnO$_2$ catalyst.
    (b) electrolysis of water.
    (c) adding MnO$_2$ catalyst to hydrogen peroxide.
    (d) adding HCl to sodium peroxide.
    (e) *all* of these reactions produce O$_2$.
11. Hydrogen is a unique element chemically because
    (a) its low atomic weight makes it very reactive.
    (b) it is only one electron away from a closed shell in both directions—gaining or losing e^-.
    (c) it is definitely metallic (group IA) but exists as a gas.
    (d) it has the lowest electronegativity of all elements.
    (e) it is unable to bond covalently.
12. The halogens are the family of elements which
    (a) lack only one electron from a closed outer shell and tend to gain that electron in bonding.
    (b) have only one electron in their outer shell and tend to lose that electron in bonding.
    (c) have closed shell configurations and tend not to react chemically.
    (d) contains nitrogen, phosphorus, arsenic, and the rest of that column.
    (e) *none* of these is true of the halogen family.

Phases of Matter

9–1 Gases

Of the three states of matter, solid, liquid, and gaseous, gases are the simplest to describe and explain. Some of the explanations of gases will also help us understand liquids, solids, and the transitions between states of matter.

Gases have several notable properties. The most important of these is their ability to expand to occupy all available space. Gases are concentrated in limited (on a cosmic scale) regions by gravitational forces, but in the small regions we normally study the distribution of the gas is uniform throughout the available space.

Gases expand to occupy a volume (V) equal to all the available space.

Four main properties must be given to describe a particular sample of gas. These properties depend upon each other, and the nature of that dependence will help us to understand gases.

First, any gas has a definite *quantity* of material. That could be measured as mass and often is in practical problems. However, the behavior of the gas actually depends on the *number* of gas molecules, not their mass. The number can be quickly and easily calculated from mass by using the molecular weight in grams, which is equal to 1 mole. We will usually refer to the number of moles of gas present and call this number n when we want to abbreviate in equations.

The quantity of gas present is conveniently stated in moles.

Second, a gas occupies a particular *volume*. That volume is simply all the available space. For gas in a container, it is the volume of the container. We will use V as the abbreviation for volume in equations.

Third, a gas has a *temperature*. We are all familiar with temperature measurements by thermometers and know that the temperature of air (a gas) varies. Actually, temperature is defined from the behavior of gases as will be described in Section 9–3. We will use T as the abbreviation for temperature in equations.

Temperature (T) and pressure (P) are also important measurable properties of gases.

Fourth, a gas exerts a *pressure*. Pressure is a force pressing against everything the gas touches. The size of that force can be measured by determining how large a force is needed to push back equally hard. We will use P as the abbreviation for pressure in equations.

One device for measuring pressure is a **barometer.** When a tube is filled with mercury and then tipped over into a cup of mercury (without letting

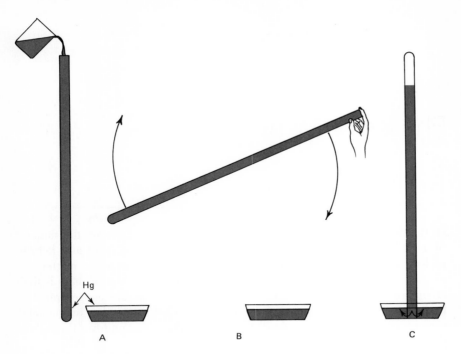

Figure 9–1 Preparation of a barometer.

    A. Fill tube with mercury.
    B. Close end and tip to place in container of mercury.
    C. Open end (under mercury) and allow mercury in the tube to seek its equilibrium level.

any air in) the height of the mercury column is a measure of the air pressure. Figure 9–1 shows preparation of a barometer. Mercury is convenient because it is so dense that the height of the column is fairly short. It also gives off very little vapor pressure (see Section 9–8) so the pressure reading does not have to be corrected for vapor pressure.

    There are other devices for measuring pressure, but the convenience of mercury columns has been used to define units of pressure. We sometimes refer to pressures as "mm Hg." That is the height, in millimeters, of the column of mercury the pressure could support. This unit has now been given the name **torr**. Its precise definition also specifies the exact gravitational field, but for practical use (on Earth) torr equals mm Hg.

    The most common gas whose pressure we measure is air. Air pressure on Earth varies with changes in altitude or weather. At sea level the average atmospheric pressure is about 760 torr. Therefore, the value 760 torr has been defined as the unit 1 **atmosphere** (atm) of pressure. Other pressure units are possible, but we will use only torr and atmospheres in this book.

Torr equals the height (in millimeters) of mercury in a barometer, and 760 torr equals 1 atm.

9–2 Gas Laws

When the quantity, volume, temperature, and pressure of gases are measured, definite relationships can be seen between the various properties. The first of these, a relationship of pressure to volume, was reported by Robert Boyle in 1660 in a book entitled "The Spring of Air." Boyle was a younger son (no responsibilities running the family estates) in a wealthy family. With the help of a hired assistant, who was very adept at building equipment, he gathered masses of data about the springlike ability of air to push back as it is compressed. Much of this data was confusing because temperature changes were complicating results, but eventually a relationship between pressure and volume was noticed and supported by selecting "good" data which fit and ignoring "bad" data. The discarding of some data without knowing why it was bad is appalling to modern scientists, but the poor control of experiments and incomplete recording of conditions in 1660 made such guesswork necessary. Boyle was then able to clear away the confusion from variations other than pressure and volume.

We now state the relationship Boyle found in the following way. *When the quantity and temperature of a gas are kept constant, the pressure is inversely proportional to the volume.* This relationship is called **Boyle's law.** It can also be stated by any one of the following three equations.

Boyle's law: P is inversely proportional to V when n and T are constant.

$$P = \frac{1}{V} \times \text{constant} \quad \text{or} \quad V = \frac{1}{P} \times \text{constant} \quad \text{or} \quad PV = 1 \times \text{constant}$$

where n and T are constant in each case.

In 1787 a second gas relationship, Charles' law, was developed by Jacque Charles in France. Figure 9–2 shows a plot of volume versus temperature similar to the data observed by Charles. All points fell on a straight line,

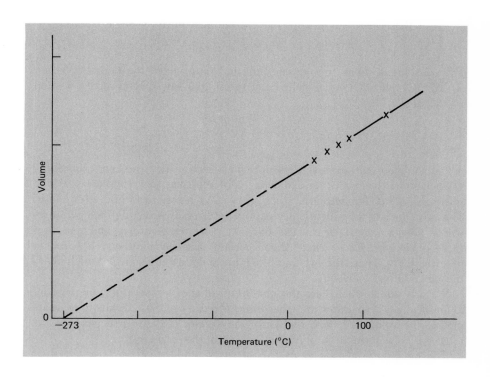

Figure 9–2 Charles' law plot (gas volume versus temperature).

but there still was volume and a continuation of the straight line at temperatures below zero on the conventional temperature scales. This problem was solved by defining a new temperature scale. On this "absolute" temperature scale the zero of temperature is the point where the gas volume would reach zero when the line in Figure 9–2 is extended. That point can never be reached in real life, but extending the line on the graph is easily done. The absolute temperature scale we use is the Kelvin scale (see Section 3–5). We must always use absolute temperatures to get simple gas relationships.

When an absolute temperature scale is used, volume and temperature are directly proportional. **Charles' law** can be stated as follows: *When the quantity and pressure of a gas are kept constant, its volume is directly proportional to absolute temperature.* It can be stated by the equation

$$V = \text{constant} \times T \quad (n \text{ and } P \text{ constant})$$

Charles' law: V is directly proportional to absolute T when n and P are constant.

T must represent the *absolute* temperature whenever it is used to describe gases in this equation and all other gas relationships described in this chapter.

Other gas relationships are

$$P = \text{constant} \times T \quad (n \text{ and } V \text{ constant})$$
$$V = \text{constant} \times n \quad (P \text{ and } T \text{ constant})$$
$$P = \text{constant} \times n \quad (V \text{ and } T \text{ constant})$$
$$n = \frac{1}{T} \times \text{constant} \quad (P \text{ and } V \text{ constant})$$

All of these gas relationships can be summed up in a single equation.

$$PV = nT \times \text{constant}$$

The constant is called the gas constant and represented by the abbreviation R. The overall equation is called the **ideal gas law** and normally written $PV = nRT$.

The gas relations are summarized in the ideal gas law, $PV = nRT$.

9–3 Ideal Gases

All of the gas relationships in the previous section have assumed "perfect" gas behavior. Real gases start to behave differently as they approach the high pressures or low temperatures where they condense to form liquids or solids. They also deviate very slightly even at the low pressures and high temperatures where they are most clearly gas like. At most temperatures and pressures these deviations are so small that "perfect" gas behavior can be used to describe them. Therefore, we find it helpful to define an ideal gas which obeys the gas relations perfectly.

We can quickly calculate the quantity, volume, temperature, or pressure of an ideal gas if we know the other three. We can use these results to estimate closely the values for most real gases, or we can compare real gases to the ideal gas to determine how nonideal the real gases are.

The deviations of real gases can often be understood in terms of their

An ideal gas would have molecules with zero space and no forces between them.

failures to meet two impossible requirements for ideal gas behavior. To be truly ideal a gas must have (1) molecules that occupy zero space and (2) absolutely zero forces between molecules.

The perfectly predictable behavior of ideal gases is used to define a "perfect" temperature scale. Most thermometers use a fluid, such as mercury, which has a fairly steady rate of expansion with temperature. If the expansion of an ideal gas was used, it would be perfectly steady in its expansion with temperature. Precise temperatures are defined from such an "ideal gas thermometer." More common temperature measuring devices are compared to the ideal gas standard and corrected for any differences.

9–4 The Kinetic Molecular Theory

The observed behavior of gases can be explained by assuming that gas molecules are in constant motion. The movement explains the ability of gases to occupy all available space. Each molecule spends some time in each available region. This theory of motion also explains liquids and solids when certain limitations are imposed.

Gas behavior is explained by the kinetic molecular theory.

The conditions needed to approach ideal gas behavior are summarized in the following points of the **kinetic molecular theory of gases.**

1. Gases have small molecules with very large average distances between molecules.
2. Molecules are in constant, random motion.
3. Collisions between molecules may transfer energy from one to the other, but the total energy (and hence the average energy) of the gas molecules is not changed.
4. The average kinetic energy (energy of motion) of the molecules is proportional to the absolute temperature.
5. The molecules of the gas exert no forces of attraction (or repulsion) on each other.

The kinetic theory leads to the gas laws as follows:

1. *Temperature.* Temperature is a measure of average kinetic energy. Kinetic energy is equal to $\frac{1}{2}$ mass $\times$ (velocity)^2. Therefore, temperature is proportional to the square of velocity and, at any given temperature, mass and (velocity)^2 are inversely proportional. The latter relationship leads directly to Graham's law, an observed relationship between molecular weight of gases and their speed in diffusion. The requirements of this temperature-energy relationship also carry over to liquids and solids and fit observed behavior in each case.

2. *Pressure.* As the moving molecules strike walls and bounce off, they transfer momentum to the walls. That exerts a force, which is the gas pressure.

3. *Boyle's law.* If volume increases, any one section of wall is hit less often. If volume decreases and the number and speed (temperature dependent) stay the same, the molecules get back to each section more often. The average force of the molecules striking the walls stays the same at a given temperature, so the pressure depends on how often each piece of wall is hit.

4. *P, T relation.* As T rises (for the same n and V), the molecules get back to any section of wall more often and hit it harder when they get there. Actually, each of those factors goes up by $T^{1/2}$ (the square root of T), but the total result gives P as directly proportional to T.

5. *Other relationships.* The variations in number of molecules to collide with the walls, frequency of reaching the walls, and intensity of collisions lead to all the other gas relationships.

9–5 Arithmetic of Gas Laws

The ideal gas relationships can be used to carry out calculations. Sometimes these calculations are quite involved. The ideal gas law, $PV = nRT$, is useful for some of these calculations, particularly those involving changes in quantity. However, we can provide enough examples to let you use (and learn to remember) the relationships while limiting ourselves to a simple "common sense" fraction method.

Most gas problems involve changes from one condition to another. In these problems we have the following information:

1. The original value of one gas property for which we must calculate the final value.
2. Both original and final values for at least one (sometimes more than one) other gas property.
3. The fact that any of the properties not mentioned in the problem remain constant. (The quantity, n, is almost always in this constant group.

We will solve the problems by taking (1) the given value of what we are calculating and multiplying it by a fraction (or fractions) obtained from (2) the set (or sets) of values of the other property (or properties). Sometimes we will also have to convert units, either to get the desired answer units or to make both parts of a fraction have the same units. The common sense part comes in when we decide how to put each fraction in. We know from the gas relations whether an increase or decrease in one property will increase or decrease the property being calculated. If the property being calculated should go up, we write the fraction with the larger value on top. If the answer should get smaller, we put the smaller value on top.

Most gas problems can be solved by multiplying the original value by a fraction (or fractions) made from the other property (or properties) that changes.

Example 9–1. If a gas occupies 25.0 liters at 1.0 atm pressure, what volume does it occupy at 0.75 atm pressure.

As P goes *down*, V must go *up*. Therefore, we use the fraction as 1.00 atm/0.75 atm.

$$25.0 \text{ liter} \times \frac{1.00}{0.75} = 33.3 \text{ liter}$$

Example 9-2. If a gas has a pressure of 740 torr at 25°C, what pressure would it have at 50°C in the same container?

As T goes *up*, P goes *up*. Therefore, we use a fraction with the 50°C temperature on top. But, all gas law problems must use *absolute* temperatures, so we must convert to Kelvins.

$$25°C = (25 + 273) \text{ K} = 298 \text{ K}$$
$$50°C = (50 + 273) \text{ K} = 323 \text{ K}$$

$$740 \text{ torr} \times \frac{323}{298} = 802 \text{ torr}$$

Example 9-3. If a gas has a pressure of 740 torr at 25°C, at what temperature would it have a pressure of 1.00 atm?

The 25°C must be converted to Kelvins.

$$25°C = (25 + 273) \text{ K} = 298 \text{ K}$$

The pressures must have the same units. Either change 740 torr to 740/760 atm or change 1 atm to 760 torr. 1.00 atm (760 torr) is larger than 740 torr, so P goes up. As P goes up, T goes up.

$$298 \text{ K} \times \frac{760}{740} = 306 \text{ K}$$

If the answer had to be in degrees Celsius, the following additional step would be needed

$$306 \text{ K} = (306 - 273)°C = 33°C$$

Example 9-4. If a gas occupies 596 cm³ at 25°C and 735 torr, what volume does it occupy at 0°C and 1.00 atm?

Two fractions are needed. P goes up (735 torr to 760 torr), so V goes down. T goes down (298 K to 273 K), so V goes down.

$$596 \text{ cm}^3 \times \frac{735 \text{ torr}}{760 \text{ torr}} \times \frac{273 \text{ K}}{298 \text{ K}} = 528 \text{ cm}^3$$

A few gas law problems have only one condition given instead of two. These can be solved directly using $PV = nRT$ if we know the value to use for the gas constant R. However, these problems can also be done by our fraction method if we know one other condition for the gas to use in making fractions. There is such a condition we should know anyhow. It is the volume of a gas at the conditions called **standard temperature and pressure (STP)**—0°C and 1.00 atm pressure. At 0°C and 1.00 atm, 1 mole of any gas occupies 22.4 liters.

Here is an example of a problem solved using that fact.

Example 9-5. What is the pressure of 10.0 g O_2 gas at 47°C in a 25.0 liter container?

All gases occupy 22.4 liters for 1 mole at standard temperature and pressure (STP—0°C and 1 atm), and this condition can be used as the "second" condition in gas problems where only one condition is given.

Using 1 mole at 0°C, 1.00 atm, and 22.4 liters as the other condition:

V goes up (22.4 liters to 25.0 liters), so P must go down. The fraction from volumes must be 22.4 liter/25.0 liter.

T goes up (273 K to 320 K), so P goes up. The fraction from temperatures must be 320 K/273 K.

n goes down (1 mole to 10.0/32.0 mole), so P goes down. The fraction from quantities must be (10.0/32.0 mole)/1 mole.

Notice that some fractions may tend to increase pressure while others decrease it. We can put each one in, one at a time, ignoring the others and using our common sense about what that factor would do if it was the only change. Then the overall calculation is

$$1.00 \text{ atm} \times \frac{22.4}{25.0} \times \frac{320}{273} \times \frac{10.0}{32.0} = 0.328 \text{ atm}$$

In testing your comprehension about gases, your instructor may choose to ask some questions that do not require any numbers at all. An example would be Example 9–3 rewritten in this form: If a gas has a pressure of 740 torr at 25°C, is the temperature higher or lower when it has a pressure of 1.00 atm in the same container?

9–6 Real Gases

No real substance follows the ideal gas law exactly. Some deviations due to the size of gas molecules can be found, but most of the deviations are caused by forces between molecules. These forces are insignificant when the molecules are far apart but increase as the separations become smaller. When the forces become large enough to hold some gas molecules together for short periods of time, the gas behaves as if there were less molecules present. That makes the observed pressures or volumes less than predicted at that temperature by the ideal gas law.

The distance-dependent forces between molecules are called Van der Waal's forces. We will use VWF as an abbreviation for these forces. VWF's are caused by distortions of the electronic distibutions. Figure 9–3 shows how molecules can be attracted to each other by inducing a slight concentration of electrons on one side of each molecule. This concentration is more easily accomplished if the molecules are large with many electrons in relatively loosely held outer shells. Figure 9–3 compares the VWF's of H_2 and I_2 to illustrate the size effect.

Large and small molecules also differ in average speed at a particular temperature. Since temperature is a measure of kinetic energy, each gas at a given temperature has the same value for $\frac{1}{2}$ mass $\times$ (average velocity)^2. As mass goes up, average velocity must go down to keep the kinetic energy constant. At room temperature (25°C) the average velocity of H_2 molecules is about 1 mile/sec whereas O_2 molecules average about $\frac{1}{4}$ mile/sec and heavier molecules go even slower.

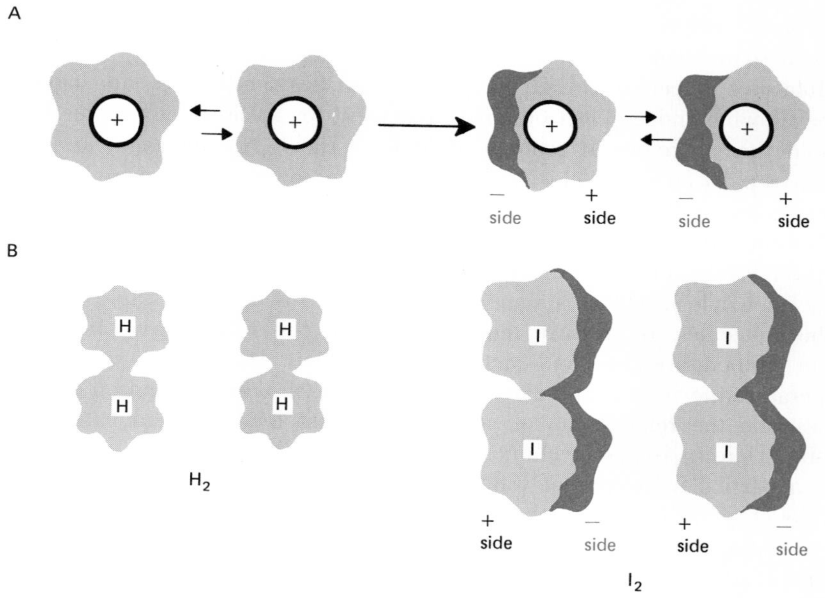

Figure 9–3 Van der Waal's forces.

A. On the left: Electron clouds of two molecules repel each other. On the right: Electrons from one molecule have been repelled toward the side away from the neighboring molecule. That leaves an excess of positive charge over negative charge on the side toward the neighbor and causes electrons on the neighbor to concentrate on the side near that positive charge. The result is a positive side of one molecule near a negative side of the second molecule. The attraction between positive and negative regions then holds the molecules together.

B. On the left: Very little distortion occurs with H_2 molecules. On the right: I_2 molecules are larger than H_2 and have many more electrons, so they are more easily distorted to set up attractions between the molecules.

Real gases deviate most sharply from ideal gas behavior under conditions of slow speeds (low T) and large electron clouds (heavy molecules) which lead to large Van der Waals forces.

The slower moving heavy molecules spend a longer time near each other during each collision. That gives more time for the electronic rearrangement needed to set up VWF's. As a result, heavy molecules are more easily affected by VWF's than light molecules. Because the heavy molecules also usually have more electrons and larger electron clouds, they exert stronger VWF's. The combination of stronger forces and more time to form them leads to a sharp, very general size effect on real gas deviations from ideal gas behavior. Small, light gases, like H_2, with no forces between molecules other than VWF's, are the most ideal, whereas heavy, large molecules give less ideal gases. Gases are also more ideal at higher temperatures because they spend less time near each other in any one collision and can break away from larger attractions. Molecules with other kinds of forces between molecules, such as the attractions between positive and negative regions of polar molecules, are less ideal than those of the same size and mass that have only VWF's.

9–7 Liquids

When forces of attraction are large enough, the molecules condense to liquid or solid phase.

If the forces of attraction are strong enough to hold molecules together against the force of their motion, a liquid phase results. When gas molecules are brought together into a group, they are said to **condense.** Liquids and solids are referred to as condensed phases.

Liquid phase does not change the average kinetic energy at a given temperature. O_2 molecules would still move at an average of $\frac{1}{4}$ mile/sec at room temperature. Forces large enough to hold molecules of that energy together are required to make O_2 a liquid at room temperature. The forces

[9–7] Liquids

between O_2 molecules are not large enough to accomplish that, so O_2 remains a gas at room temperature.

Molecules of benzene, C_6H_6, are larger and heavier than O_2 and have a shape allowing enough electronic rearrangement for VWF's (see Section 9–6) to hold them together at room temperature. The C_6H_6 molecules are still moving rapidly (average velocity about $\frac{1}{9}$ mile/sec) but going nowhere because they are constantly bouncing off neighbors or being pulled back to a neighbor.

Molecules in a liquid are held near their neighbors but are otherwise free to move. The result is a free shape but a fixed volume (essentially the volume of the molecules). There is some empty space caused by the rapid motion of the molecules. If pressure remains constant, this empty space increases as the temperature rises and increases the motion. However, at any given temperature the volume and mass are related to each other. We find it helpful to describe the relationship of volume and mass by a fraction called the density. Density is mass/volume. Since the unit of mass, gram, is defined from a particular volume of water, 1 ml at 4°C, density is often reported in units of grams per milliliter (g/ml). The density of water is exactly 1 g/ml at the temperature of maximum density (4°C) and only slightly less than 1 g/ml at all other temperatures.

Liquids have a free shape but a fixed volume. The fixed volume for a given amount leads to a definite density.

The characteristics of gases were dominated by the freedom of the molecules to move about. The characteristics of liquids are dominated by the attractive forces that hold the liquid together. Some of the notable properties of liquids caused by these attractive forces are the following.

The attractions between molecules dominate the characteristics of liquid, causing properties like surface tension, capillary rise or fall, viscosity, and slow diffusion.

1. *Surface tension.* Figure 9–4 shows that molecules on the surface of a liquid are in a less advantageous position than those in the center. As a result, there is a tension on the surface that tends to keep the surface area as small as possible. For small drops (or in the absence of gravity) this is a spherical shape. Gravity pulls large amounts of liquid down to form a flat surface, but there may still be some curvature into a slightly spherical form at the edges. A water drop in a space ship (where there is no motion through air or gravity to distort it) would form a perfect sphere because that is the shape with the least surface for a given volume.

2. *Capillary rise or fall.* The curvature at edges plus the surface tension's tendency to keep the surface area small causes capillary effects. Figure 9–5

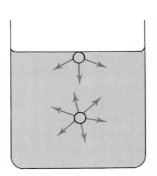

Figure 9–4 Surface tension.

Figure 9–5 Capillary fall (as by mercury in a glass container).

shows the capillary fall occurring in mercury where the attractions between liquid molecules are very large. The edges curve down. In a small tube (capillary) the edges are so close together that the center is pulled down with them. The whole column is depressed until the pressure pushing mercury up equals the force with which the curvature is pulling the center down.

Figure 9–6 shows the same effect can work in the opposite direction. H_2O is attracted to glass more strongly than to other H_2O molecules. Therefore instead of curving down at the edges, water curves up to get more molecules where they have the greatest attractive force. In a capillary, the glass pulls the water level at the edges upward, causing surface tension and the curvature to pull the center up until the gravitational force pulling the column of water down equals the force pulling the column up. This capillary rise of water also occurs in wood and other plant tissues. It is one of the mechanisms that carries water up to the leaves of trees. The attractions of H_2O to wood are very large. In tall trees the capillary action carries water about 12 times as high as air pressure could push it against a vacuum, and the trees do not have any vacuum to help draw the water up.

3. *Viscosity.* Some liquids flow more easily than others. Those that do not flow as well are called viscous liquids. Figure 9–7 shows factors that cause viscosity. The attraction of molecules to solids (similar to H_2O in the capillary rise examples) may hold those at the sides of a tube in place. If the rest of the liquid is held to those molecules, flow becomes difficult. Entanglement of the molecules may cause some viscosity, but strong attractive

Figure 9–6 Capillary rise (as by water in a glass container).

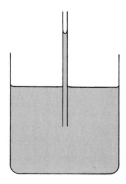

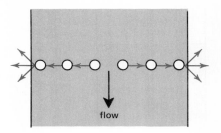

Figure 9-7 Viscosity. Attractions tend to hold the molecules in place, thus resisting flow.

forces, which must be at least partially broken to move, are an important cause of viscosity.

4. *Diffusion.* One gas can diffuse (move through) in another gas fairly quickly. Molecules of one kind can also diffuse in a liquid made up of another kind of molecules, but the process is extremely slow. The attraction between neighboring molecules leaves very little open space for the diffusing molecules to move through, so they spend most of their time going back and forth in the same place.

9-8 Evaporation and Heat of Vaporization

If a molecule on the surface of a liquid breaks away from the attractive forces, it can move off with the freedom of the gas phase. This conversion of liquid to gas is called **evaporation.** Energy must be supplied to break the liquid attractions. There are no significant forces between molecules in the gas phase, so the entire energy of attraction in liquid must be supplied when a molecule breaks free. This energy is usually supplied from the energy of motion of surrounding molecules, which lowers the average kinetic energy of molecules in the liquid, thus cooling the liquid. The human body uses this mechanism for cooling. When excess heat needs to be removed, a person sweats and thus makes water available for evaporation from the skin. The skin (and eventually body) is cooled by the evaporation. The amount of energy needed to evaporate 1 mole of liquid is called the **heat of vaporization.**

The heat of vaporization must be supplied to evaporate a liquid.

The concentration of energy on the molecule breaking loose may not make it move fast. The energy is used up to break bonds, so the escaping molecules also have low average kinetic energy. That means they have a low temperature. The situation can be understood by the analogy to a rocket shown in Figure 9-8. The rocket must be supplied with a lot of energy to escape from the Earth's gravity, but only a little energy is left as energy of motion after the rocket reaches space. The evaporating molecule also needs a lot of energy to break the bonds, and only a little is left as energy of motion after the molecule reaches the gas phase.

9-9 Vapor Pressure: An Introduction to Le Chatelier's Principle

If a liquid evaporates and the gas molecules escape from the area, the liquid will eventually disappear completely. If instead the gas molecules are kept around the liquid (by a closed container around the gas and liquid), the

Figure 9-8 Analogy of evaporation to rocket launching.

A. The rocket consumes much energy during launching to break the gravitational attraction so that very little of the available energy is left to cause motion of the rocket in orbit.

B. Much energy is consumed to break the attractions of a molecule in liquid phase so that very little of the molecule's original energy of motion remains after it has evaporated.

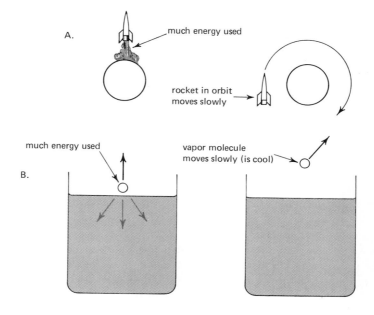

molecules may eventually return to the liquid. If they return, they will again be held by the attractive forces and release the heat of vaporization.

At any one temperature, there will be some set average rate at which molecules evaporate from a liquid surface. If that rate is faster than the rate at which molecules of that substance return to the surface, the concentration of that type of molecules in the gas phase will go up, as shown in Figure 9-9A. If we ever get so many vapor molecules that they return to the liquid faster than they evaporate, the gas concentration will drop, as shown in Figure 9-9C. In either case, the concentration will continue to change until the *balance* situation (shown in Figure 9-9B) is reached. Because pressure is a measure of how often molecules strike a surface (at a given temperature), there will be a fixed **vapor pressure** of the liquid. This balance between rates of evaporation and condensation is an example of an **equilibrium**

The vapor pressure is the pressure of the material as gas needed to give an equilibrium balance between evaporation and rate of return to the liquid.

Figure 9-9 Vapor pressure equilibrium.

A. At the condition where the pressure of the vapor is less than the equilibrium vapor pressure for that temperature liquid evaporates and the pressure of the vapor is rising.

B. At the equilibrium vapor pressure the rates of evaporation and return are equal, causing an equilibrium balance. There is no net evaporation or condensation.

C. At a pressure of the vapor greater than the equilibrium vapor pressure for that temperature vapor condenses faster than liquid evaporates and the pressure of the vapor is falling.

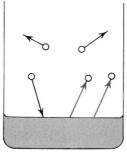 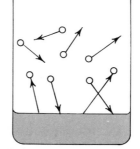 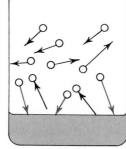

A. Liquid evaporates   B. No net rate of evaporation   C. Vapor condenses

[9-9] Vapor Pressure: An Introduction to Le Chatelier's Principle   175

situation. Equilibrium is the natural result in many situations, including many chemical reactions.

Equilibria are subject to very specific limitations, set up by the required balance between forward and reverse reactions. Equations can be set up and used to calculate the exact amounts present at equilibrium in many cases. However, there is also a nonmathematical way to predict how any equilibrium changes with changing conditions. We can predict the direction of any shift in equilibrium by using **Le Chatelier's principle.** Le Chatelier's principle can be remembered by using the crude form "nature is against us."* A more precise statement of Le Chatelier's principle is the following. *When any change is made in the conditions of an equilibrium, the equilibrium will shift in the direction to minimize the effect of the change.* We can use vapor pressure to show the meaning of Le Chatelier's principle and how to use it. We will use Le Chatelier's principle again later in this book, and it is often referred to in other subjects.

> Le Chatelier's principle: When any change is made in the conditions of an equilibrium, the equilibrium will shift in the direction to minimize the effect of the change.

The only changes that count in Le Chatelier's principle are changes in things *directly involved in the equilibrium.* In vapor pressure, the things involved are the liquid, gas pressure of the same kind of molecules, and the heat needed to convert liquid to gas. Let us change the conditions on one of those and see how it affects the equilibrium.

> The only changes that count in Le Chatelier's principle are changes in things directly involved in the equilibrium.

If we raise the temperature, we are changing the amount of heat, one of the things involved in this equilibrium. By Le Chatelier's principle, the equilibrium will work against our change. The system can use up some of the extra heat (keep the temperature from rising as much as we expected) by using it as heat of vaporization. To do that, more liquid must evaporate than the amount of vapor condensing. That means the amount in the gas phase is increased. Therefore, the vapor pressure goes up.

Actually, the increase in vapor pressure at higher temperature is caused by increased breaking of attractions, not by Le Chatelier's principle. Le Chatelier's principle only helps us see what must happen to restore the equilibrium balance. A typical variation of vapor pressure with temperature is shown in Figure 9–10. This is called a vapor pressure curve.

A second application of Le Chatelier's principle lets us predict what will happen if we compress the vapor at equilibrium into a smaller space. The reduction in volume increases the pressure of vapor, and pressure of that type of molecules is one of the things involved in the equilibrium. By Le Chatelier's principle, the equilibrium shifts in the direction opposing our change. The system can use up some of the extra pressure by condensing vapor to form more liquid. But condensing vapor releases the heat of vaporization. That will warm up the liquid and vapor and shifts us to a new position on the vapor pressure curve. This new condition cannot have as high a vapor pressure as we had immediately after we compressed the vapor. Some vapor

---

* One reviewer criticized this cliche as inappropriate in these days when we are learning to cooperate with nature. He suggested "nature tends to avoid abrupt and large changes" instead, but that statement is not complete since even small changes can cause a small shift in equilibrium. We hope you can use our crude form as a clue to nature and not an attempt to fight nature.

Figure 9-10 Vapor pressure curve.

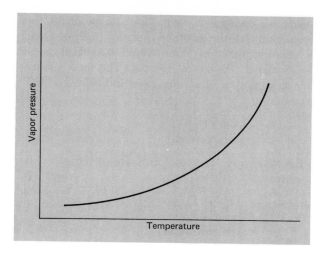

must condense to provide heat to raise the temperature. We end up with a vapor pressure between the original vapor pressure and the vapor pressure reached in the compression. That is the normal result for a Le Chatelier's principle shift. The property being changed ends up between the original condition and what it seemed to be shifting to.

The limitations of Le Chatelier's principle are shown by the case where air is pumped into a container where a liquid is at equilibrium with its vapor. At first glance we might suppose the equilibrium would shift to work against the increase in pressure. That is *not* true. *Air* pressure is *not part of the equilibrium.* The vapor pressure (pressure of the liquid type molecules) is not changed, so *nothing* at all happens to the equilibrium.

We can understand the lack of effect by looking at the balance situation. The temperature is unchanged, so the rate of escape of molecules from the liquid stays the same. The concentration of that type of molecules in the gas stays the same, so the rate of return to the liquid is unchanged. Therefore, the rates of escape and return are still equal. Air does not affect the equilibrium because it does not provide any of the molecules that could condense to be the liquid and does not change the chances of liquid molecules breaking the attractions to evaporate. In other cases it may be more difficult to see why the equilibrium is unaffected by an "outside" change, but the principle remains the same. Unless the thing being changed is part of the equilibrium, it does not count.

9-10 Partial Pressures

It is often important to keep track of the contributions from each type of molecule for purposes such as Le Chatelier's principle effects. This is usually done by using the concept of partial pressures. The ideal gas equation, $PV = nRT$, can be applied to each kind of molecule present. The volume and temperature are the same for each gas present—the volume of the total available space and the temperature of the gas. The quantity (moles) of each gas then determines the pressure that gas alone would have had in the container. That is called the partial pressure of the gas. The total quantity (moles) present is the sum of quantities of all the gases individually.

$$n_{\text{total}} = n_1 + n_2 + n_3 + \cdots$$
(sum of all gases present)

The quantity of each gas determines the partial pressure of that gas, and the total pressure in a container is the sum of the partial pressures of all the gases present.

The pressures add up in the same way. The total pressure of the gas is the sum of all the partial pressures.

$$P_{\text{total}} = P_1 + P_2 + P_3 + \cdots$$
(sum of all the partial pressures)

When a liquid is evaporating into air, the partial pressure of the liquid's vapor must always be less than or equal to the vapor pressure. That vapor mixes with the other gases above the liquid. Unless there is a breeze to carry away the vapor-laden gas, the evaporation of the liquid is limited by the rate at which the vapor can diffuse away through the other gases. As the temperature rises, the vapor pressure from the liquid makes up a larger fraction of the gas near the liquid surface, but evaporation is still limited by the rate of diffusion.

When the vapor pressure of the liquid equals or exceeds the total gas pressure above the liquid, the rate of evaporation changes drastically. Instead of diffusing through the other gases, the vapor can become 100% of the gas above the liquid and simply push the other gases out of the way. The liquid can then evaporate as fast as the heat of vaporization can be supplied. That condition is called **boiling**.

We should emphasize the fact that boiling occurs whenever the vapor pressure exceeds the gas pressure above the liquid. For water under 1.0 atm pressure, boiling occurs at 100°C. For water under more than 1.0 atm (such as an automobile's pressurized cooling system) the boiling point is higher—the temperature where the vapor pressure curve reaches that higher pressure. Under lower pressures water boils at less than 100°C. If the pressure is low enough, water can boil while it is so cold that some is freezing at the same time.

When a liquid boils, the rapid evaporation cools the surface of the liquid. If heat is being applied at the bottom of the liquid, the bottom may become several degrees hotter than the surface. Since vapor pressure goes up rapidly with temperature, the bottom of the boiling liquid may have a vapor pressure larger than the surface pressure by enough to let bubbles form. These bubbles must have enough pressure to equal the pressure above the liquid plus the

9–11 Boiling

Boiling occurs when the vapor pressure of a liquid equals or exceeds the total pressure above the liquid.

pressure applied by the depth of the liquid. Most of us look for these bubbles as a sign of boiling. If heat could be added at the surface, boiling could occur from the surface without bubble formation, but excess heating of the bottom is common.

Bubble formation actually has to overcome more than just the pressure. Even when bubbles could form, they have trouble getting started. An examination of the discussion of surface tension (Section 9–7 and Figure 9–4) may help you see why formation of new bubbles is more difficult than evaporation from an already existing surface. This difficulty can lead to **superheating** of the liquid to a temperature above the boiling point under the actual pressure. When a bubble does start to form, it then grows very rapidly as the excess heat supplies the heat of vaporization for many molecules.

In a narrow vessel like a test tube a bubble may fill the container from wall to wall and force part of the liquid out in front of the expanding bubble. With corrosive chemicals that can be dangerous. One method used to prevent superheating in chemistry laboratories is the addition of "boiling chips," which provide a surface where bubbles can form easily, without much superheating. Superheating can also occur when there is a sudden drop in pressure. When the cap is removed from a hot car radiator, the water is put in a superheated state. Bubbles formed inside the radiator may then force boiling water out and burn the person who removed the cap.

> **Superheating occurs under some conditions where difficulty in forming bubbles prevents boiling from occurring fast enough to use up the heat being supplied.**

9–12 The Critical Point

Transitions from liquid to gas or gas to liquid cannot occur at very high temperatures because liquid phase ceases to exist at high temperatures. The point beyond which a given liquid cannot exist no matter how high the pressure exerted on its vapor is called the **critical point** of that liquid. The critical point consists of a particular temperature, the **critical temperature,** and the vapor pressure at that temperature, the **critical pressure.** The critical point is the upper end of the vapor pressure curve.

The reason for the occurrence of critical points is shown in Figure 9–11. At every temperature below the critical temperature the gas phase can be *compressed* to form a liquid, which is more dense than the densest possible gas (the vapor pressure). At the critical point the gas can become as dense as the liquid, so no compression to liquid is possible. Above the critical temperature the gas can simply be compressed to denser gas, and no liquid ever appears. This material is described as a gas because it can expand to occupy all available space without having to evaporate. At high pressures such a substance is far from ideal gas behavior, but it still fits the description of a gas (compressible) better than liquid. It should be noted that liquids near their critical points are about as compressible as their very dense equilibrium vapor pressure gases, but there is still at least some difference between the two until the temperature is raised to the critical temperature.

> **Liquid phase cannot exist above the temperature of the substance's critical point.**

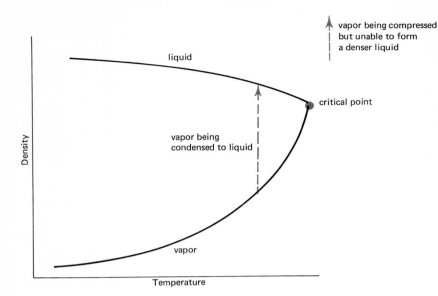

Figure 9-11 Densities of a liquid and of its vapor at equilibrium vapor pressures.

Material in solids is even less free than material in liquids. Molecules (or ions) are not only held close to each other, they are held in particular positions. The molecules cannot freely move about, and the solid cannot take on the shape of its container as a liquid can.

The arrangements in solids are those which provide stronger forces of attraction than in liquid phase. The molecules (or ions) are held in the particular positions where these stronger attractions are possible and therefore lose freedom to move to other positions and form new shapes in the way liquids do.

As temperature is raised, the molecules of the solid may break free of their rigid arrangement. If they form a liquid, an amount of energy equal to the difference between attractions in liquid and solid must be provided. The transition from solid to liquid is called melting or fusion, and the energy required is called the **heat of fusion.** Since the concentrations of molecules in both liquid and solid are fixed (at any given temperature), there is a particular temperature, the **melting point,** where the change from solid to liquid takes place. At that temperature, there is an equilibrium between the rate at which molecules escape from the solid and the rate at which they return. When heat is supplied, the equilibrium shifts to use it up, supplying heat of fusion as solid melts. If heat is removed, it is supplied by an equilibrium shift to freeze liquid and release heat of fusion. For a pure substance, all equilibrium melting and freezing occurs at one temperature, the melting point. Section 9-14 describes the one possible nonequilibrium condition that can briefly vary the temperature for freezing.

The melting point equilibrium can also be shifted by other Le Chatelier's

9-13 Transitions Involving Solids

The heat of fusion must be supplied to break up the specific arrangement of solid phase to the freer pattern of liquid.

Figure 9-12 The stable phase always has the lower vapor pressure.

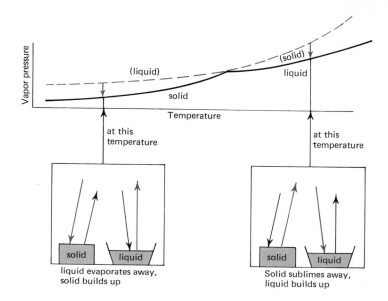

principle effects. For example, impurities, which change the concentration of molecules in the liquid, change the melting point. That shift is discussed in more detail in Chapter 10. Because the density of liquid and solid are not identical, there is also a small shift of melting point as the pressure on the material changes. Increasing pressure causes a shift toward the denser material, thus relieving the pressure slightly. This shift with pressure is noted in Section 9-15.

If a solid is converted directly to a gas, the process is called sublimation. The **heat of sublimation** is the energy required to overcome the force of attractions in the solid. The heat of sublimation must therefore equal the sum of the heats of fusion and vaporization. All of these heats vary slightly with temperature, so you will find the heats of fusion and vaporization do not quite

Figure 9-13 Equal vapor pressures at the melting point.

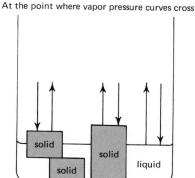

[9-13] Transitions Involving Solids 181

add up to the heat of sublimation if they are measured at different temperatures.

Solids have vapor pressure curves similar to those of liquids. At temperatures below the melting point, the vapor pressure of the solid is always less than the vapor pressure a liquid would have. At temperatures above the melting point, the liquid always has a lower vapor pressure than a solid would have. Figure 9–12 shows why the one which forms must be the one with the lower vapor pressure at that temperature. When the vapor pressure curves cross, both liquid and solid can exist together, as shown in Figure 9–13. This is the condition at the melting point. Therefore, the melting point is the temperature where the vapor pressure curves cross.

The vapor pressure of the solid (sublimation) and the vapor pressure of the liquid cross at the melting point. At other temperatures the phase (liquid or solid) that is stable has the lower vapor pressure.

9–14 Crystals and Supercooling

When a solid is heated above the melting point, it is always converted to liquid. The regular arrangement of the solid comes apart to form the less regular, fluid pattern of the liquid.

When a liquid is cooled below the melting point, it may freeze to solid. But it may not freeze. Even though the solid would have a lower vapor pressure and should eventually form, it may not. The stable, low vapor pressure solid cannot form unless the molecules are *arranged in the correct pattern*. Sometimes the molecules fail to arrange themselves in the way needed to form solid. They then remain liquid (free to move) even though the temperature is below the melting point. A liquid below the melting point temperature is called a **supercooled liquid.**

Supercooling can only occur when there is a specific pattern in the solid and difficulty in duplicating it. Many solids have specific regular patterns called **crystals.** Sometimes the patterns are hard to duplicate. We can explain the difficulties by using an example where crystals are unable to form.

Several common substances that we usually consider solids lack crystalline arrangements. Glass and rubber are in this group. When liquid sulfur is cooled quickly (by pouring it into cold water) it also forms such a "solid" with a very irregular arrangement. Such substances are called **amorphous solids.**

Liquid sulfur exists as long chains of sulfur atoms intertwined with each other and moving about. These chains form at temperatures above 160°C. Below 160°C these chains tend to break down by forming neat puckered rings of eight sulfur atoms. When sulfur is cooled slowly, these rings eventually stack together in the regular yellow crystalline arrangement called rhombic sulfur. When sulfur is cooled very quickly, the chains do not have time to form rings and to stack neatly, so supercooled liquid sulfur forms. If the supercooled sulfur is cold enough, the molecular motions are too small to move the chains from one place to another. The material then appears to be solid. Because the sulfur chains are so large and so entangled in each other, room temperature is cold enough to make the sulfur a black, apparently solid, mass that we call amorphous sulfur. This (and other amorphous solids) could also be described as a supercooled liquid. Because it has no specific

Supercooling occurs when the specific regular pattern of the crystalline solid is not available or duplicated.

There may be several different possible crystal forms for one solid, and amorphous (noncrystalline) solids may also be formed.

required locations for the molecules, it can be stretched into different shapes without breaking. Stretching tends to straighten out the long chains. When the amorphous sulfur is released, the entanglement of chains and the tendency for the chains to be bent by the bonding cause it to snap back into its earlier shape. Other rubber materials work the same way.

Amorphous solids such as the amorphous sulfur behave as very viscous liquids. There is still some liquidlike flow, although it may be very slow. For example, 900 year old glass windows in cathedrals have had enough flow to make them twice as thick at the bottom as at the top. Solids also retain some freedom for rearrangement. In sulfur, where a good crystalline solid is possible, the black amorphous form gradually converts itself to the more stable yellow solid. Bits of the yellow crystalline sulfur form at the edges and grow gradually as neighboring sulfur is directed into the proper positions.

Direction of crystal growth by the existing crystal is normal. When the crystal pattern is quite different from random nearby motion in the liquid, a supercooled liquid may persist for a long time. However, as soon as any of the crystalline solid forms, additional molecules are drawn into the correct positions by the attractive forces of the existing crystal. As the crystal grows, the heat of fusion is released and warms up the crystal and supercooled liquid. Finally the temperature reaches the melting point. At that point the rates of melting and freezing are in equilibrium balance, and no further change is seen unless additional heat is given to, or taken away from, an outside source.

Because a crystal allows the system to move to the condition of equilibrium, supercooling can only occur when no crystals are present. As soon as a crystal forms, supercooling is brought to an end. Supercooling can be eliminated by adding a crystal of the solid form to a supercooled liquid or by getting a crystal to form spontaneously. Sometimes spontaneous crystal formation can be aided by putting in something close to the desired crystal. Cloud seeding with NaI works by stimulating the formation of ice crystals. In chemical laboratories it is sometimes helpful to scrape the inside of a beaker with a glass rod. If (by random chance) you get a surface similar to the solid you want, you may start crystal growth.

There are often several different kinds of crystals possible for the same substance. For instance, sulfur exists not only as the familiar yellow solid (rhombic crystals) but also as an orange solid (monoclinic crystals). Monoclinic sulfur has a higher melting point ($119°C$) so it usually forms first as sulfur cools. However, below $95°C$ rhombic sulfur is more stable, so the orange monoclinic changes to yellow rhombic on standing. The speed of that conversion becomes slower as the temperature is lowered (hence lowering molecular motions). Thus if the monoclinic sulfur was kept very cold, it might remain monoclinic so long we would not realize another form was more stable. For changes that do not go as easily, even warm temperatures would let the less stable form continue to exist. An example is diamond. Diamonds should rearrange to a more stable crystal, graphite (the "lead" in a lead pencil), but it would take hundreds of millions of years.

If liquid sulfur was supercooled, a crystal of rhombic sulfur could be added and formation of monoclinic sulfur would be skipped. Because formation of the solid is aided by presence of other similar solid, workers in a normal, dusty laboratory may continue to get whatever solid first formed crystals there even if it is not the most stable form. There are some amusing stories of arguments between scientists about the properties of supposedly identical compounds until samples were exchanged. In a typical tale one found green crystals and the other brown. After the brown (more stable) crystals were brought into the first lab, the green crystals could never be made again. The significant scientific point behind this is that crystal structures are varied and very complex. And the growth of crystals (even under conditions where they should grow) depends on the presence of a pattern—preferably one of the crystals.

9–15 Phase Diagrams

Although the preceeding section on supercooling shows that the most stable state is not always formed, it is very useful to know whether solid, liquid, or gas should be present. This is conveniently shown by a phase diagram. Phase diagrams are graphs using temperature and pressure as coordinates and showing the points where phase changes occur. They summarize much useful information in a small space.

Figure 9–14 shows a typical phase diagram, the one for CO_2. Points of interest in this diagram are labeled. A is the critical point. B is the liquid vapor pressure curve. That can be used either to estimate the vapor pressure at a given temperature or to estimate the boiling point at a given pressure. C is the vapor pressure curve of solid CO_2 (dry ice). D is the vapor pressure

A phase diagram shows which form should be stable at each condition of temperature and pressure.

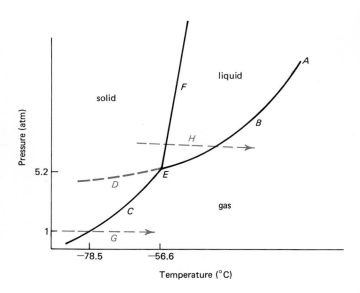

Figure 9–14 Phase diagram of carbon dioxide (CO_2).

curve of supercooled liquid CO_2. Notice that the vapor pressure of the solid (*C*) *must* be lower than the vapor pressure of liquid at temperatures below the melting point.

The two vapor pressure curves intersect at the **triple point** (*E*), the only condition of pressure and temperature where pure solid, liquid, and gas CO_2 can be present at equilibrium simultaneously. *F* is the melting point "curve." It begins at the triple point but slants toward higher temperatures as the total pressure is raised. That slant shows the Le Chatelier's principle effect that raising pressure at a given temperature may cause the equilibrium to shift toward forming more of the phase which occupies less space (the solid). The regions of pressure and temperature where CO_2 should form solid, liquid, or gas are labeled. In this diagram we notice that the triple point occurs at a high pressure. That explains why dry ice sublimes at atmospheric pressure without forming liquid, as shown by the dotted line *G*. Heating at a higher pressure (dotted line *H*) would cause it to melt to a liquid that would later evaporate.

You should be able to sketch such a phase diagram accurately enough to show all the key points (such as solid vapor pressure lower than supercooled liquid and the slant of the melting point curve). You should be able to label and identify any of the key points or areas. You should be able to describe by added lines and words what happens during specific changes. The dotted lines *D*, *G*, and *H* are examples of such "extra" information added to the basic phase diagram in our example.

Phase diagrams can be more complex. Figure 9–15 shows the more complex phase diagram of sulfur, showing the two solid crystalline forms discussed

Figure 9–15 Phase diagram of sulfur.

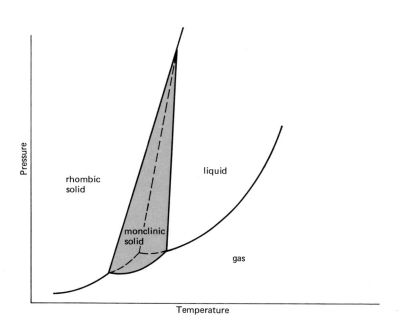

in Section 9–14. You will not be asked to duplicate complex phase diagrams, but you may find it interesting to identify the various points this phase diagram has in common with the simple diagram (vapor pressure curve, critical point, and so on), those which are doubled (two solid-liquid-gas triple points, two solid vapor pressure curves), and the unique additions (solid-solid transition, solid-solid-gas triple point).

Skills Expected at This Point

1. You should be able to state the gas law relationships including the need to use absolute scales for the quantities being varied.
2. You should be able to calculate gas law problems by using a series of factors obtained from the changing conditions plus using the molar volume at STP if only one condition is given.
3. You should be able to list the points in the kinetic theory of gases and describe the changes from those conditions which cause liquid and solid phases to form.
4. You should be able to explain properties of liquids and solids and phase transitions in terms of intermolecular forces.
5. You should be able to draw a phase diagram violating no general principles (such as relation of liquid and solid vapor pressures) and identify the important points.
6. You should be able to describe equilibrium balance and state and use Le Chatelier's principle correctly.

Exercises

1. **(a)** If the gas in a tank occupies 1.45 ft^3 at 15.2 atm, what volume would it occupy at 1.00 atm? **(b)** What volume would the same gas occupy at 735 torr?
2. If a gas has a pressure of 740 torr at 25°C, what pressure would it have in the same container at 50°C?
3. The gas in a balloon occupies 2.50 liters at 30°C. What volume would it occupy (at the same pressure) at −10°C?
4. If 10.0 g of a gas occupies 12.5 liters, what volume would 20.0 g of the same gas occupy (at the same temperature and pressure)?
5. If 10.0 g of a gas has a pressure of 720 torr, how much of the gas is needed in that same container and at the same temperature to have a pressure of 1.5 atm?
6. If a gas has a pressure of 740 torr at 27°C, what pressure would it have if it was placed in a container twice as large and heated to 127°C?
7. A balloon has a volume of 3.50 liters at 27°C and 750 torr. Assuming no gas is added or leaks out, what volume would it occupy at −123°C and 150 torr?
8. How do real gases differ from an ideal gas?
9. How do liquids differ from gases?

10. How do crystalline solids differ from liquids? How do crystalline solids differ from amorphous solids? How do amorphous solids differ from liquids?
11. **(a)** List five properties or types of behavior of liquids (which are clearly different from gases). **(b)** Describe how the attractive forces between molecules cause each of the properties in your answer to part (a).
12. Describe how Le Chatelier's principle shows what would happen in each of the following cases.
    (a) Heat is applied to an ice-water mixture.
    (b) The pressure cap on a car radiator breaks, causing a drop in gas pressure above the hot water.
    (c) Steam is pumped into a container where water and its vapor are at equilibrium.
    (d) Oxygen is pumped into a container where water and its vapor are at equilibrium.
    (e) An equilibrium mixture of N_2, H_2, and NH_3 (reaction $N_2 + 3 H_2 \longrightarrow 2 NH_3$) is compressed into a smaller space.
    (f) An equilibrium mixture of H_2, I_2, and HI (reaction $H_2 + I_2 \longrightarrow 2 HI$) is compressed into a smaller space.
13. Sketch the phase diagram of CO_2 and label clearly **(a)** the region corresponding to solid; **(b)** the region corresponding to liquid; **(c)** the region corresponding to gas; **(d)** the vapor pressure curve (of the liquid); and **(e)** the triple point.

Test Yourself

1. Which of the following was a point of the kinetic molecular theory always *applicable to real substances* (not just for ideal gases)?
   (a) The molecules have zero size.
   (b) Molecules are in constant, ceaseless motion.
   (c) No energy is transferred when molecules collide.
   (d) There are no forces between molecules.
   (e) The formula states the group of atoms that move about as a unit.
2. 25.0 g of $CaCO_3$ are decomposed by the reaction $CaCO_3 \longrightarrow CaO + CO_2$. What is the volume of CO_2 gas at STP formed by that decomposition? (At. wt. Ca 40)
   (a) 3.36 liters **(c)** 2.24 liters **(e)** 0.25 liter
   (b) 11.2 liters **(d)** 5.6 liters
3. On a day when the outside pressure is 14 lb/in.^{2}, a balloon has a gauge pressure of 4 lb/in.^{2} above the outside, for a total pressure of 18 lb/in.^{2} If it contains 36 g of gas under those conditions, how much of that same gas would it contain when the gauge pressure was 8 lb/in.^{2} (22 lb/in.^{2} total pressure)?
   (a) 8 g **(c)** 44 g **(e)** none of these
   (b) 22 g **(d)** 72 g

4. At 819 K and 380 torr pressure, 8.00 g of O_2 gas would have a volume of
   - (a) 5.60 liters
   - (b) 8.40 liters
   - (c) 22.4 liters
   - (d) 33.6 liters
   - (e) 67.2 liters
5. If a gas has a pressure of 740 torr at 25°C, what pressure would it have in the same container at 50°C?
   - (a) 2.0 atm
   - (b) 1480 torr
   - (c) 1172 torr
   - (d) 802 torr
   - (e) 370 torr
6. A balloon has a volume of 3.50 liters at 27°C and 750 torr. If no gas is added or leaks out, what volume would it occupy at -123°C and 150 torr?
   - (a) 17.5 liters
   - (b) 8.75 liters
   - (c) 1.75 liters
   - (d) 0.70 liters
   - (e) none of these
7. Choose the pair of phrases that best complete both sentences.
   (1) Liquid molecules cannot escape from the liquid to the gas phase unless we supply energy equal to the
   (2) Molecules find their way back to the liquid as fast as they escape once they have set up a maximum
   - (a) heat of vaporization, boiling point
   - (b) viscosity, vapor pressure
   - (c) evaporation, boiling point
   - (d) heat of vaporization, vapor pressure
   - (e) viscosity, critical point
8. The critical point in a phase diagram is
   - (a) the point where the vapor pressure curve reaches its highest temperature and pressure.
   - (b) the temperature at which solid changes to liquid.
   - (c) the gas pressure at which vapor at 0°C begins to condense.
   - (d) the point where gas, liquid, and solid of the same substance can exist simultaneously.
   - (e) verification of the kinetic molecular theory.
9. Choose the phrases marked (1) and (2) that best complete these sentences for a gas.
   (1) If P and n are constant,
   (2) If P and V are constant,
   - (a) (1) V decreases as T increases; (2) T increases as n increases
   - (b) (1) V decreases as T increases; (2) T decreases as n increases
   - (c) (1) V increases as T increases; (2) T increases as n increases
   - (d) (1) V increases as T increases; (2) T decreases as n increases
   - (e) (1) V increases as T decreases; (2) T increases as n decreases
10. A tank holds 50.0 kg of O_2 at 5.00 atm total pressure. How much O_2 does the same tank hold at 1.00 atm pressure at the same temperature?
    - (a) 10.0 kg
    - (b) 20.0 kg
    - (c) 100 kg
    - (d) 250 kg
    - (e) 500 kg
11. A deep sea diving vessel with a volume of 5000 liters needs a pressure of 250 atm at a temperature of 4°C. What volume of air at 1.00 atm

and 25°C is needed to fill the diving vessel to that needed 250 atm at 4°C?

(a) $5000 \times \dfrac{1}{250} \times \dfrac{4}{25}$ liters

(b) $5000 \times \dfrac{250}{1} \times \dfrac{4}{25}$ liters

(c) $5000 \times \dfrac{1}{250} \times \dfrac{298}{277}$ liters

(d) $5000 \times \dfrac{250}{1} \times \dfrac{298}{277}$ liters

(e) $5000 \times \dfrac{250}{1} \times \dfrac{277}{298}$ liters

12. If 12.0 g of gas occupies 10.0 liters, what is the mass of that gas needed to occupy 15.0 liters at the same temperature and pressure?
(a) 8.0 g
(b) 16.0 g
(c) 32.0 g
(d) 25.0 g
(e) 18.0 g

13. If 10.0 g of a gas has a pressure of 1.50 atm, how much of the gas is needed in that same container at the same temperature to have a pressure of 720 torr (mm Hg)?
(a) 15/720 g
(b) 1500/720 g
(c) 6.3 g
(d) 12.6 g
(e) 15.8 g

14. The gas in a cylinder of compressed oxygen has a pressure of 220 lb/in.^{2} at 25°C. At what temperature would the pressure in the tank reach 440 lb/in.^{2}?
(a) 50°C
(b) 273 K
(c) 298°C
(d) 323°C
(e) 546 K

15. The volume of 10.0 g of O_2 gas at 77°C and 740 torr (mm Hg) is given by the setup

(a) $10.0 \times \dfrac{22.4}{32.0} \times \dfrac{350}{273} \times \dfrac{760}{740}$ liters

(b) $10.0 \times \dfrac{22.4}{16.0} \times \dfrac{77}{273} \times \dfrac{740}{1.0}$ liters

(c) $10.0 \times \dfrac{22.4}{32.0} \times \dfrac{273}{350} \times \dfrac{1.0}{740}$ liters

(d) $10.0 \times \dfrac{22.4}{16.0} \times \dfrac{350}{273} \times \dfrac{760}{740}$ liters

(e) $10.0 \times \dfrac{16.0}{22.4} \times \dfrac{350}{77} \times \dfrac{1.0}{740}$ liters

16. On a cold day the warm air in a building has a "hot air balloon" effect which lifts some of the building's weight that normally rests upon the foundation. Calculate the "lift" provided by the hot air in the Administration Building on a day when the outside temperature is +5°F and the building contains 30,000 kg of air at 68°F and the same pressure as the air outside.

(a) $30{,}000 \text{ kg} - \dfrac{5}{68} \times 30{,}000 \text{ kg}$

(b) $30{,}000 \text{ kg} \times \dfrac{293}{258} - 30{,}000 \text{ kg}$

(c) $30{,}000 \text{ kg} \times \dfrac{68}{5} - 30{,}000 \text{ kg}$

(d) $30{,}000 \text{ kg} - \dfrac{258}{293} \times 30{,}000 \text{ kg}$

(e) none of these

17. Le Chatelier's principle states
    (a) the fact that no two electrons in an atom can have the same four quantum numbers.
    (b) the relationship between masses of two elements in a compound.
    (c) the relationship between moles and atomic weight.
    (d) the effect of gravity on an object.
    (e) the effect of changes on an equilibrium.
18. According to the kinetic theory, temperature is
    (a) a measure of the rate of reaction.
    (b) a measure of the average energy of motion.
    (c) a measure of the equilibrium constant.
    (d) a measure of the rate at which molecules strike the wall of the container.
    (e) none of these.

Note: About 90% of a class managed to get question 19 wrong on a test. They chose answer (a), which is incorrect. Think about the kinetic theory. Perhaps this trial question will help you remember the point they missed. (It is the same point being checked in question 18.)

19. If we compare three substances at atmospheric pressure and the same temperature, 25°C, at which one is a gas (air), the second is a liquid (water), and the third is a solid (sugar crystals), the kinetic molecular theory lets us know that
    (a) the gas molecules have the largest average energy of motion, the liquid molecules have a lower average energy of motion, the solid molecules have the lowest average energy of motion.
    (b) liquid and solid have the same average energies of motion, and the gas molecules have a larger average energy of motion.
    (c) the average energies of motion are the same in gas, liquid, and solid.
    (d) the gas molecules have the lowest average energy of motion, and liquid and solid have larger average energies of motion with both liquid and solid having the same average energy.
    (e) the gas molecules have the lowest average energy of motion, liquid a larger average energy of motion, and solid the largest average energy of motion.

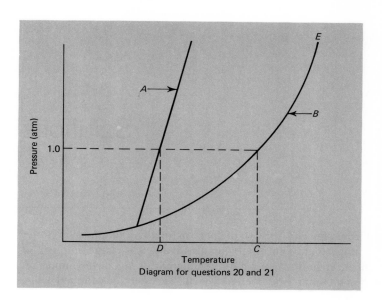

Diagram for questions 20 and 21

20. Which letter on the phase diagram shown points out the vapor pressure curve?

21. Which letter on the phase diagram shown points out the normal boiling point?

22. CO_2 is a "normal" substance, with a solid that is more dense than the liquid form. Solid CO_2 sublimes at normal atmospheric pressures (the pressure at the CO_2 triple point is about 5 atm). Sketch the phase diagram of CO_2 (showing reasonable shapes that fix the necessary relationships between the lines. Your sketch should include labeling the solid, liquid, and gas regions. Also label the axes of your graph. Then mark a large *A* beside the critical point and a large *B* beside the *dotted line* which would show the vapor pressure of the supercooled liquid.

Solutions   10

Solutions are a common and important form of matter. Solutions in water are particularly important. The ability to dissolve a very large number of different substances, including many ionic compounds, is one of the important unusual properties of water (see Section 13–2). The solutions formed when things dissolve in water are especially common and important to us, but there are also many other solutions in the world around us.

A **solution** is any dispersion of the molecules or ions of one substance throughout another on a truly molecular scale. When one substance is dispersed throughout another, it is said to **dissolve.** The simplest situation where this happens is in a mixture of gases. Since all the molecules in a gas are free to move about at random, molecules of different kinds are rapidly and completely mixed. When they become mixed, no group of molecules is set apart from the rest. The result is a perfect example of a solution. Anything that causes one group of molecules or one region of space to be set apart would prevent the existence of a single true solution. But a mixture of gases has all of the molecules moving about in all the available space. Therefore any gas is a solution if it contains two or more different kinds of molecules. Air is the most common example. It consists of a "dispersion on a molecular scale" of N_2, O_2, Ar, CO_2, H_2O vapor, and small amounts of other gases.

Liquids may also be dispersed in other liquids on a truly molecular scale, but these do not form in every case. If nothing prevents it, there is a tendency in nature for things to mix with each other. This tendency is discussed qualitatively in Chapter 11. It is covered with more mathematical detail and precision in the study of thermodynamics (specifically in the second law of thermodynamics). However, in liquid phase there are large forces between molecules that prevent the molecules from escaping to the free random motions of gas phase. Mixing in large numbers of different kinds of molecules may break up these attractions. If too much of the favorable attractions must be lost, the mixing (formation of a solution) will not occur. Appendix C includes a description of free energy and shows how both energy and tendency to mix contribute to the overall extent of changes.

The conditions where solutions could form easily between two liquids are

10–1 Nature of Solutions

A solution is a dispersion on a molecular scale.

All gas mixtures are solutions.

1. Each of the liquids has small attractive forces between molecules, so they can attract each other about as well as they can attract their own molecules; or
2. Each of the liquids has large attractive forces between molecules but can also set up large attractive forces to the other kind of molecule.

The first situation is common between nonpolar covalent molecules. The second situation can occur between strongly polar molecules, between a very polar substance and an ionic substance, or between two substances each of which can take part in hydrogen bonding (see Chapter 13). The overall result is a general rule that solutions tend to form between substances that are *alike.* Nonpolar dissolves in nonpolar, polar dissolves in polar, and hydrogen bonded dissolves in hydrogen bonded. Dissimilar materials tend *not* to form solutions. A nonpolar substance would need to break up the attractions in a polar substance to dissolve but would not give new attractions in their place, so it simply does not dissolve. This general pattern of *like dissolves in like* holds for all types of solutions, not just liquid-liquid solutions.

> Like dissolves in like; dissimilar substances do not form solutions very well.

The ability of one substance to mix with another on a molecular scale to form a solution is called **miscibility.** Two substances that can always form a solution no matter what quantities are taken are said to be **completely miscible.** Gases are always completely miscible with each other (as long as they remain gases), and some pairs of liquids are completely miscible. Other pairs of liquids cannot mix completely in all proportions. Even if the substances are dissimilar, the general tendency in nature toward mixing things will allow at least a little of one substance to dissolve in the other without requiring a significant number of attractions to be broken. The pairs of substances in these cases, where solutions form but only up to limited quantities being dissolved, are said to be **partially miscible.** When two partially miscible liquids are present in roughly equal quantities, usually neither will be able to dissolve completely into the other. In that case, two solutions will be formed, each containing the maximum possible amount of one dissolved in the other. The two solutions will be seen as two distinct, separate liquid layers. If the mixture is shaken up, the layers may be broken up into small drops, perhaps even a colloidal dispersion of one solution in the other (see Section 13–5.6). But there will still be two (and only two) types of liquid regions present, the two solutions.

> Substances that can be mixed in any proportion and still form a single solution are completely miscible.

> Substances that have a limit on how much of one can dissolve in the other are partially miscible.

As our example above shows, the condition of the maximum amount that can be dissolved in a partially miscible system is particularly important. This limiting condition is called the **solubility** of one substance in another. Partial miscibility also causes a distinction to appear between the substance being dispersed (limited to relatively small amounts by the solubility) and the substance in which it is spread out (any amount of which could go into the solution). Therefore, these are given specific names. The substance being dispersed is called the **solute,** and the substance in which it is spread out is called the **solvent.**

> The substance dissolving is the solute; the substance it is dispersed in is the solvent.

> The maximum concentration of a solute that will dissolve is the solubility.

So far we have discussed only gases and solutions of liquids in liquids.

[10–1] Nature of Solutions 193

It is also possible to have solutions of gases in solids, liquids in solids, or solids in solids. However, solutions in solids are somewhat less common than solutions with liquid solvents. Solutions where liquid water is the solvent are especially common and important. We therefore want to consider solutions of gases in liquids and solids in liquids in somewhat more detail. Although our emphasis is on solutions of gases or solids in liquids, other kinds of solutions are also important to us. Solutions of gases in solids are important in some catalysts for commercial processes, and solid-solid solutions play a vital role in the properties of the metals we use. But liquid solutions deserve our greatest attention.

10–2 Gases in Liquids

There are two basic ways in which a gas can dissolve in a liquid.

1. The gas can react chemically to form something quite different from the gaseous molecules.
2. The gas can dissolve in the same molecular form it has in gas phase.

An example of a gas reacting chemically in a liquid solvent is the case of HCl in water. HCl molecules in gas phase are covalently bonded units, somewhat polar, but with only small attractions to other molecules. In water they react to form H_3O^+ ions and Cl^- ions, which are strongly attracted to the surrounding polar water molecules (see Section 13–4). Because the reaction occurs readily and the attractions are so favorable, very large amounts of HCl can dissolve in a small amount of water. In general, gases that react chemically with water have high solubilities.

Sometimes even gases that we would write in the same molecular form in solution are still reacted to forms quite different from gas phase. An example is NH_3. Most chemists would continue to write it as NH_3 in water solution because no other formula accurately shows how it is bonded. NH_3 can form hydrogen bonds (see Section 13–3) to water both via the unshared electron pair on the nitrogen and via the hydrogen atoms of the NH_3. NH_3 is held strongly by these hydrogen bonds and is in a very different situation from an independent NH_3 molecule in gas phase. Because of the strong attractions, the solubility of NH_3 in water is very high.

In contrast to the highly soluble gases given above, all gases that do not react chemically have low solubilities in liquids. This is a natural example of the rule that like dissolves in like whereas dissimilar things do not dissolve as extensively. All gases have small intermolecular forces or they would not be gases. All liquids have large intermolecular forces or they would not be liquids. Because they are dissimilar, only small amounts of the gases will dissolve in liquids.

Solubility of gases in liquids is always low unless the gas reacts chemically with the liquid.

The amount of gas dissolved in a liquid depends directly on the pressure of that gas above the liquid. If we doubled the pressure of oxygen above water, we would double the rate at which O_2 molecules went into the water from the gas. We would then have to have twice as many O_2 molecules in the liquid (which would double their rate of escape to the gas) to establish an equilibrium where the amount of dissolved O_2 remained constant. This direct proportionality between dissolved gas and the partial pressure of that specific gas is called **Henry's law.**

Gas solubility is directly proportional to the pressure of that kind of gas over the solvent (Henry's law).

The balance between rate of addition and rate of escape of gas molecules can be upset by a change in temperature. Because the gas molecules are less strongly held than the liquid molecules, a small increase in temperature can greatly increase the rate of escape of the gas molecules. Therefore the amount of a gas like O_2 that remains dissolved in water drops sharply as the temperature rises. The best conditions for dissolving a gas in liquid are to have the liquid as cold as possible and the pressure of that gas above the liquid as high as possible.

Gas solubilities go down as temperature rises.

10–3 Solids in Liquids

Solutions of solids in liquids are very common and important. They differ sharply from solutions of gases in liquids. Molecules of solids have strong attractive forces or they would not be solids. Although they can never be completely miscible with liquids, solids may set up strong enough attractions to the liquid molecules to give high solubilities. The solubilities of solids are determined by a competition between the strong forces in the solid and other strong forces when they are in solution.

Because molecules or ions of a solid are usually held by stronger forces than the liquid molecules, they do not escape readily when the temperature is raised. Instead, it is the liquid that evaporates and escapes (leaving the solid behind) if enough heat and volume for the gas to escape into are available.

However, just as an increase in temperature causes dissolved gas to escape from the "more constrained" phase (the solution) to the freer phase (gas), a temperature increase also causes solid molecules to escape to a freer phase. In this case the "more constrained" phase is the solid itself and the "free" phase is the solution. Therefore, as the temperature rises, the solubility (the maximum possible amount) also goes up. There are occasional exceptions to this rule, just as there are exceptions to low gas solubilities caused by reactions.

Solubilities of solids in liquids normally rise as temperature rises.

In Figure 10–1 we show a typical plot of the solubility of a solid in water versus temperature. The variation of solubility with temperature is represented by a line called the **solubility curve.** Every point on that curve represents a solution with as much solid dissolved as the solubility permits at that temperature. These solutions, called **saturated solutions,** are full as far as their ability to dissolve that solid is concerned. Points below the solubility curve in Figure 10–1 represent **unsaturated solutions,** which could dissolve more of the solid.

Solutions in which the solute concentration equals the solubility are saturated.

Solutions with less solute than the solubility are unsaturated.

If more solid is added, some will dissolve if the solution is unsaturated. When the amount dissolved reaches the solubility curve, no more solid will be dissolved.

[10–3] Solids in Liquids 195

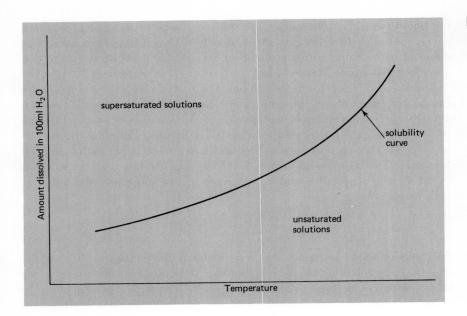

Figure 10–1 Solubility curve for a typical solid in a liquid.

If a saturated solution containing a piece of undissolved solid is cooled, the solubility drops. The extra solid in solution is removed by redepositing as solid. The presence of some solid is important, because the solid cannot redeposit unless it is guided into the correct crystal pattern. If a saturated solution with no solid present is cooled, the solid may not be able to form. In that case we move over into the region above the solubility curve labeled **supersaturated solutions.** If a solution in this region ever gets even one small speck of the solid to form, it will immediately pick up much more solid until enough has been removed to lower the solution back to the solubility curve.

All partially miscible systems will have solubility curves indicating where saturated solutions occur and regions corresponding to unsaturated and supersaturated solutions. In our practical examples at the end of this chapter we will include a problem caused by supersaturation of a solution of a gas in a liquid. Even though no crystal arrangement is required by a gas, gases may become supersaturated if the total pressure and surface tension forces prevent the formation of bubbles needed to release the excess gas rapidly.

When temperature is changed with no solid present, the solution may remain more concentrated than the solubility at the new temperature. Such solutions are supersaturated.

Gas solutes may become supersaturated when the gas pressure drops if gas bubbles are not able to form.

10–4 Concentrations

10–4.1 Molarity. We need some convenient units to state how much solute we have in various unsaturated, saturated, or supersaturated solutions. In Figure 10–1 we did this by plotting the amount dissolved in 100 ml of water, but not all of our solutions will have 100 ml of solvent or water as the solvent. If we had a saturated solution, we would have twice as much solute dissolved in 200 ml as we would in 100 ml. If we added more solvent to increase the volume of solution from 100 ml to 200 ml without dissolving any more solute, the solution would be **diluted** to an unsaturated solution. If we evaporated 50 ml of solvent from 100 ml of saturated solution, we would either cause half of the solute to come out of solution (precipitate) or if no solid was present we might **concentrate** the solution to a supersaturated solution. The units we need depend both on the amount of solute and the amount of solution.

We can write several kinds of ratios in the form

$$\frac{\text{amount of solute}}{\text{amount of solution}} \quad \text{or} \quad \frac{\text{amount of solute}}{\text{amount of solvent}}$$

Molarity, moles per liter, is a convenient unit to express solution concentrations.

We call these ratios **concentrations.** They become larger as the solution is concentrated. The most widely used units for concentration are

$$\frac{\text{moles of solute}}{\text{liters of solution}} = M$$

This unit is called **molarity** and represented by the abbreviation M. Molarity was introduced in Section 4–8, and the procedure for preparing solutions of known molarity was shown in Figure 4–4.

A 5.0 M solution (which could also be written "a 5.0 molar solution") of NaCl has 5.0 moles of NaCl in each 1 liter of solution, or would have if there was enough of that solution to make up 1 liter. If we had 25 liters of 5.0 M NaCl, there would have to be 125 moles of NaCl in solution. If we had 1.0 liter of 5.0 M NaCl, there would have to be 5.0 moles of NaCl in solution. If we had only 5.0 ml of 5.0 M NaCl, that would be only 0.0050 liter and there would be only 0.025 mole of NaCl in solution.

$$\frac{125 \text{ moles}}{25 \text{ liter}} = 5.0\ M \qquad \frac{5.0 \text{ moles}}{1.0 \text{ liter}} = 5.0\ M \qquad \frac{0.025 \text{ mole}}{0.0050 \text{ liter}} = 5.0\ M$$

Although there are many other kinds of concentration units, molarity is so generally useful that we will use it almost exclusively. We will be forced to introduce one other concentration unit in our section on colligative properties, but in every other situation molarity will be the only unit needed.

Let us consider some of the possible uses of molarity.

1. We can state the concentration of solute in a solution in terms of molarity.

    Example 10–1. What is the KNO_3 concentration when 10 g KNO_3 are dissolved in 150 ml of solution? We need the amount of solute in *moles* and the amount of solution in *liters*. We must then divide the moles by the liters. The formula weight of KNO_3 is $(1 \times 39) + (1 \times 14) + (3 \times 16) = 101$ g/mole. Then

$$\frac{10 \text{ g KNO}_3 \times \dfrac{1 \text{ mole KNO}_3}{101 \text{ g KNO}_3}}{150 \text{ ml solution} \times \dfrac{1 \text{ liter}}{1000 \text{ ml}}} = \frac{10/101}{0.150} M \text{ KNO}_3 = 0.66 \, M \text{ KNO}_3$$

(Figure 4-4 shows the procedure for preparing such a known volume of solution from a given amount of solid.)

2. We can state the concentration of molecules in a gas, of the solvent in a solution, or even of a pure substance in terms of molarity. We can calculate the number of moles in each liter for anything, whether it is a solute or not.

Example 10-2. What is the H_2O concentration in water? The formula weight of H_2O is $(2 \times 1) + (1 \times 16) = 18$, so

$$\text{moles of } H_2O = \text{g } H_2O \times \frac{1 \text{ mole } H_2O}{18 \text{ g } H_2O}$$

1 ml of H_2O weighs 1 g, 1 liter = 1000 ml, so 1 liter of H_2O weighs 1000 g.

$$\frac{1000 \text{ g } H_2O}{1 \text{ liter}} \times \frac{1 \text{ mole } H_2O}{18 \text{ g } H_2O} = 55.6 \, M \, H_2O$$

3. We can calculate the extent of dilution or concentration of a solution in terms of molarity.

Molarity can be used to calculate the results of dilutions or titrations.

Example 10-3. 75 ml of 0.55 M NaCl is diluted to 250 ml by adding water. What is the new concentration of the solution? In Figure 10-2 we show how such a dilution could be carried out accurately.

The number of moles of *NaCl* has not been changed as the water was added. The number of moles can be written as moles/liter × liters, and that is true both before and after dilution. Therefore

$$\text{moles NaCl} = \underset{\text{(before)}}{M \times \text{liters}} = \underset{\text{(after)}}{M \times \text{liters}}$$

The number of liters is the *volume*, abbreviated as V, so

$$M_1 \times V_1 = M_2 \times V_2$$

or

$$0.55 \, M \times 0.075 \text{ liters} = x \times 0.250 \text{ liters}$$

Actually, since the units of V are the same on both sides, they cancel out, so we could have left them in milliliter units instead.

$$0.55 \, M \times 75 \text{ ml} = x \times 250 \text{ ml}$$

$$x = \frac{0.55 \times 75}{250} M$$

$$= 0.165 \, M$$

Figure 10-2 Preparing a solution by dilution. [From R. H. Petrucci: *General Chemistry.* The Macmillan Company, New York, 1972.]

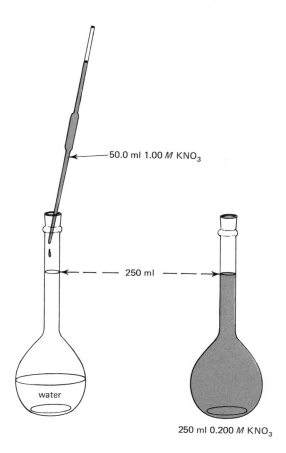

The formula $M_1 \times V_1 = M_2 \times V_2$ will be useful to us in many different problems, but we must use it only when the amount of the substance being discussed stays constant.

4. We can calculate the amount of one solution needed to react with another solution (a **titration** problem) in terms of molarity.

Example 10-4. How much $0.500\ M$ H_2SO_4 is needed to react with 27.0 ml of $0.350\ M$ NaOH? The reaction is $H_2SO_4 + 2\ NaOH \longrightarrow Na_2SO_4 + 2\ H_2O$. The equation (by its coefficients) tells us that 2 moles of NaOH react for each 1 mole of H_2SO_4. Since 2 moles NaOH is equivalent to 1 mole H_2SO_4 in this reaction, we can write the unit factor 2 NaOH/1 H_2SO_4 to change units from H_2SO_4 to NaOH in our problem.

$$\text{moles } H_2SO_4 = M \times V \text{ for } H_2SO_4$$
$$\text{moles NaOH} = M \times V \text{ for NaOH}$$

so

$$M \times V \text{ for NaOH} = 2\frac{\text{NaOH}}{\text{H}_2\text{SO}_4}(M \times V \text{ for H}_2\text{SO}_4)$$

$$0.350\ M \text{ NaOH} \times 27.0 \text{ ml} = 2\frac{\text{NaOH}}{\text{H}_2\text{SO}_4} \times 0.500\ M \text{ H}_2\text{SO}_4 \times V$$

$$V = \frac{0.350\ M \text{ NaOH}}{2 \times 0.500\ M \text{ NaOH}} \times 27.0 \text{ ml}$$

$$= 9.45 \text{ ml}$$

We can adjust quickly to any given chemical reaction between the solutions. Try the same problem except this time use the reaction equation

$$\text{H}_2\text{SO}_4 + \text{NaOH} \longrightarrow \text{NaHSO}_4 + \text{H}_2\text{O}$$

and you will get a different result.

If you are curious about other concentration units or the historical development of concepts of concentration, you may want to ask your instructor about normality, the other type of concentration unit often used to work titration problems.

10-4.2 pH. When concentrations are so low that molarity gives inconveniently small numbers, other kinds of units make values easier to write down and handle. pH is one of the units used for concentrations. When we write down the pH of a solution, we are really writing down the molarity in a way that is especially handy for small numbers. We use this to tell whether a solution is acidic, basic, or neutral.

In Section 7–6 we described acids as substances ionizing to give H^+ and bases as substances ionizing to give OH^-. Actually, small amounts of H^+ and OH^- are always present in water solutions. Even acidic solutions contain some OH^- and even basic solutions contain some H^+. Section 13–4 will describe how the ion (H^+ or OH^-) present in excess determines acidity or basicity. An excess of H^+ ions over OH^- ions makes a water solution acidic. But the concentrations of H^+ and OH^- may both be very low. In a neutral solution at 25°C, the concentrations of H^+ and OH^- are each 0.0000001 M. In basic solutions, the H^+ concentration is even smaller. These small molarities can be written in less space by using exponential numbers.

Neutral solutions contain $10^{-7}\ M\ H^+$ and $10^{-7}\ M\ OH^-$. A typical basic solution might contain $2.0 \times 10^{-9}\ M\ H^+$ and $5.0 \times 10^{-6}\ M\ OH^-$. The exponential numbers save some space, but much more time and space can be saved by defining a new way to write numbers especially fitted to this situation. Acid-base chemistry in water is so important that this new number system, pH, has been defined and is widely used. pH is a shorthand for writing molarity in exponential form. Instead of writing out the whole thing, we write down only the exponent.

In a neutral solution, the H⁺ concentration is 10^{-7} M. The exponent alone (actually called the logarithm to the base 10) would be -7. It would be easier for us to write -7 than 10^{-7} M, but it is still less convenient than a positive number. Since most of the logarithms of H⁺ concentrations are negative numbers, it was decided long ago to include one more step in the definition of the new way to write H⁺ concentrations. The sign of the logarithm is always changed. Negative numbers (which are common) become positive numbers (easy to write). The occasional positive numbers become negative. Our directions then become: take the logarithm of the H⁺ molarity and change its sign. We give the result the name pH, and we can state the definition by the equation

$$\text{pH} = -\log [\text{H}^+]$$

pH ($-\log[\text{H}^+]$) is a concentration unit particularly adapted to measurements of acidity.

For the neutral solution we have been describing, pH = 7. All acidic solutions have pH less than 7. All basic solutions have pH greater than 7. A few concentrated solutions of strong acids have negative pH, as in the case of 10 M HCl where pH = -1.

pH = 7 for neutral solutions at 25°C; pH is larger than 7 for basic solutions and smaller than 7 for acidic solutions.

For cases such as [H⁺] = 2.0×10^{-9} M, it is necessary to take the logarithm of the 2.0 as well as the logarithm of the 10^{-9}. Tables of these values are available, and they would show log 2.0 = 0.30. To make this into a pH, we must also change its sign. Therefore, pH = $-\log (2.0 \times 10^{-9})$ = $-[0.3 + (-9)] = +9 - 0.3 = 8.7$. The pH therefore becomes the negative exponent on 10 minus log of the number in front of 10.

In a water solution the concentrations of H⁺ and OH⁻ are held in a rigid relationship by the equilibrium between ionization of H₂O and recombination of H⁺ and OH⁻. As we would predict by LeChatelier's principle (Section 9–9) any increase in H⁺ concentration will cause a shift in the equilibrium resisting that increase. That shift will lower the OH⁻ concentration. The product of H⁺ concentration times OH⁻ concentration is a constant value at any given temperature (see Section 12–5 and Appendix B–2). At 25°C this relationship makes [H⁺][OH⁻] = 10^{-14} M^2. Therefore, when pH = 8.7 and [H⁺] is 2.0×10^{-9} M, [OH⁻] must be 5.0×10^{-6} M so [H⁺][OH⁻] will equal 10^{-14}. Because of this, most people never bother to write down the OH⁻ concentration. They just state the H⁺ concentration, using pH, and let the OH⁻ concentration be calculated by anyone who really needs to know it. pH is therefore the only common unit used to describe acidic or basic solutions. However, the same sort of definition ($-\log$ of molarity) can be used for other concentrations besides H⁺. This is occasionally done with OH⁻ concentration (although not nearly as often as pH is used) and the small letter p is used to indicate it is being done. Therefore, pOH = $-\log [\text{OH}^-]$ and when [OH⁻] = 5.0×10^{-6} M:

H⁺ concentration and OH⁻ concentration are related (at 25°C) by the equation [H⁺][OH⁻] = 10^{-14} M^2.

The notation p can be used to mean $-\log$ of other small numbers besides H⁺ concentrations.

$$\text{pOH} = -\log (5.0 \times 10^{-6}) = 6 - \log 5.0 = 6 - 0.7 = 5.3$$

This p notation is useful for many problems where very small numbers occur. For instance, our statement that [H⁺][OH⁻] = 10^{-14} M^2 could also be written as pH + pOH = 14. You might find it interesting and a useful drill to take four or five acidic or basic solutions, calculate pH and pOH for each, and prove for yourself that pH + pOH = 14.

10–5 Colligative Properties

10–5.1 Vapor Pressure and Raoult's Law. We have explained solubility variations by the argument that the rate of escape and rate of return of solute must be equal at the equilibrium condition represented by the solubility limit. The same sort of argument holds for the temperature variations of liquid vapor pressures, which are also limiting equilibrium conditions. Liquid vapor pressures, like rates of escape of solutes, also depend upon how many molecules are at the surface with a chance to escape. Formation of a solution changes that quantity and therefore changes the vapor pressure.

Figure 10–3 shows a simplified sketch of how formation of a solution *lowers* vapor pressure. If the available places on the surface remain constant, the number of solvent molecules there goes down as some of the places are taken by solute instead of solvent. The higher the concentration of solute, the more the vapor pressure is cut down.

If you are struggling just to get by in this course, you can skip this paragraph. It will not introduce any concepts you need right now. But this is a good place to offer the better students a little explanation of how we use models and approximations as tools to grasp concepts. The model just presented is not entirely correct. It has been shown that other factors are needed for a full explanation [K. J. Mysels: *J. Chem. Educ.*, **32**:179 (1955)]. We could give a better explanation by use of thermodynamics to describe the change in the equilibrium condition. But that would require knowledge of thermodynamic relationships (usually derived with liberal use of calculus), which are beyond the level of this course. Changes in the entropy (see Chapter 11) are much more significant than obstruction. However, the model we are using here does show many facts about the situation without requiring as much prior knowledge. It shows that solutes interfere with the loss of solvent, that the equilibrium is upset, that a change in the vapor pressure can restore the equilibrium, and that the change in vapor pressure is proportional to the solute fraction, all of which are true. By using the model we can proceed

Presence of solute interferes with the ability of solvent molecules to escape from the solution. This leads to the colligative properties.

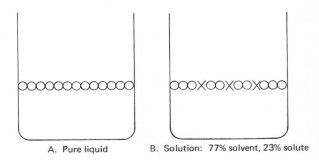

Figure 10–3 Reduction of vapor pressure by addition of a nonvolatile solute.

A. Pure liquid with 13 solvent molecules on the surface with a chance to escape.

B. A solution which is about 77% solvent and 23% solute. The number of solvent molecules on the surface with a chance to escape has been reduced to 10 because solute now occupies the other surface positions.

to a better understanding of important principles. Most of our understanding of science is based on similar models that have been put together, tested, and refined by the scientific method. We use approximations and simplifying assumptions to make models easier to understand. Sometimes our experiments show that more exact models are needed. In some areas of chemistry there have been at least three distinct types of models (in addition to the many minor refinements made on each model type) introduced progressively as our needs and knowledge became more sophisticated. The refinement and/or replacement of those models and ideas now thought to be the "best" is a continuing process. Often we are faced with a choice between old models—simple, revealing some truth, but with flaws we can see and point out—and new models—complex, closer to the truth, but still just models. In this book we use many imperfect simple models like the one here referring to obstruction of solvent escape by solute. You may go on later to understand more exact explanations and then find our explanations here amusingly simplified. But some of you will go a step further, to an understanding of why we use such simplifications even when we know they are imperfect. The latest, most sophisticated models are improving mankind's understanding of important concepts just as the simplified models improve your understanding of concepts in this course. If you really grasp both the usefulness and limitations of such models, the world of science is yours.

Vapor pressures are proportional to the fraction of molecules in the liquid that can evaporate to form each kind of vapor molecules (Raoult's law).

The amount of vapor pressure from a solution should be the vapor pressure of the pure substance times the fraction of the molecules in solution of that substance. If both solvent and solute are volatile (tend to escape to gas phase), both will have vapor pressures dependent on their fraction of the molecules. For example, a gasoline made by mixing very volatile propane, C_3H_8, with less volatile isooctane, C_8H_{18}, would have a total vapor pressure equal to the sum of the vapor pressures of the two components. Each component should have a vapor pressure equal to its fraction times the vapor pressure it has as a pure substance. As the fractions are varied the individual vapor pressures and their sum (total vapor pressure) would vary in proportion to the fractions. Figure 10–4 shows the result for such a case, solutions of two completely miscible liquids.

We see that in our gasoline example all mixtures should have vapor pressures between those of the pure propane and the pure isooctane. In cold weather, when high vapor pressure is needed to start cold engines, more propane is needed. In warm weather it would be better to have more isooctane to cut down losses by evaporation. (Isooctane also has better antiknock properties.) Oil companies make such seasonal changes in their gasolines.

Real substances do not give perfect straight lines because details of bonding forces cause complications, but the general pattern is clearly normal. This vapor pressure dependence is called **Raoult's law.** The dependence of gas solubility on pressure (Henry's law, Section 10–2) is just a special case of this general rule.

Most of the cases we are interested in are solutions of solids in liquids where another special case holds. The vapor pressure of the solute is so low

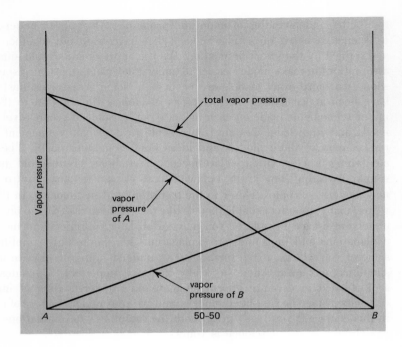

Figure 10–4 Raoult's law.

The vapor pressures of both A and B are proportional to their molecular fractions in the solution. The total vapor pressure is the sum of these two straight, molecular fraction dependent vapor pressures and is also a straight line, molecular fraction dependent variable.

that the vapor pressure of the solvent becomes the total vapor pressure. At every point of temperature, that vapor pressure is less than that of the pure solvent by the same fraction. Figure 10–5 shows the vapor pressure curves that result for a pure solvent, a solution (1), and a second solution (2) which is twice as concentrated in solute as solution 1. At every point, the vapor pressure of solution 2 is depressed twice as far (from pure solvent) as the

Solutions of nonvolatile solutes are an important type.

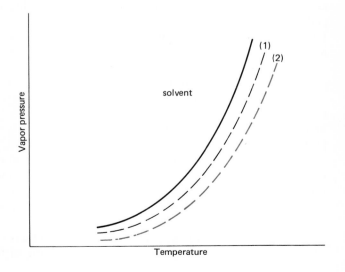

Figure 10–5 Vapor pressures.

The three curves represent the vapor pressure of a pure solvent, the vapor pressure of that same solvent from a solution (1) containing a nonvolatile solute, and the vapor pressure of that same solvent from a second solution (2) that has double the concentration of nonvolatile solute present in solution (1).

Figure 10-6 Effect of putting a solvent and solution in the same closed container.

Equal amounts of solvent and solution (shown on the left) will result in the situation shown on the right. All of the solvent will move to the solution container, forming a new solution half as concentrated as the original one.

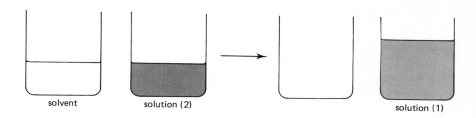

vapor pressure of solution 1 because solution 2 has twice as many solute molecules to interfere.

If we put equal amounts of pure solvent and solution 2 in a closed container, as shown in Figure 10-6, the pure solvent will continue to evaporate after the vapor pressure of the solution has been equaled or exceeded. The solvent will then start to return to solution 2 faster than it can escape while it is still escaping from the pure solvent faster than it can return. The pressure of solvent vapor in the container will be somewhere in between the equilibrium vapor pressures of the solution and of the pure solvent. As this happens, the pure solvent will evaporate away, and the solution will be diluted. The vapor pressure of the solution will rise as it becomes more dilute, but it will always be less than the vapor pressure of the pure solvent. Therefore the transfer of solvent will continue until the container of pure solvent is empty. If we started with equal amounts of solvent and solution 2, the solution will have been diluted to half the original concentration, which makes it now the concentration of the solution 1 shown in Figure 10-5.

In Figure 10-7, we consider a similar case starting with equal amounts of solution 1 and solution 2. Again the portion with the higher vapor pressure [the less concentrated solution 1 in this case] loses solvent by evaporation, and solvent is transferred to the more concentrated solution 2. As solvent is transferred, solution 1 becomes more concentrated (causing its vapor pressure to drop) and solution 2 becomes more dilute (causing its vapor pressure to rise). Eventually, the two solutions reach the same concentration, the vapor pressures become equal, and the transfer of solvent stops. If we begin with equal amounts of the two solutions, the final concentrations are half-way between the two original concentrations—something we could call solution $1\frac{1}{2}$ if you like. That happens when solution 1 is reduced to $\frac{2}{3}$ of its original volume and solution 2 is diluted to $1\frac{1}{3}$ times its original volume.

Figure 10-7 Effect of putting two solutions of different concentrations in the same closed container.

Equal amounts of two solutions (solution 2 twice as concentrated as solution 1) lead to transfer of enough solvent (in this case $\frac{1}{3}$ of the original total) from solution 1 to solution 2 to make both solutions reach the same new concentration, as shown on the right.

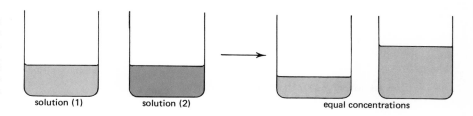

10-5.2 Molality. The reduction of vapor pressure (and all of the other phenomena we will discuss that are related to it) depends on the fraction of molecules that are solute or solvent. Our concentration unit of molarity does not give such a fraction. It compares amount of solute to volume, not to molecules of solvent or total molecules. Therefore, another kind of concentration unit is used for colligative property problems. It is called **molality** and abbreviated with a small *m* or the word **molal**. It is defined so there is very little difference between molarity and molality for dilute solutions in water. We simply use the moles of solute per *1000 g solvent* instead of per liter of solution. For almost pure water, the two are nearly identical. As solutions in water are made very concentrated, or when something other than water is used as the solvent, molality becomes different from molarity.

Solutions can be made up to known molality, as shown in Figure 10–8A,

All colligative properties depend upon the relative number of species, and molality, moles solute per 1000 g solvent, is a convenient concentration unit to use in these cases.

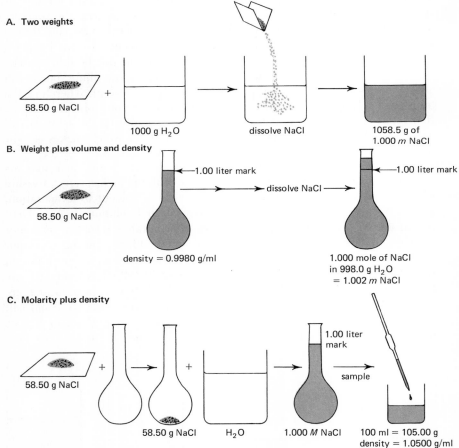

Figure 10–8 Methods of determining molality.

206 Solutions [10]

by weighing both the solute and solvent. The same thing is accomplished, as shown in Figure 10–8B, by adding a known weight of solute to a measured volume of solvent, provided the density of the solvent is known. However, in most cases the solutions are already made up to known values of molarity. Molality for such a solution can be calculated if the density of the solution can be measured, as shown in Figure 10–8C. We will do that calculation as an example.

Example 10–5. Calculate the molality of a 1.00 M NaCl solution in water, given that the density is 1.0500 g/ml.

$$1.0500 \frac{\text{g soln}}{\text{ml soln}} \times \frac{1000 \text{ ml soln}}{1 \text{ liter soln}} = 1050.0 \times \frac{\text{g soln}}{1 \text{ liter soln}}$$

in that solution there are

$$\frac{1.00 \text{ mole NaCl}}{1 \text{ liter soln}} \times \frac{58.5 \text{ g NaCl}}{1 \text{ mole NaCl}} = 58.5 \text{ g NaCl/liter soln}$$

therefore, there are

$$1050.0 \text{ g} - 58.5 \text{ g} = 991.5 \text{ g H}_2\text{O/liter soln}$$

$$\frac{1 \text{ mole NaCl}}{991.5 \text{ g H}_2\text{O}} \times \frac{\frac{1000}{991.5}}{\frac{1000}{991.5}} = \frac{\frac{1000}{991.5} \text{ mole NaCl}}{1000 \text{ g H}_2\text{O}} = 1.009 \; m \text{ NaCl}$$

10–5.3 Boiling and Freezing Point Changes.

Colligative properties include lowering of vapor pressure, freezing point depression, boiling point elevation, and osmosis.

Changing the fraction of solvent molecules changes the likelihood of their escaping to a crystal surface as well as their likelihood of escaping to gas phase. Therefore, the temperature at which both solid and liquid are present (melting point) must be lower to cut back the rate of escape from the solid by as much as the solution cuts the rate of escape from the liquid. The melting point must still occur where the vapor pressure of the solid is just equal to the vapor pressure of the liquid. Since the solid usually stays the same (pure frozen solvent) while the solution has a lower vapor pressure than pure solvent, that crossover occurs at a lower temperature. The resulting changes in both vapor pressure and melting point curves for a water solution are shown by dotted lines in Figure 10–9. The solid lines represent the phase diagram of pure water.

As the vapor pressure is depressed further (as more concentrated solutions are formed), the vapor pressure crossover and hence the melting point is also pushed farther down. Since the vapor pressure drop is related to the fractions of molecules in solution, the *concentration in molality* can be used to calculate the change in melting point. Each solvent will have a characteristic amount by which the melting point of a 1 m solution will be pushed down. This should depend only on the relative shapes of the liquid and solid vapor pressure curves. If we get very far from the original melting point, these shapes may vary enough to change the characteristic effect of each 1 m of concentration.

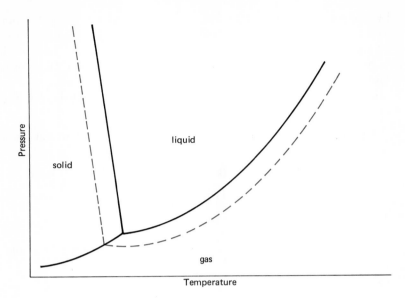

Figure 10–9 Phase diagram for water and for a water solution: solid lines, water; colored lines, solution.

We will avoid that problem by staying with fairly dilute solutions, and in that dilute range we will call the fraction

$$\frac{\text{drop in melting point in } °C}{\text{molality of solution}} = K_{fp}$$

the **molal freezing point depression constant,** K_{fp}. In water K_{fp} has a value of $1.86°C/m$.

The freezing point of a solution is

$$\text{freezing point of pure solvent} - K_{fp} \times \text{molality}$$

Therefore, a $0.65\ m$ solution of CH_3OH in water has

$$\text{freezing point} = 0°C - 1.86°C/m \times 0.65\ m$$
$$= 0°C - 1.21°C = -1.21°C$$

Similar calculations can be done for other solvents if we know *both* their freezing points as pure solvents and their molal freezing point depression constants. For example, given that benzene freezes at $+5.50°C$ and has a freezing point depression constant of $4.90°C/m$, we can calculate the freezing point of $0.65\ m\ CH_3OH$ in benzene.

$$\text{freezing point} = +5.50°C - 4.90°C/m \times 0.65\ m$$
$$= +5.50°C - 3.18°C$$
$$= +2.32°C$$

A similar constant, the **molal boiling point elevation constant,** K_{bp} has been defined to show how much the temperature must be raised to get the vapor pressure to equal the total pressure and allow boiling. Figure 10–10

Figure 10–10 How lowered vapor pressure of solutions causes boiling point elevation and freezing point depression.

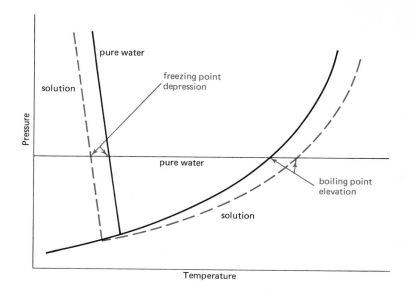

shows why boiling point must go up as a solution becomes concentrated while freezing point goes down. K_{bp} for water is $0.52°C/m$. A $1.0\ m$ solution of glucose in water at 1 atm pressure would have the following boiling point.

$$\text{boiling point} = \text{bp of pure solvent} + K_{bp} \times \text{molality}$$
$$= 100°C + 0.52°C/m \times 1.0\ m$$
$$= 100°C + 0.52°C = 100.52°C$$

If the total pressure was not 1 atm, both pure water and the solution would have different boiling points, but the difference between them would stay about the same. Therefore, under a total pressure where water boils at $98.50°C$, the same $1.0\ m$ solution has

$$\text{bp} = 98.50°C + 0.52°C/m \times 1.0\ m$$
$$= 98.50°C + 0.52°C = 99.02°C$$

10–5.4 Effect of Ionization. Because of variations in the curve shapes, boiling and freezing point changes vary from the expected values as the boiling and freezing points get far away from normal values. These changes remain small and can be ignored when we limit ourselves to dilute solutions, but there is another factor causing large changes even in dilute solutions. That is the breakup of molecules into separate ions.

All colligative properties (vapor pressure, boiling and freezing point changes, and osmosis) depend upon the *relative number* of solvent and solute particles. When an ionic substance (like NaCl) ionizes, it forms several ions. Each of these ions is a separate piece able to get in the way of vaporization or related things. Therefore, a solution of 1.0 mole NaCl in 1000 g H_2O

Individual ions each count separately in determining relative numbers for colligative properties.

is 1.0 m in NaCl, but because all the NaCl is ionized to Na^+ ions and Cl^- ions it is 2.0 m in total dissolved particles. Therefore, its freezing point is

$$fp = 0°C - 1.86°C/m \times 2.0\ m \text{ ions}$$
$$= 0°C - 3.72°C = -3.72°C$$

Similarly, a 1.0 m $CaCl_2$ solution in water, where each $CaCl_2$ forms one Ca^{2+} ion and two Cl^- ions, is 3.0 m in ions and

$$fp = 0°C - 1.86°C/m \times 3.0\ m \text{ ions}$$
$$= 0°C - 5.58°C = -5.58°C$$

If a substance is only partially ionized, the concentration of separate species actually formed determines the effective molality that must be used to calculate colligative properties. Let us consider the example of a 0.1 molal solution of $NaHSO_4$ in water.

If it existed as $NaHSO_4$ molecules in solution, it would have

$$fp = 0°C - 1.86°C/m \times 0.1\ m = 0°C - 0.186°C = -0.186°C$$

If it ionized to Na^+ ions and HSO_4^- ions, it would have

$$fp = 0°C - 1.86°C/m \times 0.2\ m \text{ ions} = 0°C - 0.372°C = -0.372°C$$

If it ionized to Na^+ ions, H^+ ions, and SO_4^{2-} ions, it would have

$$fp = 0°C - 1.86°C/m \times 0.3\ m \text{ ions} = 0°C - 0.558°C = -0.558°C$$

The actual situation is that it ionizes completely to Na^+ and HSO_4^- and then 29.2% of the HSO_4^- ionizes further to H^+ and SO_4^{2-}. The final concentrations are therefore 0.1 m Na^+, 0.0708 m HSO_4^-, 0.0292 m H^+, and 0.0292 m SO_4^{2-}. The overall total is then 0.2292 m ions and

$$fp = 0°C - 1.86°C/m \times 0.2292\ m = 0°C - 0.426°C = -0.426°C$$

The important point is that only the total concentration of all dissolved species matters to the colligative properties. There can be several different ions with different concentrations of each, but only the total concentration of everything added together (all ions plus any remaining nonionized molecules) matters.

10–5.5 Osmosis. If a solution and the pure solvent (or two solutions of different concentrations) are separated by something through which the solvent can pass but the solute cannot, the flow is slower from the more concentrated solution where solute gets in the way. The resulting net flow from the less concentrated to the more concentrated solution is called **osmosis**. Actually, there are many kinds of porous membranes, that is, membranes with small holes that let solvents through but are not large enough for many solutes to pass. Ions in water solutions are surrounded by water molecules, making them quite bulky and unable to pass through small holes.

When the tendency to escape to gas phase is lowered for a solution compared to pure solvent, the rate of return can also be cut down by having

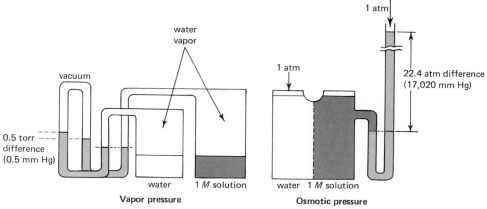

Figure 10-11 A comparison of osmotic pressure to vapor pressure differences for a 1 M solution at 25°C.

a lower vapor pressure in the gas phase. A 1 m solution in water at room temperature has a vapor pressure that is smaller than that of pure water by less than 0.5 mm Hg or 1/1000 atm. That lowers the concentration of molecules with a chance to return to the solution by enough to restore the balance.

In osmosis, it is not possible to decrease the rate of flow from pure solvent to solution because the pure solvent just stays pure solvent. The concentration of solvent molecules stays the same as long as any liquid at all remains. If the flow from pure solvent to solution cannot be slowed, the only way to restore a balance (no net flow going on) is to increase the flow from the solution to the pure solvent. That can be done by applying pressure to the solution. The amount of pressure needed is called the **osmotic pressure,** and these osmotic pressures are *very, very large*. The same 1 m solution at room temperature, which has a vapor pressure less than 1/1000 atm smaller than pure water, exerts an osmotic pressure of about 22.4 atm when compared to pure water.

Osmotic pressures are much larger than vapor pressure differences.

Figure 10-11 shows how much larger osmotic pressures are than vapor pressure differences. These large osmotic pressures are usually never reached in natural osmosis, such as water transport through the walls of plant roots. If balance is never reached, a steady flow continues from the less concentrated to the more concentrated solution. In experiments where the pressure in the solution is allowed to build up, the porous membrane barrier may break under the pressure long before the true osmotic pressure is reached. The same thing can happen to plant or animal cells if an osmotic imbalance is set up when the diffusing solvent is trapped within limited available space.

If the osmotic pressure is never reached, a steady flow continues from the less concentrated to the more concentrated solution.

10-6 Transfer of Water and Nutrients in Plants and Digestion

The walls of plant roots and the linings of digestive systems are porous membranes that permit osmosis. The possible osmotic pressures are not reached, so a steady flow of water occurs from the less concentrated solution (ground water in the case of plants) to the more concentrated solution (inner fluids). These membranes allow some solutes to pass through with the water. Other solutes are tied up by the membranes on the outside and then released inside—a process called active transport because it forces certain substances through the membrane. As long as there is some solute present that will not pass through and remains in excess on one side of the barrier, water flow will continue. Those solutes able to pass through the membrane may also be carried along by the same osmotic flow delivering water across the barrier. Some of the nutrients absorbed by plants from soil are carried in this way; others require active transport. All animal life also depends upon a net transfer of food across a porous barrier.

Nutrients and water are carried into plants and animals by osmosis.

All land plants and animals depend upon the net osmotic flow bringing in water. The eventual loss of water by evaporation must be balanced by the gain through osmosis. The process requires greater solution concentrations inside than outside. If that situation is reversed, the living system is in serious trouble. If plants grow in the cracks between patio blocks or bricks, they can be killed off and prevented from reappearing by applying salt. As the ground water around the plant roots becomes very concentrated in dissolved salt, the solution inside the roots becomes the less concentrated instead of the more concentrated. The osmotic flow is then reversed, and water is steadily removed from the plant. The plant becomes dehydrated and dies.

Land plants and animals must have a net osmotic inward flow to replace water losses by evaporation.

The same thing can happen to animals. If a person drank salty ocean water, the solution in his intestines could reach a higher molality than the blood on the other side of the porous barrier. Even though some of the salts are transported across the barrier to reestablish the original difference, so much water would be lost by osmosis before then that the person would become at least violently ill from dehydration and might not survive.

Sometimes reversals of osmotic flow can be caused by human error. Several years ago there was a tragic situation when a nurse mixed up the hospital formula supply with salt instead of sugar. Salt is not a poison, but the combination of a lower molecular weight than sugar plus the doubling of the molality by the ionization of NaCl made the solution concentrated enough to reverse the ordinary osmosis. Several babies died from the dehydration. Plants are also accidental victims sometimes. In Section 13–5.4, we will describe how irrigation can lead to concentration of salts in the irrigated soil. If the concentration becomes high enough, desirable plants are unable to maintain an adequate inward osmotic water flow for survival.

Obviously some plants and animals can maintain a proper osmotic balance in more concentrated solutions than others. The oceans support extensive life. Actually, cells completely surrounded by water must avoid the buildup of osmotic pressure, which could rupture their cell walls. They might escape by simply being so porous that solution flows back out whenever pressure builds up, but that would also carry out the dissolved nutrients and cell proteins the cell lives on. Most cells adjust to the particular concentration they are in. Some species require certain solution concentrations around them to survive, and very few cells can survive a rapid change in concentration.

Water plants and animals must maintain an osmotic equilibrium with the surrounding solution.

Placing cells in distilled water is one method of breaking them open so the internal contents can be analyzed, and a rapid change from concentrated to dilute solution could have the same effect.

Sometimes differences in ability to tolerate salt determine natural selection processes. Large numbers of the oceanic animal species depend upon brackish estuary areas for breeding. Their young develop in the salt marshes and other areas too salty for land or fresh water species but not salty enough for ocean species. They then move out and adjust to the sea as they become mature enough to compete more successfully. What appear to be useless estuaries and salt marsh areas are irreplaceable resources because of this. The breeding migrations to fresh water by salmon and the seagoing steelhead trout are also classic examples of the advantage some species can gain if they can adjust to a changing osmotic environment.

10-7 Some Problems Illustrating Solution Phenomena

Bends in divers and some fish kills are caused by N_2 supersaturation.

Human activities often upset long established natural balances. By ignorance or lack of concern we sometimes set up solution phenomena that we do not want. Problems caused by exceeding the solubility of N_2 in water are examples. When human divers work in pressurized suits and breathe compressed air, they set up an equilibrium between the N_2 pressure in the compressed air and the N_2 dissolved in their blood and other body fluids. As expected from Henry's law, more N_2 dissolves at the higher pressures. As they return to the surface and pressure is released, some of the dissolved N_2 must be removed. If the pressure drop is rapid, the lungs are not able to remove the excess N_2 fast enough. The amount of N_2 in the blood then becomes a supersaturated solution. If it becomes sufficiently supersaturated, bubbles may form and block blood circulation. This condition is called the bends and may be serious or fatal. It is treated by putting the person back under enough pressure to exceed the N_2 pressure in the gas bubbles and make them collapse. Then a gradual release of the pressure allows the dissolved N_2 to escape slowly without reaching a level of supersaturation great enough to form bubbles.

Fish can also suffer from the same problem. Large kills of migrating steelhead trout have been traced to N_2 supersaturation of rivers caused by flow of spring floodwaters over dam spillways. The general principles affecting gas solubilities show us how this supersaturation can occur. As the water comes down over the spillway, its momentum carries it down deep into the water below the dam, as shown in Figure 10–12. Large amounts of trapped air bubbles are carried down with it and compressed by the water pressure. Less than 10 ft of water depth is needed to compress the air enough for the N_2 pressure alone to be greater than the entire air pressure above the water, and the deep water becomes saturated with the concentration of N_2 matching this high pressure. The deep water may also be quite cool, particularly if part of it comes from output of electrical generators. These generating units take water in from far below the lake surface above the dam, and it is much cooler than surface water (see Section 13–5.7). The combination of high

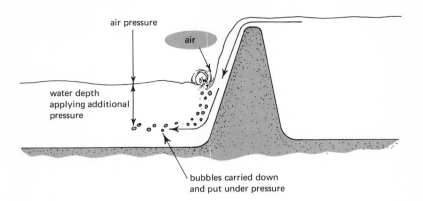

Figure 10–12 Nitrogen entrapment by spillways.

Air bubbles are carried down with the water flow to depths where they are under a higher pressure—the sum of atmospheric pressure plus the pressure of the water depth. By Henry's law the solubility of gases is greater in water touching these high pressure bubbles than in water in equilibrium with air at normal atmospheric pressure.

pressures and cold temperatures makes much more N_2 dissolve than would go into solution at the surface. Fish swimming in the turbulent deep water below the dam gradually have their blood saturated to the same N_2 concentration as the water. When they then swim up to some surface area, which is also usually warmer, the N_2 in their blood can form N_2 gas bubbles with pressures greater than the total pressure on the fish. These bubbles then form and expand rapidly. The sickening fish is unable to swim back to deep water where he could recover, so eventually he dies. Because of this phenomenon, officials of the Army Corps of Engineers would be quite interested in any new ideas about how to keep small young trout weighing about 0.1 kg (or $\frac{1}{4}$ lb) from the bottom of their huge dams (which weigh perhaps 100 million kg).

An even more classic example of human interference with solution phenomena on a large scale is occurring in Great Salt Lake in Utah. Great Salt Lake has been a saturated salt solution much of the time. In summer as water evaporates from the lake it becomes saturated in NaCl and excess NaCl goes out of solution. In winter, water evaporation is slowed and, as more water flows in, some of the NaCl can be redissolved. However, as it gets cold, the lake becomes saturated in Na_2SO_4 (which forms the solid hydrate $Na_2SO_4 \cdot 10\ H_2O$); this compound has a solubility that varies much more with temperature than NaCl does.

The continuous precipitation and redissolving of these salts has made the lake very shallow, only about 3 ft deep for each mile from the shore. The lake, shown by the map in Figure 10–13, is surrounded by mountains and divided by Promontory Point, which sticks well down into the lake from the north. The first transcontinental railroad was completed at Promontory Point in 1869. Long ago, as railroads were modernized, the route was shortened by building trestles across the two shallow arms of the lake to Promontory Point. In the 1950s, the trestle on the west arm needed to be replaced, and this was done by building a gravel fill across that arm of the lake. At that point, the laws of solutions began to operate in a new and different way on Great Salt Lake.

As long as the lake was open (even with a trestle across it) storms could

Figure 10-13 Railroad route across Great Salt Lake.

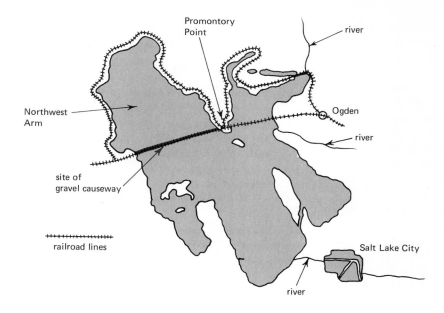

A gravel barrier has caused concentration and precipitation of salt in one part of Great Salt Lake, leaving the rest of the lake less concentrated and making the lake level drop.

stir up the lake and mix the salt solution from one area to another. When the gravel fill was completed this mixing was cut off. This is the first of the major factors upsetting the lake's balance. Salt solution could flow through the gravel, but turbulent mixing was no longer possible between the northwest arm and the rest of the lake. Both the northwest arm and the rest of the lake could be mixed by storms, but they were separated from each other by the gravel fill where only gentle flow (with no mixing) could take place.

At that point geography entered the picture as the second major factor affecting the lake's balance. The gentle flow through the gravel is always in one direction—into the northwest arm. The area around the northwest arm is very dry, so there is a high evaporation rate and little water flow in from streams. The main rivers flowing into the lake are on the northeast corner, on the east side, and on the southeast corner. Some water entering at those points must eventually flow to the northwest arm to make up for the evaporation there. As the water flows through the gravel, it carries dissolved salt with it.

In the absence of mixing, the only way the salt could be returned across the gravel barrier would be by diffusion. The third major factor in the situation is the fact that diffusion in liquids is much too slow to help at all. Even the very limited mixing possible inside the gravel would do more good than diffusion, which is limited to about 1 ft/year.

All these factors led to a series of logical results. First, as saturated salt solution moved into the northwest arm and then water evaporated away, large amounts of salt precipitated in the northwest arm. This salt was not redissolved in the usual yearly cycle because the fresh water from the rivers was not mixed into the northwest arm. The flows into the northwest arm were always

[10-7] Some Problems Illustrating Solution Phenomena

salt solution from the main lake. As a result, during the first 5 years after completion of the gravel fill about 10^9 metric tons of salt were precipitated in the northwest arm of Great Salt Lake.

This transfer of salt to the northwest arm led to a second effect—dilution of the main body of the lake. As fresh water entered from the rivers and salt solution moved to the northwest arm, the main lake was diluted to an unsaturated solution. For a while excess salt on the lake floor could dissolve to keep the solution saturated, but eventually this failed to maintain the concentration. As the main body of the lake became diluted, Raoult's law (Section 10–5.1) set in to produce a third effect. The vapor pressure of the diluted solution was higher, so water began to evaporate from the lake more quickly. There was no corresponding drop in evaporation from the northwest arm because the solution was already saturated; additional solid salt formed but the concentration (and evaporation rate) stayed the same. Therefore, the lake as a whole had an increased evaporation rate.

During the first few years after completion of the causeway, the level of the lake dropped. The lake level had been falling slowly for at least 80 years, and it seemed that the new causeway would speed the drop. The increased evaporation rate would accentuate an already existing problem. The evaporation rate was large enough to cause the lake to lose water faster than it was being brought in by rivers and rain. Therefore, the lake began to shrink. Eventually the lake level would drop enough to cut the surface area down to the point where average evaporation would just equal the average rate at which water comes in. However, lowering the lake level exposes new areas of very salty land. Eventually rainfall will wash out much of the salt, but in the meantime the salty soil prevents any plant growth because of the unfavorable balance of osmotic forces (Sections 10–5.5 and 10–6). The lack of plant growth leaves the areas around the lake very susceptible to erosion by wind, and dust storms have become an irritation to the surrounding areas, which includes most of the population of Utah.

However, the balance creating the lake level is complex. As the evaporation from the lake increases, the average rainfall and snowfall in the lake's drainage basin also rises. The effect on the amount of moisture entering or leaving the basin is not clear. In fact, the lake level, which had been falling from 1872 to 1963, has now begun rising. By 1973 it had reached the highest levels in 40 years. The high water level has forced closure of a road causeway to a state park on an island in the lake and covered most of a beach used by tourists. We simply do not know enough about weather to say how the railroad causeway has contributed to or lessened the rise in precipitation.

We can say with certainty that man-made changes have now divided Great Salt Lake into three distinctly different regions. First, Willard Bay at the northeast end of the lake has become a freshwater lake because it is cut off from the main body of the lake and it receives fresh water from the Bear River. We have not made any earlier mention of this change because the resulting small fresh water body is considered good. Second, the northwest arm is saturated salt solution with a large amount of solidified salt on the bottom. By 1973 the estimate of salt newly precipitated in the northwest arm since the causeway was built exceeded 2 billion tons. That is about one quarter of the salt in the lake's waters. Third, the main body of the lake has become unsaturated but still remains quite salty.

Although water plants like algae can adjust to much higher salt concentrations than land plants (which must have a steady inward osmosis of water), the difference between the northwest arm and the rest of the lake has already caused evolution to produce significantly different algae growths in the two areas. The northwest arm has a rose color whereas the main body of the lake has lighter greenish algae.

The adjustments necessary for industrial plants using the lake as a source of salts may be more serious. Three salt producers and a magnesium and chlorine plant are located on the main body of the lake. These firms have sued the railroad for endangering their salt supply. The railroad has been joined as defendant by an industrial firm using the northwest arm to produce both salt and potassium sulfate for fertilizer. They argue that the concentrated northwest arm is an industrial boon. It has been estimated that a 1500-ft long opening, costing $12 million, would allow the lake to restore the earlier equilibrium between the northwest arm and main body. By a later estimate a 500-ft opening would allow the main body to maintain 90% of the concentration of the northwest arm. But a recent study indicated that the migration of salt north is no longer very rapid. There is some doubt that any action will be taken.

In summary, the sequence of events was

1. The gravel fill stopped mixing by storms.
2. The geography led to a steady flow into the northwest arm.
3. Diffusion in liquids is too slow to return the excess salt carried into the northwest arm.
4. Large amounts of salt precipitated from the saturated northwest arm.
5. The main body of the lake became unsaturated.
6. Raoult's law increased the rate of evaporation from the less concentrated main body of the lake; the evaporation from the northwest arm stayed constant as the solution remained concentrated.
7. The level of the lake dropped to compensate for the greater evaporation rate.
8. The newly exposed land was too salty for plants and subject to dust storms.
9. Increased rainfall has now raised the lake level, but the separation into regions of dissimilar salt concentrations remains.

This situation occurred because a large engineering project was undertaken without adequate prior thought about the nature of salt solutions and possible environmental consequences. The actual situation is complex and not completely understood even afterward. It was certainly not well enough understood at the time the causeway was constructed.

Skills Expected at This Point

1. You should be able to define and/or use correctly the terms solution, dissolve, completely miscible, partially miscible, solute, solvent, concentrate, dilute, concentration, solubility, saturated, unsaturated, supersaturated, molarity, molality, pH, Henry's law, Raoult's law, colligative properties, freezing point depression, boiling point elevation, and osmosis.
2. You should be able to apply the principle "like dissolves in like" to predict whether a solubility might be large or is certain to be small.
3. You should be able to identify and list the conditions that favor solubility of gases.
4. You should know most solids become more soluble at higher temperatures.
5. You should be able to calculate concentrations in molarity and use them to work dilution problems and titration problems.
6. You should be able to identify a solution as acidic, basic, or neutral if you are told its pH.
7. You should be able to calculate molality, recognize how ionization changes the effective molality, and calculate freezing point depressions or boiling point elevation when given the necessary data.
8. You should be able to describe the very large pressures due to osmosis and the role osmosis plays in living things.

Exercises

1. It is possible to get (at most) 5.50 moles of NaCl to dissolve in 1 liter of solution (in water) at a particular temperature. Take that information plus the conditions in a case in which 11.7 g NaCl (0.20 mole) and 95.0 g H_2O were used to make 100 ml of solution (at the same temperature) and describe the situations correctly using *as many of the terms listed in statement 1 of the "skills expected at this point"* as you can get information about from this data. You should be able to use at least 12 of the 22 terms listed. You will notice we could not even write down the question without using at least one of the terms (solution).
2. CH_3OH is a covalent liquid that forms hydrogen bonds readily and is not very polar, H_2O forms hydrogen bonds readily and is quite polar, C_6H_{14} is a covalent liquid that does not form hydrogen bonds and is not very polar, NaCl is an ionic solid, and I_2 is a nonpolar covalent solid. Which pairs are most likely to allow highly concentrated solutions to form?
3. Draw the solubility curve for a typical solid in a liquid and indicate clearly the region of supersaturated solutions.
4. **(a)** List the factors favoring higher solubility for gases. **(b)** Show how bubbles carried down underwater by the water flowing over dam spillways are put under some of the solubility favoring conditions you listed in part (a).

5. Draw a graph showing (a) solid lines representing the phase diagram of water. (b) On the same graph, show dotted lines representing the vapor pressure and melting point curves for a water solution. (c) Draw a line across at the 1 atm pressure condition and label clearly the places (include arrows to point out these places) that show how the boiling point elevation and freezing point depression occur.
6. Calculate the molarity of each of the following solutions:
   (a) 1.5 moles of Na_2SO_4 in 500 ml
   (b) 58.5 g NaCl in 250 ml
   (c) 8.50 g $Mg(NO_3)_2$ in 1.50 liters
7. Calculate the molality of each of the following solutions:
   (a) 1.5 moles of Na_2SO_4 in 400 g H_2O
   (b) 149 g KCl in 2500 g H_2O
8. Given five solutions that have, respectively, pH = 5.5, pH = 12.5, pH = 4.0, pH = 7.0, and pH = −1:
   (a) Which of them are acidic?
   (b) Which of them are basic?
   (c) Which of them are neutral?
9. (a) Calculate the H^+ concentration in a pH = 9.0 solution.
   (b) Calculate the OH^- concentration of the same solution.
10. Given K_{fp} for H_2O is $1.86°C/m$, calculate the freezing points of:
    (a) A 1.50 m solution of the covalent compound glucose, $C_6H_{12}O_6$, in water.
    (b) A 0.75 m solution of $Mg(NO_3)_2$, which ionizes completely to form Mg^{2+} and NO_3^- in water.
11. (a) A thin film separating the vapors above a solution and the pure solvent would be stretched a little by a pressure difference if the solvent vapor was the only gas present on each side. Which direction would it be pushed and why?
    (b) If the same thin film was porous (allowed osmosis) and was used down in the liquid to separate the solution from the pure solvent, it would soon be stretched by a pressure difference. Which direction and why?
    (c) In part (a), the film would stretch but not break, but in part (b) it would keep stretching until it did break. Why does that difference in behavior happen?
12. Why does salty soil kill plants?
13. Why does distilled water make cells burst?
14. What part does Raoult's law play in the accelerated drying up of Great Salt Lake?
15. Why can underwater plants adjust to higher salt concentrations than land plants?
16. What volume of 0.500 M $Mg(OH)_2$ is needed to react with 55.5 ml of 0.750 M HCl to make $MgCl_2$ and H_2O?
17. Propose a solution to one of the practical problems brought up in this chapter.

Test Yourself

1. Give brief definitions for each of the following terms:
   (a) dissolve
   (b) saturated solution
   (c) solubility
   (d) molarity
   (e) freezing point depression
   (f) solute
2. (a) State briefly how the solubility of a gas in a liquid normally varies when (1) the temperature is raised and (2) the pressure of that gas is increased.
   (b) State briefly how the solubility of a solid in a liquid normally varies when (1) the temperature is raised and (2) more of the solid is added.
3. What is the pH of $10^{-4}\ M$ HCl solution?
4. If 250 ml of 0.550 M NaCl solution is concentrated to 75 ml by evaporating water, what is the concentration of the new solution? What kind of solubility condition might prevent the NaCl concentration from reaching this new value when the water is evaporated, leaving the NaCl behind?
5. The addition of a nonvolatile solute to water will
   (a) increase the boiling point.
   (b) increase the freezing point.
   (c) increase the conductivity in all cases.
   (d) increase the vapor pressure of water at a specific temperature.
   (e) none of these.
6. What is the molarity of a solution of $CuSO_4$ that contains 24.95 g of $CuSO_4 \cdot 5H_2O$ in 250 ml of solution? (At. wts.: Cu, 63.5; S, 32.0)
   (a) 24.95 M
   (b) 1 M
   (c) 0.1 M
   (d) 0.2 M
   (e) 0.4 M
7. Which of the following is the boiling point at standard pressure of a solution of 90 g of glucose (mol. wt. = 180, covalent compound) in 250 g of water? ($K_{bp} = 0.52°C/m$)
   (a) 100.00°C
   (b) 101.04°C
   (c) 102.08°C
   (d) 105.20°C
   (e) None of these
8. Which of the following could cause the greatest freezing point depression if 1 mole of each compound were dissolved separately in 1000 g of water?
   (a) $HClO_3$ (an acid)
   (b) Na_2SO_4 (a salt)
   (c) $NaNO_3$ (a salt)
   (d) $C_{12}H_{22}O_{11}$ (sugar)
   (e) HCl (an acid)
9. Suppose you used $H_2C_2O_4 \cdot 2H_2O$ to standardize your base in lab, and you found that 35.0 ml of your NaOH was required to neutralize 0.598 g of $H_2C_2O_4 \cdot 2H_2O$. What is the molarity of the NaOH solution? (*Hint:* The equation for the reaction has a $H_2C_2O_4 \cdot 2H_2O$ reacting with 2NaOH.)
   (a) 0.100 M
   (b) 0.135 M
   (c) 0.270 M
   (d) 0.543 M
   (e) 0.355 M
10. The pH of a solution of sodium hydroxide that is 0.001 M will be
    (a) 3
    (b) 5
    (c) 9
    (d) 11
    (e) Impossible to calculate

11. Which of the following would be most likely to form a solution readily?
    (a) An ionic substance in a nonpolar solvent
    (b) A covalent substance in a nonpolar solvent
    (c) A covalent substance in a molten salt
    (d) A highly hydrogen bonded substance in a non-polar, nonhydrogen bonded solvent
    (e) A high melting solid in a gas
12. Proper manipulation of a solid such as salt and a liquid such as water may produce a supersaturated solution because
    (a) extra dissolved solid is trapped in the hydrogen bonded structure as the liquid cools.
    (b) supersaturated solutions always form when excess solid is added to the liquid.
    (c) the solid cannot precipitate until the correct crystal structure is provided or occurs by chance.
    (d) salt reduces the solubility of nitrogen and causes bubbles to form.
    (e) all of these answers are correct.

Mixing and Pollution by Disorder

11–1 Spontaneous Changes and Irreversible Changes

Everyday experience shows us that some things tend to happen naturally while other things do not. Those changes that can occur without outside interference are called **spontaneous** changes. We use knowledge about different kinds of spontaneous changes to predict what will happen in future situations, and we use those predictions to avoid dangers or accomplish useful tasks.

For example, we know that heavy objects spontaneously move toward the center of the earth unless something actively prevents that movement by getting in the way. As objects move toward the center of the earth they release energy. We can use that knowledge to avoid dangers. When we park a car on a slope, we know it could spontaneously move down the hill. Therefore, we prevent the change by using brakes, wheels turned against the curb, or even wood blocks behind the wheels. But we can also use our knowledge to accomplish tasks. If we find our car on the hillside will not start, we might get it to a service station down the hill by releasing the brakes and other restraints and allowing the spontaneous movement downhill. In the process we convert potential energy (gravitational) into an equal amount of kinetic energy (motion). To stop at the service station we have to get rid of the kinetic energy, normally by using brakes to convert this energy to heat which escapes. Throughout the process the total energy remains constant (law of conservation of energy). Energy is only converted from one form to another.

But not all energy conversions are spontaneous. Gravitational energy can spontaneously convert to kinetic energy. If the car rolled down one hill and then up another hill, the kinetic energy could be spontaneously converted back to gravitational energy (although the small amount converted to heat by friction would not be converted back to gravitational energy). The conversion between gravitational and kinetic energy is a **reversible** process—it can go in either direction. But energy converted to heat by the brakes cannot be converted back spontaneously. The heat escapes to the surroundings and is not available to be converted back to kinetic or gravitational energy. That process is **irreversible**—it goes in only one direction. The heat is always lost from the brakes to the surroundings. We never see a car absorbing heat into its brakes, converting that energy to motion, and thus climbing hills.

Some conversions of energy from one form to another occur spontaneously.

Energy conversions may be reversible or irreversible.

Spontaneous changes involve irreversible steps.

Figure 11-1 illustrates these examples of spontaneous, reversible, and irreversible changes.

It is a simple fact of nature that spontaneous changes involve irreversible changes. There may be some reversible steps along the way (such as conversion of gravitational to kinetic energy), but eventually an "escape" of energy (like the loss of heat) will occur that is irreversible. In Appendix C-1 we will consider the relationship of energy changes to spontaneous processes. We will also give some attention in Section 12-2 to when (how fast) restraints are overcome to allow spontaneous processes to occur. But for now we will concentrate on the "escape" type process that is involved in irreversible changes. This "escape" does not convert the heat from the car brakes into

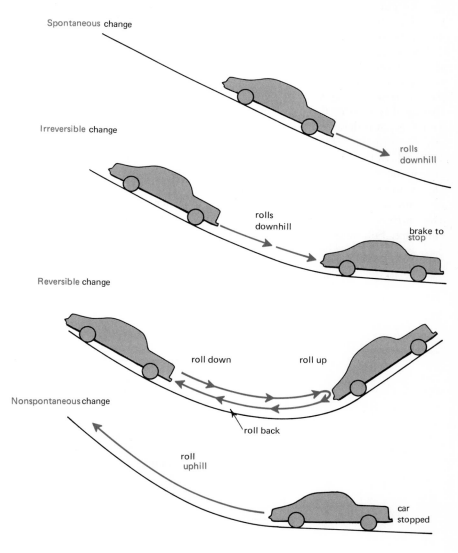

Figure 11-1 Types of changes.

[11-1] Spontaneous Changes and Irreversible Changes 223

any other energy form, and we will see in other examples that it does not even have to involve energy escaping.

11-2 Entropy: the Universal Tendency to Disorder

There is a tendency in nature for things to mix. Mixing processes are spontaneous even when there is no change in the amounts of energy present in various forms. Mixing processes lead to disorder—nature is always less completely orderly in its arrangement after mixing. The amount of disorder is related to a property called **entropy.** A perfectly regular arrangement would have zero entropy, all real (somewhat disordered) materials have a positive entropy, and entropy increases during spontaneous changes where no energy changes occur. Let us consider some examples of entropy increases during spontaneous mixing processes.

Figure 11-2 shows the mixing of two gases, 400 cm^3 of N_2 and 100 cm^3 of O_2. Even though there is no change in the temperature, pressure, or chemical indentities (molecules), the gases will mix if the barrier between them is removed. They form a gaseous solution, as described in Chapter 10. During this mixing the gases become more disordered. Entropy increases. There is considerable disorder at the start—each N_2 molecule could be anywhere in the 400 cm^3 and each O_2 molecule could be anywhere in the 100 cm^3. But no N_2 molecules are in the 100 cm^3 of O_2 and no O_2 molecules are in the 400 cm^3 of N_2. That orderliness (no N_2 in one region and no O_2 in another region) is lost when the gases mix. After mixing, each N_2 molecule and each O_2 molecule could be anywhere in the entire 500 cm^3. The entropy increases by the increase in uncertainty about where the molecules are. Solutions, where substances are mixed on a molecular scale, are

There is a property called entropy that is related to the natural tendency toward mixing and disorder.

Entropy increases as mixing occurs and disorder is increased.

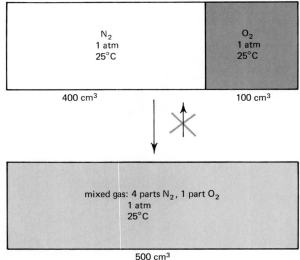

Figure 11-2 Spontaneous mixing of gases.

Figure 11-3 Spontaneous flow of heat.

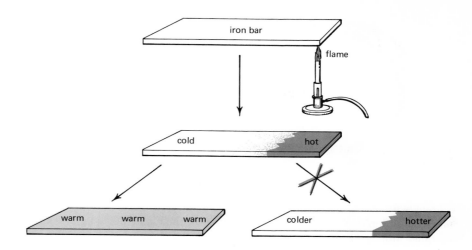

Mixing is spontaneous only in the direction in which entropy increases.

higher in entropy than the separate pure substances from which they are formed.

The mixed gas in Figure 11-2 is approximately the same as normal air. We can easily prove by experiment that the N_2 and O_2 will mix as shown. We also know from everyday experience that the opposite change, separation of air into pure N_2 and pure O_2, does not occur spontaneously. That separation process would require a decrease in entropy because it would make the locations of molecules more orderly. When no energy change occurs, only processes that increase entropy are spontaneous.

Figure 11-3 shows a kind of mixing in which the matter stays essentially where it was. Only heat moves. We all know the heat will flow spontaneously from the hot end of the bar to the cold end and not in the opposite direction. When one end of the bar is hotter than the other end, any particular unit of heat energy could be located anywhere in the bar, but it is more likely to be in the hot end. As the heat flows, this orderliness (the heat units more likely to be in one place) is lost. When the temperature of the bar reaches the same throughout, we have reached the state of maximum disorder (equal chance of finding a given unit of heat anywhere in the bar). That is therefore a state of higher entropy than the original state, and the spontaneous process (with no changes of the energy from heat to any other form) is again the one with increasing entropy.

The irreversible change shown in Figure 11-1 clearly involves an entropy increase of the heat transfer type. Heat is lost spontaneously from the hot brakes to the surroundings.

Figure 11-4 shows a mixing of arrangements in space. O_2 gas can be detected and its amount in air measured because it is the only major component in air that has molecules acting like small magnets. A strong magnet can cause all the O_2 molecules to point in the same direction, as shown in Figure 11-4. The principle is the same as that which causes a compass needle (a small magnet) to point toward the Earth's north magnetic pole. The

[11-2] Entropy: the Universal Tendency to Disorder 225

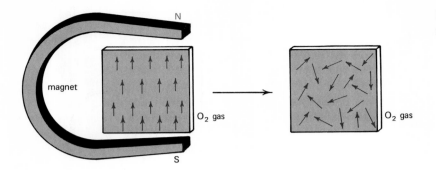

Figure 11–4 Spontaneous mixing of arrangements.

arrangement of O₂ molecules with the same magnetic direction is clearly somewhat ordered. When the magnet is removed, the molecules spontaneously rearrange into a disordered arrangement with higher entropy.

There are two notable points in this example. First, the distributions of both molecules and heat energy could be (and were) at maximum disorder while the entropy still had a way to increase through another kind of disorder. Second, the spontaneous tendency to increase entropy can be strong enough to overcome some forces. Even though the Earth itself has a magnetic field, that is not enough to force the oxygen molecules at room temperature into the orderly arrangement of one magnetic direction. The Earth's magnetic field is not strong enough to overcome the tendency toward increased entropy.

The spontaneous process in Figure 11–1 which we labeled a reversible change is in fact an example of mixing arrangements. There is an entropy increase in going from the fixed position and energy form (potential energy) of a parked car to the pendulumlike movement shown, and the process is not completely reversible. Outside effort would be needed to catch and stop the car at its original position. All other spontaneous changes can also be shown, on careful examination, to involve some entropy increase.

All spontaneous changes involve some entropy increase.

Although all of our examples in this section have shown entropy increasing, it is also possible to carry out processes where entropy decreases for the material being studied. Processes that decrease entropy (such as stacking molecules in an orderly arrangement) simply cannot be spontaneous without some energy change. As a matter of fact, the energy change needed to cause a decrease in entropy will always cause an increase in entropy somewhere else (such as flow of heat from a hot to cold object). That entropy increase is always as large or larger than the entropy decrease.

A spontaneous process may cause an entropy decrease in a limited region, but it must cause an entropy increase for the universe as a whole.

11–3 Natural Degradation of Energy

The spontaneous tendency toward mixing could ultimately lead to uniformly mixed matter all at the same temperature. At present we are still far from that situation. There are sharp differences in concentrations of energy, and many substances are separated into parts with a high degree of order. Life as we know it requires some precisely ordered arrangements of matter and

> Low entropy conditions are achieved in living things by harnessing the spontaneous energy flow from hot to cold.

continual conversions of energy. We do well to remember that once that order of matter and convertibility of energy are lost, with an increase in entropy, they cannot be spontaneously recovered.

The highly organized (and therefore relatively low entropy) state of matter in life is achieved in nature by harnessing the spontaneous entropy increasing flow of energy from "hot" to "cold." Large amounts of energy are released as heat in the Sun through a series of nuclear reactions (see Sections 15–3, 15–6, and 15–7). As the Sun became very hot, the spontaneous tendency for that heat to flow to cooler material eventually caused heat to be lost as fast as it was formed. The means by which the heat is lost is through conversion into light radiated away (sunlight). That sunlight is the low entropy starting form from which we get the spontaneous reactions sustaining life on earth.

To understand why sunlight is a relatively low entropy form of energy, we must understand the relationship between the wave and quantum natures of light. Light behaves like waves, with each particular wavelength (distance from the top of one wave to another) corresponding to a different color. The longest wavelengths of visible light are red, then orange, yellow, green, blue, and the shortest wavelengths are violet. White light is a mixture of all colors. There are also light wavelengths longer and shorter than those we can see. Methods of measuring wavelengths are described in basic physics books. The points of interest to us here are the following. (The first two of these were also needed to reach the conclusions of Section 6–2 about atomic energy levels.) (1) Light appears to carry energy in definite blocks (called quanta or photons). (2) The size of the energy blocks is related to the wavelength. The shortest wavelengths have the highest energy per quanta. The actual energy is given by the formula $E = hc/\lambda$ (E is energy of the quanta, h is a constant of nature called Planck's constant after Max Planck who proposed this relationship, c is the speed of light in a vacuum, and λ is the wavelength). (3) There is an upper limit on how fast quanta of a particular wavelength can be emitted from one point.

> A given amount of energy in light has higher entropy when it is in longer wavelength quanta.

> The upper limit on how fast quanta of each wavelength can be emitted forces hot objects to emit some short wavelength, low entropy quanta.

To reach maximum entropy, the energy in light should be divided into as many quanta as possible. Each of those quanta can then be emitted in any direction and be reflected, absorbed, or perhaps diffracted (bent in direction) by things in its path. The uncertainty about the fate of each of the quanta contributes to the entropy. If a cool object has only a small amount of energy to radiate, it can emit quanta of the few longest wavelengths fast enough to remove all the energy being radiated. But a hot object like the sun must radiate energy so fast that it also uses the shorter wavelengths. At the shorter wavelengths, the energy is contained in less quanta than the same amount of energy would require at longer wavelengths. If energy can be converted to longer wavelengths, the entropy associated with it increases. That change to longer wavelengths is accomplished on Earth.

Figure 11–5 shows the actual distributions of energy given off at each wavelength for a cool object and a hot object; the dotted line shows the distribution for equal numbers of quanta at each wavelength. Figure 11–6

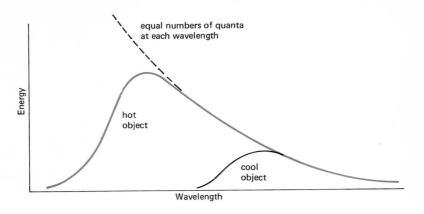

Figure 11–5 Energy emitted at each wavelength.

The dotted line represents the energy emitted at each wavelength if equal numbers of quanta are emitted in each equivalent range of wavelength. The black line represents emission from a cool object and the colored line represents emission from a hot object. Each matches the dotted line for the long wavelength (low energy quanta) emissions where the maximum number of quanta is being emitted at each wavelength. The hotter the object, the shorter the wavelength reached before the curve falls away from the dotted line.

shows how each point on the Earth's surface can radiate many more quanta than it absorbs. The rate at which quanta can be emitted at any one wavelength is limited, and that limitation plus the random choice of direction causes the rate of possible emissions to be the same in all equal sized solid angles. Points on the Earth can emit into space both day and night and at directions in a much larger solid angle than the angles in which they receive sunlight. Therefore, the Earth can stay at the same overall heat content while converting high energy quanta into a much larger number of low energy (long wavelength) quanta. Figures 11–7 and 11–8 show how this spontaneous change is harnessed to do work in sustaining life. In this process we obtain energy at low entropy from the Sun, a very hot nuclear reactor located a safe 93 million miles away, and dispose of that energy at high entropy to the vastness of outer space. While that energy is on Earth, a brief stay compared to the age of the atoms involved in life, the energy is stored in chemical form, used to cause nonspontaneous entropy decreasing changes, and eventually released by spontaneous, entropy increasing reactions.

The Earth can emit energy as fast as it receives energy from the Sun and emit the energy in longer wavelength, higher entropy forms.

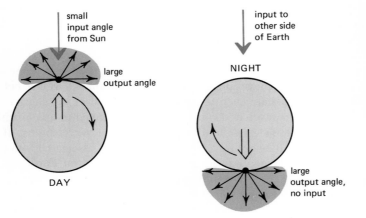

Figure 11–6 Potential input and output of quanta by earth.

The number of quanta that could be emitted by Earth is much larger than the number that could be received in the small angle of incidence from the Sun.

228   Mixing and Pollution by Disorder   [11]

Figure 11–7 Entropy increase on Earth.

[Adapted from H. A. Bent: Haste makes waste, pollution and entropy, *Chemistry,* 44(9):6–14(1971). Copyright 1971 by the American Chemical Society. Used by permission of the copyright owner.]

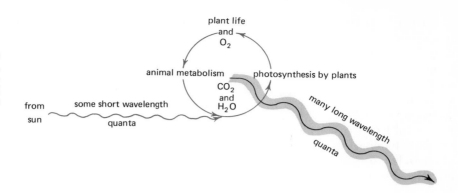

Figure 11–8 A natural energy chain.

[Adapted from H. A. Bent: Haste makes waste, pollution and entropy, *Chemistry,* 44(9):6–14(1971). Copyright 1971 by the American Chemical Society. Used by permission of the copyright owner.]

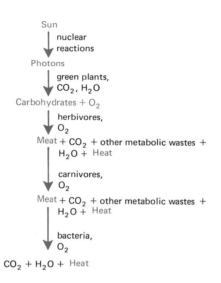

11–4 Disorder Produced by Human Activity

Human activities may degrade energy in wasteful ways.

11–4.1 Artificial Degradation of Energy. Many human activities have the effect of degrading energy forms. For example, concentrated sources of chemical energy (fuels) used to power transportation devices (cars, trains, airplanes) are degraded through friction and air turbulence to widely diffused heat. That degradation of energy is often not directly harmful, but it can be wasteful. The waste is significant because similar energy degradation is necessary to support life, and the total input and output of degradable energy is limited.

One of the most widely used artificial degradations of energy is in the production of electric power. Some power is produced in hydroelectric facilities where gravitational potential energy is utilized. That energy would be degraded to heat naturally without directly supporting life. We will see in Chapter 13 that recovery of this energy for power may interfere with the natural delivery of oxygen needed by life in streams. However, most power generation involves direct conversion of concentrated, stored energy into power and waste heat. The fraction of the stored energy that can be converted into useful power is limited by certain physical facts. The precise limitations can be determined (see any standard text on thermodynamics or elementary physics), but for our present purposes it is sufficient to note that the limitations are related to the amount of degradation (entropy increase) possible for the waste heat. That means the waste heat must be disposed of at a lower temperature (and therefore higher entropy) than the temperature at which the energy can first be put in as heat. As a result, cool objects that can accept waste heat become as important as the sources to provide the energy. It is the *flow* of energy from low entropy (hot) to high entropy (cold) that allows conversion of some energy into the work which produces the desired energy form, electric power.

Cool objects to accept heat are as important as energy sources for processes such as generation of electricity which depend on flow of energy.

Burning fossil fuels (coal, oil, natural gas) and operation of nuclear reactors increase the total input of energy above that received from the Sun. If that additional energy can be radiated into space in a high entropy form, it will increase the capacity of Earth to support entropy decreasing processes such as the organization of living things. However, there are serious problems. The rate of radiation into space increases only as the radiating object becomes hotter. Life on Earth is adapted to the normal temperatures on Earth, and temperature increases may upset the conditions necessary for particular plants and animals. Although the Earth as a whole is not warmed much by the amounts of energy released by man, some areas of the Earth, such as rivers used as convenient receivers of waste heat, are damaged by human use of energy.

Artificial releases of energy may upset the normal temperature pattern of the Earth.

11-4.2 Rearrangements of Materials. Human activities use the flow of energy toward high entropy to create changes in our environment. Our environment has three types of components that are important to people. We have relatively young and very complex cultural components, complex biological components which had their beginnings several billion years ago, and somewhat simpler physical components originating about 5 billion years ago. Changes in the biological and cultural environments are often accomplished by manipulating the simpler physical environment. We will concentrate this discussion on changes in the physical environment.

Production of goods involves arrangement of materials in particular patterns and places. That creation of order decreases entropy, which can only be achieved at the expense of some other increase in entropy. Degradation of energy is the usual source of the entropy increase. The net effect is the concentration of some material while energy and (in most cases) other materials are dispersed.

Production of goods also leads to high entropy byproducts or wastes.

For example, the arrangement of iron into steel girders (for construction of buildings or bridges) is accomplished at considerable costs in dispersal of energy and materials. Concentrated chemical potential energy in the form of oil, coal, and oxygen is converted into dispersed heat. In the process, large amounts of matter are also moved from concentrated forms to dispersed forms that have high entropy and cannot easily be reconcentrated. Oil based fuels burned in bulldozers, power shovels, trucks, trains, and boats are converted to carbon dioxide and water vapors mixed into the air. Coal, ore, and limestone are dug from deposits where they were quite concentrated and changed to new forms. Although the iron is concentrated in the process, other substances such as the limestone are made less pure by tying up unwanted substances. Everything but the desired iron is thrown away.

The disposal of unwanted materials creates major problems. Debris from mining and discarded solid impurities from iron and steel manufacture are often left in useless and bothersome piles. The inconvenience of being blocked off from possible working space or more useful materials (such as topsoil) makes solid waste seem worse than liquid or gaseous wastes. But wastes in liquid form (usually substances dissolved in water) or gaseous form also cause problems—water and air pollution.

11–5 Wasting Entropy Increases

Modern technology has given man control over large scale events involving tremendous entropy changes. Many human activities are carried out in ways that are wasteful of precious increases in entropy. We need to recognize that the available entropy increase from the sunlight is essentially constant and other sources of entropy increase are both relatively limited and potentially upsetting to the natural Earth we are adapted to survive on. As the scale of human technology grows larger, the consequences of our wastefulness become increasingly serious.

Some of our worst waste of entropy increase occurs in waste disposal methods. When only small amounts of wastes were produced, the convenient disposal was dispersal. Wastes allowed to accumulate in one place were offensive, but the same wastes spread over a large area could be diluted so much that no one was bothered or offended by them. Often those wastes even became involved in other spontaneous reactions (such as plant growth) and were converted back to more desirable forms. But there are limits on these natural spontaneous processes. All of them are driven by the entropy increase of the absorption of sunlight and emission of radiation to space. Man can

Man can increase the production of wastes beyond the capacity of nature to dispose of them.

increase the production of wastes beyond the capacity of nature to dispose of them, and dilution ceases to be helpful when the waste levels become high enough to be bothersome and offensive everywhere.

As we approach that sad condition of universal pollution, the facts of entropy become important. There is a greater entropy increase in wide dispersal of wastes than in accumulating them in a small space. Therefore, if it becomes necessary to remove the offensive wastes, it costs more work to get them back after they have become dispersed. Allowing them to become dispersed is wasteful of entropy increase. To conserve the limited entropy increase of Earth for other useful tasks, such as sustaining life or concentrating useful substances, we need to control our wastes in exactly the opposite way from our forefathers. Instead of dilution into our rivers and air, we must trap the dangerous pollutants before they become diluted.

Wastes can be recovered more efficiently if they are not allowed to disperse.

The necessary evolution of an entropy conserving attitude is shown by the development of coal burning techniques. As coal furnaces were improved from simple early forms to more efficient designs for electric power plants and other large scale uses, two of the problems were getting the coal to burn completely and getting rid of the ash. The pulverized coal method (which was adopted in the 1920s and carefully developed ever since) burns coal quite completely. One of its "advantages" was the fact that almost half of the ash went up the smokestack. Power companies did not have to truck that ash away, and the smoke problems of dirt and lung damage to people downwind were not charged to the companies. Now as companies are increasingly forced to trap the airborne ash, it is more expensive to handle than ash left in the furnace. Thoughtful engineers are now trying to move the solution back nearer the source by considering other (nonpulverized coal) furnace designs that would not create such smoke problems. They must catch up with the 40 years of technical development which went into pulverized coal, but the basic scientific principles of entropy assure them that their approach is the correct one.

Entropy conserving attitudes lead to changes in our technology.

Entropy increase is also wasted by excessive uses of energy to accomplish a given task. Transportation provides good examples. Walking a mile requires an entropy increase—conversion of the chemical potential energy of food to diffuse heat. Running 10 miles by a roundabout route to reach the same point would require a greater entropy increase and could be considered slightly wasteful. But the man who flies from Los Angeles to New York instead of walking a mile to work requires a tremendously greater entropy increase. The ability to travel from Los Angeles to New York may be important enough to the cultural component of our environment to be worth the cost in entropy, but we should take care not to spend such large quantities unless they are needed. Particularly expensive transportation, such as supersonic transports, may be too wasteful to be tolerated. And the most serious present problem involves a relatively simple and cheap form of transport, the automobile. Single passengers in vehicles designed for six and long commuting distances from suburbs to cities are clearly wasteful, and every large city suffers noticeably from the consequences of this waste. Car pools and public transportation can help, but a reduction in frequency and distance of travel would be even better and should be a key objective of long term planning.

Transportation (cars and planes) and electric heat are particularly wasteful of entropy increase.

Some forms of waste are hard to recognize. A common example is electric heating. Some well meaning people have suggested cleaning up the air by converting from "dirty" coal, oil, or gas heat to "clean" electric heat. Actually, the electric heat is about three times as "dirty" as coal, oil, or gas. Heat is a relatively high entropy form of energy that can be obtained by complete, spontaneous conversion of chemical (coal, oil, or gas) or electrical energy. But electricity is not so easily obtained. Most electricity is produced by burning coal and recovering only a fraction of the energy as work which can be used to make electricity. Only about one third of the heat energy is converted to electrical energy. The other two thirds of the heat is taken away by cooling waters in the power plant. The entropy increase as that two thirds of the heat flows from high to low temperature makes the process possible.

Therefore, when someone uses electric heat, the pollutants from burning coal are simply released somewhere else (at the power plant) in about three times as large a quantity as would have been necessary. There is also the additional problem from the heat released in the power plant cooling water, which may upset the environment. All of this is obvious to anyone who uses the available knowledge of science to analyze the situation. And that is a key reason why we need a large number of citizens with an understanding of the nature of science and a smaller number of active scientists to expand our knowledge continually and make that knowledge available when needed.

Reference Reading

Although detailed principles of thermodynamics are covered more completely in texts on that subject, students inclined toward a qualitative picture with minimum mathematics can find some interesting and accurate discussions in an unusual place—a science fiction magazine. Isaac Asimov discusses entropy in "Order! Order!" in the February, 1961, issue of *The Magazine of Fantasy and Science Fiction* and relates entropy to the evolution of life in "The Modern Demonology" in the January, 1962, issue of the same magazine.

Skills Expected at This Point

1. You are expected to be able to describe (with examples) the natural tendency toward disorder and associate the term entropy with it.
2. You are expected to know and describe the relationship between entropy and spontaneous or nonspontaneous changes.
3. You should be able to recognize and/or suggest natural and artificial examples of processes where entropy increases are used to bring about changes.
4. You should identify examples of waste in entropy increase, including those involved in pollution.

Exercises

1. (a) Write down three examples of spontaneous changes from everyday experience. (b) Write down two examples of changes that are not spontaneous but are frequently carried out.

2. Entropy increases during mixing. Explain briefly what entropy is and how mixing causes it to increase.
3. Show how each example in your answer to exercise 1 has an increase in entropy connected with it. In part 1b examples, remember to consider the "outside" changes used to cause the nonspontaneous changes.
4. You are supposed to learn some chemistry in this course. Consider the process of your studying and learning chemistry and answer the following questions about it.
   (a) Is your mental state about chemistry going to be at higher or lower entropy when you have organized an understanding of the facts and their relationships to each other?
   (b) Is the process spontaneous?
   (c) Is work required to achieve the learning?
   (d) Is there an entropy increase required to achieve the learning? If so, describe it. (*Hint:* Do you ever work up a sweat when concentrating hard?)
5. If we have equal amounts of energy in two beams of light, one red (long wavelengths) and the other blue (short wavelengths), which of the following is true?
   (a) There are more quanta of red light than blue.
   (b) There are equal numbers of quanta of red and blue.
   (c) There are more quanta of blue light than red.
6. Which gives off energy at lower entropy, an iceberg or a fire? Why?
7. Consider your own activities of the past week. What one activity (or type of activity) do you think was most unnecessarily wasteful of entropy increase?
8. Consider the activities of your community as a whole. What type of activity do you think could be most easily changed to significantly reduce waste of entropy increase?

Test Yourself

1. List three examples of spontaneous mixing processes.
2. What does entropy measure in the universe? In the absence of energy changes, does entropy always stay constant? If entropy changes, what is the direction of the change? When energy changes are used to force a process to go in a nonspontaneous way, is entropy increased or decreased? What must then happen in some other part of the universe?
3. Which of the following would be the best policy (assuming each could be achieved)?
   (a) No entropy increasing processes should be allowed to occur on Earth.
   (b) Artificial processes should be carried out for useful changes but with an effort to avoid those that are very wasteful of entropy increase.
   (c) All available energy should be put to work as quickly as possible to rearrange materials into products.

(d) Only naturally occurring processes should be allowed.

(e) Everything should be converted to the highest possible entropy state.

4. Which of the following best conserves entropy increase on Earth? Assume each is supplied enough energy to maintain the same temperature as the others.
   (a) An uninsulated house heated by an open fire (100% of the energy converted to heat in the house) and losing half of that heat through the walls and the other half through the ventilation needed to bring in fresh air.
   (b) A house heated by an open fire (as in part a) but carefully insulated so only 10% of the heat loss is through the walls and 90% through the needed ventilation.
   (c) An uninsulated house (like the one in part a) heated by a gas furnace that converts 90% of the energy to heat in the house (eventually lost through the walls) and 10% to heat lost up the chimney.
   (d) A carefully insulated house (like the one in part b) heated by an oil furnace that gets three quarters of the heat to the house (later lost through the walls) and loses one quarter up the chimney.
   (e) A carefully insulated house (like the one in part b) heated by electric heat that gets 100% of the electricity converted to heat in the house (later lost through the walls).

5. Which of the following does *not* reduce the waste of entropy increase?
   (a) Forming car pools instead of driving individually.
   (b) Removing ash from smoke before it goes out a smokestack.
   (c) Forcing water to flow over a dam spillway instead of down the original watercourse.
   (d) Keeping tin cans and glass in separate waste containers instead of combined with other trash.
   (e) All of these reduce waste of entropy increase.

Kinetics and Equilibrium (Qualitative Basis and Principles)

12–1 Coverage in This Book

Chemical equilibrium is one of the most important topics in chemistry. The conditions of equilibrium can be used as a powerful tool to analyze many important processes, ranging from the ability of humans to expel carbon dioxide in breathing to the formation of sedimentary rocks in the oceans. Unfortunately, use of equilibrium also leads to the most complex mathematical problems encountered by beginning chemistry students and is therefore the most difficult subject for many students.

In this book we are attempting to offer important concepts with the necessary mathematical background, but without emphasizing the mathematics. Equilibrium is a difficult subject to cover in this way because an emphasis on the mathematical applications is needed to show the true strength and usefulness of this tool. To assist you and your instructor in choosing the extent of coverage appropriate to your course objectives, we have divided this subject into three parts. Your instructor may cover all parts in detail, or some parts may be omitted, or you may be asked to read some parts only to see what can be done without being asked to learn to do it yourself.

This chapter derives a qualitative basis for the principles of equilibrium. The approach is via a brief consideration of kinetics, the rate at which reactions occur. Kinetics is also a very important subject but, when studied in detail, even more complex than equilibrium. The very simplified analogies to reaction rates in this chapter will point out a number of very important and relatively easily learned principles of kinetics. The concept of equilibrium and its mathematical form (the law of mass action) can then be developed through qualitative relationships to the principles of kinetics.

Appendix B at the back of the book is concerned with the practical matter of how equilibrium is applied to real situations. This appendix emphasizes the forms used for particular situations and has only a few sample problems chosen for their mathematical simplicity.

Appendix C discusses the thermodynamic basis for equilibrium. The rather limited discussion here is intended to provide some insight into the existence of this much more precise approach to changes and insight into some of the consequences of its principles. Appendix C also considers further examples, which are not as mathematically simple as those in Appendix B. However, emphasis is on problem solving methods that make these problems simple and on determining when the simplifications will work. The coverage in Appendix C is only intended as an introduction. It is presumed that anyone wanting to really master the thermodynamic principles will proceed to a book on thermodynamics. Similarly, anyone wanting to master complex equilibrium problems (such as buffering systems in blood) will need additional practice which is available through various chemistry problem books.

12-2 Factors Needed for Reaction

12-2.1 Reactants.

In Chapter 11 we discussed some spontaneous processes, such as mixing of gases and flow of heat. But these processes do not always occur, even though they are spontaneous. Barriers may prevent them from occurring. Gases can be kept separate by container walls, and the flow of heat from hot to cold can at least be slowed down by insulation, such as is used in the walls of houses. The processes that actually occur are those which are both spontaneous and able to occur by a path avoiding prohibitive barriers.

Chemical reactions are processes subject to the same limitations as other processes. They can occur only if the conditions are proper for reaction. There are several necessary conditions, the first of which must be suitable **reactants,** materials that can react with each other. There must be a favorable potential for interaction between these reactants to make the process spontaneous. We will come back to a further discussion of the requirements for spontaneous reaction in Appendix C-1, but at this point let us concentrate on what is needed to set up strong interactions.

Reactants should have the potential for spontaneous reaction.

One kind of interaction between different substances is the formation of ionic compounds from two very dissimilar elements. Although the attraction between positive and negative ions plays a crucial role in the overall favorable bonding, it is the sharp difference between the parts, one attracting electrons strongly and the other attracting its last one or more electrons rather weakly, that makes the formation of ions possible. The bonding that results is very different from the interactions possible between the identical atoms of a single element.

Elements with sharply different attractions for electrons have the greatest potential for partnership in forming ionic bonding.

When the parts are sufficiently different, with one tending to gain electrons and the other to lose electrons, the ionic bonding has a greater bonding force. The reaction to reach the more strongly bonded position is then a spontaneous process, with the extra energy of bonding lost as heat or some other energy form. The change to the stronger bonding can be compared to the spontaneous fall of a rock toward the earth, in which the extra energy of being closer to the center of gravity is lost as energy of motion, air turbulence, heat, and vibrations as the rock strikes the ground. Figure 12-1 shows these spontaneous processes. Substances that can release energy to other forms by reacting are usually suitable reactants (although we will have to modify that condition

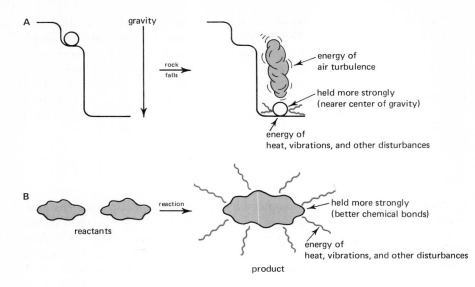

Figure 12–1 Spontaneous processes.

A. Conversion of gravitational potential energy to other forms.

B. Conversion of chemical potential energy to other forms.

for reaction to a more precise standard in Appendix C–1. For ionic reactions, the most dissimilar substances tend to be the best reactants.

Covalent bonding occurs between atoms that are closer in their strength of attractions for electrons than those which form ionic bonds. The bonding depends upon these atoms being similar enough to share in common the electrons forming the bonds. The sharing may not be equal; in fact some preference toward the more electronegative atoms may be necessary in order to provide a bonding advantage for compounds greater than the completely equal sharing between atoms in covalently bonded elements. But the covalent bonding of dissimilar atoms depends upon having enough similarity to share effectively their common interest, the bonding electrons. If they can share electrons, they may be suitable reactants for covalent compound formation. If they do not share electrons effectively, the covalent bonding is lost. If, at the same time, they are not dissimilar enough to form favorable ionic bonding, the bonding available as separate elements may be stronger than the bonding in a compound and these atoms will not react.

The conditions discussed for suitable reactants, sharply dissimilar for ionic reactions or capable of sharing a common interest for covalent bonding, can be illustrated quite well by a social analogy, relationships between men and women. Interaction between a couple can lead to a favorable bonding reaction if strong attractions can be established. Sometimes the attractions can be built on sharp differences, which are accentuated by the contrasts between the two. The sharply polarized roles resulting may seem so totally different to outside observers that there would be no basis for interaction. But the attraction is to the strongly opposite character of the other party. It is based not on being equivalent but on emphasizing the differences. These are sharp, spicy relationships that can be considered analogous to the polar nature of ionic

Covalent bonding requires partners similar enough to share effectively.

bonding. We can note that ionic substances have (when concentrated) an irritating effect on unprotected body tissues. Polarized roles in a couple can also have irritating effects on many people. But if the differences are strong and natural, their emphasis can contribute a forceful vigor. It can only work for complementary pairs with wide natural differences. But if it fits, as the French would say, "viva la difference!"

Sharing type relationships, analogous to covalent bonding, are more common among couples (and more generally approved of by our society). Strong bonds can be formed based on common interests and sharing. But it is not necessary or desirable to remove all differences to get effective sharing. In fact, it is the clear remaining difference which provides the advantage for a couple (analogous to a compound) over sharing and common interests among a group of boys or a group of girls (analogous to two dissimilar elements). And there clearly is an advantage great enough to induce reactions. Platonic (brotherly and sisterly) relationships where response to the difference has been inhibited cannot form the same sort of bond as relationships with both sharing and a clear natural difference. The strongest bonds utilize both effective sharing and the advantages available through responding to the dissimilarity.

The earth's population serves as one clear evidence of the suitability of men and women as reactants. The variety and quantities of chemical compounds are evidence of reactants being suitable for many chemical reactions.

12–2.2 Contact: Rate Dependence on Concentration.

It is not enough to have reactants capable of a reaction. Absolutely no reaction takes place until the reactants are brought together. Bonds cannot form between atoms unless the atoms (or the molecules containing them) come in direct physical contact. Such physical contact requires two things, a fluid to permit movement of at least one species to the other, and the event of all the necessary parts reaching the same point together. The needed ability to move will have important effects on reactions involving solids, such as some reactions involved in iron and steel manufacturing which are described in Section 16–6.2. For the present let us assume our reactants are gases, liquids, or substances dissolved in liquids. The contact requirement for reaction is then bringing together all the necessary reactants.

Contact between the reactants is necessary for reaction.

If the reactants come together frequently, there will be many chances for reaction. If they come together only rarely, there will be few chances for reaction if the time for reaction is the same. To get the same amount of reaction in the second case would take a longer time. Therefore the **rate of reaction** (how fast it occurs) depends upon the frequency with which the reactants come together. The rate of reaction can be stated as the fraction

$$\text{rate} = \frac{\text{amount reacting}}{\text{time for reacting}}$$

The chance of the reactants coming together depends on their chances of being at the same spot. Each reactant has a chance of being at a given spot determined by its concentration, not just upon the total quantity. Putting the same number of molecules in half the space will increase the chance of being at a given spot just as much as doubling the number of molecules while the space remains constant. The chance of getting two reactants together will depend on both their concentrations. This dependence on concentration can be illustrated by returning to our analogy to reactions between men and women.

If we took all the unmarried young men and women (let us say ages 18 to 25) in a whole state, we would have a large group of "reactants" with a high potential for reaction in the process known as courtship and marriage. The state marriage license bureau would assure us that the process was proceeding at a fairly consistent rate. But this large group would be spread over a large area. The concentration would be fairly low. As a result, the number of contacts that could start the reaction would be limited and be mostly between young men and women from the same area. Some of these would not find a suitable matching "reactant" among the readily available contacts and either be left out of the reaction or react very slowly.

Increasing the concentration of young men and women would speed up the contacts and increase the rate of reaction. That could be done by putting in more reactants. Opening a new Army base could raise the rate of reaction of local girls. But a more likely solution would be to confine the available young men and women in a smaller space. In this example, let us assume that 10% of the eligible men and women enroll in the State University. The *quantity* on the university campus would be less than in the whole state, but the *concentration* would be much higher. The number of contacts each girl made with young men would *rise* with the concentration change, not fall with the drop in total quantity of men.

But a large university campus is still too big to allow meaningful contacts between all the possible combinations. A large fraction of the most "reactive" pairings must miss each other. Some campus socialites try to correct this by increasing contacts further. One method used to do this is to schedule parties. Then the concentration of young men and women (under conditions where contacts are possible) becomes really high. The rate of reaction rises. Actually, even in the low concentration situation of men and women over the whole state, most contacts and reactions occur in social settings where temporary local concentrations increase the rate. The advantages of higher concentrations are well recognized in our social patterns.

When we talk about reaction rate depending on concentrations, we must keep in mind that all the necessary reactant concentrations play a role. The chance of getting together is given by *multiplying* the concentrations of all of the involved reactants by each other. If two reactants must come together, the chance is [concentration$_1$][concentration$_2$]. If three reactants must come together the chance is [concentration$_1$][concentration$_2$][concentration$_3$]. If three species must come together, but two of them are the same (number 1), the chance is [concentration$_1$][concentration$_1$][concentration$_2$]. These dependences can be shown by a few simple examples.

> Reaction rates depend upon concentrations.

> The concentrations of all necessary reactants must be multiplied to get the overall chance for contact.

First, all reactants must be represented to have any reaction. For example, a dorm with 500 men and no women has no man-woman contacts at the dorm.

$$\text{rate} = \frac{500 \text{ men}}{\text{dorm}} \frac{0 \text{ women}}{\text{dorm}} = 0$$

Squeezing 1000 men into the dorm will increase their concentration but will not help.

$$\text{rate} = \frac{1000 \text{ men}}{\text{dorm}} \frac{0 \text{ women}}{\text{dorm}} = 0$$

But if the 500-man dorm shares a cafeteria with a dorm of 250 girls, there will be contacts in the cafeteria fitting the following rate:

$$\text{rate} = \frac{500 \text{ men}}{\text{cafeteria}} \frac{250 \text{ women}}{\text{cafeteria}} = 125{,}000 \text{ chances for contacts}$$

(Please pardon the units. We do not have the conversion factor to change to contacts/hour.) Then, if 1000 men are squeezed into the dorm, the rate will rise with the increase in concentration.

$$\text{rate} = \frac{1000 \text{ men}}{\text{cafeteria}} \frac{250 \text{ women}}{\text{cafeteria}} = 250{,}000 \text{ chances for contacts}$$

Similarly, doubling the girls to 500 would also double the rate

$$\text{rate} = \frac{1000 \text{ men}}{\text{cafeteria}} \frac{500 \text{ women}}{\text{cafeteria}} = 500{,}000 \text{ chances for contacts}$$

Notice that doubling both concentrations raised the rate by 4 times (2×2).

We should note here that the concentrations involved in reaction rates are only those needed at the key point in the reaction. For example, marriage could be thought to require three factors: a man, a woman, and a preacher (or other party to perform the wedding). But the concentrations of preachers could be halved or doubled without affecting the reaction rate. The preacher simply is not around at the really decisive step in the reaction. The key step is referred to as the **rate determining step.** Many real chemical reactions have rate determining steps that do not depend on the same concentrations we would expect from the overall reactants.

12-2.3 Activation Energy. So far we have considered the needs for things that could react and for getting the reactants together. But the substances also must be *suitably reactive* at the particular time and place of the contact. The requirements for suitable reactivity are enough energy (the main factor) and proper shape and position arrangements (the steric factors to be covered in Section 12-2.4). Most molecules are not suitably reactive. This has the effect of lowering the number of contacts that really matter or, taking another point of view, lowering the effective concentrations of reactants. Since the ability to react at a given time depends mostly on the energy of the molecules, we find it convenient to classify the reactants as active (those with enough energy to react) and inactive (those with too little energy to react). Only contacts between active reactants matter. The minimum energy required to be active is called the **activation energy** and represented by the symbol E_{act}.

Only those reactants with a certain minimum energy called the activation energy can react.

We can illustrate the problem by an example in our man-woman relationships analogy. Suppose 6 men and 6 women are invited to a party. There should by $6 \times 6 = 36$ chances for contacts. But one of the men is Poor Joe, who had a rough time the previous night. He was up all night studying for his chemistry quiz. That plus a cold have him exhausted, and his spirits weren't helped any when he couldn't stay awake during the quiz and blew the whole thing. He arrives at the party, ends up in a chair, and has nothing to offer except blank, glassy eyeballs all evening. He is below the activation energy for effective reaction. The chances for contacts are now down to $5 \times 6 = 30$. If Poor Joe is joined in his stupor by Frank and Eddie, who drink themselves under the table in the first hour, and the group includes Charlie, who has needed vitamins or Geritol or something for a long time, and Phil, who has been permanently deactivated by making the drug scene, the girls could have a pretty dull evening.

For most real chemical reactions the fraction of active molecules is even lower than the one out of six men remaining active in the example. As a result the rate of reaction is very sensitive to changes in the fraction of molecules that have the activation energy. Energy is clearly the most important factor in determining which chemical reactions are fast and which are slow. We will come back to this energy dependence in more detail in Sections 12–3.2 and 12–3.3.

12–2.4 Steric Factors.

Reaction rates can also be affected by geometry. Figure 12–2 illustrates a simple example where a change in arrangement affects the chance for reaction. The reaction whose rate is being considered is $2\ HI \longrightarrow H_2 + I_2$. These shape and position dependent effects are called **steric factors**. Steric factors may involve difficulty in getting at the point where reaction must occur. Specific shape requirements play an important role in limiting biochemical reactions in living cells to specific desired reactions.

Steric factors are shape or geometry effects that affect the chance for reactions to occur.

The common example of steric effects on man-woman interactions is female beauty. We will try for a little originality here and use the less common reverse for our analogy example. In looking over the field at a social get-together, the women might find a handsome, broad shouldered, robust masculine figure more interesting than the average male present. That increases the chances for effective contact (more girls are "active" when he is around) and makes conditions more favorable for reaction. This factor might not lead to a reaction, particularly if he has become overimpressed with his assets from the ease

Figure 12-2 Steric effect.

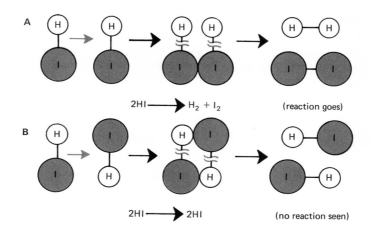

with which he gets attention. But the point still holds: some shapes are more suitable for reaction than others. That is a steric factor.

12-2.5 Rate Constant. At any given temperature, the energy factor (fraction of molecules with the activation energy), steric factor, and the factors for converting the concentrations into frequency of contacts can be grouped together into a single constant value, the probability for reaction (per unit time) for concentrations at unity. This constant is called the **rate constant** and represented by k. The reaction rate is therefore: rate $= k$(concentration terms)

The reaction probability factors, other than concentration, are grouped into the rate constant, k.

12-3 Changing Reaction Rates

12-3.1 Reaction Order and Concentration and Surface Effects. In Section 12-2.2 we showed how changes in concentration affect reaction rates. It is quite clear that any reaction can be speeded up by increasing the concentration of the reactants. (It is equally clear that the reaction rate slows down as the reactants are used up and the concentrations drop.) However, some concentrations affect the reaction rate more than others. For example, if the reaction $4\,HCl + O_2 \longrightarrow 2\,H_2O + 2\,Cl_2$ depended on all the reactants in its rate determining step, it would be much more sensitive to changes in HCl concentration than O_2 concentration. The rate of the reaction would be (where square brackets means concentration).

$$\text{rate} = k[\text{HCl}][\text{HCl}][\text{HCl}][\text{HCl}][O_2]$$

or

$$\text{rate} = [\text{HCl}]^4[O_2]$$

Writing in the exponential form ($[\text{HCl}]^4$) is a convenient way of expressing the concentration dependence. The exponent, 4 in the case of HCl, is especially important and is used to determine what is called the **order** of the reaction. If the exponent is one, the reaction is **first order,** an exponent of two is **second order,** and so on. We can speak of the order in one reactant (its exponent) or the overall order of the reaction (the sum of all the exponents). The HCl plus O_2 reaction, if as described above, would be fourth order in HCl, first order in O_2, and fifth order overall.

Orders do not have to be whole numbers. For instance, the reaction $H_2 + Br_2 \longrightarrow 2\,HBr$ occurs with a rate determining step that is not $H_2 + Br_2$. The complex reaction results in rate $= k[H_2][Br_2]^{1/2}$, a dependence on $[H_2]$ and on the square root of $[Br_2]$. The reaction is therefore first order in H_2, $\frac{1}{2}$ order in Br_2 and $\frac{3}{2}$ order overall.

In every case, the higher the order of a given reactant, the more dependent the reaction is on the concentration of that reactant.

In reactions involving solids, concentration is usually not the important factor. If the solid is a pure substance, it always has the same concentration. However, only those species on the surface of the solid can react with the surrounding gas or liquid. Therefore, the reaction rate depends upon the amount of surface on the solid, not on concentration of the solid. For example, the rate at which NaCl can dissolve in water would be given by

$$\text{rate} = kS$$

where S equals surface area of the solid. If the salt was broken up into a fine powder instead of large salt crystals, the surface area would be increased and it would dissolve in water more rapidly.

Surface can also be important in reactions between substances dissolved in two different, immiscible liquids.

> The order of a reaction for any given substance is the exponent needed in the rate expression for the concentration term of that substance, and overall order is the sum of all the exponents.

12-3.2 Temperature Effects.

Reaction rates can also be changed by changing the temperature. Temperature is a measure of the average energy of molecules (Section 9-4) and energy is the most important factor determining whether reactions are fast or slow (Section 12-2.3), so a dependence of rate on temperature seems reasonable. But the rate of reaction does not depend upon the *average* energy; it depends upon the *fraction* of molecules having a certain minimum energy, the activation energy. Therefore reaction rate does not change in the same proportion as the temperature change. To understand how reaction rates vary with temperature we must know something about the energies of individual molecules and the size of the activation energies for reactions.

At any given temperature, the molecules of a substance will have a wide variety of energies of motion. Some will have little or no energy whereas others have very high energies, but the molecules of the whole substance average out to an average energy proportional to the absolute temperature. In the collisions going on between molecules, some of the molecules near average energy will lose energy and become low energy molecules and others will pick up energy to become high energy molecules. At the same time the

Figure 12–3 Distribution of energies.

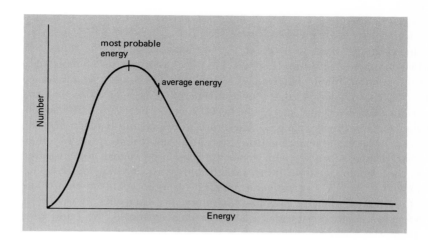

The distribution of molecular energies fits a curve determined by statistical laws.

low energy molecules are more likely to gain energy than lose it in collisions and the high energy molecules are more likely to lose energy. The result is gains and losses of energy by individual molecules but a constant distribution of molecular energies dictated by the average energy and the statistical laws of random chance. (To fit statistical chance a large number of units are needed, and any sample of material large enough to see and work with has an appropriately large number of molecular units.)

Figure 12–3 shows the energy distribution that occurs because of the statistical random chance variations. Figure 12–4 shows the same graph for one temperature, T_1, and the distribution for a higher temperature, T_2. In each case, the number of molecules at any given energy rises from none at zero energy to a maximum (at an energy near the average energy, but actually only two thirds as much as the average energy) and then drops with increasing energy. As the energy goes up, the number of molecules with that energy approaches closer to zero but never actually gets down to zero. Even at very

Figure 12–4 Temperature dependence of energy distribution (T_2 is a higher temperature than T_1).

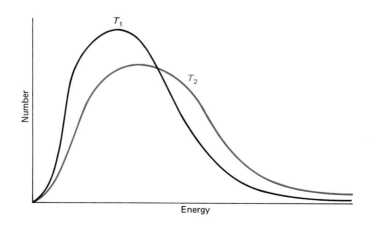

[12–3] Changing Reaction Rates 245

high energies, there is some chance (but very small) for a molecule of that energy to be present in the distribution.

At this point we need to consider how various values for the minimum energy requirement (activation energy) will affect the fraction of molecules capable of reaction. Figure 12–5 shows several possible choices for activation energies. If, like choice A, the activation energy is zero or close to zero, all molecules can react and there will be little or no effect on that fraction when the temperature is changed. There are a few chemical reactions like that, but only between very rare (low concentration) species made by other reactions. For any substance with concentrations large enough for measurement in the common quantity ranges, collisions at room temperature occur so quickly that such a zero activation energy reaction would be completed in less time than the fastest possible laboratory measurements. It would be over before we could be sure the reactants had been mixed. We know that most chemical reactions, such as burning of fuel, are not that rapid.

If the activation energy was similar to the average energy, as in choice B of Figure 12–5, the fraction that could react would change with a temperature change. But the fraction reacting would still be so large that the reaction would be much faster than the chemical reactions we find and study. Common reactions must have high activation energies and therefore very low fractions of molecules that can react. If the activation energy is too high, as in choice D, the reactive fraction is so small that millions of years or more would be required for significant amounts of reaction. Choice C, where the activation energy is 100 times as large as the average energy, is in the range of real reactions, fast enough to be found and slow enough to be measured.

Figure 12–6 shows how the fraction of molecules exceeding the activation energy changes as the temperature changes. A relatively small temperature increase, enough to increase the average energy by about 3.3%, causes a large increase in the fraction that can react. The exact percentage of the increase depends upon the temperature and the activation energy for the particular case. For the chemical reactions most commonly observed, the reaction rate is usually about *doubled* for each *10° rise in temperature* (using Kelvin sized degrees). The 3.3% increase in energy shown in Figure 12–6 was chosen to fit a 10° temperature increase from room temperature (about 25°C).

Real reactions usually have activation energies much higher than the average molecular energies.

Figure 12–5 Effect of different activation energies.

Activation energy choices: A, very low activation energy; B, activation energy equal to average energy at T_1; C, activation energy equal to 100 times average energy at T_1; D, activation energy equal to 300 times average energy at T_1.

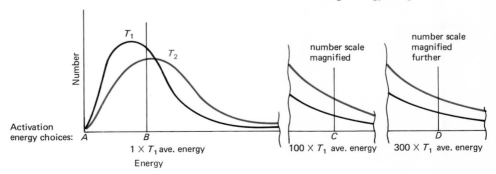

Figure 12-6 Effect of temperature increase on reaction rate.

At temperatures near room temperature a 10K rise in temperature is about a 3.3% rise in average energy, but this may cause a 100% increase in the fraction of molecules with more than the minimum energy needed for reaction.

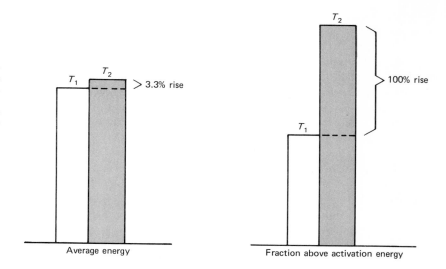

Because the fraction of active molecules is small, reactions are very sensitive to temperature changes and energy is the most important single factor in reaction rates.

Measured on the absolute (Kelvin) scale needed to relate temperatures to average energy, the increase is from 298 K to 308 K. Figure 12-6 shows how this would double the rate of a typical reaction by doubling the molecules with the activation energy. The case shown would fit only if the reaction required only one active molecule (a first order reaction). But the general principle is also true for other reactions. There is approximately doubling of reaction rate for each 10° because the fraction of active molecules goes up faster than the average energy.

Although increasing temperature makes any reaction go faster, it may not increase the amount of product produced. There may be other reactions using up the product (see Section 12-4), and those reactions will also be speeded by heating.

Catalysts can speed reactions by lowering the activation energy.

12-3.3 Catalysts. Reaction rates can also be increased by addition of catalysts. Catalysts are substances that make possible another reaction path which has a lower activation energy. The fraction of molecules that can react is increased, and therefore the reaction becomes faster.

A catalyst must play an active role in the reaction, but (since only the reactants are changed by the reaction) the catalyst must reappear in its original form when the reaction is completed.

We can illustrate the role of a catalyst by an example from our analogy to reactions between men and women. Low lights and soft music are a simple catalyst. The reaction does not change. It still requires reactants (man and woman), contact, and suitable reactivity. But the requirements for suitable reactivity have changed. If all other factors are suitable, the fraction energetic enough to react is increased by lowering the resistance to reaction. The activation energy has been lowered. If a reaction does occur, the catalyst is still available to assist later reactions. The same dark room and records could be used again. If a catalyst is effective, it may be a determining factor in setting the reaction rate. The rate of man-woman interactions might depend upon the concentrations of men, of women, and the concentration of secluded spots to act as catalysts.

Reactions can also be speeded by increasing the chance of reactants getting together. Sometimes surfaces act in that way by absorbing reactants and allowing them to move around on the surface. The adsorption on the surface concentrates the reactants and may therefore increase the chance of all necessary reactants getting together. These same surfaces may also lower the activation energy for reaction. Many important commercial processes use solid catalysts which work through such surface effects. The role of surfaces in getting reactants together, often under circumstances that also lower the activation energy, could be compared to the role played by matchmakers in man-woman reactions.

12-3.4 Inhibitors.

Substances that interact strongly with reactants do not always speed reactions. Good catalysts interact strongly to tie reactants up in forms that can go ahead and complete the reaction more easily than the original reaction process. But other similar materials may interact strongly to tie up reactants in forms that do not go on to complete the reaction. That has the effect of lowering the concentration of free reactants or of increasing the activation energy to include the additional energy necessary to break up the nonreactive combinations. The reaction is then slowed. Substances with this slowing effect are called inhibitors. They are essentially negative catalysts. We use many inhibitors to slow reactions we do not want. Examples are paint to protect surfaces from weathering reactions, rust inhibitors for car radiators and other applications, and food preservatives. In the case of food preservatives we are probably not tying up the reactants but instead interfering with the agents (such as bacteria) that could have caused other reactions.

An example of an inhibitor in man-woman relations would be an engagement that eventually broke up. While the engagement was on the reactants were removed from the possibility for other reactions or at least the activation energy for other reactions should be higher. That slows the rate of reactions reaching completion.

Inhibitors slow reactions by tying up reactants in nonreactive forms.

Some processes in nature can only occur in one direction. But other processes, including many chemical reactions, can occur in both a given way and the reverse of the first way. These processes are called **reversible** processes. A chemical reaction occurring in the direction it is written is called a **forward reaction.** The reaction going in the opposite direction is called a **back reaction.**

In man-woman relationships, engagement would be a forward reaction and

12-4 Back Reactions

Many chemical reactions are reversible, allowing back reactions to occur.

a broken engagement would be the back reaction paired with engagement. Similarly marriage would be a forward reaction matched with divorce as the back reaction. In the case of marriage, the process is supposed to go only one way (be **irreversible**), but it is clear that at least some of the marriages in our society have been reversible. Chemical reactions that seem quite final and complete also turn out to be at least slightly reversible upon close examination, so the remaining sections in this chapter (which fit only reversible processes) will be applicable to all of chemistry.

Back reactions are subject to the same types of constraints as forward reactions.

Back reactions are subject to the same limitations as forward reactions. They depend upon the concentrations of their reactants (which happen to be the products of the forward reactions). For example, the rate for broken engagements would depend upon the concentration of engaged couples who might break up. It might also depend upon other concentrations if they are involved in the key step. The concentrations of unattached men and women offering alternative choices might have an effect in some cases.

Back reactions also have an activation energy. The activation energies for the forward and back reactions will be related to each other. Whatever path was best (lowest activation energy) in the forward direction will also be best for the back reaction, and the activation energies will differ only by the amount of the energy difference of the original state (reactants) and the final state (products). Figure 12-7 shows this relationship between activation energies.

A catalyst must lower the activation energy for the back reaction by the same amount that it lowers the activation energy for the forward reaction.

Figure 12-7 also shows the effect of a catalyst on the activation energies. Back reactions can be catalyzed and, in fact, any catalyst that lowers the activation energy for reaction in one direction must lower the activation energy for the reverse reaction by the same amount. For example, low lights and soft music might catalyze a back reaction broken engagement (by aiding response to other unattached men and women) by as much as it catalyzed the forward reaction. If the product has a real advantage (more favorable binding energy), the back reaction will have a higher activation energy than the forward reaction had and back reaction may be very slow. But the fact remains: anything that helps the forward reaction helps the back reaction too.

Figure 12-7 Activation energies of forward and back reactions.

Black line: minimum energy path for uncatalyzed reaction. Colored line: minimum energy path for catalyzed reaction.

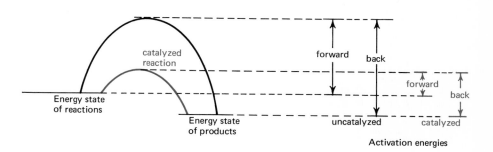

[12-4] Back Reactions 249

If a forward reaction and its back reaction are allowed to continue long enough, they will eventually reach a condition where they balance each other. Product is formed and used up at the same rate and there is no more net change. This condition is called an equilibrium. Equilibrium occurs when the concentrations of reactants and products are in the correct proportions to make the forward and back reaction rates equal. In Section 9–9 we discussed the direction of shifts in equilibrium with changing conditions; now we need to discuss the precise equilibrium condition.

Figure 12–8 shows how a reaction will naturally adjust its concentrations to reach the equilibrium condition. Figure 12–8 also shows that neither forward nor reverse reaction stops when equilibrium is reached; the condition is a balance between the reactions. The faster reaction in terms of rate constant must have lower concentrations of its reactants to slow it enough to just balance the other reaction. Since the faster reaction in terms of rate constant is the one with the lower activation energy, that means the material (reactants or products) with the higher energy state (less binding) must go to lower concentrations. Most of the material is therefore in the low energy (more strongly bonded) state when equilibrium is established.

When equilibrium is reached, the rates of the forward and back reactions must be equal,

$$\text{rate}_{(forward)} = \text{rate}_{(back)}$$

12–5 Equilibrium

Chemical equilibrium is established when the rates of forward and back reactions are equal.

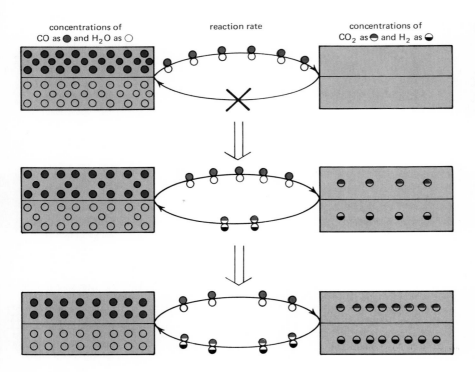

Figure 12–8 Reaching equilibrium.

An idealized diagram of the approach to equilibrium in the reaction of carbon monoxide (CO) with water vapor (H_2O) to form carbon dioxide (CO_2) and hydrogen (H_2).

We know that the rates of the forward and reverse reactions are given by their rate constants times the important concentration terms. Giving the rate constants the names k_1 and k_2 so we can tell them apart, we get

$$\text{rate}_{(\text{forward})} = k_1(\text{concentration terms forward reaction})$$
$$\text{rate}_{(\text{back})} = k_2(\text{concentration terms back reaction})$$

Because the rates are equal at equilibrium, we can see that the rate constant times concentrations expressions are also equal.

$$k_1(\text{concentration terms forward reaction})$$
$$= k_2(\text{concentration terms back reaction})$$

We can rearrange this equation so that all of the constants are on one side and all the concentrations are on the other side. Dividing both sides by k_2 and by the concentration terms for forward reaction gives us

$$\frac{k_1}{k_2} = \frac{(\text{concentration terms back reaction})}{(\text{concentration terms forward reaction})}$$

The ratio k_1/k_2 is a constant at any given temperature; k_1/k_2 is given a new symbol, K, and a new name, the **equilibrium constant.** The ratio of concentration terms becomes a simple, predictable combination of concentrations. The rule for predicting that combination of concentrations is called the law of mass action.

12–6 Law of Mass Action

Reaction rates depend upon concentrations in many complex ways, but the dependence of the equilibrium constant upon concentrations is always simple. A few examples will show that this is true and give us some understanding of why it is true.

First, let us consider a reaction that depends simply on the concentrations of reactants for the forward reaction and on the concentrations of products for the back reaction. Assume that the reaction of hydrogen and iodine is such a reaction.

$$H_2 + I_2 \rightleftharpoons 2\,HI$$
$$\text{rate}_{(\text{forward})} = k_1[H_2][I_2]$$
$$\text{rate}_{(\text{back})} = k_2[HI][HI] = k_2[HI]^2$$

at equilibrium

$$k_1[H_2][I_2] = k_2[HI]^2$$

and

$$\frac{k_1}{k_2} = \frac{[HI]^2}{[H_2][I_2]}$$

so

$$K = \frac{[HI]^2}{[H_2][I_2]}$$

The equilibrium constant, K, equals the concentration of the product, HI, raised to the exponential power 2 (which is its coefficient in the balanced equation) divided by the concentrations of the reactants, H_2 and I_2, with each having the exponential power 1 (which is its coefficient in the balanced equation). That relationship is the law of mass action.

The **law of mass action** is

$$K = \frac{\text{concentrations of products, each with their coefficient as exponent}}{\text{concentrations of reactants, each with their coefficients as exponent}}$$

Law of mass action: At equilibrium the ratio of the concentrations of products, each with their coefficient from the balanced chemical equation as exponent, over the concentrations of reactants, each with their coefficients as exponents, is a constant (at any fixed temperature) called the equilibrium constant, K.

If the same reaction was speeded by a catalyst, the rate would change and also depend upon the catalyst. Let us assume that a metal surface acts as a catalyst and S represents the amount of metal surface. The forward reaction rate is then (using k_1' for the new, larger rate constant).

$$\text{rate}_1' = k_1'[H_2][I_2] \times S$$

But the back reaction will also be speeded by the same surface acting as a catalyst. Using k_2' for the new larger rate constant, the back reaction rate is then

$$\text{rate}_2' = k_2'[HI]^2 \times S$$

at equilibrium

$$k_1'[H][I] \times S = k_2'[HI]^2 \times S$$

and

$$\frac{k_1'}{k_2'} = \frac{[HI]^2 \times S}{[H_2][I_2] \times S} = \frac{[HI]^2}{[H_2][I_2]}$$

so

$$K = \frac{[HI]^2}{[H_2][I_2]}$$

The law of mass action still holds. The S term cancels out because it affects both forward and reverse reactions. In fact anything that affects one will affect the other. The reactants and products are the only things that can affect the two reactions differently and therefore they are the only concentrations remaining when temperature is constant. (Since heat is also a reactant or product in real reactions, changes in T, which is the concentration of heat, will affect the equilibrium. The T effects are included in the constant, K, and cause K to have a different value at each different temperature.)

The law of mass action holds for any equilibrium condition.

Even unusual dependences of reaction rate on concentrations of reactants will not affect the law of mass action. If the reaction

$$4\ \text{HCl} + \text{O}_2 \rightleftharpoons 2\ \text{H}_2\text{O} + 2\ \text{Cl}_2$$

had a rate dependence

$$\text{rate}_1 = k_1 [\text{HCl}]^2 [\text{O}_2]$$

the back reaction would have a rate dependence

$$\text{rate}_2 = k_2 \frac{[\text{H}_2\text{O}]^2 [\text{Cl}_2]^2}{[\text{HCl}]^2}$$

and the equilibrium constant would be

$$K = \frac{[\text{H}_2\text{O}]^2 [\text{Cl}_2]^2}{[\text{HCl}]^4 [\text{O}_2]}$$

as expected from the law of mass action. The above statements can be verified by considering the rate determining step $2\ \text{HCl} + \text{O}_2 \longrightarrow$ (unit of $2\ \text{HCl} + \text{O}_2$), and a separate fast equilibrium reaction, (unit of $2\ \text{HCl} + \text{O}_2$) + $2\ \text{HCl} \longrightarrow 2\ \text{H}_2\text{O} + 2\ \text{Cl}_2$. The first reaction gives

$$K_1 = \frac{[(\text{unit of } 2\ \text{HCl} + \text{O}_2)]}{[\text{HCl}]^2 [\text{O}_2]}$$

The second reaction gives

$$K_2 = \frac{[\text{H}_2\text{O}]^2 [\text{Cl}_2]^2}{[\text{HCl}][(\text{unit of } 2\ \text{HCl} + \text{O}_2)]}$$

The K_2 equation can be rearranged to

$$[(\text{unit of } 2\ \text{HCl} + \text{O}_2)] = \frac{[\text{H}_2\text{O}]^2 [\text{Cl}_2]^2}{[\text{HCl}]^2 \times K_2}$$

When that is substituted for $[(\text{unit of } 2\ \text{HCl} + \text{O}_2)]$ in the K_1 equation, we get

$$K_1 = \frac{[\text{H}_2\text{O}]^2 [\text{Cl}_2]^2}{[\text{HCl}]^2 \times K_2} \bigg/ [\text{HCl}]^2 [\text{O}_2]$$

$$= \frac{[\text{H}_2\text{O}]^2 [\text{Cl}_2]^2}{[\text{HCl}]^2 \times K_2 \times [\text{HCl}]^2 [\text{O}_2]} = \frac{[\text{H}_2\text{O}]^2 [\text{Cl}_2]^2}{K_2 \times [\text{HCl}]^4 [\text{O}_2]}$$

multiplying both sides by K_2

$$K_1 K_2 = K = \frac{[\text{H}_2\text{O}]^2 [\text{Cl}_2]^2}{[\text{HCl}]^4 [\text{O}_2]}$$

The above exercise is unnecessary if we use the law of mass action, which is always true. We can simply say that from the balanced equation $4\text{ HCl} + O_2 \rightleftharpoons 2\text{ H}_2\text{O} + 2\text{ Cl}_2$ the law of mass action tells us that

$$K = \frac{[\text{H}_2\text{O}]^2[\text{Cl}_2]^2}{[\text{HCl}]^4[\text{O}_2]}$$

Remember

> The law of mass action is
>
> $K = \dfrac{\text{concentrations of products, each with their coefficient as exponent}}{\text{concentrations of reactants, each with their coefficients as exponent}}$

Skills Expected at This Point

1. You should be able to list the factors required for reactions to occur, define activation energy, and point out energy as the most important limiting factor in most real reactions.
2. You should be able to define and use the terms rate constant, order of the reaction, and catalyst and be able to list the ways a reaction can be made faster or slower.
3. You should be able to describe the balance between opposing reactions at equilibrium.
4. You should be able to state the law of mass action and use it to write the equilibrium constant expression for any reaction.

Exercises

1. List three types of conditions that must be satisfied before two molecules can react.
2. What effect could each of the following be expected to have on the rate of a typical gas phase reaction which is first order in A, first order in B, and second order overall?
   (a) a 10°C temperature rise
   (b) doubling A (B constant)
   (c) tripling B (A constant)
   (d) compressing the gases to one fourth the original volume
   (e) doubling the pressure by adding a third gas, helium
3. Given the following data (in each case, everything else is held constant)

   double X—reaction rate doubles
   double Y—reaction rate quadruples
   double anything else (other than X or Y)—no change

   Write the expression for the rate of reaction and identify the terms.
4. Write the equilibrium constant expression for the equilibrium reaction,

$$PCl_3(g) + Cl_2(g) \rightleftharpoons PCl_5(g) + \text{heat},$$ and state how the equilibrium and the equilibrium constant would be affected by each of the following:

Change	Effect on Equilibrium	Effect on K
Temperature rise		
Adding Cl_2		
Compressing the gases		
Adding H_2		

5. Write equilibrium constant expressions for the following reactions. (*Note:* All products and reactants are listed, but the equations may not be balanced. (aq) means in water solution.)
   (a) $H_2(g) + Cl_2(g) \rightleftharpoons 2\ HCl(g)$
   (b) $H_2(g) + O_2(g) \rightleftharpoons H_2O(g)$
   (c) $H_2S(g) + O_2(g) \rightleftharpoons H_2O(g) + SO_2(g)$
   (d) $S_8(g) + O_2(g) \rightleftharpoons SO_2(g)$
   (e) $5\ Fe^{2+}(aq) + MnO_4^-(aq) + 8\ H^+(aq) \rightleftharpoons 5\ Fe^{3+}(aq) + Mn^{2+}(aq) + 4\ H_2O$

Test Yourself

1. Given a reaction with rate $= k[NO]^2[O_2]$
   (a) List *all* the facts that this statement can tell you about the order of the reaction
   (b) What does the letter k in the equation represent?
   (c) What does a large value for k do to the reaction?
   (d) What is the most important factor in determining the size of the value for k?
2. State the law of mass action briefly and show what it means by using an example (one reaction).
3. (a) What is a catalyst? (b) How do most catalysts work?
4. At room temperature, a $10°C$ rise in temperature is about a 3% temperature rise. For most observed chemical reactions, a $10°C$ rise in temperature will
   (a) not affect the reaction rate.
   (b) slow the rate of reaction.
   (c) raise the reaction rate by less than 3%.
   (d) raise the reaction rate by 3%.
   (e) raise the reaction rate by much more than 3%.
5. The equilibrium constant expression

$$K = \frac{[A]^2[B]^3}{[C][D]}$$

fits the equation:
(a) $2A + 3B \rightleftharpoons C + D$
(b) $C + D \rightleftharpoons 2A + 3B$
(c) $C + D \rightleftharpoons A + B$
(d) $A + B \rightleftharpoons C + D$
(e) $CD \rightleftharpoons A + B$

6. Complete and balance the equation for the gas phase reaction of H_2S with O_2 to form H_2O and SO_2. Use your result to identify which of the following is the correct equilibrium constant expression for the reaction.

(a) $K = \dfrac{[H_2O]^2[SO_2]}{[H_2S][O_2]^2}$

(b) $K = \dfrac{[H_2O]^2[SO_2]^2}{[H_2S]^2[O_2]^3}$

(c) $K = \dfrac{[H_2O][SO_2]}{[H_2S][O_2]}$

(d) $K = \dfrac{[O_2][H_2S]}{[H_2O][SO_2]}$

Part III

Common Things in the World Around Us

Water

13-1 Abundance and Transport Cycle

Water is a common and important compound.

Water interacts with its surroundings as it moves about in the water transport cycle.

Human activities interfere with the normal water transport cycle.

Water is the most important single chemical compound. It covers about three fourths of the Earth's surface as rivers, lakes, and oceans. It covers substantial amounts of land as snow and ice. It also permeates soil and rock formations and is an important component of the atmosphere. The unique properties of water are a dominant factor in determining the nature of the Earth's crust and atmosphere. All life as we know it is part of this water-dominated system.

Movement is very important to the physical and chemical effects of water on the Earth's crust and life. Under the pressure and temperature conditions of Earth, water exists in all three phases: solid, liquid, and gas. As we shall note in Section 13-3, that is so only because of some of the unique characteristics of the water molecule. The conversion of water into the gaseous state and then back to liquid or solid starts a transport cycle that moves tremendous quantities of water. As it moves, water interacts with its surroundings. It supports life, both by direct use and by carrying vital substances. It releases much energy of motion. It dissolves many substances and moves tremendous quantities of material, dissolved and not dissolved, to new locations. It changes the climate in some areas. The chemical nature of water determines the physical characteristics that control these effects.

Sometimes man is not satisfied with the natural water transport cycle. Desert areas would produce for man if more water was supplied. Low areas could be used by man if excess water was kept out. The energy of water motion could be used to do work. As changes are made or contemplated, it becomes increasingly important for us to understand the properties of water and its transport system. A thorough understanding would require extensive study in other areas, such as geology. Our study with references to some chemistry should provide the basic insights needed to approach this subject, and any later studies of it, intelligently.

13-2 Unusual Properties of Water

Many thousands of compounds have solid forms that are more dense than their liquid form. Therefore, as a "normal" material like a parafin wax melts, the solid remains at the bottom of the vessel. Water is the *only* common compound that behaves differently. Ice is less dense than liquid water and therefore rises to float at the top. This is only one of a group of properties in which water is uniquely different. We will list some of the more important ways in which water is unique and then consider the explanation and consequences of this uniqueness.

1. *Ice floats on water.* Normal solids are more dense than their corresponding liquid and therefore sink.
2. *Liquid water has a maximum density at 4°C.* Below that temperature, the liquid density falls as temperature falls. Normal liquids always increase in density as temperature falls.
3. *The melting point and normal boiling point of water are surprisingly high* for a compound with such a low molecular weight (18). By contrast, CH_4, (mol. wt. 16) has a melting point of $-182.5°C$ and a normal boiling point of $-161.5°C$, and these values agree with general trends much better than the values of $0°C$ and $100°C$ for H_2O do.
4. *Melting, warming and evaporating water all require unusually large amounts of heat* compared to other compounds. For example, H_2O requires 540 cal for each gram vaporized at its normal boiling point of $100°C$, whereas SO_2 requires only 94.9 cal for each gram vaporized at its normal boiling point of $-10°C$.
5. *Water is an effective solvent* for many ionic compounds that do not dissolve in most other covalently bonded liquids.
6. *Water is a relatively viscous liquid,* resistant to flow. Although some compounds with large molecules are much more viscous than water, most compounds with small molecules flow more easily and rapidly than water. It is much easier to pump gasoline through a small pipe than it is to pump water through the same pipe at the same speed.

The above properties lead to many other interesting characteristics of water, some of which will come up in our discussion later in this chapter or to which references have already been made in Chapter 10. Since we see more water than all other liquids combined, many of these characteristics seem completely natural to us. Sometimes it is hard for us to recognize anything unusual, and therefore it is hard to appreciate how the entire nature of our world depends upon water *not* following the same patterns as other compounds. Let us try to sharpen our sensitivity to the importance of the unique character of water by explaining how the well-known process of ice skating can happen. To get everything in proper perspective, we should start with a simple statement of fact that you may find hard to believe.

Ice is not very slippery.

An arctic Eskimo would be able to accept that fact better than the rest of us. He would know that in extremely cold weather things do not slide easily on the solidly frozen ice. It is only as the ice warms to temperatures nearer the melting point that sharp runners can slide smoothly over ice.

As the temperature rises very close to the melting point, even a dull object

---

Water is the only common compound whose solid is less dense than its liquid.

Liquid water has a maximum density at 4°C.

The melting and boiling points of water are surprisingly high.

Melting, warming, and evaporating water all require unusually large amounts of heat.

Water is an effective solvent for ionic as well as covalent compounds.

Water is a relatively viscous liquid.

Ice is slippery because a layer of liquid water is formed when pressure is applied.

like the bottom of a shoe or boot slides easily. If the Sun melts the surface to leave a thin film of water on top of the ice, even a wide, light object like a wood shaving or piece of paper can slide freely about, pushed by the gentlest of breezes. Actually, the last example is what is really happening in each case—the objects are all sliding on thin layers of *liquid water*, not solid ice. The layer of water beneath an ice skate runner forms because of the special density relationship of ice to water. When it becomes so cold that nothing can cause the water layer to form, the easy sliding characteristic disappears.

Let us examine the reasons for formation of a water layer on ice under a sharp skate runner (or under a dull boot) and contrast that with what would happen to a "normal" material. For any compound, if we know both the temperature and the pressure, we can say definitely whether it must exist as a liquid, solid, or gas, or if it is on a borderline where two of these phases can exist simultaneously. The results can be summarized in a phase diagram like the one for benzene shown in Figure 13-1. (See Section 9-15 for a discussion of phase diagrams.)

In this typical phase diagram for a "normal" compound we see that the line for liquid-solid transition slopes toward the liquid region as the pressure goes up. The slope has been greatly exaggerated in the figure to make it more noticeable. If we take a sample of benzene at low pressure and a temperature just above the freezing point and raise the pressure while carefully holding the temperature constant, we will eventually cross the liquid-solid transition line as shown in Figure 13-1. The benzene must then exist as a solid instead of a liquid. Therefore the liquid benzene will be solidified by application of pressure. In this process, the benzene becomes more dense and occupies less space. The conversion of liquid to solid thus allows the benzene to give way somewhat as the pressure is applied, as expected from Le Chatelier's principle (Section 9-9).

For water the effect is exactly the opposite, as shown in Figure 13-2.

Figure 13-1 Phase diagram of benzene.

The arrow in part B of the figure shows the shift from liquid to solid phase that occurs when liquid benzene near the melting point is put under increased pressure.

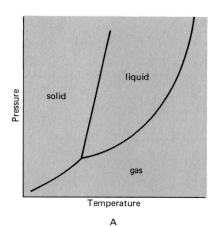
A

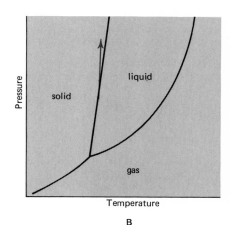
B

[13-2] Unusual Properties of Water 261

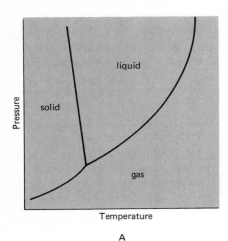 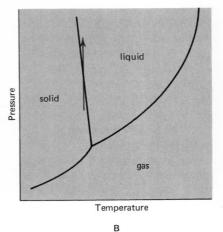

A B

Figure 13-2 Phase diagram of water and ice skating effect.

Because liquid water is denser than ice, the melting point line slopes in the opposite direction from "normal" phase diagrams. As a result, increasing pressure can cause the solid form (ice) to melt if the solid was near the melting temperature. The arrow in part *B* of the figure shows this shift. The liquid that forms when pressure is applied causes slipperiness. The pressure-induced slipperiness is the ice skating effect.

Because water ice is less dense than liquid water, the only way water could "give way somewhat" as pressure is applied would be to convert solid water (ice) to liquid water, which would occupy less space. That is exactly what happens. To fit this trend, the liquid-solid transition line must slope toward the *solid* region as pressure rises. This trend, which is greatly exaggerated in Figure 13-2 to make it more noticeable, occurs for water alone among common compounds.

Ice skating works because applied pressure can convert some ice into a slippery liquid layer. For moving objects, friction also aids slipperiness substantially by warming the surface. The heating by friction is probably the more important factor causing moving objects to slide in almost all cases. The pressure effect, however, can be important in allowing a nonmoving object to begin sliding and in lowering the friction needed to continue sliding. The greater the pressure (from greater weight or less surface—like under a sharp skate blade), the lower the temperature at which the liquid can still form. When the water cannot be formed, the ice surface loses its special slipperiness.

The phase diagram of water is different from those of normal compounds.

Heating by friction helps form the slippery liquid layer on ice.

13-3 Hydrogen Bonding and the Structure of Water

We represent water with the formula H_2O and the Lewis dot diagram structure

$$H\!:\!\overset{..}{\underset{..}{O}}\!:$$
$$H$$

In this structure we expect to find four pairs of electrons in the outer shell around oxygen, two of which are used for covalent bonds to hydrogens and two of which are "lone pairs" not involved in bonding. These four pairs of electrons could be arranged in a tetrahedral pattern around the oxygen for maximum separation of negative charges. However, presence of protons

Water molecules are polar.

bonded to two of the four pairs causes an electrical imbalance in the molecule. This makes water molecules *polar* molecules with one region (including the two protons) having a net positive charge, and the other side of the molecule (including the two lone pairs) having a net negative charge. Because of this, the two pairs used in bonds are less strongly repelled from each other (because they are somewhat neutralized by the attached positively charged hydrogen nuclei) than the lone pairs are. This distorts the tetrahedral arrangement somewhat, forcing the hydrogens closer together, at an angle of about 105° instead of the regular tetrahedral angle of 109°. This structure of individual H_2O molecules is a good picture of water as a gas, where the molecules are far apart from each other.

In liquid or solid phases the H_2O tetrahedral geometry sets up strong forces that lead to the unique properties of water. The oxygen is surrounded by four regions of charge, two ending in positively charged hydrogen nuclei and the other two ending in the unshared electron pairs. As H_2O molecules come together, a positive region on one molecule can be attracted to a lone pair on another molecule. The hydrogen nucleus then ends up between two pairs of electrons, the pair originally bonding it to oxygen and the lone pair from the neighboring oxygen. This sharing of a hydrogen nucleus between electron pairs is called a **hydrogen bond.** It is somewhat comparable to the sharing of an electron pair between two positive electrical charges (the atomic nuclei) in a covalent bond. The energy released when these hydrogen bonds form accounts for the very high heat of vaporization of water and the high normal boiling point. The molecules can simply be held together in liquid phase against higher temperatures which provide motions violent enough to tear less strongly bonded molecules apart.

Hydrogen bonding and the geometry of H_2O molecules explain the unusual properties of water.

As the temperature of liquid water is lowered, the molecules arrange themselves into a more regular pattern so more hydrogen bonds can be formed. The formation of these extra hydrogen bonds releases additional heat. That extra heat explains why water requires more heat when it is warmed and releases more heat per degree on cooling than liquids without hydrogen bonding. Additional heat can be released when enough hydrogen bonds are formed to fix the molecules into a rigid solid structure. That explains the high heat required to melt ice (or released when water freezes) and the relatively high temperature at which freezing occurs. At all temperatures, liquid water is therefore not really individual H_2O molecules but a mixture of polymers, $(H_2O)_x$, held together by hydrogen bonds. The lower the temperature, the higher the average value for x becomes until finally all H_2O's are connected in one huge unit, which is a solid.

Liquid water is a group of hydrogen bonded aggregates (or polymers) which becomes bonded into larger units as the temperature is lowered.

To get the maximum number of hydrogen bonds in solid phase, it is necessary to arrange the molecules in a particular geometrical pattern. If this pattern became completely regular (which could only happen for ice at a very low temperature), each oxygen would be surrounded by four hydrogens, two from its original bonds and two from hydrogen bonds to other H_2O units. Each hydrogen would be in a hydrogen bond between two oxygens. The geometry around each oxygen would be tetrahedral, and the geometry around each hydrogen would be linear—in each case giving the maximum possible separation of the groups of electrons.

If we begin with one hydrogen sticking straight out of the paper toward us, the oxygen behind it has three other hydrogens attached, each on bonds angling into the paper, as shown in Figure 13–3, and each leading to another oxygen farther along in those directions. Each of these oxygens has three more hydrogens attached (and oxygens farther along in those directions). If we set one of these in each case as being aimed directly into the sheet, the other two (plus the one connecting to the original oxygen) angle up out of the sheet toward us, as shown in Figure 13–4. Each of the oxygens at the end of these angles has the same kind of arrangement as the very first oxygen—one hydrogen straight up and three angling down into the paper (one of which is the hydrogen we just put in to attach that oxygen). In each case, one of these hydrogens angling down into the paper will lead to a new oxygen, which is also reached by a hydrogen bond from one of the other oxygens we have already put down in Figure 13–4. The result is the series of puckered hexagons shown in Figure 13–5. This pattern can continue to be built up indefinitely into a three-dimensional array.

The structure is made slightly more complex by the fact that each hydrogen atom is slightly closer to one of the oxygen atoms (the one to which it is covalently bonded in an H_2O unit) than to the other oxygen atom (to which it is bonded by the hydrogen bonding). But that does not change the fundamental arrangement—tetrahedral around each oxygen with a hydrogen (near or somewhat farther away) in each of those four directions and another oxygen farther away in each of the same four directions. Putting this structure together is an exercise in solid geometry that may not be clear to all of the students reading this section. But the net result is very simple—the best arrangement for hydrogen bonding leaves an *empty hole* in the center of each hexagonal ring. The consequence of that result should be simple and clear to everyone. The highly hydrogen-bonded structure (ice) needs more space than a structure where molecules can move around to fill up the holes (liquid water). The

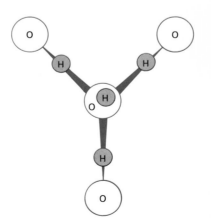

Ice is a hydrogen bonded unit that requires empty holes to line up all the hydrogen bonds correctly.

Figure 13–3 Tetrahedral bonding pattern around a single oxygen atom in solid water (ice).

264   Water   [13]

Figure 13-4 Tetrahedral bonding patterns around four neighboring oxygen atoms in solid water (ice).

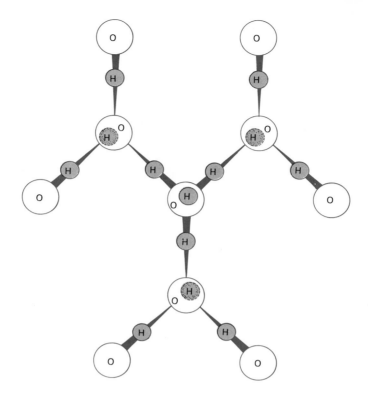

At 4°C (the temperature of maximum density for water) the decrease in volume as hydrogen bonds break upon heating (which is the dominant factor when water is heated at less than 4°C) and the increase in volume due to increased molecular motion (which dominates when water above 4°C is heated) are equal.

result is a solid less dense than the liquid, something that never happens for compounds without this special need to leave holes in the solid structure.

As ice is heated up from the very low temperatures where hydrogen bonding is essentially complete, increased molecular vibrations cause some of the hydrogen bonds to break, but enough remain (or are formed back) to hold the overall solid structure (including the vacant holes) together. The increased molecular vibrations also push the atoms farther apart, thus lowering the density of the ice as the temperature rises. Eventually so many hydrogen bonds are broken that the solid structure collapses into a group of large molecular aggregates that can allow some flow (liquid behavior). Even though the extent of molecular vibrations increases as the temperature rises and the less regular arrangement of the liquid requires some extra space compared to a closely packed solid, these factors are far outweighed by the reduction in space because some of the holes in the hydrogen bonded structure can now be filled up. The result is the observed sharp increase in density when the solid is melted.

The liquid water formed as ice melts still has very extensive hydrogen bonding. Breaking only about 15% of the original hydrogen bonds is enough to allow the solid structure to collapse. As the liquid water is heated, additional hydrogen bonds break and more of the puckered hexagons collapse. For a while, the resulting loss in volume is greater than the increase in volume

[13-3] Hydrogen Bonding and the Structure of Water 265

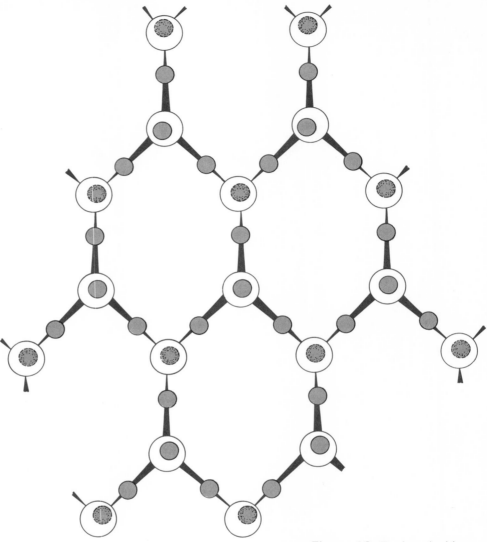

Figure 13-5 Interlocking tetrahedral bonding of a region of solid water (ice) showing the "holes" left in the center of "hexagons."

due to increased molecular motions. In this range, from 0 to 4°C, the density of the liquid water increases with increasing temperature. At 4°C, the expansion due to increased molecular motion just equals the contraction due to hydrogen bond breaking, and water passes through its point of maximum density. Above 4°C the increasing molecular motion with increasing temperature becomes the larger factor, and the density decreases with increasing temperature (as it does at *all* temperatures for other liquids).

At all temperatures, liquid water has considerable hydrogen bonding. The resulting large aggregates are bulky and tend to be interlocked. As a result they flow less easily than small molecules that are not hydrogen bonded, and the viscosity of water results.

Under all conditions each H_2O molecule has the definite areas of positive and negative electrical charge concentration mentioned in our structure. These charged areas can become associated around positive or negative ions. The energy from these interactions is crucial in allowing H_2O to dissolve ionic substances. Sometimes these associated H_2O molecules are kept with the ions in solid salts as well as in solution. The resulting salts including water are called hydrates.

13–4 Acids and Bases

One of the most important associations between ions and water is the combination with protons (H^+ ions) to form the species responsible for what we call **acid behavior.** The lone pairs of electrons on oxygen in H_2O are readily available and can form strong covalent bonds with free protons. We know these bonds are strong because they are the same sort of covalent sharing that already holds two other protons to the oxygen in each H_2O unit. The resulting H_3O^+ unit, which we call a **hydronium ion,** can then be surrounded by additional H_2O molecules with their negative regions attracted toward the positive charge. It has been suggested that $H_9O_4^+$ or some other formula with even more H_2O's added to H^+ would better represent what would really happen if a free H^+ was placed in water. For convenience, we will simply call this ion H^+ (even though we know there are always H_2O's attached in water) or sometimes H_3O^+ to show we know at least one (and probably more) H_2O is attached.

In water, H^+ ions always have one or more H_2O molecules attached.

Free protons are not available in nature in the sorts of materials we normally experience; they exist only in cosmic rays or other gaseous material that has been subjected to high energy bombardment, such as could be done in a mass spectrometer. But the H_3O^+ unit is so highly favored that it can often form by "stealing" an H^+ away from another chemical bond. For example, HCl molecules (which are covalently bonded in gaseous HCl) can be broken up by H_2O in the reaction

$$HCl + H_2O \longrightarrow H_3O^+ + Cl^-$$

The dot diagrams of these substances show how the proton merely shifts from one pair of electrons to another and is never a free H^+ ion.

$$H\!:\!\ddot{\underset{H}{O}}\!: \;+\; H\!:\!\ddot{\underset{\cdot\cdot}{Cl}}\!: \;\longrightarrow\; H\!:\!\ddot{\underset{H}{O}}\!:\!H^+ \;+\; :\!\ddot{\underset{\cdot\cdot}{Cl}}\!:^-$$

Some other hydrogen containing molecules can also be broken up to produce the same H_3O^+-type species.

$$H_2SO_4 + H_2O \longrightarrow H_3O^+ + HSO_4^-$$
$$HNO_3 + H_2O \longrightarrow H_3O^+ + NO_3^-$$

Any compound that produces these H_3O^+ ions in water is an acid. As noted in Section 10–4.2, any water solution with more H_3O^+ ions than water would form by itself is acidic. If the H_3O^+ ions form very readily, with essentially every molecule forming an H_3O^+, the material is called a **strong acid.** If only a few of the molecules at a time form H_3O^+ ions, the water solution does not exhibit the acid properties (which are really the properties of H_3O^+ in water) as extensively, and the material is called a **weak acid.** An example is acetic acid, which is only slightly reacted to H_3O^+ and $^-OOCCH_3$ when added to pure water.

$$HOOCCH_3 + H_2O \longrightarrow H_3O^+ + {}^-OOCCH_3$$

The species made from H^+ and water, which we sometimes call H_3O^+, the hydronium ion, is responsible for acid behavior in water.

Acids are strong acids if all the molecules (or ions) react with water to give H_3O^+; they are weak acids if H_3O^+ is produced but the reaction is incomplete.

another interesting weak acid is H_2O itself, which can form H_3O^+ by the reaction

$$H_2O + H_2O \longrightarrow H_3O^+ + OH^-$$

This reaction is never very extensive because the reverse reaction, $H_3O^+ + OH^- \longrightarrow 2\,H_2O$, efficiently removes most of the H_3O^+ ions. A solution is **acidic** only if there are more H_3O^+ ions than OH^- ions.

The ability of OH^- to destroy the acid character of a water solution gives solutions with more OH^- than H_3O^+ ions an opposite kind of behavior we call **basic.** Basic solutions and acidic solutions react with each other (forming H_2O) in a way that neutralizes both of them. If we get a solution with equal numbers of H_3O^+ and OH^- ions, it is said to be **neutral.** Neutral (the borderline between acidic solutions and basic solutions in water) is simply the condition equivalent to having pure water, because the ionization of water must always give equal numbers of H_3O^+ and OH^- ions.

Base behavior in water leads to OH^- ions in solution.

A water solution with more H_3O^+ ions than OH^- ions is acidic; one with more OH^- than H_3O^+ is basic; one with equal numbers of H_3O^+ and OH^- is neutral.

Anything that can produce OH^- ions in water is called a **base.** Using terms comparable to our terms, strong acid and weak acid, substances that produce OH^- extensively and efficiently are called **strong bases,** whereas those giving only limited reaction to form OH^- are **weak bases.**

Strong bases produce OH^- efficiently in water; weak bases are only partially reacted with water to give OH^-.

Acids and bases were originally defined as those things which gave H^+ (acids) or OH^- (bases) directly. (That is the Arrhenius theory of acids and bases—see Section 7–5.2.) A more general definition was needed to include things that produced an excess of H_3O^+ or OH^- indirectly through interaction with water. The definition chosen was that acids include anything that can act as an H^+ donor in reaction (for example, $NH_4^+ + H_2O \longrightarrow NH_3 + H_3O^+$) and bases include anything that can act as an H^+ acceptor in reaction (for example, $CH_3COO^- + H_2O \longrightarrow CH_3COOH + OH^-$). This theory is identified as the **Brønsted-Lowry theory** of acids and bases. Reactions like the examples shown, where an ion reacts with water to produce an acidic or basic solution, are **hydrolysis** reactions and are discussed in Appendix B.

Anything that serves as an H^+ donor is an acid; anything that serves as an H^+ acceptor is a base (Brønsted-Lowry theory).

When H_3O^+ or OH^- are produced by the reaction of some other ion with water (instead of direct ionization to H^+ or OH^-), the process is called hydrolysis.

We should note that even a weak acid can react completely to give up H^+ if a good base is available to accept the protons. Their incomplete reaction when alone in water (or when in acidic solution) occurs because the other ion they form (such as CH_3COO^- from the weak acid CH_3COOH) is a stronger base by hydrolysis than H_2O is as a proton acceptor to form H_3O^+. Therefore,

Lewis acids can accept electron pairs; Lewis bases can donate electron pairs.

some of the weak acid is regenerated in the unreacted form. A similar thing happens for weak bases to make their reaction to OH⁻ incomplete.

An even more general definition, the Lewis theory of acids and bases, was reached by noting that the sources (donors) of H^+ we call acids are able to accept a share in a pair of electrons in reaction and bases (sources of OH^- or H^+ acceptors) are able to provide a previously unshared pair of electrons to be shared by the acid. Therefore, anything that can *accept an electron pair* to form a bond is called a **Lewis acid** and anything that can *donate an electron pair* to form a bond is called a **Lewis base.** This concept will be useful to us later in discussing some organic chemistry and some of the complexing phenomena vital to life. Table 13–1 lists some examples of Lewis acids showing how they act as electron pair acceptors. Table 13–2 lists some examples of Lewis bases showing how they act as electron pair donors.

Table 13–1 Lewis Acids

Acid	Reaction*
HCl	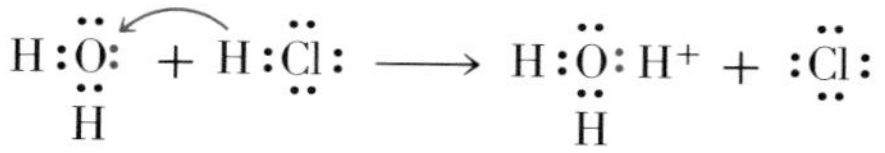 H:Ö:H + H:Cl: ⟶ H:Ö:H⁺ + :Cl:
H_3O^+	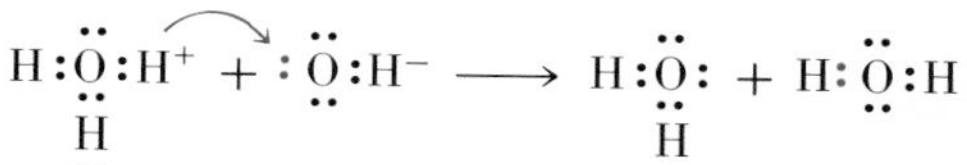 H:Ö:H⁺ + :Ö:H⁻ ⟶ H:Ö: + H:Ö:H
NH_4^+	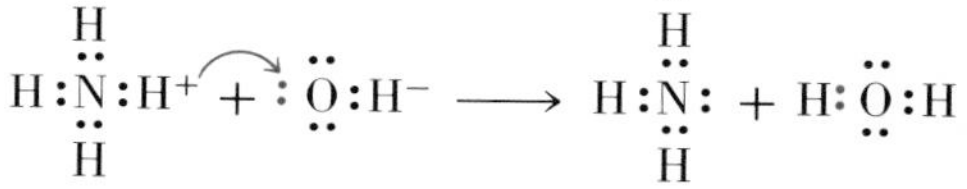 H:N̈:H⁺ + :Ö:H⁻ ⟶ H:N̈: + H:Ö:H
SO^{2+}	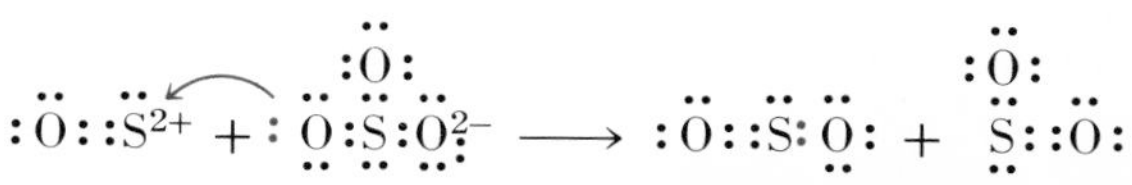 :Ö::S²⁺ + :Ö:S:Ö:²⁻ ⟶ :Ö::S:Ö: + :S::Ö:

*The electron pair being accepted is colored.

Table 13–2 Lewis Bases

Base	Reaction*
OH⁻	 H:Ö:⁻ + H:Cl: ⟶ H:Ö:H + :Cl:⁻
CH_3COO^-	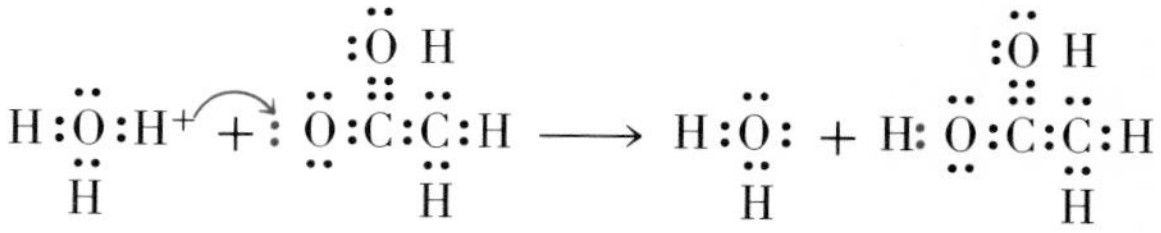 H:Ö:H⁺ + :Ö:C:C:H ⟶ H:Ö: + H:Ö:C:C:H
NH_3	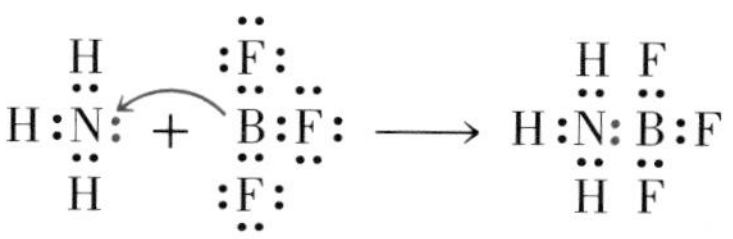 H:N̈: + B:F: ⟶ H:N̈:B:F

*The electron pair being donated is colored.

In general, nonmetals and oxides of nonmetals will hold electrons quite strongly. Therefore, they act as electron acceptors and are often good sources of H^+. The H^+ can be taken away from the bond electrons (which are strongly held by the other atom) and hydrogen compounds of nonmetals and nonmetal oxides are therefore acidic. The nonmetal oxides can usually react with water to form acidic hydrogen containing compounds. The strongest acids will be the ones that hold electrons away from the hydrogen completely, thus leaving the hydrogen free to move as H^+ to a different pair of electrons it can share more effectively.

The strongest acids are the ones that can attract electrons away from hydrogen most strongly.

By contrast, metals give up electrons quite easily. Therefore, they act as electron donors and are often good sources of OH^-. Metal oxides can react with water to form hydrogen containing compounds in which the oxygens have excess electrons received from the metal. These negatively charged oxygens cannot pull electrons away from hydrogen very effectively to release H^+, so the oxygen and hydrogen remain strongly covalently bonded and it is the whole unit, OH^-, that separates from the metal ion. The strongest bases will be the ones where the metal gives up its outer electrons to oxygen most completely.

The strongest bases are the ones where electrons can be given up most readily.

As the oxidation state of a metal goes up, it becomes harder to remove the electrons. As a result, low oxidation state metal ions give up electrons easily and their oxides act as bases in water solution, but high oxidation state compounds of the same elements cannot give up the electrons as easily and may act like the nonmetals do (their oxides are acidic in water solution). This phenomena was discussed in Section 8–6 and Figure 8–14. There are also intermediate states that may have some acidic behavior and some basic behavior. The dominant factor continues to be the question of whether the substance accepts or gives up electrons readily.

As the oxidation state of an element goes up, it holds electrons more strongly and therefore becomes more acidic.

13–5 Natural and Man-Made Uses of Water

13–5.1 Flooding. Like all other living things, man is dependent upon an adequate supply of water. He also needs an adequate supply of food, and all of his food sources depend on adequate water. Some of the best locations for satisfying these needs have been in the flood plains of rivers. The water in the surrounding soil supported plant life through the driest periods, and fresh water was available from the river for man. When normal seasonal variations in the water transport cycle caused floods, man would retreat to higher ground and return afterward. During the flood, water spread out over a large area and moved slowly at the edges. In these calm backwaters soil washed away from upstream areas and suspended in the turbulent, fast moving streams was able to settle out and build up additional layers of fresh soil and new nutrients. This made the bottom lands an especially productive and desirable location for man. Later these flat bottom areas also proved to be easy places to build man's structures and transportation systems.

Floods are a natural part of the water transport cycle.

Because of the advantages of the low areas, it has become advantageous

Protection from flooding allows greater use of the desirable flood plain lands.

Levees raise the level of flood waters by preventing their release over flood plains and prevent silt deposit on flood plains.

to men to use them extensively. As the investment in structures and the potential gain from possible crops increased, man began to resist the natural water cycle. Figure 13–6 shows some of the changes often made by man. Barriers (levees) were built to keep the river out of the flood plains so man would not have to retreat from the seasonal floods. Sometimes these barriers fail and tremendous amounts of human effort are lost as the water covers places man did not expect it to reach.

When the barriers succeed, the water transport cycle is changed. The calm backwaters are eliminated and the deposition of new soil onto the flood plain stops. Floods become more violent as the full force of the water is concentrated into a smaller space without relief from flows out onto the plain. Soils remain suspended in the turbulent main flow and are carried farther downstream. To assist in removal of the water, man relieves constraints slowing flow by straightening the river's course. The more rapid flow continues in dry periods causing lower water levels in the river and in the water table of the soils nearby. The more rapidly flowing stream has a very different ecological system, lacking some of the slow flowing areas in which some fish breed and developing a muddier water that is unfavorable to other species.

River straightening causes faster flow, leading to muddy water and ecological problems.

13–5.2 Water Pollution.

Muddy rivers are a less desirable water source for man as well as fish, but there are more serious problems. As human settlements built up around rivers, the rivers were used both as a water source and as a universal disposal system. This merely moves the wastes to new locations, where they may or may not be degraded by natural processes. Wastes dumped into the water can make sharp changes in the chemical makeup of the water solutions, and they can spread human diseases. These are serious problems that can make further use of the water difficult or impossible, and the problems are complicated by the variety of pollutants. Later users may have difficulty recognizing all of the problems they should take care of before using the water. Some of the common problems of water pollution will be discussed in Chapters 14 and 16.

People have used rivers as a universal disposal system, creating polluted waters which cause problems for later users.

13–5.3 Wells and Hard Water.

One common solution to the problem of surface water pollution is use of ground waters drawn from wells. Water from deep wells is usually free from human disease bacteria, but the same long path through soil and rocks that frees it from bacteria and silt allows it to pick up dissolved minerals. Various salts can be dissolved to give **hard water** from wells, but the most general problem is the "temporary hardness" due to $CaCO_3$ dissolved through an acid-base reaction.

Water from deep wells is free from diseases but loaded with dissolved minerals.

The process begins when CO_2 in the air is dissolved in rainfall. The nonmetal oxide CO_2 forms the weak acid H_2CO_3. The acid solution reacts with $CaCO_3$ in limestone formations converting it into Ca^{2+} and HCO_3^- ions in solution. The equations involved are

Acid-base reactions possible because of CO_2 dissolved in rainfall dissolve rock to make hard water. Heating regenerates the rock and CO_2.

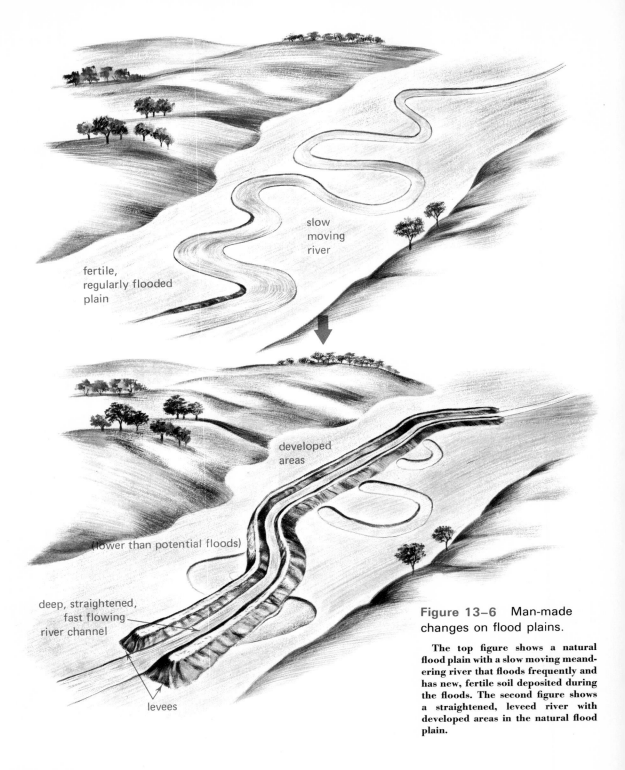

Figure 13-6 Man-made changes on flood plains.

The top figure shows a natural flood plain with a slow moving meandering river that floods frequently and has new, fertile soil deposited during the floods. The second figure shows a straightened, leveed river with developed areas in the natural flood plain.

$$CO_2 + H_2O \rightleftharpoons H_2CO_3$$
$$H_2CO_3 \rightleftharpoons H^+ + HCO_3^-$$
$$H^+ + CaCO_3 \rightleftharpoons Ca^{2+} + HCO_3^-$$

This is called **temporary hardness** because the $CaCO_3$ precipitates out when the solution is heated. The whole series of reactions is reversed with $CaCO_3$ solid forming and CO_2 gas being driven out of the solution.

13-5.4 Irrigation and Salting Effects.

Man directs large amounts of water to use by plants and animals, particularly in irrigating crops in areas otherwise too dry to grow them.

Evaporation of water from plants and soil is an important source of atmospheric water in the water transport cycle. Evaporation from the oceans is an even more important source of atmospheric water. In the normal cycle, water picks up materials on its way back to the oceans. Solids eventually settle out, but dissolved substances remain in the oceans. Volatile materials like O_2, N_2, and CO_2 may also escape from the surface, so an equilibrium balance is set up between additions and escape. However, because of the definite regions of positive and negative charge concentration in H_2O molecules, water also dissolves many ionic salts. These are not volatile. They remain behind as the H_2O evaporates and, as a result, the oceans (and any other spot where water flows in and is removed only by evaporation) become quite concentrated with ionic salts.

Water moves salts which are left behind when the water evaporates.

Irrigated farm lands are no exception to the above pattern of evaporative salt concentration. Unless some water is purposely "wasted" so it can carry excess salts away, the soil will gradually become more concentrated in ionic salts. The buildup of salts will be especially rapid when hard water from wells or water from an especially salty river is used. The rapid evaporation of soil water in the dry areas where irrigation is done also speeds the salt buildup. Some salts may be incorporated into plants and taken away in crops, but the nature of areas where irrigation is most useful almost assures that a net salt increase will occur. The effects of this salt buildup on plants were discussed in Section 10-6.

Irrigation may concentrate salts in farm land.

13-5.5 Erosion.

Both air and water are able to move solid materials about as they move from place to place. As we have noted, water can move material either as chunks of solid (plant materials, silt, sand, or even large rocks) or as dissolved material. Removal of material from a location is called erosion.

Erosion by dissolving the material is particularly important in flow of ground water. As we noted above in Section 13-5.3, rain water is made slightly acidic by dissolved CO_2 and is therefore able to dissolve $CaCO_3$ from limestone formations. As water seeps through rock formations, channels are dissolved away, and eventually a path for a flowing stream is opened up.

If the water source ever dries up or recedes, the opening is left as a cave. The process of cave building by underground streams is going on constantly under almost every part of the earth. If these ground waters pass through a space where the water can evaporate away, their dissolved solid burden is redeposited, as in stalactites, stalagmites, and other deposits in caves. Figure 13–7 shows these steps in cave formation. Even in those caves now dry enough for easy human exploration, the eroding water is usually present, as in the rivers within Mammoth Cave in Kentucky and the waterfall inside Ruby Falls Cave in Chattanooga.

Surface erosion is mostly by movement of undissolved chunks. The relatively viscous, hydrogen bonded nature of water contributes significantly to this erosion. Because water does not flow completely freely, it is often unable

Underground water flows produce caves.

Figure 13–7 Cave formation.

Rain water absorbs CO_2 from the atmosphere to form an acidic solution. As the acidic rainwater flows through the ground, it dissolves $CaCO_3$ and forms channels through the rock. As new channels form or grow, some old channels are left as dry caves. Dripping water in the dry caves may form stalactites and stalagmites.

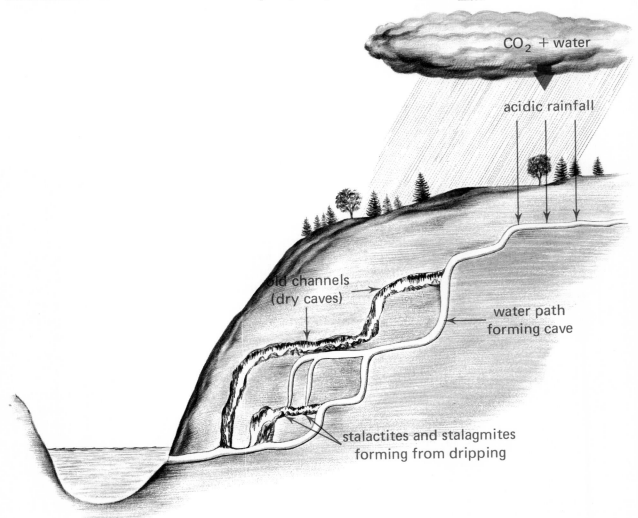

274 Water [13]

Surface erosion moves large amounts of suspended solids.

The smallest particles tend to remain suspended the longest.

Farming assists the breakdown of soil into small particles subject to erosion.

to maintain a smooth, relatively frictionless flow. The flow becomes turbulent and this turbulence helps tear chunks of material loose and mix them into the water stream. When the flow slows and the turbulence ceases, the chunks can settle back out of the water, as occurs on natural flood plains. Those solids that are broken up into the smallest particles by the turbulent water remain suspended the longest, and somewhat larger particles can remain suspended for a longer average time in water than would be the case in a less viscous liquid.

Rates of soil erosion are often increased by farming techniques. When soil is broken up, the normal pathway for ground water to be carried to the surface and possible evaporation is interrupted, and more water is left in the soil for use by plants. But that same loose surface soil is much more easily washed away by water during rainstorms. It also more easily reaches the small sizes that can be carried further by water.

13-5.6 Colloids. We know that fine particles do not settle out as quickly as large suspended particles. The ultimate fine particles would be actual individual molecules or ions dissolved as a true solution, which is significantly different in nature and behavior from suspended particles. But undissolved aggregates containing from about a thousand to a billion atoms are also small enough to be significantly different in behavior from large suspended particles. The suspensions of these fine aggregates are called **colloidal dispersions,** and their properties will help us understand suspensions in general and how they persist.

Colloidal particle dimensions are often 1 mμ to 200 mμ.

Dimensions of colloidal particles have often been stated in units of millimicrons (mμ), which are 10^{-6} mm or 10^{-7} cm. The modern (SI) name for mμ is nanometers (nm). Dimensions in the range 1 nm to 200 nm are common, but some colloidal particle dimensions may fall outside that range. We will limit our discussion to colloids in water, although other colloidal dispersions, like solids or liquids in gases (smoke, dust, aerosols), are also important. Moreover, any suitably fine dispersion can be classified as colloidal, even including gases in liquids (foams) or solids (pumice rock, floating soap) where it is hard to talk about the dispersed "particle."

Colloids can be formed by dispersion or condensation.

Colloids can be formed either by dispersion (breaking down larger particles) or condensation (growth from smaller units). Dispersion can sometimes be accomplished by grinding, and some colloids are formed by grinding solids in a liquid so the particles are kept separated. There is an obvious analogy to the way turbulent water and stream rocks can break up soil into fine silt. The ultimate particle size in each case depends on the efficiency of grinding.

Hydrophilic colloids disperse spontaneously in the peptization process. They have strong attractions to water, are very viscous, readily form gels, and coagulate reversibly in concentrated ionic solutions.

Some solid substances such as gelatin, glue, and starches disperse spontaneously to form colloids in water. This process is called **peptization.** These particles are already of colloidal size. They also have a strong attraction for water and are referred to as **hydrophilic** (water loving) colloids. These attractions tie up water molecules (which already have considerable aggregation of their own due to hydrogen bonds) and make very viscous colloidal systems. The same substances also readily form gels, colloidal dispersions of liquid in solid. Because of their strong attractions to water, these colloids do not recombine (coagulate) to larger aggregates unless conditions are changed quite drastically. Addition of concentrated ionic salt solution will coagulate hydrophilic colloids, but the process is reversible. Addition of water will disperse them as colloids again, as shown in Figure 13–8.

Some other dispersed colloids have little or no affinity for water. These are called **hydrophobic** (water hating) colloids and are usually inorganic elements or compounds. They have low viscosities, never form gels, and are easily and irreversibly coagulated by addition of ionic solutions. Until ions are added, they avoid recombination because all of their dispersed particles have a net electrical charge of the same sign, either positive or negative. This provides a small repulsive force to keep the particles separated. The existence of these charges can be shown by putting positive and negative electrodes into the colloidal dispersion. The colloid will coagulate on the electrode with the electrical charge opposite to the colloidal particles. Figure 13–9 shows the behavior of a hydrophobic colloid.

In each of the above classes, colloids were stabilized by some force that prevented the particles from associating to form larger aggregates which would eventually separate out. This kind of separation can sometimes be aided by a substance that is neither the colloid nor the dispersing phase (which is always water in our limited examples). Milk is an example of such a colloidal dispersion familiar to all of us. Milk is a colloidal dispersion of butterfat in a water solution. Normally it is not a very good colloid because the butterfat drops are too large and too easily separated. This can be corrected by breaking them down to smaller drops in the form we call homogenized milk, which separates very slowly or not at all. However, when milk does separate, it forms cream, not butterfat. The water phase cannot be completely removed without further steps like churning (and even then some water remains in the butter). The link between fat and water is through a protein material, casein, which is attracted to both and forms a connecting layer on the boundary between fat and water, as shown in Figure 13–10. The material stabilizing the colloid is called an **emulsifying agent.** The colloids consisting of one liquid dispersed in another liquid where this phenomenon is most common are called emulsions. Soaps and detergents work by acting as emulsifying agents or wetting agents. Wetting agents also assist formation of colloids by lowering surface tension and thus easing the separation of particles into water dispersed colloids.

Hydrophobic colloids have little affinity for water, all their particles have an electrical charge of the same sign, and they are easily and irreversibly coagulated by ionic solutions.

Emulsifying agents help separate colloidal particles by surrounding them and being attracted to both the colloid and the dispersing phase.

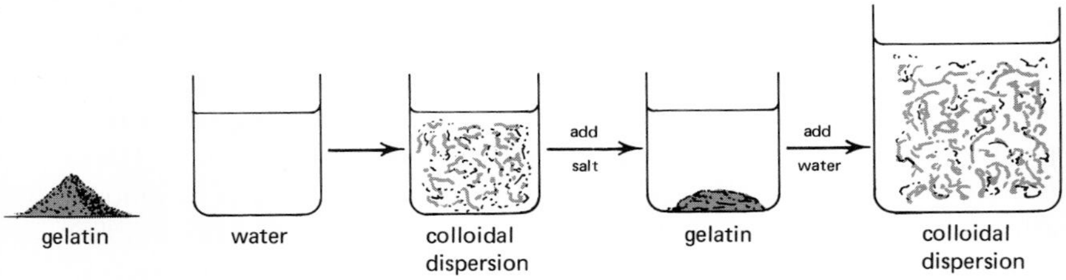

Figure 13–8 Hydrophilic colloids

Hydrophilic colloids, such as gelatin, may spontaneously disperse in water. They can be coagulated by adding ionic salts and then reform the colloid if the salt is removed or diluted.

Figure 13-9 Hydrophobic colloids

A gold colloid can be formed by chemical reaction between solutions. All hydrophobic colloids have particles with an electrical charge, as can be shown by their attraction to one electrode in an electrolysis cell. When the hydrophobic colloid is broken up by adding an ionic salt or attracting the colloidal particles to an electrode, the colloid cannot be regenerated by any simple method such as addition of water.

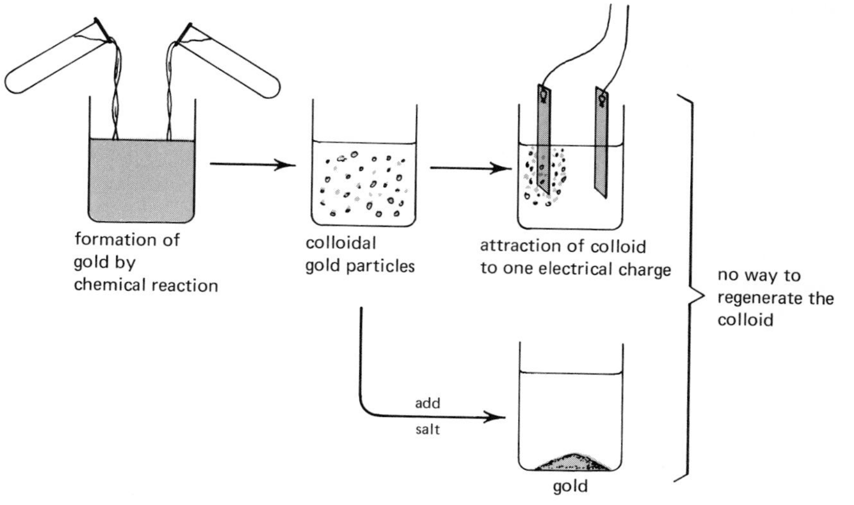

formation of gold by chemical reaction → colloidal gold particles → attraction of colloid to one electrical charge → no way to regenerate the colloid

add salt → gold

Molecular motion can be seen in the Brownian motion of colloidal particles which is caused by their being struck by the surrounding molecules.

When particles can be kept as small as colloidal size, they remain suspended almost indefinitely. That is because the light particles are pushed about and kept "stirred up" by the motion of the surrounding molecules. This motion can be seen when a colloid is observed through a microscope and is called **Brownian motion.** Equivalent stirring for larger particles can be provided only by having larger forces of motion, such as water turbulence. The larger the particles, the greater the turbulence required to maintain the suspension. However, some of the factors that stabilize colloids—attractions to water,

Figure 13-10 Emulsifying agent.

The emulsifying agent (casein in the case shown) interacts strongly with both liquid phases. By forming a coating around one phase (the butterfat) the emulsifying agent prevents the recombination of small drops and thus prevents the formation of separate layers of liquid (butterfat and water phase).

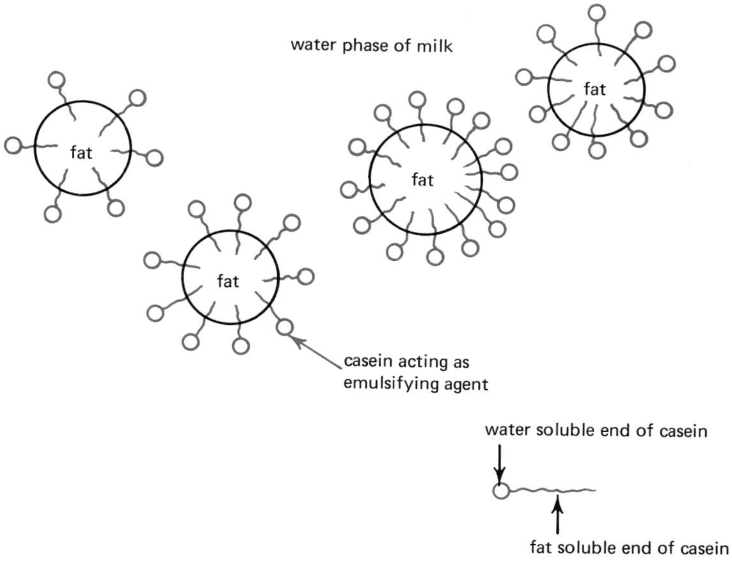

water phase of milk

fat

casein acting as emulsifying agent

water soluble end of casein
fat soluble end of casein

[13-5] Natural and Man-Made Uses of Water

tendency to pick up net charges of a particular kind, and coating by substances attracted to both the colloid and liquid—can also help keep other particles smaller and more easily suspended. The structure of water molecules make them particularly effective at setting up strong attractions directly to particles such as soil or to things adsorbed on the particles, and the ability of water to dissolve ionic salts leads to possible formation of charged particles by selective adsorption of certain ions.

13–5.7 Effects of Lakes and Dams. Most suspended solids will settle out of water when they reach a point with no (or very slow) flow. If this point is an ocean, even true colloids would tend to coagulate in the presence of ionic salts. The deposition of deltas at river mouths is an important feature of our world, but the deposition of solids in fresh water lakes is often even more important to us. We use lake waters for a wide range of purposes varying from recreation to drinking water. We also create artificial lakes by building dams. The ability to store water is often a major reason for building dams.

Building dams to create lakes reduces the extent of floods and increases water availability for irrigation and other uses in dry periods. It is the natural and normal fate of lakes eventually to fill with solids. As this happens, the lakes undergo natural changes that make their waters less desirable to man, as we will discuss further in Section 14–6. The filling in of lakes, natural or artificial, is generally considered undesirable, but human activities tend to speed up the process. Erosion due to farming fills natural lakes more rapidly than natural erosion would. Many artificial lakes fill even more rapidly because they are located on large (and often relatively turbulent) streams.

Lakes gradually fill up with solids, destroying their usefulness as lakes, and human activities speed up the filling of lakes.

The density properties of water have some interesting effects on lakes, particularly in areas where freezing occurs. Because ice is less dense than water, only the surface water freezes in lakes with appreciable depth. When the water temperature reaches 4°C, further cooling of the surface makes the surface waters less dense than the bottom waters. Therefore, the colder water remains on top until it eventually freezes. In the absence of mixing, the bottom waters are not cooled any further than the maximum density temperature of 4°C. The unfrozen deep water permits winter survival of many fish and other aquatic life. In summer, the surface is warmed and becomes less dense than the deep waters because it is warmer than the maximum density temperature of 4°C. Many of the most desirable types of fish thrive in the cool bottom waters which have less seasonal temperature variation. In a stream where the water is mixed, the water might become too warm for them to survive.

Deep lakes continue to be liquid at the bottom and support life when the surface freezes.

Lakes that freeze and thaw will reach a point twice each year when all the water is approximately 4°C and therefore the density is the same throughout. At that time extensive mixing can take place. Dissolved oxygen in the surface water can be carried to the bottom, and materials on the bottom such as decaying dead vegetation can be carried to the top. These periods are sometimes described as times when the lake turns over.

Lake water can mix freely throughout when the surface water and bottom water is all at 4°C.

278 Water [13]

Water used in power generation at dams is usually from the deeper water levels with less temperature variations.

The difference in temperature patterns for surface and deep waters has some effects on the water released from dams when power is being generated. The inlets to the generators are usually fairly deep in the reservoir to assure access to water and allow the lake level to be varied as water is stored and released. Therefore, the water going through the generators is cool deep water even when the surface water is quite warm. This can change conditions significantly in the stream below the dam. The area below Table Rock Dam on the White River in southern Missouri gets enough cold water from the power plant to become significantly colder than other rivers or the surface regions of lakes in the area. As a result, it also becomes the most favorable spot in many miles for trout, which require cool water.

Bodies of water affect temperatures and rainfall nearby.

Large bodies of water like lakes and oceans have a strong moderating influence on temperatures. Because of the large amount of heat released as water cools and freezes, cold winter air may be warmed considerably as it passes over water. This protects temperature-sensitive plants, including some desirable fruit plants on the downwind side of the lake, from the severest freezes. The same downwind areas also receive extra moisture from evaporation off the lake and therefore have heavier snows (or rainfall) than other surrounding areas. Dams forming large lakes may cause noticeable changes in the temperatures and rainfall in nearby areas.

13-5.8 Water as a Coolant.

There are many human activities where removal of heat is an important step. Sometimes the heat is simply allowed to dissipate into surrounding air, but that removal is limited in speed and in the quantity of heat easily handled. For rapid removal of large quantities of heat, water is almost always the coolant selected. Water is the most abundant liquid on earth. It is usually available in large quantities at little cost. As a liquid, it is easily transported to the item to be cooled, a substantial mass can be receiving heat from the hot object at one time, and mixing brings additional cool water to the object continuously. In addition, there are very high heat requirements to melt, warm, or evaporate water because of the hydrogen bonding which must be broken up. This gives water a very high capacity for accepting and storing heat energy. The combination of abundance, convenience, and heat capacity makes water the best coolant choice in all cases where it is not eliminated by some unfavorable limiting condition (such as temperatures where water is no longer liquid).

Water is widely used as a coolant, and thermal pollution of waters can be a problem.

When water is used to remove really large quantities of heat, **thermal pollution** of the water output may result. Processes such as condensation of steam in steam turbine electric power plants or formation of steel sheets from molten metal in water-cooled continuous process units are most efficient when the coolest available water is used. Then less water needs to be circulated, and, in the case of power plants, a more complete conversion of heat into electricity is possible. Therefore, water is taken from the coolest available source (often the bottom of a river or lake). When it is returned at a higher temperature, the body of water may become too warm for survival of the natural fish and plants. Even if conditions remain acceptable in temperature, the transfer of cool bottom waters to a warm surface layer may cause a transport of material from bottom to top where normally the difference in density would not allow it. As we will describe more fully in Section 14–6, that can have a disastrous effect on some lakes.

The alternative to return of heated water is transfer of the heat to something else. This is normally done by cooling towers in which the evaporation of some of the water cools off the remainder. This introduces large amounts of water vapor into the atmosphere in warm air (heated by exposure to the warm water). When this is cooled off by contact with cooler air, mists or fog may form. In wintertime, the water vapor may condense out on cool surfaces and freeze. The resulting ice is hazardous to traffic, and the weight of ice may break wires and other structures. In summary, use of water as a coolant does not end the heat disposal problem. The water may be available and cheap by our usual methods of accounting, but it is rarely or never free. We must be careful to include the secondary effects on our surroundings when we compute the real costs.

Skills Expected at This Point

1. If you are asked to identify the unusual properties of water, you should be able to list the six specific points given in Section 13–2.
2. You should be able to describe the structure of an individual H_2O molecule and describe how the polar character leads to hydrogen bonding and to the effectiveness of water as a solvent for ionic compounds.
3. You should be able to describe a hydrogen bond and how hydrogen bonding leads to the density characteristics, high energy of transitions, and viscosity of water.
4. You should be able to define the terms acidic, basic, neutral, strong acid, weak acid, strong base, weak base, Lewis acid, Lewis base, and hydrolysis.
5. You should be able to describe a normal water transport cycle including flooding and list several problems caused by human activity which causes abnormal conditions.
6. You should be able to describe hard water and the chemistry leading to temporary hardness.
7. You should know what a colloid is, the approximate size of colloidal particles, and the types of factors present that keep the particles separated in the various types of colloids and emulsions.
8. You should be able to explain the extensive use of water as a coolant.

Exercises

1. List the most noticeable unusual properties of water.
2. Draw the phase diagram for water and identify **(a)** the region where ice is stable; **(b)** the vapor pressure curve for water; **(c)** the melting

point curve; and **(d)** the feature that shows that ice is less dense than water.
3. Use the phase diagram for water and about 15 to 20 words to describe how ice skating happens.
4. Describe the structures of **(a)** an individual water molecule; **(b)** a "perfect" ice crystal; **(c)** liquid water at about 2°C.
5. What is a hydrogen bond?
6. Why does ice form a structure such that it floats on water?
7. Why does water have a maximum density at 4°C?
8. How does water interact with a salt like NaCl to dissolve it?
9. Define the following terms: **(a)** strong acid, **(b)** weak acid, **(c)** Lewis acid, **(d)** basic, **(e)** hydrolysis.
10. What terms are used to describe the following reactions?
    (a) $NH_3 + BF_3 \longrightarrow NH_3BF_3$
    (*Hint:* Write out the dot diagram structure if you don't recognize this.)
    (b) $NH_4^+ + H_2O \longrightarrow NH_3 + H_3O^+$
11. Why is H_2SO_4 a stronger acid than H_2SO_3?
12. Describe the normal water transport cycle.
13. Describe the chemical reactions causing temporary hardness in water.
14. A farmer's field has been badly eroded by water. Describe how **(a)** the water moves materials away by two different means; **(b)** plowing increased the erosion; **(c)** the erosion will affect the nearby lake downstream from his farm.
15. What is a colloid?
16. What factors could help colloids to form? Describe the properties of hydrophilic colloids. Describe the properties of hydrophobic colloids.
17. Why is water used extensively as a coolant?
18. A power plant has been built on the shore of a deep lake. It is particularly important that it have access to cold water for cooling in the summer.
    (a) Describe the temperature profile of the lake in summer and where, based on that information, the plant engineer will choose to draw the cooling water from.
    (b) When this water is put back on the surface of the lake as warm water, what will the effect be on the rate of water evaporation and the climate nearby?

Test Yourself

1. **(a)** How are water molecules associated into groups (in liquid and solid)? **(b)** To what character of water molecules are these bonds mostly due? **(c)** The same bonds cause water to have high melting point and boiling point. What other property of water is unusually high?
2. How does the Brønsted-Lowry theory of acids and bases describe acids and bases upon ionization in water?
3. **(a)** What type do compounds containing the combination of hydrogen, oxygen, and metals tend to be? **(b)** What tendency do electrons of a

metal atom have? **(c)** Upon ionization, what sort of ions do these compounds produce?
4. List one advantage gained when levees are built along flood-prone rivers, and list two different kinds of problems the levees may cause.
5. Write the chemical reactions that lead to rainwater dissolving $CaCO_3$ rock. Use these same equations to show why heating water with "temporary hardness" causes the rock to reform.
6. List three properties of a hydrophobic colloid.
7. All but one of the following was listed as one of the "unusual properties of water" given in your chapter. Which one does *not* fit as an "unusual property of water"?
   (a) Liquid water has a maximum density at 4°C.
   (b) Water expands as it is heated above 4°C.
   (c) Water is less dense as a solid than as a liquid.
   (d) Water is an effective solvent for ionic as well as covalent compounds.
   (e) Water is relatively viscous liquid.
8. Which of the following properties of water is *not* "unusual" when compared to the observed trends in other compounds?
   (a) Ice floats on water.
   (b) Water melts at 0°C and boils at 100°C (at 1 atm pressure).
   (c) The heat of vaporization of water is 540 cal/g (at the normal boiling point).
   (d) Water has a maximum density at 4°C.
   (e) Liquid water has a surface tension that would draw it into a spherical (ball) shape in a spaceship under weightlessness.
9. Water contracts as it is heated from 0°C to 4°C because
   (a) increasing strength of hydrogen bonds puts the atoms closer together.
   (b) evaporation of some of the water reduces the volume of water remaining.
   (c) the hydrogens are absorbed by the oxygen in nuclear reactions.
   (d) the molecules become increasingly polar and are therefore held together more strongly.
   (e) collapse of empty holes in the structure as hydrogen bonds break causes more shrinkage than the expansion caused by extra vibration.
10. Choose the three phrases that best complete the sentences.
    1. Water molecules in ice are forced to take certain positions by
    2. The arrangement of the water molecules in ice
    3. The "unusual property" caused is that

    (a) polar bonds; holds the molecules together strongly; ice can sublime
    (b) hydrogen bonds; leaves empty holes; ice floats
    (c) Van der Waals attractions; holds the molecules together strongly; the melting point of ice is unusually high
    (d) hydrogen bonds; restricts free movement; ice is a solid
    (e) polar bonds; leaves empty holes; ice has a high viscosity

11. The effectiveness of water as a solvent for ionic as well as covalent substances is due to
    (a) the polar water molecules.
    (b) the hydrogen bonding in water.
    (c) the density pattern of water, which leaves empty spaces for solutes.
    (d) the high heat of vaporization of water.
    (e) none of these.
12. Two important true facts about H^+ ions in water are
    (a) H^+ is always bonded to H_2O molecules to form H_3O^+ (or even larger units) instead of existing as free H^+; H^+ ions are responsible for acidic behavior.
    (b) H^+ ions react with H_2O to form H_2 gas; some H^+ ions are formed by reaction of H_2O with itself.
    (c) H^+ ions react with H_2O to form H_2 gas; H^+ ions are responsible for acidic behavior.
    (d) some H^+ ions are formed by reaction of H_2O with itself; compounds that ionize to form H^+ in water are called salts.
    (e) H^+ is always bonded to H_2O molecules to form H_3O^+ (or even larger units) instead of existing as free H^+; compounds that ionize to form H^+ in water are called salts.
13. Which of the following should be most strongly *basic* according to the Lewis theory of acids and bases?
    (a) CO_2
    (b) SiO_2
    (c) SnO_2
    (d) PbO_2
    (e) none could be basic
14. Which of the following should be most strongly acidic?
    (a) $HClO$
    (b) $HClO_2$
    (c) $HClO_3$
    (d) $HClO_4$
    (e) All should be equally strong acids

Nutrient Materials

14–1 Limiting Nutrients

Plant or animal growth is limited by the required item which is used up first. Among the many different elements required for growth, carbon, hydrogen, and oxygen are particularly important. Plants require H_2O and CO_2 plus light for an energy source, and animals require H_2O, O_2, and carbon containing energy sources. Plant growth (and the animal life dependent upon it) may be limited by the amount of H_2O or CO_2 available, but often both of these (plus light) are so readily available that other less abundant substances become the limiting factors. The **limiting factor** in growth will always be whichever required item is *used up first*.

Growth is limited by the nutrient used up first.

The limiting factor may be an item such as H_2O, which is needed in very large amounts and is available in only moderate amounts (as is true in areas of limited rainfall). Or it may be an item such as copper, which is needed only in traces but which is almost totally absent in some locations. But most often the limiting material will be a nutrient that is both necessary in fairly large amounts and not very readily available. Nitrogen, phosphorus, and potassium are the three elements used in large quantity that are most often in short supply. These important nutrient materials are the major ingredients in fertilizers. We will study the two most important of these growth limiting, short supply elements, nitrogen and phosphorus. Although they are not *always* the growth limiting factors, their influence will help us understand how natural evolution or catastrophic disruptions of growth patterns can occur. We will also need reference to the chemistry of nitrogen and phosphorus to understand proteins and biochemical processes in Part IV, The Chemistry of Life.

Nitrogen, phosphorus, and potassium are often limiting nutrients.

14–2 Nitrogen

Nitrogen is less abundant than the neighboring elements of the periodic table, carbon and oxygen (see Sections 15–2.2 and 15–8), but it is still a relatively common element. The thing that makes it a common limiting factor to plant growth is the inability of most plants to use its common form, N_2 gas. The diatomic N_2 molecules have a triple covalent bond, which is exceptionally strong. For each mole of N_2 molecules formed from nitrogen atoms, 225 kcal

The very strong bond in N_2 often prevents its use as a nutrient.

of energy can be released. Oxygen, nitrogen's neighbor in the periodic table, also forms diatomic molecules, but only 118 kcal of energy can be released for each mole of O_2 molecules formed from oxygen atoms. The difference in bond strengths is much greater than we would expect between the triple bond of N_2 and the expected double bond of O_2.

The molecular orbital theory of bonding provides a good explanation of this difference, as described in Section 8–1. For present purposes, we can simply note the sharp difference in behavior of O_2 and N_2. O_2 is very reactive because its bonds can be broken and replaced by two new bonds to other atoms on each oxygen. Each of these bonds is almost as strong as the oxygen–oxygen bond that was broken. For example, each new hydrogen–oxygen bond provides about 110 kcal/mole, almost as much as the 118 kcal/mole for the oxygen–oxygen bond.

By contrast, N_2 is very nonreactive because its existing triple bond is about as strong as all the new bonds it could form put together. For example, new hydrogen–nitrogen bonds each provide about 93 kcal/mole, much less than the 225 kcal/mole for the nitrogen–nitrogen bond. Even though three hydrogen–nitrogen bonds can be formed on each nitrogen, most of the bonding energy is used up to replace the lost nitrogen–nitrogen bonds and hydrogen–hydrogen bonds (which provide 104.2 kcal/mole of H_2). At room temperature only 11 kcal of energy is released for each mole of NH_3 gas formed from N_2 and H_2, whereas over 57 kcal is released for each mole of H_2O gas formed from O_2 and H_2.

The lower energy release explains the nonreactivity of N_2 only partially. Reactions of N_2 are also very slow unless temperatures are high. As noted in Section 12–2.3, energy is the most important factor in rates of reaction. Even reactions like $H_2 + O_2$ require some energy (from high temperatures) to start the breakdown of existing bonds so new ones can begin to form. The bonds in N_2 are so strong that much higher temperatures are usually needed to get it to react in a reasonable length of time. Living systems have difficulty providing conditions where N_2 can react. Exceptionally effective catalysts are needed to get significant reaction rates at the temperatures where life can survive.

The most common oxidation states of nitrogen are −3 and +5.

Although nitrogen can form covalently bonded combinations (particularly with oxygen) in which we would assign it other oxidation states, the most common oxidation states are −3 and +5. These result if nitrogen either gains enough electrons to reach the closed shell configuration of neon or loses enough to reach the helium closed shell. It is not actually possible to remove the five outer electrons in chemical bonding, so the +5 state and all other positive oxidation states are actually reached by covalent bonding in which the nitrogen tends to complete its outer shell. When all of those covalent bonds are to atoms more electronegative than nitrogen, the nitrogen atom is credited with none of the electrons in the calculation of oxidation states. Hence the +5 state is reached if all the outer electrons are being so shared. We can see that a +5 state for nitrogen atom, in which it really has a partial share of eight outer electrons, is not at all like a +5 ion, in which it would truly have to be left with an empty second shell. After the second shell is completed to eight electrons, there are no other energy levels low enough to accept electrons very strongly. Therefore, nitrogen never has more than the neon configuration (two inner electrons plus eight in the shell involved in bonding) in any situation.

The electronic configuration of nitrogen leads to unusual bonding situations. The most important of these in living things is the type of bonding arrangement present in ammonia, NH_3. The nitrogen has completed its outermost shell by forming three covalent bonds. As shown in the dot diagram, two of the original five electrons in the outer shell were not needed to form ordinary covalent bonds and are left as an unshared electron pair. This structure

$$\begin{array}{c} H \\ \cdot\cdot \\ :N:H \\ \cdot\cdot \\ H \end{array}$$

is intermediate between the structures formed by bonding hydrogen around carbon and around oxygen.

$$\begin{array}{cc} H & \\ \cdot\cdot & \cdot\cdot \\ H:C:H \quad & H:O: \\ \cdot\cdot & \cdot\cdot \\ H & H \end{array}$$

Carbon is able to form a tremendous variety of compounds because of its abilities to form chains and rings, to have various branching combinations because of the several groups bonded on each carbon, to be involved in single, double, or triple covalent bonds, and to have geometry that makes molecules differ sometimes when the order of attached groups is changed (even though the groups themselves are the same). Chapter 17, Carbon and Its Compounds, describes these bonding characteristics in more detail. Although nitrogen is not as good as carbon at forming chains with itself, it is close to carbon in both size and electronegativity and can be bonded onto (or into) the carbon chains and rings. The capacity for three bonds is enough to let it bond in almost all the ways carbon does with a lone pair replacing one of the bonds. The structures where nitrogen is linked to a carbon chain with only single covalent bonds (plus the one lone pair) are the most common.

Lone pairs of electrons were a key feature in setting up the hydrogen bonding and resulting unique properties of H_2O. Ammonia (NH_3) and the related compounds where one (or more) hydrogen has been replaced by a carbon chain can also have hydrogen bonding. The nitrogen differs from oxygen by having (1) less electronegativity, (2) more covalent bonding, and (3) less lone pairs to form hydrogen bonds. Hydrogen bonding is important in NH_3 but less extensive than in H_2O. Because nitrogen is less electronegative, NH_3 is less able to donate H^+ (act as an acid) than H_2O. Nitrogen's lower electronegativity compared to oxygen also makes NH_3 able to share

Nitrogen can have both a variety of bonding like carbon and effective hydrogen bonding like oxygen.

NH₃ and related compounds are basic and tend to form positive ions.

its lone pair of electrons (act as a base) more readily than the lone pairs on H_2O. Therefore NH_3 is a base which tends to pick up protons to form NH_4^+ ions. These characteristics carry over to the compounds in which nitrogen is bound to carbon chains or rings. The lone pair on nitrogen acts as a *base*, tending to pick up H^+ to form *positive ions*. As an extreme case, nitrogen may even bond to a fourth carbon chain and still form an ion of $+1$ charge. Formation of positive ions or hydrogen bonding through the nitrogen greatly increases the water solubility of the covalent carbon chain compounds by making them more like water.

Large amounts of N_2 are converted to NH_3 by the commercial Haber process. N_2 and H_2 are reacted at high temperatures (to increase rate of reaction) and pressures (to increase yield) in the presence of catalysts. Some of the NH_3 is converted to positive oxidation states of nitrogen before use, and essentially all of it is converted to a positive state by soil bacteria when it is applied as fertilizer. This occurs because of the abundance of oxygen, which can displace nitrogen from its negative oxidation state. One possible reaction of O_2 with NH_3 forms H_2O and N_2. The very strong bond in N_2 makes this reaction energetically favorable, but it is undesirable when we are seeking forms of nitrogen usable by plants or animals. By use of catalysts we can selectively bypass N_2 formation and get nitrogen oxides instead. Soil bacteria are also able to react NH_3 to positive oxidation states instead of producing N_2. Positive oxidation states of nitrogen are not easily formed from N_2, but direct formation of some nitrogen oxides in car engines has become a factor in production of smog.

Nitrogen oxides with nitrogen oxidation states of $+1$, $+2$, $+3$, $+4$, and $+5$ are known. The simplest formulas of the oxides are N_2O, NO, N_2O_3, NO_2, and N_2O_5. When NH_3 reacts with O_2, there is only one nitrogen atom in each reacting NH_3. Therefore, N_2O, N_2O_3, and N_2O_5 cannot be formed unless two reacting molecules get together. If two nitrogens get together, the very stable N_2 molecules are usually formed. Therefore, NO and NO_2 are the important oxides formed when NH_3 is reacted with O_2. This seems reasonable since these are the simplest oxide molecules that could form, and simple units allow higher entropy (greater disorder), which is favorable as noted in Chapter 11.

There are a number of nitrogen oxide molecules and negative ions with properties determined by the odd electron arrangement of nitrogen.

But NO and NO_2 are actually very unusual molecules. Nitrogen is an odd numbered element, having seven electrons. When it bonds to an oxygen with eight electrons (or to two oxygens with eight electrons each) the resulting molecule has an *odd number* of electrons. NO and NO_2 are called **odd molecules** because of their odd number of electrons. The dot diagrams for NO shows that there must either be an unshared electron (and incomplete outer shell) on nitrogen or on oxygen or a resonance involving some of both. This unpaired electron makes the molecule very reactive. One possible reaction

$$\cdot \ddot{N} :: \ddot{O} : \qquad : \ddot{N} :: \ddot{O} \cdot$$

is to combine with another identical molecule to form a **dimer** (unit made from two of the basic parts). The dimer N_2O_2 has a very satisfactory dot diagram, but it does not form readily.

$$:\ddot{O}::\ddot{N}:\ddot{N}::\ddot{O}:$$

The best explanation of this failure to dimerize would require use of molecular orbital theory plus the entropy factor favoring the smaller NO units. A second possible reaction would be combination with any other reactive molecule present. If NO is released in air, the most reactive other molecule present is O_2. NO reacts rapidly in air to form NO_2, a brownish gas. The dimerization of NO_2 to N_2O_4 occurs more readily than dimerization of NO. Both NO_2 and N_2O_4 are formed, and an equilibrium is set up between dimerization of NO_2 and decomposition of N_2O_4.

In the presence of excess O_2, the nitrogen should be oxidized to the highest possible oxidation state, $+5$. If any other oxidation state was also going to be favored, we would expect it to be the $+3$ state, corresponding to the filled $2s$ subshell configuration. Both of these states can be formed with simple units containing only one nitrogen atom if they are *negative ions* instead of neutral molecules. Both the NO_2^- ions ($+3$ state) and NO_3^- ions ($+5$ state) are given extra stability by resonance, which is shown by the existence of equivalent favorable dot diagram structures. The nitrate form, NO_3^-, is by far the most common.

Nitrate, NO_3^-, is the common form in which nitrogen occurs in compounds.

$$\left[:\ddot{O}:\ddot{N}::\ddot{O}:\right]^{1-} \longleftrightarrow \left[:\ddot{O}::\ddot{N}:\ddot{O}:\right]^{1-}$$

$$\left[\begin{array}{c}:\ddot{O}:\\:\ddot{O}:\ddot{N}::\ddot{O}:\end{array}\right]^{-} \longleftrightarrow \left[\begin{array}{c}:\ddot{O}:\\:\ddot{O}::\ddot{N}:\ddot{O}:\end{array}\right]^{-} \longleftrightarrow \left[\begin{array}{c}:\ddot{O}\\:\ddot{O}:\ddot{N}:\ddot{O}:\end{array}\right]^{-}$$

Nitrate salts are very soluble in water and can be taken in and utilized by plants. Nitrite, NO_2^-, is not common in nature. Its use as a synthetic food preservative has led to some concern. Under some conditions it may form nitrogen containing biological materials that could have a relationship to cancer. Nitrite is also formed during the "natural" food preservation process of smoking meats, and it has been suggested that a diet with very heavy consumption of smoked meats may be inadvisable. Nitrate in great excess can also upset biological balances, apparently by being converted to excess NH_4^+. Cattle have been killed by excess nitrate, an interesting paradox since controlled supplements of nitrogen compounds can aid their growth.

14–3 Phosphorus

Phosphorus is in the same chemical family as nitrogen, but it has one more electronic shell. Chemistry of some nitrogen and phosphorus halides was used in Section 8–5 to illustrate the effects of the extra shell. Phosphorus is less abundant than nitrogen, but it does not form a nonreactive diatomic molecule like N_2. Phosphorus atoms are larger than nitrogen atoms, so the π bonding formed from parallel p orbitals is less effective in phosphorus. Therefore, multiple bonding on phosphorus (although it is possible) provides much less extra bond strength over single bonding than was the case in nitrogen, and the bond in a P_2 molecule would not be particularly strong. Instead, phosphorus forms single bonds to other phosphorus atoms and builds up a three-dimensional structure to provide complete bonding.

Elemental phosphorus is singly bonded as giant molecules (red phosphorus) or P_4 units (white phosphorus), which are very reactive.

The most stable arrangement is called **red phosphorus.** It consists of complex giant molecules. When phosphorus is formed, the atoms usually come together in a simpler but less stable form called **white phosphorus.** White phosphorus consists of P_4 molecules, and it can be melted or evaporated without disrupting these P_4 units. Red phosphorus burns readily in oxygen to form phosphorus oxides. The P_4 structure of white phosphorus burns even more readily. It must be covered to prevent spontaneous ignition in air. Each phosphorus atom in the P_4 unit has three single bonds and one lone pair, much like the structure around nitrogen in NH_3, but the bond directions are held much closer together than a normal tetrahedral arrangement with one lone pair. These strained bond directions can be relieved by inserting an oxygen (with the usual "bent" geometry of the two bonds on each oxygen) into each of the six edges to form P_4O_6. An additional oxygen can bond on each lone pair by a coordinate covalent bond to form P_4O_{10}. Figure 14–1 shows the structures of P_4, P_4O_6, and P_4O_{10}.

Phosphorus can form other compounds such as PH_3 (comparable to NH_3), but the readily formed positive oxidation states are much more common and important. As expected from the general trend in the periodic table (see Sections 7–9 and 8–5), phosphorus is less electronegative than nitrogen and tends to form positive oxidation states more readily than nitrogen. As was the case with nitrogen, the $+5$ state (P_4O_{10} and its derivatives) is the usual product in the presence of excess oxygen. Again, by forming negative ions, simpler stable species can be formed. However, the phosphorus atom is enough larger than nitrogen that more bonding groups of electrons are accommodated. Also, the multiple bonding character of the resonance in an NO_3^- type structure would not bond as effectively around the larger phosphorus atom

Figure 14–1 Structures of P_4, P_4O_6, and P_4O_{10}.

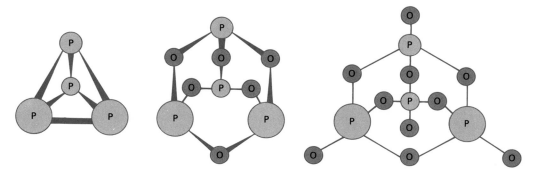

because of the poorer π bonding of the more separated p orbitals. Therefore, instead of the structures involving resonance bonding found for NO_2^- and NO_3^-, all bonds to the phosphorus are single bonds* and the stable ions formed are PO_3^{3-} and PO_4^{3-}. The common +5 form, the phosphates, exist in other forms besides the orthophosphate, PO_4^{3-}.

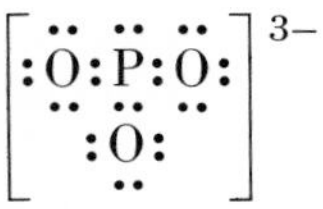

phosphite 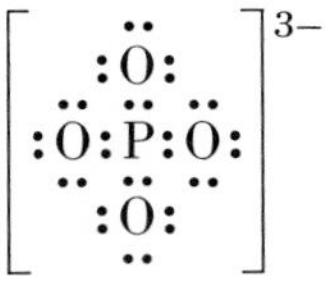   orthophosphate

Upon heating forms containing hydrogen and orthophosphate, water can be driven off and two phosphorus atoms become linked via a shared oxygen. The products formed by linking two or more phosphates in this way are called **polyphosphates**. The $P_2O_7^{4-}$ unit formed when one oxygen "bridge" links two phosphorus atoms is called pyrophosphate. If a second oxygen "bridge" forms to link a third phosphorus atom, we get a $P_3O_{10}^{5-}$ unit called triphosphate.

Phosphorus forms highly charged negative ions containing oxygen including oxygen-bridged polyphosphates.

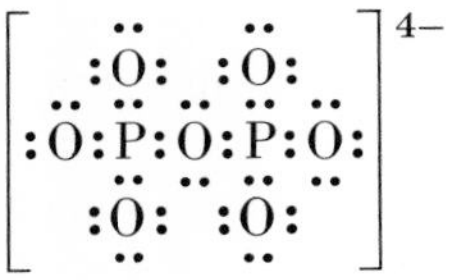

pyrophosphate 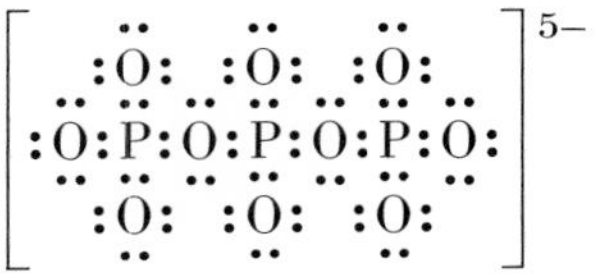   triphosphate

If the formation of oxygen bridges continues long enough, the "extra" oxygen at the end of each chain becomes insignificant compared to the large number of PO_3^- units that have been tacked on. The resulting compounds, called metaphosphates, are written with the formula PO_3^- (like $NaPO_3$, sodium metaphosphate) but the PO_3^- is not at all like nitrate, NO_3^-. NO_3^- ions are individual ion units, whereas each PO_3^- is just one part of a long polyphosphate chain.

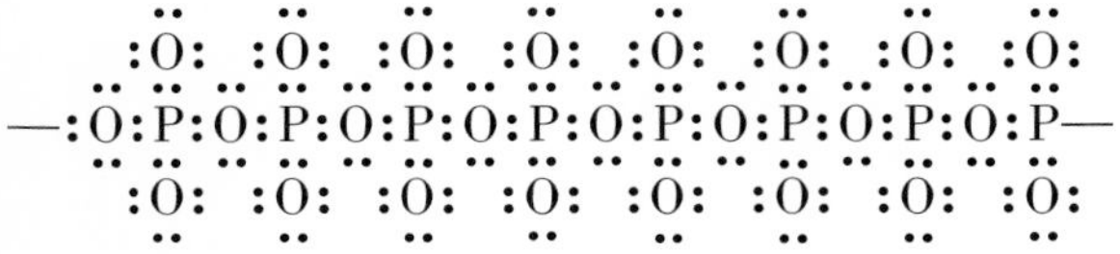

part of a metaphosphate
(charge of -1 for
each PO_3 unit)

---

*There is also some double bond character, utilizing the d electron level available in phosphorus. Therefore some books will show PO_4^{3-} with 3 single bonds and one double bond. We choose not to show this relatively weak double bond contribution.

The importance of these polyphosphates will be shown in Part IV, The Chemistry of Life. One of the specific compounds mentioned will be adenosine triphosphate, mentioned in Chapter 21 in the discussion of energy storage and use. Energy can be stored by forming polyphosphates. When the oxygen bridge is broken (but replaced by another oxygen atom so there is no reduction in the number of bonds), the negative charges can move away from each other and the stored energy is released. All energy in biological systems (such as energy used for muscle movements) works through this polyphosphate mechanism. Other crucial biochemicals, such as genetic material, also involve phosphates.

Insolubility of many phosphates limits the supply of phosphorus available for growth.

Because phosphates occur as the highly charged -3 ions (or the even more complex polyphosphates) instead of a simple -1 species like nitrate, they tend to form insoluble solids, particularly with ions having charges greater than $+1$. $Ca_3(PO_4)_2$ is a particularly common and important insoluble phosphate. It is the main constituent in bones. Since plants take in their nutrients by osmosis through porous membranes, these insoluble phosphates are not available to them. Only phosphorus in the form of *soluble* orthophosphates is useful to plants. These can be produced by acid treatment of insoluble orthophosphates. Phosphoric acid, H_3PO_4, is an acid that ionizes in three steps, ranging from one moderately strong acid ionization to a very weak acid ionization.

$$H_3PO_4 \rightleftharpoons H^+ + H_2PO_4^-$$
$$H_2PO_4^- \rightleftharpoons H^+ + HPO_4^{2-}$$
$$HPO_4^- \rightleftharpoons H^+ + PO_4^{3-}$$

H_3PO_4 is a moderately strong acid, salts of $H_2PO_4^-$ are weakly acidic, but salts of HPO_4^- usually act as bases, accepting H^+ from solutions rather than donating H^+ when the solution is near neutral. The K_a values for the three ionizations are

$$K_1(H_3PO_4) = 7.5 \times 10^{-3}$$
$$K_2(H_2PO_4^-) = 6.2 \times 10^{-8}$$
$$K_3(HPO_4^{2-}) = 4.8 \times 10^{-13}$$

If $Ca_3(PO_4)_2$ is treated with acid, it can be converted into the soluble $Ca(H_2PO_4)_2$ which provides usable phosphate to plants. The usual methods of preparation for commercial phosphate fertilizer are

$$Ca_3(PO_4)_2 + 2\ H_2SO_4 + 4\ H_2O \longrightarrow Ca(H_2PO_4)_2 + 2\ (CaSO_4 \cdot 2H_2O)$$
(mixture sold as superphosphate of lime)

$$Ca_3(PO_4)_2 + 4\ H_3PO_4 \longrightarrow 3\ Ca(H_2PO_4)_2$$
(triple superphosphate of lime)

Acid converts orthophosphate to a soluble form, $H_2PO_4^-$ ions.

Both $Ca_3(PO_4)_2$ and $CaHPO_4$ are quite insoluble; they can be used as fertilizers, but the phosphorus is not available to plants until it is acidified to the $H_2PO_4^-$ form. This occurs naturally when the fertilizer is exposed to rainfall because dissolved CO_2 makes rainfall acidic (see Section 13-5.3).

The release of phosphorus by this method is slow, so the amount of available phosphorus (and hence plant growth) may be quite limited.

14–4 Fertilizer and Food Production

When more of the limiting nutrient becomes available, growth increases until the supply of this nutrient or some other becomes exhausted and a new growth limit is reached. Water and CO_2 are returned for reuse by transport through the atmosphere. Water is often the limiting factor for growth of land plants, so irrigation (and also cultivation techniques that cut down water evaporation) can cause large increases in yield. As production is increased by irrigation or as crops are harvested, some other limiting growth factor is likely.

Under natural conditions, that is, when plants grow and die in the same spot, the other nutrient materials are returned for reuse. When crops are removed, the solids incorporated into them are depleted from the soil and may become limiting. Nitrogen (in forms other than N_2) often becomes the limiting factor. Phosphorus (in soluble form) and potassium also become depleted at times, and a large number of other elements occasionally become a growth limiting factor. In these cases, growth can be increased by **fertilizing** with added amounts of the needed substance or substances. There are two basic ways in which the necessary nutrients can be restored. The natural cycle for return of nutrients can be completed by bringing back the nutrient containing wastes from animals (or people) who used the harvested crops. These materials (such as manure) are **natural fertilizers.** The other method is to supply specific chemical **artificial fertilizers** to make up the shortages.

Artificial fertilizers can be much more efficient than natural fertilizers if they can be selected to fit the specific needs of the soils and plants. For best results, a chemical analysis of the soil should be performed first. These analyses can be complex because many elements, like phosphorus, have sharp differences between the amount present and the amount that is usable. Often farmers simply add the fertilizers that they know from experience are usually needed—nitrogen alone for crops like corn which need a lot of it, or nitrogen, phosphorus, and potassium.

This practice can lead to costly mistakes or excesses. Excessive fertilizer is useless if the supply of carbon dioxide, water, or light becomes the limiting factor. In corn belt areas with good rainfall, greater use of fertilizer has had to be matched by closer planting to increase efficiency of light utilization and other factors. Concentration on the three major fertilizer nutrients can also lead to shortages of other necessary elements. Some fertilizer production has been converted from $(NH_4)_2SO_4$ to NH_4NO_3 to increase the amount of usable nitrogen per pound, but taking the sulfate out is not always wise. Sulfur is also required for growth and is removed in harvested crops, and, like phosphates, much of the sulfate in nature is in insoluble form. Some soils need the sulfur in the relatively soluble $(NH_4)_2SO_4$ more than they need the nitrogen. Artificial fertilizers can aid food production tremendously, but the

Artificial fertilizers are selected to increase the limiting nutrients, and knowledge is needed for best use.

great chemical complexity of soils and living things makes continuous study and careful use important.

Natural fertilizers may avoid some problems by returning small amounts of the necessary trace elements (including things like copper and boron) even if we are not informed enough to know they are necessary. However, it is also possible for natural processes to separate the various nutrients and let us return the ones that are already abundant and not the ones that are needed. Natural fertilizers can also sometimes spread diseases. In parts of Asia where human wastes are used extensively for fertilizer, vegetables may cause severe intestinal illnesses for anyone but the natives who have already developed immunity. Some animal diseases are spread similarly.

Our "civilized" solution is to discharge human wastes, treated or untreated, into lakes and streams. In these waters the wastes do the same thing they would do on land—they act as natural fertilizers. In places where the amounts are large because of accumulation from cities, human wastes can have a tremendous effect on the waters. Concentration of animals also creates large amounts of natural fertilizers. When these can be spread and allowed to go into soils, they are useful as fertilizers. However, in areas with freezing weather there are times of year when they cannot get into the soil. When thaws and heavy water runoffs occur, much of the manure spread in winter is carried off into streams and lakes where it fertilizes the water.

Natural fertilizers such as human and animal wastes are rich sources of limiting nutrients.

14–5 Fertilization of Waters

Plants growing in water may have their growth limited by the available CO_2, light, or a nutrient substance in short supply. Lakes and streams formed from rainwater with little dissolved solids will have very little capacity to support plant growth. Any of a large number of elements may be so totally absent that the growth of the plants that need them is prevented. In these quite pure waters plenty of dissolved CO_2 (which is absorbed from air) is available, and light is available in any surface body of water. Therefore, the dissolved salts are the limiting growth factor.

As water proceeds toward the oceans it picks up more dissolved salts. The ability to support growth then increases, and the variety of nutrients that are limiting is cut down as all become available in at least moderate amounts. The most common limiting factors then become nitrogen (other than N_2), which is needed in quite large amounts, phosphorus, which is needed in smaller amounts but is not very available in soluble forms, and (as growth becomes very great) CO_2. In lakes the nitrogen or phosphorus may be almost completely tied up in plants (and fish) at the peak of the growing season. When the plants die the remains settle to the bottom and decay, thus releasing the nutrients. Since the decay is on the bottom (or in the silt on the bottom) whereas most growth occurs in the surface regions where the necessary light is most available, stirring of nutrients from the bottom to the surface becomes an important factor in fertilizing a lake.

Oceans follow patterns similar to lakes except that most soluble salts are available in excess. In the oceans nitrogen and phosphorus become almost exclusively the limiting nutrients, and exchange between the bottom and surface becomes especially important. Light is absorbed as it passes through water, so all plant growth must occur within a surface layer that is, at most, about 100 m thick. In deep ocean areas, the dense bottom waters have almost zero mixing with the less dense (usually warmer) surface water. Nutrients that settle to the bottom are essentially lost forever. Because of this the central areas of the world's oceans have little nitrogen and phosphorus and cannot support much plant growth. It has been said that 90% of the world's ocean surfaces are biological deserts, supporting life but not very abundantly.

Nutrients in water systems tend to be concentrated in sediments or bottom waters.

The shallower areas near shore are quite different. Mixing from bottom to surface returns enough nitrogen and phosphorus to support more life in these areas. Even more striking is the extreme fertility of those few areas where natural patterns bring the nutrient-rich bottom waters up to the surface. The tremendous plant growth in these areas supports small animals and larger fish in fantastic numbers. The area of upwelling currents off the west coast of Peru produces one of the world's richest fishing areas, and other upwelling areas have similar effects.

Stirring and upwelling currents stimulate growth in water by returning nutrients to the surface waters.

The massive fertilizing effect of such an upwelling current can be illustrated by a side effect. Sea birds eat millions of tons of the abundant small fish near Peru and Ecuador. It has been estimated that these birds deposit about a million tons of bird droppings per year on the offshore islands they use for nesting. The thick layers of droppings are rich in usable nitrogen and phosphorus, so men are now "harvesting" these droppings by bulldozers for use as fertilizers in South America. Some of the most interesting ideas for long term increases in the world's food supply involve the creation of similar regions of upwelling as byproducts of other human activities. In particular, the waste heat disposal problems mentioned in Section 13–5.8 could be solved by using hot water to cause an upwelling current offshore.

14–6 Secondary Effects of Fertilizer

It is ironic that the plant growth explosion considered thrilling in the oceans is considered a disaster when it happens in our lakes. The clear, quite pure water of our biologically least productive lakes is considered a greater asset than the possible food production. Our handling of nutrient materials can transform lakes and streams from clear water to unwanted gardens. The process of enrichment of nutrients in lakes, called eutrophication, is the natural fate of lakes, but human activity may speed the process up.

Our clearest lakes tend to be near the water source, fairly deep, and geologically recent. The water flowing into them has not had much chance to dissolve solids, and the lakes have not received much washed-in silt. Because they are deep they stratify with a thermocline, a sharp temperature variation, separating a surface layer (warm in summer, perhaps freezing in winter) and a denser body of deep water (about the same temperature year around). There is very little mixing between these layers, so nutrients carried to the bottom in silt and plant or animal remains is not easily returned to the surface layer where plants would grow.

As nutrients reach the lake, plants and fish develop, but only in low populations. Plants are eaten by small "forager" fish which are in turn eaten by larger "scavenger" fish. Because the populations are low, enough dissolved oxygen reaches all levels of the lake to support fish and other animal life. The larger fish, including varieties such as trout which are desirable to people, thrive particularly in the cool, relatively constant-temperature deep water.

As lakes age, they tend to fill up with silt and build up more nutrients. As they become shallower, they also allow more nutrients to be stirred from the bottom to the surface, and they may lose their stratification and change over to different types of fish such as carp. As the lake continues to fill in with silt, its normal fate would be to become a swamp, which is very productive biologically, and then fill in to form dry land. All of the changes are slow enough to allow a gradual shift to new life forms fitting the new conditions.

Fertilization of water can lead to eutrophication of lakes.

If the surface waters become rich in nutrients while the lake is still stratified by thermocline formation, the deep water may undergo a drastic change. Plant growth near the surface, particularly of the blue-green algae, may become great enough to cut off all light to plants farther down, which then die. When very large amounts of these dead plants settle to the bottom, the bacteria decaying them may consume oxygen faster than it can be supplied to the deep water across the thermocline barrier. The deep water then becomes depleted in oxygen. Bacteria can reduce the oxygen concentration far below the level needed by fish. That is the natural result any time the nutrients for animal life (food supplies) are available out of proportion to the oxygen. This changes the entire chemical nature of the deep water from an oxidizing environment (excess oxygen present) to a reducing environment.

Biological oxygen demand greater than the oxygen supply damages water quality.

The interaction of the water with the bottom silts then becomes completely different. Some normally insoluble compounds begin to dissolve in the water with bad results. An example is iron, normally present as Fe^{3+}, which forms compounds of low solubility. The positive regions of H_2O molecules cannot provide attractive forces to negative ions that are competitive with the attractions to Fe^{3+}, so Fe^{3+} compounds are not enough like water to dissolve well. But when a reducing environment converts the Fe^{3+} to Fe^{2+}, the difference in attractive forces is less and the solubility is greater, as expected from the "like dissolves in like" principle (Section 10–1). This dissolved iron is considered *very* undesirable in its flavor effect on drinking water. And it appears in the bottom water which is normally the most desirable as water supply. If a city shifts its water inlets from a lake to shallow water to avoid dissolved iron, other problems arise. The water is then hot in summer, the inlet may freeze in winter, and the water is loaded with algae which are hard to filter out.

The lack of oxygen also kills all the fish in the layer below the thermocline. That removes the most desirable game fish, and it may also upset the balance between "forager" fish and "scavenger" fish, particularly if human activity made the change so abrupt the fish could not adjust. If the "forager" fish have a great increase in food supply while most of the deep dwelling "scavenger" fish die from lack of oxygen, a population explosion occurs. Eventually unfavorable conditions cause large groups of the forager fish to die, and tons of dead fish may be washed up on the shores of the lake. The growth of algae may also form a thick soup or scum on the lake's surface.

The results of eutrophication are undesirable when people want to use a lake for water supply and recreation, so it is unfortunate that human activity tends to accelerate the process where many people are present. Since eutrophication is basically the result of applying nitrogen and phosphorus fertilizers, we cause it when we add these substances to the lake's water sources. The biggest source is via the natural fertilizers, human and animal wastes. Even complete degradation of sewage does not remove the basic nutrient elements. Water downstream from city sewage treatment plants is greatly enriched. Other sources contribute, such as the washing off of manure which was spread on frozen ground. Washing off of artificial fertilizers can also add to the total, but this is usually not much of a problem. They are usually applied at times and in ways that keep most of the fertilizer in the soil.

Human wastes can fertilize plant growth or deplete oxygen in water directly by serving as animal nutrient.

Nutrients added in forms not used as fertilizers can also add to eutrophication. The principal example is the phosphates used in detergents to react with some of the components of hard water that would otherwise interfere with the detergent action. Soluble orthophosphates improve the efficiency of detergents in water. Therefore they have been added to many detergents, particularly the best heavy duty ones. In 1970 it was estimated that half of all the usable phosphates going into our lakes and streams came from detergents. A study of the causes of eutrophication of Lake Mendota in Madison, Wisconsin, showed about half of all phosphates entering the lake came from a single dairy plant which used heavy duty industrial detergent to clean equipment.

There is much furor to stop the eutrophication of lakes, but not all the proposed solutions are sensible ones. Some lakes would be undergoing eutrophication naturally without human interference. Reversing the trend in such cases is probably impossible. Where the process can be stopped, the nature of the nutrient materials must be kept in mind. At Lake Tahoe on the California–Nevada border, nutrient materials are being chemically removed from sewage output after the ordinary treatment has been completed. One part of this plant converts the nitrogen to NH_3 and then takes it out of the water and blows it out into the air. A little thought about hydrogen bonding and like-in-like solubility characteristics shows that the NH_3 will dissolve in water droplets and be brought back to the lake in rainfall or even spray droplets from the tower in which it is separated from the water. Because of insolubility, phosphates can be removed more easily than nitrogen compounds, so phosphate removal is (quite properly) getting most of the attention in antieutrophication efforts. The same Lake Tahoe plant that tries to remove nitrogen is doing a very good job of removing the phosphorus.

Phosphorus can be removed from water more easily than nitrogen.

Even when the problems are clearly known, as is the case with eutrophication, there is need for some care not to get carried away with "simple" solutions to very complex problems. In 1969 and 1970 there was a very large scale movement to substitute NTA (nitriloacetic acid) for phosphates as a water conditioner in detergents. Although this might have been a wise move, the pressures to do it rapidly forced detergent makers to begin the changeover before they could consider and test the possible effects of large amounts of NTA on sewage plants and the environment. Some of the possible heavy metal poisoning problems from this substitution are discussed in Section 16–12. Research results late in 1970 showed that serious side effects from NTA were possible, so the government banned its use in detergents.

Perhaps the saddest chapter in the NTA story came in the announcement by a major detergent manufacturer that they would no longer consider NTA regardless of the results of further study. Public confidence in it had been undermined by its being banned. The implication seems to be that pressure from environmental groups could get companies to adopt a product with a possible incompletely tested potential to poison us, but no amount of scientific evidence could restore its use later even if it turned out to be safe and helpful. A reliance on the scientific method, with the patience to test carefully and use the conclusions (also with care), would lead to better results. In the case of NTA, testing on a very large scale would be possible now because Canada and Sweden have allowed it to be put into widespread use. The aftereffects of this use should be carefully studied for several years so that NTA could be withdrawn if hazards appear or could be approved for wider use if it proves safe.

14–7 Human Effluents and Biological Oxygen Demand

The damaging effects of eutrophication occur because concentration of dead plant material increases the demand for oxygen to more than natural processes can supply. The region where this happens then becomes chemically different from the "normal" situation in our oxygen-rich planet. Very similar effects can be caused anywhere else where the local demand for oxygen becomes very high. This frequently occurs in rivers because of our practices in handling wastes. The biological need for oxygen to degrade materials is called **BOD** or **biological oxygen demand.**

Our concentration of people into cities and our collection of wastes for disposal into water creates points of unusually high biological oxygen demand. Human wastes are incompletely used energy sources as well as being natural fertilizers. Bacteria complete the oxidation of the wastes, and they normally consume oxygen in the process. When wastes enter streams in moderate amounts and at different locations, enough oxygen can be supplied from the air to complete the oxidation. Fast flowing "white water" sections of rivers are particularly effective at getting oxygen into the water if it is needed. Under those conditions the main problem caused by the wastes is their support of bacterial life, including disease-causing bacteria.

When wastes go into a lake, as often happens to sewage from summer cabins, it pollutes the lake with disease bacteria, fertilizes the waters, and also increases the demand for oxygen in the deep water where decay takes place and oxygen supply is already subject to depletion. When the collected wastes of a city go into a lake or river, the need for additional oxygen can become very large. The oxygen concentration may remain below the level required by fish for many miles downstream. A river then becomes a system of plants and bacteria that can survive at low oxygen concentrations, with no higher animal life present.

These damaging effects can be lessened by sewage treatment. The effectiveness of the treatments used varies considerably. **Primary** (or one stage) **treatment** is simply filtering solid materials out of the sewage. No chemical changes are made, and bacteria and much of their food supply continue to go into the output. These outputs still have a very large biological oxygen demand. In small amounts, a somewhat better treatment can be obtained by holding the material for a while in a septic tank. That allows bacteria to oxidize most of the remaining food supplies before the sewage escapes. The oxidation can be done either by aerobic bacteria, which use oxygen, or by anaerobic bacteria, which live in the absence of oxygen. Anaerobic bacteria use up part of the available energy in the low oxygen regions, forming products like NH_3 and H_2S, which are often odoriferous and objectionable to humans. As oxygen becomes available farther along the sewage path, aerobic bacteria complete the process.

To accomplish a similar bacterial degradation (called **secondary treatment**) on a city's sewage requires some mechanical help. The oxygen requirements are so high that air is bubbled through tanks in which the bacteria are oxidizing wastes. Since only the oxygen in the air is being used, some sewage plants are now separating the oxygen from air and blowing pure oxygen into the tanks. Smaller tanks can then accomplish the job. These tanks are covered so the escaping gases can be trapped and recycled to avoid wasting the excess oxygen. The sewage flow carries the growing bacteria toward the outlet, and large amounts of bacteria-filled sludge form and are removed. The efficiency of the plant is improved by adding some of this sludge to the input so a good supply of the most appropriate types of bacteria is available from the start. This recycled sludge, which has evolved to the best types of bacteria for that kind of sewage, is called activated sludge and the process is called the **activated sludge process.** The sludge produced in these plants is an excellent natural fertilizer, but to prevent spreading disease it must be treated to kill the bacteria before use to fertilize crops like vegetables for human consumption.

Sewage treatment by the activated sludge process can lessen biological oxygen demand.

Secondary treatment leaves much of the nutrients in solution as soluble nitrates and phosphates. Processes to remove these nutrients are being set up to prevent eutrophication problems. These processes are called **tertiary treatment.** The plant at Lake Tahoe mentioned in the preceding section (Section 14–5) is a pioneer effort at tertiary treatment. Its successful portion (the phosphorous removal) may be widely adopted.

14-8 Oxygen Demand of Industrial Effluents

Industrial activity can cause large, concentrated biological oxygen demands.

The rate of oxygen supply to water is related to water roughness.

Because it is the excessive availability of food nutrients that leads to oxygen depletion problems in water, any good nutrient source usable by bacteria or other animal life can contribute to the problem. When plants grow and are used by animals in the same area, there is a balance between oxygen released by the plants and used by the animals. But when plants grow (releasing oxygen) in one place and are degraded in another, degradation must be limited by the rate at which oxygen can be supplied. Industrial activities often result in the movement of large amounts of material into water. This can cause a number of different pollution problems, but we will concern ourselves here only with effluents that create a biological oxygen demand in water.

Discarded remains from food processing are an obvious food source for animal life. Some sections of the Snake River in Idaho are seriously affected by output from potato processing plants. These materials are not poisonous in any way, but they can kill off fish in the river by using up the dissolved oxygen. The problem becomes serious because (1) so much oxidizable material is put in at one spot by a plant effluent and (2) biological oxygen demand ends up being put in faster than the river can pick up oxygen.

Building dams contributes to the problem by lowering the rate at which the water can pick up oxygen. Free flowing streams with rapids areas have much more air mixed into the water than the surfaces of calm lakes, so the exposure to oxygen and the rate at which it can be dissolved is reduced when dams form lakes. In the case of the Snake River, the "white water" flow in rugged Hell's Canyon restores the river's oxygen between the Idaho potato processing area and the lower portions of the river. Most other rivers have already passed their last "white water" sections before reaching the industrial areas, so they may never recover from oxygen depletion. The Cuyahoga River in Cleveland is an example of a river completely uninhabitable to fish and unable to recover its depleted oxygen.

Actually, most industrial effluents are less obvious than potato starch and peelings as food sources. One of the most important degradable items is wood. The size of the industrial effluent problem related to wood can be seen from the following facts. In 1970 a modified wood pulp mill went into operation offering the possibility of a "great improvement" in paper production. The amount of each tree recovered as useful product could be raised from about 45% to 52%. That means that the unused portions have been (and will continue to be) about equal to the total production of paper, even in the *most efficient* plants. Much of this waste is bark burned as fuel at the paper mills or small branches left in the forests and burned as "slash," but the amounts coming out in the plant effluent are also large.

Treatment areas, somewhat analogous to municipal sewage treatment, can separate or degrade some of this from the water before it reaches the rivers, but the harsh chemicals used to make wood pulp create a very difficult environment for the degrading bacteria. The treatment may be effectively limited to a settling pond to remove solid, the equivalent of mere primary treatment of sewage. The potential polluting effects of the effluents that are not removed by the treatment is staggering. In some cases the natural streams can handle the BOD load, but care and a consciousness of the chemistry going on are very important when any unnatural large bulk of material is handled. Anything that can serve as either a plant or animal nutrient could eventually upset the chemical balance.

Skills Expected at This Point

1. You should be able to list the principal needs for plant growth or for animal growth and the key role played by limiting nutrients.
2. You should be able to describe the N_2 molecule and the other bonding patterns around nitrogen and relate these structures to the biological significance and limited supply of usable nitrogen.
3. You should know the structures of elemental phosphorus and of the phosphate ions and be able to contrast their behavior to N_2 and NO_3^- and explain the differences.
4. You should be able to describe preparation and use of fertilizers, recognize wastes as a fertilizer, and be able to explain how fertilizers can be harmful in some water systems.
5. You should recognize the term "biological oxygen demand" and its abbreviation BOD and be able to use the term in explanations of water pollution problems.
6. You should be able to describe eutrophication and its consequences with reference to the limitations on oxygen supply.

Exercises

1. For each of the following situations, what one or two things are most likely to be the growth limiting factor? Give a very brief explanation for your choice in each case.
   (a) Plant growth in a desert.
   (b) Corn growth in a field that has been farmed for many years.
   (c) Fish growth below the thermocline of a eutrophied lake.
   (d) Plant growth in the open ocean.
   (e) Plant growth in a mountain lake.
   (f) Bacterial growth in a river.
2. Compare the structures and reactivity of elemental nitrogen and elemental phosphorus.
3. Why are NO_3^- and PO_4^{3-} common forms on earth?
4. List three main causes of excessive BOD in water.
5. Draw three dot diagrams showing polyphosphates with different numbers of oxygen bridges.
6. (a) Describe the sequence of events in natural eutrophication.
   (b) Describe the sequence of events in artifically induced eutrophication.
   (c) Compare the opportunities for evolution of species adapted to the new situations in part (a) and (b).
7. List the main types of natural and artificial plant fertilizers.

8. Describe the activated sludge process, its purposes, and its limitations.
9. Outline the waste disposal program which should be expected from a large paper mill. Be reasonable—do not ask for elimination of "problems" that would not cause damage.

Test Yourself

1. Phosphorus is limited in availability as a nutrient because
   (a) the usable forms of phosphorus evaporate into the air.
   (b) the common form, P_4, is so strongly bonded that few plants or animals can use it.
   (c) the common form, PO_3^-, has been concentrated in the oceans.
   (d) most phosphorus is in the PO_4^{3-} form but plants can only use the PH_3 form.
   (e) the common forms of phosphorus have low water solubility.
2. Nitrogen is limited in availability as a nutrient because
   (a) the usable forms of nitrogen evaporate into the air.
   (b) the common form, N_2, is so strongly bonded that few plants or animals can use it.
   (c) the common form, NO_3^-, has been concentrated in the oceans.
   (d) most nitrogen compounds are very insoluble in water.
   (e) none of these.
3. A limiting nutrient is
   (a) something absent during plant or animal growth.
   (b) something abundant that is used in growth.
   (c) something present in small quantities that is used in small quantities in growth.
   (d) the substance, abundant or rare, that is used up first in growth.
   (e) the substance used in the largest quantities in growth.
4. The chemistry of N_2 molecules has a strong influence on growth because
   (a) N_2 is an essential nutrient.
   (b) N_2 is in very short supply on earth.
   (c) N_2 has an exceptionally strong chemical bond that cannot be broken by most plants and animals.
   (d) nitrogen has an odd number of electrons and does not form favorable covalent bonds with other elements.
   (e) reaction with N_2 would use up the limited available O_2, which the plants and animals must have for other purposes.
5. The output from a good activated sludge sewage treatment plant is generally safe to the environment except for one of the following problems. Identify the one associated with properly treated sewage output.
   (a) eutrophication of deep lakes.
   (b) depletion of dissolved O_2 in rivers due to the high BOD.
   (c) poisoning plants and fish downstream.

(d) destroying biological productivity in the oceans.
   (e) creating salt and bacteria levels that prevent treatment and reuse of the water by downstream cities.
6. The most biologically productive regions of the oceans are (choose two answers)
   (a) at river mouths where nutrients enter the sea.
   (b) in regions of upwelling where bottom waters reach the surface.
   (c) in deep ocean areas where pollutants have settled out.
   (d) in areas near shore where some material from the bottom is stirred to the top.
   (e) near polar regions where the surface water is colder than bottom water.
7. An important step in the preparation of phosphate fertilizers is
   (a) reaction with acid to make the phosphate more soluble.
   (b) oxidation of phosphite minerals to phosphate.
   (c) hydrolysis of metaphosphate to orthophosphate.
   (d) conversion of PO_4^{3-} to PO_3^{-}
   (e) none of these.
8. Ocean areas with extremely rich fishing (such as the area off Peru and Ecuador) are characterized by
   (a) incoming ocean currents, which concentrate fish in the area.
   (b) upwelling currents, which bring up nitrates and phosphates from the bottom.
   (c) warm water temperatures, which increase the rate of growth.
   (d) deep water, which allows plants to be spread out in a larger volume.
   (e) sheltered breeding areas.
9. Eutrophication of a lake leads to fish kills because of
   (a) nitrate poisoning.
   (b) starvation due to shortage of food.
   (c) inability to move in the viscous solution.
   (d) shortage of oxygen in the water.
   (e) all of these.
10. Eutrophication (over enrichment) of a lake is
    (a) a natural process that can neither be speeded up nor slowed down.
    (b) an unnatural process caused by human interference with natural balances.
    (c) a slow natural process that can be speeded up by human activities.
    (d) a desirable change that should be encouraged to increase fishing productivity.
    (e) an undesirable change that kills off plant life by "salting out" effects.
11. Activated sludge treatment plants remove BOD more efficiently than rivers because
    (a) the concentrated activity raises the temperature and therefore increases the activity of the bacteria.
    (b) the material is spread out too much in the rivers so bacteria cannot find it.

(c) the treatment plant has stagnant ponds in which bacteria grow while the river current prevents growth.
(d) O_2 is pumped into the treatment plant while the river runs out of O_2 to be used by the bacteria.
(e) the noxious gases produced by the bacteria can be removed in the treatment plant whereas they build up and kill the bacteria in rivers.

Nuclear Stability and Chemical Abundance

15-1 Radioactivity

In 1896 a Frenchman, Henri Becquerel, found that uranium compounds emitted radiation that could pass through paper or other thin sheets of solid and affect a photographic plate. Substances that emit such radiations are said to be **radioactive.** Interest in this new phenomena of **radioactivity** (emitting radiations) spread rapidly among scientists. The experiments they devised led to rapid advancement of our knowledge in many areas, including atomic structure (Chapter 5). We now know a great deal about radioactivity, both as it occurs naturally and how it can be produced artificially. Study of this information is useful in itself, but it can also explain the origin and distribution of elements and why some are common and others rare on the surface of the earth.

Natural radioactivity was quickly shown to produce several different types of emissions. These could be separated (see Figure 15–1) by passing them through an electrical field. Some of the particles were deflected in the direction expected for particles with positive electrical charges. These electrically positive emissions had a very intense effect on their target (such as a photographic plate) and could penetrate only thin barriers. They were given the name **alpha particles** and represented by the symbol alpha (α), first letter of the Greek alphabet. We now know they are actually He^{2+} ions moving very rapidly. A second type of emissions, which were named **beta particles** from the second Greek letter beta (β), were deflected in the opposite direction. We now know they are actually electrons. A third type of emission was unaffected by electrical fields. They were named **gamma rays** from the third Greek letter, gamma (γ). They are electromagnetic waves similar to radio waves or light rays. The terms α, β, and γ have proven a useful shorthand for describing radioactive emissions and have continued to be used, even though we now know their real identities.

Emissions observed in natural radioactivity are principally α particles ($_2^4He$ nuclei), β particles (electrons), and γ rays (electromagnetic radiation).

When a **radioisotope** (an isotope that is radioactive) emits radiation, the nucleus of the atom changes. When an α particle is emitted, the nucleus loses two units of positive charge and about 4 atomic mass units (amu). The products (α plus what is left of the nucleus) have the same total positive charge as the original nucleus. Because this positive charge is caused by the

Figure 15–1 Effect of an electrical field on radioactive emissions.

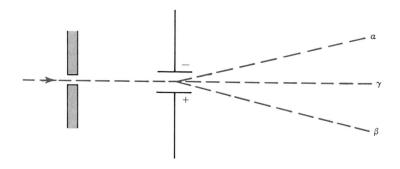

protons in the original nucleus, we can see that protons are not created or destroyed during α emission. Neutrons are also not created or destroyed during α emission. The sum of protons plus neutrons (which is the mass number, written above and to the left of the chemical symbol when we write the symbol for a particular isotope) is therefore the same for the original nucleus and the sum of its products. These facts are summarized in the following **nuclear equation** for α emission by the isotope, uranium 238 ($^{238}_{92}U$)

$$^{238}_{92}U \longrightarrow {}^{234}_{90}Th + {}^{4}_{2}He$$

Any reaction where one atomic nucleus breaks up into two or more products is a **radioactive decay,** and this one is an example of **α decay,** which is shown in Figure 15–2.

Nuclear equations can be balanced by balancing sums of mass numbers and charges.

Nuclear equations (for decays or any other kind of nuclear reaction) can be balanced by making the superscripts (mass numbers) add to the same total on each side and making the subscripts (charges), add to the same total on each side. If we do not know what one of the products (or reactants) is, we simply figure out the mass number and charge and that tells us what it must be. Therefore, if we knew $^{238}_{92}U$ underwent α decay, we could figure out the other product as follows:

mass number = 238 (mass no. of ^{238}U)
$\qquad$ = x (mass no. of other product) + 4 (mass no. of α)
$\qquad$ 238 = x + 4
$\qquad$ x = 234

charge = 92 (at. no. of $^{238}_{92}U$)
$\qquad$ = y (at. no. of other product) + 2 (at. no. of α)
$\qquad$ 92 = y + 2
$\qquad$ y = 90

Figure 15–2 α decay.

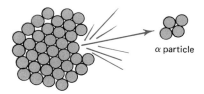

α particle

The atomic number of thorium is 90, so the other product must be $^{234}_{90}$Th. We will use nuclear equations throughout this chapter, but the important point for us to note now is that the **mass number** and **total charge** are the things kept constant. The number of protons and the number of neutrons also stayed constant during α decay, but we will see that is not true during β decay.

The $^{234}_{90}$Th formed in the example above is also radioactive, but by β decay instead of α decay. When we write down the nuclear equation for the decay of $^{234}_{90}$Th, we must have values for mass number and charge of the β particle (an electron). The electron has no neutrons or protons, so its mass number (sum of protons and neutrons) is zero. Its charge is -1. Therefore, our equation is

$$^{234}_{90}\text{Th} \longrightarrow ? + ^{0}_{-1}e$$

The other product must have mass number 234 and charge (which is also the atomic number for any nucleus) of 91 for the equation to balance. Therefore, it is $^{234}_{91}$Pa. We see that the overall effect of β decay is to raise the atomic number by one. Figure 15-3 shows a β decay.

The two decays described above are actually only the first two steps in a long series. Each product formed by α or β decay is itself radioactive until finally the **stable** (not radioactive) **isotope** $^{206}_{82}$Pb is reached. In this series from $^{238}_{92}$U to $^{206}_{82}$Pb a total of 32 units must be lost from the mass number. That happens in a total of eight α decays. Those eight α particles lost would reduce the charge by $8 \times 2 = 16$. A final charge of $92 - 16 = 76$ would be obtained if there were only α decays. The actual final charge of 82 is reached by having a total of six β decays, each of which raises the charge by one. Therefore, we need exactly eight α decays and six β decays to get from $^{238}_{92}$U to $^{206}_{82}$Pb.

Actually, it is common to have one or more γ rays produced as well as the α or β particles in each of the decay steps. The gamma rays occur when the original decay creates an **excited state** of the product. When the excited state rearranges to the ground state (the lowest energy arrangement for that kind of nucleus), the extra energy appears as a gamma ray. Therefore, the series of reactions produces α and β particles and γ rays and all three could be found in the natural radioactivity of uranium minerals. Figure 15-4 illustrates γ decay.

There are two other long decay series in nature, $^{232}_{90}$Th to $^{208}_{82}$Pb and $^{235}_{92}$U to $^{207}_{82}$Pb. There should be one more so all mass numbers could fit into one of the series, but this fourth series is not found in natural materials on earth. Actually, it probably did occur when the earth was first formed, but all of

The number of α and β decays in a decay chain can be derived from the total mass and charge changes.

Natural decay chains lead to the heaviest stable isotopes.

Figure 15-3 β decay.

Figure 15–4 γ decay.

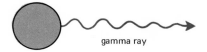
gamma ray

Half-life is a useful measure of decay rates.

the isotopes in that series were used up by radioactive decay many millions of years ago.

Decay is a random chance process, but each isotope has its own individual probability for decay. When that probability is high, more of the nuclei decay in a given time than would have if the probability was low. Laws of chance and statistics can be used to calculate a mathematical statement of the decay probability (called the decay constant) for each isotope from measurements of the decay rate. The statement of decay probability can be used to calculate what the extent of decay will be during any time period, but there is a simpler approach available to those wanting less mathematical complexity. No matter how much of a particular isotope we start with, the random chance of decay will use up half of it in a certain set period of time we call the **half-life**. If we start with twice as much, twice as many decays take place at each point in time, but the time when half of the original group is gone is still the same. When we have half as much, the time for half of that amount to decay is still one half-life.

$^{238}_{92}U$ has a half-life of 4.5×10^9 years, and measurements of the ratios and amounts of uranium and lead isotopes lead to the estimate that somewhere near half of all the $^{238}_{92}U$ originally present on Earth has decayed to $^{206}_{82}Pb$. These measurements are the basis for an estimate of the age of the Earth as about 5×10^9 years. Because a lot of $^{238}_{92}U$ has not yet decayed, we still have it present to start its decay chain. Similarly, $^{232}_{90}Th$ is still here to start its decay chain because its half-life is even longer, 1.4×10^{10} years. $^{235}_{92}U$ has a shorter half-life, 7.1×10^8 years, so most of it has already decayed. In the first 7.1×10^8 years, half of the original $^{235}_{92}U$ decayed. In the next 7.1×10^8 years, half of the remaining half decayed, bringing the amount left down to one fourth of the original. In the next half-life, the $^{235}_{92}U$ was reduced to one eighth of the original. In 5.0×10^9 years, there would be $(7)(7.1 \times 10^8)$ years or seven half-lives for $^{235}_{92}U$ decay. That would reduce the $^{235}_{92}U$ to $(\frac{1}{2})^7$ of the original or $\frac{1}{128}$. Therefore, $^{235}_{92}U$ (which is the desired form of uranium for nuclear reactors) is quite rare, but there is still enough around for us to find it and the decay chain it starts. Figure 15–5 shows the decay of $^{235}_{92}U$ through four half-lives.

The longest half-life of any isotope in the "missing" decay series is 2.2×10^6 years for $^{237}_{93}Np$. In 5.0×10^9 years there would be 2270 half-lives for $^{237}_{93}Np$ decay. The amount remaining is therefore so small that no one ever managed to find it in nature. Measurement of its half-life and other members of its decay chain have been made using material produced artificially by man.

There are some natural radioactivities besides those in the decay chains. The most important radioisotope not in a decay chain is $^{40}_{19}K$. It has two ways

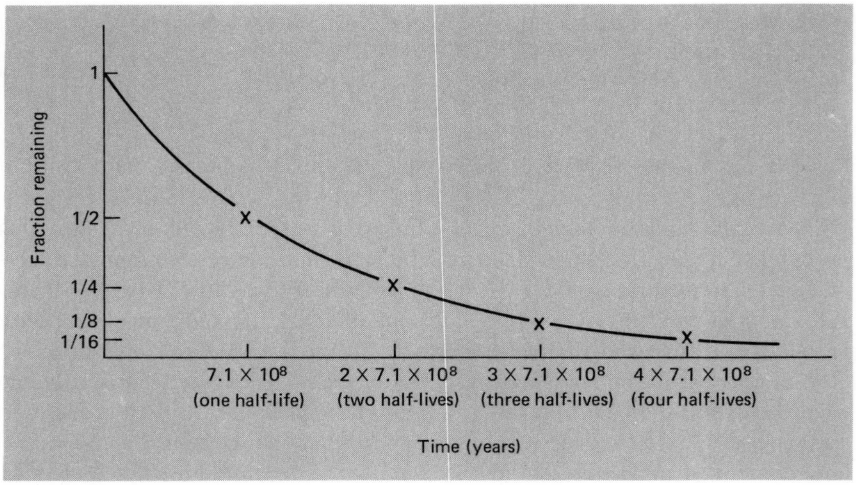

Figure 15–5 Decay of $^{235}_{92}$U showing half-life.

to decay. Some $^{40}_{19}$K nuclei undergo β decay to form $^{40}_{20}$Ca, but others undergo **electron capture.** In electron capture the nucleus reacts with one of the electrons in the atom and has its nuclear charge (and atomic number) reduced by one. The two "decay" reactions of $^{40}_{19}$K are shown in the following equations.

β decay

$$^{40}_{19}\text{K} \longrightarrow {}^{40}_{20}\text{Ca} + {}^{0}_{-1}e$$

electron capture

$$^{40}_{19}\text{K} + {}^{0}_{-1}e \longrightarrow {}^{40}_{18}\text{Ar}$$

When electron capture does occur, the captured electron is almost always a $1s$ electron because they are closest to the nucleus. Electrons in higher energy levels can then move to fill that vacant position and to fill the others left open each time one moves to fill the previous vacancy. The energy released as these electrons shift is evidence that electron capture has occurred. Although the process is chemically very interesting because much of the energy is often released by ejecting other electrons (called an Auger process) to form high charged positive ions, it was very hard to identify. Figure 15–6 shows an electron capture and Auger process. Over 40 years passed between identification of α, β, and γ decay and the identification of electron capture by Alvarez in 1938. We will use ec as an abbreviation for electron capture in this chapter.

One of the most important forms of radioactivity is fission. In fission, the nucleus breaks up into two (or more) large pieces instead of just emitting a small particle or ray. This also occurs naturally, but it was not detected until the late 1930s. Even then, it was found from experiments on artificial

Electron capture is often followed by Auger processes giving large positive charges.

Electron capture, positron emission, fission, and neutron emissions are forms of radioactivity.

Figure 15–6 Electron capture and Auger process.

When the first electron (from the innermost shell) is captured by the nucleus, the Auger process follows. An electron from the second shell moves to fill the vacant first shell position, and the energy released causes another electron from the second shell to be emitted from the atom. Similar pairs of electrons from the third shell have one filling a second shell vacancy while the other is emitted from the atom. The process could continue if the atom had more than three shells containing electrons.

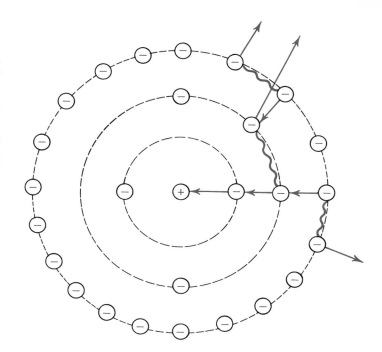

Because of their similarity, electron emission is called beta minus (β^-) decay and positron emission is called beta plus (β^+) decay.

radioactivity before it was found to occur in natural radioactive decay. We will discuss fission in Section 15–5.

In addition to the five types of radioactivity occurring naturally on Earth, (α, β, γ, ec, and fission), there are several other types of decay that can be artificially induced. The two most important of these are neutron emission, which is necessary for operation of nuclear reactors, and positron emission, which is important to our understanding of the following sections of this chapter.

A positron is a positive electron. It is identical to an electron in every way except its charge, which is $+1$ instead of -1. Positrons do not last long in our world because a positron can react with an ordinary electron in a way that destroys them both by converting them into energy (as γ rays). Isotopes like $^{40}_{19}K$ cannot form positrons because there is not enough energy available in the $^{40}_{19}K$ decay to equal the energy needed to make a positron, but many artificial radioisotopes have plenty of decay energy to make a positron and a need to decrease their nuclear charge. This decrease can be accomplished by emitting a positron, just as some other radioisotopes emit electrons. In fact, the processes are so similar they are *both* called β decay. To keep them straight we now add the sign of the emitted β particle, $-$(electron) or $+$(positron), as a superscript on the β. Therefore, all the β decays we mentioned before this paragraph should have been called β^- decays. We will use that system throughout the rest of this chapter. Examples of the two types of β decay are

β^- decay
$$^{14}_{6}\text{C} \longrightarrow {}^{14}_{7}\text{N} + {}^{0}_{-1}e$$

β^+ decay
$$^{11}_{6}\text{C} \longrightarrow {}^{11}_{5}\text{B} + {}^{0}_{+1}e$$

15–2 Nature's Preferences

15-2.1 The elements about us. The world about us is made up of 90 naturally occurring chemical elements plus a few extras that man has managed to make out of the originals. Each of these elements has its own particular characteristics which we can study and compare with the other elements. But the elements are far from equal in importance for practical purposes. Some elements are important because their properties are especially useful, but sheer abundance is also important in practical considerations. On the surface of the earth almost half of all matter available to us is the element oxygen. The chemistry of oxygen is therefore a vital and inescapable part of almost everything around us. On the other hand, rhenium, element 75, has chemical properties that are more interesting than oxygen in some ways, but it is so rare that most people will never have to deal with it. In fact, one must pay a fairly high price for the privilege of being intrigued by rhenium chemistry. To study elements 43, 61, or any above 92, you must actually make the element to have any to study at all.

Like the chemical behavior, the abundances of elements are not random and can be related to structural patterns. Individual elements in nature are mixtures of all the isotopes of that element that exist. The total abundance of an element depends upon which isotopes exist and upon their relative abundance. We must therefore ask two questions. What isotopes exist, and what isotopes are abundant? The answers to these questions show us the structural patterns of elements.

15-2.2 Even Numbers. When we calculate the number of protons and the number of neutrons in each of those isotopes that exist in nature, we find a definite preference for even numbers over odd numbers. Even numbered elements (even protons) have more natural isotopes than odd numbered elements, and in each group there are more isotopes with even numbers of neutrons than odd. If we count only the stable isotopes, we get the distribution in Table 15–1.

This striking preference for even numbers goes even farther than the number of isotopes. Although there are about four times as many isotopes of even numbered elements as there are of odd, the total abundance on earth of even elements is ten times as great as the abundance of odd elements. Isotopes with even numbers of neutrons are also noticeably more abundant.

Nature prefers even numbers of protons and even numbers of neutrons.

Table 15-1

		Number of neutrons	
		Even	Odd
Number of protons	Even	164	55
	Odd	50	4

Among the ten most abundant elements on Earth, three have only even n isotopes and among the other seven (with 17 stable even n isotopes and seven stable odd n isotopes) only magnesium with 10.11% has as much as 5% abundance of the odd n isotope.

The observed preference for even numbers of protons and neutrons indicates a tendency for them to exist in pairs. This tendency is operating separately for both protons and neutrons but not for the total of the two, as shown by the bias against odd-odd isotopes where the sum would be even. The preference for even protons is about equal to the preference for even neutrons, as shown by the similar numbers of even proton-odd neutron and odd proton-even neutron isotopes. Therefore, protons and neutrons must follow rules independent of each other but each showing the same sort of preference for grouping in pairs.

15-2.3 Balance of Protons and Neutrons.

Among light elements, nature prefers equal numbers of protons and neutrons.

Although protons and neutrons are independent in their preference for even numbers, they do seem to be related to each other in total numbers. Protons are never found in groups without at least one neutron also included, and the reverse is also true. In fact, among the light elements the number of protons and the number of neutrons in stable isotopes is always equal or nearly equal. Since protons and neutrons follow the same sort of rules, we conclude that the preference for same numbers of protons and neutrons means that they follow exactly the same rules. Apparently protons and neutrons are being "stacked" in, and both stacks must be about the same "height." These stacks are energy levels similar to the various shells and subshells available for electrons, and the height is the number of levels filled.

This balancing of the stacks would not work, however, unless there was some way to get from one stack to the other. Apparently if the neutron stack is higher a neutron can "fall off" to the lower proton stack. If the proton stack is higher, the reverse can happen. But electrical charge must be conserved. When a neutron becomes a proton, a balancing negative charge, such as an electron, must also be formed. When a proton becomes a neutron either the negative charge of an electron must also disappear or a charge like an electron, but positive, must be formed. All of these things do happen. Neutron-rich isotopes emit β^- rays, which are electrons, and proton-rich isotopes absorb electrons (electron capture decay) or emit positrons (β^+ decay), which have the same mass as electrons but a positive charge. Unbalanced isotopes could also reach proton-neutron balance by emitting their extra neutrons or protons, but that is less common.

15-2.4 Coulomb Repulsion and the Curve of Stability.

Since electrical charges of the same sign repel each other, protons repel each other. As the number of protons in a nucleus increases, the repulsion becomes very large. This coulombic (electrical) repulsion affects the energy levels for protons in nuclei, making the energy levels higher (less favorable) as the number of protons rises. Neutrons have no such repulsions for each other. Therefore, in any nucleus containing more than one proton, neutron energy levels are lower than those of protons.

For light elements the difference due to coulomb repulsion is small. As the number of protons rises, the difference increases until "stacks" with the same number of protons and neutrons are not the same "height" in energy. Then we may have a proton converted to a neutron to give "stacks" of more even "height" even though they are now unbalanced in favor of more neutrons than protons. This is what we find in nature. Above element number 20, all stable isotopes have *more neutrons* than protons, with the deviation steadily increasing as the atomic number rises.

This is shown by Figure 15-7. The stable isotopes are plotted in terms of their number of protons and number of neutrons. A smooth curve drawn through the region of stable isotopes is called the curve of stability. This curve represents a balancing between protons and neutrons, not in number but in energy levels. Therefore, everything above the curve has higher proton

Repulsions between protons force the curve of stability to bend toward an excess of neutrons as the atomic number rises.

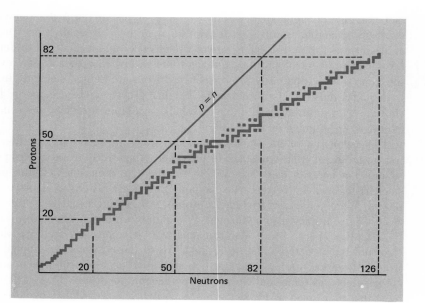

Figure 15-7 The stable isotopes.

The curve of stability would be the smooth curve passing through the center of these stable isotopes at each mass number region.

312 Nuclear Stability and Chemical Abundance [15]

energy levels than neutron energy levels. If the difference is great enough, the element may be radioactive by electron capture or β^+ emission. Everything below the curve has higher neutron energy levels and may be radioactive by β^- emission. Stable isotopes are those on or near the curve of stability, where the energy levels are close enough that no conversions between protons and neutrons occur. At very large mass numbers and atomic numbers, the curve of stability stops. Beyond this point there is a continuation of the curve where isotopes are not β^- or β^+ radioactive, but the coulomb repulsions become so large that the isotopes are radioactive in other ways, usually by α emission but sometimes by fission.

15-2.5 Magic Numbers and Nuclear Shells.

Although all even numbers are favored for protons and neutrons in general, there are a few numbers that are especially favored. The numbers 2, 8, 20, 50, 82, and 126 are so highly favored that they are called "magic numbers" for protons or neutrons. The number 28 is sometimes also included in this list. Evidence of the preference for these numbers shows up both in the large number of stable isotopes having "magic numbers" of protons or neutrons and in the high abundance of these isotopes in nature. Here is some evidence of the preference for "magic numbers."

> The numbers 2, 8, 20, 50, 82, and 126 are called magic numbers because they are the favored closed shell values for protons and for neutrons.

1. The two heaviest stable isotopes, ^{208}Pb and ^{209}Bi, both have 126 neutrons.
2. All three of the natural radioactive decay chains end at lead (82 protons).
3. There are more naturally occurring isotopes with 82 neutrons or 50 neutrons than any other numbers.
4. The element with the largest number (ten) of stable isotopes is tin (50 protons).
5. The stable isotopes of calcium (20 protons) also cover an unusually large range including one with 20 neutrons and one with 28 neutrons.
6. The most abundant isotope on earth is ^{16}O (eight protons, eight neutrons).
7. The isotope ^{4}He (two protons, two neutrons) is so stable it is a common product of radioactive decay (α particle). Also, there are no stable isotopes at all with mass of 5 or 8. Those that do not decay by β^- or β^+ break up to give ^{4}He. This leads to important complications in the operation of stars and to low abundances of elements 3, 4, and 5.

The observed "magic numbers" for protons and neutrons are similar to the "closed shell" arrangements for electrons. Each is a point where everything is held very strongly but an extra particle (electron, proton, or neutron) could not be held nearly as well. With the historical background of chemistry as a guide, a pattern of four quantum numbers, selection rules, and relative orders of energy levels has been developed. These quantum numbers and energy levels are very useful and do predict some magnetic characteristics of nuclei in addition to fitting the "magic numbers." However, some fine details of nuclear structure are hard to fit into this simple energy level approach. It is good to remember that development of this model for nuclei was very strongly influenced by earlier conclusions about electron arrangement in atoms, which we think we understand quite well. It would be naive of us to expect all of it to match nuclei neatly, since the forces within nuclei are much greater than those on electrons in atoms. Critical assessment of the models requires a good physics background and some very hard thinking.

15–3 Conversion of "Mass" to Energy

Every one of the radioactive decay processes we have been describing has one more product. Energy is produced. The amount of energy released is directly proportional to a reduction of the **rest mass** by the equation $E = mc^2$ where E is energy, m is mass, and c is the speed of light in vacuum. This well known equation, derived from Einstein's special theory of relativity, is constantly misinterpreted by almost everyone. They conclude, falsely, that mass can be converted into energy and therefore things lose mass as more energy is made available.

Actually, the theory of relativity tells us that rest mass (which can be converted to energy) and actual mass of an object are not the same thing. When we weigh objects, they are usually at rest (no motion) and we are measuring the rest mass. But the theory of relativity predicts that the same object will increase in mass as it is made to move faster and faster. This predicted increase has been proven actually to occur. Therefore the mass of an object has two parts, its rest mass and an increase caused by its motion. $E = mc^2$ simply tells how much increase in mass we get when a given amount of energy is used to cause motion.

The rest mass is a fixed amount for any given object (like a particular nucleus). The mass due to energy is variable because energy can be transferred from one object to another. When energy is transferred, one object gains mass and something else loses the same amount of mass. The total mass of the universe remains the same. When one (or more) object is changed into another object, it must change its rest mass to the correct fixed amount for the new object. Once again, the total mass of the universe must remain constant. Therefore, if the amount in the fixed form (rest mass) goes down, the amount in the transferable form (energy) must go up by the same amount.

As energy is released in any reaction, the *rest mass* of the material is reduced. The total mass is constant.

In a radioactive decay, like $^{238}_{92}U \longrightarrow {}^{234}_{90}Th + {}^{4}_{2}He$, the products must be formed with the same actual mass as the reactant originally had. However, since the sum of the rest masses of $^{234}_{90}Th$ and $^{4}_{2}He$ is less than the rest mass of $^{238}_{92}U$, the amount of real mass present in the form of energy must be greater. This energy is used to make the products fly apart from each other. As they move apart, they strike other atoms or molecules and transfer away energy (and therefore mass). We can see and measure this release of energy. If we later measure the masses of the thorium and helium (now at rest), we find they are lighter than the uranium was and "mass has been converted to energy." Actually, rest mass of the uranium was converted to energy-mass of whatever ended up with the released energy. Total mass stayed the same. For a more quantitative and meaningful picture of these interesting phenomena you should study modern physics.

The tendency of radioactive decay (and other reactions in nature) to release heat indicates a tendency to have as much of the mass as possible in the freely transferable energy form instead of the fixed rest mass form. Decay must occur in that direction because any excess energy can be used as motion of the products away from each other. It is usually physically impossible to go in the other direction with rest mass increasing, because the products cannot end up with negative energies of motion. When two or more reactants come together, the energy of the motion that brought them together can be partially used for an increase in rest mass, but such reactions are less common than energy releasing reactions.

15–4 Artificial Radioactivity

Nuclear reactions are used to produce artificial radioactivity.

Study of natural radioactivity led scientists to develop methods for making artificial radioisotopes, many of which are quite useful. These are produced by causing nuclear reactions to occur. In addition to the new radioisotopes, these studies led to the discovery of artifically induced nuclear fission and to development of nuclear energy sources.

Perhaps the simplest of the induced nuclear reactions are photonuclear reactions very similar to natural radioactive decay. Decay can occur when a nucleus forms products having a lower total rest mass, therefore allowing some energy to be released. If the products have a higher total rest mass than the reactant, no decay is possible. However, if the reactant nucleus first absorbs a block of energy (such as a γ ray) to form an excited state, it would then have the rest mass of that excited state instead of the normal (ground state) form. The excited state would have a rest mass equal to the original rest mass of the nucleus plus the mass corresponding to the energy absorbed (and therefore not available as motion). If that is higher than the sum of the product rest masses, the excited state can "decay" to products just like any ordinary radioactive decay. The sequence is absorption of energy to increase the rest mass followed by breakup to products. An example is

$$^{39}_{19}K + ^{0}_{0}\gamma \longrightarrow ^{38}_{18}Ar + ^{1}_{1}H$$

Massive particles can also be added to a nucleus to form an intermediate that is either stabilized by emitting energy or breaks up into new products. Most nuclear reactions fall in this category. Some examples are

$$^{27}_{13}Al + ^{4}_{2}He \longrightarrow ^{30}_{15}P + ^{1}_{0}n$$
$$^{23}_{11}Na + ^{1}_{0}n \longrightarrow ^{24}_{11}Na + ^{0}_{0}\gamma$$

The reaction producing $^{30}_{15}P$ was done by Joliet and Curie using α particles from natural radioactivity. The $^{30}_{15}P$ product was particularly interesting because it is a β^+ emitter, thus making available artificially a kind of radioactivity not found in nature. β^+ emitters have proven especially useful for "mapping" problems. The positrons do not get very far (at least in liquids or solids) before finding an electron and being annihilated. As the electron and positron destroy each other, their rest masses must be converted into energy. For conservation of momentum, this energy must appear as two identical γ rays traveling in exactly opposite directions as shown in Figure 15–8. Each γ particle has an energy equivalent to the rest mass of an electron (which is also the rest mass of a positron) 0.51 million electron volts (Mev). If a series of counters sensitive to γ rays of that energy are arranged around a sample, with electronics that can tell which two of the counters detect rays simultaneously, we can know that the β^+ emission occurred somewhere on the straight line between those two detectors. If we make a large number of these straight line determinations, we can determine the entire region where the β^+ emitter is present. Figure 15–9 shows how this could be used to locate a brain tumor before surgery. A compound is selected that can be made using a β^+ emitter and that is picked up much more rapidly by the tumor than normal tissues. Very little is needed, so even compounds that are poisons in larger quantities can be used. One example is the use of BF_3 containing the β^+ emitting radioisotope $^{18}_{9}F$.

β^+ emitters are particularly useful because the positron annihilation identifies the location of the decay.

Natural radioisotopes are not always useful or available to cause desired nuclear reactions, so **particle accelerators** have been built to induce reactions. Any positive or negative ion can be accelerated to high energies by electrical fields and aimed at a target. Protons, deuterons (^2_1H nuclei), and $^4_2He^{2+}$ ions are the most common choices. The biggest problem with these nuclear reactions is that positively charged nuclei repel each other electrically. Therefore, the accelerated particle must overcome this coulombic (electrical) repulsion to reach the other nucleus and react. The repulsion has two effects. One sets a minimum energy necessary to penetrate instead of just bouncing off. The second tends to make particles bend away and miss if they are aimed a little off dead-center to start with. As a result, bombarding nuclei with other

Reactions of charged particles are made more difficult by coulomb barriers.

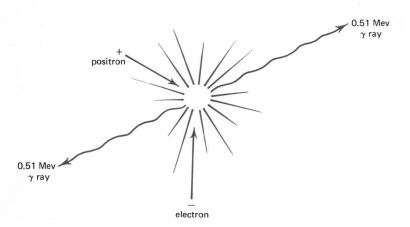

Figure 15–8 Annihilation of positrons.

Figure 15-9 Mapping by use of positron emitters.

The figure shows a brain tumor region that has absorbed radioactive BF_3 and is emitting positrons via the decay of $^{18}_{9}F$. Each emission leads to an annihilation nearby, which creates the pair of simultaneous 0.51 Mev γ rays going in opposite directions. Detection of these pairs of γ rays in the surrounding detectors allows identification of the region where the emissions were occurring.

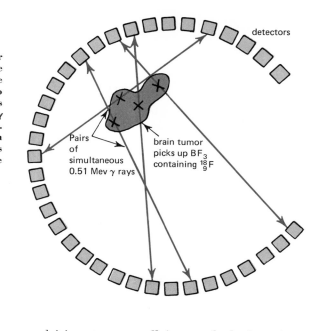

Neutrons can react without coulomb barriers.

nuclei is not a very efficient method of causing nuclear reactions (although it is often the only way to get the desired reaction).

In the early 1930s the neutron was discovered via reactions similar to $^{27}_{13}Al + ^{4}_{2}He \longrightarrow ^{30}_{15}P + ^{1}_{0}n$. The availability of a nuclear particle without a positive charge was welcomed by Enrico Fermi and others as a way to get around the coulomb barrier problem. Many new artificial radioisotopes were produced and studied. Some examples of neutron induced reactions are

$$^{107}_{47}Ag + ^{1}_{0}n \longrightarrow ^{108}_{47}Ag + ^{0}_{0}\gamma$$

$$^{27}_{13}Al + ^{1}_{0}n \longrightarrow ^{24}_{11}Na + ^{4}_{2}He$$

$$^{32}_{16}S + ^{1}_{0}n \longrightarrow ^{32}_{15}P + ^{1}_{1}H$$

$$^{19}_{9}F + ^{1}_{0}n \longrightarrow ^{18}_{9}F + 2\,^{1}_{0}n$$

Early workers obtained their neutrons from other nuclear reactions. They were always fairly high in energy because they carried energy away from the reaction in which they were formed. For the last of the reactions above, high energy was needed so one neutron going in could excite the nucleus enough to make two neutrons break loose, but no excess energy was needed for the other reactions. There is no coulomb barrier to overcome.

Neutron reactions are increased as the neutrons are slowed down (moderated).

Fermi and his coworkers studied neutron reactions with silver (an element with a nuclear charge much too large to allow efficient reactions with positive particles), and they found that lowering the neutron energies (called thermalization) by allowing them to bounce around in paraffin wax increased the efficiency of reaction. The wave nature of matter and the relationship between energy and wavelength are important in the explanation of this phenomenon.

The phenomenon itself is vitally important in the design and operation of nuclear reactors.

In addition, Fermi and his coworkers investigated neutron reactions with uranium and identified many new radioactive substances. They expected these to be isotopes of elements near uranium, but the chemical behaviors often did not fit known elements near uranium. These "misfits" were eventually resolved into two extremely important discoveries. First, nuclear fission was identified by showing that some of the new radioisotopes were elements much lighter than uranium. Second, one "misfit" was identified as a previously unknown element beyond uranium in the periodic table.

15-5 Fission and Nuclear Power

Figure 15-10 shows a "typical" fission of uranium. Addition of one neutron to $^{235}_{92}$U makes an unstable excited $^{236}_{92}$U which then breaks up into two large pieces plus a few neutrons. A large amount of energy is released. The large products vary from one fission to the next. The number of neutrons released also varies but averages about three. This neutron-induced fission to two main pieces is not the only kind of fission possible. Some radioisotopes decay by spontaneous fission, and there is some evidence that occasionally nuclei fission to form three large pieces instead of two. However, neutron-induced fission is the key to practical use of nuclear energy. It can be done with reasonably available materials ($^{235}_{92}$U present in nature or $^{239}_{94}$Pu or $^{233}_{92}$U which can be made fairly easily from $^{238}_{92}$U and $^{232}_{90}$Th). It can be "turned on" when we want it by making neutrons available. It releases a lot of energy. And it produces the neutrons needed to continue the reaction faster than it uses

Neutrons can induce fission of some isotopes.

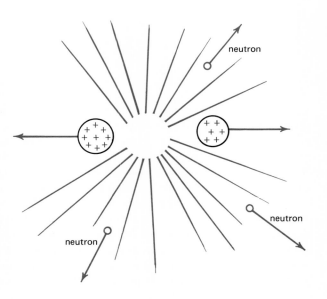

Figure 15-10 A typical uranium fission.

An energy producing chain reaction can be set up with neutrons inducing fission which releases more neutrons to induce more fission.

them. The only problem with neutron-induced fission is that the neutrons normally escape and react with nuclei that cannot fission.

By concentrating fissionable material (like $^{235}_{92}U$) in one region, we increase the chance of some neutrons causing additional fission. However, the very high energy neutrons released in fission tend to miss the nearby nuclei and escape from the region of fissionable material before reacting. In 1934, Fermi discovered that a neutron's chance of reacting increased as it lost energy. Fermi later used this knowledge to devise a method to increase the number of neutrons causing fission and thus keep a continuous series of fissions going. He alternated regions of concentrated fissionable material with **moderators,** substances that allowed the neutrons to lose energy in collisions but did not use many neutrons in nonfission nuclear reactions. The result was a **nuclear reactor.**

The need for a moderator is a key feature in the design of nuclear reactors. Both the methods of controlling nuclear reactors and the practical use of nuclear energy depend upon the conditions necessary for moderating the energy of the neutrons.

Because of the time lag between production of neutrons in one fission and the reaction of those neutrons to cause new fissions, speedup or slowdown of the fission rate is not instantaneous. There is also a delay in neutron availability because a few of the neutrons are produced by radioactive decay. Reactors are actually designed so any increases or decreases stay within the small fraction coming from radioactive decays. The fission rate changes are therefore slow enough to be easily controlled by slow mechanical changes. These changes are made by pushing in or pulling out **control rods.** Control rods are simply very *bad* moderators. Good moderators allow neutrons to lose energy without undergoing nuclear reactions. Control rods are made of materials neutrons react with readily. The control rods are positioned so that the combination of geometry, moderator, and control rods let exactly one neutron from each fission get back to cause another fission. To increase reactor power, the control rods are withdrawn slightly. Then each one billion fissions may lead to one billion plus one or one billion plus two new fissions, and the rate of fissions gradually increases. A gradual reduction in reactor power can be accomplished by pushing the control rods in slightly.

The fission chain reaction can be controlled in a nuclear reactor because of the need for moderation to be efficient and because some neutrons formed by radioactive decay are delayed in appearance.

If the control rods are pushed in quickly, the number of neutrons getting back to cause new fissions drops far below one per fission and the reactor is rapidly turned off. All reactors do this automatically whenever any sign of malfunction is detected.

If the control rods were pulled completely out, the rate of fission would increase rapidly. As long as some of the delayed neutrons from radioactive decay are needed to reach a rate of one new fission from each original fission, the rate of increase is held down by the time lag for radioactive decay and the reactor can be controlled. When the control rods are removed far enough for the rate of fission to increase even without the delayed neutrons from decay, the increase is so fast that no external change, like moving the control rods, could be done quickly enough to stop the increase. The reactor is then out of control and undergoing a **power excursion.** The heat produced during a power excursion may damage the reactor, but it also tends to *turn the reactor off*.

During a power excursion, the moderator in a reactor losses effectiveness. If water is the moderator, it may be converted to steam. That reduces the amount of moderator inside the reactor. A larger fraction of the neutrons then leave the region of the reactor before reaching the low energies where they are most effective at causing fission. Even if the moderator does not vaporize, more collisions would be needed for the same energy loss when the moderator is hot, so more neutrons are lost before reaching low energy.

When the moderator loses effectiveness, less of the neutrons cause new fissions. Eventually a point is reached where so few neutrons cause fissions that the fission rate goes down instead of up. The power level of the reactor then drops rapidly, turning the reactor off. Many reactors (particularly small ones) are designed so any power excursion automatically shuts the reactor down without damaging any parts. These reactors can be "pulsed" (an intentional power excursion) to give brief periods of very high neutron production for experiments. The total time from the start of the "pulse" to reactor shutdown is usually about $\frac{1}{100}$ sec.

Loss of moderator efficiency during a power excursion may turn a reactor off.

Shutdown of a reactor by inserting the control rods can also be very rapid and, as this writer can testify, quite exciting. Controlling instruments detected something irregular while I was standing on the platform above a large operating nuclear reactor. The resulting automatic shutdown included the blasting into place of control rods by release of compressed gas. That sounded like an explosion. It was accompanied by gas bubbling up through the water around the reactor, by red lights and sirens going off throughout the building, and by a loud clang of the thick metal door slamming as part of the automatic sealing of all exits (sealing me and the reactor crew inside). At a time like that it is comforting to know that most reactors could not even come close to exploding, but I still would not recommend the experience for anyone with heart trouble.

Reactors are equipped with several automatic safety shutoff controls.

Actually, no nuclear reactor (however large it may be) would ever be expected to concentrate its fissionable isotope effectively enough to restart a chain reaction without the moderator. Restating that in every day language: *it is impossible for a nuclear reactor to become an atom bomb*. But that does not mean that nuclear reactors are harmless. A very large nuclear reactor could generate so much heat in a power excursion that the internal parts would melt. In a water-moderated reactor that would expose reactive metal to hot steam. That would generate more heat, which might melt a hole in the containing vessel. A hole in the top of the vessel would let out the dangerous radioactive products from fission. However, the very heavy, dense nuclear fuels would almost certainly melt their way out the bottom instead. As the radioactive materials continued to release heat, the molten mass would then follow what has been described in the United States as the "Chinese syndrome." The heavy mass would open its own hole by melting rock and would move off toward the center of the earth (generally toward China).

Although nuclear reactors cannot explode like atomic bombs, large reactors could melt and cause safety problems.

Design of nuclear power reactors is intended to prevent this rather unpredictable and uncontrolable series of events from ever happening. Several separate systems of controls are used, all of which would be set off by any real problem. When any one of these controls reports a problem, the reactor is automatically shut off. The shutdown incident described above was an example of a "false alarm" shutdown. The actual failure was an electrical problem in one of five separate control systems.

15-6 Binding Energy and Fusion

Nuclear fission releases a great deal of energy because the products of fission are held together with greater attractive force than the original nucleus. That is easily understood because we can see that there should be an energy advantage in taking the positively charged protons away from each other. However, breaking nuclei up into smaller, less positively charged parts does not always give an energy advantage. Among light nuclei (those with only a few protons and neutrons) there is a very large energy advantage in going the opposite direction, putting nuclei together to make a new, larger nucleus. This process is called **fusion.** It occurs in stars like our sun and is the principal source of energy in the universe.

It is clear that there is a large natural force that tends to hold protons and neutrons together in a nucleus. In the normal scientific manner theories have been proposed to explain this force. You would find these ideas (exchange forces, mesons, and nuclear wave functions) described in some physics courses. However, for our purposes, the important part is not why there are such forces. We need to consider how these forces interact with other forces to result in the chemical elements found in our world.

There are a number of conflicting forces at work in atomic nuclei.

1. The attractive force between protons and neutrons tends to hold the nucleus together
2. The electrical repulsion between protons tends to break the nucleus apart
3. The tendency for protons and neutrons to be in the lowest (most strongly held) levels makes the proton-neutron ratio stay at or near the curve of stability
4. The preference for even numbers of protons or neutrons gives all even proton–even neutron combinations an energy advantage
5. The preference for the "magic numbers" gives isotopes with those numbers an extra energy advantage

There are some simplifications that help us see the important general trends established by the nuclear forces.
1. We only find the most stable isotopes in nature. The others have disappeared through radioactive decay. Therefore, we will consider the bonding forces only on those isotopes that fall *on the curve of stability*. Then we can ignore the effects of item 3 in the list above.

2. Differences in binding forces will be easier to see if they are stated as an amount *per nucleon* (proton or neutron) instead of the total for the whole nucleus. Then we can clearly see the difference possible if the same number of protons and neutrons were put into larger or smaller nuclear groups. Figure 15-11 shows a graph of the quantity

$$\frac{\text{binding energy of the nucleus}}{\text{number of nucleons in the nucleus}}$$

versus the number of nucleons (the mass number of the isotope) for those isotopes on the curve of stability. It shows us several interesting and important facts.

First, when there are only a few nucleons, the binding energy not only goes up when extra neutrons or protons are added, it goes up *faster* than the number of nucleons goes up. That occurs because the number of possible nucleon-nucleon bonds goes up rapidly. With two nucleons, there is only one possible bond, or one half bond per nucleon. When a third nucleon is added, it can bond to each of the other two. That brings the number of bonds up to three or one bond per nucleon. When a fourth nucleon is added and bonds to each of the other three, the number of bonds reaches six, or one and one-half bonds per nucleon. This increasing bonding per nucleon continues as more nucleons are added. Eventually the number of bonds per nucleon levels off when each one already has as many nucleons bonded as can get close enough to form bonds.

Second, when the number of protons becomes very large, the binding energy per nucleon goes down as more nucleons are added. The repulsion between protons becomes more important as the number of protons goes up.

Third, there is a region of maximum binding per nucleon (actually occurring

There is a region of maximum binding energy for nuclei and a series of high points in stability for even-even nuclei compared to their neighbors.

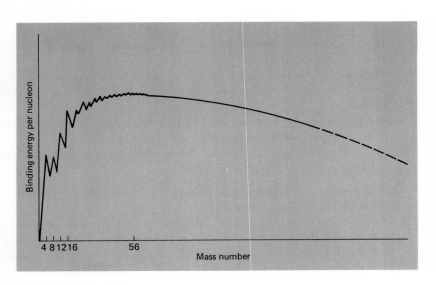

Figure 15-11 The binding energy curve.

At each mass number the binding energy per nucleon is given for the isotope of that mass number that has the greatest binding energy.

around iron and nickel). Elements below that point can release energy by fusion; elements above that point can release energy by fission.

Fourth, there are some isotopes favored over others in the same region, particularly among the light elements. In the region where the curve of stability follows equal numbers of protons and neutrons, there is a big advantage for both protons and neutrons to be present in even numbers. Therefore, there are peaks in binding energy per nucleon at four nucleons (two protons and two neutrons), eight nucleons (four protons and four neutrons) 12 nucleons (six protons and six neutrons), 16 nucleons (eight protons and eight neutrons), and at each fourth number thereafter. When the curve of stability gets away from equal numbers of protons and neutrons, the advantage of being even-even in protons and neutrons may be cancelled out by another isotope being closer to the curve of stability, so the "bumps" become less noticeable.

Fifth, the special preference for "magic numbers" has some effects. The most noticeable one is the special preference for ^4_2He (two protons and two neutrons) over the even but not "magic" number four (four protons and four neutrons) in ^8_4Be. In a mass region where most isotopes can release energy by fusion, ^8_4Be can (and does) fission to form two ^4_2He plus energy. Since ^8_4Be is the most strongly bonded isotope between mass number four and mass number 12, that whole region of elements is affected by the instability of ^8_4Be. The important consequences are discussed in the following sections on Operation of Stars (Section 15–7) and Abundance of Elements (Section 15–8).

15–7 Operation of Stars

Fusion of light elements to form other elements nearer the maximum binding energy is the energy source in stars.

We know that most of the energy in stars is produced by fusion of light elements, particularly hydrogen. The operation of stars, the history of the universe, and the abundances of elements can all be explained in a logical manner from the curve of binding energy per nucleon shown in Figure 15–11 and the assumption that everything began as hydrogen.* Hydrogen nuclei (protons) could release energy in fusion, but the repulsion of positive charges for each other tends to keep the hydrogen nuclei apart. This repulsion can be overcome if the nuclei are brought together with very high energies. These high energies can be supplied if the hydrogen is very, very hot. That is accomplished by the formation of a star. When a very large mass of hydrogen is drawn together in one small region, the hydrogen is heated by the energy from gravity pulling the mass together. (If a steel ball was dropped on pavement from a ten story building, the force of the blow on the pavement would heat up the pavement and ball somewhat in a similar conversion of gravitational energy to heat.) When the hydrogen gets hot enough (and very dense), fusion of hydrogen atoms starts to occur often enough to replace gravitational contraction as the source of the star's energy.

* If instead everything began as neutrons, we would get the same result. The neutrons would eventually decay to protons plus electrons (β^- particles), the components present in hydrogen.

Fusion can provide the energy emitted by the star for many billions of years as the ^1_1H is converted to ^4_2He (plus some β^+ decays). The ^4_2He does not undergo fusion at the temperatures where hydrogen fusion is occurring. A higher temperature is needed to overcome the larger repulsions of the $+2$ helium nuclei from other nuclei. The fusion of ^1_1H and ^4_2He also fails because the isotopes of mass number five break apart to form the especially stable ^4_2He plus a proton or neutron. Therefore, the ^1_1H tends to "burn up" in fusion before any ^4_2He begins to react. Our sun is in this stage, converting hydrogen to helium.

Eventually, as much ^4_2He is formed and as gravity concentrates the helium in the center with the lighter hydrogen outside, the center of the star becomes mostly helium. When the center runs out of hydrogen for fusion, the central part again begins to contract from gravitational forces. It then becomes much denser and hotter until finally ^4_2He fusion begins. In almost pure helium the expected reaction would be $^4_2He + ^4_2He \longrightarrow ^8_4Be$, but that does not happen because any ^8_4Be comes right back apart to reform ^4_2He. However, if three ^4_2He nuclei can be brought together at once, $^{12}_6C$ can be successfully formed. In this process, the star "skips" the region of beryllium. Once some $^{12}_6C$ is formed, it can react with ^4_2He to form $^{16}_8O$, which can react with ^4_2He to form $^{20}_{10}Ne$ and so on up the line by jumps of four in mass number. This sequence is called the **main fusion sequence.**

The main fusion sequence produces ^4_2He, $^{12}_6C$, $^{16}_8O$, $^{20}_{10}Ne$, $^{24}_{12}Mg$, $^{28}_{14}Si$, $^{32}_{16}S$ and other energetically favored isotopes.

Energy released by the helium fusion reactions keeps the stars operating and stable until much of the material has been converted to the region of maximum binding energy per nucleon, around iron and nickel. Further fusion reactions then release no energy, and the center again begins to contract from gravitational forces. As the center becomes very hot and very dense, a tremendous variety of nuclear reactions occur including fusions to very heavy nuclei, fission to different light nuclei, and generation of neutrons that react with the isotopes present. This brief, violent period corresponds to the supernova explosion of the star in which a small, very dense center forms while the rest of the star is blown away into space.

Supernova explosions convert main sequence isotopes to heavy elements and other light isotopes besides the most favored ones.

Our solar system appears to have been formed by the gravitational drawing together of some of the residue from an old supernova. The elements present are the ones (and in the amounts) formed by the sequence of star operations that we have just described.

15–8 Abundance of the Elements

In the universe as a whole, the most abundant element by far is hydrogen. As noted in Section 15–8, the process seems to have started with only hydrogen present. Even when the center of a star goes through a supernova, a lot of the original hydrogen is still on the outside, unreacted, and is blown away along with part of the center. The material from which our solar system formed is mostly hydrogen.

As various masses of material gathered together in our solar system, the light hydrogen remained at the outer edges of each mass. Eventually the largest center of gravitational force (the Sun) stripped most of that loosely held hydrogen away from the other bodies. Therefore, the Sun is mostly hydrogen while the Earth has almost no free hydrogen. The only hydrogen kept was the hydrogen chemically combined with other elements to form heavier molecules (principally H_2O) which Earth's gravity could hold more effectively.

The other abundant elements in the universe are those formed in the series of fusion reactions from hydrogen to iron and nickel. The ones nearest to hydrogen (helium, carbon, and oxygen) are the most abundant. The reaction never reaches completion before stars explode as supernova, and the simplest reactions are the most likely ones to occur before the end comes.

Elements beyond iron and nickel and those falling between the especially stable isotopes in the main fusion sequence can only form in secondary reactions. These occur mainly during supernova, and the amounts formed are quite limited, particularly if the product is far removed from the abundant isotopes formed in the main fusion process. Therefore we find the following patterns:

1. Elements heavier than nickel are less common, and their abundances continue to drop as they become heavier.
2. Elements like nitrogen are less abundant than those with isotopes formed directly by fusion, but they are reasonably abundant because they form extensively from their abundant near-neighbors. They follow the abundance pattern of the main fusion products fairly closely (decreasing from carbon to nickel).
3. Lithium, beryllium, and boron, whose isotopes should have formed fairly extensively from ^8_4Be if it had been part of the fusion sequence, are all much less common than any other light elements.

Abundance of lithium, beryllium, and boron is held down by the absence of ^8_4Be, and heavy elements become rarer as they get farther from the region of maximum bonding.

There are a few additional factors that show up in the abundances of elements, such as extra abundance of oxygen because it has the special stability from the "magic number" eight, but the trends described here summarize the major points. Abundances are discussed further in Section 16–1. Figure 16–3 shows a graph of the relative abundances of elements on which the patterns we have just described show clearly.

We have already noted the difference between the abundance of hydrogen in the universe and its much lower abundance on earth. The abundances of other elements within our access are also different than in the universe as a whole. Elements that are found extensively as gases (N_2 and all the noble gases) have been at least partially stripped from Earth by the Sun's gravity. Elements found as very dense materials (particularly iron, nickel, and some other metals) tend to be concentrated in the core of the Earth, beyond our reach. By contrast, oxygen (which is quite abundant in the universe as a whole) has been concentrated in the crust of the Earth by forming various oxides which are solid (or liquid) resistant to stripping by the Sun, but not dense enough to sink to the Earth's core.

The crust of the Earth has concentrated those elements that were light enough (meaning not the most dense) not to sink to the core but not gases that could be stripped off by the Sun.

Because of its abundance and concentration on the Earth's crust, oxygen dominates the chemistry of our world. A similar oxygen dominance is virtually certain, for the same reasons, on any other planet. Occasionally we find a science fiction story describing a planet with an F_2 atmosphere instead of O_2 as an oxidant or with NH_3 seas instead of H_2O. The facts of nuclear stability and operation of stars make it clear that such tales are purely fiction, not science. On the other hand, an atmosphere or ice caps of CO_2 are completely reasonable possibilities, and a lack of O_2 (or only small amounts) is probable on planets where no plant life exists.

Skills Expected at This Point

1. You should be able to define and use the terms radioactive, radioactivity, natural radioactivity, artificial radioactivity, emission, α particle, β particle, γ ray, radioisotope, nuclear equation, decay, mass number, total charge, stable isotope, excited state, decay chain, half-life, electron capture, fission, β^-, β^+, neutron emission, fusion, coulomb repulsion, light nuclei, heavy nuclei, curve of stability, magic numbers, rest mass, total mass, particle accelerator, moderator, thermalization or moderation, nuclear reactor, chain reaction, control rods, power excursion, shutdown, binding energy, main fusion sequence, supernova, abundance, and gravitational stripping.
2. You should be able to write nuclear equations to describe radioactive decays and able to balance and complete a nuclear equation with only one item missing.
3. You should be able to calculate the fraction remaining after any whole number of half-lives.
4. You should be able to list the factors showing nature's preferences in nuclear stability, sketch the curve of stability and use it to predict the kinds of decay in different regions, and know the magic numbers.
5. You should be able to list the necessary parts in a nuclear reactor and describe neutron-induced fission, the relationship between the energy of neutrons and their efficiency in nuclear reactions, and the situation of balance in neutrons necessary in an operating nuclear reactor.
6. You should be able to draw a crude sketch of the binding energy curve and describe how both fission and fusion release energy.
7. You should be able to describe the main fusion sequence in stars, the production of other elements in supernova explosions, and the resulting low abundances of lithium, beryllium, boron, and the heavy elements.

Exercises

1. Complete nuclear equations for the following:
    (a) α decay of $^{214}_{84}Po$
    (b) β^- decay of $^{210}_{83}Bi$
    (c) electron capture by $^{40}_{19}K$
    (d) $^{9}_{4}Be + ^{4}_{2}He \longrightarrow ^{1}_{0}n +$
    (e) $^{27}_{13}Al + ^{1}_{0}n \longrightarrow ^{4}_{2}He +$

2. If you had 1 g of a radioisotope with a half-life of 7.00 days (a) how much would be left after four weeks (28.00 days)? (b) At what time would $\frac{1}{8}$ g be left?
3. List all the known types of radioactive decay.
4. List the observed preferences in nature for nuclear stability.
5. List the values of the magic numbers.
6. $^{14}_{7}$N is stable ($^{14}_{6}$C decays to $^{14}_{7}$N), $^{18}_{8}$O is stable ($^{18}_{9}$F decays to $^{18}_{8}$O), and $^{22}_{10}$Ne is stable ($^{22}_{11}$Na decays to $^{22}_{10}$Ne).
   (a) What general trend causes $^{14}_{7}$N to be stable?
   (b) What other general trend works against $^{18}_{9}$F and $^{22}_{11}$Na being stable?
   (c) What other special factor (not present in the $^{22}_{11}$Na to $^{22}_{10}$Ne case) might also help $^{18}_{8}$O to be stable instead of $^{18}_{9}$F?
7. Sketch the curve of stability and clearly lable the region where β^- decaying isotopes are found.
8. Most commercial radioisotopes are now made using neutrons from nuclear reactors. Explain why this works better than charged particle accelerators (which were available sooner).
9. How can the effectiveness of a stream of neutrons for causing nuclear reactions be increased?
10. Use a very crude diagram to show the parts of a nuclear reactor and what each part does.
11. What are the key features of nuclear fission that make control of a nuclear reactor possible?
12. What steps are taken to guard against nuclear reactor accidents?
13. Sketch the curve of binding energy per nucleon for the most stable isotopes (those along the curve of stability) and use it to point out (a) why fission releases energy; (b) why fusion releases energy? (c) why $^{8}_{4}$Be is not stable.
14. Use the element synthesis in stars to explain (a) why lithium, beryllium, and boron are rare; (b) why $^{32}_{16}$S is more common than $^{34}_{16}$S (which fits the curve of stability about as well); (c) why elements of the first transition series are more abundant than those in the second and third transition series.
15. Use the general abundance of the elements and the concentration effects during formation of the solar system to explain (a) why the center of the earth is mostly iron and nickel; (b) why oxygen is so abundant on earth; (c) why there is less nitrogen on earth than phosphorus.

Test Yourself

1. List the five types of naturally occurring radioactive decay.
2. $^{3}_{1}$H has a half-life of 12.46 years. How long must a sample of $^{3}_{1}$H be stored to reach the point where only one fourth of the original $^{3}_{1}$H is left? (The other three fourths will have formed $^{3}_{2}$He by the decay reaction.)

3. What property of the neutron makes it especially useful for producing artificial radioactivity?
4. What is the function of the moderator in a nuclear reactor?
5. Which of the following is *not* a complete, correct nuclear equation?
   (a) $^{234}_{91}\text{Pa} \longrightarrow\ ^{234}_{92}\text{U} + ^{\ \ 0}_{-1}e$
   (b) $^{39}_{19}\text{K} + ^{0}_{0}\gamma \longrightarrow\ ^{38}_{17}\text{Cl} + ^{1}_{1}\text{H}$
   (c) $^{235}_{92}\text{U} \longrightarrow\ ^{140}_{56}\text{Ba} + ^{90}_{36}\text{Kr} + 5\,^{1}_{0}n$
   (d) $^{235}_{92}\text{U} + ^{1}_{0}n \longrightarrow\ ^{92}_{38}\text{Sr} + ^{142}_{54}\text{Xe} + 2\,^{1}_{0}n$
   (e) $^{9}_{4}\text{Be} + ^{4}_{2}\text{He} \longrightarrow\ ^{12}_{6}\text{C} + ^{1}_{0}n$
6. $^{12}_{6}\text{C}$, $^{16}_{8}\text{O}$, $^{20}_{10}\text{Ne}$, $^{24}_{12}\text{Mg}$, $^{28}_{14}\text{Si}$, $^{32}_{16}\text{S}$, $^{36}_{18}\text{Ar}$, and $^{40}_{20}\text{Ca}$ are more abundant than most other isotopes of similar mass because
   (a) they fit more closely on the curve of stability.
   (b) all isotopes are formed steadily in stars but the natural preference for even numbers gives the even-even ones a slight preference.
   (c) all isotopes are formed about equally in stars but many decay radioactively to the especially stable even-even isotopes.
   (d) they are formed extensively both by fusion of light isotopes and by fission of heavy isotopes.
   (e) they are the only ones formed in the main fusion sequence in the center of stars whereas the others are filled in by less important side reactions and supernova explosion reactions.
7. Choose the three phrases that best complete the two sentences.
   (1) The graph of the curve of stability as shown here has the coordinates named in (phrase A) and (phrase B)
   (2) Isotopes in the region marked by the letter C decay by (phrase C)

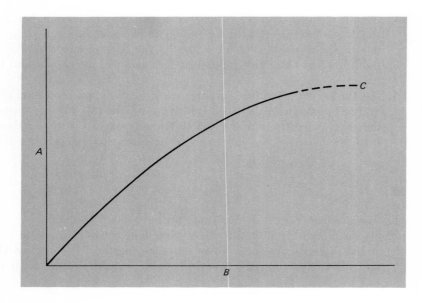

328   Nuclear Stability and Chemical Abundance   [15]

(a) (A) binding energy per nucleon, (B) mass number, (C) β^+ or ec
(b) (A) neutrons, (B) mass number, (C) α or β^-
(c) (A) protons, (B) mass number, (C) β^+ or ec
(d) (A) protons, (B) neutrons, (C) α or fission
(e) (A) nuclear charge, (B) neutrons, (C) α, β, or γ

8. Which of the following is *not* an important factor in determining which isotopes are stable?
   (a) The mass number should be an even number.
   (b) The number of protons should be an even number.
   (c) For light elements, protons should equal neutrons.
   (d) For heavy elements, there should be more neutrons than protons.
   (e) Certain "magic numbers" of protons or neutrons are preferred.

9. Which of the following is *not* a factor in causing the presence on the earth's surface of the particular elements we find to be abundant and the lower abundance of others?
   (a) Hydrogen is the starting material for synthesis of all other elements.
   (b) The main fusion sequence in stars does not produce significant amounts of elements much beyond iron and nickel.
   (c) The instability of ^8_4Be causes the main fusion sequence in stars to skip from helium to carbon.
   (d) Our sun is operating on conversion of hydrogen to helium.
   (e) Condensed materials (mostly compounds) of low density were concentrated on the Earth's surface.

10. Choose the answer that lists the products (in order) needed to complete the following nuclear equations

    $$^{40}_{19}\text{K} \longrightarrow {}^{40}_{20}\text{Ca} +$$
    $$^{210}_{84}\text{Po} \longrightarrow {}^{206}_{82}\text{Pb} +$$
    $$^{137}_{54}\text{Xe} \longrightarrow {}^{136}_{54}\text{Xe} +$$

    (a) ^1_1H, ^4_2He, ^1_0n (d) ^4_2He, $^0_{-1}e$, ^1_1H
    (b) ^1_0n, ^1_1H, ^4_2He (e) $^0_{-1}e$, $^0_{-1}e$, ^4_2He
    (c) $^0_{-1}e$, ^4_2He, ^1_0n

11. What general trend causes $^{14}_7\text{N}$ to be stable?
    (a) There is a preference in nature for even numbers of protons plus neutrons.
    (b) There is a preference in nature for equal numbers of protons and neutrons.
    (c) There is a preference in nature for certain "magic" numbers of protons and neutrons.
    (d) $^{14}_7\text{N}$ is the isotope nearest to the observed atomic weight of nitrogen, which is 14.
    (e) Repulsions between protons make the curve of stability bend toward excess neutrons as the number of protons goes up.

12. Although only one answer to question 11 explains why $^{14}_7\text{N}$ is stable, three of the other four statements are also factors affecting nuclear

stability. Which one of the statements is not a factor affecting nuclear stability?

13. Which of the following lists gives the important components of a nuclear reactor?
    (a) Fissionable fuel, control rods, igniter
    (b) Fissionable fuel, heat source
    (c) Fissionable fuel, moderator, igniter
    (d) Fissionable fuel, accelerator, magnetic field, output wires
    (e) Fissionable fuel, moderator, control rods

14. A *large* nuclear power reactor could, if allowed to undergo a power excursion, be hazardous because
    (a) it would explode as an atomic bomb.
    (b) the continued high output would cause thermal pollution of cooling waters.
    (c) it would melt and release radioactivity.
    (d) the level of fission would increase enough to penetrate the shielding with heavy doses of γ rays.
    (e) None of these. It would safely turn itself off.

15. Beryllium is the only even numbered element whose most common isotope ($^{9}_{4}Be$) has an odd number of neutrons. Why is $^{9}_{4}Be$ more common than $^{8}_{4}Be$?
    (a) Five is an especially favored number of neutrons.
    (b) Coulomb repulsion makes it necessary for beryllium to have more neutrons than protons.
    (c) $^{8}_{4}Be$ decays to two $^{4}_{2}He$ because $^{4}_{2}He$ has a "magic number" of protons and neutrons.
    (d) $^{9}_{4}Be$ is the product formed from one of the observed natural decay chains.
    (e) None of these.

16. Oxygen is the most abundant element on earth because
    (a) $^{16}_{8}O$ has a magic number of protons and neutrons.
    (b) $^{16}_{8}O$ has both even numbers and equal numbers of protons and neutrons.
    (c) $^{16}_{8}O$ is formed in the main fusion sequence of stars.
    (d) oxygen forms compounds of the right density to be concentrated on the Earth's surface.
    (e) all of these.

16 Use of Metals

16-1 Occurrence of Metals

There are a large number of metallic elements.

The metallic elements in the second row of the periodic table and in the third row through the transition elements are abundant in the universe.

Metals are those elements that tend to lose electrons and form positive ions. When they are not chemically bonded in compounds, they may form metallic crystals in which positive ions are surrounded by a sea of electrons (see Section 8–4 and Section 7–2). The general trend of shell effects makes elements increasingly metallic as the number of shells increases (See section 8–5). That makes the total number of metallic elements very large, because most of the relatively rare heavy elements are metals. Figure 16–1 (which shows the same information as Figure 8–4) shows the very large number of different metallic elements.

Although the large number of metallic elements provides a variety of properties that may fit particular needs, many of the metals are too rare for any large scale uses. Nuclear stability factors and the synthesis of elements in stars (see Sections 15–6, 15–7, and 15–8) give a strong preference to formation of the lighter elements. Figure 16–2 shows the 25 most abundant elements in the universe as a whole. The low abundance of lithium, beryllium, and boron is a conspicuous gap, since all other light elements are very abundant. That gap has been explained through the instability of $^{8}_{4}Be$ (see Section 15–6) and the resulting skipping of $^{8}_{4}Be$ (and the other lithium, beryllium, and boron isotopes that could eventually form from $^{8}_{4}Be$) in the main fusion cycle in stars (see Sections 15–7 and 15–8). The elements beyond the maximum binding energy point (iron and nickel, see Section 15–6) were also not formed as extensively in stars (see Sections 15–7 and 15–8). Throughout the entire range, even numbered elements are favored over odd numbered elements because of the nuclear preferences for even numbers (see Section 15–2). That shows up as odd numbered elements scandium and vanadium fail to make the most abundant 25. The general trend toward decreasing abundance as the elements increase in atomic number has an exception in the area of maximum binding energy around iron and nickel. The abundances in that area are about as high as those of elements with atomic numbers 10 units lower. Therefore the odd elements near iron and nickel make the list of the 25 most abundant elements and the even numbered elements (iron and nickel) are among the top 15 in abundance.

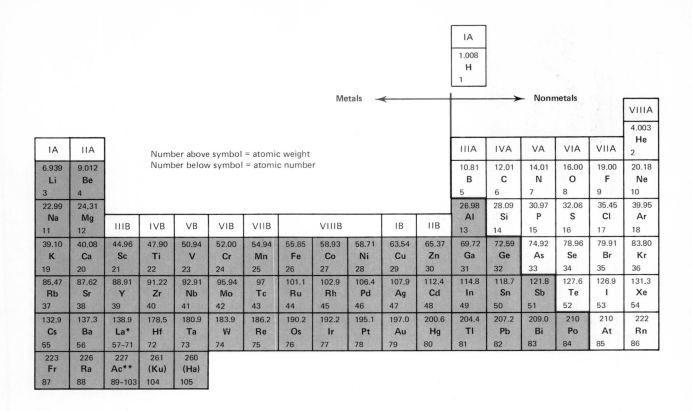

Figure 16–1 Metals and nonmetals.

Exact values for abundance of elements are hard to measure, so there are disagreements between sources about the best values. Among the rarer elements, there are even disagreements about the order of elements by abundance. Figure 16-3 shows a graph of the relative abundances from one of the books sometimes used as a text on this subject in geology courses, *Distribution of the Elements in Our Planets* by Louis Ahrens (McGraw-Hill Book Company, 1965, in paperback).

The total abundance of elements in the universe and the abundance of elements available for use on Earth are significantly different. Figure 16–4 shows the 19 most abundant elements in the accessible regions of the Earth's crust, oceans, and atmosphere. When we compare Figure 16–4 to Figure 16–2, we see that abundances in the universe influence abundances on Earth.

But some other factors also show up. The highly reactive (and less dense) metals in the first two columns of the periodic table are favored over some of the less reactive, very dense transition metals. The noble gases (helium, neon, argon) are also less abundant on Earth.

The abundant elements within our reach are those that formed light (but not gaseous) compounds able to float to the surface of the Earth's dense molten metal (mostly iron and nickel) core. Even the water of our oceans and gases of our atmosphere came from these other light compounds (often released by volcanic activity). The light compounds formed were those of the most reactive metals (particularly if they were also light metals) with the most reactive nonmetals (particularly the lightest ones). The only very reactive, low density elements left out are the ones with particularly low abundance in the universe because they are odd numbered (rubidium, cesium) or in the group skipped because of $^{8}_{4}Be$ instability (lithium, beryllium, boron).

The tendency to concentrate light compounds on the Earth's surface is even more apparent when we show only the ten most abundant elements, as in Figure 16–5. We find only oxygen (light, very reactive, even numbered,

Elements of low density and high chemical reactivity have been concentrated on the surface of the earth.

Figure 16–2 The 25 most abundant elements in the universe.

Number above symbol = atomic weight
Number below symbol = atomic number

[16-1] Occurrence of Metals

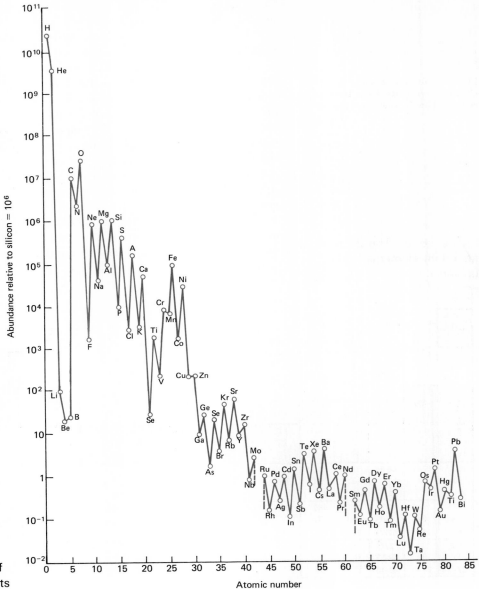

Figure 16-3 Diagram of the abundance of elements in the universe.

The vertical scale is logarithmic and the estimated abundances are relative to 10^6 atoms of silicon. [From *Distribution of the Elements in Our Planets* by Louis Ahrens. Copyright 1965 by McGraw-Hill Book Company. Used with permission of McGraw-Hill Book Company.]

IA																	VIIIA
1.008 H 1	IIA											IIIA	IVA	VA	VIA	VIIA	4.003 He 2
6.939 Li 3	9.012 Be 4		Number above symbol = atomic weight									10.81 B 5	12.01 C 6	14.01 N 7	16.00 O 8	19.00 F 9	20.18 Ne 10
22.99 Na 11	24.31 Mg 12	IIIB	IVB	VB	VIB	VIIB		VIIIB		IB	IIB	26.98 Al 13	28.09 Si 14	30.97 P 15	32.06 S 16	35.45 Cl 17	39.95 Ar 18
39.10 K 19	40.08 Ca 20	44.96 Sc 21	47.90 Ti 22	50.94 V 23	52.00 Cr 24	54.94 Mn 25	55.85 Fe 26	58.93 Co 27	58.71 Ni 28	63.54 Cu 29	65.37 Zn 30	69.72 Ga 31	72.59 Ge 32	74.92 As 33	78.96 Se 34	79.91 Br 35	83.80 Kr 36
85.47 Rb 37	87.62 Sr 38	88.91 Y 39	91.22 Zr 40	92.91 Nb 41	95.94 Mo 42	97 Tc 43	101.1 Ru 44	102.9 Rh 45	106.4 Pd 46	107.9 Ag 47	112.4 Cd 48	114.8 In 49	118.7 Sn 50	121.8 Sb 51	127.6 Te 52	126.9 I 53	131.3 Xe 54
132.9 Cs 55	137.3 Ba 56	138.9 La* 57–71	178.5 Hf 72	180.9 Ta 73	183.9 W 74	186.2 Re 75	190.2 Os 76	192.2 Ir 77	195.1 Pt 78	197.0 Au 79	200.6 Hg 80	204.4 Tl 81	207.2 Pb 82	209.0 Bi 83	210 Po 84	210 At 85	222 Rn 86
223 Fr 87	226 Ra 88	227 Ac** 89–103	261 (Ku) 104	260 (Ha) 105													

*Lanthanide series	140.1 Ce 58	140.9 Pr 59	144.2 Nd 60	147 Pm 61	150.4 Sm 62	152.0 Eu 63	157.3 Gd 64	158.9 Tb 65	162.5 Dy 66	164.9 Ho 67	167.3 Er 68	168.9 Tm 69	173.0 Yb 70	175.0 Lu 71
**Actinide series	232.0 Th 90	231 Pa 91	238.0 U 92	237 Np 93	242 Pu 94	243 Am 95	247 Cm 96	247 Bk 97	249 Cf 98	251 Es 99	254 Fm 100	253 Md 101	256 No 102	254 Lr 103

Figure 16–4 The 19 most abundant elements on Earth.

The most abundant elements on the Earth's surface are oxygen and the relatively abundant metals and nonmetals which react with oxygen most readily to form nongaseous compounds.

Metals in the crust of the Earth got there in the form of compounds.

and therefore by far the most favored nonmetal) and those fairly abundant light elements that are best at reacting with oxygen to make light compounds. They are the three metals in the lightest row with any abundant metals, four metals in the next lightest row (including the two least dense and most reactive), two nonmetals (hydrogen and silicon) (which are quite reactive with oxygen), and oxygen itself. It is also worth noting that less metallic elements, such as carbon, are less favored. Their tendency to form gaseous oxides rather than liquids or solids let the Sun's gravity pull more of these nonmetals away from the earth.

The metals on Earth exist in the forms and locations caused by their chemical characteristics. Large amounts of unreacted metal exist in the Earth, particularly iron and nickel. However, those metallic elements present in abundance are dense materials that are concentrated in the Earth's core (see Section 15–8). The metal that ended up on the Earth's crust where we can reach it was in the less dense form of various chemical compounds. The rocks and soil of Earth consist mainly of compounds of the most common metallic and nonmetallic elements. Less common metals (and nonmetals) are also

IA																	VIIIA
1.008 H 1	IIA											IIIA	IVA	VA	VIA	VIIA	4.003 He 2
6.939 Li 3	9.012 Be 4		Number above symbol = atomic weight Number below symbol = atomic number									10.81 B 5	12.01 C 6	14.01 N 7	16.00 O 8	19.00 F 9	20.18 Ne 10
22.99 Na 11	24.31 Mg 12	IIIB	IVB	VB	VIB	VIIB		VIIIB		IB	IIB	26.98 Al 13	28.09 Si 14	30.97 P 15	32.06 S 16	35.45 Cl 17	39.95 Ar 18
39.10 K 19	40.08 Ca 20	44.96 Sc 21	47.90 Ti 22	50.94 V 23	52.00 Cr 24	54.94 Mn 25	55.85 Fe 26	58.93 Co 27	58.71 Ni 28	63.54 Cu 29	65.37 Zn 30	69.72 Ga 31	72.59 Ge 32	74.92 As 33	78.96 Se 34	79.91 Br 35	83.80 Kr 36
85.47 Rb 37	87.62 Sr 38	88.91 Y 39	91.22 Zr 40	92.91 Nb 41	95.94 Mo 42	97 Tc 43	101.1 Ru 44	102.9 Rh 45	106.4 Pd 46	107.9 Ag 47	112.4 Cd 48	114.8 In 49	118.7 Sn 50	121.8 Sb 51	127.6 Te 52	126.9 I 53	131.3 Xe 54
132.9 Cs 55	137.3 Ba 56	138.9 La* 57–71	178.5 Hf 72	180.9 Ta 73	183.9 W 74	186.2 Re 75	190.2 Os 76	192.2 Ir 77	195.1 Pt 78	197.0 Au 79	200.6 Hg 80	204.4 Tl 81	207.2 Pb 82	209.0 Bi 83	210 Po 84	210 At 85	222 Rn 86
223 Fr 87	226 Ra 88	227 Ac** 89–103	261 (Ku) 104	260 (Ha) 105													

*Lanthanide series		140.1 Ce 58	140.9 Pr 59	144.2 Nd 60	147 Pm 61	150.4 Sm 62	152.0 Eu 63	157.3 Gd 64	158.9 Tb 65	162.5 Dy 66	164.9 Ho 67	167.3 Er 68	168.9 Tm 69	173.0 Yb 70	175.0 Lu 71
**Actinide series		232.0 Th 90	231 Pa 91	238.0 U 92	237 Np 93	242 Pu 94	243 Am 95	247 Cm 96	247 Bk 97	251 Cf 98	254 Es 99	253 Fm 100	256 Md 101	254 No 102	257 Lr 103

Figure 16–5 The ten most abundant elements on Earth.

included in smaller amounts. Light elements are favored by both their greater abundance (see Sections 15–7 and 15–8) and the frequently lower densities of their compounds.

The factors just discussed can also help us understand why other elements are rare. For example, gold is rare on Earth because its high atomic number and odd number give it a low abundance in the universe and its high density and low reactivity keep it from forming light compounds that would reach the Earth's surface.

In the millions of years they have been in the Earth's crust, the metals may have taken part in various chemical reactions. Some of the changes they may have undergone are as follows.

1. The less reactive metals may have been displaced from compounds and appear as free metal. Gold (although rare) is an example of a quite unreactive metal sometimes found in the metallic form. Copper also existed as free metal, but the fraction of it in that form was less than gold because copper is more reactive chemically than gold. Those metals that are substantially more reactive than copper cannot be changed into

Less reactive metals may be found as free metal which was displaced from compounds naturally.

Elements present as soluble compounds have been concentrated in the oceans.

The least soluble compounds of the most common substances precipitate from oceans to form minerals.

Veins of ore may occur at boundaries between minerals.

Mining consists of location and recovery of desired elements in desirable chemical form.

Transportation and disposal problems with impurities in low grade ores give premium value to the richest ores and sometimes justify enrichment processes.

the free metal form by the chemical reactions normally occurring, so they are always found in compounds.

2. Water soluble compounds may be dissolved. The water transport cycle (see Section 13–1) circulates water which carries away the most soluble materials. Elements that form soluble compounds, such as sodium and magnesium, are thus transported to and concentrated in the oceans.
3. Dissolved materials may be precipitated out as especially insoluble substances. As the dissolved materials are concentrated in oceans, some substances may exceed their solubilities and precipitate out. As concentrations and temperatures change, the precipitates may vary. This allows buildup of various minerals, each differing in elemental composition.
4. Dissolved materials may be carried to locations where they react. They are then concentrated at the reaction site. The buildup of particular elements at the boundary between different minerals creates "veins" of rich ore.

Mining operations consist of the location and recovery of material particularly rich in the desired substance. That means both that it should be rich in the desired element and that the element should be in a chemical form that can be easily converted to the free metal. The sources will be minerals in which the desired material was precipitated or simply remained, veins where natural processes have concentrated the desired material, or occasionally (as in the case of magnesium) the oceans or other salt water from which the material can be chemically removed.

In most cases, much undesirable material will be in the mined ore along with the commercially valuable material. If the unnecessary material must be transported and processed with the valuable part, costs are increased and disposal problems are created at the processing points. That provides strong economic pressure toward selection of the richest ores. Sometimes this "selection" can be continued by treatments to enrich the ore before it leaves the mining site. That saves transportation. If the enrichment process leaves the unwanted materials in their original form, they may be put back into the mining site and also cut down on pollution.

Transition Metal Chemistry

Transition metals form a variety of oxidation states by involving none, some, or all of their d electrons.

Our discussion in Chapter 8 on General Trends in Chemical Behavior included the alkali metal family (Li, Na, K, Rb, Cs). Some of these elements are common and important on earth, but there is very little to their chemistry except the ability to lose one electron. The second column of the periodic table, the alkaline earth metals (Be, Mg, Ca, Sr, Ba), are also quite important, but dull. They lose two electrons. On the other hand, the transition metals are both abundant enough to be important (at least in the first row of transition elements) and varied enough in their chemistry to be quite interesting.

In Section 8–7 we discussed the effects of the d level electrons and changing nuclear charge on the oxidation states of transition metals. You should *read*

Section 8–7 again now to review the energy level situation in the first transition series and the multiple oxidation states possible because of the availability of d electrons.

Some transition metals form all of the oxidation states expected. An example is vanadium, which forms V^{2+}, V^{3+}, V^{4+}, and V^{5+}. Other transition metals have "missing" oxidation states. An example is chromium, which forms Cr^{2+}, Cr^{3+}, and Cr^{6+} in water but not the Cr^{4+} or Cr^{5+} that might be expected. These missing oxidation states are thought to be formed and exist very briefly in some chemical reactions. We do not find them because they react quickly to form other products which are more energetically favored in nature. They can react by **disproportionation reactions** to form more stable oxidation states, one product being a higher oxidation state and the other being a lower oxidation state. An example would be the reaction $3\,Cr^{4+} \longrightarrow 2\,Cr^{3+} + Cr^{6+} +$ heat. Stronger bonding forces exist (in water) for the chromium as two Cr^{3+} and one Cr^{6+} than as three Cr^{4+}, so the reaction proceeds spontaneously to form the more strongly bonded products and release heat.

Some possible oxidation states disproportionate to other higher and lower oxidation states.

Disproportionation reactions do not always yield two positive oxidation states. Cu^+ disproportionates in water to yield Cu^{2+} plus copper metal (oxidation state = 0). The only requirement for disproportionation is that the products have a more favorable total bonding energy than the reactants. Cu^{2+} has much more than double the forces of attractions to H_2O than Cu^+ has. Cu^{2+} has double the charge of Cu^+, is a smaller ion so the H_2O molecules can have their lone pairs of electrons held closer to the positive charge, and (because of the stronger forces) Cu^{2+} ions get more H_2O molecules to establish bonding forces than Cu^+ ions. In the case of copper in water, the extra forces are large enough to make up for the extra energy required to form one $+2$ ion instead of two $+1$ ions. Therefore, the disproportionation occurs. However, when H_2O is not present the extra attractive forces to Cu^{2+} ions do not make up for the extra energy required to form a $+2$ ion instead of two $+1$ ions. Therefore, Cu^+ can continue to exist and is, in fact, very common in solids. The disproportionation of Cu^+ in water occurs because of an especially favorable situation of one of the products. It is the strong interaction of Cu^{2+} with water that causes the reaction.

Disproportionation occurs when the products have a more favorable total bonding energy than the reactants.

Every disproportionation reaction must have a product that is especially favored for some reason. If it did not, electrical repulsions would tend to keep the electrical charges distributed as evenly as possible. There clearly are some arrangements in nature that are strongly favored over others. One of the particularly favored arrangements is the Cr^{3+} oxidation state. It is the preference for $+3$ that causes the disproportionation of Cr^{4+} and Cr^{5+}. It also makes it very easy to oxidize Cr^{2+} to Cr^{3+}. That is unusual behavior for a transition element, where $+2$ is usually the most common and favored oxidation state.

The special preference for Cr^{3+} occurs because of the strongly bonded **complex ions** it can form. Complex ions can be formed by all positive ions, but some form with much stronger bonding than others. The most favored

Positive ions can form complex ions, some of which have more favorable bonding forces than others.

complex ions cause some of the oxidation state patterns. Their strong chemical bonding also leads to many other important reactions. Complex ion formation with iron is the mechanism used by the human body to transport and store needed oxygen. Good complexing metals also play an important role in certain vitamins and in other processes vital to life. Therefore, it is important for us to understand the basic principles that cause complex ion formation and why some complexes are more stable than others. We will find that the same electronic configurations that gave transition metals a variety of oxidation states give them a special central importance in this vital subject of complex ions.

Complex formation plays a vital role in living things.

16-3 Complex Ion Formation

Some details of complex ion formation can be best understood by using more complex bonding models than we have been using in this book. However, many of the most important features can be understood from the electronic orbitals and sharing of electron pairs we described in Chapter 6 and Chapter 7. We will limit our explanations to using those principles and accept the occasional errors that could be avoided by more sophisticated models.

Positive ions form complexes with negative ions or polar molecules by forming the bonds we described as **coordinate covalent bonds** in Section 7-8. The metal ions have unfilled, low lying energy levels in which they can accept pairs of electrons. Molecules or ions that have **lone pairs** of electrons in their outermost shells can share their lone pairs with the metal, thus forming bonds. Those substances that can provide an unshared pair of electrons and form a bond are called **ligands.** All of these ligands act as electron pair donors, which makes them Lewis acids according to the definitions of acids and bases in Section 13-4. The metals act as Lewis bases, electron pair acceptors.

Complex ions form by coordinate covalent bonds in which the positive metal ion (or sometimes even a neutral transition metal atom) accepts electron pairs from ligands.

The ligands act as Lewis acids and the metals as Lewis bases during complex formation.

The important features needed by the metal to form good complex ions are *empty orbitals* to accept electron pairs and *strong attractions* for electrons in those orbitals. The transition elements are in a very good position to provide both. They have more empty, low lying orbitals than the elements to their right because they have available d electron levels. On the other hand, they can provide stronger attractions than the alkali or alkaline earth metals. The transition metals have larger nuclear charges than alkali or alkaline earth metals to attract electrons to the same shells and energy levels. Therefore transition metals tend to form the strongest complexes.

To form strongly bonded complexes, a metal must have empty orbitals and strong attractions for electrons in those orbitals.

Transition metals tend to form strong complexes.

In some cases, it is not even necessary for the metal to form a positive ion before complexing ligands. Transition metals with partially filled d levels can accept electron pairs into strongly held levels even without removing any of the electrons originally on the neutral atoms. Carbon monoxide is a very good ligand. It can form complexes called carbonyls with nickel, iron, and chromium. The complex with nickel is particularly easy to form. The $Ni(CO)_4$ complex, nickel tetracarbonyl, which forms is gaseous (and quite poisonous). Formation of $Ni(CO)_4$ is used commercially to separate nickel from impurities, which are left behind as solids while nickel is converted to gaseous $Ni(CO)_4$ and carried away.

The most favorable complexes are formed with positive ions (which help attract the lone pairs of ligands). The best complex formers have the maximum usable low lying empty orbitals near the largest available nuclear charges. If the attractions are small, only two ligands will be attracted. These two ligands will be as far apart as possible. An example of that type of complex ion is Cu^{1+} plus two H_2O ligands. The ligands are written as part of the formula of the complex ion product, in this case as $[Cu(H_2O)_2]^{1+}$. When the forces are somewhat larger, four ligands can be held in the tetrahedral arrangement we discussed in Section 7–13. An example is $[Zn(NH_3)_4]^{2+}$ formed from Zn^{2+} and four NH_3 molecules. The tetrahedral arrangement of four ligands is the maximum possible if the metal is limited to accepting the eight electrons that would give it a closed shell in its outer shell. Normally that outer shell is completely empty before addition of ligands because all of the outer electrons were lost in forming the positive ion (see Section 7–7).

Some metals, particularly the transition metals, can go beyond the tetrahedral four-ligand arrangement. They do that by using low lying d electron levels. The particularly stable arrangement possible using d levels is the octahedral arrangement (see Section 7–13) of six ligands; $[Cr(H_2O)_6]^{3+}$ is an example. The octahedral arrangement is so favorable that it is sometimes achieved even when some of the electron pairs in that arrangement around the metal atom do not form bonds to ligands. The most common examples are square planar arrangements of four ligands. In such cases there are also two lone pairs on the metal, one above and one below the square planar shape (see Section 7–13). An example of a square planar complex is $[Ni(CN)_4]^{2-}$ (formed from Ni^{2+} and four CN^-).

We find it convenient to use a symbolism called valence bond hybrid orbitals to distinguish between different ligand arrangements. In this symbolism we list the number and types of atomic orbitals that would have to be combined to form the arrangement used in the bonding. This combination of orbitals into hybrids is one model used to explain chemical bonding. It is neither the most sophisticated nor the least sophisticated model available. It is often used as a compromise to give fairly good explanations in a moderate amount of space. It is used particularly heavily by organic chemists and by chemists discussing complex ions.

Arrangements in complex ions are named from the hybrid orbitals which give their bonding.

The hybrid orbital descriptions of the complexes we used as examples are sp for the *linear* two ligand $[Cu(H_2O)_2]^+$ complex, sp^3 for the *tetrahedral* $[Zn(NH_3)_4]^{2+}$ complex, d^2sp^3 for the *octahedral* $[Cr(H_2O)_6]^{3+}$ complex, and dsp^2 for the *square planar* $[Ni(CN)_4]^{2-}$ complex. sp^2, a trigonal planar arrangement (see Section 7–13) is also a common hybrid orbital arrangement, appearing frequently in organic chemistry.

Common kinds of complexes are sp (linear, small attractions), sp^3 (tetrahedral, larger attractions), d^2sp^3 (octahedral, very strongly favored), and dsp^2 (square planar).

The hybrid orbital symbolism is useful because it lets us go back and figure out which orbitals must be empty for a metal to form strong complexes. We will use diagrams to represent the electron positions on the metal atom.

Figure 16–6 shows the energy levels in a chromium atom, a Cr^{3+} ion,

Figure 16-6 Favorable complexing of Cr^{3+}.

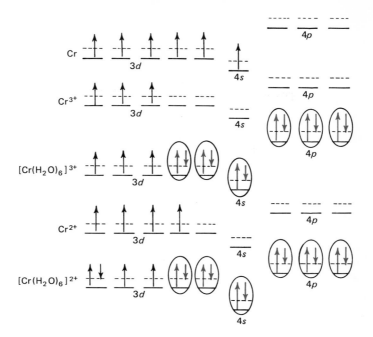

a Cr^{3+} complex ion, a Cr^{2+} ion, and a Cr^{2+} complex ion. In each case the 18 strongly held $1s$, $2s$, $2p$, $3s$, and $3p$ electrons are not shown. The circles show the empty energy levels that accepted electron pairs in the d^2sp^3 complex ions. Each electron is represented by an arrow. The arrow points up if the electron has the spin direction most favored by external magnetic fields (half-filled subshell effect, Section 6–7). The arrow points down if the electron has the less favored spin. Each orbital, capable of holding two electrons, is represented by a solid line for the energy of the electron with favored spin and a dotted line for the energy of the electron with less favored spin.

Since the ligands must bring in electrons in *pairs* (their lone pairs), the metal atom must have *completely empty* orbitals to accept them. To form a d^2sp^3 complex, the metal atom must have *exactly* two d orbitals, one s orbital, and three p orbitals available. It cannot use any extra d orbitals, even if they are available and would hold the electrons at lower energy than the p orbitals. There must be three p orbitals, two d orbitals, and one s orbital to get the favorable shape and bonding of the octahedral complex.

We see that Cr^{3+} has the necessary orbitals available. It can form the maximum bonding of a d^2sp^3 complex without requiring any energy to rearrange its electrons. On the other hand, Cr^{2+} has one electron blocking a needed d orbital. That electron can be moved to pair up in another d orbital, as shown in Figure 16–6. However, energy must be supplied to put that electron in the less favored spin. The formation of a d^2sp^3 complex is so advantageous that the extra electron in Cr^{2+} does pair up in spin, but the product is less stable by the energy needed to pair one set of spins. The

[16-3] Complex Ion Formation

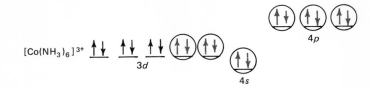

Figure 16-7 Favorable complexing in $[Co(NH_3)_6]^{3+}$.

availability of one electron less strongly held than the others (because of the unfavorable spin) also explains why Cr^{2+} can react easily to form Cr^{3+}.

The most favored complexes are those, like Cr^{3+}, that can form without rearranging electrons or those with larger nuclear charges that can rearrange the electrons so that all bonding is done in low lying levels. An example of the latter is $[Co(NH_3)_6]^{3+}$. Figure 16-7 shows the arrangement in $[Co(NH_3)_6]^{3+}$. In this and all following figures, only a single line is used to represent each orbital although, in fact, each orbital has two levels to represent the two different spins.

The less favored complexes in which higher energy levels must be used can also form, particularly when a very good ligand is available. The higher level may be used as a place to put electrons from the metal so lower levels are available for the complexing. Or the upper levels can be used to form the hybrid orbitals for the complex. Figure 16-8 shows the two possible arrangements that could allow an octahedral complex of Co^{2+}. The first arrangement, using the low lying d levels for complexing, is called an **inner sphere complex**. The second arrangement, using the d levels of the fourth shell instead of $3d$ for complexing is called an **outer sphere complex**.

Figure 16-9 shows the arrangements for Zn^{2+} when it forms a tetrahedral, four-ligand complex and when it forms an outer sphere, octahedral, six-ligand complex. In this case, the octahedral complex forms if the ligand can set up particularly strong bonds. The Cu^{2+} complex with H_2O is such a case.

Complexes where there is the maximum nuclear charge for either no needed rearrangement of electrons or rearrangement filling the low lying levels are particularly favored.

Complexes needing d levels may use inner d levels, if available, to form inner sphere complexes, or they may use outer d levels to form outer sphere complexes.

Figure 16-8 Octahedral complexes of Co^{2+}: uses of higher energy levels.

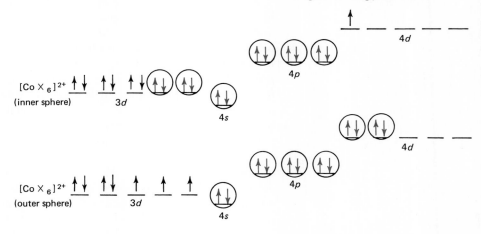

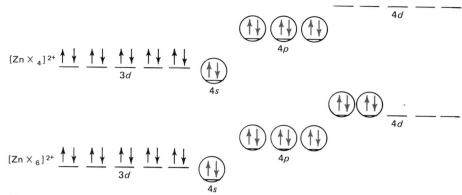

Figure 16-9 Complexes of Zn^{2+}.

The especially favored bonding of Cu^{2+} in H_2O which causes disproportionation of Cu^+ (see Section 16–2) is formation of $[Cu(H_2O)_6]^{2+}$. That is an outer sphere complex which is distorted from the normal octahedral shape. Four of the water molecules on $[Cu(H_2O)_6]^{2+}$ are known to be held closer to the copper than the other two.

From consideration of the energy levels alone, we might expect outer sphere complexes to form only in cases like zinc where there are no openings in the lower d levels. However, nature is not that simple. Even in the Co^{3+} case, an outer sphere complex can form with certain ligands. Figure 16–10 shows the arrangement of $[CoF_6]^{3-}$, which is an outer sphere complex. The powerful forces involved in forming complexes cause these variations in type. Those forces and variations are beyond our understanding here, but the existence of even outer sphere complexes does help to emphasize an important point we can understand. There is a *very strong tendency in nature to go to d^2sp^3 complexes,* whenever good ligands and metals with available d electron levels get together.

Octahedral complexes are strongly favored.

Figure 16–11 shows the contrast between the two types of four-ligand complexes. When one d orbital can be made available, the square planar dsp^2 complexes may form. When the low lying d orbitals are full, the tetrahedral sp^3 complexes are usually favored.

The complexes formed in nature are most often those that use up all of the readily available low lying electron energy levels.

Figure 16–10 An outer sphere Co^{3+} complex.

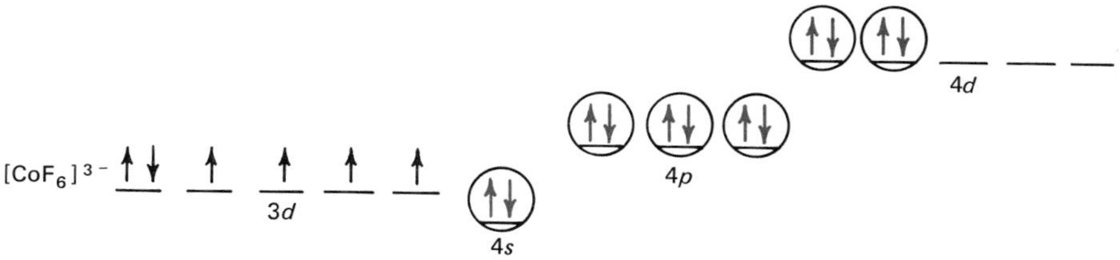

[16–3] Complex Ion Formation

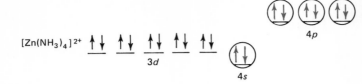

Figure 16-11 Two types of four-ligand complexes.

16-4 Methods of Ore Enrichment

Sometimes ores can be enriched by duplicating the common natural method of moving material—dissolving it in water. This is particularly useful when a rare substance is being removed from a large quantity of other material. The process is called **leaching**. It is often used commercially in copper mining. Copper is an odd-numbered element and therefore not very abundant (see Sections 15-2 and 15-8). However, copper is often present in ores as copper oxides, which can be recovered by leaching. Copper oxides react as bases with acidic solutions, allowing the copper to be dissolved. Low grade copper ores (often less than 1% copper) are pushed into position so that rainwater accumulates and slowly seeps through the ore. The rainwater is weakly acidic because of dissolved carbon dioxide (CO_2), so it dissolves copper oxides. The other substances in the ore are less soluble in acidic water, so they are left behind. After seeping through many piles of ore, the water is processed to recover the copper. Figure 16-12 shows recovery of copper by leaching.

Another process useful for separating small amounts of desired ore from

Leaching dissolves small amounts of desired material, such as copper oxides, out of a large mass of impurities.

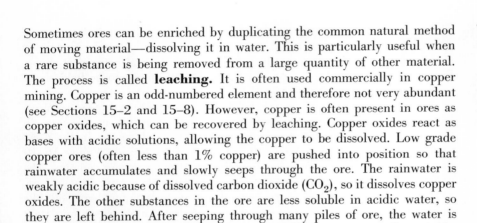

Figure 16-12 Recovery of copper by leaching.

Rainwater made acidic by dissolved carbon dioxide is allowed to flow through loose piles of low grade copper ore. Copper oxide dissolves in the acidic solution and is carried out of the ore pile in the water. After several exposures to ore piles, the rainwater containing dissolved copper oxide is collected for chemical processing to recover the copper.

Flotation lifts ore components out in a foam while impurities settle out in water.

Mechanical separations can separate large quantities of ore from large quantities of impurities by using a difference in a physical characteristic like density.

Magnetic iron ores can be mechanically enriched by use of a magnetic field.

a large bulk is **flotation.** If the desired ore can be made to adsorb small bubbles of air, it can be floated to the surface in a foam and skimmed off while the unwanted material settles to the bottom of a tank of water. Usually some material, such as an oil, is added which is preferentially adsorbed by the desired ore particles. This coating then helps the particles pick up bubbles and rise in the foam as air is pumped through the water. Figure 16–13 shows a flotation process.

Flotation is essentially a method of mechanically sorting materials. Other **mechanical separations** are also possible. One of the oldest of these is gold panning. Rocks were swirled in a shallow pan of water. Light pieces washed over the edge while heavy pieces of gold remained in the bottom. More refined devices are possible to separate dense and light components of ores. Mechanical crushing to small pieces is needed to make such separations effective. Crushing is also necessary for flotation and aids leaching.

An important variation of mechanical separation uses magnetism instead of density. Most (but not all) iron ores are magnetic. The magnetic iron ores can be separated from nonmagnetic impurities by passing crushed ore through a magnetic field. The magnetic ore can be deflected into a bin, as shown in Figure 16–14, while the nonmagnetic material continues undeflected and is discarded. Iron is a very common element (see Section 15–8), so a fairly

Figure 16–13 Flotation.

Crushed ore is placed in water and exposed to oil or a foam forming compound. Air bubbles forced into the water tend to cling to the ore particles because they are preferentially coated by the foam forming compound. The air bubbles then cause the ore to float upward and become entrapped in the foam forming at the surface. The foam is removed and enriched ore is recovered from the foam.

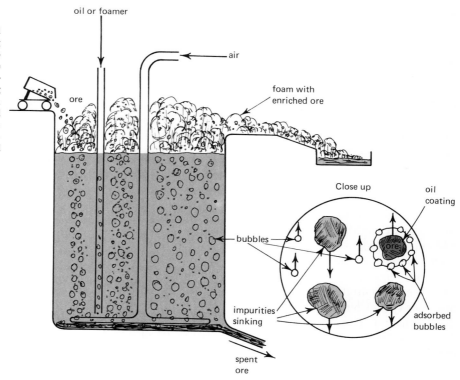

[16–4] Methods of Ore Enrichment

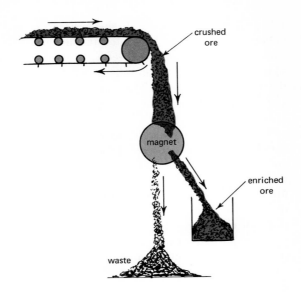

Figure 16-14 Magnetic separation of iron ore.

The iron ore is deflected by the magnetic field, whereas the nonmagnetic impurities are unaffected by the magnetic field.

large fraction of the material being passed through the magnetic field is usually ore to be kept. Mechanical separations (density or magnetic) are most effective when both the portions to be kept and to be discarded are fairly large. That is sharply different from leaching and flotation, which are usually used to separate small amounts from a large batch.

Leaching used chemical reactions to separate small amounts from a large batch. The amounts are limited by solubilities of minerals in water. On the other hand, concentrated chemical reagents can permit us to do **chemical separations** in which large amounts of material are separated from impurities. These separations are often done at centralized plants instead of at the mining site because of the large amounts of chemicals and equipment needed. The Bayer process, used in aluminum production, is an important commercial example of chemical separation on an ore (see Section 16–5.2).

Chemical separation can handle large quantities by using concentrated reagents and using differences in chemical reactions.

16–5 Production of Aluminum

16–5.1 Selection of Ore. Aluminum is the third most abundant element in the Earth's crust. That abundance is surprising because aluminum is an odd-numbered element and therefore less favored in nuclear structure than even-numbered elements (see Section 15–2). The other four of the five most abundant elements in Earth's crust are all even-numbered (O, Si, Fe, and Ca). The concentration of aluminum in the earth's crust is caused by the low density and high chemical reactivity of aluminum. Aluminum is also a light element in the range formed most extensively by stars (see Section 15–7). Aluminum formed stable chemical compounds that had the appropriate low density to float to the surface before the Earth solidified. The sixth, seventh, and eighth most abundant elements in the Earth's crust (sodium, potassium, and magnesium) are similar cases of moderately abundant light metallic elements concentrated in the crust by their reactivity and low density (see Section 16–1).

Aluminum is an important useful metal, particularly in uses where its low density is helpful. However, the same high reactivity that helped concentrate aluminum in the Earth's crust makes it very difficult to get the aluminum out and converted into metallic form. Most of the aluminum on Earth is in the form of aluminum silicates. These common compounds of the three most common elements (oxygen, silicon, and aluminum) exist in many various forms. The clay materials in soil are aluminum silicates familiar to most of us. No practical method has ever been found to separate the aluminum from the undesirable silicon component and produce aluminum metal from aluminum silicates. Therefore, the great majority of the aluminum on earth is in compounds that are worthless as a source of aluminum metal.

Aluminum is very abundant, but most of it is in aluminum silicates which are not usable.

A very small fraction of the aluminum on earth is present as the oxide, Al_2O_3. All commercial production of aluminum is from this chemically more manageable source. The raw Al_2O_3 is in an ore called bauxite. Even though the aluminum is not chemically combined with silicon in this ore, there is usually some silicon and some iron (the second and fourth most abundant elements) present along with various other impurities. These impurities would interfere with the production of aluminum metal, so they are removed by a chemical separation called the Bayer process.

Practical production of aluminum begins from bauxite ore (Al_2O_3) available in limited quantities.

Al_2O_3 is purified by the Bayer process.

16–5.2 Bayer Process. The Bayer Process, shown in Figure 16–15 takes advantage of a chemical property in which aluminum acts differently from its common impurities. Aluminum oxide is **amphoteric,** capable of

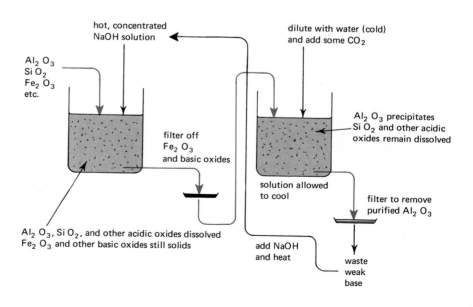

Figure 16–15 The Bayer process.

In the first stage aluminum oxide, silicon oxide, and other acidic oxides are dissolved while iron oxides and other basic oxides remain as solids. After those solid impurities are filtered off, water and carbon dioxide are added to weaken the base. The aluminum oxide then reprecipitates while the more acidic oxides remain in solution. The purified aluminum oxide is removed by filtering. The remaining waste solution of weak base can be regenerated by heating, which drives off water and carbon dioxide. Small amounts of sodium hydroxide may be needed to bring the regenerated base back to the strength needed to dissolve more aluminum oxide ore.

[16–5] Production of Aluminum

acting as either an acid or a base (see Section 8–8). Because silicon is more nonmetallic than aluminum, silicon oxide acts only in the acidic way characteristic of nonmetal oxides (see Section 13–4). Iron, which is farther than aluminum from the borderline between metals and nonmetals, has oxides that act only in the basic way characteristic of metal oxides. Iron cannot form the very high oxidation states that are acidic even for metals (see Sections 8–7 and 8–8).

The first step in the Bayer process is dissolving of acidic oxides in a hot, strongly basic solution of NaOH in water. The acidic oxides (including those of aluminum and silicon) form negative ions that dissolve in water. The actual reactions may involve loss of an H^+ from one of the water molecules associated with the oxide, as shown in Figure 16–16. Basic oxides (such as those of iron and other metals, which are far from the borderline and not present in very high oxidation states) remain undissolved. They cannot give off H^+ to form negative ions, and whatever solubility they have to form positive ions plus OH^- is sharply reduced by a Le Chatelier's principle shift. The large concentration of OH^-, which is a product of dissolving basic oxides in water, shifts the solubility equilibrium toward keeping the basic oxides as solids.

The separation of Al_2O_3 from basic oxides like Fe_2O_3 occurs because the basic oxides cannot react like acids in strong base.

After the undissolved solids are filtered off, the basic solution of Al_2O_3 and other acidic oxides is treated to remove the aluminum. That could be done by making the solution acidic. An acidic solution would dissolve basic oxides, including Al_2O_3, but not acidic oxides like silicon oxide. However, that is not economical because large amounts of acid and base would be used up in neutralization and could not be recovered for reuse. Instead, a sort of halfway measure is used in which the basic solution is only weakened. Because the amphoteric aluminum has only a weakly acidic oxide, it must have a strongly basic solution to dissolve. As the solution becomes less strongly basic, the Al_2O_3 precipitates out. The more strongly acidic oxides remain dissolved as negative ions under the same conditions. The usual conditions to weaken the NaOH solution are addition of some water to dilute it and sometimes addition of some CO_2 to neutralize part of the NaOH. The solution is also cooled, which lowers the solubility of Al_2O_3 and helps cause its precipitation. When the leftover solution is reheated, some H_2O and CO_2 can be driven off. Addition of only a small amount of NaOH can then bring

The separation of Al_2O_3 from acidic oxides like SiO_2 occurs because the less acidic Al_2O_3 precipitates out in weak, cold base where acidic oxides remain dissolved.

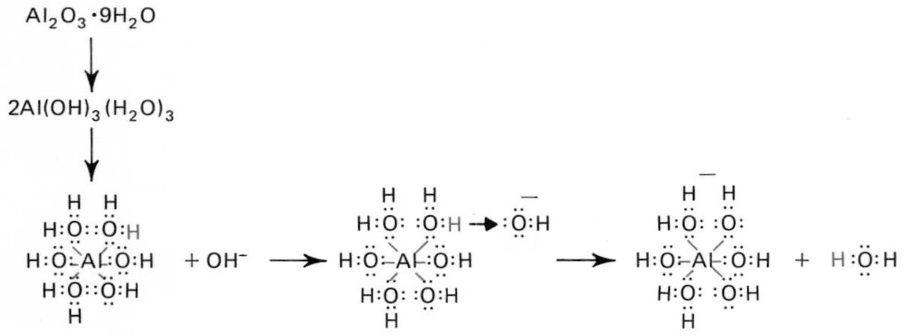

Figure 16–16 Aluminum oxide acting as an acid.

it back to the concentration needed to dissolve more ore in step 1 of the Bayer process.

In summary, the Bayer process works because Al_2O_3 is not acidic enough to stay dissolved in cold, dilute NaOH. Impurities either never dissolve or else remain dissolved.

The Bayer process also changes the crystal structure of Al_2O_3. The bauxite ore is much harder to dissolve than the purified Al_2O_3 formed in the Bayer process. However, when Al_2O_3 from the Bayer process is stored for a long time, it gradually becomes more and more difficult to dissolve. This shows that the Al_2O_3 from the Bayer process (a crystal which forms quickly during precipitation) is gradually changing back to the crystalline form in bauxite (a more complicated but more stable arrangement).

16–5.3 Oxidation, Reduction, and the Reduction Step for Aluminum.

Conversion of aluminum compounds to aluminum metal is difficult. No practical way has been found to get aluminum metal from aluminum silicates, and even purified Al_2O_3 yields aluminum only with difficulty. The problem is caused by the great chemical reactivity of aluminum.

Reactions in which an element forms a positive oxidation state in a compound (see Sections 7–7 and 8–2) are called oxidations. They are often what happens in reaction with oxygen (such as $4\ Al + 3\ O_2 \longrightarrow 2\ Al_2O_3$), but any reaction in which electrons are lost by the element we are talking about is an oxidation, as noted in Section 8–2. In order to form any metal from compounds in which the metal has a positive oxidation state (see Section 7–7), we need the opposite kind of reaction—a gain of electrons. In Section 8–2 we defined any reaction in which electrons are gained as a reduction. Our most common reactions of this type, formation of a metal from a compound, cause a reduction in sample weight. But any reaction in which electrons are gained is called a reduction. Production of any metal from compounds must have a **reduction step** somewhere in the process.

Production of metal from a compound is a reduction step.

Aluminum is one of the most easily oxidized elements, giving up electrons easily to form Al^{3+} in compounds. Those reactions which go most vigorously in one direction in nature are also those which go with greatest difficulty in the reverse directions. Therefore Al^{3+} in compounds is one of the hardest things in nature to reduce. In fact, there is no cheap, readily available substance that can react chemically with aluminum compounds to produce aluminum metal. In this case (and with other very reactive metals) we resort to a simple alternative. If we cannot find a chemical to give electrons for the reduction step, we *give electrons directly* to the reaction by *passing an electric current* through the substance. This process is called **electrolysis.**

No cheap, available chemical can produce aluminum from aluminum compounds, so electrolysis is used.

16–5.4 Hall Process.

Electrolysis will cause the reactions that go most easily among the substances through which the electric current is passed. But there is no reaction at all unless the electric current can flow. Those limitations cause problems in electrolysis of Al_2O_3. When Al_2O_3 is dissolved in strong base (or acid), ions are present that can move, carrying the electric current. But those solutions contain things more easily reduced than Al^{3+}, particularly the H^+ in H_2O. Therefore, no aluminum metal is formed from electrolysis of solutions in water. When pure Al_2O_3 is used, it is a solid substance (very high melting point) which has no ions free to move about. No electric current flows through the Al_2O_3, so no reaction is possible.

The problems of electrolysis of Al_2O_3 have been solved in the **Hall process** by finding an ionic salt of aluminum that can be melted so that electric current will flow through it. The salt used is Na_3AlF_6. Aluminum is the substance most easily reduced in Na_3AlF_6, so aluminum metal is the reduction product formed. Na_3AlF_6 exists in small amounts naturally and was given the name cryolyte. (Its systematic chemical name should be sodium hexafluoroaluminate, but it is usually just called cryolyte.) In addition to its ability to conduct electricity, molten cryolyte is able to dissolve Al_2O_3.

Aluminum is produced by the Hall process, electrolysis of Al_2O_3 in molten cryolyte, Na_3AlF_6.

In the solution of Al_2O_3 in cryolyte, aluminum and oxygen are the elements that react during electrolysis. Electricity enters and leaves the solution through electrodes. The electrode with a negative electrical charge is called the cathode. At the cathode aluminum metal is formed by addition of electrons to Al^{3+} from the solution. The other (positively charged) electrode is called the anode. At the anode combined oxygen (in forms equivalent to O^{2-}) loses electrons and could combine to form O_2 molecules. In practice, the best available material for the electrodes is carbon. Most of the oxygen reacts with the carbon anode to form CO_2 gas. That reaction gradually uses up the carbon, and eventually the worn out anode must be replaced with a fresh one. As aluminum and oxygen are removed by reaction, more Al_2O_3 can be added to the solution. Except for small losses by side reactions, the cryolyte is not used up and therefore does not need to be replaced. Figure 16–17 shows how aluminum is produced by electrolysis in the Hall process.

Aluminum is formed at the cathode (negative electrode) while O_2, which reacts with the carbon electrode to produce CO_2, is formed at the anode (positive electrode).

The Hall process uses a lot of electricity. Three electrons must flow through the cathode for each aluminum metal atom formed. Therefore, 3 moles of electrons must flow through the electrolysis for each 1 mole of aluminum metal. Since the atomic weight of aluminum is only 27, only 27 g of aluminum are produced for each 3 moles of electrons, or 9 g of aluminum per mole of electrons. In some electrolysis processes, only a very small voltage is needed (see Section 16–10) so many electrolyses can be run at once along the same line of flowing electric current. In the Hall process, the number of electrolyses on the same line is kept fairly low by the requirement of a fairly large voltage drop (at least 3 volts) across each electrolysis. In terms of the common units for electricity (which includes the voltage, the amount of current flowing, and how long it flows), 1 kilowatt hour of electricity could produce about 125 g of aluminum. That makes electricity important enough to the total cost of aluminum to cause location of all plants in areas where electricity is cheap. Aluminum reduction plants also often use "interruptable" power, which means they must shut down whenever other demands for electricity rise. Other important economic factors in the Hall process are the large investment in

Major expenses in aluminum production are for the large amount of electricity and for replacement of carbon anodes.

Figure 16-17 The Hall process.

Electric current is supplied through carbon electrodes. The current is conducted through a molten salt solution, Na_3AlF_6, with dissolved Al_2O_3. Aluminum metal is produced at the cathode while much of the oxygen produced at the anode reacts with the carbon anode to form CO_2 gas.

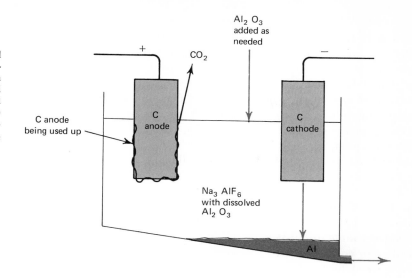

equipment, cost of the frequently replaced carbon anodes, and the labor needed to exchange anodes.

The Hall process requires fairly pure Al_2O_3 to avoid difficulties with the electrolysis. That is one reason for the chemical purification of the raw material by the Bayer process. However, this prepurification effort also helps produce a fairly pure metal product. Electrolysis also tends to be a clean type of reduction step that does not introduce other impurities. Aluminum can be used for most purposes essentially as it comes from the Hall process, with only forming into shapes needed. The following sections on iron and copper will show that later purification requirements are common with other metals.

Aluminum can often be used directly in the purity produced by the Hall process.

16-6 Production of Iron

16-6.1 Effects of Ready Availability. Iron is the fourth most abundant element in the Earth's crust. Unlike aluminum, almost any compound of iron could be used to produce iron metal. The abundant iron sources create economic pressures for very large scale and very cheap handling at every stage in the production of iron metal and its particularly useful form, steel. Since plenty of iron ores are available, those which can be processed in the cheapest way are able to supply all of the demand. Those which are quite suitable in every way except for slightly higher costs become totally worthless. Because of the large amounts of material being handled, transportation costs become an important factor.

Iron is present abundantly in chemically processable forms.

The best iron ores are iron oxides. Iron oxides exist naturally in large amounts, mixed with other minerals. When these impurity minerals are transported with the iron oxides, they add to the expenses. Therefore, ore with a higher iron oxide content is worth considerably more at the mine. The cheapest way to reach the richer ores is usually roughly stripping away

Competitive iron ores must be cheap in every way, including minimum transportation and minimum impurities.

the less rich materials, including any soil or trees over the ore. When the richest ores have been used up, an area may be left with large amounts of iron oxide in poorer ores but be unable to continue production because of prohibitively higher shipping costs. That problem had been developing in the Mesabi region of Minnesota, causing a severe depression for the area. Sometimes mechanical separation using magnetic deflection can enrich ores by enough to pay more than the cost of separation in reduced shipping charges. Ore processing plants of that type are bringing renewed economic activity to the Mesabi region. There is never a case where chemical separation to really purified iron oxide would pay economically.

16-6.2 Blast Furnace. Most iron is reduced by chemical reaction in a blast furnace. Figure 16–18 shows the operation of a blast furnace. Carbon is being oxidized while iron is being reduced in the blast furnace. The carbon is obtained from coal. The coal is converted to a form called coke (fairly pure solid carbon) by heating it to drive off everything that can evaporate. (The material driven off is the source of many useful chemicals.) The coal, like the iron ore, is handled in bulk as cheaply as possible, which encourages rough and sometimes destructive mining techniques.

In a blast furnace, the carbon (as coke) and iron ore cannot react directly. They are both solids and are therefore unable to mix and react on a molecular scale. That problem is solved by allowing a limited amount of air to enter and burn the coke. When there is a shortage of oxygen, carbon burns to

Ore is usually reduced to iron in a blast furnace.

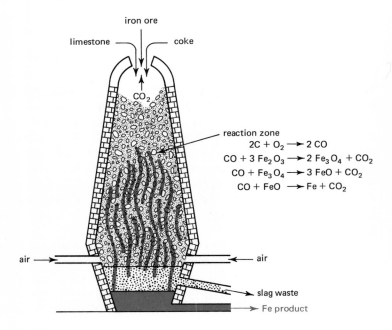

Figure 16–18 Blast furnace.

Raw material (iron ore, coke, and limestone) are added at the top. Air enters near the bottom and flows upward. Carbon dioxide plus other product gases and unreacted air components are exhausted at the top. In the reaction zone coke reacts with oxygen to form carbon monoxide, and the carbon monoxide reduces iron oxides to iron metal. The iron metal melts and flows to the bottom, where it is drawn off as the product. Slag, formed from the limestone and impurities, also melts and flows downward. The slag is less dense than iron, so it forms a layer floating on top of the iron and can be removed separately.

The active reducing agent in a blast furnace is CO formed by the reaction of O_2 with coke.

form CO (carbon monoxide) gas instead of CO_2 (carbon dioxide) gas. The CO gas can move to the various particles of solid iron oxide and react. In each reaction a CO picks up an additional oxygen to become CO_2. Eventually all of the oxygen is removed from the iron oxide and iron metal is left as the product. Equations for some of the reactions taking place are

$$2\,C + O_2 \longrightarrow 2\,CO$$
$$CO + 3\,Fe_2O_3 \longrightarrow CO_2 + 2\,Fe_3O_4$$
$$CO + Fe_3O_4 \longrightarrow CO_2 + 3\,FeO$$
$$CO + FeO \longrightarrow CO_2 + Fe$$

The reaction of carbon with O_2 from the air generates enough heat to melt the iron metal. As it forms it melts and runs down to the bottom where it is collected.

Acidic oxide impurities are removed from the blast furnace by reaction with CaO (formed from limestone, $CaCO_3$) to form slag.

Because the iron ore put into blast furnaces is far from pure, there are many impurities to dispose of. One of the most common and troublesome impurities is silicon. Much of the silicon and other elements that form acidic oxides can be removed during the iron reduction in the blast furnace. This is done by adding limestone (a cheap, readily available rock). The limestone ($CaCO_3$) breaks down to CaO (a basic oxide) and CO_2 when heated in the blast furnace. The CaO reacts with acidic oxides like SiO_2 to form compounds that settle to the bottom as a molten rock mixture called slag. The slag is less dense than molten iron, so it floats on top of the iron. Batches of molten iron and slag are drawn off as necessary to keep the levels about constant. Iron ore, coke, and limestone are continually added at the top. It is neither necessary nor desirable to stop the continuous operation of the blast furnace for anything but major repairs or maintainence.

16-6.3 Purification of Iron.

The molten iron drawn from a blast furnace is called **blast furnace iron.** If it was allowed to cool and solidify in blocks called "pigs," it would be called pig iron. That is rarely allowed to happen in modern steel mills because it would waste heat. Heat would have to be replaced later by burning fuel to melt the iron again.

Blast furnace iron contains many impurities, present in varying and generally unknown quantities. There is always a substantial amount of carbon (often 3 to 5%) plus elemental silicon, phosphorus, sulfur, and other elements which were reduced by reaction with CO before they could be completely tied up in the slag. These impurities have marked effects on the properties of the metal. Often the effects are bad. Even when the effects are good, it is necessary to control the amounts of added material. Therefore, the blast furnace iron is treated to remove (as nearly as possible) everything but iron. Additives are then put back in controlled amounts to convert the pure iron into steel or other desired products.

Blast furnace iron is purified by burning out impurities.

One of the methods for removing impurities is to burn them off. Carbon and some other impurities are oxidized more easily than iron, so they can be removed first. However, if oxygen is supplied after the impurities are gone, the iron can also burn, forming iron oxide. That would put us back to the stage before the blast furnace, so some kind of control over the supply of oxygen for burning is needed.

The earliest method for burning off impurities was the **Bessemer process.** Although it is almost never used now, it will introduce us to the type of purifications needed. In the Bessemer process molten blast furnace iron is placed in a large ladle with holes in the bottom. Air is pumped through the holes and allowed to bubble up through the iron, and oxygen in the air reacts. As long as there is carbon present, most of the oxygen is used to form CO. As the hot gases leave the ladle, they mix with air. The CO then burns with the oxygen in that air to form CO_2. The combustion of CO to CO_2 can be seen as a blue flame.

Figure 16-19 shows a Bessemer Process ladle during a "blow." After about 15 min in a typical batch, the carbon in the iron is used up. The formation of CO then stops and the blue flame goes out. That is taken as a signal to turn off the air pump and pour the finished iron out of the ladle. This method is fairly effective at removing carbon without oxidizing much iron, but it does not assure or measure the extent of removal of other impurities. The product is generally of low quality with unpredictable variations from one batch to the next.

In order to get more complete removal of impurities and hence better quality, the steel industry converted almost completely to the **open hearth process,** shown in Figure 16-20. Instead of bubbling air through the iron, it was passed over the surface of a shallow pool of molten iron. As impurities

The Bessemer process is blowing air through the iron until the carbon is gone, but it does not give good control of other impurities.

The open hearth process gives better control by a slower surface reaction with air and by reaction of oxide impurities with the acidic or basic lining.

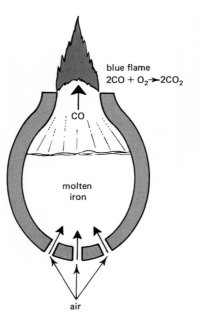

Figure 16-19 The Bessemer process.

Air is blown through molten iron. As long as carbon is present, some CO forms. That CO burns as it enters the air above the vessel. (The actual flame is blue.) When that flame goes out, the molten iron is poured out of the vessel and considered finished product.

Figure 16-20 The open hearth process.

Air and fuel are preheated by passing over hot bricks. The fuel then burns and the hot gases pass over a shallow pool of molten iron. Impurities are oxidized by contact with the hot, oxygen-containing gases and then removed by reaction of the oxide products with the lining of the hearth. The lining may be either basic or acidic, depending upon which impurities are present in the iron being processed. The exhaust gases heat a brickwork, and flow is reversed regularly to conserve heat. When analysis shows the iron is sufficiently pure, a plug is removed and the molten iron is allowed to flow out as the product.

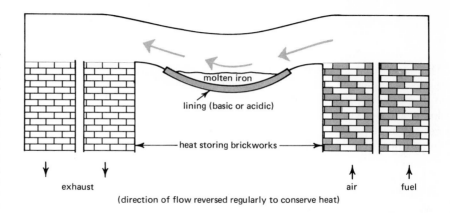

were converted to their oxides, they could be removed by reaction with the brick lining of the pool. If the iron came from ore where the main impurities formed basic oxides, the lining could be made from acidic oxides to react and remove the impurities. An open hearth with acidic lining is called an acidic open hearth process. However, in most cases the main impurities (silicon, phosphorus, and sulfur) form acidic oxides. These are removed by selecting a basic lining. That is called a **basic open hearth process.**

The open hearth process is quite slow—typically 8 hr per batch. Fuel is burned in the air over the iron pool to keep it molten. In order to cut down fuel consumption, large heat adsorbing brickworks are used at entrance and exhaust and the direction of flow is reversed at regular intervals. The slow progress also allows time for samples of the iron to be analyzed. The process can then be stopped when analysis shows all impurities have been taken care of, not just the carbon.

The basic oxygen process is an open hearth with basic lining plus pure O_2 injected onto the surface to speed reaction.

The **basic oxygen process** is a modern variation of the open hearth process. It is now the most widely used process. The basic oxygen process uses a shallow pool and a basic lining, just like the common open hearth process. It differs in that the process is speeded up by blowing a stream of pure oxygen across the hot surface. The oxygen is inserted through a "lance," as shown in Figure 16-21. The pure oxygen speeds the reaction considerably

Figure 16-21 The basic oxygen process.

The open hearth type process is speeded by injecting pure oxygen gas onto the molten iron surface.

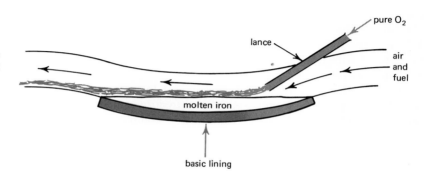

[16-6] Production of Iron 355

from the pace when the partially oxygen depleted air of the burned air-fuel mixture is the only source of oxygen. Rapid modern analytical methods allow the same sort of sample analysis control as the open hearth process.

The speed of modern tools may make other fast, high quality methods possible. One method that has been under study involves a stream of molten iron dropping through air (sort of a reverse Bessemer process). A computer makes evaluations of analyses and directs adjustments in the size of the stream to get the best results.

In addition to these methods for upgrading blast furnace iron, high quality iron can be produced by avoiding the blast furnace completely. A significant and growing fraction of steel is produced by electrolytic reduction. The process is expensive, but the low impurity levels make it attractive for some special quality products. It also has some advantages for small plants (serving local markets) in areas where electricity is cheap.

16–7 Steel

Iron by itself is not a very useful metal. It is only when certain impurities are added in carefully controlled amounts that iron takes on the properties of strength and toughness we need in most uses. These combinations are extremely varied and complex. A full course in metallurgy would be needed to understand even the basic categories, and a lifetime of study would be needed to approach a full understanding of those combinations we call steel. We will limit ourselves here to discussing a few of the basic points involved in steel. That should give us a reasonable layman's understanding of what steel is and some understanding of the variety of its forms and behavior.

Pure iron is soft and not very useful.

Steel is basically a combination of iron and carbon. The combination is much stronger than pure iron. The strength and other properties of steel vary greatly as the proportions of iron and carbon change and as the arrangement (which can be affected by heating and cooling) changes. These changes can best be understood by starting with an extreme case. Iron and carbon can and do form a compound with the formula Fe_3C, called cementite. Cementite is an extremely hard substance. Its behavior would be more like a diamond than like an iron bar. It if was struck, it could not bend like a nail. If it was hit hard enough to change its shape, it would shatter. Because of its hardness, cementite would be very strong but so brittle it would be useless.

At the other extreme, pure iron is so soft that it lacks the strength to be useful. However, the softness lets it bend into new shapes without breaking. Therefore, it can withstand blows better than the brittle cementite.

Steel is an intermediate form that has some of the hardness of cementite to hold its shape and some of the softness of iron to resist blows without brittleness. This compromise is produced in steel by forming a crystal pattern that is more interconnected than pure iron but less completely bonded than cementite. The ability of carbon to be dispersed in iron and to form strong interactions (as shown by the ability to form the compound cementite) makes several intermediate crystal types possible. One of these crystal types gives

Figure 16-22 Effects of cooling pattern on iron-carbon mixtures.

(A) A slow cooled mixture that separated into soft, almost pure iron and chunks of carbon. (B) A tempered mixture (steel). The iron and carbon solution has solidified to a steel, which is strong but not too brittle for use. (C) A quick cooled mixture. The product has frozen while it still contains much cementite in an iron and cementite solution. The cementite makes the product rigid and brittle.

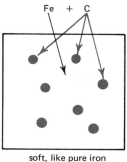 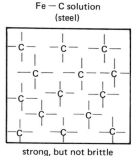 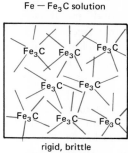

soft, like pure iron　　strong, but not brittle　　rigid, brittle
A. Slow cooled　　B. Tempered　　C. Quick cooled

us the desirable properties of steel. Steelmaking is the art of creating the proportions and conditions which form the desirable crystal type.

The proper proportions of carbon and iron are not sufficient to form steel. They must also have the proper arrangement. If a solution of carbon in iron is cooled slowly, the solubility of carbon drops and chunks of solid carbon separate out. The result, shown in Fig. 16-22A is fairly pure iron with a few carbon chunks that have no strong interaction with the iron. The substance acts like pure iron and is too soft for most uses. When the iron-carbon mixture is heated to a molten state, the carbon dissolves and reacts to form Fe_3C. The reaction uses up heat as shown by this equation.

Carbon is lost from steel by becoming insoluble in the iron and forming separate chunks when molten steel is cooled slowly.

$$\text{heat} + 3\,Fe + C \rightleftharpoons Fe_3C$$

An equilibrium is set up between reactants and products, and the concentration of heat (temperature) affects the equilibrium as expected from Le Chatelier's principle (see Section 9-9). At high temperatures the mixture is mostly Fe_3C; at lower temperatures it is mostly a solution of carbon in iron. If the very hot solution (most of the carbon in Fe_3C) is cooled quickly, the Fe_3C is frozen in before it has time to break down to iron and carbon and separate out as chunks of carbon in iron. The very strong, rigid Fe_3C interacts with the surrounding iron, making the whole crystal hard and brittle, as shown in Figure 16-22C.

At high temperatures, iron and carbon form cementite, Fe_3C, which makes steel brittle if it is frozen in by quick cooling.

Steel is formed by reaching the intermediate condition shown in Figure 16-22B. Dissolved carbon (plus some Fe_3C) interacts with the iron to form a fairly hard structure, but the forces are not large enough to cause a complete (brittle) resistance to changes in shape. The result is a metal both strong (hard) and tough (able to resist blows). Hardness also increases as the concentration of carbon in the steel is increased.

Steel is a solution of carbon in iron, strong but not brittle.

16-8　Tempering Steel

The arrangement in steel is reached by the process called **tempering.** Molten steel is allowed to solidify and cool slowly until most of the Fe_3C has broken down to iron and carbon. If it was allowed to continue to cool slowly, the carbon would then begin to separate out and gather into chunks. (Even a "solid" metal has enough internal motion when quite hot to allow the carbon some freedom to move around and group together.) That is prevented by "quenching" the steel by rapid cooling, usually by placing it in or spraying it with water. Thus tempering consists of slow cooling at high temperatures (to avoid excess Fe_3C and brittleness) followed by quick cooling through the lower temperatures (to avoid carbon separation and softness).

When we understand the need for tempering steel, we can understand how fire can ruin steel which seems unaffected. If steel beams in a building become very hot in a fire and then cool slowly, they will allow the carbon to separate out. The resulting soft iron bars will lack strength. They may sag. The structure could collapse under its own weight when the rearrangement is completed—perhaps several days after the fire. On the other hand, if steel beams got hot enough for extensive Fe_3C formation and were then cooled quickly by water from a fire hose, they might become very brittle. They might then shatter when they cannot adjust to applied forces.

The differences in tempering characteristics can also be used to gain better steel performance. The most common example is **case hardening** of things like roller bearings which will be subjected to intense wear. A roller made from normal tempered steel or even a fairly soft iron can be given an extremely hard coating as shown in Figure 16–23. The roller is heated and then given a light coating of powdered carbon. The carbon dissolves in the surface. The whole process must be done quickly. The carbon is not allowed time to mix throughout the roller, but enough time is allowed to form a very high concentration of Fe_3C and dissolved carbon in the outer layer. The roller is then cooled quickly. The surface is thus frozen in the very hard, brittle form with much cementite. However, since most of the roller is still a softer steel, the roller as a whole can absorb blows without letting the surface crack. We have effectively converted a single part into two very different types of steel and gotten the advantages of each type.

Tempering is slow cooling by enough to remove cementite followed by fast cooling (quenching) to prevent carbon separation.

Hard surfaces can be made by case hardening done by adding carbon to the surface while it is hot and then quickly cooling the part.

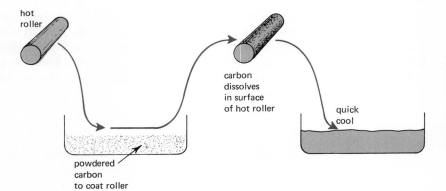

Figure 16–23 Case hardening.

A hard surfaced roller bearing is formed by passing a hot steel roller through powdered carbon. Some carbon dissolves in the surface of the hot roller. When the roller is then cooled quickly, the surface of the roller is tempered to a hard, high carbon steel whereas the center of the roller remains a softer, low carbon steel.

16–9 Steel Alloys

Carbon is only one of many elements that can be added to iron to form desirable solutions. By careful selection of the type and amount of additives, thousands of different types of steel can be produced with varying properties. The specific combinations of two or more elements are called alloys. Commercial steel is identified by number codes to indicate which alloy is being used.

Some of the most common materials added to iron are other transition metals. Manganese is used very extensively and tends to add toughness. Vanadium is also widely used. Chromium and nickel are particularly important because they help protect the steel from damaging chemical corrosion. Steels with more than the minimum amounts of chromium and nickel needed for good corrosion resistance (about 18% chromium and 8% nickel is needed) are called **stainless steels**. Stainless steels, like regular steels, are produced in a large number of different alloys tailored for particular uses.

There are many steel alloys, including the stainless steels that have enough chromium and nickel to resist corrosion.

16–10 Production of Copper

Aluminum and iron are the two most abundant metals in the earth's crust and are very important to us commercially. However, many other metals are also useful and important, even though less abundant. Several of these are also considerably less reactive than aluminum or iron, thus leading to quite different processing. We are going to discuss one of the most important of these low abundance, low reactivity metals, copper.

Copper was the metal first used extensively by man. Because of its low reactivity, some copper existed as free metal which could be found and used. When the need for metal exceeded that limited supply, methods were found to produce copper simply from ores. Like steel, copper is useful in thousands of different alloy combinations with other metals. The two most important groups of copper alloys are copper–zinc alloys, which are called brass, and copper–tin alloys, which are called bronze.

The largest present uses of copper are in electrical wiring and soft flexible tubing. Copper is valuable for electrical uses because it is a particularly good conductor of electricity. Only silver, which is too rare and expensive to compete, can conduct electricity better than copper. Copper used in wiring needs to be quite pure, so our copper production methods must include a good purification step.

The limited amount of copper on earth is mostly in the form of very insoluble oxides (CuO or Cu_2O) or sulfides (CuS or Cu_2S). When it is obtained as a sulfide, the sulfur is removed by **roasting** the ore. Roasting is simply heating the ore in air. Many metals that are mined as sulfides can be converted to oxides by roasting. An example is conversion of ZnS to ZnO. The oxide product can then be reduced to metal by methods similar to those used for aluminum and iron. However, some of the less reactive metals go directly to metal during roasting. Although some copper oxide forms, most of the copper sulfide is converted directly to copper by reactions such as

$$CuS + O_2 \longrightarrow Cu + SO_2$$

Copper is also used in many alloys, particularly brass (copper–zinc) and bronze (copper–tin).

Whatever CuO or Cu_2O is formed can be easily reduced to copper metal. The copper oxide in molten copper metal obtained from roasting can be converted to copper by simply stirring the molten mass with green wood sticks. The sticks are partially oxidized as the copper oxide is reduced. Copper oxide ores can also be converted to copper metal by very mild reducing agents.

The copper produced directly by the simple reduction reactions is called blister copper. It is not pure enough for many uses. It can be purified by **electrorefining.**

Figure 16–24 shows the purification of copper by electrorefining. At the impure copper, which is the positive electrode (anode), the main reaction is

$$Cu \longrightarrow Cu^{2+} + 2\,e^-$$

At the same time, any metallic impurities that are *more reactive* than copper also react to form ions in solution. The *less reactive* impurities are left behind. Eventually, as shown in the closeup at the left side of Figure 16–24, the copper (and reactive impurities) are reacted away from around the less reactive bits. Finally, the unreacted piece drops off and falls to join the **anode sludge** shown in Figure 16–24. This sludge is later processed to recover the silver, gold, and other valuable metals contained in it.

At the cathode, pure copper is produced by the reaction

$$Cu^{2+} + 2\,e^- \longrightarrow Cu$$

The more reactive metals in the solution stay there. There are plenty of Cu^{2+} ions to accept all the available electrons, so none are left for reactions with the less easily reduced ions from more reactive metals.

Since the reactions are identical at the two electrodes, almost zero voltage is needed. A very large number of electrolyses can be run in series on one

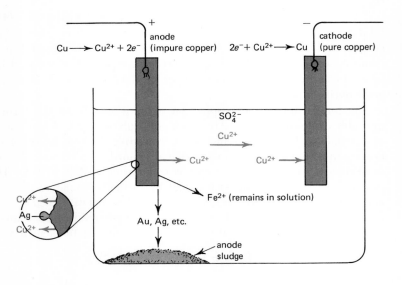

Copper sulfides can be converted to copper by roasting (heating in air), and copper oxides can be reduced by weak reducing agents such as green sticks used to stir copper–copper oxide mixtures.

Copper can be purified for modern uses like wiring by electrorefining.

In electrorefining, less reactive metals are left unreacted and fall off into an anode sludge, whereas more reactive metals go into solution and cannot come back out.

Figure 16–24
Electrorefining of copper.

Copper from the impure anode reacts to form Cu^{2+} ions, which later react at the pure copper cathode to form copper metal. Reactive impurities such as iron react to go into solution but do not plate out on the cathode. Less reactive impurities are left behind, as shown by the closeup of the region around some silver. When the copper around the silver has all been removed by reaction, the silver will fall to the bottom to become part of the anode sludge.

line, so the cost of electricity is much lower than in the production of aluminum by electrolysis.

16–11 Side Effects of Metal Production

16–11.1 Sulfur Problems.

Production of large quantities of metals leads to a number of side effects. Some of the most serious of these involve the release of sulfur.

Lead, zinc, and copper are mined mostly as sulfides. Some other metals are also found as sulfides because the very insoluble sulfides tend to be concentrated into recoverable ores. When these ores are roasted in smelters, large quantities of SO_2 are produced. SO_2 is both a gas (so it can be moved around in the air) and an acidic oxide. When it comes in contact with water, it produces sulfurous acid, H_2SO_3. It can then be oxidized to form the even stronger acid, sulfuric acid, H_2SO_4. Plant and animal tissues are among the places where the SO_2 can find water and form acid. The acid formed is very irritating to humans and can damage plants. It is one of the most dangerous air pollution problems. In addition to metal smelting, SO_2 is often produced when coal or other fuels are burned. Coal is usually the worst in causing SO_2 problems because some elemental sulfur is present in most coal deposits.

SO_2 released in roasting sulfide ores damages plants and animals by forming acid in their tissues.

The SO_2 from metal smelters is often particularly bad because of poor wind circulation patterns. Mining is often done in hilly or mountainous areas because ore can be located and reached from the surface there. A nearby metal smelter is usually located in a valley. The walls of the valley tend to hold in the fumes, letting them build up to become a significant hazard. The destruction of plant life in these areas shows the damage being done, and people in the area may also suffer damage to their health. An example of such a valley can be seen when one drives through Kellogg and Smelterville, Idaho, on Interstate 90. The desolation of the area is especially striking when compared to the lush timber on all the surrounding mountains and valleys. Some of the treeless areas may be due to large forest fires in the past, but similar areas far from the smelter develop grass and shrubs (often huckleberry bushes). Some of the hillsides near the smelter have so little plant life that they appear to be bare brown dirt even in the middle of the summer growing season.

The geography of smelter locations and low quality of product possible from smelter SO_2 make pollution control difficult.

The solution to SO_2 emissions is to trap the gas, make acid out of it (by oxidizing it so H_2SO_4 is obtained), and use the acid for some chemical process. Such a plant has been built at Kellogg, Idaho, but it has run into problems. Removal of H_2S from natural gas (to prevent SO_2 formation when the gas is burned) is being done on a large scale and very efficiently in nearby western Canada. The cheap sulfur produced has taken all of the acid markets in the area. The smelting plant produces a poorer grade of acid. The impure metal ores make a poor starting material, and the whole process suffers from the many impurities present. As a result, acid from the smelter cannot find buyers at any price. They cannot produce acid and then dump it in the river. (It would kill everything and pollute downstream lakes and rivers.) Therefore, the acid plant sits idle while SO_2 goes off into the air. Presumably a solution will be found that removes the SO_2, but the situation does show that pollution control problems are often harder to handle in practice than on paper.

Another sulfur problem comes as a side effect of coal mining (much of which is done for use in iron and steel production). Sulfur sitting in deposits with coal is nonreactive with ground water flowing through the coal bed. However, when the coal is mined, the sulfur is exposed to air and can be oxidized. Water flowing through the mine can then react with the SO_2 and further oxidation can convert it to an H_2SO_4 solution. The water which flows out of old abandoned coal mines is therefore very acidic. This **acid mine drainage** is often a serious water pollution problem.

Exposure of coal mines to air leads to acidic pollution of streams.

16–11.2 Smoke and Heat. Metal production, particularly iron and steel, also causes problems with smoke and heat. Blast furnaces and the coking ovens where coal is converted to coke for the blast furnaces are particularly likely sources of smoke, and the total amount of gases given off during iron and steel processing is so huge it can affect the atmosphere somewhat even when spread over a large region.

Smoke and heat from steel mills damage their environments.

Heat disposal can also be a problem. Steel plants heat the air around them. They also heat up rivers by using large volumes of cooling water. A change of a few degrees in the temperature of a river may be enough to upset its biological balances.

16–12 Poisons Associated with Metal Use

16–12.1 Lead. Almost any substance could be poisonous if present in excessive quantities. Some metallic elements (particularly a few of the heaviest ones) can be poisonous even in fairly small amounts. Our very extensive use of various metals has made some of these elements available (in biologically usable forms) in greater quantities than their normal rare occurrences in nature. Lead and mercury are the two which most frequently cause problems, but there are others that could become problems if used carelessly.

Lead can cause poisoning when its concentrations are raised above normal levels, and acidic solutions allow its transfer into people via the digestive tract.

In discussing metal poisoning problems, we should note that *none* of these elements is a poison *at the concentrations occurring in nature*. It is only at *excessive* concentrations (higher than living things have had a chance to adjust to) that they kill plants or animals. An example of this concentration dependence is seen in the case of copper. Small amounts of copper are absolutely necessary for many living organisms. Since copper is relatively rare, copper deficiency can be a problem in some areas, causing sick or dying cattle or other difficulties. However, copper in concentrated solutions (a very unnatural situation) is an effective poison for almost everything. It is used to clean out bacteria or algae from tanks by killing every living thing there.

Lead is a moderately available element, easily produced from its ores, and with a wide variety of uses. Unfortunately, compounds of lead tend to react with acidic solutions (such as gastric juices in human stomachs) and dissolve.

Therefore, when chips of paint containing lead are eaten by small children, the lead is digested and can wreak havoc with many complex body functions. Although this could kill a child, it more frequently leads to symptoms of general sickness and lethargy. Permanent damage can be done to the person's physical health, and substantial mental retardation can result.

Lead compounds have also been used as a coating on pottery to give a shiny gloss. If lead glazed pottery which was not fired at a high enough temperature is exposed to acidic solutions, such as orange juice, the lead can dissolve and cause serious lead poisoning. Pottery made in the United States and most other countries avoids lead glazing to prevent accidental poisonings. However, some old pottery or imported pieces may be dangerous, particularly if they come from "cottage industries" where small amounts are produced at home by well meaning individuals unaware of possible dangers.

The largest input of lead into our environment is in vapor form from automobiles. Lead can form bonds to hydrocarbon groups. When four such bonds are formed to a single lead atom, we get compounds called lead tetraalkyls. These compounds improve the performance of gasolines. Millions of pounds are used every year. This lead is not put directly into the digestive system like the above examples, so it may not cause lead poisoning. However, it is in a form chemically bonded to carbon. It is therefore very much like the substances incorporated into living tissues, and it does get spread around in the air. Trees have been found to be concentrating lead in their wood in the past few years when leaded gasolines have become common. Even though it is not absorbed in concentrations large enough to cause lead poisoning, many people are deeply concerned about possible long term effects of this known poison on plants, animals, and people.

The ability of living organisms to incorporate lead and other poisonous heavy metals can serve as a way to remove them from our environment. The bacteria in sewage treatment plants (see Section 14–7) are very effective at tying up heavy metals. The lead content in sewage plant sludge is usually quite high. That puts the lead in solid form where it may not be put back into later biologically usable forms. This removal of lead (and other metals) prevents the lead from going on to become part of the water supply of downstream communities. However, this protection against poisoning is being threatened by a very surprising group of people—those trying to protect the environment.

Many of the efforts to prevent over-enrichment of waters (see Section 14–6) have centered on removal of phosphates. Although phosphates could be removed chemically as precipitates, treatment of all sewage would be fairly expensive. Most of the attention has therefore centered on cutting down use of phosphates. The biggest use of phosphates has been in laundry detergents, and great pressure has been applied to find substitutes. One function of phosphates is to tie up "hardness" ions (mostly Ca^{2+} and Mg^{2+}) which would otherwise interfere with the detergent. The same thing could be done with a complexing agent.

Nitrioacetic acid, which is called NTA, is a complexing agent that has actually been put into use as a substitute for phosphates. Unfortunately, good complexing agents like NTA tie up the small amount of heavy metals as well or better than they tie up the hardness ions. If NTA was used extensively, it could carry much or all of the lead and other heavy metals through the sewage plants. The bacteria would be unable to tie up the complexed metal ions. Even if over 99% of the NTA was used up by bacteria in the sewage plant (which would be a remarkably effective operation), there would be plenty left to carry all of the lead and other heavy metals through. We would then be left with lead dissolved *in water* instead of in a solid sludge.

No one knows whether or not natural processes would eventually remove the lead before the water reached the inlet of a downstream city's water supply. That sort of uncertainty should not be answered by an experiment on 200 million people. We need money to buy some research and, even more, public patience to buy the time to do the research before proceeding. The government has recognized this by a ban (at least temporarily) on NTA in detergents. However, NTA has been allowed in Canada and Sweden with no bad effects known as of this writing. Research on its effects there would help determine whether it would be safe in regions of greater population density, such as the eastern regions of the United States.

> Lead is removed from sewage by sewage disposal plants, but complexing agents like NTA in the sewage could interfere, causing possible poison hazards.

16–12.2 Mercury. The uncertain fate of metals in rivers, lakes, and ocean has been brought to public attention by mercury, another heavy metal known to be poisonous. Mercury is uniquely useful because it is the only metal that is a liquid at room temperature. It is an excellent solvent for other metals, as expected from the "like dissolves in like" principle (see Section 10–1). It has been used extensively in electrolysis processes (in which some mercury reacts and is lost), is used to make electrical contacts in devices like "silent" switches, and is used extensively in scientific laboratories to measure pressures or pump gases around in closed containers.

The hazards from liquid mercury are fairly well known from experience in scientific laboratories. Although there is some low maximum level of mercury that can be excreted with no damage, everything above that level accumulates over an entire lifetime. Such substances are called **cumulative poisons.** Therefore, ten small exposures to mercury would add up to a final poisoning equal to one exposure to all of the mercury at once. Symptoms of mercury poisoning include shakiness, loss of teeth and hair, and sharply increased susceptibility to heart attacks. However, all of this can be avoided by care in handling it. Mercury has a very low vapor pressure at room temperature, and if left standing it becomes covered by dust, water, or other substances that prevent the mercury from evaporating. The two conditions that make mercury dangerous are

> Mercury is a cumulative poison.

Heat. As soon as you heat mercury you raise the vapor pressure to dangerous levels. A glassblower who puts his flame on a spot with a small drop of mercury can evaporate more in $\frac{1}{10}$ sec than would evaporate from

> Metallic mercury is dangerous when heated or when fresh clean surfaces are created by a spill.

a pool of cool mercury in years. Never heat anything you know has had mercury on it.

Fresh clean surfaces. When mercury is spilled, it breaks up into thousands of small drops providing a very large area of clean surface. If left that way, it can fill a whole room with enough mercury vapor to give people a steady buildup of cumulative poisoning. If you ever spill mercury, get it *dirty* as soon as possible. In a sink, cover it with water. On the floor, cover it with dust. The very best dust to use is sulfur, which gradually reacts with it to form a safe compound.

The knowledge of lab safety considerations left many scientists confident that mercury would not be much of a problem in the environment. We could be concerned about silent switches if they get warm during use, but if the mercury is well enclosed and the vapor reaches cool surfaces before escaping, the hazard should be small. The mercury lost in processes such as electrolysis is reacted to compounds. Most mercury compounds are insoluble, so they should end up in the mud at the bottom of rivers, lakes, or ocean and never reappear.

Unfortunately, it does not work that way. Bacteria can convert mercury into mercury alkyls with mercury–carbon bonds. Those compounds can be incorporated into worms or other organisms in the mud. Apparently that also occurs naturally with the fairly small amounts of mercury in soil and mud. As these organisms are eaten by higher organisms, we get a **food chain concentration.** One animal eats 10 times its weight of lower organisms, using most of the material to generate energy. However, it cannot use any of the mercury alkyls to generate energy, so it ends up with 10 times as high a concentration of mercury alkyls.

Figure 16–25 shows the effects of concentration in the food chain. After four or five steps up the food chain, the mercury level can begin to approach the maximum that can be handled without beginning the cumulative poison

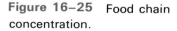

Mercury in mud is incorporated in organisms and then concentrated by the food chain.

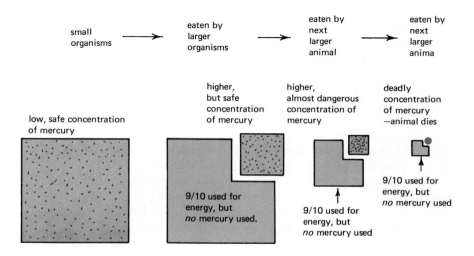

Figure 16–25 Food chain concentration.

Unusable materials, such as mercury alkyls or DDT, become more concentrated as they advance up a food chain. This example shows four levels of organisms and assumes each uses nine tenths of the incoming food supply for energy and one tenth for growth.

[16–12] Poisons Associated with Metal Use   365

effects. Man is at the end of the food chains. In areas where the availability of mercury in mud has been sharply increased, some foods (particularly those reached by a large number of steps in the food chain) may become dangerous to man. The lower species will continue to thrive in good health, with low mercury levels. Only the highest species are threatened.

Food chain concentration can occur with any substance that is biologically incorporated but not usable anywhere. DDT and other chlorinated hydrocarbon pesticides are notable examples.

16-12.3 Cyanide. Materials used in processing metals may sometimes create poisoning problems as bad or worse than those from metals. A notable example is cyanide. Sometimes one metal is deposited on the surface of another metal to obtain better properties. When this is done by an electrolysis, it is called electroplating. A common example is electroplating chromium (corrosion resistant but expensive) on steel (cheap). The biggest problem is getting the different metals to bond solidly together. Solutions containing CN^- ion, a very deadly poison, often help good bonding form in electroplating. CN^- left in worn out electroplating solutions must be chemically destroyed before the solution is discarded into a sewage system or river. This author recalls a time while he was in high school when an employee at a local electroplating plant released a solution before reacting the cyanide. The next day dead fish were piled 8 ft high on a dam 5 miles downstream. About 5 miles further downstream the fish kill stopped as natural chemical reactions eliminated the cyanide. Fortunately, no city or town was using that portion of the river for a water supply.

Some chemicals used in metal processing, like cyanides, are dangerous poisons.

Our extensive use of metals causes substantial problems in handling the materials when we are done with them. These problems begin with the physical removal of wastes from the site where they were used. From the beginning we suffer from disorganization. We have many different kinds of waste, and we often do not know exactly what each waste item is made of. Usually we put all wastes together to simplify hauling the heavy and bulky materials away. We end up with a conglomeration we know almost nothing about, a high entropy condition (see Chapter 11).

In choosing metals, we often select ones like aluminum or stainless steels that resist corrosion. They last longer in use, but they also last longer in junk. If they simply remained as junk, we could eventually convert all our usable space into piles of assorted, unidentified wastes.

Eventually, we must recover and reuse the material from our wastes. They could be looked upon as a rich deposit of ore, ready to be mined. They are richer in metals than many ores used commercially. However, they are not as uniform. In order to use them we must restore some of the order we lost in our first handling of them as wastes. That is an entropy decreasing process.

16-13 Metal Disposal and Recycling

Disposal problems encourage disorganization of material which complicates recycling efforts.

As noted in Chapter 11, such processes are nonspontaneous. They require work.

One method of restoring order is hand sorting. It has been with us for a long time in the form of the junk man, scavenging through discards searching for items of value. The method is very inefficient, costing much labor and not reducing the waste accumulation significantly. The more recent version of this method, the volunteer clean-up group, does more good because a larger fraction of the waste is sorted out for recovery. But the low value of the extra recovered material makes it hard for such groups to be economically viable.

Mechanical sorting can be better, particularly in the one case where a sharp difference in behavior can be used. Iron parts can be separated by magnetic means similar to the magnetic mechanical separation of ores described in Section 16–4. The result is a mixture of the many different kinds of steel. This can be melted and reused. Large iron scrap, such as automobiles, can be melted with it for reuse. However, we still have to restore one more kind of order to use it effectively. We have to analyze each batch and find what we have.

The amount of iron recovered in scrap is already fairly large and can be increased. Steps are also being taken to recover other materials, such as aluminum in cans, by getting people to return them before they become mixed into the waste conglomeration. These are steps in the right direction, but we still have a long way to go to raise our technology of handling wastes to the same level as our technology of preparing metals for use.

Skills Expected at This Point

1. You should be able to list the ten most abundant elements on earth and to list low density, high reactivity, and reasonably high cosmic abundance as factors leading to abundance on the earth's surface.
2. You should be able to point out the variety of transition metal oxidation states and the high relative abundance of the first transition series elements.
3. You should be able to explain the importance of complexes by pointing out examples in living things, define the term ligand and describe the bonding of ligands to metals by coordinate covalent bonds, and list four different common types of complexes, giving the number of ligands and shape in each case.
4. You should be able to describe minerals as the concentrations of particular substances caused by their different properties and the natural processes on earth.
5. You should be able to list the tasks and purposes of mining operations and the common methods for ore enrichment.
6. You should recognize the abundance of iron and be able to explain why it has greater *effective* abundance than aluminum or copper.
7. You should be able to describe the production of iron in a blast furnace

and describe the basic oxygen process and identify it as the main commercial steel process.
8. You should be able to identify steel as being different from iron and list some ways in which it is different.
9. You should be able to list the steps in the process of tempering steel and describe what tempering achieves.
10. You should be able to define the term alloy and recognize that many are in common use.
11. You should be able to describe copper as a metal that is easily obtained from its ores but fairly rare in nature.
12. You should be able to explain sulfur air pollution in terms of SO_2 and acidity.
13. You should be able to identify lead and mercury as poisons and be able to describe the concentration of poisons by food chains.

Exercises

1. One of the requirements for an element to be abundant on the Earth's surface is that it be fairly abundant in the universe as a whole. List two other factors that contribute to an element's abundance on the Earth's surface.
2. Describe the trends in oxidation states found among transition metals. (Note that you are responsible for the material in Chapter 8 that you were referred to in this chapter.)
3. (a) Define a disproportionation reaction. (b) Under what conditions do disproportionation reactions occur?
4. Why is the subject of complex ions important?
5. What is a ligand?
6. List the common shapes of complexes and the hybrid orbital name for each of those shapes.
7. Why does Cr^{3+} form more stable complexes than each of the following? (a) Cr^{2+} (b) V^{3+} (c) Mn^{3+}
8. (a) List three ways in which natural processes separate elements. (b) List four ways in which ores can be artificially enriched. (c) What pair of processes (one from your first list and one from your second) are most alike? What pair are most different and in what ways are they so different?
9. What are the best ores for (a) aluminum, (b) iron, (c) copper.
10. Why does silicon fail to reprecipitate when aluminum does in the Bayer process?
11. List all the factors that seem likely to be *important* factors in the cost of producing aluminum.
12. List all the factors that seem likely to be *important* factors in the cost of producing steel.
13. List all the factors that seem likely to be *important* factors in the cost of producing copper.

14. Some cost factors are more important with one metal than with another.
    (a) Pick one factor in aluminum production which seems particularly important there but much less important in iron or copper production.
    (b) Pick one factor in steel production that seems particularly important there but much less important in aluminum and copper production.
    (c) Pick one factor in copper production that seems particularly important there but much less important in aluminum and steel production.
15. Describe carefully (including a sketch if it helps) a common method for reduction of (a) aluminum, (b) iron, (c) copper.
16. Why is carbon removed from blast furnace iron when it must eventually be put back to make steel?
17. Describe the steps in tempering steel and what occurs in each step.
18. What differences would you expect in the properties of "high carbon steel" and "low carbon steel"?
19. What is an alloy?
20. Metal production is often accompanied by sulfur releases. List explanations for (a) why this is considered a problem; (b) why it continues to happen; (c) what control measures are needed; (d) problems encountered in control and how they are likely to add to costs.
21. What chemical condition tends to make lead compounds digestible poisons?
22. Why are predator fish (which eat smaller fish) more subject to mercury poisoning than seaweed or fish that eat only seaweed?
23. Would deposits on containers serve a useful technological function even if the containers were to be used only as raw materials instead of being reused as containers? Why?

Test Yourself

1. List the factors necessary for an element to have high abundance on Earth.
2. The hybrid orbital combination used in forming square planar complexes is
   (a) sp (c) sp^3 (e) d^2sp^3
   (b) sp^2 (d) dsp^2
3. Why is good aluminum ore harder to find than good iron ore?
   (a) Aluminum is less abundant on earth than iron.
   (b) Iron compounds are more effectively concentrated in the Earth's crust by density factors.
   (c) Iron has greater chemical differences from other common elements than aluminum has.
   (d) Most aluminum is tied up in aluminosilicates which are chemically unsuitable for commercial processing.
   (e) None of these.

4. Choose the two phrases that best answer both sentences. Name the process currently most important for making steel. Of what process is this a modern improvement?
   (a) basic oxygen process, open hearth process
   (b) leaching process, blast furnace process
   (c) open hearth process, Bessemer process
   (d) Hall process, Bayer process
   (e) blast furnace process, tempering process
5. The electrolysis of Al_2O_3 in cryolyte (Hall process) is an example of
   (a) mechanical ore enrichment
   (b) chemical ore enrichment
   (c) alloying the metal
   (d) a reduction step
   (e) refining metal to remove minor impurities
6. Silicon (as silicates) does not precipitate with the purified Al_2O_3 in the Bayer process because
   (a) silicon is more acidic, so it remains in solution at the conditions where Al_2O_3 precipitates.
   (b) silicon was left behind in an earlier step of the Bayer process.
   (c) silicon was never present in the material being processed.
   (d) silicon compounds are always more soluble than aluminum compounds.
   (e) none of these.
7. Which of the following is an ore enrichment method?
   (a) blast furnace   (c) oxidation   (e) tempering
   (b) Hall process   (d) leaching
8. Which of the following is a reduction step in metallurgy?
   (a) blast furnace   (c) oxidation   (e) tempering
   (b) Bayer process   (d) leaching
9. Choose the words that best answer both questions. The strength of steel is due to the effect of what substance with the iron?
   Name the careful heat treatment that provides the proper combination of strength without too much brittleness.
   (a) cementite, case hardening
   (b) carbon, fire polishing
   (c) cementite, fire polishing
   (d) carbon, tempering
   (e) carbon monoxide, blast furnace

Part IV

Chemistry of Life

17 Carbon and Its Compounds

17–1 Bonding with Carbon

Organic chemistry, the chemistry of the compounds found associated with life or closely related to those associated with life, is the chemistry of carbon compounds. Carbon is an abundant element (see Sections 15–7 and 15–8), but its importance to life is not caused by abundance. Carbon is not even among the ten most abundant elements on the Earth's surface. And most of the carbon in forms meeting the criteria of liquid or solid compounds of low density (criteria needed to cause concentration on the Earth's surface) are carbonates, which are not "organic" chemicals.

The importance of carbon in organic chemistry is due to the chemical bonding possible with carbon. The bonding pattern of carbon has three outstanding features compared to other elements. First, it permits a large number (four) of the strong regular covalent bonds—bonds with one electron contributed from each atom. Second, it permits multiple bonding—more than one pair of electrons in a bond. Third, carbon can form chains by bonding to other carbon atoms.

The electronic configuration of a carbon atom is $1s^2 2s^2 2p^2$. The four electrons in the outer shell ($2s^2 2p^2$) can be shown by the dot diagram

Carbon has the ability to form four bonds and forms strong covalent bonds.

That is exactly halfway between two different closed shell configurations. We expect carbon to reach a closed shell in bonding. To reach a closed shell it must either gain four electrons or lose four electrons. Carbon actually does this by forming four covalent bonds. It picks up a share of four electrons in the covalent bonds, and it shares four electrons of its own to form the covalent bonds.

Whether we count carbon's gains or contributions, the number of bonds must be *four*. We will be able to check carbon's bonding in various cases by simply remembering that there must *always* be four bonding pairs of electrons on each carbon. Four pairs is the maximum for any element that does not exceed a closed shell arrangement. Cases where more than four bonds form are limited to the coordinate covalent bonding in complexes (see Section 16–3) plus a very few cases of regular covalent bonding. Only elements with d electron levels available in their outer shells (or next to outer shell) can exceed four bonds. Therefore, carbon (with four bonding pairs) always has the maximum number of bonding electrons possible for its row in the periodic table.

As carbon forms its four bonds, it can easily form bonds to other carbon atoms. The bonds being formed by carbon are truly covalent. Because carbon is right in the middle of the row, it is neither metallic (needing to lose electrons) nor strongly electronegative (needing to gain extra electrons taken from something else). Carbon does not need to react with an "opposite" that has a much greater tendency to gain or lose electrons. Therefore, bonds between carbon atoms form readily and are not easily broken apart by reactions with other elements.

The four pairs of electrons forming bonds around a carbon atom can be in several different patterns. The most likely of these patterns is the tetrahedral pattern described in Section 7–13, which gives the maximum separation of the four groups of electrons. That pattern is the one resulting from an overall combination of the $2s$ electronic orbital and the three $2p$ orbitals into an sp^3 hybrid orbital that can hold four pairs of electrons. This sp^3 hybrid orbital is the name from the valence bond model of bonding described for complex ions in Section 16–3. In that section we indicated that the valence bond hybrid orbital names are used extensively in organic chemistry. We will use that system from here on. Under the hybrid orbital system we can restate the second sentence of this paragraph as follows: sp^3 bonding is the most likely pattern around carbon. The sp^3 hybrid orbital is shown in Figure 17–1.

It is not necessary for all of the s and p orbitals in the outer shell to combine into a hybrid orbital. The $2s$ orbital and two $2p$ orbitals may combine to form an sp^2 hybrid orbital, which can hold three pairs of electrons. The sp^2 orbital will form three bonds. The fourth bond to that carbon must then be a bond using the one remaining $2p$ orbital. Figure 17–2 shows an sp^2 hybrid orbital and an sp^2 hybrid orbital plus the remaining p orbital. Because the sp^2 hybrid includes the concentrations in the particular directions of two p orbitals, it is concentrated along the plane defined by their two directions. The remaining p orbital is therefore the one not in that plane. That is why

Four single bonds in tetrahedral sp^3 bonding is the most likely pattern around carbon.

Top view   Side view

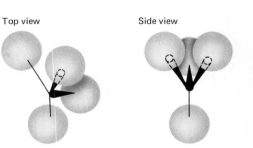

Figure 17–1 sp^3 hybrid orbital.

The shape is tetrahedral. The direction from the nucleus toward each lobe is shown by a line in black. When that direction includes angling into or out of the page, the end of the line closest to our viewing position is shown thicker than the more distant parts of that line.

Figure 17-2 sp^2 hybrid orbital and $sp^2 + p$ orbitals.

The sp^2 orbital is trigonal (planar). The remaining p orbital has its axis perpendicular to the plane of the sp^2 orbital. The direction from the nucleus toward each lobe is shown by a line in black. When that direction includes angling into or out of the page, the end of the line closest to our viewing position is shown thicker than the more distant parts of that line.

Carbon can also double bond with an sp^2 hybrid orbital plus one p orbital or have a triple bond or two double bonds by using an sp hybrid orbital plus two p orbitals.

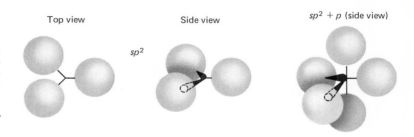

sp^3 bonding holds four other atoms to the carbon by single covalent bonds in a tetrahedral arrangement.

Figure 17-3 sp hybrid orbital and two p orbitals.

The sp orbital is shown in black. The p orbitals (each a pair of lobes) are shown in color. Each p orbital has an axis perpendicular to the sp orbital and perpendicular to the other p orbital axis. The axis of the sp orbital is shown by a black line and each p orbital axis is shown by a colored line.

it must be perpendicular to the plane of the sp^2 hybrid, as shown in Figure 17-2.

A hybrid orbital is also sometimes formed from the $2s$ orbital and one $2p$ orbital. This sp hybrid orbital can hold two pairs of electrons and will form two bonds. It is concentrated along one direction (the direction of the one $2p$ orbital being used), as shown in Figure 17-3. The other two bonds to the carbon must then involve the two remaining $2p$ orbitals. As shown in Figure 17-3, they must be both perpendicular to the sp orbital and perpendicular to each other.

You may wonder why, if an s and a p orbital can form a hybrid, the two $2p$ orbitals do not form a hybrid. Actually, in bonding p orbitals behave as if they do form a hybrid. Figure 17-4 shows the doughnut shape that would result if p orbitals form a hybrid orbital. The $2p$ hybrid "doughnut" would be around the sp hybrid as shown. You can think of bonding by sp hybrids either as shown in Figure 17-3 or as in Figure 17-4. The important points are that the sp part is a straight line shape and that something sticking out to the sides must be used for the other two bonds. We will refer back to the shapes of these bonding arrangements when we discuss carbon compounds in Section 17-2.

When carbon bonds in an sp^3 arrangement, four other atoms can bond to the one carbon atom. Each of the four lobes of the sp^3 hybrid can hold a pair of electrons, one from the carbon atom, one from another atom. Each of the atoms being bonded also provides an orbital lobe pointing toward the carbon atom. The electron pair is shared by the overlap between the lobes. The attraction of the electrons toward an atomic nucleus makes each lobe point directly toward the atom to which it forms a bond. Therefore, each of the four bonded atoms must be directly out in the line along which the

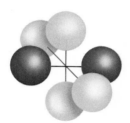

[17-1] Bonding with Carbon 375

Figure 17-4 *sp* hybrid orbital and p^2 hybrid orbital.

The *sp* orbital is shown in black. The doughnut shape shown in color is the hybrid orbital formed from two *p* orbitals. The plane of the doughnut shape is perpendicular to the axis of the *sp* orbital. The axis of the *sp* orbital is shown by a black line.

bonding lobe is concentrated, as shown in Figure 17-5. The forces causing the sp^3 shape around carbon therefore tend to move the bonded atoms into the same shape arrangement as the sp^3 shape, as shown in Figure 17-5. Since no single atom can be in two places at once, there must be four bonded atoms, one on each sp^3 lobe.

The bonding just described has one pair of electrons in each bond, with the electrons concentrated around the line drawn between the nuclei of the two atoms being bonded. That line between nuclei is called the **bond axis**. A bond centered on the bond axis is called a **sigma bond** (Greek letter sigma, σ). Only one pair of electrons can be held in a σ bond, so these are covalent **single bonds.**

However, carbon can also form multiple covalent bonds. That is done by forming the one allowed σ bond and then adding bonds to *p* orbitals. When the carbon atom has an sp^2 hybrid plus one *p* orbital, a second bond can be formed by exchange between the *p* orbital on carbon and a *p* orbital on the other atom, as shown in Figure 17-6. This is a **double bond.** To be effective in bonding, the exchange must have the *p* orbitals as close to each other as possible. Carbon atoms (with only two shells of electrons) are quite small, which helps the bonding between *p* orbitals. The bonding between *p* orbitals is also helped by lining the orbitals up parallel to each other (the closest possible arrangement). The advantage of that parallel arrangement is so strong that the atoms are held in that position. The rigidity of this bonding will be an important feature of our later discussion of double bonding. The sharing of electrons between *p* orbitals gives a concentration out away from the bond axis. It is called a **pi bond** (Greek letter pi, π).

When a carbon atom has an *sp* hybrid plus two 2*p* orbitals, a **triple bond** can be formed involving, in one direction, one σ bond and two π bonds. Or the carbon can have two double bonds, one on each side. In every case

Each of the four lobes of an sp^3 hybrid orbital shares an electron pair with an orbital lobe from another atom with the electrons in a σ bond concentrated along the bond axis.

Double bonds are formed from one σ bond involving a lobe of a hybrid orbital plus one π bond involving exchange away from the bond axis between parallel *p* orbitals.

The necessity for the *p* orbitals in a π bond to be parallel holds double bonds rigid with all of the atoms bonded by sp^2 hybrid orbital σ bonds held in trigonal arrangements in the same plane.

Triple bonds are formed from one σ bond involving a lobe of an *sp* hybrid orbital plus two π bonds involving *p* orbitals.

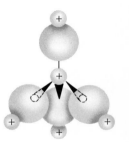

Figure 17-5 Bonding to a sp^3 hybrid orbital.

The direction from the central atom nucleus toward each lobe is shown by a line in black. When that direction includes angling into or out of the page, the end of the line closest to our viewing position is shown thicker than the more distant parts of that line.

Figure 17-6 A double bond.

One pair of electrons is in the π bond represented by the two regions at the top and bottom of the figure.

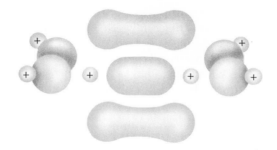

where there is anything but single bonds, there must be π bonds involved. The concentration of electrons away from the bond axis in π bonds will have an important influence on organic reactions, to be discussed in Chapter 18, particularly Section 18-2.

We are going to show bonding structure by a shorthand in all the organic chemistry sections. Hydrogens or other atoms or groups which always bond through a single covalent bond will be written next to the atom they are bonded to

$$CH_4 \quad \text{for} \quad H:\overset{..}{\underset{..}{C}}:H.$$
$$\phantom{CH_4 \quad \text{for} \quad }H$$

A shorthand for organic structures has symbols for hydrogen and other groups that are always singly bonded next to the carbon holding them and uses one dash per bond for other bonding.

All other bonds will be shown by writing one dash (—) for each *pair* of electrons being shared. Therefore, a single bond between carbons is C—C (as in CH_3—CH_3), a double bond is C=C (as in CH_2=CH_2), and a triple bond is C≡C (as in CH≡CH). In our exercises and sample test items we refer to this shorthand as the CH_3—CH_3 type shorthand.

17-2 Hydrocarbons

Hydrocarbons are compounds containing only hydrogen and carbon.

Carbon can form chains or rings by bonding to other carbon atoms.

Most compounds of carbon associated with life contain hydrogen as well as carbon. The remains of living things are often naturally converted to compounds (in oil, natural gas, and parts of coal) that contain only carbon and hydrogen. These compounds are called hydrocarbons. Hydrocarbons are important to us as fuels and as raw materials for chemical reactions. They also provide a basic background for nomenclature and an understanding of other organic chemical compounds. We want to progress to an understanding of some chemical processes involved in life, and the study of hydrocarbons is the best first step we can make toward that goal.

Because carbon can form bonds to other carbon atoms, very large and complex organic molecules can be assembled. The possible carbon–carbon linkages fall into two main groups. The carbons can be arranged in chains, or the carbons can link around to form rings (ending up where they started). Although the presence of rings makes some difference, most carbon rings behave chemically almost like carbon chains. Compounds involving carbon chains or acting similarly are called **aliphatic compounds**.* However, there is a group of carbon ring compounds that behave differently. These compounds involve resonance (see Section 7–12) of the especially favorable kind found in benzene, C_6H_6 (see Section 7–12 and Section 17–3.7). Because these compounds behave so very differently chemically, they are classified as a separate group called **aromatic compounds.**

Because so many different organic compounds are possible, we group them together in families with similar bonding and behavior. These families are called **homologous series.** By learning homologous series and typical reactions of each series instead of studying each individual organic compound, you can greatly increase the amount of understanding you gain for your effort in organic chemistry. That does not make organic chemistry simple or remove the need for memorization of facts, but it can make organic chemistry manageable and make the facts learned rewarding.

The simplest homologous series (and the one we will use as the reference to compare other series with) is the **alkanes.** The alkanes are hydrocarbons with only single bonds. The simplest alkane has one carbon atom with four hydrogen atoms attached, CH_4. The next member of the same series has two carbons. Each of those carbons has three hydrogen atoms bonded, with the fourth bond used to link to the other carbon atom. The formula is C_2H_6. The next member of the alkane series has three carbons. The two at the ends each have three hydrogen atoms and one bond to the center carbon atom. The one in the middle has two hydrogen atoms and two bonds to other carbon atoms. The formula is C_3H_8. These three alkanes are shown in Figure 17–7. Notice that the chains must be bent, not straight, to fit the tetrahedral shape of sp^3 bonding. We will go on to other alkanes in Section 17–3 where we will use this series as the basis for naming organic compounds.

Aliphatic hydrocarbons are those with carbon chains or which behave similarly to those with carbon chains.

Organic chemistry can be studied more efficiently by grouping related compounds into families called homologous series.

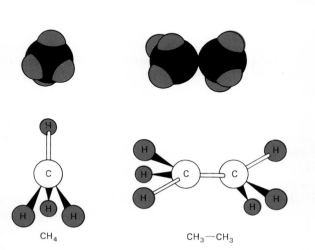

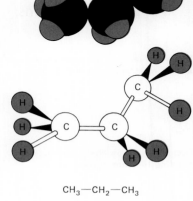

Figure 17–7 The three simplest alkanes.

Carbon atoms are shown in black and hydrogen atoms are shown in color. The stick figures shown below each model allow enough extra space (not present in the real molecules) to make the arrangement easier to see.

---

*The word aliphatic is actually derived from a Greek word for fats. Fats are aliphatic, but the word is now used to describe all carbon chain type compounds including many that are not fats.

Alkanes, which have all bonds as single bonds, are described as saturated hydrocarbons because they have the maximum number of possible bonding groups.

Alkenes, with a double bond, and alkynes, with a triple bond, are described as unsaturated hydrocarbons because additional groups could be bonded if the double or triple bonds were replaced by single bonds.

The alkanes are called **saturated** hydrocarbons. By having only single bonds, they have the maximum possible number of atoms bonded to the carbons. Any hydrocarbons containing double or triple bonds are called **unsaturated** hydrocarbons. More atoms could be bonded to their carbons if the double or triple bonds were replaced by single bonds.

Those hydrocarbons which contain one double bond (with all other bonds single bonds) are called **alkenes.** Since hydrogen can only form single bonds, there must be at least two carbons in an alkene so the double bond can form between the carbons. The simplest alkene has two carbons, each of which has two hydrogen atoms attached and the other two bonds used in the double bond to the other carbon. The formula is C_2H_4. The next alkene has three carbons. One has two hydrogen atoms attached and two bonds used for the double bond to the next carbon atom. That second carbon atom has two bonds used for the double bond, one bond used for a single bond to the third carbon atom, and only one bond left to attach a single hydrogen atom. The last carbon atom has three hydrogen atoms attached and one bond used for the single bond to the center carbon. The formula is C_3H_6. Another alkene would have four carbons as follows: One carbon with two hydrogens and a double bond to the next carbon; the second carbon with a double bond, a single bond to the third carbon and one hydrogen, the third carbon with two hydrogens and two single bonds to carbons, and the fourth carbon with three hydrogens and one single bond to the third carbon. The formula is C_4H_8. These three alkenes are shown in Figure 17–8.

Figure 17–8 The three simplest alkenes.

Carbon atoms are shown in black and hydrogen atoms are shown in color. The stick figures shown below each model allow enough extra space (not present in the real molecules) to make the arrangement easier to see.

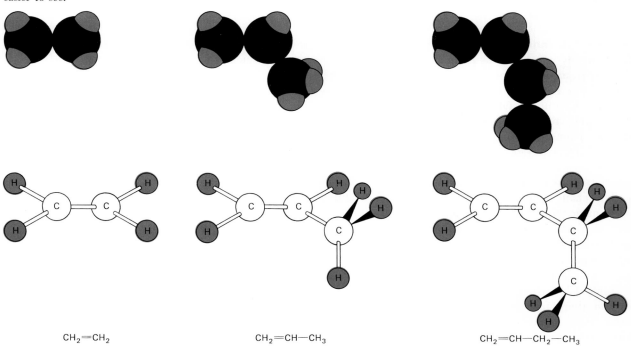

$CH_2{=}CH_2$ $CH_2{=}CH{-}CH_3$ $CH_2{=}CH{-}CH_2{-}CH_3$

[17–2] Hydrocarbons

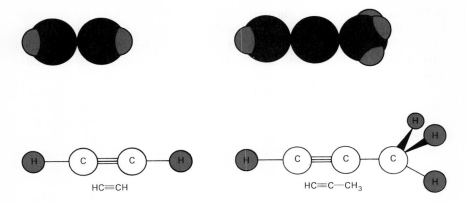

Figure 17-9 The two simplest alkynes.

Carbon atoms are shown in black and hydrogen atoms are shown in color. The stick figures shown below each model allow enough extra space (not present in the real molecules) to make the arrangement easier to see.

Actually, there are four different alkenes with the formula C_4H_8, but different arrangements of atoms. These different arrangements are called **isomers**. We will discuss them in Section 17–3 where we will also provide a naming system to tell them apart.

A third series of aliphatic hydrocarbons would have one triple bond and all the rest single bonds. This series is called the alkynes. Once again at least two carbons are needed. The first two alkynes, C_2H_2 and C_3H_4, are shown in Figure 17–9.

Additional series will be introduced as we proceed in Sections 17–3 and 17–4 and in later chapters.

17–3.1 Alkanes. The three homologous series described in Section 17–2 have already introduced us to an important feature of organic nomenclature. The series names alkanes, alkenes, and alkynes differed only in the endings used. Those endings are used to identify the class of compound. The ending -ane means saturated (all single bonds), -ene means one carbon–carbon double bond, and -yne means one carbon–carbon triple bond. The endings are put on root names providing further information. All three of these series are made from chains of carbon atoms. The chain groupings are called alkyl compounds. The name alkyl is shortened to the root name alk- to mean carbon chain type hydrocarbon compounds. Therefore, alkanes are alk- (meaning carbon chain hydrocarbon compounds) -anes (meaning all single bonds).

To name individual alkanes, alkenes, or alkynes we need root names telling us which carbon chain they contain. We are going to get those root names by learning the names for the most common alkanes. Then by using the same root names with the endings -ene or -yne, we can also name the alkenes and alkynes with the same carbon chains.

Table 17–1 lists the names of the alkanes containing a single chain of carbon atoms from one to ten carbon atoms long. We could continue this list to much higher numbers, but the first ten are the most frequently used and a representative sample of how the naming system works.

17–3 Naming Hydrocarbons and Isomers

Hydrocarbons are named by using root names to indicate the carbon chain and the endings -ane, -ene, or -yne to indicate the single or multiple bonding.

Table 17-1

Formula	Arrangement	Name
CH_4	CH_4	methane
C_2H_6	CH_3-CH_3	ethane
C_3H_8	$CH_3-CH_2-CH_3$	propane
C_4H_{10}	$CH_3-(CH_2)_2-CH_3$ or $(CH_3-CH_2-CH_2-CH_3)$	butane
C_5H_{12}	$CH_3-(CH_2)_3-CH_3$	pentane
C_6H_{14}	$CH_3-(CH_2)_4-CH_3$	hexane
C_7H_{16}	$CH_3-(CH_2)_5-CH_3$	heptane
C_8H_{18}	$CH_3-(CH_2)_6-CH_3$	octane
C_9H_{20}	$CH_3-(CH_2)_7-CH_3$	nonane
$C_{10}H_{22}$	$CH_3-(CH_2)_8-CH_3$	decane

The names of the first ten alkanes without branching are methane (one carbon), ethane (two carbons), propane (three carbons), butane (four carbons), pentane (five carbons), hexane (six carbons), heptane (seven carbons), octane (eight carbons), nonane (nine carbons), and decane (ten carbons).

17-3.2 Branching Isomers. There is only one possible arrangement with the formula CH_4. There are also only one each with the formulas C_2H_6 and C_3H_8. But starting with C_4H_{10}, there are two or more arrangements possible for each formula. These alternate arrangements with the same formulas are called **isomers.** The isomers of alkanes occur because of **branching** of the carbon chain. The C_4H_{10} molecules with the arrangement $CH_3-CH_2-CH_2-CH_3$ have a single chain of connected carbon atoms. But the C_4H_{10} isomer with the arrangement

$$CH_3-\underset{\underset{CH_3}{|}}{CH}-CH_3 \quad \text{or} \quad CH_3-CH(CH_3)-CH_3$$

has a branching position at the center carbon. The carbon chain begun at the left continues on two branches at the center carbon atom.

Both of the C_4H_{10} isomers are alkanes with four carbon atoms. The root name "but-" (pronounced "beaut") is used to identify groups of four-carbon atoms, so both C_4H_{10} isomers are butanes. We need a naming system that can tell them apart. A very old system did that by presuming single chains were "normal" and branched chains were the isomers. $CH_3-CH_2-CH_2-CH_3$ was called normal-butane (abbreviated n-butane) and $CH_3-CH(CH_3)-CH_3$ was called isobutane (abbreviated i-butane).

You may see those names used, but it is really a very poor system. Only two names are possible and from C_5H_{12} on there are more than two isomers to be named. It would be better to use a systematic pattern of names that identifies the parts present and their arrangement. Systematic naming patterns have been established and many of the names we will describe fit the current official system established by the International Union of Pure and Applied Chemistry (IUPAC). However, some very common substances have old-fashioned, less systematic names that are used more frequently than the official names. Often these "common" names are much shorter than the systematic names. The most important reference for locating information about chemical

Systematic organic nomenclature and a detailed compromise using some common names plus the systematic names exist.

research results is *Chemical Abstracts*. *Chemical Abstracts* has chosen to use a mixture of systematic names plus some of the well-established common names. We will also use common and IUPAC names, but we will not attempt to cover all the types of names needed by *Chemical Abstracts*. We will try to emphasize the IUPAC names of important classes but also mention the common names of a few important compounds.

A very much more detailed description of nomenclature appears at the front of the subject index of *Chemical Abstracts* for January–June, 1962 [*Chemical Abstracts*, **65,** subject index Jan.–June pp. 1N-69N (1962)]. It includes both more detailed descriptions of systematic rules and a much larger number of important common names than we list here.

The systematic names are set up to handle many different compounds, not just hydrocarbons. The rules of this system that affect alkane names are

1. The compounds are named from the *longest* single carbon atom chain present.
2. Any branches are named as substituent groups attached to the main chain.
3. The position where the substituent group is attached is named by a number obtained by counting the number of carbons from the end of the chain to the substituent group. If the number is different, depending on which end we start counting from, we start from the end that gives us the lowest number (or numbers).

Branching isomers are named from their longest carbon chain, with branch chains named as alkyl groups and located (when necessary) by numbering the carbons from the end of the chain.

Therefore, $CH_3-CH_2-CH_2-CH_3$ is named butane. We do not list the positions of attached hydrogens. We assume only hydrogen atoms are attached unless we have a name stating a substituent group, which then replaces a hydrogen. However, $CH_3-CH(CH_3)-CH_3$ is not named as a butane. It is named as a propane because its longest carbon atom chain is three carbons long. It has a CH_3 group attached to the number 2 carbon in its propane chain.

Carbon chains attached to the main chain can be named from the root name of the alkane with the same chain. Groups like CH_3- or CH_3-CH_2- are called alkyl radicals and represented by the ending -yl. Therefore CH_3- is methyl, CH_3-CH_2- is ethyl, and so on. $CH_3-CH(CH_3)-CH_3$ is therefore 2-methylpropane.

Alkyl groups are named like the corresponding alkane except for the ending -yl instead of -ane.

Numbering the carbon atoms is not necessary with 2-methylpropane because it is the only possible methylpropane. However, as the carbon chains grow longer, numbering becomes necessary. For instance, methylpentanes could appear in these arrangements:

$$CH_3-\overset{\overset{\displaystyle CH_3}{|}}{CH}-CH_2-CH_2-CH_3 \qquad CH_3-CH_2-\overset{\overset{\displaystyle CH_3}{|}}{CH}-CH_2-CH_3$$

$$CH_3-CH_2-CH_2-\overset{\overset{\displaystyle CH_3}{|}}{CH}-CH_3$$

Counting to locate position can be done from either end of the main carbon chain, and when two molecules can be counted out to the same name, they are in fact identical molecules. The name with the lowest numbers is used.

At first glance, we might expect their names to be 2-methylpentane, 3-methylpentane, and 4-methylpentane. The first two are correct, but there is no such thing as 4-methylpentane. We are supposed to name it using the lowest possible numbers. We get the lowest number in the last case by counting from the right, as shown here.

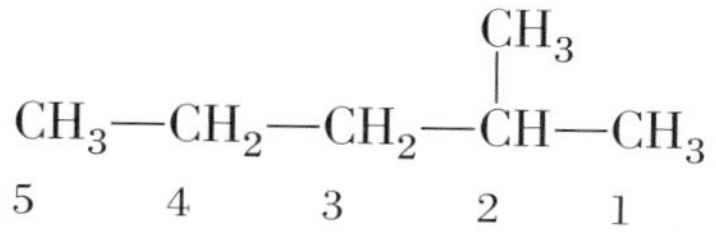

$$CH_3-CH_2-CH_2-\underset{1}{\overset{\overset{\displaystyle CH_3}{|}}{CH}}-CH_3$$
$$54321$$

Its name is then 2-methylpentane, the same as the first methylpentane we named. In fact, the two *are* the same compound. The molecule does not know which end we are writing at the left on the paper, and molecular motions flop it around from one side to the other anyhow. The rules for systematic nomenclature can often help us recognize cases like this where two ways of writing the structure both represent the same thing.

When several identical substituents are present, one carbon number is listed for the position of each substituent.

Sometimes we need to name a compound that has the same substituent group used several times. It may be on several different carbons, or it may be used twice on the same carbon. In our system, we will simply list the numbers of the carbons, listing one number twice if two groups are attached there. We will also indicate the number of each group type used by putting a prefix on that name. The prefixes used are the same ones used in inorganic nomenclature of compounds between nonmetals. The most common ones are di- (2), tri- (3), tetra- (4), penta- (5), and hexa- (6).

The prefixes di- (2), tri- (3), tetra- (4), penta- (5), and hexa- (6) are used to tell how many of a particular substituent or type of multiple bond are present in each molecule.

The compound

$$CH_3-\underset{\underset{\displaystyle CH_3}{|}}{\overset{\overset{\displaystyle CH_3}{|}}{C}}-CH_2-\overset{\overset{\displaystyle CH_3}{|}}{CH}-CH_3$$

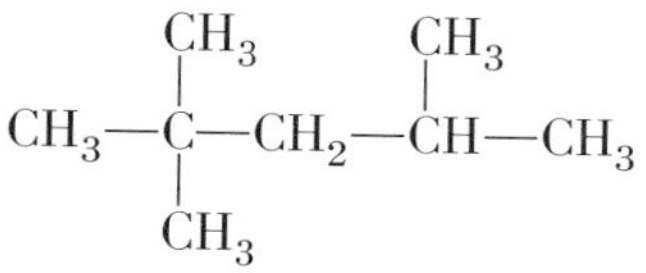

is a very desirable component of gasoline. It has the old-fashioned name "isooctane," although that name leaves the reader totally uninformed about which of the 16 branched octanes it is. It happens to be the most important of the 16 because it is the standard for defining "octane rating" of gasoline. It has the value of 100 on the octane scale of antiknocking property. (The other reference point for the octane scale is zero for hexane, a compound that causes considerable knocking in engines.) Its systematic name (which tells us exactly which of the octanes we mean) is 2,2,4-trimethylpentane. It is not likely that any gas stations are going to start quoting the "2,2,4-trimethylpentane rating" of their gasolines, but now that you understand how systematic naming of alkanes works you can use such names to identify hydrocarbons exactly.

17–3.3 Alkenes and Dienes.

Once the names for alkanes have been established, alkenes can be named easily by substituting the -ene ending for the -ane ending. Therefore, C_2H_6 is ethane so C_2H_4 is ethene. C_3H_8 is propane so C_3H_6 is propene. C_4H_{10} in the isomer without branching is butane, so the C_4H_8 without branching is butene. These first three alkenes also have some old (nonsystematic) names with -yl- inserted before the -ene ending, but we will use only the systematic names in our discussions.

The alkenes will also have branching isomers corresponding to the branching isomers of the alkanes. For example, 2-methylpropane has a counterpart alkene, 2-methylpropene. However, alkenes also have another kind of isomer. We need to list the location of the double bond when it is in a chain of four or more carbons. Differences in positions on the same chains will cause **position isomers.** The butene without branching has two position isomers, $CH_2\!=\!CH\!-\!CH_2\!-\!CH_3$ and $CH_3\!-\!CH\!=\!CH\!-\!CH_3$. ($CH_3\!-\!CH_2\!-\!CH\!=\!CH_2$ is not a third isomer because it is identical with $CH_2\!=\!CH\!-\!CH_2\!-\!CH_3$, just turned around.) These positions are identified by numbering the carbons, just like substituent locations are numbered. The lowest numbers are again preferred. They are found by counting from the nearest end of the chain to the carbon where the double bond begins. Therefore, $CH_2\!=\!CH\!-\!CH_2\!-\!CH_3$ is 1-butene and $CH_3\!-\!CH\!=\!CH\!-\!CH_3$ is 2-butene.

In position isomers differing in positions of their multiple bonds, the location of double or triple bonds is given by the number of the multiple bonded carbon atom that is closest to the end of the chain from which numbering begins.

When we had more than one substituent group, we indicated the number (by prefixes) and locations (by carbon number) of each one. The same thing can be done with double bonds. Two methyl substituents on the same molecule made it a dimethyl. Similarly, two -ene groups (double bonds) on the same molecule make it a -diene. Dienes are a new homologous series. They do not fit the alkene definition of having *one* double bond. However, alkenes and dienes are obviously closely related to each other. Examples of dienes are 1,3-butadiene ($CH_2\!=\!CH\!-\!CH\!=\!CH_2$) or 1,2-butadiene ($CH_2\!=\!C\!=\!CH\!-\!CH_3$). Note that an extra vowel (an a) was inserted when the di prefix of -diene made a hard-to-pronounce double consonant if put directly after the root name.

When branching and double (or triple) bonding occur in the same molecule, we always take care of the double bond first in naming and then do whatever is needed to name the substituent groups. We even choose a shorter carbon chain, if necessary, to make the double bond appear in the basic chain. Therefore,

Whenever possible, all multiple bonds are included in the main chain used for naming, even if that makes the chain shorter than some other chain that does not include the multiple bonds.

$$CH_3\!-\!CH_2\!-\!\underset{\underset{CH_2}{\|}}{C}\!-\!CH_2\!-\!CH_2\!-\!CH_3 \text{ or}$$

$$CH_2\!=\!\underset{\underset{CH_3}{|}}{\overset{CH_2}{C}}\!-\!CH_2\!-\!CH_2\!-\!CH_3 \text{ or } CH_2\!=\!C(CH_2CH_3)\!-\!CH_2\!-\!CH_2\!-\!CH_3$$

is named as a pentene (the longest chain including the double bond) instead of as a hexane. It is 2-ethyl-1-pentene. We prefer to name the ethyl group as a branching chain instead of naming the CH_2 group double bonded to carbon as a branching chain.

17-3.4 *Cis* and *trans* isomers. We have already mentioned three isomers of butene, the branching isomer 2-methylpropene and the two position isomers, 1-butene and 2-butene. Actually, there are four isomers. 2-Butene exists in two different forms called *cis* and *trans* isomers. Those isomers exist because the two other groups (besides the double bond) are not identical on either end of the double bond, and the double bond holds those dissimilar groups in a rigid arrangement. The rigid arrangement of double bonds was pointed out in Section 17-1. Figure 17-10 shows how that rigidity causes two isomers of 2-butene. The isomer in which the two methyl groups are on the same side of the molecule is called a **cis isomer.** The isomer where the methyl groups are on opposite sides (and therefore farther apart) is called a **trans isomer.** These isomers can be named *cis*-2-butene and *trans*-2-butene. As molecules become more complex, *cis-trans* isomerism becomes more common. *cis-trans* isomerism is also very important biologically. Many biological reactions require a particular shape in the reactants, and *cis* and *trans* isomers are quite different in shape. Natural rubber is a series of *cis* isomers, with the linkages to a continuing carbon chain always both on the same side of each double bond. In making synthetic rubber, we find it wise to mimic nature and also select the *cis* isomers.

> When neither end of a double bond has two identical groups bonded on the same carbon, the rigid double bond causes *cis-trans* isomers.

17-3.5 Alkynes. Alkynes can also be named by substituting their ending, -yne, for the -ane ending on the alkane with the same carbon chain. The position of the triple bond can be located by numbering carbons in the same way they were numbered in alkenes. Therefore, the simplest alkynes are ethyne (CH≡CH), propyne (CH≡C—CH₃), 1-butyne (CH≡C—CH₂—CH₃), and 2-butyne (CH₃—C≡C—CH₃). Ethyne has an old fashioned name **(acetylene)** which is used essentially all the time. Acetylene is one of the common names worth learning. It is so common that other alkynes are sometimes even named as if they were ethyne with a substituent group (or groups). For example, propyne is often called methyl acetylene.

There are less isomers of alkynes than there are of alkenes. Branching isomers are impossible at the carbons that have triple bonds. With three of the carbon atom's four bonds used in the triple bond, there can only be one other bond. Two other bonds to different carbons would be needed for

> There are less isomers of alkynes than of alkenes because of the simpler arrangement around carbons with triple bonds.

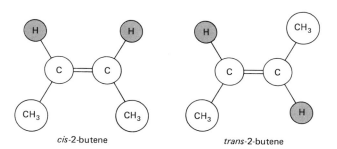

Figure 17-10 *cis* and *trans* isomers of 2-butene (CH₃—CH=CH—CH₃).

branching. There are also no *cis-trans* isomers, since the straight line shape of the bonding and small number of bonded groups eliminates the possibility for differences.

Sometimes both double and triple bonds appear in the same molecule. That causes a naming problem because the carbon numbers for location placed at the front of the name would not tell which number located the double bond and which located the triple bond. In these cases (and all other complicated naming situations) we simply move the numbers to positions right next to the part of the name they refer to. For example, $CH_3-C\equiv C-CH_2-CH=CH-CH_2CH_3$ is named from its octane-like carbon chain. Its name is oct-5-ene-2-yne. Many systematic organic names can be understood by realizing that the information is put where it tells us the facts.

17-3.6 Optical Isomers. There is one more important class of isomers, called optical isomers. Optical isomers are so similar to each other that their properties such as density, melting point, and boiling point are identical. They do have different effects on polarized light. They are named d or l isomers.

Figure 17-11 shows the optical isomers of 3-methylhexane. One of these isomers will rotate plane polarized light right and the other will rotate plane polarized light left. We usually just refer to a mixture of the two isomers as *dl*-3-methylhexane. The *dl* part of the name says that both the d and l forms are present.

Optical isomers occur when four different groups are attached to the same carbon. When there are only two or three different groups, one arrangement of the groups can be rotated to become the same as any other arrangement (except for those cases where double bonds have *cis-trans* isomers). When there are four groups on the same carbon, different arrangements can be rotated to the same arrangement only if at least two of the four groups are identical. If all four are different, there are two different orders of arrangement. These arrangements are **mirror images** of each other—exact reversals of order like the exact reversal of left for right you see in a mirror. The examples

Optical isomers occur when there are four different groups bonded on the same carbon. They occur because there are two different orders of arranging the four groups which form mirror images of each other.

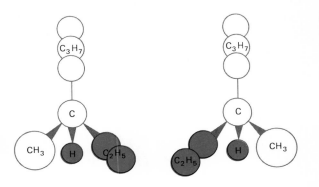

Figure 17-11 Optical isomers.

of mirror images which you look at most often are your right and left hands. They are very similar, but no amount of rotating can make a right hand become a left hand. Optical isomers are similar "right-hand" or "left-hand" pairs.

Biochemical reactions often lead specifically to only one of the optical isomers.

Optical isomers are important in biological systems. Biochemical reactions often produce products put together in the specific order of one of the optical isomers. For example, the most common sugar produced by plants is *d*-glucose. Its optical isomer, *l*-glucose, is either totally nonexistant in nature or at least so rare that it is unimportant. We will defer all further discussion of optical isomers to Section 19–3 and later sections where the biologically important optical isomers are described.

17–3.7 Aromatic Hydrocarbons. We are going to concentrate most of our attention on the aliphatic hydrocarbons and their derivatives. The other main class of hydrocarbons, aromatic hydrocarbons, is sufficiently important for us to learn the names of a few of the most important compounds.

The basic aromatic unit is a six-membered ring of carbon atoms. The simplest aromatic hydrocarbon, C_6H_6, is called **benzene.** That is not the IUPAC name, but it is so common it is even used to name related compounds. The ring is flat, made from sp^2 bonding on each carbon. The remaining p orbitals (one on each carbon) are involved in π bonding with all the other carbons in the ring. That results in the resonance structure shown in Figure 17–12 and described in Section 7–12 and Figure 7–16. The aromatic ring is often written in the form with alternating single and double bonds shown in Figure 17–13, but it cannot actually contain single and double bonds.

Figure 17–12 Resonance in benzene.

The π bond electron ring above the ring of carbon atoms is shown in solid color, whereas the π bond ring below the ring of carbon atoms is shown in a less intense shade.

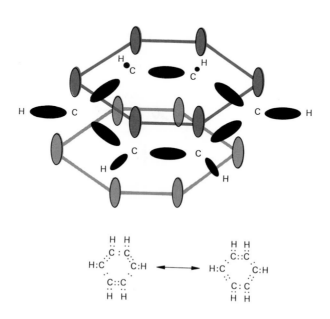

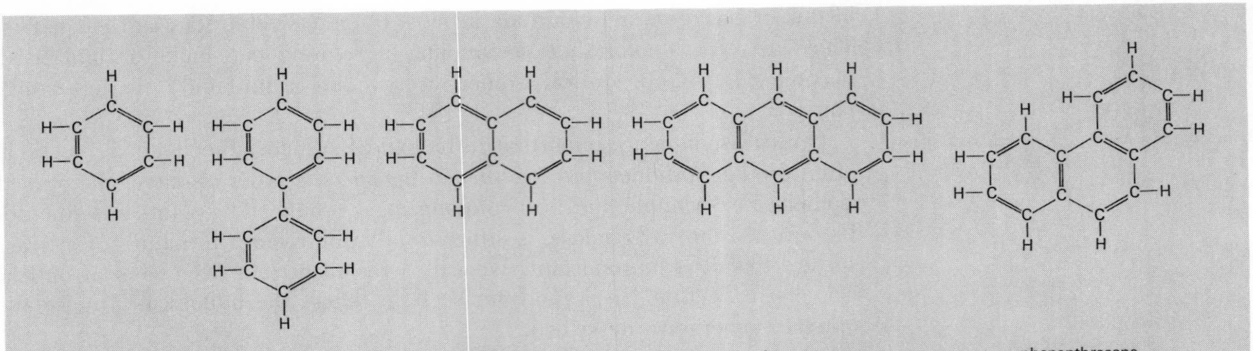

Figure 17-13 Aromatic rings.

Five of the simplest and most common types of aromatic rings are shown. Each could form derivatives by substitution of other groups for hydrogens.

It must exist as the resonance sharing shown in Figure 17-12 with a pair of π **cloud** rings above and below the carbon atom ring.

The six-membered aromatic rings can be linked together either by bonds between two independent aromatic rings or by interlocked rings that expand the resonance sharing to more than one ring. Figure 17-13 shows some of the possible arrangements and their names. Remember, in each case the actual bonding is a resonance sharing, not the single and double bonds used in Figure 17-13 to make it easier to write them down. Figure 17-14 shows the actual π bonding resonance of naphthalene, the simplest case with interlocking rings.

The hydrogens on aromatic rings can be replaced by other substituents. If these are aliphatic hydrocarbon chains, they can be named just as branching chains were on aliphatic hydrocarbons. The simplest such cases are shown in Figure 17-15. You will notice that methylbenzene is one of those compounds with a common name, **toluene,** which is used by everyone.

If there are two substituents on one ring, we need a carbon numbering system. Since there is no "end" of the ring, we begin counting from one

Aromatic hydrocarbons have π cloud rings of resonance sharing above and below the carbon atom rings.

Carbon numbering for location of substituents on aromatic rings is done starting from a carbon holding one of the substituents.

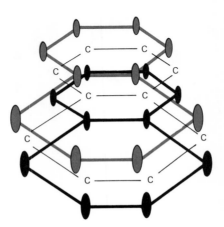

Figure 17-14 Resonance in naphthalene.

The π bond electron rings above the rings of carbon atoms are shown in color, whereas the π bond rings below the rings of carbon atoms are shown in black.

388   Carbon and Its Compounds   [17]

Figure 17–15 Introduction of alkyl side chains.

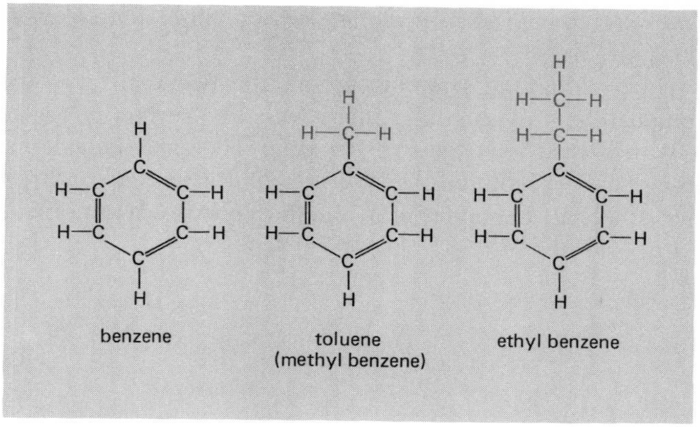

benzene   *toluene (methyl benzene)*   *ethyl benzene*

of the substituents. Figure 17–16 shows the dimethylbenzenes. Once again, they have a common name, **xylene.** The common names also include a system for saying how far apart the substituents are. Ortho (abbreviation *o*-) means "on neighboring carbons" (1,2-substitution), meta (abbreviation *m*-) means "separated by one intervening carbon" (1,3-substitution), and para (abbreviation *p*-) means "as far apart as possible" (1,4-substitution).

The examples listed above should be enough for you to recognize the names given to some of the simpler aromatic hydrocarbons. We will leave the rest of aromatic nomenclature for your later study if and when you need it. You should refer to an organic chemistry textbook if you need to know more about aromatic compounds.

Nonaromatic rings can have any number of carbon atoms equal or larger than three, but five or six carbon atom rings are the most common because they best fit the shape of bonding around carbon.

17–3.8 Nonaromatic Ring Compounds. Carbon atoms can also form rings that do not have the special resonance condition of aromatic compounds. These essentially aliphatic compounds with rings (sometimes referred to as alicyclic compounds) may be saturated (all single bonds) or unsaturated (some double bonds or, very rarely, a triple bond). The rings may be as small as three carbon atoms or as large as the maximum number of carbon atoms

Figure 17–16 The dimethylbenzenes.

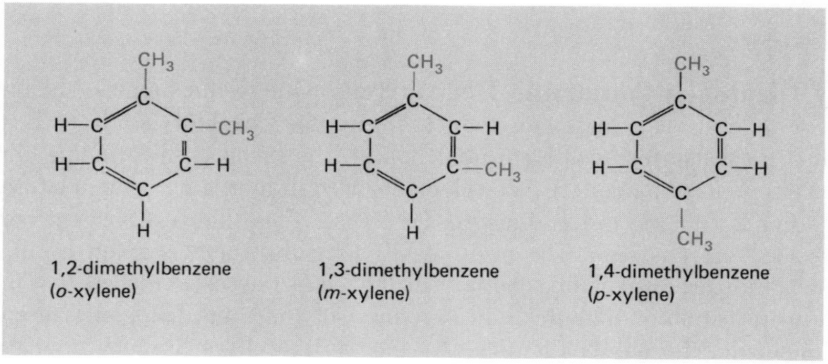

1,2-dimethylbenzene (*o*-xylene)   1,3-dimethylbenzene (*m*-xylene)   1,4-dimethylbenzene (*p*-xylene)

[17–3] Naming Hydrocarbons and Isomers

available. When there are enough carbon atoms, two or more rings may be formed in the same molecule.

Because of the shape of bonding around carbon, the two really favored (and therefore common) kinds of rings are those with five carbon atoms in the ring and those with six carbon atoms in the ring. They are named just like aliphatic hydrocarbons except for the addition of the prefix cyclo- to indicate the presence of a ring. Examples of common **cyclic hydrocarbons** (those with rings) are

The prefix cyclo- is used to indicate a cyclic, but not aromatic, hydrocarbon.

cyclopentane

$$\begin{array}{c} CH_2 \\ CH_2 \quad CH_2 \\ | \quad\quad | \\ CH_2 - CH_2 \end{array}$$

cyclohexane

$$\begin{array}{c} CH_2 - CH_2 \\ /\quad\quad\quad \backslash \\ CH_2 \quad\quad CH_2 \\ \backslash\quad\quad\quad / \\ CH_2 - CH_2 \end{array}$$

cyclohexene

$$\begin{array}{c} CH = CH \\ /\quad\quad\quad \backslash \\ CH_2 \quad\quad CH_2 \\ \backslash\quad\quad\quad / \\ CH_2 - CH_2 \end{array}$$

methylcyclopentane

$$\begin{array}{c} CH_3 \\ | \\ CH \\ /\quad\;\; \backslash \\ CH_2 \quad CH_2 \\ | \quad\quad | \\ CH_2 - CH_2 \end{array}$$

So far we have limited our discussion to compounds containing only carbon and hydrogen. We named basic chains of hydrocarbons and then named the **substituents** tacked on whenever a hydrogen was replaced by anything else. Those substituents do not have to be limited to hydrocarbon chains. We can also replace hydrogens by one atom or a group of atoms containing elements other than carbon or hydrogen. If we do so, we can name and locate these nonhydrocarbon substituents in almost the same way we named branched hydrocarbons. The names (and in fact the compounds themselves) are derived from the hydrocarbons with the same arrangement (except for the hydrogen-substituent exchange). These compounds, which are based on hydrocarbons but contain at least one other element, are called **hydrocarbon derivatives**.

17–4 Hydrocarbon Derivatives

Hydrocarbon derivatives have other elements bonded to carbon in place of some hydrogens of hydrocarbons.

17–4.1 Halogen Containing Derivatives. One of the simplest substitutions is to replace a hydrogen (which forms one covalent bond) with a halogen (which also forms one covalent bond). The halogen substituents are named from the root names for those elements used in inorganic nomenclature (see Section 2–6) plus the ending -o. Therefore, F is fluoro, Cl is chloro, Br is bromo, and I is iodo. The position of each substituent is given by the number of the carbon it is on, counted in the same way as branched hydrocarbons were counted. If more than one atom of the same halogen is used in the molecule, the number is given by the same prefixes (di-, tri-, and so

Halogen containing hydrocarbon derivatives are named like the corresponding hydrocarbons with the number and positions of halogens given using fluoro-, chloro-, bromo-, or iodo- prefixes and the same numbering and locating system as branched hydrocarbon naming.

on) used in hydrocarbon names. Some examples of halogenated hydrocarbon derivatives are

CH_3Cl	chloromethane
$CHCl_3$	trichloromethane (common name: *chloroform*)
$CH_2Br—CHBr—CH_3$	1,2-dibromopropane
$CHCl_2—CHI—CH=CH_2$	4,4-dichloro-3-iodo-1-butene

(The double bond gets first preference in choosing how to make the numbers smallest.)

17-4.2 Oxygen Containing Derivatives. Since oxygen is the most abundant element on the Earth's surface (and is very reactive), the most common and important hydrocarbon derivatives are those containing oxygen. Oxygen can form two bonds, so there are several ways in which oxygen can be bonded to a hydrocarbon chain. Sometimes oxygen can be bonded in different ways in the same molecule. Some particular combinations of oxygen bonding are especially important biologically and will be discussed in later chapters. At this point we will introduce and name the types of oxygen containing hydrocarbon derivatives without considering the combinations involving several different types in the same molecule.

The most important hydrocarbon derivatives are those containing oxygen.

Alcohols. The simplest way to bond oxygen to a hydrocarbon in place of hydrogen is to substitute an OH unit for a hydrogen atom. One of the oxygen's bonds is used to bond the hydrogen in the OH unit, so only one remains to form a bond to carbon. Replacement of hydrogen (one bond) by OH (one bond) leaves the rest of the hydrocarbon unchanged. Hydrocarbon derivatives containing one oxygen as an OH group are called **alcohols**.

Alcohols contain OH groups and are named by dropping the -e from alkane names and adding -ol.

Alcohols are named by dropping the final -e from the name of the corresponding alkane and adding -ol. Therefore, the following are systematic names for alcohols.

$CH_3—OH$	methanol
$CH_3—CH_2—OH$	ethanol
$CH_3—CH_2—CH_2—OH$	1-propanol
$CH_3—\underset{\underset{OH}{\vert}}{CH}—CH_3$	2-propanol

or

$CH_3—CHOH—CH_3$

$CH_3—CH_2—OH$ also has an old fashioned name, **ethyl alcohol.** It is a poison to humans, but when used in moderation its strong biological effects merely cause mild paralysis instead of death. Since that paralysis strikes first at brain centers related to inhibitions in behavior, alcoholic beverages are popular drugs among people who want to escape from their own inhibitions. Other alcohols also have biological effects. Methanol (sometimes called wood alcohol) is known to be a poison dangerous to humans. Methanol is thought to be less poisonous than ethanol, but it also affects different areas. Sometimes methanol has been substituted for ethanol in illegal alcoholic beverages. The methanol affects vision before it affects inhibitions. When taken in excess (by people trying to reach a state of repressed inhibitions), methanol has been known to cause permanent blindness (or in greater excess, death). It should be noted that the same amount of ethanol would kill a person, not just blind him. Fortunately, not many people manage to overindulge themselves that badly on ethanol, although there are some fatalities. In Chapter 22 we will discuss poisons and drugs (and the need for moderation) somewhat further.

Ethanol and methanol can be hazardous to human beings.

It is possible to have more than one OH group on a molecule. A molecule with two OH groups is a diol (di, two; ol, alcohol). Two simple diols are 1,2-ethandiol (CH_2OH-CH_2OH) and 1,2-propandiol ($CH_2OH-CHOH-CH_3$). Diols have an old fashioned name, **glycols.** The old name for 1,2-ethandiol is **ethylene glycol** and 1,2-propandiol is **propylene glycol.** They are the two main constituents sold as permanent antifreeze.

Diols (two OH groups) are important commercially and the simplest triol, glycerol, is the key building block in animal fats.

There is an even more common triol (three OH groups) in nature. 1,2,3-propantriol ($CH_2OH-CHOH-CH_2OH$) is called by the common name **glycerol.** In later chapters we will see that it is a key building block in animal fats. Glycerol is the simplest possible triol. It is usually not possible to put two OH groups on the same carbon because such a compound would decompose by losing water.

Sugars, starches, and cellulose have even larger numbers of OH groups, but their structures are also complicated by other substituents besides OH groups.

Aldehydes and ketones. A second way for oxygen to bond to hydrocarbons is by a double bond between carbon and oxygen. An oxygen with a double bond to the hydrocarbon will not have any other group bonded to it. An oxygen with a double bond to a hydrocarbon is called a **carbonyl** grouping

An oxygen with a double bond to carbon forms a carbonyl group.

$$\overset{\overset{\displaystyle O}{\|}}{-C-} \quad \text{or} \quad -CO-$$

Carbonyl groups are important in nature. Carbonyl groups also have an important influence on chemical reactions.

In chemical reactions there is a noticeable difference between carbonyl groups in which the carbon has a hydrogen atom bonded to it and those which have only carbon atoms bonded to the carbon of the carbonyl. Therefore, separate names are given to the class of carbonyl compounds with a hydrogen on the carbonyl carbon and the class of carbonyl compounds with only carbons on the carbonyl carbon.

Carbonyl compounds with a hydrogen on the carbonyl are called **aldehydes.** Their systematic names are like alcohol names except for the ending, which is -al instead of -ol. Some examples of aldehydes are

$$CH_3—CHO \qquad \text{ethanal}$$
$$CH_3—CH_2—CHO \qquad \text{propanal}$$
$$CH_3—CH(CH_3)—CH_2—CHO \qquad \text{3-methylbutanal (counting starts from the aldehyde end)}$$

Methanal (HCHO) is usually called by the old fashioned name **formaldehyde.** Formaldehyde is used as an embalming agent (kills bacteria).

Carbonyl compounds with two carbon atoms attached to the carbon of the carbonyl are called **ketones.** Their systematic names use the same root name as alcohols (the alkane name with the final -e taken off) with the ending -one. If more than one position is possible, the position of the ketone group is given by the number of carbons from the nearest end of the chain. The simplest ketone, propanone

$$CH_3—\overset{\overset{\displaystyle O}{\|}}{C}—CH_3 \quad \text{or} \quad CH_3—CO—CH_3$$

is almost always called by the common name, **acetone.** Other ketones are sometimes called by old fashioned names based on the alkyl groups attached to the ketone carbonyl. Therefore butanone, $CH_3—CO—CH_2—CH_3$, is called by the old fashioned name methylethylketone. Acetone and methylethylketone are widely used as solvents for other organic compounds. For example, acetone is used as a solvent and then evaporated away in the manufacture of cellulose acetate fibers used to make acetate fabrics.

Carboxylic acids. So far we have described oxygen containing hydrocarbon derivatives with OH groups (alcohols) or with carbonyl groups (aldehydes and ketones). The two kinds of bonding can exist in the same molecule. If they exist with the OH group bonded to the carbon of the carbonyl group, we get an especially important class of compounds. The grouping

$$—\overset{\overset{\displaystyle O}{\|}}{C}—OH$$

is called a **carboxyl group.** It is particularly important because of its ability to act as an acid, serving as a source of H^+ ions. There are a few other kinds of organic acids, but most acid organic compounds are **carboxylic acids** with the —COOH grouping.

Systematic naming of carboxylic acids is based on the same root name as other oxygen containing hydrocarbon derivatives, the alkane name without the -e on the end. The ending is -oic acid. Some of the simple carboxylic acids are still called by old fashioned names from the same old system that

---

Compounds with a hydrogen on the carbon of a carbonyl group are called aldehydes; compounds with two carbon atoms bonded to the carbon of a carbonyl group are called ketones.

Carbonyl compounds are named by dropping the -e from alkane names and adding -al for aldehydes or -one for ketones.

The grouping of an oxygen double bonded to a carbon that also has an OH group attached is called a carboxyl group. Compounds with carboxyl groups are carboxylic acids, the most common type of organic acids.

the name formaldehyde came from. Some examples of organic acids and their names are

HCOOH	methanoic acid* (common name: *formic acid*)
CH$_3$—COOH	ethanoic acid (common name: *acetic acid*)
CH$_3$—CH$_2$—COOH	propanoic acid
CH$_3$—CH(CH$_3$)—CH$_2$—COOH	3-methylbutanoic acid

Salts of organic acids are named by changing the -ic acid ending to -ate. That is the same as the way inorganic acids like nitric acid are changed to name salts. Some examples of salts are

> Carboxylic acids are named by dropping the -e from the corresponding hydrocarbon name and adding -oic acid. Names of salts of the carboxylic acids end in -oate.

Na$^+$⁻OOC—CH$_3$ or NaOOC—CH$_3$	sodium ethanoate (common name: sodium acetate)
K$^+$⁻OOC—CH$_2$—CH$_2$—CH$_2$—CH$_3$ or KOOC—(CH$_2$)$_3$—CH$_3$	potassium pentanoate

Ethers. When an oxygen atom has a single bond to one hydrocarbon chain, the other bond does not have to be to a hydrogen atom. It can be to a carbon in a second hydrocarbon chain. This arrangement (—C—O—C—) is called an **ether.** The insertion of an oxygen between carbons in the chain breaks up the relationship to alkane names. Instead, the —O— group is named by the word ether and the groups attached on each side are named like branch chains in hydrocarbon naming. Some examples of ethers and their names are

> Compounds with an oxygen that has two hydrocarbon chains bonded to it are called ethers. They are named by naming the two attached alkyl groups like branched chains on the oxygen, which is indicated by the word ether.

CH$_3$—O—CH$_3$	dimethyl ether
CH$_3$—O—CH$_2$—CH$_3$	methyl ethyl ether
CH$_3$—CH$_2$—O—CH$_2$—CH$_3$	diethyl ethyl

Diethyl ether is used as an anaesthetic for surgery and is sometimes called simply **ether.**

Esters. If an oxygen is attached to a hydrocarbon chain on one side and to the carbon of a carbonyl group on the other side, we have a compound of the class called an ester. Esters are very common in nature. Animal fats are esters. Esters that are volatile enough to smell have pleasant odors. Perfumes and other pleasant odors in nature are often esters.

> Compounds with an oxygen bonded to a hydrocarbon chain on one side and to a carbonyl carbon on the other side are called esters. They are very common in nature. Esters are named like acid salts with the alkyl group as the positive part.

An ester has a structure like a carboxylic acid except for substitution of an alkyl group for the acid hydrogen. Esters are named like salts of the carboxylic acids, using the name of the alkyl group where the metal ion's

---

*You may notice that HCOOH would also fit the brief definition we gave for aldehydes. The acid behavior dominates, so HCOOH is named as an acid (formic acid or methanoic acid), not as an aldehyde. We left the qualification "unless there is also an OH group on the carbon" out of the aldehyde definition to keep it simple. We will continue to leave out such minor points, which explain only a few exceptions, in all our definitions.

name would go in a salt name. Some examples of esters and their names are

$CH_3-COO-CH_3$ methyl ethanoate (old name: methyl acetate)

$CH_3-COO-CH_2-CH_2-CH_2-CH_3$ butyl ethanoate

$CH_3-CH_2-CH_2-COO-CH_2CH_3$ ethyl butanoate

Acid anhydrides. If an oxygen is attached to two carbons of carbonyls, we have a compound of the class called **acid anhydrides.** They are named from the name of the related acid, dropping the word acid and adding the word anhydride. An example is

$$CH_3-\overset{\overset{O}{\|}}{C}-O-\overset{\overset{O}{\|}}{C}-CH_3 \quad \text{ethanoic anhydride (common name: acetic anhydride)}$$

Compounds with an oxygen bonded to two carbonyl carbons are called acid anhydrides. They are named by replacing acid by anhydride in the name of the corresponding carboxylic acid.

Acid anhydrides are very dangerous on contact with the skin.

You may use acetic anhydride in a laboratory experiment. If so, you will be told it is very hazardous if allowed to touch your skin. The danger occurs because the anhydride is not acidic and behaves almost like a hydrocarbon. Acids would be kept out on the surface of the skin, but the acid anhydride tends to go right through the skin to the tissues below. There each anhydride molecule reacts with water to form two carboxylic acid molecules. That acid irritates or destroys the tissues, and it cannot get out through the skin any better than acids on the outside can get in. Therefore, acid anhydride spilled on the skin must be rinsed off immediately or it becomes equivalent to injecting acid through the skin with a needle. No amount of rinsing can remove it once it gets inside and begins to sting.

Hydrocarbon derivatives containing $-NH_2$ are called amines. They are named by the prefix amino- used with hydrocarbon naming in the same way that halogen substituents were named. The hydrogens on $-NH_2$ can be replaced by other groups to form substituted amines.

17-4.3 Amines. There are several different classes of hydrocarbon derivatives involving nitrogen. The most important of these are the amines. Amines have an $-NH_2$ group substituted for a hydrogen. (Compounds with two carbon chains bonded to $-NH-$ or three carbon chains bonded to $-N\!\!<$ are also amines.) They are called by the name amino-. Some examples of amines are

CH_3-NH_2 aminomethane
(can be written CH_3NH_2)

$$CH_3-\overset{\overset{NH_2}{|}}{CH}-CH_2-CH_3 \quad \text{2-aminobutane}$$
(can be written $CH_3-CH(NH_2)-CH_2-CH_3$)

Amino acids, compounds which are both amines and acids, are of great natural importance.

Amines are particularly important in compounds that are also carboxylic acids. The amino acids are the building blocks of proteins (see Section 19-3). In

Section 18–5 we will also describe the use of amines as building blocks in the manufacture of the synthetic polymer nylon.

17-4.4 Other Substituents. We have limited ourselves to the hydrocarbon derivatives involving halogens (which can substitute simply for hydrogen), oxygen (the most abundant and important hydrocarbon derivatives), and nitrogen as amines (important in the chemistry of life). Other hydrocarbon derivatives can be formed involving other elements. Some of these are also formed in living organisms or are of importance in other ways. We will illustrate the variety of other substituents and their properties by an example involving sulfur.

Sulfur is in the same family as oxygen and can also form two bonds. It can form essentially the same variety of bonding arrangements as oxygen.

The first bonding pattern we discussed for oxygen in hydrocarbon derivatives was OH groups in alcohols. The SH group can also bond to hydrocarbon chains in place of a hydrogen. The family of compounds formed are called **mercaptans,** or (more systematically) thiols. The mercaptans have strong biological interactions, particularly in terms of smell. We mentioned that occasionally people overindulge in alcohol (ethanol) and kill themselves. Mercaptans could also cause illness if used in large amounts. However, we can state with complete confidence that no human being will ever voluntarily be exposed to too much mercaptan. CH_3—CH_2—CH_2—CH_2—SH (named butyl mercaptan or, systematically 1-butanthiol) is the active odor ingredient in skunk oil, and many other mercaptans smell far worse.

This last example has been inserted to illustrate the facts that other classes of derivatives are possible, that some have unique characteristics, and that simply being able to make certain organic compounds does not mean you would necessarily want to have them around.

Many other hydrocarbon derivatives are possible.

Compounds with SH groups in place of the OH group of alcohols have powerful and unpleasant odors.

1. If you are asked to list the important bonding characteristics of carbon, you should be able to list the ability to form four bonds, the ability to bond to other carbon atoms, and the ability to form either single or multiple covalent bonds.
2. You should be able to list aliphatic and aromatic as the main categories of hydrocarbons and describe each type briefly, including pointing out the resonance in aromatic compounds.
3. You should be able to define and use the terms homologous series, alkane, alkene, alkyne, saturated hydrocarbon, unsaturated hydrocarbon, branching isomer, position isomer, alkyl group, diene, substituent, alcohol, aldehyde, ketone, carbonyl group, carboxyl group, carboxylic acid, ether, ester, acid anhydride (in the organic chemistry sense), and amine.
4. You should be able to recognize and identify by name each structure class or condition described by a term listed in expected skill 3, and you should

Skills Expected at This Point

be able to write down the characteristic structure when given the name of any class of compounds appearing in expected skill 3.

5. You should be able to write out the structure (using the CH_3—CH_3 type shorthand) for any hydrocarbon or hydrocarbon derivative not exceeding a ten-carbon chain in one of the classes listed in expected skill 3 if you are given its systematic chemical name.

6. You should be able to write the systematic chemical name for any hydrocarbon or hydrocarbon derivative not exceeding a ten carbon chain in one of the classes listed in expected skill 3 if you are given its structure in the CH_3—CH_3 type shorthand.

Exercises

1. Describe three significantly different ways in which a carbon atom can use its four bonds.
2. List the hybrid orbitals on carbon used to form the σ bonds in each of the following:
   (a) CH_4
   (b) CH_2=CH_2
   (c) CH≡CH
3. (a) What is the nature of a π bond of the sort found in alkenes?
   (b) What is the nature of a π bond of the sort found in aromatic compounds?
4. What is a homologous series?
5. Give the systematic names for the following:
   (a) CH_3—CH_2—CH_2—CH_3

   (b) CH_3—$\underset{\underset{CH_3}{|}}{C}$=$CH_2$

   (c) CH_3—$\underset{\underset{CH_2—CH_3}{|}}{\overset{\overset{CH_2—CH_3}{|}}{C}}$—$CH_2$—$CH_3$

   (d) CH_3—CH_2—$\underset{\underset{CH_3}{|}}{\overset{\overset{CH_3}{|}}{C}}$—$CH$—$\underset{\underset{}{}}{\overset{\overset{CH_3}{|}}{C}}$—$CH_3$

   (e) benzene ring structure

6. Write the structures (using the same type of structure writing as those shown in problem 5) for the following:
   (a) 1,3-butadiene
   (b) 2,2-dimethyloctane
   (c) 1-butyne
   (d) 3-ethyl-2-pentene
   (e) ethane
7. List all the branching isomers of heptane, C_7H_{16}. How many are there?
8. List all the isomers of C_5H_{10}. (There is one common ring isomer, cyclopentane

   $CH_2\overset{\overset{CH_2—CH_2}{}}{\underset{\underset{CH_2—CH_2}{}}{|}}$

   You should be able to identify all of the isomers that do not involve rings.)

9. Give examples of each of the following kinds of isomers: (a) branching isomers; (b) *cis-trans* isomers; (c) optical isomers.
10. What *class* of hydrocarbon derivatives does each of the following belong to?
    (a) CH_3Cl
    (b) $CH_3-CH(CH_3)-CH_2OH$
    (c) CH_3-CH_2-CHO
    (d) $CH_3-CO-CH_2-CH_2-CH_3$
    (e) $HCOOH$
    (f) $CH_3-CH_2-COO-CH_3$
    (g) $CH_3-CH_2-CH_2-O-CH_2-CH_3$
    (h) $CH_3-CH_2-CH(NH_2)-CH_3$
11. Give the systematic names of the compounds shown in problem 10.
12. Write the structures for:
    (a) ethanol
    (b) glycerol (1,2,3-propantriol)
    (c) 3-pentanone
    (d) 2-methylbutanoic acid
    (e) diethylether

Test Yourself

1. Write structures for the following compounds. (*Do not* give empirical formulas, that is, $C_{11}H_{23}Cl$.) Be sure to show all atoms. *Example:* Propane is

    $CH_3CH_2CH_3$ or H—C(H)(H)—C(H)(H)—C(H)(H)—H *not* C_3H_8 or C—C—C

    (a) formaldehyde (methanal)
    (b) acetylene
    (c) cyclohexene
    (d) 1,2,4-tribromobenzene
    (e) 2-propanone
    (f) propanoic anhydride
    (g) methylpropane
    (h) diethyl ether
2. Why are there more organic compounds than all the other compounds put together?
3. Give the class name derived from the presence of the functional group for each of the following compounds. *Example:* CH_3-O-CH_3, ether.
    (a) CH_3-CH_2-CHO
    (b) $CH_3-COO-CH_3$
    (c) CH_3-COOH
    (d) $CH_3-C\equiv CH$
    (e) CH_4
    (f) $CH_3-CH_2-NH_2$
    (g) $CH_3-CO-CH_3$
    (h) $CH_3-CH_2-CH_2-OH$
4. Which of the following compounds have a mirror-image isomer?

(a) 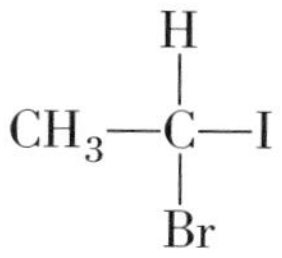 CH$_3$—C(H)(Br)—I

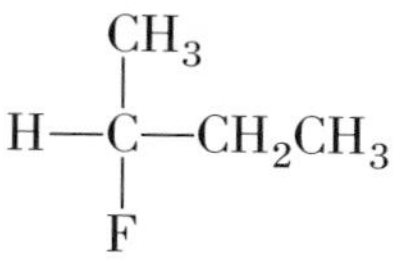

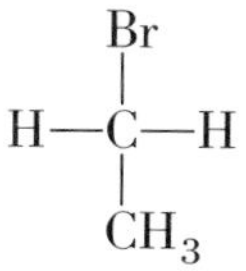

(b) H—C(Br)(CH$_3$)—H

(c) H—C(H)(H)—Cl

(d) H—C(CH$_3$)(F)—CH$_2$CH$_3$

(e) CH$_3$CH=CHCH$_3$

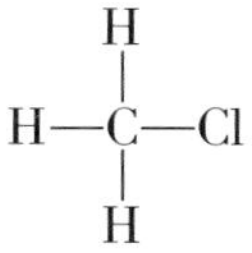

5. CH$_3$—CH$_2$—CH$_2$—CH$_2$—CHO is
 (a) an ester
 (b) a ketone
 (c) an aldehyde
 (d) an ether
 (e) 1-pentanol

6. 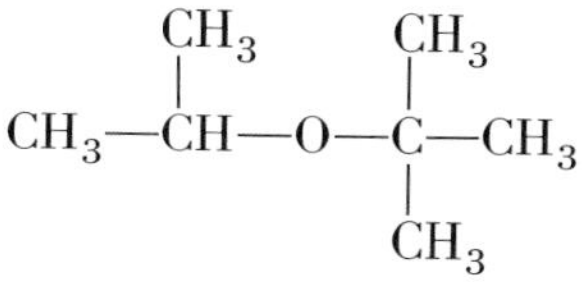 CH$_3$—CH(CH$_3$)—O—C(CH$_3$)(CH$_3$)—CH$_3$ is

 (a) an ester
 (b) a ketone
 (c) an aldehyde
 (d) an ether
 (e) isooctane

7. Which of the following is *not* an important fact about bonding at carbon which contributes to the variety of organic compounds?
 (a) Carbon can form four covalent bonds.
 (b) Carbon can bond to other carbon atoms.
 (c) Carbon can form strong multiple bonds.
 (d) Double bonds between carbons have a shape causing extra possible isomers.
 (e) Triple bonds between carbons have a shape causing extra possible isomers.

8. Choose the systematic name for CH$_3$—C(CH$_3$)(CH$_3$)—CH=CH$_2$

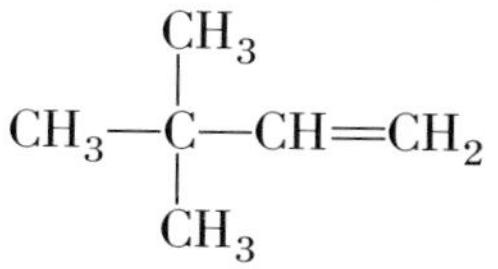

 (a) hexane
 (b) hexene
 (c) 2,2-dimethyl butane
 (d) 3,3-dimethyl 1-butene
 (e) None of these

9. Choose the systematic name for CH$_3$—C(CH$_3$)(CH$_3$)—CH$_2$—CH(CH$_3$)—CH$_3$

[17–4] Hydrocarbon Derivatives

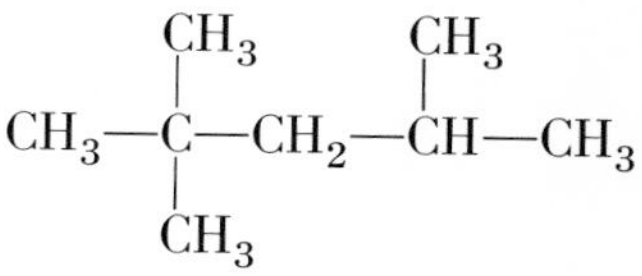

(a) octane
(b) isooctane
(c) pentane
(d) 2,4-dimethylpentane
(e) 2,2,4-trimethylpentane

For questions 10 through 16. Given this group of structures
(a) $CH_3-CH_2-O-CH_3$
(b) $O=CH-O-CH_2-CH_3$
(c) $CH_3-CO-CH_2-CH_3$
(d) $CH_3-CH_2-CH_2-CH_2-CHO$
(e) $CH_2OH-CHOH-CH_2-CH_3$
(f) $CH_3-CH_2-CH_2-CHOH-CH_3$
(g) $CH_2=CH-CH_2-CHNH_2-CH_3$
(h) 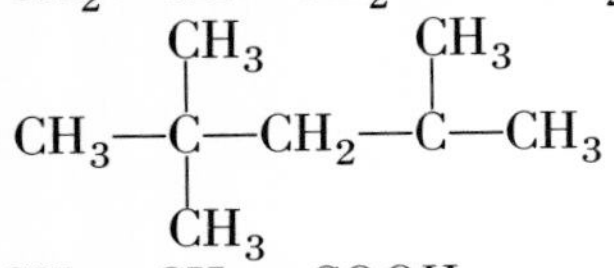 $CH_3-\underset{\underset{CH_3}{|}}{\overset{\overset{CH_3}{|}}{C}}-CH_2-\overset{\overset{CH_3}{|}}{C}-CH_3$
(i) CH_3-CH_2-COOH
(j) none fits

10. Which of them is an alkene?
11. Which of them is an ester?
12. Which of them is a ketone?
13. Which of them is an ether?
14. Which of them is 3-butanone?
15. Which of them is 2-pentanol?
16. Which of them is an amine?

18 Organic Reactions

18–1 Reactions of Alkanes

18–1.1 Low Reactivity of Alkanes.
In our discussion of organic compounds and nomenclature, we began with the fairly simple situation of alkanes, with only carbon and hydrogen and only single bonds. We then moved on to the more complicated arrangements of unsaturated hydrocarbons and hydrocarbon derivatives. We will find the same sequence helpful in discussing organic reactions. The reactions of alkanes are quite limited and fairly simple. We can then proceed to the more reactive unsaturated hydrocarbons and hydrocarbon derivatives and their more complicated reactions.

The very low reactivity of alkanes is caused by their saturated bonding. There are already four other atoms bonded to each carbon, and four is the maximum number of bonds possible for carbon. Therefore, nothing can happen unless one or more of the existing chemical bonds is broken. Breaking bonds is difficult, so alkanes often simply do not react at all when other compounds would react easily. When alkanes do react, it is under quite strenuously reactive conditions. We will describe those conditions and the types of reactions here, but you should keep in mind the fact that the reaction conditions are usually quite severe. The first and foremost point for you to learn about alkane reactions is that *alkanes do not react well* when compared with other compounds.

Alkanes do not react well compared to other organic compounds because a bond must always be broken before another bond can form.

18–1.2 Substitution Reactions.
One of the easiest reactions to picture for an alkane is the replacement of one bonded group by another. An example is the **substitution reaction** of a halogen atom (with one single covalent bond) into a position from which a hydrogen atom (with one single covalent bond) is removed. Bonding a halogen atom to carbon is easy. The hard part is managing to remove the hydrogen atom. It simply does not happen at room temperature energies. Substitution can occur with the help of either high temperatures or of light energy via a special reaction mechanism we will describe in Section 18–3.4. In either case, one atom of a diatomic halogen molecule bonds to the hydrogen being removed while the other halogen atom bonds to the alkane.

In substitution reactions something is bonded to carbon in place of a hydrogen and something else must bond to the hydrogen being removed.

In this and following chapters we will want to write equations describing various reactions. We can show that these reactions fit any member of a homologous series if we can leave the description in general terms rather than using any one exact molecular structure. We will do this by using the symbol R— to represent any alkyl radical. Under that method, any alkane can be represented by R—H (an alkyl radical bonded to hydrogen). The general equation for reaction of Cl_2 with an alkane is then

$$R—H + Cl_2 \longrightarrow R—Cl + HCl$$

Reactions can be written in a general form by using R to represent any radical.

Similar general reactions could be written using F_2, Br_2, or I_2.

There are two other likely variations of the general reaction substituting one bonded group for another. One of these is to have the incoming atom form bonds to both the carbon and to the hydrogen whose bond to the carbon is being broken. That is often called an **insertion reaction** since something is being inserted between two atoms which had been bonded together. The atom (or group) being inserted must be able to form two bonds. The obvious common reactive substance fitting that requirement is oxygen. Although the process of insertion is probably quite complex, oxygen can sometimes be inserted into carbon–hydrogen bonds to form alcohols.

An insertion reaction is a substitution where both of the new bonds are formed to the same atom, which must be something like oxygen which can form two bonds.

The other likely variation is substitution for a carbon–carbon bond instead of a carbon–hydrogen bond. Carbon–carbon bond breaking is sometimes called a displacement reaction. This type of displacement reaction creates new, shorter carbon chains.

Substitution at a carbon–carbon bond is called displacement.

18–1.3 Cracking. Like other substitution reactions, displacement reactions require new bonding for both groups obtained from the broken bond. However, alkyl chain fragments do not necessarily have to add a bond to a new atom to restore stable bonding. They can also lose a hydrogen from a neighboring carbon and form a double bond. Here are some examples of such double bond formations.

Alkyl free radicals can restore stable bonding by losing a hydrogen (or transferring it to a neighboring alkyl radical) and forming a double bond.

$$R—CH_2—CH_2 \cdot \longrightarrow R—CH{=}CH_2 + H\cdot$$
$$R—CH_2—\overset{\cdot}{C}H—CH_2—R' \longrightarrow R—CH{=}CH—CH_2—R' + H\cdot$$
$$R—CH_2—\underset{\cdot}{C}H—CH_2—R' \longrightarrow R—CH_2—CH{=}CH—R' + H\cdot$$

The single dots represent one unshared electron. The fragments with one such unshared electron are called **free radicals.** Free radicals are discussed further in Section 18–3.4. If another unshared electron can be made available on the next atom, a π bond can be set up as shown in Figure 18–1.

Unshared electrons are represented in equations by single dots.

The hydrogen atom released during formation of a double bond from an alkyl radical can form a bond with something else. It could bond to another alkyl radical, as shown in this reaction.

$$R\cdot + H\cdot \longrightarrow R—H$$

It is possible to have a bond broken in an alkane with a double bond forming in one alkyl radical fragment and the released hydrogen atom combining with

Figure 18-1 π bond.

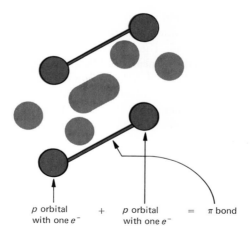

p orbital with one e^- + p orbital with one e^- = π bond

At high temperatures hydrocarbons undergo cracking to form shorter chains (half alkane or H_2 and half alkene) by radical hydrogen transfers.

the other fragment. This process is called **cracking.** It occurs at very high temperatures and is an important process in oil refining.

When a carbon–carbon bond is broken in cracking, the products are two shorter hydrocarbon chains, one an alkane and one an alkene. In our economy the volatile short carbon chain hydrocarbons in gasoline are worth more than the long chain hydrocarbons in thick oils and tars. Cracking long chain hydrocarbons lets us increase the proportion of crude oil usable as gasoline. It also gives us a large supply of alkenes which can be separated and used in chemical reactions, as described in Sections 18–2 and 18–5.

When a carbon–hydrogen bond is broken in cracking, the products must be H_2 and an alkene. H_2 byproduct from cracking is used for various chemical processes, particularly the production of NH_3 for fertilizers by the Haber process, as mentioned in Section 14–2.

18–1.4 Combustion. The one reaction of alkanes with which most of us are familiar is combustion. All hydrocarbons, including alkanes, can be burned in oxygen (usually provided from air) to form CO_2 and H_2O. An example is the reaction $CH_4 + O_2 \longrightarrow CO_2 + 2\,H_2O$. Like other alkane reactions, combustion does not take place until some mechanism gets the process started, and high energies are usually required. However, once combustion starts, enough heat is released in most cases to keep the reaction going by providing a large number of highly energetic molecules for further reaction.

18–2 Reactions of Alkenes

Alkenes are much more reactive than alkanes. Because of their double bond, they can react without breaking anything completely away. Instead, the double bond can become a single bond while another single bond is formed at each of the two carbons. Therefore, it is possible to add two groups to the molecule without taking anything out. The process is called an **addition reaction**.

Halogens can be added across a double bond. The reaction goes so easily that it is used as a simple test for unsaturated hydrocarbons. Bromine, which has an easily visible reddish brown color, can be used up by addition to double bonds. Br_2 (usually as a solution in a nonreactive solvent) is simply shaken with an unknown hydrocarbon at room temperature. If double bonds are present, the bromine color disappears as it is used up by the addition reaction.

Alkenes are quite reactive by addition reactions because no groups need to be broken away to let others bond.

$$R-CH=CH-R + Br_2 \longrightarrow R-\underset{\underset{Br}{|}}{CH}-\underset{\underset{Br}{|}}{CH}-R$$

Most of the commercial addition reactions with halogens are chlorination reactions. Some of the products are pesticides. Since the alkenes become saturated during the addition reactions, the products are sometimes like alkanes—not very reactive. Therefore, chlorinated hydrocarbons are subject to much discussion and criticism as persistent poisons.

Addition to a double bond can occur with different atoms added to the two carbons. Some chlorinated hydrocarbons are produced by adding HCl to double bonds instead of adding Cl_2. When nonidentical groups are added to a double bond, the groups add on in a particular order. We will have some further comments on that when we discuss mechanisms in Section 18–3.

The H and OH groups from an H_2O molecule can also be added across a double bond to form an alcohol. That reaction does not go as readily as addition of halogen molecules, but it is a possible reaction. This reaction will be discussed further in Section 18–3.

Halogens and polar molecules like HCl react readily by addition, whereas H_2O also reacts by addition but less readily.

It is also possible for one atom to form new bonds to both of the carbon atoms of a double bond. Oxygen atoms can do that, forming two new single bonds and a three atom ring. That reaction is carried out commercially (under controlled conditions) with ethene or propene to give products described by

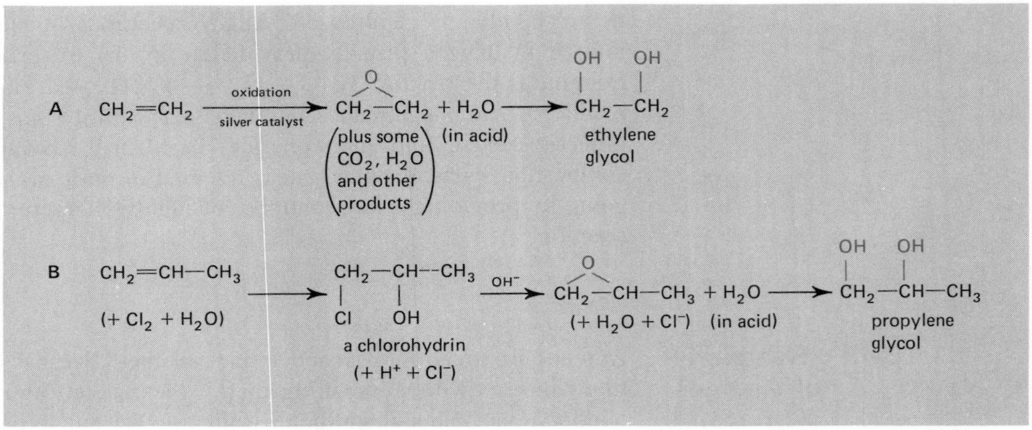

Figure 18–2 Oxidation of alkenes and formation of diols.

(A) A direct (catalyzed) oxidation. (B) Formation of the oxide via the chlorohydrin process.

Oxygen can bond to both carbons of a carbon–carbon double bond and form a ring.

the old-fashioned names ethylene oxide and propylene oxide. When water is added to those compounds, they can be converted to the diols, ethylene glycol (1,2-ethandiol) and propylene glycol (1,2-propandiol), which are the main components of permanent antifreeze. Ethene can be directly oxidized by air in the presence of silver catalyst. The sequence of reactions is shown in Figure 18–2A. The comparable reaction of propene is so inefficient that a different route must be used commercially. This chlorohydrin process (which also works for ethene) involves the steps shown in Figure 18–2B.

18–3 Mechanisms of Reactions

18–3.1 Electrophilic Reagents.

The outstanding feature of double bonds (which are reactive) compared to single bonds (which are much less reactive) is the π **bond.** The π bond gives a concentration of electrons in two regions not along the bond axis, but paralleling the axis, as shown in Figure 18–3. There are no bonded atoms in the directions of those π bond electron concentrations. Therefore, the π bonds are open to attack by potential reactants. The reactivity of double bonds is caused by this openness to attack of the π bond electrons.

The reagents that are most effective at attacking double bonds are the ones that most effectively interact with the electron concentrations of the π bonds. Therefore, the best reagents are the ones with strong attractions for electrons. They are called **electrophilic reagents,** meaning electron loving reagents.

If we consider the addition of H_2O to a double bond, it is obvious that one part of the H_2O is more electrophilic than the other part. Water molecules by themselves have two unshared pairs of electrons on the oxygen. The positive parts of the water (the H atoms) are able to interact with π bond electrons better than the oxygen atoms can, but while bonded to oxygen the hydrogen atoms cannot set up very good interactions with the π bond electrons. The electrons in the unshared pairs and the electrons in the π bond repel each other. The interaction can be improved by ionization of the H_2O to form H^+ (which we know is really H_3O^+ or some larger grouping as described in Section 13–4) and OH^-. The H^+ can then act as the electrophilic reagent and begin the interaction with the double bond.

However, water does not ionize very extensively to H^+ and OH^-. Therefore,

Addition reactions occur by electrophilic (electron loving) reagents attacking the concentration of electrons in the π bond portion of the double bond.

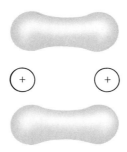

Figure 18–3 Regions of electron concentration in a π bond.

there are very few H⁺ ions in water to act as the electrophilic reagent attacking double bonds. As a result, a typical alkene like ethene can be mixed with water without getting the addition reaction. To get any reaction, we must increase the availability of H⁺ ions to serve as the electrophilic reagent. That can be done by adding concentrated H_2SO_4 to the water. It is also necessary to heat the reaction to speed the addition of H_2O to double bonds so product can be formed at a reasonably good rate. Under those conditions other reactions also occur, leading to carbon char, formed when partial oxidation leaves the carbon portion uncombined, and various other products.

$$CH_2{=}CH_2 + H_2O \xrightarrow[\text{heat}]{H_2SO_4} CH_3{-}CH_2{-}OH$$

(plus carbon char and other products)

Since the H⁺ ion was the attacking electrophilic reagent in the addition of H_2O to an alkene, we should be able to do better with a reactant that is better at supplying H⁺ ions. The donation of H⁺ ions is acidic behavior (see Section 13–4), so we might expect good acids to react well with double bonds.

HCl is a strong acid in water. The same ability to provide H⁺ ions makes it effective at addition to a double bond. Therefore, HCl addition can be carried out at much milder conditions than H_2O addition. HCl addition to ethene can be carried out at conditions where there is no charring or other complicating reactions.

Strong acids like HCl are better sources of H⁺ (an electrophilic reagent) than H_2O is, so H_2O addition needs added acid to start the reaction.

$$CH_2{=}CH_2 + HCl \longrightarrow CH_3{-}CH_2Cl$$

The reaction clearly goes first by addition of an H⁺ ion followed by reaction of the Cl⁻ ion. The half-way point of this reaction is a positive ion, a carbonium ion, which will be discussed in more detail in Section 18–3.2.

Before continuing our discussion of H⁺ additions to a double bond, we should emphasize that H⁺ is not the only electrophilic reagent. Halogen molecules can also be very effective electrophilic reagents. To perform this function they must be polarized into essentially a positive ion and a negative ion. The positive ion is then the electrophilic reagent that attacks a double bond. The reaction can be pictured as follows:

$$CH_2{=}CH_2 + Br_2 \longrightarrow CH_2{=}CH_2 + Br^{+\delta}{-}Br^{-\delta} \longrightarrow \overset{Br}{\underset{|}{CH_2}}{-}\overset{+}{CH_2} + Br^-$$
$$\downarrow$$
$$CH_2Br{-}CH_2Br$$

(where δ is less than the full charge of an electron)

The reaction of halogens with alkenes is interesting because it goes so well. We might expect halogens to react poorly because they certainly are not as good as acids like HCl at providing positive ions. However, halogen molecules are easily polarized into positive and negative parts. What is needed is a region of positive or negative electrical charge to apply electrical forces and cause the halogen molecule to become polarized. An alkene molecule has such a region of charge—the cloud of electron concentration in the π bond.

As the pi bond approaches a halogen molecule, it polarizes the halogen molecule, making the closer halogen atom become positive and the one farther away negative. The positive halogen atom can then react with the π bond, followed by reaction of the negative halogen to give one halogen on each of the two carbons. In our earlier examples of electrophilic reagents reacting with double bonds, we talked about the electrophilic reagent attacking the double bond. In reactions with halogens, we could turn the statement around. The double bond attacks the halogen molecule. The attack works because it makes the halogen electrophilic. All reactions with alkenes involve electrophilic reagents reacting with the π bond.

Addition of halogen molecules goes well because halogens are easily polarized to positive (electrophilic) and negative regions. The polarization can be caused by approach of the alkene π bond electrons to the halogen molecule.

18–3.2 Carbonium Ions. When an H^+ ion adds to a double bond, it forms a positive ion, which we call a carbonium ion. The carbonium ion is thought to exist only briefly and then react to form a stable molecule. It can either react with a negative ion, or it can react with a molecule in a way that releases a positive ion. The reaction with HCl is an example of carbonium ion reaction with a negative ion.

$$CH_2{=}CH_2 + H^+ + Cl^- \longrightarrow CH_3{-}\overset{+}{C}H_2 + Cl^- \longrightarrow CH_3{-}CH_2Cl$$

In the addition of H_2O, we had to get extra H^+ ions from an added acid (H_2SO_4), so OH^- ions are not very available to complete the reaction. Therefore, the carbonium ion reacts with the neutral H_2O molecules to "steal" OH^-, leaving H^+. In this case, the carbonium ion attacks one of the lone pairs of electrons on H_2O, and then an H^+ is lost.

Addition of a positive electrophilic reagent to one carbon of a double bond leaves the positive charge on the other carbon originally in the double bond, making it a carbonium ion which goes on to complete the reaction.

$$CH_2{=}CH_2 + H^+ + H_2O \longrightarrow CH_3{-}\overset{+}{C}H_2 + \overset{H}{\underset{|}{O}}{-}H \longrightarrow$$
$$CH_3{-}CH_2{-}OH + H^+$$

In either case, the fact that the carbonium ion forms first lets the H^+ ion bond to the carbon atom of its choice. Whenever there is a difference, that makes the bonding always occur in the same way—the way that puts the electrophilic reagent in the best place. An example of this specific position bonding is shown in the reaction of HCl with propene.

$$CH_2{=}CH{-}CH_3 + H^+ + Cl^- \longrightarrow CH_3{-}\overset{+}{C}H{-}CH_3 + Cl^- \longrightarrow$$
$$CH_3{-}CHCl{-}CH_3$$

The electrophilic reagent adds to the carbon of the double bond which has the most hydrogens bonded to it.

The hydrogen atom is always bonded to the end carbon and the chlorine atom to the center carbon. Since the electrophilic reagent is often H^+ and there are other hydrogen atoms on the alkene, this position selection can be remembered by a short cliche. The principle (known as Markonikov's rule and stated more properly in organic chemistry books) fits the statement "those that have get more." The H^+ will go on whichever carbon already has the larger number of hydrogen atoms. The rule works because the bonds formed are strongest when the carbon atom has less of its bonds to other carbon atoms.

18-3.3 Catalysts. At this point we can introduce an interesting (and important) insight into the difference in reactivity of polarizable and non-polarizable molecules. Addition of H_2 across a double bond is an important type of reaction. It converts an alkene to an alkane. But H_2 molecules are not as reactive as halogen molecules. The hydrogen can form strong bonds, stronger than those formed by halogen atoms in addition. But the strong bonds are not enough to make the reaction happen. To react with a double bond we must have an electrophilic reagent, and H_2 is very poor as an electrophilic reagent. Hydrogen, with only two electrons, cannot be distorted into positive and negative regions by the approaching alkene π bond as easily as halogen molecules can. Therefore very extreme conditions are needed to cause H_2 to add to double bonds.

But catalysts are known that allow the reaction of H_2 with double bonds at low or moderate temperatures. There are some metals, such as nickel, which interact with H_2. These metals can accept electrons from ligands when they form complexes. We made special note of the ability of nickel to accept electrons, even as neutral nickel atoms instead of positive ions. The formation of $Ni(CO)_4$ from neutral nickel and CO was described in Section 16-3. If the atoms on the surface of nickel metal also accept extra electrons, H_2 molecules may be adsorbed on the surface with their electrons held by the nickel. The hydrogen (with its electrons partly taken by the nickel) must then be in a form somewhat like H^+ ions. Notice that the hydrogen is not being polarized into positive and negative portions as halogens are during reaction. Both hydrogens are being made positive as the nickel pulls their electrons away. In that form, the hydrogen can attack a double bond. Alkene molecules can also be adsorbed on the nickel surface, as the electrons in their π bonds are held by the nickel. If adsorbed hydrogen and adsorbed alkenes come together, they can react.

Hydrogenation (addition of hydrogen to a double bond) does not go readily because the H_2 is neither polar nor very polarizable, but catalysts can set up strong forces that make the hydrogen more reactive.

The above picture is tremendously oversimplified, but it does contain some basic truths. Nickel metal does set up forces with hydrogen. Nickel can adsorb alkenes on its surface. And nickel does cause hydrogen and alkenes to react with each other. Nickel is a **catalyst** for hydrogenation (the addition of hydrogen to double bonds). Catalysts are discussed in Section 12-3.3. In this case the catalyst (nickel) made the hydrogen more electrophilic and therefore more reactive with double bonds.

18-3.4 Free Radicals. When a carbonium ion is formed by addition of an electrophilic reagent to a double bond, both of the electrons from the π bond are taken to form the new bond to the electrophilic reagent. The other carbon is then left with the positive charge and the ability to accept a pair of electrons while forming a bond. You may have noticed that we always wrote our carbonium ion structures showing the positive charge on the former double bond carbon which did *not* get the bond to the electrophilic reagent. Sometimes we get organic reactions which do not go via this ionic route involving electrons in pairs. Instead we may get one electron from the hydrocarbon combining with one electron from a reactant to form a new covalent bond.

If we bring a reactant with one unshared electron up to a double bond, it can react to form a bond to one carbon and leave one unshared electron on the neighboring carbon. We will represent a single unshared electron by a dot, as we did earlier in Section 18–1.3. We then represent a typical reaction of this type, addition of a hydrogen atom to the double bond of ethene, as follows:

$$H\cdot + CH_2{=}CH_2 \longrightarrow CH_3{-}CH_2\cdot$$

The product of the reaction is an alkyl radical, in this case, ethyl radical. But it is not a radical bonded as a side chain on a main hydrocarbon chain. It is unbonded and free. The unshared electron is not used as one of two forming a covalent bond. It is left as an unpaired electron in an orbital that could hold two electrons.

Any unbonded radical with an unpaired electron is called a free radical. The stable molecule NO, discussed in Section 14–2, meets the definition of a free radical. Free radicals are very reactive chemically. The unpaired electron can form a bond to almost anything. In fact that is what the hydrogen atom in our example does when it reacts with ethene. The hydrogen atom is a free radical. It reacts with ethene, even though it cannot form a really stable product by doing so. The product free radical will also react with almost anything. The product of its reaction may be another free radical. The process may continue for many steps, with each step using one free radical and making a new one. This sequence is called a **chain reaction.** Free radical chain reactions are an important type of organic reaction mechanism.

Free radicals can be formed by bond breaking as well as by addition to double bonds. Reactions of alkanes, where a bond must be broken to get a reaction, are examples of free radical formation by bond breaking. We will discuss one of the reactions of alkanes, the substitution reaction with bromine, to illustrate free radical formation by bond breaking and to illustrate a chain reaction.

The equation for the substitution reaction of an alkane with bromine is

$$R{-}H + Br_2 \longrightarrow R{-}Br + HBr$$

The reaction does not go very well. At room temperature the R—H bond does not get broken, so no reaction can take place. However, we can get the reaction at high temperatures. At high temperatures there is enough energy to break bonds and start a reaction. Actually, at the temperatures used for

Free radicals are very reactive and may generate other free radicals when they react.

this reaction there still is not enough energy to break significant numbers of R—H bonds. But there is enough energy to break Br_2 bonds (which are weaker), thus forming bromine atom free radicals. The reaction is

$$Br_2 \longrightarrow 2\ Br\cdot$$

The bromine atoms are able to react with alkane molecules by **abstraction reactions.** They pull off hydrogen atoms, leaving a free radical behind.

$$R{-}H + Br\cdot \longrightarrow HBr + R\cdot$$

The alkyl free radical can then react with Br_2, pulling off one bromine atom.

$$Br_2 + R\cdot \longrightarrow R{-}Br + Br\cdot$$

These last two reactions form a chain reaction cycle which converts R—H and Br_2 to R—Br and HBr. That is the way the substitution reaction takes place.

The free radicals can also undergo other reactions. Some of these, such as the reaction of an alkyl radical to abstract a hydrogen from another alkane, just give back the same group of radicals and molecules. They are called chain transfer reactions. We will choose to ignore the kinds of situations where they have any importance. Much more important to us are the **termination reactions.** Whenever two free radicals get together, they can combine. An example is $R\cdot + Br\cdot \longrightarrow R{-}Br$. Termination reactions use up free radicals.

When an alkane-bromine mixture is heated, free radicals are generated by bond breaking. Reactions that create free radicals where there were none are called **initiation reactions.** The concentration of free radicals then rises until they are being used up by termination reactions at the same rate they are being formed by initiation reactions. We then get the chain reaction going at the rate caused by the equilibrium free radical concentration. At that concentration the initiation and termination reactions are in balance. If we lower the temperature the initiation reaction slows or stops. The termination reactions then use up the remaining free radicals and the whole chain reaction stops. The total set of the main steps and some minor steps in the reaction $RH + Br_2$ are shown in Figure 18–4.

When we understand the free radical chain mechanism, we can understand how to cause the substitution reaction at lower temperatures. All we need

Free radicals may be produced in initiation reactions when bonds are broken by heat or light energy. They react by a chain reaction series of abstraction or addition reactions, and they are removed by termination reactions with other free radicals.

Initiation: Br_2 + energy $\longrightarrow$ $Br\cdot + Br\cdot$

Chain: $Br\cdot + R{-}H \xrightarrow{\text{and}} R{-}Br + H\cdot$ or $Br\cdot + R{-}H \xrightarrow{\text{and}} H{-}Br + R\cdot$

$H\cdot + Br_2 \longrightarrow H{-}Br + Br\cdot$ or $R\cdot + Br_2 \longrightarrow R{-}Br + Br\cdot$

Termination: $Br\cdot + Br\cdot \longrightarrow Br_2$
or $H\cdot + Br\cdot \longrightarrow HBr$
or $H\cdot + H\cdot \longrightarrow H_2$ } each of these requires another molecule to carry off the excess energy

or $R\cdot + Br\cdot \longrightarrow R{-}Br$
or $R\cdot + H\cdot \longrightarrow R{-}H$ } the excess energy can be distributed among the several bonds in these products

Figure 18–4 Chain reaction of bromine with an alkane.

is a low temperature initiation reaction to produce free radicals. Our example of the reaction of bromine with an alkane was chosen because there is a way to initiate the reaction at room temperature. Br_2 molecules can absorb light and use the light energy to break apart into bromine atoms. When a strong light is directed at a bromine–alkane mixture, free radicals (bromine atoms) are generated and the chain reaction begins. However, if the light is turned off, the termination reactions eliminate the free radicals and the reaction stops. Because of its dependence on the light energy, this reaction can literally be turned on and off by a switch. Some of you may get a chance to run this light switch controlled reaction in the laboratory.

We will discuss some commercially important free radical chain reactions in Section 18–5.2, addition polymers.

18–3.5 Electron Shifts. In our discussions of reaction mechanisms we have talked about the use of electrons from one source, such as a π bond, to form another bond. Sometimes organic chemists follow the reaction mechanism by drawing a series of arrows for electrons moving around from one place to another. It is actually an oversimplification to talk about particular

Figure 18–5 Some examples of use of arrow representations of organic reaction mechanisms.

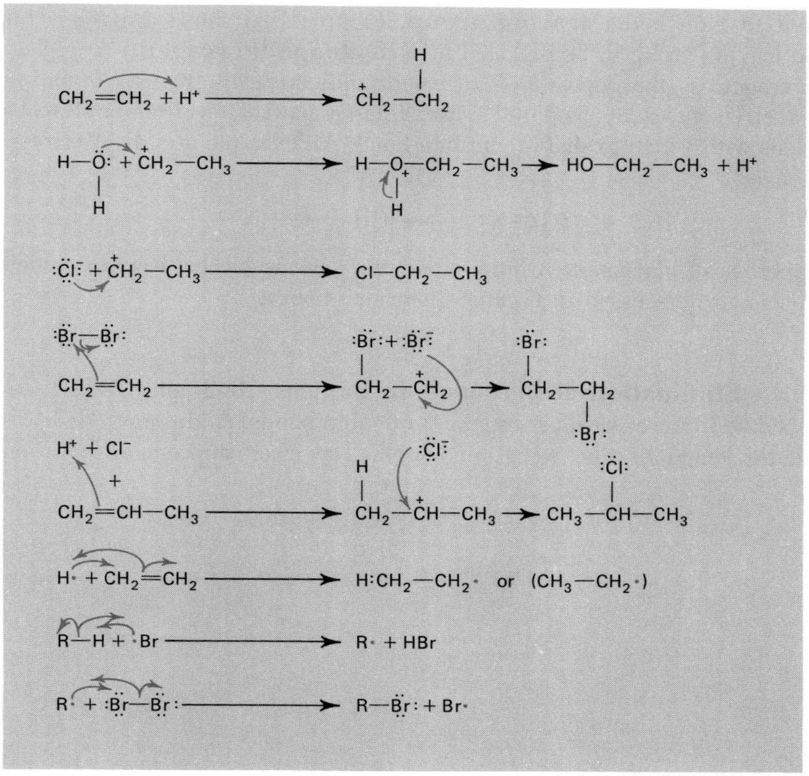

electrons moving from one place to another. Other chemists sometimes make fun of organic chemists for their use of arrows and shoving electrons around. But organic chemists use this model for very sensible reasons. It is fairly simple and it works. Figure 18–5 shows arrow representations for some of the reactions we have described. Single arrows mean a *pair* of electrons moved as shown. A double arrow in opposite directions means one electron from a pair went one way (usually to form a bond with a free radical) and one electron went the other way (to form a free radical at that position). We will continue to use this picture method with arrows whenever it seems useful to explain an important point.

Description of organic reaction mechanisms by using arrows to follow electron shifts is convenient and helpful.

18–4 Reactions Involving Hydrocarbon Derivatives

18–4.1 Displacement Reactions. In Section 18–1 we indicated that substitution of a different group for hydrogen on an alkane is difficult. Substitution for any other group on a saturated hydrocarbon is also difficult because a bond must still be broken in the reactions. We will call the displacement of one nonhydrogen group by another a **displacement reaction,** but it is really just another form of substitution reaction. Other groups may be easier to replace than hydrogen.

One of the conditions that would make a group more reactive than hydrogen is for it to be a better **leaving group.** Figure 18–6 shows two ways for a good leaving group to be replaced by a **nucleophilic reagent.** A nucleophilic reagent is the opposite of an electrophilic reagent. Instead of trying to find electrons, a nucleophilic reagent has a pair of electrons and seeks a nucleus with which to share them in a bond. An example of a displacement reaction is

$$CH_3I + {}^-{:}C{\equiv}N{:} \longrightarrow CH_3{-}C{\equiv}N + I^-$$

Displacement reactions, substitution for a group other than hydrogen, go better if the group being replaced is a good leaving group and the attack is by a nucleophilic reagent.

Numerous displacement reactions and the various mechanisms by which they occur are discussed in organic chemistry books.

18–4.2 Elimination Reactions. In Sections 18–2 and 18–3.1 we described addition of various reagents to double bonds. Addition of HCl was one of the reactions that went fairly well. The exact opposite of addition

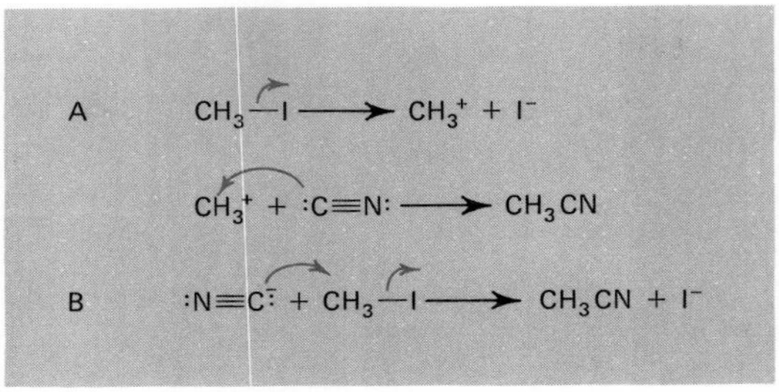

A $CH_3{-}I \longrightarrow CH_3^+ + I^-$

$CH_3^+ + {:}C{\equiv}N{:} \longrightarrow CH_3CN$

B ${:}N{\equiv}C{:}^- + CH_3{-}I \longrightarrow CH_3CN + I^-$

Figure 18–6 Displacement reactions.

In each case a leaving group (the I^- ion) takes the electrons from the bond with it. The attacking reagent (the CN^- ion) is a nucleophilic reagent with a lone pair of electrons available for sharing.

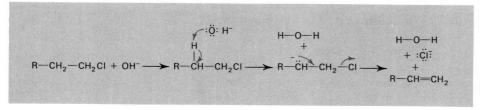

Figure 18-7 Elimination of HCl (in base).

Elimination reactions are the opposite of addition reactions and also occur by an ionic mechanism.

reactions can also occur. A molecule can be formed by taking one group apiece from two adjacent carbon atoms. The two carbon atoms then become double bonded instead of single bonded. That process is called an **elimination reaction.** HCl (which was a good reactant for addition) is one of the molecules that is eliminated (removed during an elimination reaction) quite readily. The reason for that is fairly simple. Elimination, like addition, goes via ionic mechanisms, so polar molecules can take part more readily than nonpolar molecules. An example of an elimination reaction is

$$R-CH_2-CH_2Cl \longrightarrow R-CH=CH_2 + HCl$$

In an elimination reaction a π bond must form between the carbons. Figure 18-7 shows the arrow type picture of this reaction, which usually occurs in the presence of base. Any hydrogen halide molecule can be eliminated fairly readily by this reaction.

Nonpolar molecules like H_2 or halogen molecules do not eliminate as readily as hydrogen halides. That fact can be used to obtain particular products. For example, $CH_2=CHCl$ (common name: vinyl chloride) can be obtained from an elimination reaction on CH_2Cl-CH_2Cl. The HCl eliminates better than Cl_2 or H_2, so $CH_2=CHCl$ is the product. $CH_2=CHCl$ is used extensively to manufacture a commercial polymer, polyvinyl chloride (see Section 18-5.2). Starting from common raw materials, $CH_2=CHCl$ can be prepared in this two step reaction.

$$CH_2=CH_2 + Cl_2 \xrightarrow{\text{addition}} CH_2Cl-CH_2Cl \xrightarrow{\text{elimination}} CH_2=CHCl + HCl$$

Alcohols should also be able to react by elimination to form alkenes. An example would be $CH_3-CH_2OH \longrightarrow CH_2=CH_2 + H_2O$. But removal of a hydrogen from carbon is fairly difficult. Before the extreme conditions needed for elimination are reached, alcohols find another source of hydrogen atoms that are easier to remove. The resulting reaction, condensation, is discussed in Section 18-4.3. Organic chemistry is full of examples like this, where a reaction is possible but there is another even better reaction possible.

In Section 18-3.3 we described how a catalyst could set up forces that would increase the rate of H_2 addition to a double bond. The reverse of that reaction, elimination of H_2, should also be possible. Like the other examples, it occurs under conditions similar to those that cause the opposite reaction. Direct addition of H_2 to double bonds would not occur under reasonable conditions because H_2 is not polar and is not polarizable. Similarly, H_2

elimination does not occur directly at moderate conditions. However, a catalyst that can set up a strong attraction for hydrogen can assist the elimination of H_2.

The catalyst aids elimination of H_2 in much the same way that catalysts assist the addition of H_2. Sometimes the catalyst for reaction in one direction differs from that for reaction in the other direction, although strong forces must be involved in each case. However, in other cases the same catalyst works in either direction. Metals in the nickel family in the periodic table (particularly palladium and platinum) are used commercially for both hydrogenation (addition of H_2) in some cases and dehydrogenation (elimination of H_2) in other cases. Such dual uses are understandable once we recognize the principle that opposite reactions often go via the same intermediate steps.

Dehydrogenation requires a catalyst and works with the same catalysts used for hydrogenation.

18-4.3 Condensation Reactions.

There is an available source on alcohols of hydrogen atoms that are easier to remove than the hydrogens on carbon. The hydrogens on oxygen are easier to remove. Therefore, an H_2O is removed by taking the OH from one alcohol molecule and a hydrogen atom from a neighboring alcohol molecule. During the process the oxygen atom that is not removed forms a bond to the carbon which loses OH. The two carbon chains are combined together. This combination of chains while H_2O is removed is called a **condensation reaction.** In Section 18-5.1 we will show a condensation where another small molecule (HCl) is removed while groups are linked, but most condensations involve removal of H_2O. The product is an ether, as in this reaction

$$CH_3-CH_2-OH + CH_3-CH_2-OH \longrightarrow CH_3-CH_2-O-CH_2-CH_3 + H_2O$$

Condensation reactions, removal of a small molecule such as H_2O while two groups are linked, occur by an ionic mechanism.

Once again, the actual mechanism of the reaction is ionic. Figure 18-8 shows a simplified version of how acid conditions can cause the formation of an ether by condensation. More favorable conditions for ether formation are obtained by reacting the alcohol with H_2SO_4 to form a different, more reactive compound. Figure 18-8 shows how that is done in the actual commercial preparation of diethyl ether for an anasthetic or other uses.

Since the removal of water goes by an ionic mechanism, it works best when OH^- and H^+ can form most readily. Alcohol hydrogens are a slightly better source of H^+ than alkyl hydrogens, but alcohols are still not a good source of H^+. The removal of water would go better with a more acidic compound which could supply H^+ better. The obvious organic chemical to choose for acidity is a carboxylic acid. Condensation reactions between alcohols and carboxylic acids (to form esters) do go more readily than condensation reactions between two alcohol molecules. They can also go by reactions similar to the ether formation reaction shown in Figure 18-8 or by a more direct reaction not requiring the intermediate. It is also possible to eliminate H^+ from one carboxylic acid and OH^- from another carboxylic

Alcohols condense to ethers instead of having elimination reactions because the hydrogen on an alcohol is easier to remove than hydrogen on a carbon, and esters form more easily than ethers by condensation because the hydrogen on a carboxylic acid is easier to remove than hydrogen on an alcohol.

Figure 18-8 Mechanism of ether formation.

The reaction shown is the commercial method for producing ether.

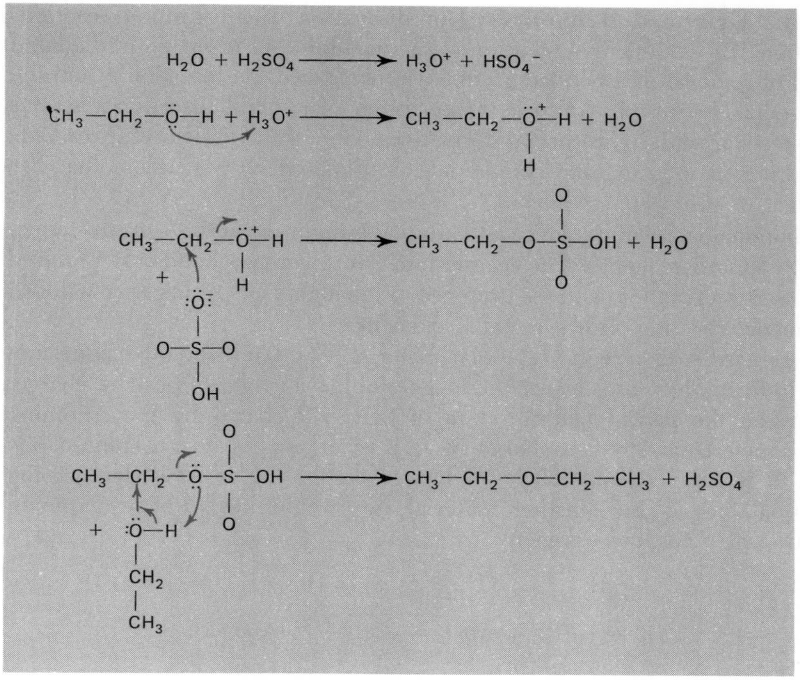

$$H_2O + H_2SO_4 \longrightarrow H_3O^+ + HSO_4^-$$

$$CH_3-CH_2-\ddot{O}-H + H_3O^+ \longrightarrow CH_3-CH_2-\overset{+}{\underset{H}{\ddot{O}}}-H + H_2O$$

$$CH_3-CH_2-\overset{+}{\underset{H}{\ddot{O}}}-H \longrightarrow CH_3-CH_2-O-\underset{O}{\overset{O}{\underset{\|}{S}}}-OH + H_2O$$

(with $:\ddot{O}^-\!-\!S(=O)_2\!-\!OH$ attacking)

$$CH_3-CH_2-O-\underset{O}{\overset{O}{\underset{\|}{S}}}-OH + :\ddot{O}-H\!-\!CH_2\!-\!CH_3 \longrightarrow CH_3-CH_2-O-CH_2-CH_3 + H_2SO_4$$

acid molecule to form an acid anhydride by a direct condensation reaction very similar to the direct formation of esters. Figure 18–9 shows the direct mechanism for acid anhydride formation by condensation in an acid.

Condensation reactions can also be carried out using amines as one reactant. Condensations between amines and carboxylic acids are very important in living things and in commercial reactions like the production of nylon. Nylon is discussed in Section 18–5.1.

Figure 18-9 Acid catalyzed condensation to an acid anhydride.

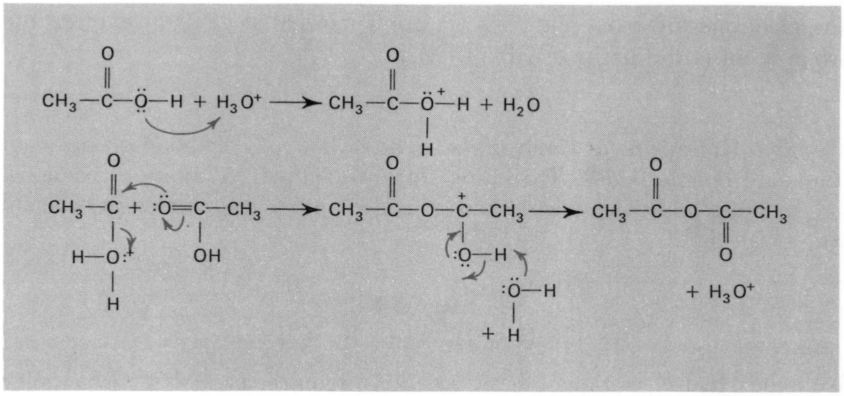

$$CH_3-\underset{\|}{\overset{O}{C}}-\ddot{O}-H + H_3O^+ \longrightarrow CH_3-\underset{\|}{\overset{O}{C}}-\overset{+}{\underset{H}{\ddot{O}}}-H + H_2O$$

$$CH_3-\underset{H-\overset{+}{O}-H}{\overset{O}{\underset{\|}{C}}} + :\ddot{O}=\underset{OH}{\overset{}{C}}-CH_3 \longrightarrow CH_3-\underset{\|}{\overset{O}{C}}-O-\overset{+}{\underset{:\ddot{O}-H}{C}}-CH_3 \longrightarrow CH_3-\underset{\|}{\overset{O}{C}}-O-\underset{\|}{\overset{O}{C}}-CH_3 + H_3O^+$$

[18–4] Reactions Involving Hydrocarbon Derivatives

18-4.4 Grignard Reagents.
Our discussion of substitution reactions in Section 18–1.2 included mention of the possibility of insertion into a bond instead of removal of a bonded group. We mentioned the insertion of oxygen in particular, because it can form the necessary two bonds. Insertion reactions can also occur with hydrocarbon derivatives. One of the most useful reagents for making various organic compounds is prepared by a reaction that has the effect of insertion.

Carbon–halogen bonds can react with magnesium metal to form substances called Grignard reagents. The magnesium can form two bonds, as required in an insertion reaction, and the presence of the highly electronegative halogen atom favors reaction with a metallic element.

Grignard reagents are made by insertion of magnesium into carbon–halogen bonds.

A Grignard reagent is a very polar material. The tendency for magnesium atoms to form Mg^{2+} ions leaves the magnesium as a center of positive charge. Even when the partial neutralization of that $+2$ charge by the adjoining halide ion is considered, a charge of $+1$ is left on the magnesium–halide group. It is balanced out by a -1 charge on the adjoining carbon. Using a chloroalkane as the starting material, we would have this example of formation of a Grignard reagent.

$$\begin{array}{c} H \\ | \\ R-C-Cl \\ | \\ R \end{array} + Mg \longrightarrow \begin{array}{c} H \\ | \\ R-C^-\ ^+MgCl \\ | \\ R \end{array}$$

The carbon with a negative charge is able to react with other organic or inorganic compounds. Because it has an ionic charge, it can react rapidly and effectively by ionic mechanisms. The negatively charged carbon is a nucleophilic reagent and can react by displacement reactions like those described in Section 18–4.1. In each case, the negative carbon of the Grignard reagent will react with the positive part of the molecule it reacts with. Some examples are

The positive charge on magnesium in a Grignard reagent makes the carbon negative and therefore a nucleophilic reagent which reacts with the positive part of other molecules.

$$R-MgX + H_2O \longrightarrow R-H + MgXOH$$
$$R-MgX + CO_2 \longrightarrow R-COO^-\ ^+MgX$$

(The electronegative oxygens on CO_2 are somewhat negative, therefore the carbon atom is the positive part of CO_2.)

18-4.5 Reactions at Carbonyls.
The carbon–oxygen bonds in carbonyl groups are double bonds. Therefore, they are subject to addition reactions. For example, addition of hydrogen to a carbonyl is sometimes possible to form an alcohol.

$$\begin{array}{c} O \\ \| \\ R-C-R \end{array} + H_2 \longrightarrow \begin{array}{c} OH \\ | \\ R-CH-R \end{array}$$

Like other hydrogenations, these reactions require a catalyst or a polar

Carbonyl groups can react by addition reactions and also react with nucleophilic reagents like Grignard reagents because the carbonyl is polarized with the carbon positive.

reactant. However, unlike additions to carbon–carbon double bonds, the reagent does not have to be electrophilic.

Electrons are not shared equally in a carbon–oxygen double bond. The more electronegative oxygen has somewhat more than its share of the electrons. As a result, the oxygen is somewhat negative and the carbon of the carbonyl is somewhat positive. The electron concentration around oxygen can be attacked by an electrophilic reagent, although the reaction may not go as well as attack on an alkene because electrons are harder to remove from oxygen. However, the positive carbonyl carbon is available for a new kind of attack. It can accept electrons from a nucleophilic reagent. When the carbon accepts a pair of electrons, it can release the electrons in its π bond to oxygen and still have the necessary four bonds. In the process, the electronegative oxygen picks up a full negative charge. That single bonded O^- may later pick up an H^+ ion and become an alcohol OH group.

One group of the known nucleophilic reagents are the Grignard reagents we just described in Section 18-4.4. Various specific Grignard reagents can be added to carbonyl groups to make specific desired alcohols. An example of such a synthesis of an alcohol is shown in Figure 18-10.

There are many other possible reactions of carbonyl groups. The important points for us to recognize about carbonyls are (1) that they are reactive because of their double bonds and (2) the internal polarization of the carbonyl bond opens up additional reactions not available to alkenes.

18-4.6 Other Organic Reactions.

We have hardly scratched the surface of the possible organic reactions. The time needed for real coverage

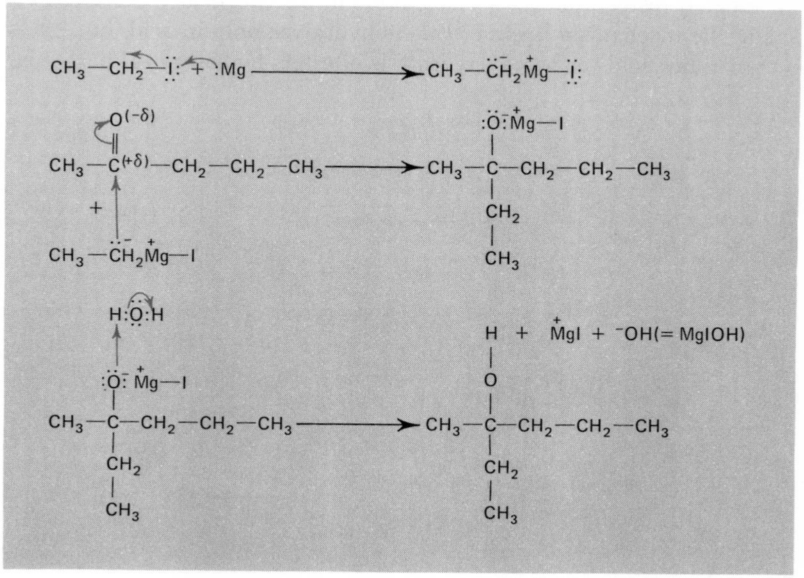

Figure 18-10 Use of a Grignard reagent in an alcohol synthesis.

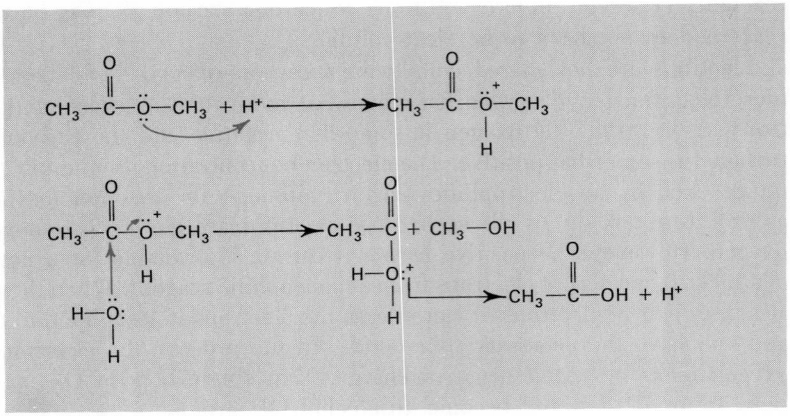

Figure 18–11 Hydrolysis of an ester in acid.

of organic reactions is simply not available in this course. Therefore, we must refer you to an organic chemistry text for more detail. However, we have been able to introduce the most important general features of organic reactions here. Important types of reactions are substitution reactions (including those we called displacement reactions), addition reactions, and elimination reactions (including those we called condensation reactions). Most reactions go by *ionic* mechanisms. Some go by free radical mechanisms. Often there are pairs of reactions that are opposites, like HCl addition and HCl elimination.

The list of opposites could be increased if we listed more reactions. For example, one of the most important reactions we listed was condensation, H_2O removal while units are linked together. Its opposite is one of the most important reactions we have not yet listed, **hydrolysis.** H_2O can be added while an ether or ester linkage is being broken, as shown in Figure 18–11. Once again, the mechanism is ionic. Ethers hydrolyze only in acid, but esters hydrolyze in either acids or bases. Hydrolysis of esters by bases is of particular

Hydrolysis is the opposite of condensation reactions and also goes by an ionic mechanism.

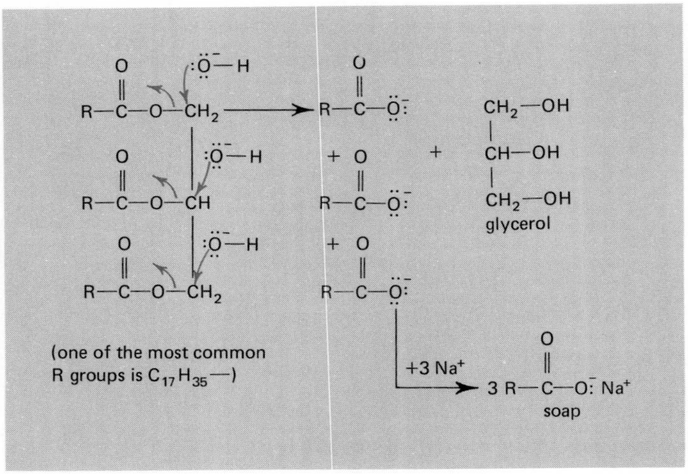

Figure 18–12 Hydrolysis of an ester in base.

The reaction shown is hydrolysis of a common triglyceride fat. The products of this reaction are soap and glycerol.

418   Organic Reactions   [18]

interest because that is the reaction used in manufacture of soap. Figure 18–12 shows the manufacture of soap from animal fats.

That last reaction is a good example of how we have selected the limited number of organic reactions presented here. In addition to bringing up the most general principles, we have concentrated on reactions significant in life and in those processes we use on a large scale commercially.

18–5 Polymers

18–5.1 Nylon, a Condensation Polymer. We will discuss one more general class of reactions, formation of polymers. The reactions are actually just repetitions of types we have already discussed. We will now discuss them as carried out under conditions that produce very large molecules. These polymers meet our two criteria for selecting areas to cover. They are large scale commercial processes, and they are important to life. Polymers play a key role in biochemistry, and there are close resemblances between some biopolymers and some commercial polymers.

In our discussion of organic reactions we mentioned two main processes that put several pieces together into larger groupings. Those processes were condensation reactions and addition reactions. Each process can be used to build up the large molecules we call polymers.

Condensation polymers include the important biopolymers plus nylon and some other commercial polymers.

Condensation reactions link molecules together while H_2O is being removed. Condensation reactions are very common biologically. Life exists in water systems, and the products of water elimination are present as starches, cellulose, fats, proteins, and even the triphosphate unit which is used as an energy transfer agent in living systems. Of those natural materials, cellulose and proteins have the largest molecules. They are natural polymers. They also have interesting properties as fibers and as structural materials. Cellulose is responsible for the strength in wood. Hair, fingernails, and skin are forms of protein.

Nylon and protein are both amine-carboxylic acid condensation polymers.

Protein is held together by amide linkages from condensation reactions between amines and carboxylic acids. Nylon was discovered by a chemist who was investigating similar condensations between amines and carboxylic acids. He found that the product of his reaction could be drawn out into strong, thin fibers somewhat similar to hair. This strong flexible material could be made in large quantities with consistent quality. It led to development of the synthetic fiber industry. Nylon was the first of a series of synthetic fiber products.

The usual commercial nylon is nylon 66. It is produced by the reaction of a six-carbon diacid with a six-carbon diamine as shown.

$$NH_2-CH_2-CH_2-CH_2-CH_2-CH_2-CH_2-NH_2 + HO-\overset{O}{\underset{\|}{C}}-CH_2-CH_2-CH_2-CH_2-\overset{O}{\underset{\|}{C}}-OH$$

$$\downarrow$$

$$-(NH-CH_2-CH_2-CH_2-CH_2-CH_2-CH_2-NH-\overset{O}{\underset{\|}{C}}-CH_2-CH_2-CH_2-CH_2-\overset{O}{\underset{\|}{C}})_x$$

where x is usually between 40 and 100.

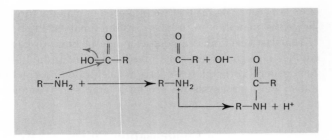

Figure 18–13 A possible mechanism for the condensations in polymerization of nylon.

The condensation between amines and carboxylic acids can occur under the conditions of high temperature and pressure used in the commercial process. The amines are quite basic and able to use their lone pair of electrons on nitrogen as a nucleophilic reagent. The carbon on the carboxylic acid is a somewhat positive carbonyl carbon subject to attack by a nucleophilic reagent, as described in Section 18–4.5. Figure 18–13 shows how the reaction might proceed. Actual reactions between acids and amines require some kind of activation by catalysts, high temperatures, or formation of more reactive intermediates.

The nylon 66 product is useful because it can be drawn into particularly strong strands. That ability depends on details of the structure of the nylon product, and upon the polymer chains being lined up as the nylon is pulled into a fiber.

Actually, other diacids and diamines can also react to form nylons. Some of these actually react more readily than the commercial mixture. However, they do not all give such good strong strands as nylon 66. An example of

Figure 18–14 Steps in formation of a nylon by condensations releasing HCl.

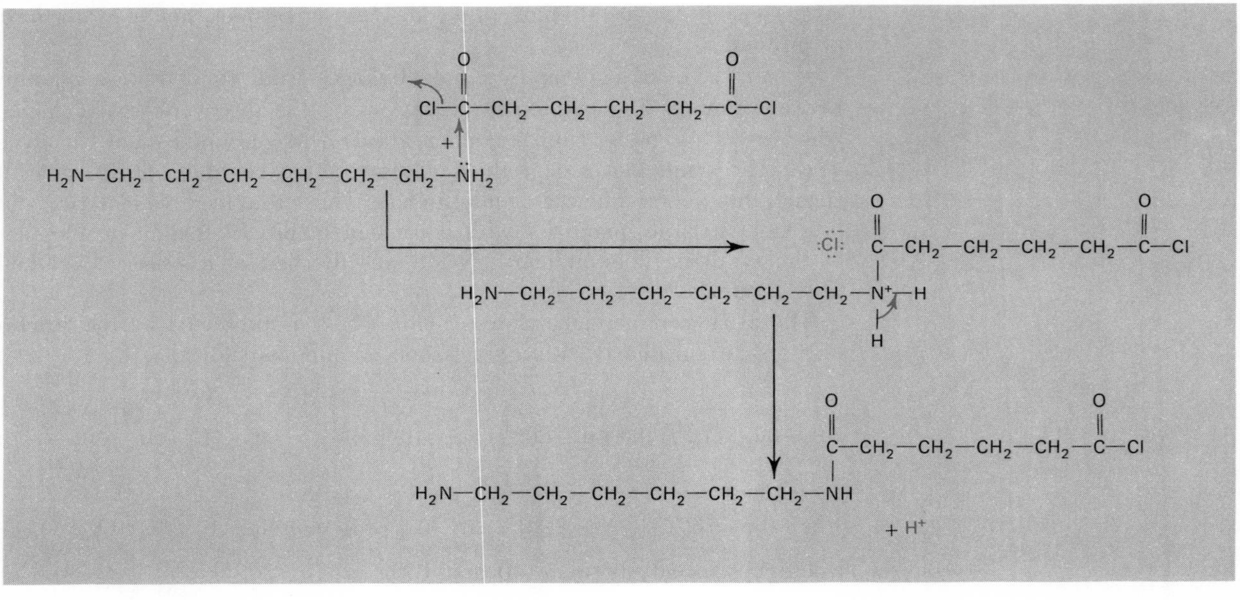

Nylon can be made in several forms and from several sets of reactants.

another nylon is nylon 106. It is made by the reaction of a ten-carbon diacid with a six-carbon diamine.

Nylons can also be produced by condensation reactions that do not release water. Nylon 66 can be made by the reaction of the six-carbon diacid chloride with the six-carbon diamine. (The diacid chloride is a derivative in which the OH groups of the diacid have been replaced by chlorine atoms.) HCl is released in that reaction, as shown in Figure 18–14. Cl^- is a better leaving group than OH^-, so attack on the carbonyl carbon by the nitrogen lone pair is more effective. This reaction is not commercially practical because the acid chloride is more expensive than the carboxylic acid. However, it is an interesting laboratory experiment since it can be done without the elaborate high temperature, high pressure equipment required for the commercial process.

Some of the other commercial synthetic fibers are also condensation polymers. Examples are Dacron and Orlon.

Addition polymers like polyethylene, polypropylene, and polyvinyl chloride are formed by free radical addition chain polymerization.

18–5.2 Addition Polymers. The other main process we mentioned that links organic molecules together is addition. Addition reactions are used to make polymers by free radical chain reactions. An example is the formation of polyethylene. These reactions need a source of free radicals for an initiation step. They are usually carried out in the presence of catalysts. If we presume initiation of the reaction by hydrogen atoms, we get the reaction sequence shown in Figure 18–15.

Addition chain polymerization can occur at any multiple bond. If there are groups other than hydrogen attached to the double bond, the product has somewhat different chains. That causes somewhat different behavior of the plastic polymer product. An example is polypropylene, formed by chain polymerization of propene in the reactions shown in Figure 18–16.

Polymers differ in properties because of their differences in molecular structure.

The CH_3 groups sticking out from the main chain give polypropylene noticeably different properties from polyethylene. We can grossly oversimplify the difference by noting that the CH_3 groups might become entangled with neighboring chains, thus making the molecules harder to pull apart. Polypropylene makes much stronger (harder to pull apart) ropes than polyethylene does. The real effect of the side chains is to force a different alignment (crystal structure) of neighboring chains.

Figure 18–15 An addition chain polymerization.

In each step a free radical reacts with an ethene molecule to form a new, larger free radical. The process continues with similar reactions to form long chain free radicals. A very long carbon chain forms before the process is stopped by some reaction that converts the free radical to a stable molecule.

$$H\cdot + CH_2{=}CH_2 \longrightarrow CH_3{-}CH_2\cdot$$
$$CH_3{-}CH_2\cdot + CH_2{=}CH_2 \longrightarrow CH_3{-}CH_2{-}CH_2{-}CH_2\cdot$$
$$CH_3{-}CH_2{-}CH_2{-}CH_2\cdot + CH_2{=}CH_2 \longrightarrow CH_3{-}CH_2{-}CH_2{-}CH_2{-}CH_2{-}CH_2\cdot$$
$$CH_3{-}CH_2{-}CH_2{-}CH_2{-}CH_2{-}CH_2\cdot + CH_2{=}CH_2 \longrightarrow CH_3{-}CH_2{-}CH_2{-}CH_2{-}CH_2{-}CH_2{-}CH_2{-}CH_2\cdot$$

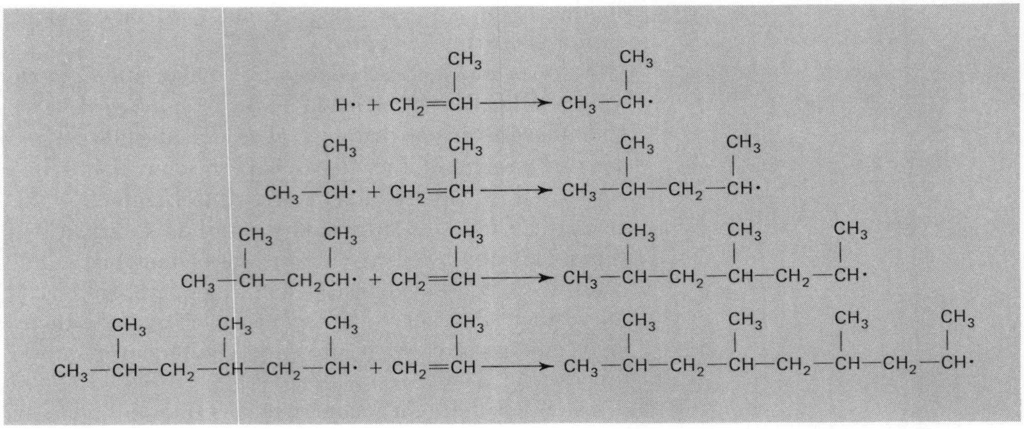

$$H\cdot + CH_2{=}CH(CH_3) \longrightarrow CH_3{-}CH(CH_3)\cdot$$

$$CH_3{-}CH(CH_3)\cdot + CH_2{=}CH(CH_3) \longrightarrow CH_3{-}CH(CH_3){-}CH_2{-}CH(CH_3)\cdot$$

$$CH_3{-}CH(CH_3){-}CH_2CH(CH_3)\cdot + CH_2{=}CH(CH_3) \longrightarrow CH_3{-}CH(CH_3){-}CH_2{-}CH(CH_3){-}CH_2{-}CH(CH_3)\cdot$$

$$CH_3{-}CH(CH_3){-}CH_2{-}CH(CH_3){-}CH_2{-}CH(CH_3)\cdot + CH_2{=}CH(CH_3) \longrightarrow CH_3{-}CH(CH_3){-}CH_2{-}CH(CH_3){-}CH_2{-}CH(CH_3){-}CH_2{-}CH(CH_3)\cdot$$

Figure 18–16 Chain polymerization of propene.

Addition chain reactions across the double bonds leave one methyl side chain for each two carbons in the main polymer chain. The process continues with similar reactions to form long chain free radicals, which are eventually terminated to give long carbon chain polymers.

Addition chain polymerization can also occur at carbon–carbon double bonds of hydrocarbon derivatives. Chloroethene (old fashioned name vinyl chloride) is polymerized to the widely used plastic polyvinyl chloride as shown in Figure 18–17. Once again, the differences in the polymer chain cause differences in the behavior of the plastic. Polyvinyl chloride fits some uses, such as automobile seat covers, better than polyethylene.

In some cases addition reactions occur across a pair of double bonds in a 1,3-diene instead of across one double bond. This is called 1,4 addition because it forms bonds on the first and fourth carbon atoms. 1,4 addition chain polymerization occurs in the formation of natural rubber, as shown in Figure 18–18. The raw material, 2-methyl-1,3-butadiene (old name: isoprene), forces the bonding groups to approach from the side away from the bulky CH_3 group. There is originally a resonance of π bonding among the four carbons of the butadiene unit. As bonds are formed at each end, the two double bonds are reduced to one double bond located between the center

Addition polymerization can also occur by 1,4 addition to 1,3-dienes.

Figure 18–17 Formation of polyvinyl chloride.

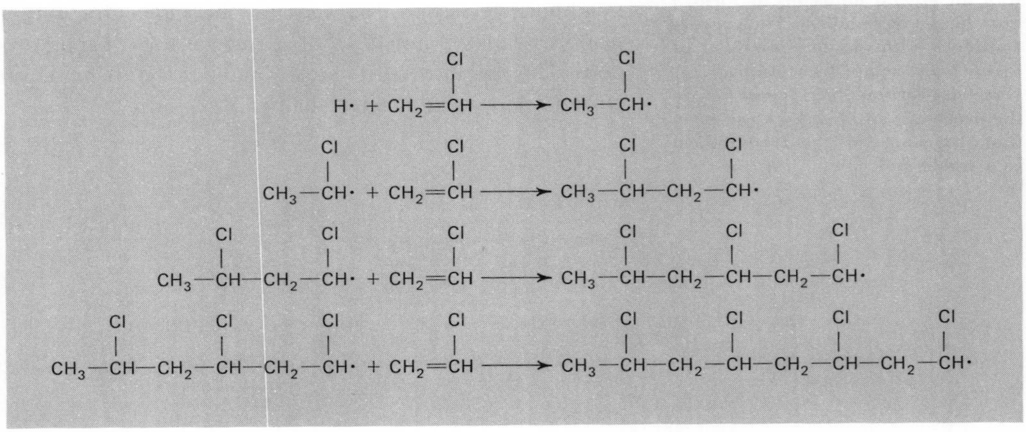

$$H\cdot + CH_2{=}CH(Cl) \longrightarrow CH_3{-}CH(Cl)\cdot$$

$$CH_3{-}CH(Cl)\cdot + CH_2{=}CH(Cl) \longrightarrow CH_3{-}CH(Cl){-}CH_2{-}CH(Cl)\cdot$$

$$CH_3{-}CH(Cl){-}CH_2{-}CH(Cl)\cdot + CH_2{=}CH(Cl) \longrightarrow CH_3{-}CH(Cl){-}CH_2{-}CH(Cl){-}CH_2{-}CH(Cl)\cdot$$

$$CH_3{-}CH(Cl){-}CH_2{-}CH(Cl){-}CH_2{-}CH(Cl)\cdot + CH_2{=}CH(Cl) \longrightarrow CH_3{-}CH(Cl){-}CH_2{-}CH(Cl){-}CH_2{-}CH(Cl){-}CH_2{-}CH(Cl)\cdot$$

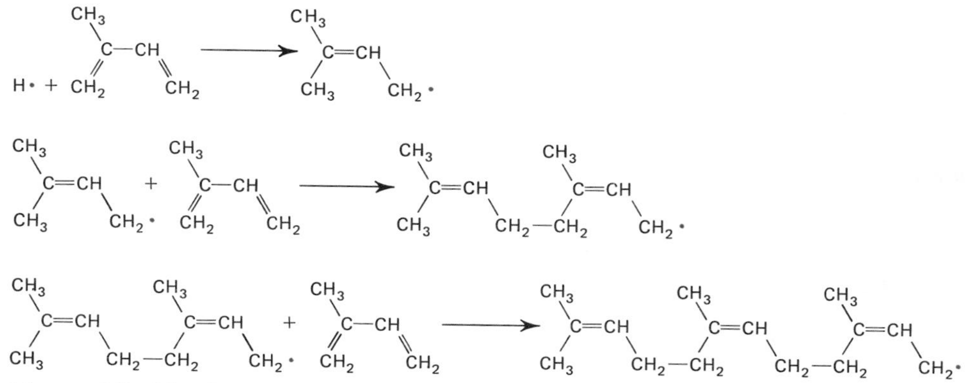

Figure 18–18 Formation of natural rubber.

All of the double bonds in the product polymer have the main polymer chain entering and leaving the bond on the same side—a *cis* configuration.

Natural rubber is a 1,4 addition polymer with *cis* double bonds. *cis*-polybutadiene is a synthetic rubber which also avoids the nonelastic *trans* arrangement.

two carbons. That 2-butene type arrangement is capable of existing as *cis* or *trans* isomers. Since both new bonds are formed by groups coming from the same side (the side away from the CH_3 group), only the *cis* isomer is formed. Natural rubber is a polymer with *cis* double bonds every fourth carbon along its chains. That *cis* arrangement is necessary for the properties of rubber. The *trans* arrangements of the double bonds can be formed under some conditions. The product, called gutta-percha, is described as a horny substance, not elastic.

We now know how to use certain catalysts to make 1,3-butadiene polymerize into the *cis* pattern like natural rubber. The product, called *cis*-polybutadiene or by trade names like Diene or Tuf-syn, is actually superior to the natural rubber containing CH_3 side groups. Figure 18–19 shows the formation of *cis*-polybutadiene.

In the above examples we have shown that many different polymers can be formed, that the two main classes of polymers are condensation polymers and addition polymers, that the characteristics of polymers are often useful and important, and that the characteristics depend on the structure of the polymer chains. Additional discussions on natural polymers will be included in later sections of this book.

Figure 18–19 Formation of *cis*-polybutadiene.

A catalyst is required to direct the bonding into the *cis* configuration.

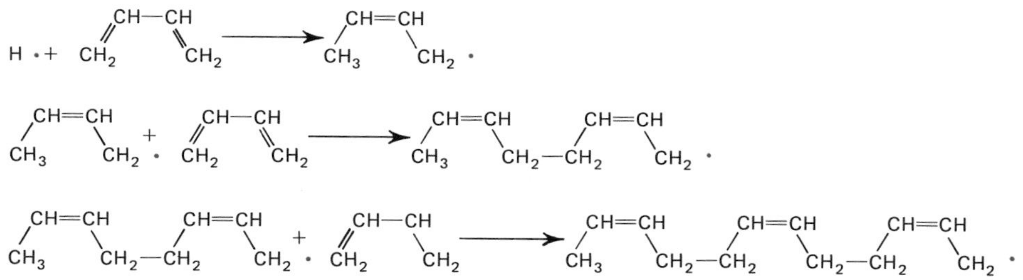

(catalyst needed to maintain the cis arrangement)

[18–5] Polymers 423

Skills Expected at This Point

1. You should be able to recognize and describe the classes of reactions we called substitution reactions, insertion reactions, addition reactions, elimination reactions, condensation reactions, and displacement reactions.
2. You should be able to describe the low reactivity of alkanes, the much higher reactivity of alkenes, and be able to represent reactions in a generalized way by using R to represent any radical, and you should understand that symbolism when used by others.
3. You should be able to define electrophilic reagent and nucleophilic reagent and explain how polar molecules make good electrophiles and react effectively in addition reactions.
4. You should be able to state that halogens are effective in addition reactions and explain that their polarizability is the cause of that reactivity.
5. You should be able to identify and describe carbonium ions and free radicals and recognize the single dot symbol used to represent the unshared electron of a free radical.
6. You should be able to write equations for the formation of ethers and the formation of esters by condensation reactions.
7. You should be able to describe carbonyls as reactive because of the double bond and able to react in ways not possible for alkenes because of the polar carbonyl bond.
8. You should be able to describe the importance of polymers, list condensation polymers and addition polymers as the main classes of polymers, and describe the preparations of nylon and polyethylene.

Exercises

1. Why are alkanes so unreactive?
2. What characteristic is needed for a reagent to react readily with alkenes?
3. How does the reaction of alkenes with halogens (like Br_2) differ from the reaction of alkenes with hydrogen halides (like HBr)?
4. What is a carbonium ion? What is a free radical?
5. Describe the role of a catalyst in hydrogenation of alkenes.
6. The reaction of H_2 with Cl_2 to form HCl occurs as a free radical chain reaction. Compare the reaction to the free radical chain reactions described in this chapter and list your conclusions about the following.
    (a) How could the $H_2 + Cl_2$ reaction be initiated?
    (b) What are the equations for the chain reaction steps which produce HCl?
    (c) What termination reactions could occur in the $H_2 + Cl_2$ reaction?
7. Write the equation for an HCl elimination reaction.
8. Write equations for condensation reactions to form (a) an ether, (b) an ester, (c) a polymer.
9. Devise a series of reactions to make the alcohol 3-ethyl-3-pentanol from the following starting materials: H_2O, ethene, HCl, magnesium, and 3-pentanone. (*Hint:* Make and use a Grignard reagent.)

10. Which would you expect to undergo hydrolysis in acid solution more easily, an ether or an ester? Why?
11. Synthetic rubber is a hydrocarbon polymer made by addition chain polymerization. Polyethylene is also a hydrocarbon polymer made by addition chain polymerization, but it is not like rubber. What feature of rubber does polyethylene lack in its polymerization process?

Test Yourself

1. Which of the following is an *addition* polymer?
   (a) rubber
   (b) nylon
   (c) protein
   (d) starch
   (e) none of these

2. Which of the following lists shows the structures of the raw materials used commercially to make nylon 66?

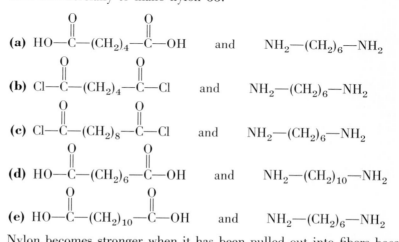

3. Nylon becomes stronger when it has been pulled out into fibers because
   (a) additional amide type bonds form making larger molecules.
   (b) chains are lined up so that covalent bonds in the chains hold the fiber together.
   (c) the chains become more entangled around themselves.
   (d) fibers are the only arrangement in which nylon can solidify.
   (e) dyes are added to the fibers and serve as a glue to hold it together.

4. Which answer best finishes each sentence?
   Ethene and propene can each be polymerized to form long chain hydrocarbon polymers. The main difference between the products (polyethylene from ethene, polypropylene from propene) is (1)
   This difference leads to the result that polypropylene is (2)
   (a) (1) less frequent double bonds in polypropylene (2) less reactive
   (b) (1) more cross linkages in polyethylene (2) weaker
   (c) (1) rings in the polypropylene structures (2) softer
   (d) (1) more atoms in each polypropylene chain (2) harder
   (e) (1) CH_3 side chains on the polypropylene chain (2) stronger

5. Which of the following shows the kind of linkage present in nylon and proteins?

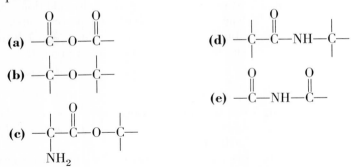

6. Alkenes are more reactive than alkanes because
   (a) electrophilic reagents can attack π bond electrons.
   (b) nucleophilic reagents can displace good leaving groups.
   (c) isomerism gives twice as many possible products.
   (d) the carbon–carbon bonds are weaker.
   (e) none of these.

7. Both natural and synthetic rubber are characterized by
   (a) being long chain alkanes.
   (b) peptide bonds.
   (c) formation by condensation polymerization.
   (d) ring structures.
   (e) *cis* double bonds in the chain.

19 Biological Building Blocks

19–1 Biopolymers

In Section 18–5.1 we noted that polymers play a key role in biochemistry and that there are close resemblances between some biopolymers and some commercial polymers. Condensation reactions lead to many biologically important linkages. We began our discussion of polymers with nylon, a material patterned after the bonding in the natural polymers we call proteins. The formation of nylon was a fairly simple example of a condensation polymer.

The important biopolymers are condensation polymers.

The important biopolymers are condensation polymers. We are going to begin our discussion of them with proteins, the materials bonded with the same kind of linkages as nylon. But we are going to find that proteins are not simple examples of condensation polymers. They will be discussed first because their complexity and variety make proteins a logical and necessary factor in the discussion of other important biopolymers. Some other biopolymers are important because they are involved in making proteins. Still other biopolymers are important energy reservoirs, which are made and used up by reactions involving proteins. All of these other important biopolymers are also condensation polymers.

Proteins have unique importance which depends on the reactive groups present, their arrangement, and the polymer structure.

Since proteins are so central to our discussion of biopolymers, we must give some thought to the factors that make proteins unique in nature. Important factors will include the reactive groups present on the component parts of the proteins, the arrangement of the groups, and the structure of the biopolymers that form. The structure of the biopolymers will be discussed at several different levels of complexity, all of which are important.

19–2 Reactivity of Organic Acids and Amines

The condensation reaction in proteins (and in nylon) is between an amine grouping ($-NH_2$) and a carboxylic acid grouping ($-COOH$). In Section 18–5.1 we mentioned that production of nylon by direct reaction of an amine with a carboxylic acid requires high temperatures and pressures. We also noted that nylon could be formed under more gentle conditions if an acid chloride was substituted for the carboxylic acid. The acid chloride is equivalent to the acid (the same segment is incorporated into the polymer), but the acid chloride is a more reactive form. The acid chloride amounted to an activated carboxylic acid. The reaction between amines and carboxylic acids always requires some type of activation of the carboxylic acid group, either by high temperatures or by some chemical activation. In biological systems high temperatures cannot be used, so the carboxylic acid must be activated by other means. Synthesis of proteins is carried out by reactions utilizing enzyme catalysts and activated carboxylic acids.

An amine has a lone pair of electrons on the nitrogen which could be used to form another bond if something else will accept a share of the electron pair. This ability to act as an electron pair donor in forming a coordinate covalent bond makes the amine a **Lewis base** (see Section 13–4). Lewis bases can act as ligands in complex ion formation (see Section 16–3) or as nucleophilic reagents in organic reactions (see Section 18–4). In water (the liquid surrounding proteins in living systems) an amine may pick up an H^+ ion and form a positive ion.

Carboxylic acids also react readily to form ions. The carboxylic acid can donate an H^+ ion. In this process the H^+ must be bonded to electrons on another molecule. Therefore, the carboxylic acid acts as an electrophilic reagent when it donates H^+. The negative ion left behind when H^+ is removed from a carboxylic acid is able to act as an electron pair donor and is therefore a Lewis base and a nucleophilic reagent.

Amine and carboxylic acid groups provide reactive sites.

19–3 Amino Acids

Nylon is usually made by reacting one compound, which has two amine groups, with another compound, which has two carboxylic acid groups. Proteins are made by reacting compounds in which each molecule contains both an amine group and an acid group. These compounds are called **amino acids.**

Many amino acids are possible, but only a few are biologically important on Earth. All of the biologically important amino acids share three key characteristics.

First, the important amino acids have the amine and acid groups as *close to each other* as possible. The acidic and basic groups affect each other strongly, and the molecules with the amine and the carboxyl groups on the same carbon are the most reactive and are the most biologically significant molecules. These are sometimes called **α-amino acids** (based on an old fashioned naming system).

Second, the important amino acids have a hydrogen atom bonded to the carbon that has the amine and the carboxyl groups. Hydrogen atoms are small. Therefore they let other molecules approach the carbon closely. That increases the chance of reaction at the amine or carboxyl groups. Once again we see that the most reactive molecules are the biologically significant ones.

Third, the important amino acids all have the same order of arrangement of —NH_2, —COOH, —H, and the other group (which we represent by —R). This point is not as easily shown as the first two points. It is also not as

The biologically important amino acids have an amine, a carboxyl group, and a hydrogen atom bonded to the same carbon.

easily explained. We will begin by setting up a diagram to show the arrangement that occurs. (At the end of Section 19-4 we note that these compounds may actually occur in the form called a zwitterion which has —NH_3^+, —COO^-, —H, and —R as the four groups. We are ignoring that internal self-ionization in the discussion and formulas we present.)

Figure 19-1 shows the arrangement of groups in an amino acid. Unless the R group is H, NH_2 or COOH (for convenience the bonds will no longer be shown), there is optical isomerism possible for the amino acid. The carbon bonded to NH_2, COOH, H, and R has four nonidentical groups bonded to it. Therefore, there are two optical isomers of each amino acid corresponding to the two arrangements shown in Figure 19-1. However, only one of the two isomers occurs in proteins, and it is the same order arrangement in each amino acid. That one isomeric arrangement appears to have been selected early in the development of life on Earth and consistently reproduced in later living systems.

There is a standard way of writing and naming optical isomers. The molecule is put in the position with carbon–carbon bonds bending backward above and below the carbon which has four different groups. The other two bonds then point left and right and also stick forward toward the viewer.

Figure 19-1 The optical isomers of an amino acid.

The top two figures are drawn according to the system used for setting up a shorthand for optical isomers. The view is from the side toward which the —H and —NH_2 groups point and is from halfway between the H and NH_2 groups (bonds not written for convenience). Therefore, the H and NH_2 groups point left and right and somewhat toward the viewer whereas the other two bonds point up and down and somewhat away from the viewer. The figures are repeated with the view from another direction such that the COOH groups are pointing straight up and the other three bonds form a "three legged stool" arrangement. The compound shown here on the left (the one with the NH_2 group pointed left) is the L-amino acid. The compound shown here on the right is the D-amino acid.

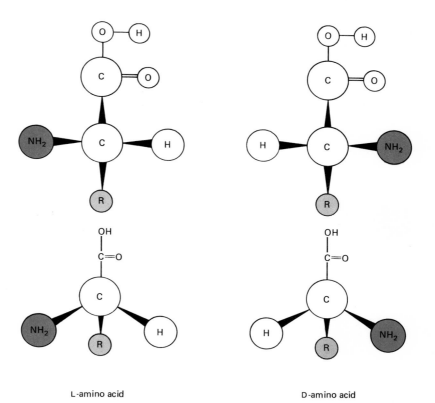

L-amino acid   D-amino acid

[19-3] Amino Acids 429

If there is a COOH group or carbonyl (C=O) group on the carbon, it is put in position above (and behind) the central carbon if possible. Figure 19–2 shows this arrangement for the two optical isomers of lactic acid, a fairly simple (and biologically important) compound with optical isomerism. If we ignore the bending of bonds away from or toward the viewer, we can simplify the picture into a flat sketch showing groups above, below, to the left, and to the right of the central carbon. These two-dimensional sketches are used widely by scientists to show optical isomers. We have to remember that the groups above and below the central carbon are the ones angling away from us and the groups to the left and right are angling toward us. Most scientists find this convention much easier to learn than three-dimensional drawing.

The left, right, above, and below convention has another advantage besides saving readers from scientists' efforts at drawing. They let us set up a naming system. When one of the groups on the carbon is a hydrogen, only one of the groups on the left or right is a larger (not hydrogen) group. In that case we can name the isomer by stating whether the large (not hydrogen) group is to the left or right as we look at the molecule in the conventional way. This is done by using the capital letter L if it is to the left and capital D if it is to the right. The letters are taken from the Latin words *levo* (meaning left) and *dextro* (meaning right). As shown in Figure 19–2, L-lactic acid is the isomer with the OH group to the left and D-lactic acid is the isomer with the OH group to the right.

Amino acids can also be named using the D and L symbols. In this case, an L-amino acid has the NH_2 group to the left and a D-amino acid has the NH_2 group to the right. Figure 19–3 shows the arrangements of D- and L-alanine (note how similar they are to D- and L-lactic acid) and of D and L isomers of a generalized amino acid where R represents any radical grouping.

Optical isomers can be written and named in a standard way, and all the amino acids used by higher life forms except one (glycine with no optical isomerism) are L-amino acids.

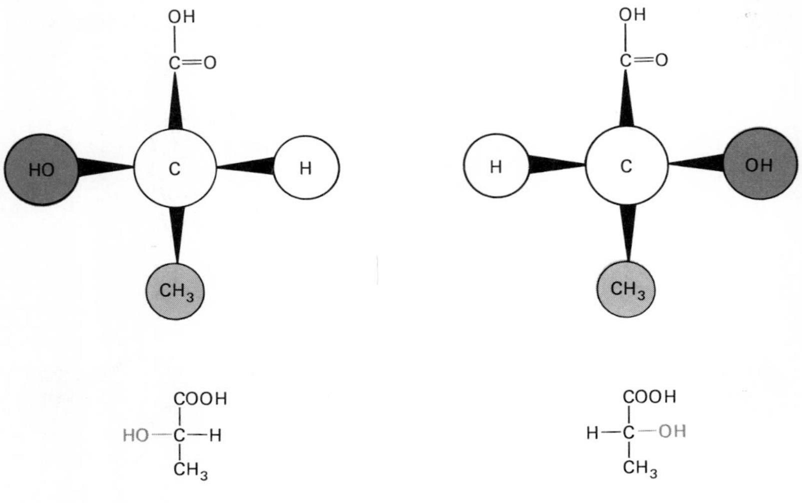

Figure 19–2 D- and L-lactic acid.

The compound shown on the left (with the OH group pointing left) is L-lactic acid. The compound shown on the right is D-lactic acid. In each case the compounds are shown once (above) in the perspective drawing with the same point of view as the upper drawings in Figure 19–1. Below those perspective drawings the same two compounds are shown in the two-dimensional shorthand form used to represent that same point of view.

L-lactic acid

D-lactic acid

Figure 19-3 D- and L-amino acids.

The formulas shown here are in the shorthand forms using the standard approach, the same approach as was used in the lower figures in Figure 19-2. L-Alanine has the NH$_2$ group to the left when written in this standard form and D-alanine has the NH$_2$ group to the right. Every other amino acid follows the same pattern with the L isomer, which occurs in life on Earth, having the NH$_2$ group to the left when written in standard form. The generalized D- and L-amino acids allow any singly bonding group to be used as R except H, NH$_2$, or COOH. If one of those three is R, there are no optical isomers because there are not four different groups on the same carbon atom. The only common amino acid without optical isomers is glycine, where the R group is H.

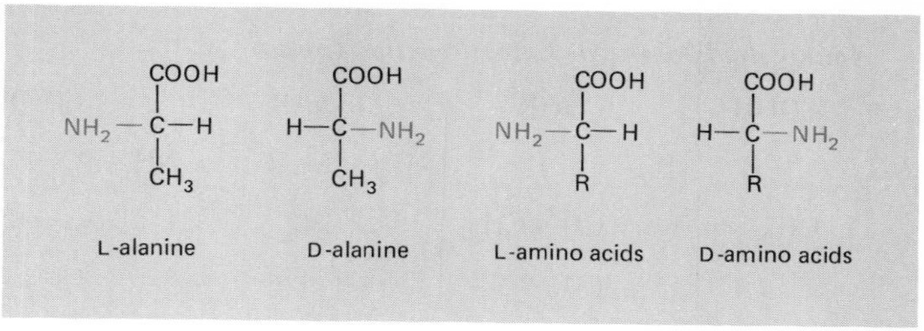

On Earth, only the **L-forms** of amino acids are important. If amino acids are chemically synthesized, we get equal amounts of D and L forms. D-Amino acids are found in limited uses in a few bacteria. But all higher forms of life on Earth make and use only L-amino acids plus glycine, which has no optical isomers because the R group is H.

There have been many wild and ridiculous science fiction stories about life on other planets based on fluorine instead of oxygen or other scientifically unsound ideas. But there is one very sensible possible variation of life on other planets. It could be based on D-amino acids instead of L-amino acids. There seems to be no compelling reason for the natural selection of L-amino acids on Earth other than the fact it happened. Life here evolved using L-amino acids. All later life continued to use L-amino acids and make more L-amino acids. If the first amino acid molecules used in life had happened to be a D-amino acid, the whole pattern might have evolved using only D-amino acids. If a human being was marooned on a D-amino acid planet, he would starve of protein deficiency even though he ate plenty of protein-rich foods.

19-4 Structure and Reactivity of Naturally Important Amino Acids

Amino acids form reactive ions in acidic or basic solutions and may even form internally ionic zwitterions in neutral solutions.

There are 20 amino acids used in making natural proteins. Table 19-1 lists these 20 plus one other, which is essentially formed within the protein polymer in some cases. We will group these amino acids according to the reactive groups they contain. We will also list their names and some abbreviations for their names. You will not want to memorize all this information, but it will serve as a useful reference to check when differences in the behavior of amino acids are discussed later. You can use the name or abbreviation to look up the structure, and that can explain the behavior.

These amino acids contain a number of very reactive groups, but the most reactive are the adjacent amine and carboxylic groups. We will make special note here of two of the ways in which these groups can be reactive.

First, the amine and carboxyl groups are very polar and easily form ions. In acid, the NH$_2$ picks up H$^+$ making the amino acid into a positive ion. In base, the carboxyl group loses H$^+$ making the amino acid into a negative ion. Even in neutral solution the amino acid may be ionic. The carboxyl group can lose H$^+$ by transferring it to the basic NH$_2$ group. The result is a neutral amino acid molecule that has internally a negative "ion" in one part of the

Table 19-1
Amino Acids and Substituent Groups

Names, abbreviations, and structures are given for 21 important amino acids.

Amino Acids with No Extra Reactive Groups

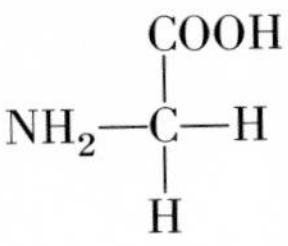

$$\begin{array}{c} \text{COOH} \\ | \\ \text{NH}_2-\text{C}-\text{H} \\ | \\ \text{H} \end{array}$$
glycine (Gly)

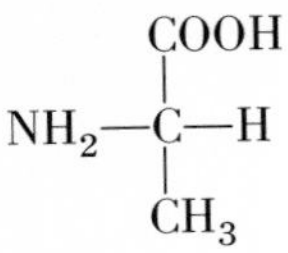

$$\begin{array}{c} \text{COOH} \\ | \\ \text{NH}_2-\text{C}-\text{H} \\ | \\ \text{CH}_3 \end{array}$$
alanine (Ala)

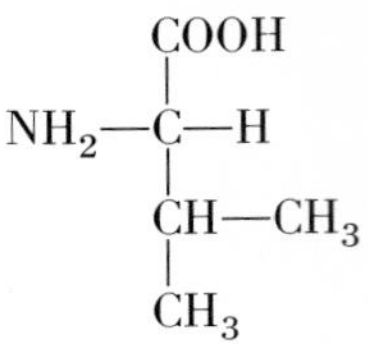

$$\begin{array}{c} \text{COOH} \\ | \\ \text{NH}_2-\text{C}-\text{H} \\ | \\ \text{CH}-\text{CH}_3 \\ | \\ \text{CH}_3 \end{array}$$
valine (Val)

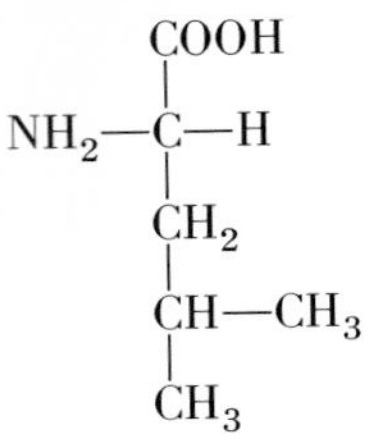

$$\begin{array}{c} \text{COOH} \\ | \\ \text{NH}_2-\text{C}-\text{H} \\ | \\ \text{CH}_2 \\ | \\ \text{CH}-\text{CH}_3 \\ | \\ \text{CH}_3 \end{array}$$
leucine (Leu)

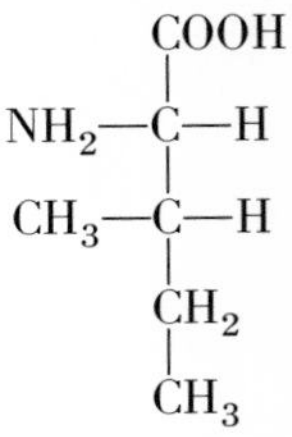

$$\begin{array}{c} \text{COOH} \\ | \\ \text{NH}_2-\text{C}-\text{H} \\ | \\ \text{CH}_3-\text{C}-\text{H} \\ | \\ \text{CH}_2 \\ | \\ \text{CH}_3 \end{array}$$
isoleucine (Ileu)

Amino Acids with an Extra Reactive Group

Extra OH group

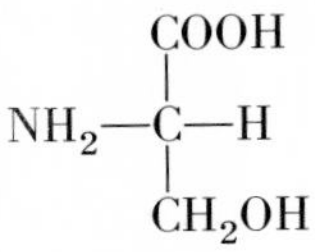

$$\begin{array}{c} \text{COOH} \\ | \\ \text{NH}_2-\text{C}-\text{H} \\ | \\ \text{CH}_2\text{OH} \end{array}$$
serine (Ser)

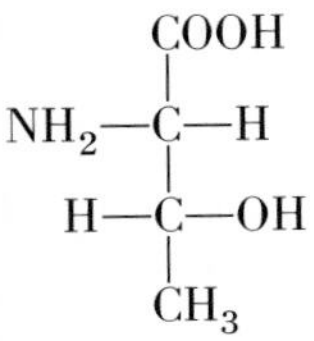

$$\begin{array}{c} \text{COOH} \\ | \\ \text{NH}_2-\text{C}-\text{H} \\ | \\ \text{H}-\text{C}-\text{OH} \\ | \\ \text{CH}_3 \end{array}$$
threonine (Thr)

Extra carboxyl group

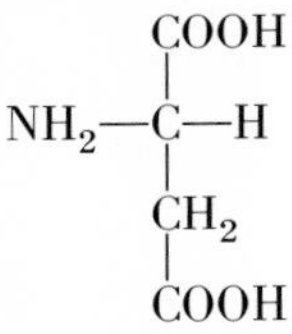

$$\begin{array}{c} \text{COOH} \\ | \\ \text{NH}_2-\text{C}-\text{H} \\ | \\ \text{CH}_2 \\ | \\ \text{COOH} \end{array}$$
aspartic acid (Asp)

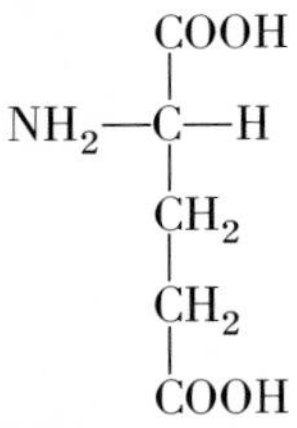

$$\begin{array}{c} \text{COOH} \\ | \\ \text{NH}_2-\text{C}-\text{H} \\ | \\ \text{CH}_2 \\ | \\ \text{CH}_2 \\ | \\ \text{COOH} \end{array}$$
glutamic acid (Glu)

Extra amine group or amine group derivatives

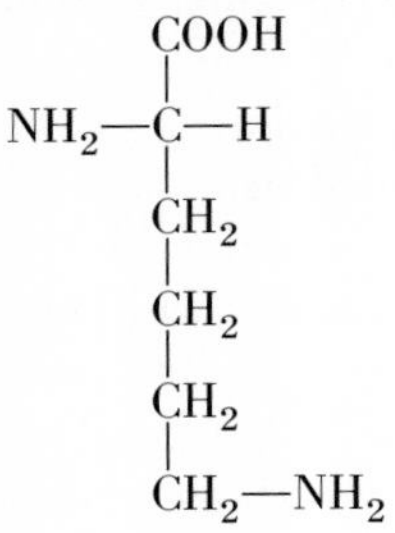

$$\begin{array}{c} \text{COOH} \\ | \\ \text{NH}_2-\text{C}-\text{H} \\ | \\ \text{CH}_2 \\ | \\ \text{CH}_2 \\ | \\ \text{CH}_2 \\ | \\ \text{CH}_2-\text{NH}_2 \end{array}$$
lysine (Lys)

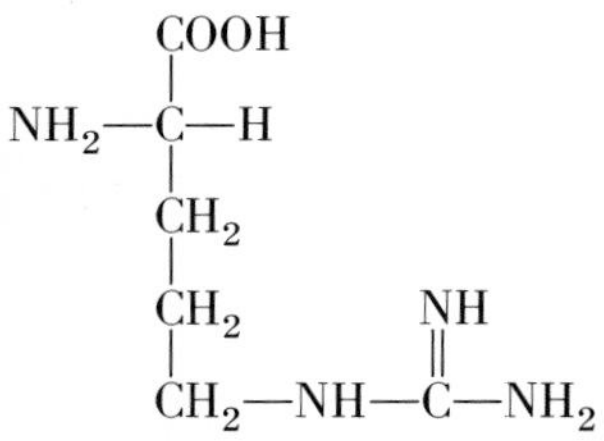

$$\begin{array}{c} \text{COOH} \\ | \\ \text{NH}_2-\text{C}-\text{H} \\ | \\ \text{CH}_2 \\ | \\ \text{CH}_2 \\ | \\ \text{CH}_2-\text{NH}-\overset{\overset{\text{NH}}{\|}}{\text{C}}-\text{NH}_2 \end{array}$$
arginine (Arg)

Table 19–1 (Continued)   Extra amide group (NH_2 on a carbonyl carbon)

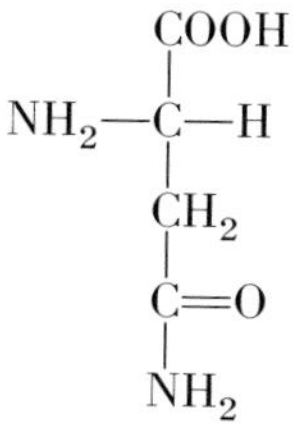

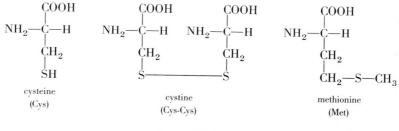

asparagine (Asp·NH_2)

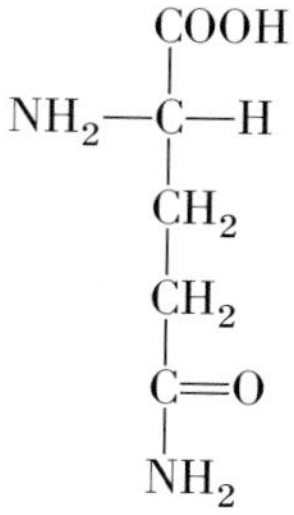

glutamine (Glu·NH_2)

Sulfur containing compounds

cysteine (Cys)

cystine (Cys-Cys)
(not one of the 20 basic amino acids)

methionine (Met)

Amino Acids Containing Rings

Aromatic hydrocarbon rings

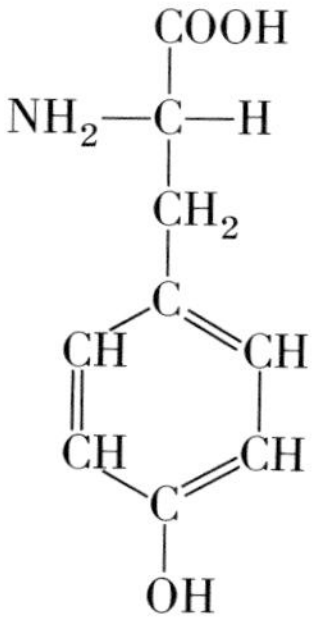

tyrosine (Tyr)

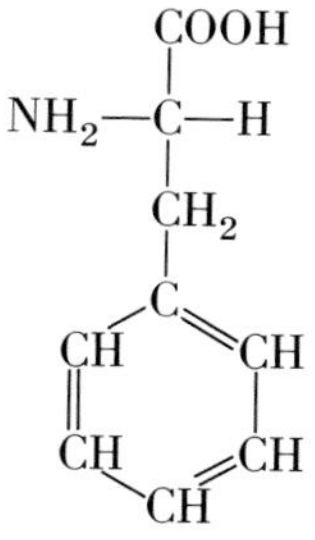

phenylalanine (Phe)

Heterocyclic rings (see Section 19–8 for a discussion of heterocyclic compounds)

Table 19–1 (Continued)

tryptophan (Try)   proline (Pro)   histidine (His)

molecule and a positive "ion" in another part of the molecule. This species is called a zwitterion. Zwitterions are not particularly reactive, so we will not write our formulas in the form showing these charges. But the tendency to form zwitterions emphasizes the presence of two dissimilar reaction sites on each amino acid molecule.

One of the likely reactions of amino acids is for the amine part of one amino acid to be attracted to the carboxyl part of the neighboring amino acid. If the carboxylic acid portion has been activated (which we represent by substitution of an activating group, X, for the OH), a condensation reaction may form a bond between the two amino acids. An HX unit is lost in that condensation reaction, but the overall process also includes the substitution for X for OH in the formation of the activated carboxylic acid. Therefore, the overall process links two amino acids while an H_2O unit is lost.

The condensation between the amine on one amino acid and the activated carboxylic acid on another forms a peptide linkage.

The amide bond that is formed is called a **peptide linkage** and is the basis for the formation of proteins.

19–5 Polypeptides

When a large number of peptide links are formed, amino acids can form long chain polymers called polypeptides. The bonding pattern in polypeptides is shown in Figure 19–4. Proteins are polypeptides. Sometimes people call "short" peptide chains with up to 50 peptide bonds polypeptides while calling longer peptide chains proteins. Actually all proteins are polypeptides, so separation into "long" and "short" groups is unnecessary.

Figure 19-4 A polypeptide.

Each of the peptide bonds between the NH and C=O groups shown in color links one more amino acid onto the chain.

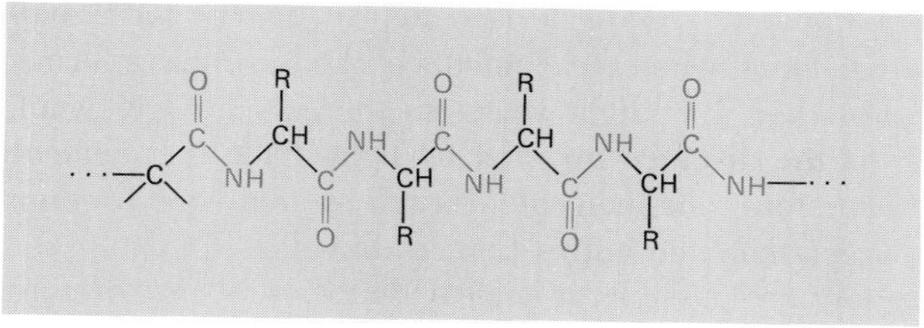

No matter how large (or small) the R groups are, the polypeptide chain has units linked in the sequence CHR, C=O, NH, CHR, C=O, NH, and so on. The sequence of bonded atoms —C—C—N—C—C—N— and so on in the chain is sometimes referred to as the **backbone** of the polypeptide. We will use that term in our discussion of protein structure. We will describe the shape of the backbone. That will not be the entire shape of the protein because R groups stick out away from the backbone, but it will tell us the shape we are starting from.

19-6 Structure of Proteins

19-6.1 Primary Structure.

Proteins are polypeptides made from the 20 common amino acids listed in Section 19-4. There are many different proteins because there are many different polypeptides possible. Each protein has a particular structural pattern which determines its properties. We will describe here the types of structural patterns which are important in determining the nature of proteins.

The first characteristic we must consider about a protein is the pattern of amino acids polymerized into the chain. Proteins differ in the number of amino acids in one chain and in the relative amounts of the 20 common amino acids present. But they also differ in the order in which the amino acids are placed. This arrangement of the amino acids in a particular order on the polypeptide chain is called the **primary structure** of the protein.

All proteins are polypeptides, with each having a particular amino acid sequence called the primary structure of that protein.

The number of possible different proteins is huge.

Because there are so many possible orders of amino acids, the number of possible proteins is huge. The molecular weights of proteins vary from several thousand to several million. If we assume an average molecular weight of 60,000 per polypeptide chain, each protein would contain about 500 amino acid units. At each of those 500 amino acid units any one of the 20 common amino acids could be present. Therefore there are 20 different choices for the first amino acid in the chain. For each of those there are 20 choices for the second amino acid, so there are 20 times 20, or 400, choices for the combination of the first two amino acids. There are $20 \times 20 \times 20$ choices for the first three amino acids. There are 20^4 ($20 \times 20 \times 20 \times 20$) choices for the first four, 20^5 for the first five, and so on. Therefore, there are 20^{500} choices for the order of amino acids in a 500-amino acid protein. That number is so large that the best we can do to explain it is to refer to units from astronomy. If there was just *one* protein molecule of each kind with 500 amino acids in the chain, they would fill solidly a cube approximately 10^{189} light years on each side. Light given off when the elements of the Earth were formed has had time to travel less than 10^{10} light years, so the cube's edges would be 10^{179} times as long as the distance traveled by light since the elements of Earth were formed. Therefore, one molecule of each protein would require a space tremendously larger than the entire known universe.

Because so many proteins are possible, it is clear that those which do form are only a tiny fraction of those possible. The duplication of large numbers of identical proteins by plants and animals requires specific guidance into particular amino acid sequences. We will discuss that guidance further in Section 19–11.

It should be noted that a polypeptide chain has a specific direction as well as a specific order. One end of the chain terminates at a NH_2 group which has not formed a peptide bond. The other end of the chain terminates at a COOH group which has not formed a peptide bond. In listing the order of amino acids the listing is normally begun from the end terminating with a NH_2 group. This is sometimes called the **N-terminal end.** Counting from that end, the groups along the backbone are in the order

$$NH_2-CHR-\overset{O}{\underset{\|}{C}}-NH-CHR-\overset{O}{\underset{\|}{C}}-NH-CHR-\overset{O}{\underset{\|}{C}}- \quad \text{and so on.}$$

Listing of the order of amino acid groups in a polypeptide is begun from the N-terminal end.

19–6.2 Secondary Structure. Proteins do not exist simply as long chains. The chains are folded back around themselves in particular patterns. Certain very regular patterns are favored over others because of the forces set up between the various amino acid groupings in the primary polymer chain. These regular patterns are called the **secondary structure** of the protein.

There are two main kinds of secondary structure, the alpha helix and the pleated sheet. The first of these to be found was the alpha helix.

Alpha helix structures form in such a way that there is a maximum amount of hydrogen bonding between NH and $C=O$ groups on the polypeptide chain. The alpha helix structure consists of a chain of amino acid units wound into a spiral which is held together by hydrogen bonds between a $C=O$ group and the NH group of the third amino acid residue on along the chain. The helix could be coiled in either the "right handed" or "left handed" directions shown in Figure 19–5. As in the case of amino acids where only L-amino acids are important in most life on Earth, there is a preference for one direction of the helix. Starting with the N-terminal end at the bottom, life on Earth uses right handed helices. If you take your right hand, as the helix moves up in the direction of your thumb, the coils curve as they rise with the direction of curves like those you can form with the fingers of your right hand. Figure 19–5 identifies the right handed helix. You may want to compare that picture to your right hand so you can later use your hand as a reminder of the helix direction.

The backbone of the protein can be arranged in secondary structure patterns of an alpha helix, pleated sheets, or combinations of those connected by less regular patterns.

Figure 19-5 Alpha helix.

On the left side of the figure a left-handed helix is shown. Left handed helices are not normally present in proteins. The right-hand side of the figure shows the common, right-handed alpha helix found in proteins. The dotted lines represent the hydrogen bonds between each NH group and a C=O group on the next lower turn of the helix.

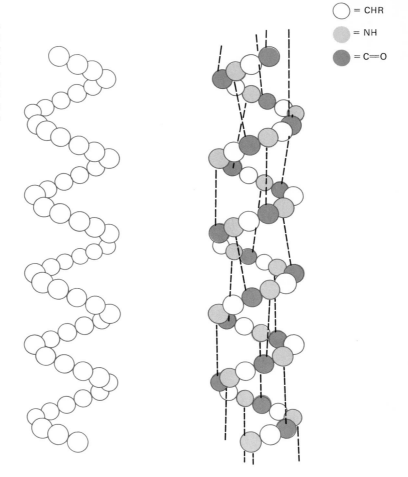

Pleated sheet and alpha helix structures give favorable hydrogen bonding.

Extensive hydrogen bonding can also be accomplished between two protein chains in the pleated sheet arrangement shown in Figure 19-6. In the pleated sheet arrangement two protein chains are paired. The sequences (NH—CHR—C=O) of these chains are lined up in opposite directions. In that arrangement, every NH group on one chain can be placed beside a C=O group on the other chain while every C=O group on the first chain is also beside a NH group on the second chain. Geometry prevents alignment of all of these pairs in positions permitting hydrogen bonding, but it is possible to permit half of the pairs to hydrogen bond. The remaining groups in each protein chain are held back away from the other chain by the tetrahedral bond angles. Note that each of these chains has another set of C=O and NH groups in the portions held back and that these groups are in exactly the arrangement needed to hydrogen bond to another protein chain in the same way. If a large number of parallel protein chains become bonded into

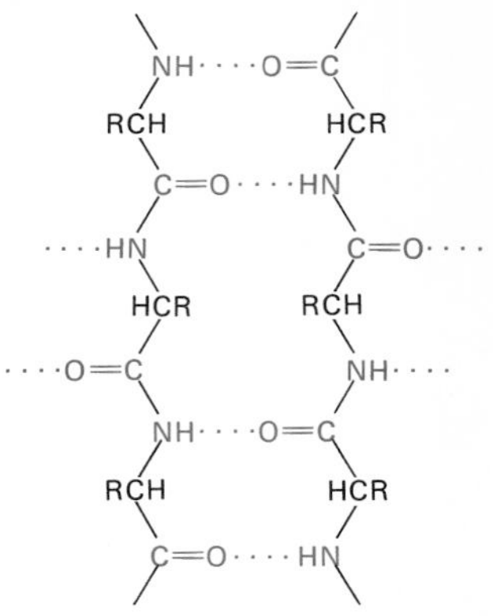

Figure 19-6 A pleated sheet structure in protein.

The dotted lines represent the hydrogen bonds. Additional chains can be hydrogen bonded on each side to build up a sheet structure.

one unit, they form a sheet of protein material. These sheets are pleated by the various forces between neighboring groups.

19-6.3 Tertiary and Quaternary Structures. The hydrogen bonds that form the secondary structures are not the only forces possible between units in polypeptide chains. A chain can be linked to itself in various **tertiary structure** patterns by various forces. These forces can include **ionic bonding** between the extra amine and carboxylic acid groups of some of the amino acids, **hydrogen bonding** between some amino acid R groups and main chain C=O and NH_2 groups, **covalent bonding** between cysteine units by forming the disulfide bridge of cystine (see Section 19-4 for structures), **dipole-dipole interactions,** or **hydrophobic interactions.** Figure 19-7 shows examples of these interactions along a polypeptide chain.

The most important of these interactions seem to be the hydrophobic interactions. In Section 10-1 we discussed the tendency for like to dissolve in like. The nonpolar R groups on some amino acids are not at all like the polar water around the protein in cells. Therefore these hydrophobic (water

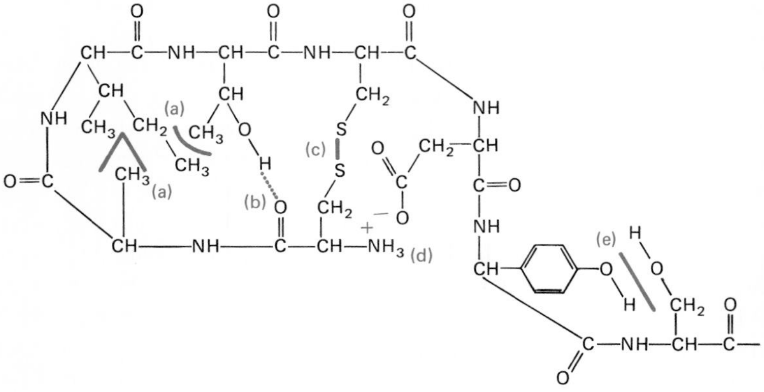

Figure 19-7 Types of bond forces between protein segments.

The most important type of force holding protein chains in particular patterns of tertiary or quaternary structure are hydrophobic attractions, in which water-repelling hydrocarbon portions of side groups are held to each other, forming "oil drop" like regions. These forces, marked (a), are caused by the tendency of the CH_3 group from an alanine to be surrounded by the $CH(CH_3)$—CH_2—CH_3 group from a nearby isoleucine. This "oil drop" region also attracts the CH_3 portion of the CHOH—CH_3 group of a threonine. The threonine group also takes part in another common type of bond, a hydrogen bond (b) to a carbonyl position in the chain. The next bond shown (c) is an actual chemical bond formed between the sulfurs of cysteine groups. Groups that exist with ionic charges can be held together by ionic attractions such as that shown in (d). When fully ionic sites do not exist, there may be attractions between polar side groups such as the OH groups shown bonding in (e). The bonds shown are only a sampling to illustrate the important general types.

Ionic bonding, hydrogen bonding, covalent bonding by disulfide bridges, dipole-dipole interactions, and hydrophobic interactions cause tertiary or quaternary structure.

Hydrophobic interactions are the most important cause of protein shapes, producing "oil drop" shapes with "oily" centers surrounded by charges.

hating) R groups tend to "dissolve" in each other by grouping together in the center of the protein. On the other hand, the polar R groups on some other amino acids are hydrophilic (water loving) and tend to stick out from the protein into the surrounding water. The result is a pattern where the inside of the protein is filled with "oily" R groups while the outside is covered with groups forming electrical charges. The protein becomes essentially an oil drop covered with charges. This is sometimes called the **oil drop theory** of protein structure. The protein forms the shape that best fits this oil drop model.

You may want to review Section 13–5.6 on colloids to compare the discussion of hydrophilic and hydrophobic behavior there with the hydrophilic and hydrophobic behavior in proteins.

All the structure discussed so far has been concerned with formation of the protein and its folding upon itself. But after a protein is completely folded into the most energetically favored shape it may still be able to set up forces (similar to those causing tertiary structure) to parts of other proteins. These forces can cause several proteins to group together in a particular way. The resulting pattern is called **quaternary structure.** Hemoglobin, the component of blood that carries oxygen, is a protein made from four polypeptide chains held together in a specific quaternary structure.

19–7 Nucleic Acids

Nucleic acids have equal numbers of three major components, inorganic phosphate, a five carbon sugar, and a pyrimidine or purine base.

Living cells contain another type of biopolymer closely related to proteins but not made from amino acids. These biopolymers were named nucleic acids because they were first isolated from cell nuclei.

Nucleic acids have three major components. Upon hydrolysis they can be broken down into equal numbers of inorganic phosphate units, sugar units (each of which is a five-carbon atom sugar), and a complex organic base. The organic bases are all heterocyclic compounds of the classes called pyrimidines or purines. There are five common organic bases used in natural nucleic acids, and their structures and bonding create a code that is the basis for heredity.

19–8 Heterocyclic Compounds

When we discussed hydrocarbons we included aromatic ring compounds (Section 17–3.7) and nonaromatic ring compounds (Section 17–3.8). In each case the rings were made of carbon atoms bonded to each other. Rings can also be formed in which both carbon atoms and one or more other kind of atoms are bonded to each other in a ring. A simple example is ethylene oxide

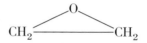

mentioned in Section 18–2. Ethylene oxide is a cyclic ether. Any compound with more than one kind of element in the ring itself is called a **heterocyclic**

compound. Only the atoms in the ring count. Those bonded to ring atoms (such as hydrogens bonded to the carbons of a ring) are not in the ring.

The usual angles of chemical bonds make five- or six-atom rings the most common. Most of these atoms are usually carbon, but there may be more than one atom of another element in the ring. An example of a six-atom ring including two noncarbon atoms is 1,4-dioxane

$$\begin{array}{c} CH_2 - CH_2 \\ O \qquad\quad O \\ CH_2 - CH_2 \end{array}$$

The most important biological heterocyclic compounds have carbon and nitrogen atoms in the rings. The nitrogen may be present as NH groups in place of the CH_2 group that would otherwise have been present in a saturated hydrocarbon ring. The amino acid proline has such a saturated ring (see Section 19–4).

In other cases nitrogen with no hydrogens bonded to them are present in place of a CH group of an unsaturated ring. The nitrogen then has one single bond and one double bond in the ring or the equivalent to that bonding in resonance-stabilized structures such as six-membered aromatic rings. In other cases the nitrogen may be in the ring as an NH unit (with only single bonds) while there is a double bond somewhere else on the ring.

Both the amino acids tryptophan and histidine (see Section 19–4) have NH units in five-membered rings that are unsaturated (double bonds elsewhere in the ring). Histidine also has nitrogen with a single and a double bond in a five-membered ring. The simplest example of nitrogen in a six-membered aromatic ring is pyridine, where nitrogen replaces one CH unit of benzene. The pyridine ring is

$$\begin{array}{c} CH - CH \\ CH \qquad\quad N \\ CH = CH \end{array}$$

Aromatic or nonaromatic five- or six-membered rings containing carbon and nitrogen are particularly important heterocyclic compounds.

Another heterocyclic aromatic compound is pyrimidine, in which two CH groups of benzene (at the first and third positions) have been replaced with nitrogen. The structure of pyrimidine is shown in Figure 19–8. Purine is a derivative of pyrimidine in which the fourth and fifth carbons of the pyrimidine ring are also part of a five-membered ring with two nitrogens and one more double bond, as shown in Figure 19–8.

The important natural derivatives of purine and pyrimidine are formed by substituting OH or NH_2 groups for some of the hydrogens. In thymine a CH_3 group is also substituted. When OH groups are substituted the product

19–9 The Pyrimidine and Purine Bases

There are five biologically important natural pyrimidine and purine bases.

Figure 19-8 Pyrimidine and purine.

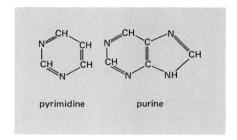

Figure 19-9 The biologically important pyrimidines and purines.

The structures are written in two forms for those having enol and keto forms. The enol forms (on the left in each case) show the relationship to pyrimidine or purine, whereas the keto form (shown on the right) is more stable and likely to be the form present most of the time. Adenine has only the one structure shown.

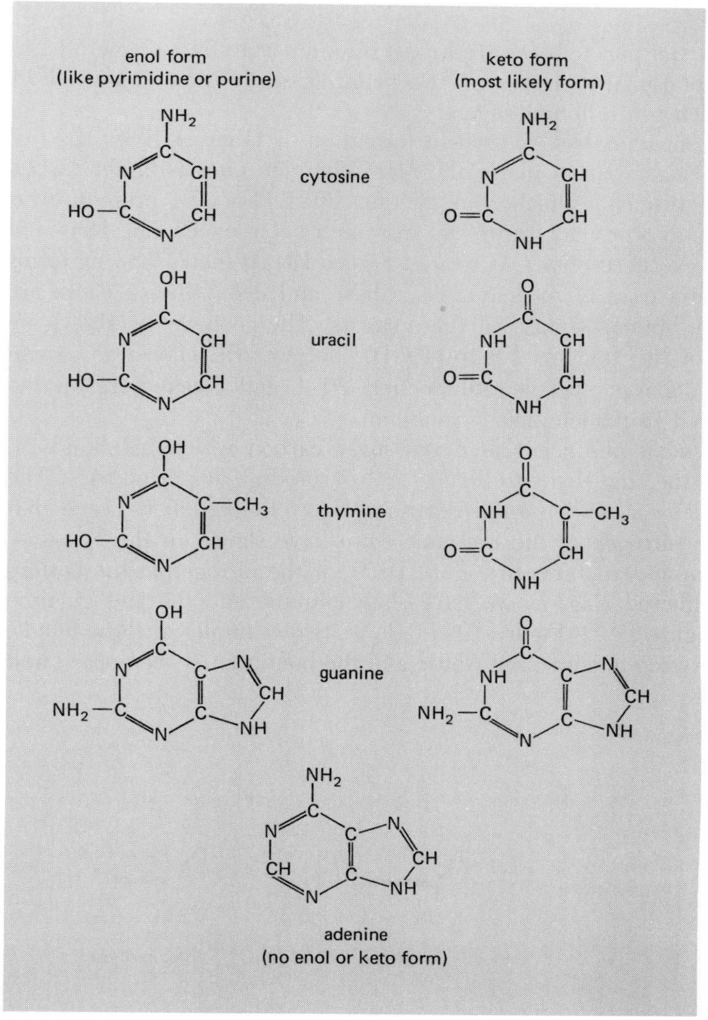

[19-9] The Pyrimidine and Purine Bases 441

is an enol—a double bond (ene) next to an alcohol (ol). The enol can rearrange to a keto (ketone) form by transferring the hydrogen from the oxygen to a neighboring atom. Figure 19–9 shows the five important purine and pyrimidine bases. The enol forms are shown to indicate the close relationship to pyrimidine (or purine) and the keto forms are shown to indicate the likely actual structure.

The resulting carbonyl (C=O) and amine (NH_2) groups are important because they provide sites for later interactions by these bases. The ring structures are important because they cause very rigid arrangements which make each base react most effectively with a specific one of the other bases which best fits its shape and available bonding sites.

The pyrimidine and purine bases are linked into important biopolymer chains by sugars and phosphates. There are two main classes of these chains, RNA and DNA, which use different sugars.

Each of the sugar molecules used in formation of DNA or RNA has five carbon atoms. Sugars (particularly the more common ones with six carbon atoms) will be discussed further in Section 20–1. For the present, it is sufficient for us to note that sugars can exist in a heterocyclic ring form and that they have several reactive OH groups. Figure 19–10 shows the ring forms of the two sugars used in nucleic acids, ribose and deoxyribose. There are actually specific optical isomers of these sugars. The geometry is shown by the side view of the rings in Figure 19–10, but we will choose to ignore the details of sugar geometry until Section 20–1 and concentrate on the bonding involved in nucleic acid formation.

In every segment of nucleic acid the sugar carbon which has bonds to two oxygens in the ring structure bonds to a pyrimidine or purine base. The bond is formed by a condensation reaction (H_2O eliminated) between that carbon and the nitrogen at the bottom, as we have shown in the cytosine, uracil, and thymine structures in Figure 19–9, or the nitrogen at the bottom of the five-membered ring, as we have shown in the adenine and guanine structures in Figure 19–9. Figure 19–11 shows two examples of these bonds, the bonding between adenine and ribose and the bonding between uracil and ribose.

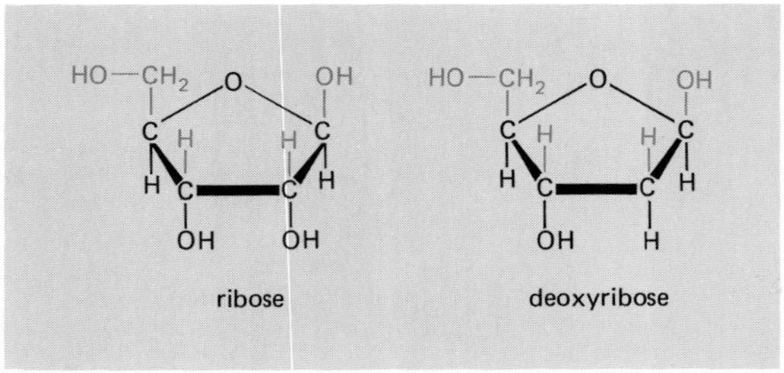

19–10 DNA and RNA

DNA and RNA are polymers of the pyrimidine and purine bases plus sugars and phosphate.

Figure 19–10 Ribose and deoxyribose.

Figure 19-11 Adenosine and uridine.

Adenosine is the condensation product (H_2O removed) of adenine plus ribose. Uridine is the condensation product (H_2O removed) of uracil and ribose.

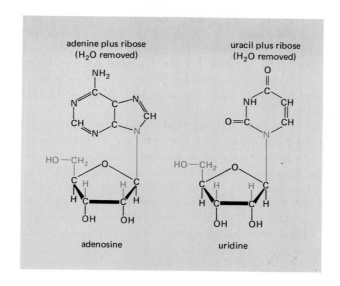

The resulting sugar-base unit still has OH units that can take part in further reactions. They can react by condensation-like reactions with phosphate units as shown in Figure 19-12. As shown in our two examples, the OH group reacting with the phosphate varies. The numbers indicated by the primed notation (3′ or 5′) indicate the carbon atom of the sugar to which the phosphate is attached.

The base-sugar-phosphate units shown in Figure 19-12 are called nucleotides. Nucleotides can continue to react by further condensation reactions.

Figure 19-12 Examples of nucleotides.

Condensation reactions between a pyrimidine or purine base, ribose or deoxyribose, and phosphate lead to base-sugar-phosphate units called nucleotides.

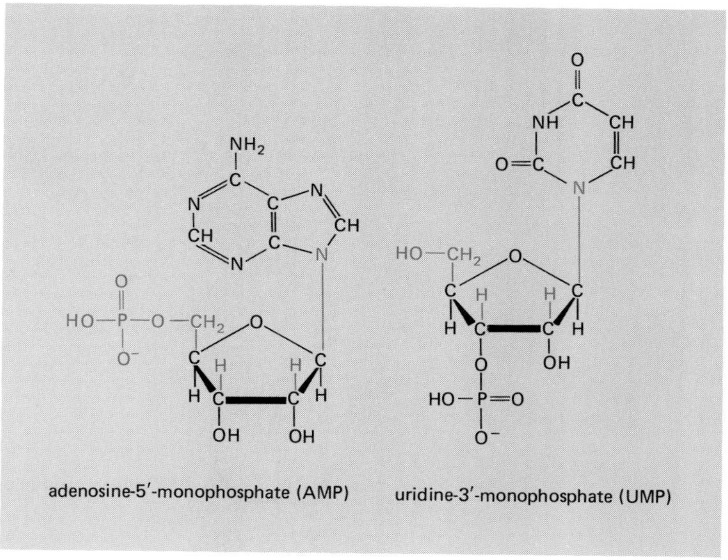

[19-10] DNA and RNA 443

These reactions could be with any other condensation reaction reagent, such as hydrogen phosphate units. Figure 19–13 shows the series of reactions building up a triphosphate unit on adenosine. The product is called adenosine triphosphate or ATP. Energy is required for the reaction because of the concentration of negative charges. This reaction is one of the key steps in storage and later recovery of energy by living organisms. Much of Chapter 21 will be spent discussing ATP.

When the monophosphate on the base-sugar-phosphate unit links to the sugar of a second base-sugar-phosphate unit, the units are combined into a chain that can continue to grow by the same kind of reaction. That is the reaction which forms the biopolymers called nucleic acids. These chains are acidic because of the phosphate units. Those containing ribose are called **ribonucleic acids** or **RNA**. Those containing deoxyribose are called **deoxyribonucleic** acids or **DNA**. Both RNA and DNA contain adenine, guanine, and cytosine. RNA also contains uracil but no thymine. DNA contains thymine but no uracil.

DNA, RNA, and protein are related to each other in the following way.

Figure 19–13 Formation of adenosine triphosphate (ATP).

Adenosine monophosphate (AMP) is converted to adenosine diphosphate (ADP) by condensation with a phosphate species. ADP can then be converted to adenosine triphosphate (ATP) by another similar condensation with a phosphate.

444 Biological Building Blocks [19]

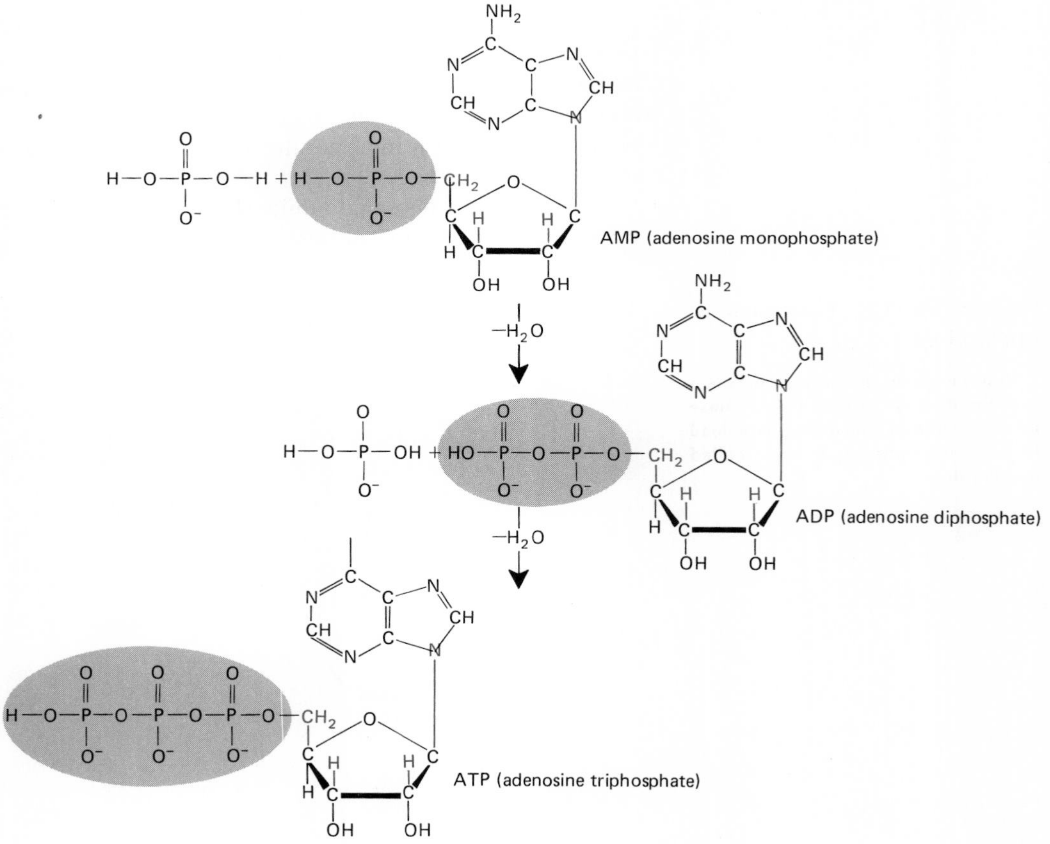

Figure 19-14 Hydrogen bonding between DNA chains.

Because of their geometry, cytosine always bonds to guanine and adenine always bonds to thymine. The cytosine-guanine bonds are stronger because they form three hydrogen bonds whereas adenine-thymine pairs form only two hydrogen bonds.

DNA exists as a double helix with chains running in opposite directions held together by hydrogen bonding between specific base pairs.

Using the symbols C (cytosine), G (guanine), A (adenine), T (thymine), and U (uracil) the base pairing is C≡G (with three hydrogen bonds) and A=T (with two hydrogen bonds) in DNA and C≡G (three hydrogen bonds) and A=U (two hydrogen bonds) in RNA.

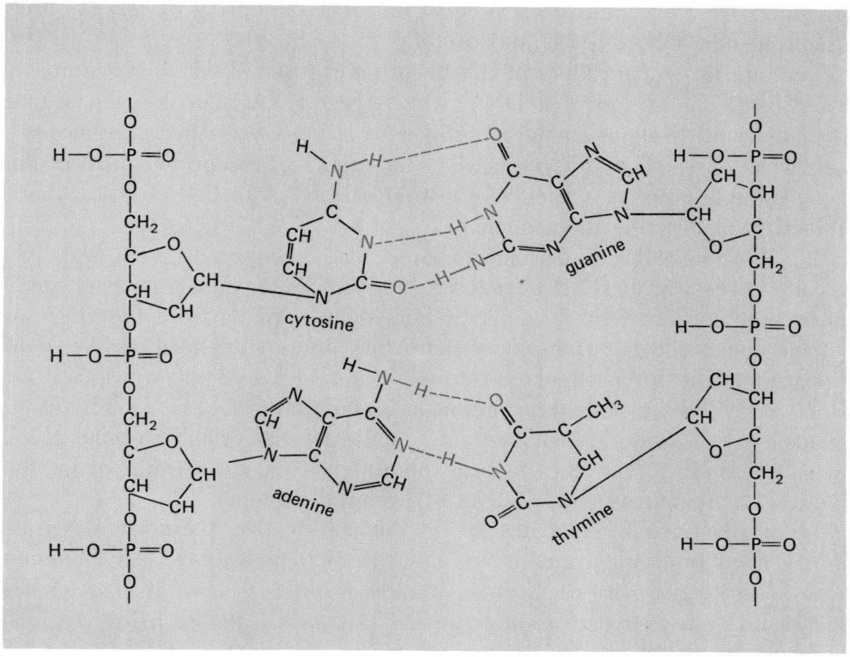

DNA is used to make RNA and RNA is used to make protein. To allow growth of new cells, it is also necessary to make new DNA. The structure of DNA allows it to both duplicate itself and produce RNA. DNA remains principally in the nucleus of living cells whereas RNA can be found in many different parts of each cell.

DNA has been shown to exist as a pair of coiled helical chains. The story of the discovery of the DNA structure, written in an interesting form by James Watson (*The Double Helix*, Signet, 1968), gives an insight into competition in science. Watson and Francis Crick used the earlier discovery of the alpha helix structure of proteins by Linus Pauling for ideas to race against Pauling to discover the structure of DNA.

The two chains run in opposite directions and have a specific crosslinkage between the bases, as shown in Figure 19-14. The chains are held together by **hydrogen bonding** between pairs of bases. The hydrogen bonds are shown as dotted lines in Figure 19-14. Because of the shapes and available sites for hydrogen bonding, every cytosine (C) unit bonds to a guanine (G) unit. There are three hydrogen bonds holding the guanine and cytosine together. Every adenine (A) unit bonds to a thymine (T) unit. The adenine-thymine pairs have only two hydrogen bonds and are therefore more easily split than the cytosine-guanine bonds. To remind us of this difference in bond strength, we will write the units as C≡G for cytosine-guanine with three hydrogen bonds and A=T for adenine-thymine with two hydrogen bonds.

In Section 19-11 we will describe how this specific pairing of bases is used in making new DNA, RNA, and proteins.

There are three main types of RNA molecules, each of which is essentially copied from one strand of a DNA. The largest RNA molecules are found associated with proteins in the ribosomes of living cells. The ribosomes are small parts of a cell that serve as the "factories" where proteins are made. We will simply note the existence of **ribosomal RNA,** called **rRNA** without attempting to describe them further.

Three major forms of RNA are rRNA, used in the ribosomes which are the cell's "protein factories", mRNA, used to provide the directions for protein synthesis, and tRNA, used to bring in particular amino acids.

The RNA molecules that are most characteristic of each particular organism are called **messenger RNA** or **mRNA.** The mRNA molecules are long single chains used to carry the specific directions for proteins from the DNA in the cell nucleus to the ribosomes where the proteins are formed. As such, they are specific for particular proteins. The rRNA and the transfer RNA (which we will discuss next) are not specific and can in fact be used by other organisms. For example, a human cell could use bacterial ribosomal RNA to make proteins. It is the message on the mRNA that would make the products be human proteins instead of bacterial proteins.

mRNA exists as long strands formed from the complementary base units to a single strand of DNA.

The smallest RNA molecules are in the class called **transfer RNA** or **tRNA.** Their molecular weights are about 25,000 to 40,000. tRNA molecules do not remain as single long strands as mRNA does. Instead tRNA coils around itself so that a single strand can form "double helix" regions similar to those formed by the two separate strands in DNA. tRNA molecules form a particular pattern of helical sections and connecting loops that has been described as a cloverleaf structure.

tRNA can coil upon itself with the base pairs aligning it in a "cloverleaf" structure that can bond a particular amino acid at one end and has a code set (anticodon) of three unpaired bases at the opposite side.

Figure 19-15 shows a schematic cloverleaf structure of tRNA with the two particularly important features noted. Each tRNA can bond a particular amino acid on the "stem" of the cloverleaf, and each tRNA has a particular set of three bases at the opposite side of the molecule from the stem that carry a particular code. Recent x-ray crystallographic work has shown tRNA to be more L shaped than shown in Figure 19-15. The actual shape may have particular details required to position the tRNA for reaction, but that would not affect the two main features we have shown—a site for the amino acid and a site carrying a three-base code.

As we will see in Section 19-12, each tRNA must have a set of three bases in its code positions which are the complement (the other member of the pair that can hydrogen bond) to the bases on the mRNA providing directions for the protein synthesis. Each set of three bases on the mRNA is called a **codon.** The reverse set on the tRNA molecules are called **anticodons.** There is a type of tRNA for each of 61 anticodon combinations and, as described in Section 19-12, each of these has a specific amino acid associated with it. That amino acid is the only one which can bond at the amino acid bonding position on its associated tRNA's. (There are also three anticodon combinations that stop the protein synthesis, as noted in Section 19-12.)

The coiled sections of tRNA are held together by hydrogen bonding similar to that in DNA. There are C≡G bonds (three hydrogen bonds) just like

Figure 19-15 tRNA

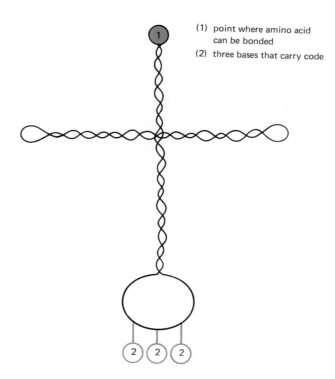

(1) point where amino acid can be bonded
(2) three bases that carry code

those in DNA. However, instead of A=T bonds, there are adenine-uracil bonds (with two hydrogen bonds) in RNA. We will represent them as A=U bonds. There is generally no thymine in RNA. Wherever thymine would have appeared if DNA was duplicated, uracil is substituted when RNA is formed. That substitution occurs in all RNA regardless of whether the RNA coils like tRNA or remains uncoiled like mRNA. DNA and RNA therefore give the following products when completely broken apart by hydrolysis.

DNA	*RNA*
phosphoric acid	phosphoric acid
deoxyribose	ribose
cytosine	cytosine
guanine	guanine
adenine	adenine
thymine	uracil

19–11 Replication and Heredity

The arrangement of bases along a DNA or RNA makes up a code pattern. The definite pairing of the bases by hydrogen bonding gives a way to read the code and use it. This code determines the unique set of proteins making an organism what it is. The code can be duplicated and passed on to offspring. That is the molecular basis for heredity.

The hereditary patterns are contained in DNA double helices and passed on to new cells or to offspring by replication (duplicating) of DNA. That is done by breaking the helix apart and building new strands on each of the two parts. Since the bases only hydrogen bond in particular pairs (C≡G and A=T), each new strand matches the one that was taken away when the double helix was split apart. The result is two new double helical DNA units. The process is shown in simplified form in Figure 19–16.

The actual process of DNA replication is complex. It probably involves different enzymes that react to open up the double helix, guide new units to their proper places, and link the new units onto the growing polymer chain. An enzyme was discovered that was thought to do this, but later evidence showed that it could not adequately explain the whole process. As of this writing biochemists continue to search for further details of DNA replication.

We can note a few important chemical facts about DNA replication without attempting to understand the details of the process. First, it is clear that the DNA double helix must be separated before new groups can be added. That is fairly easily accomplished because the hydrogen bonds holding the chains to each other are much weaker than the covalent bonds holding each chain together. Second, the accuracy of the pattern depends upon the pairing in C≡G and A=T base pairs. That occurs because there are several hydrogen

Replication of DNA is achieved by separating the two strands and having each pick up the parts for a new partner, which is always identical to the old one because of the limitation to C≡G and A=T pairing.

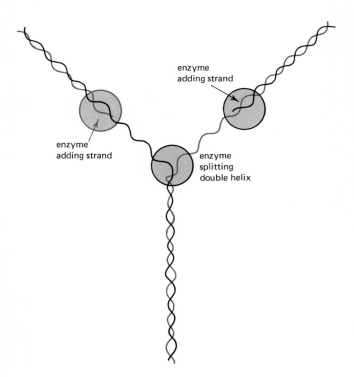

Figure 19–16 Duplication of DNA.

Hydrogen bonding is weak enough to allow separation of DNA strands, but the total bonding with two or three hydrogen bonds per pair is strong enough to cause accurate selection of the pairs that fit.

Formation of RNA is similar to DNA replication except a single RNA chain is formed which does not remain with the DNA, and a U (on the RNA chain) is paired with each DNA chain A instead of the T that would have paired with A in DNA replication.

bonds (two or three) in each pairing. The total bonding from the two or three hydrogen bonds is enough to force selection of the right base. Because of the rigid shapes of the bases, a "false" choice of the matching base would not be able to give two or more hydrogen bonds. Therefore errors in transferring the base sequence to new DNA are kept very rare by the selectivity of the hydrogen bonding.

Formation of RNA from DNA is very similar to DNA replication. The differences are that only one strand of the DNA is matched by complementary bases and that the bases are bound to ribose instead of deoxyribose. Pairing gives the RNA chain a C wherever the DNA has a G, a G wherever the DNA had a C, an A wherever the DNA had a T, and a U wherever the DNA had an A. Notice the use of U (uracil) in RNA instead of T (thymine). Different DNA chains make different RNA's. tRNA is produced with just the right order to allow coiling on itself in a double helix. The sequences of some tRNAs are now known, and the crystal structure of the tRNA for phenylalanine has been determined.

The DNA controls the order of bases and therefore determines which RNAs form. When the complete set of DNAs of an organism are passed on to a descendant, that descendant will have the same combination of RNAs. That passing on of patterns, particularly patterns for mRNAs, is hereditary transfer of genes.

19–12 Synthesis of Proteins

Proteins are manufactured in ribosomes by bringing in tRNAs with amino acids so that the tRNA's anticodons pair with three-base codon segments of the mRNA.

The three base codons on mRNA include messages to start protein synthesis, stop synthesis, or direct addition of particular amino acids, with as many as six codons for the same amino acid.

Proteins are manufactured on the ribosomes of cells by RNA. The ribosomes serve as factories that bring the components together and make them react. The mRNA serves as the set of directions to be used. You can think of the mRNA as a blueprint or (in an even better fit to its actual long chain structure) as a punched tape being fed into a teletype and printing out instructions. The tRNA serves as the assembly line workers. tRNAs pick up amino acids (the parts needed to make protein), bring them to the ribosomes, and put them into the protein when told to do so by the directions.

The directions are given by codons made up of groups of three consecutive bases on the mRNA chains. Since there are four bases possible at each position, there are $4 \times 4 \times 4$ or 64 possible codons. Those 64 codons direct selection from among the 20 amino acids used to make proteins. They also include directions to start or stop formation of a protein. Because there are more codons than needed, many of the amino acids are selected by more than one codon. All 64 codon "words" have now been identified. That identification was one of the most spectacular triumphs in the history of science. Three of the 64 codons stop the growth of the protein. One starts protein formation, but if it occurs in the mRNA chain after the protein formation is already started that codon simply selects a particular amino acid to be added. The other 60 select particular amino acids to be added, with as many as six codons selecting the same amino acid. A single mRNA may direct the formation of several proteins. Whenever a "stop" codon is reached, one protein comes off. The mRNA then continues to pass through the ribosome and be read until the next "start" code is reached and a new protein begins to form.

The mRNA is read by tRNA molecules being hydrogen bonded to the

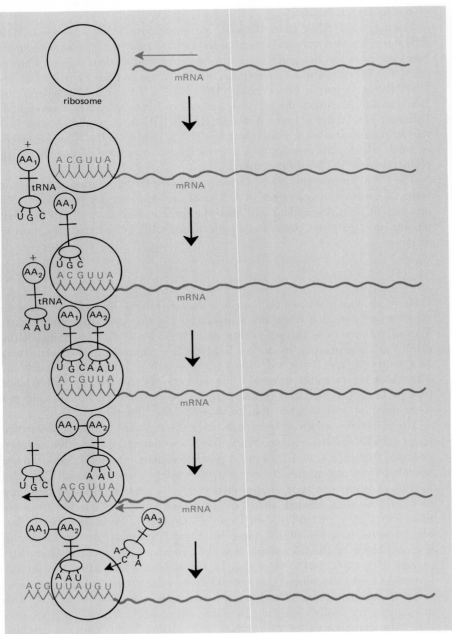

Figure 19–17 Synthesis of proteins.

An mRNA strand reaches a ribosome. At the ribosome two of the three base codon "words" on the mRNA are "read" by attaching tRNA molecules with the correct anticodons. Amino acids of the two adjoining tRNA's then react, and the first of the tRNA's is lost. The mRNA then moves over three base units, another tRNA is matched to the next codon, and the process is repeated. The mRNA continues to move through the ribosome until the synthesis is completed.

450 Biological Building Blocks [19]

mRNA. Each group of three bases on the mRNA picks up a tRNA with the proper set of three bases in the "code" position on the tRNA. That set must be the opposite of the mRNA group so the favorable C≡G and A=U hydrogen bonding can form. The overall process is complex, including steps that release the energy needed to cause the reactions. However, the general sequence of reaction is shown (in simplified form) in Figure 19-17. Two tRNA units are bound to the mRNA simultaneously on consecutive groups of three bases. A bond then forms between the amino acids at the other end of the tRNAs. The units then move over, another tRNA with attached amino acid bonds to the next set of three bases, and the process is repeated. As the protein comes out of the ribosome, it may begin to fold into its secondary and tertiary structure while its chain is still being completed by further reactions in the ribosome.

19–13 Functions of Proteins

Many structural proteins are fibrous materials made up of cable-like groupings of alpha helixes.

Proteins carry out a variety of functions in living organisms. Some proteins are structural materials. Other proteins are enzymes, antibodies, or hormones. They are used in necessary chemical reactions. Some proteins have both important structural properties and take part in important reactions.

Many structural proteins are fibrous materials. Hair, skin, fingernails, feathers, and connective tissues and tendons are made of fibrous structural proteins. These fibrous materials are composed of protein alpha helixes which are often wound about each other in small groups to form strands like cables. Figure 19–18 shows typical fibrous protein "cables" made by winding alpha helix chains around other alpha helix chains.

Figure 19–18 Fibrous protein structures.

A strong fiber is built up by a cable-like winding of protein alpha helices. The resulting helix strands may then be further wound with other multihelix strands in another cable arrangement. This intertwining of strands can build up a large and very strong fiber. (A) The intertwining of three alpha helix proteins to form a strong fibril. This structure is thought to exist in collagen which is used in connective tissues. (B) A more complex strand in which nine alpha helix proteins are wound around two central alpha helix proteins. That structure is thought to occur in keratin, the protein in hair.

A

Side view

Top view

B

Muscle proteins make up another class which has important structural properties. The contractile proteins that are used in muscular work have structural patterns that can be changed somewhat in reversible ways. These proteins have sites that are involved in chemical reactions with energy sources, and these reactions cause the structural pattern to change, thus doing work. This process is described in Section 21–6.

Many other proteins exist in patterns with no structural importance other than having the structure needed to permit certain chemical reactions. These proteins are called globular proteins because they lack the definite single directional pattern of fibrous proteins. Globular proteins include the following:

1. *Enzymes.* Enzymes are catalysts that bring about specific necessary reactions of the living organism.
2. *Hemoglobin and myoglobin.* These proteins are used to transport and store oxygen for use in the organism.
3. *Hormones.* Some important cell processes are controlled by regulatory proteins called hormones.
4. *Antibodies.* Proteins are manufactured that interact with and protect the organisms from foreign substances (such as those introduced by infections).
5. *Nucleoproteins.* Globular proteins are involved in controlling the reactions described in Sections 19–11 and 19–12.

Chemically reactive proteins tend to be globular arrangements leaving certain available reactive sites and limiting the reactants that can reach those sites.

Although the globular proteins have a tremendous variety of structures and functions, it is fair to generalize them as arrangements that make accessible particular chemically reactive sites or combinations of chemically reactive sites. They may also have particular shapes that prevent access to the reaction sites by substances other than the desired reactants. It is the detailed secondary and tertiary structures (see Section 19–6) that give the globular proteins their ability to function. Quaternary structures may also be important, particularly when forces align potential reactants in a specific way on the globular protein.

Globular proteins fold into particular secondary and tertiary structures as they come off the ribosomes in which they are formed. Conditions such as exposure to acids or bases, heating, reducing agents, organic solvents, detergents, or x rays or strong light sometimes break up the secondary and tertiary structures of proteins. Such conditions may then set up attractions to neighboring molecules which prevent the protein from folding back into the original structure. Because the structure was necessary to give the protein its specific reactivity, a change in structure destroys the protein's ability to function and gives it an entirely different nature. This process is called **denaturation** of the protein. Notice that the protein chain (primary structure) is not affected. Changes in the secondary and tertiary structure are enough to destroy the natural biological activity of the protein.

When the secondary and tertiary structures of proteins are changed by denaturation, the protein loses its function even though the primary protein chain is unchanged.

The very regular structure of fibrous proteins leaves little opportunity to expose "active sites" or limit access to particular reactants. Therefore, the proteins involved in reactions are globular instead of fibrous. Muscle proteins, which have both structural functions and involvement in reactions, contain

Muscular proteins include both fibrous and globular portions.

both globular portions and fibrous portions. The energy supplying substances react with the globular portions during muscle contraction. That causes a rearrangement of the fibrous portions relative to each other that causes contraction and does muscular work. The overall process is quite complex, but the general features of the process are described in Section 21–6.

Skills Expected at This Point

1. You should be able to state the number of amino acids used in proteins and describe the characteristics shared by those naturally important amino acids, including the functional groups and their positions and the optical isomer occurring (when there are optical isomers).
2. You should be able to describe the peptide linkage.
3. You should be able to describe primary, secondary, tertiary, and quaternary structures of proteins and the types of bonding forces that lead to protein structure.
4. You should be able to describe the structures and functions of DNA and the three categories of RNA and list the bases involved and how they pair in coding.
5. You should be able to describe (in general terms) the processes leading to heredity, replication of DNA, production of RNA, and production of proteins.
6. You should be able to describe the functions of fibrous and globular proteins, explain how their structures fit their functions, and describe how and why denaturation can occur.

Exercises

1. List three important classes of biopolymers.
2. Draw the structure of serine according to the standard convention described in this chapter and explain how this indicates its true shape and arrangement. The R group on serine is CH_2OH.
3. (a) List four amino acids that are most likely to be directed in toward the center of a protein helix and the sort of forces likely to cause them to point inward. (b) List four amino acids that are most likely to be on the outside of a folded protein and the sort of forces likely to cause them to point outward.
4. Describe a peptide linkage. Include a structure to show what you mean.
5. List the features of an alpha helix.
6. What is the difference between tertiary and quaternary structure of proteins?
7. Define a heterocyclic compound.
8. List the five biologically important purine and pyrimidine bases, the classes of compounds they appear in, and how they pair.
9. List the components of DNA and how they are linked (their sequence).

10. How does DNA direct the correct parts into place when more DNA or RNA is formed?
11. List the three classes of RNA and the function of each.
12. Why are enzymes globular proteins?

Test Yourself

1. Name the configuration of the amino acids found in most proteins on earth.
2. Draw the structure of the tripeptide composed of glutamic acid, valine, and tyrosine linked in the order given. The R groups are

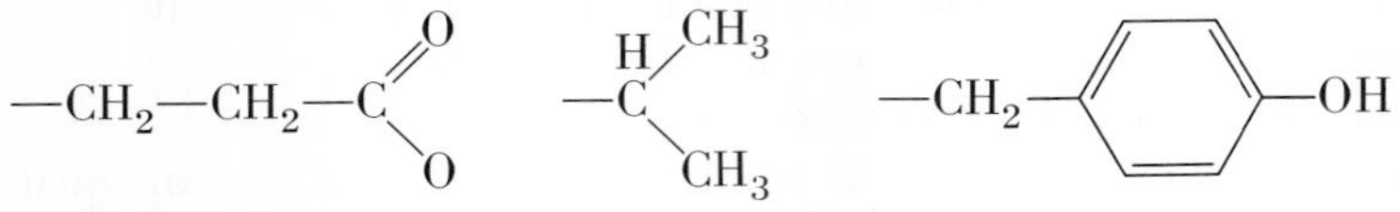

$$-CH_2-CH_2-C\overset{O}{\underset{O}{\diagup}} \qquad -C\overset{H}{\underset{CH_3}{\diagdown}}\overset{CH_3}{\diagup} \qquad -CH_2-\!\!\bigcirc\!\!-OH$$

Glu   Val   Tyr

3. Which of the following shows the kind of linkage present in nylon and proteins?

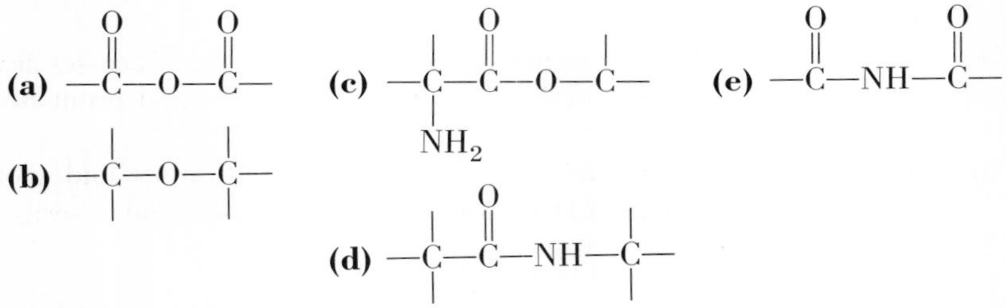

   (a) $-\overset{O}{\overset{\|}{C}}-O-\overset{O}{\overset{\|}{C}}-$

   (b) $-\overset{|}{\underset{|}{C}}-O-\overset{|}{\underset{|}{C}}-$

   (c) $-\overset{|}{\underset{|}{C}}-\overset{O}{\overset{\|}{C}}-O-\overset{|}{\underset{|}{C}}-$ with NH_2

   (d) $-\overset{|}{\underset{|}{C}}-\overset{O}{\overset{\|}{C}}-NH-\overset{|}{\underset{|}{C}}-$

   (e) $-\overset{O}{\overset{\|}{C}}-NH-\overset{O}{\overset{\|}{C}}-$

4. The double helix structure in DNA may be compared to what in proteins?
   (a) primary sequence   (d) hydrogen bond
   (b) pleated sheet   (e) hydrophobic bond
   (c) disulfide bond

5. Which of the following types of bonding is believed to be the *most* important in determining the three-dimensional structure of proteins?
   (a) hydrogen bonding   (d) hydrophobic bonding
   (b) electrostatic bonding   (e) disulfide bonds
   (c) hydrophilic bonding

6. The following structure represents what two base pairs?

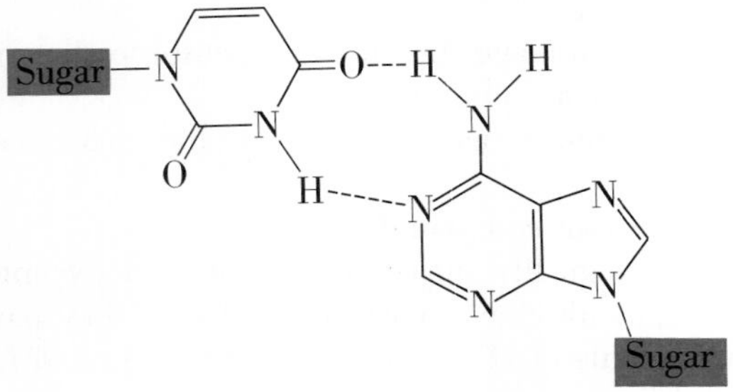

(a) adenine-uracil (d) uracil-guanine
(b) thymine-guanine (e) cytosine-guanine
(c) thymine-adenine

7. The number of different tetrapeptides that can be formed from linking one molecule each of glycine, alanine, arginine and lysine together in every possible combination is
   (a) 6   (d) 24
   (b) 4   (e) 64
   (c) 12

8. If the average nucleotide has a residue weight in mRNA of 333, what *approximate* molecular weight mRNA is needed to code for a protein 200 amino acids long?
   (a) 66,600   (d) 133,200
   (b) 200,000   (e) none of these is within 1000
   (c) 100,000

9. Which of the following is *not* necessary for protein synthesis at the time and place where synthesis occurs?
   (a) tRNA   (d) DNA
   (b) ribosomes   (e) amino acids
   (c) mRNA

10. The *anticodons* UUU, CCC, GGG, and AAA exist on tRNAs which carry the amino acids lysine, glycine, proline, and phenylalanine, respectively. In a complete cell-free system capable of protein synthesis, the artificial mRNA "poly A" will lead to the production of
    (a) polylysine   (d) polyphenylalanine
    (b) polyglycine   (e) none of these
    (c) polyproline

Energy Reservoirs

20-1 Sugars

20-1.1 Energy Storage Molecules. In Chapter 19 we discussed the structure of proteins and of the DNA and RNA used to direct protein synthesis. The specific amino acid patterns in the resulting proteins and the forces that establish secondary, tertiary, and quaternary structure lead to proteins with both specific shapes and some potentially reactive sites. The reactions at proteins (and at DNA and RNA) are vital to life on Earth. But other types of molecules, neither proteins nor nucleic acids nor the compounds from which proteins and nucleic acids are assembled, are also required in life. Many of these molecules are used to store the energy needed for the biochemical reactions in the living organism.

The reactions of life organize matter in regular patterns, an entropy decreasing process. As we noted in Chapter 11 and Figure 11-7, an entropy increasing reaction is needed to drive the entropy decreasing reactions. The flow of energy from low entropy sunlight to high entropy long wavelength radiations into space is the driving force used by life on Earth. But very few biochemical reactions can occur at times and places where sunlight is being adsorbed and high entropy light is being emitted. Therefore, the energy from sunlight must be stored in a form that can be put to use when and where it is needed later. There are a number of types of compounds involved in the energy storage process. One of the principal types is the class of compounds we call **sugars.** Sugars are the simplest of the **carbohydrate** compounds and the building blocks from which more complex carbohydrates are assembled.

Sugars are made of carbon, hydrogen, and oxygen. The hydrogen and oxygen are always present in the same ratio as in water, two hydrogens to each oxygen. We will consider here some sugars containing three carbons, five carbons, or six carbons plus some of the other carbohydrates formed by combinations of these sugars.

20-1.2 Monosaccharides, the Simple Sugars. Two sugars that contain three carbon atoms are possible. One of these is an aldehyde and the

Energy from sunlight is stored until needed in the form of compounds such as carbohydrates.

Sugars are the building blocks of carbohydrates.

other is a ketone. Each of them has one OH group on each carbon other than the carbonyl carbon. The formulas are shown in Figure 20-1. The aldehyde sugar (2,3-dihydroxypropanal) is known by the common name glyceraldehyde. The ending -ose is used to identify sugars. Glyceraldehyde is a **triose** (three-carbon sugar) and also an **aldose** (an aldehyde sugar). The other triose (1,3-dihydroxypropanone) is known by the common name dihydroxyacetone and is a **ketose** (a ketone sugar). Most of the important common five- and six-carbon atom sugars (**pentoses** and **hexoses**) we will discuss are considered to be aldoses, somewhat comparable to glyceraldehyde. One of the common hexoses is considered to be a ketose, somewhat comparable to dihydroxyacetone. We will see that none of these more complex sugars exists primarily in a form with an aldehyde or ketone structure, but the comparison to the simple forms of the trioses helps us better understand the origin of the more complex structures of the pentoses and hexoses.

The structure of glyceraldehyde is complicated by the presence of an asymmetric carbon atom. When the two possible optical isomers are written according to the convention described in Section 19-3 and Figure 19-2, we get the structures shown in Figure 20-2 and the names L-glyceraldehyde and D-glyceraldehyde. These structures and names are important because all sugars with more than three carbons are named by reference to the glyceraldehydes. When pentose or hexose sugars are written in the form showing an aldehyde or ketone group, the end of the chain away from the aldehyde (or farthest from the ketone in a ketose) has an asymmetric carbon with an OH group and a hydrogen atom adjacent to the terminal CH_2OH grouping. That arrangement matches the situation in glyceraldehyde. If the OH group is in the position written on the left in standard form (as in L-glyceraldehyde), the sugar is named as a L-sugar. If the OH group is on the right as written in standard form (as in D-glyceraldehyde), the sugar is named as a D-sugar. These arrangements are shown in Figure 20-3.

In Section 19-3 we noted that only one optical isomer of each amino acid was produced and used by most life on Earth. That was always the L-amino acid. The important common sugars on earth include several optical isomers differing in their arrangements around the asymmetric carbon atoms nearer to the carbonyl group. But at the "lower end," which is named by comparison to glyceraldehyde, the situation is similar to the amino acids. There is only

---

Sugar names end in -ose including names for the classes of monosaccharides indicating the type carbonyl (aldose and ketose) or the number of carbons (triose, pentose, and hexose).

L and D names are assigned to the sugar isomers which match the L and D isomers of glyceraldehyde at the two carbons farthest from the carbonyl.

All common sugars on Earth are the D isomers.

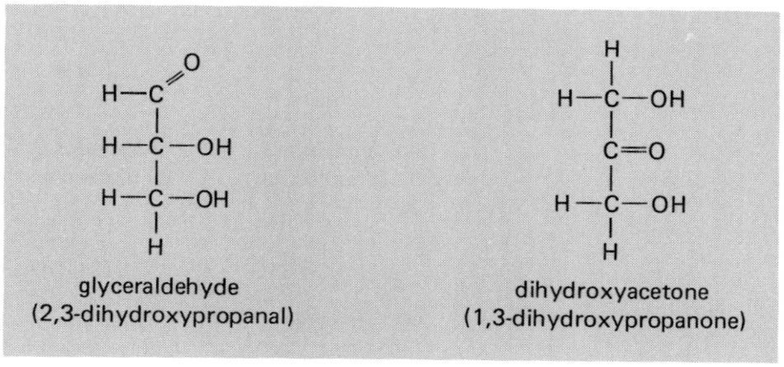

Figure 20-1 The trioses.

The aldehyde form of three-carbon sugar, shown on the left, is named glyceraldehyde. Its systematic name would be 2,3-dihydroxypropanal. The ketone triose, shown on the right, is named dihydroxyacetone. Its systematic name would be 1,3-dihydroxypropanone.

glyceraldehyde
(2,3-dihydroxypropanal)

dihydroxyacetone
(1,3-dihydroxypropanone)

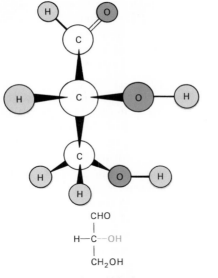

D-glyceraldehyde

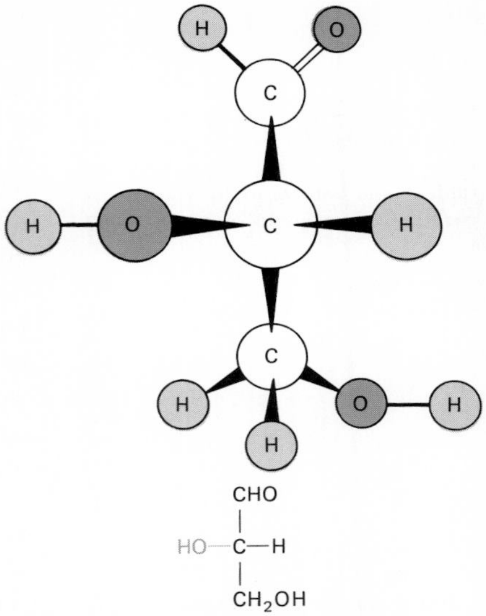

L-glyceraldehyde

Figure 20–2 L- and D-glyceraldehydes.

In each case the compound is shown in perspective (thicker bonds toward the viewer) above and in the conventional two-dimensional shorthand below. The D isomer has the OH group to the right (when written in standard form), and the L isomer has the OH group to the left (when written in standard form).

one isomer for each of the sugars and all of the common sugars have the same isomer—in this case always the D-sugar.

Although the trioses (or to be more exact, phosphate esters of the trioses) are important in the reaction sequences by which sugars are made and utilized biochemically, the principal sugars found in life are all pentoses, hexoses, or combinations of pentoses or hexoses with other compounds or with other identical sugar molecules. There is a very simple reason for this special preference for five- or six-carbon atom molecules. Five and six are the numbers of atoms needed to form ring structures that are particularly stable (see Section 17–3.8). All common pentoses and hexoses actually exist principally as five- or six-atom ring structures.

Figure 20–4 shows ring structures formed by six of the most common and important sugars. The dark lines indicate bonds on the side of the ring toward us as we view from one side and slightly above. The groups and bonds to groups above the ring are shown in color. In each case an oxygen bridge is formed between the OH position second from the noncarbonyl end (the position from which the D or L names are assigned to sugars) to the carbon

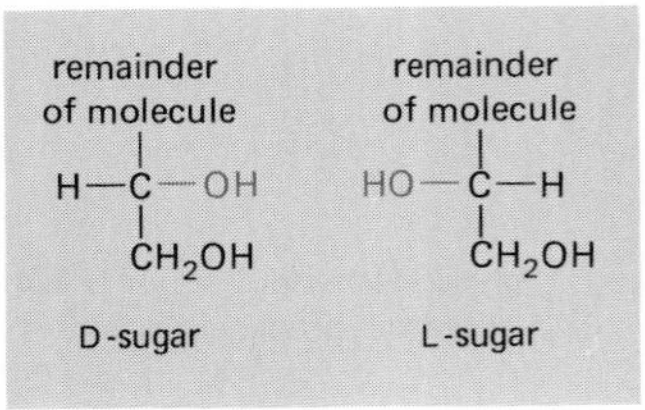

Five- and six-carbon atom sugars, which exist principally in ring forms, are naturally favored.

D-sugar   L-sugar

Figure 20–3 L- and D-sugars.

458   Energy Reservoirs   [20]

Figure 20-4 Sugar ring structures.

These figures are drawn with a side (and slightly above) view of the rings. The bonds in the ring on the side closer to the viewer are shown as thicker lines. The groups and the bonds to groups attached from *above* the ring are shown in color, whereas the groups and bonds to groups attached from *below* the ring are shown in black. Both α and β forms are shown for D-glucose. Each of the others is shown only in the α ring form, but each can also exist in a β form.

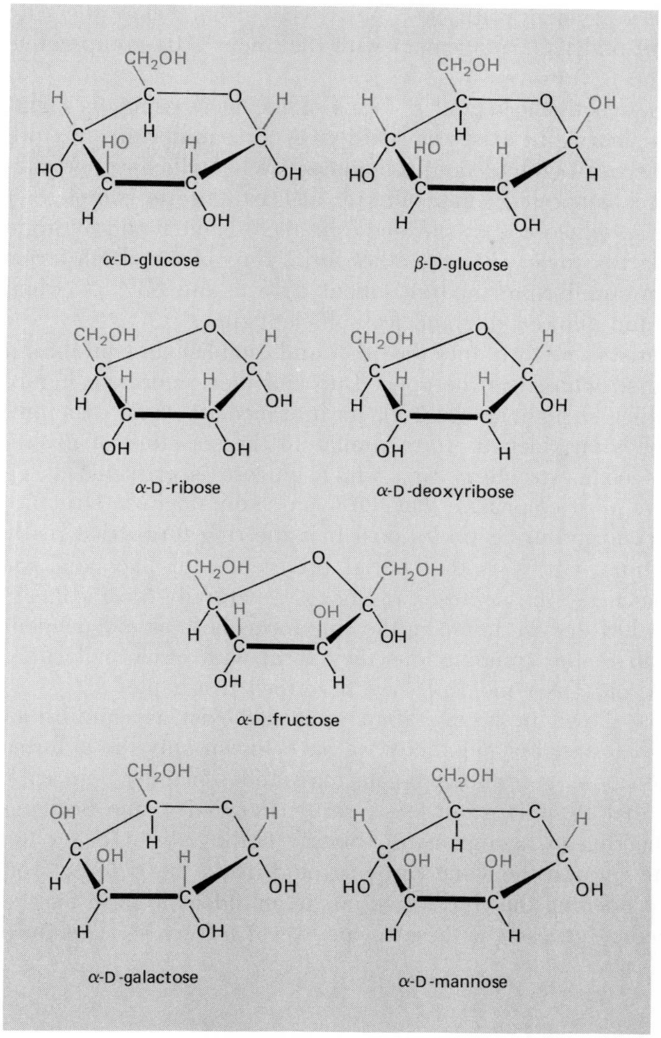

α-D-glucose

β-D-glucose

α-D-ribose

α-D-deoxyribose

α-D-fructose

α-D-galactose

α-D-mannose

that would have been a carbonyl if rings did not form. When the oxygen bridge forms, the hydrogen atom from that OH is transferred to the carbonyl oxygen, forming an OH group. Depending upon which side the bridging oxygen approached from, that OH group may be either on the side of the ring with the groups on the right side of the usual way of writing optical isomers (see Section 19-3) or on the side of the ring with the groups written on the left side of the standard notation.

We have drawn all ring structures so the "right side" (from the standard notation described in Section 19-3) groups are below the ring and the "left side" groups are those shown in color that are above the ring. When shown

[20-1] Sugars 459

in this form, the arrangement with the "new" OH group below the ring is called the α **form** and the arrangement with the "new" OH group above the ring is called the β **form**.

Figure 20–4 shows both the α and β forms of the most common sugar, D-glucose. D-Glucose normally exists in solution as an equilibrium mixture of the two ring isomers, α-D-glucose and β-D-glucose. When glucose is crystallized from water, it is converted completely (or at least close to completely) to α-D-glucose. When D-glucose is crystallized from the organic liquid pyridine (C_5H_5N), it is similarly converted to the other form, β-D-glucose. Both forms are converted to an equilibrium mixture (about 37% α and 63% β) when dissolved in water and allowed to stand for sufficient time.

The ability to convert one form into the other and establish an equilibrium shows that the ring structures can be opened up and then reformed. Figure 20–5 shows how this is thought to occur via an intermediate chain structure. The chain structure is an aldehyde form similar to the structure of glyceraldehyde. It is this chain structure from which glucose is classified as an aldose. However, we must emphasize that the open chain aldehyde structure is only a momentary form during transitions. It is the ring forms that really exist most of the time, and it is the special preference for five- and six-membered rings that make the pentoses and hexoses naturally favored molecules. Pentoses and hexoses are found in the ring forms both as independent molecules and in all of the combinations they form with other molecules, including the RNA and DNA molecules we described in Chapter 19.

All of the sugars shown in Figure 20–4 normally exist as equilibrium mixtures of α and β forms. For simplicity we have shown only the α forms of D-ribose, D-deoxyribose, D-fructose, D-galactose, and D-mannose. In each case you will find that the OH group on a carbon bonded to the bridging oxygen is below the ring (α arrangement). Simply trading that OH for the hydrogen above the ring on the same carbon would make the β form. You will also notice that opening the ring would create an aldehyde form for the momentary intermediate in each of these sugars except D-fructose. Therefore

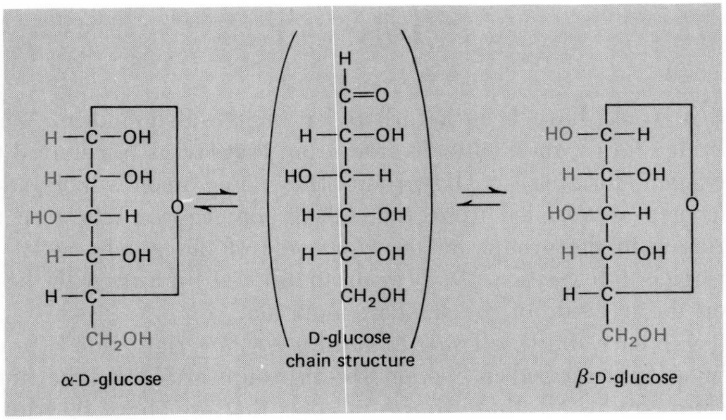

Sugar rings can form in two arrangements which are named α and β rings.

α and β rings can be interconverted via formation of an open chain form.

Figure 20–5 Conversion between α and β forms.

The α and β forms and the open chain intermediate form are all shown in the standard shorthand for easy comparison. The convention requires viewing each carbon as if the attached carbons were bent backward. That requires a twisting of the bottom carbon that would be impossible in the real ring structure molecules, which are held rigid by the ring. The groups that are actually above the ring as pictured in Figure 20–4 are shown in color to help indicate the actual arrangement in the ring structures.

all of these sugars are classified as aldoses except D-fructose, which would open to a ketone chain and is therefore a ketose.

You may have noticed that D-deoxyribose is not quite equivalent to the other sugars we have shown. Deoxyribose has one carbon with no bonded oxygen atoms (two hydrogen atoms bonded instead of the usual hydrogen and OH group). That slight difference has obviously been used effectively in the genetic code where a master code on DNA is able to form similar, but not quite identical, working material in the form of RNA. Other compounds that are closely related to sugars (but not within the pattern of the standard sugars) are also biologically important. Some of these are sugar derivatives which will be discussed in Section 21–4 on metabolism. Another closely related compound that is sometimes called a modified sugar is ascorbic acid (vitamin C), which will be discussed in Section 22–6. Ascorbic acid is a partially oxidized hexose that exists in a ring structure very similar to sugars.

20–1.3 Disaccharides.

Monosaccharides can link in particular patterns by condensation to form disaccharides and larger units.

The sugar compounds we have described so far are called monosaccharides, meaning single sugars. Those single sugar molecules are able to link together by condensation reactions. If two sugar molecules are linked, the product is called a disaccharide, meaning two sugars. When three monosaccharides are linked, the product is a trisaccharide. Chains formed by linking many monosaccharides are called polysaccharides. We will begin a study of these linkages by examining the simplest type, the disaccharides.

The disaccharide most familiar to us is sucrose, the form of sugar sold commercially as granulated sugar. Sucrose is formed by a condensation reaction between D-glucose and D-fructose. The product can be crystallized and sold in solid form, which is easier to ship and handle than the syrups obtained from other sugars. Sucrose can be obtained from sugar beets, sugar cane, and from other plant sources that are not as good for commercial sugar production.

The structure of sucrose is shown in Figure 20–6. The condensation reaction involves the OH groups from each molecule which are on the carbon bonded to the bridging oxygen. That is the OH which determines whether the sugar is in the α or β form. Sucrose is formed by an oxygen bridge between β-D-fructose and α-D-glucose. The fact that the condensation linkage is favored with one particular set of ring forms is typical of bonding between sugars. We will see that a change in the ring forms can change the nature of the product and that such a change causes the difference between starch and cellulose.

Another common disaccharide is lactose, which is found in milk. Lactose is formed from D-galactose and D-glucose. In this case only one of the sugars (galactose) bonds via the OH group which determines α or β form. That OH group is said to be on the number 1 carbon—the carbon at the "aldehyde" end of the sugar. The glucose does not bond through its number 1 carbon.

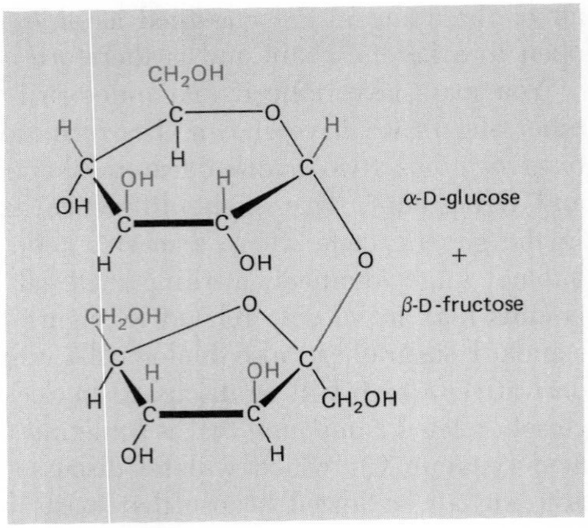

Figure 20-6 Sucrose

Bonds in the rings are shown thicker when they are closer to the viewer, groups and bonds below the rings are shown in black, bonds above the rings are shown in color, and groups above the rings are shown in color except for the bridging oxygen (which is below one ring and above the other).

Instead, the OH group on the number 4 carbon (counting from the number 1 carbon "end") forms the bond. That OH group is on the α side of the ring—below the ring as we have been writing the rings. It bonds to galactose which is in the β form. The glucose is in the α form when this bond forms. The result is the structure of β-D-galactose-α-D-glucose shown in Figure 20-7. The 1,4 bonding (carbon 1 on one ring to carbon 4 on the other) is the most common linkage found between sugars.

A third common disaccharide is maltose, which is shown in Figure 20-8. Maltose is formed as a product of hydrolysis of starch, and it shows us the basic structure of starch. Maltose consists of two units linked by condensation, but it only has one kind of monosaccharide. Both of the units are D-glucose. The linkage is a 1,4 bond similar to that in lactose. Both glucose units are in the α form. However, since the OH at the number 4 carbon is also α, one of the rings must be turned over (as shown in Figure 20-8). Sometimes it is convenient to write the structures showing sharply bent bonds so the sugar can be shown in the same arrangement every time and be more easily recognized. Such a structure (which is less like the real molecules) is shown

Sucrose is a glucose-fructose disaccharide, lactose is a glucose-galactose disaccharide, and maltose is a glucose-glucose disaccharide.

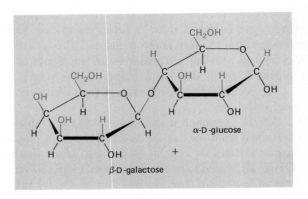

Figure 20-7 Lactose (drawn according to the same conventions as used in Figure 20-6).

Figure 20-8 Maltose (drawn according to the same conventions as used in Figure 20-6).

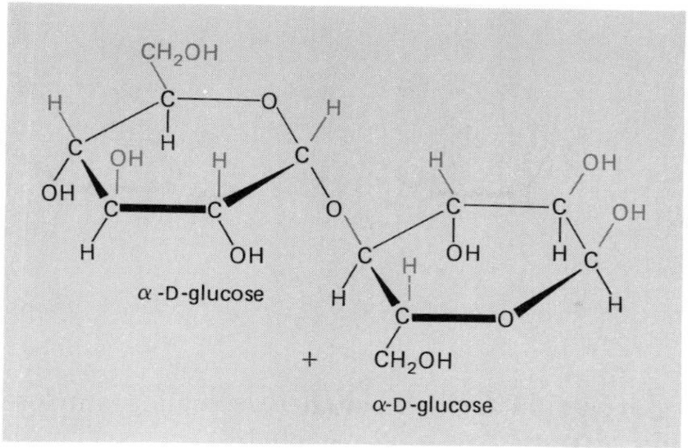

α-D-glucose + α-D-glucose

Figure 20-9 Maltose (twisted view).

Drawn according to the same conventions as used in Figures 20-4 and 20-6. The bond through the bridging oxygen has been twisted so both rings can be shown from the same side—the view used in Figure 20-4.

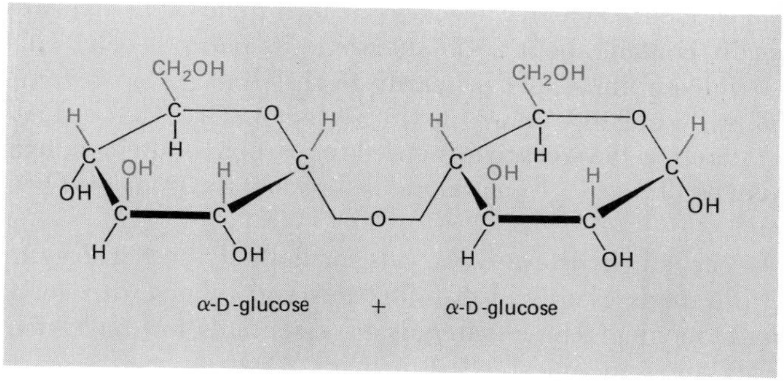

α-D-glucose + α-D-glucose

in Figure 20-9. We will use such structures to describe starch and cellulose structures, but that is only for writing convenience. Real chemical bonds do not bend around corners.

20-2 Starches and Cellulose

Starches are polymers of α glucose rings linked by 1,4 linkages or by 1,4 and 1,6 linkages.

20-2.1 Starches. Although the monosaccharides are the compounds formed in energy storing reactions and utilized to provide energy for later biochemical needs, the bulk of the actual energy storage in plants and animals is in the form of more complex molecules. One major form of energy storage is in the polysaccharides we call starches and cellulose. All starches and cellulose are polymers of glucose, but there is a sharp, biologically significant difference in the way they are linked to form the polymers.

Starches contain two types of linkages between glucose rings, the 1,4 linkage described for maltose and a 1,6 linkage. The 1,6 linkages are less common, but they can provide branching in the polymer chains by allowing a single glucose unit to be bonded to three others—one at carbon number 1, one at carbon number 4, and one at carbon number 6, which is the CH_2OH group not in the ring. All of the glucose units in starches are in the α form.

Figure 20-10 shows a portion of the simplest form of starch, a single chain

[20-2] Starches and Cellulose   463

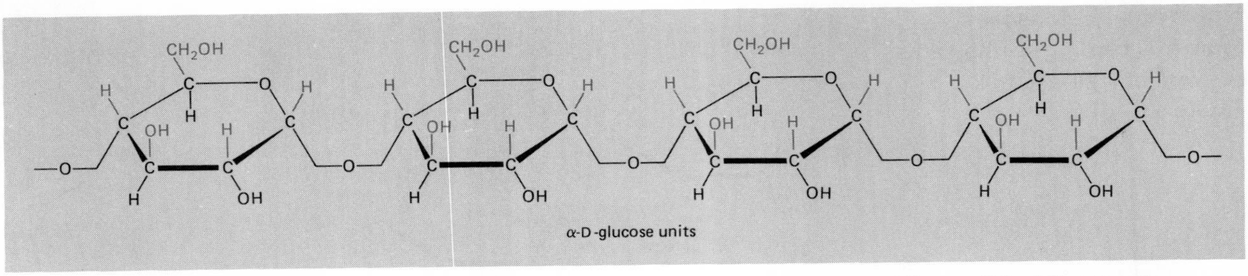

[Figure: Amylose starch structure showing α-D-glucose units]

α-D-glucose units

of 1,4 linkages between α-D-glucose units. Such starches are called amylose and typically occur with about 300 glucose units per chain.

Figure 20–11 shows a portion of the branched starch formed when some 1,6 linkages are also formed. Such structures occur in starches called amylopectins, which typically contain about 2000 glucose units per molecule. This same pattern of α-D-glucose units held primarily by 1,4 linkages with some 1,6 linkages providing branching is present in glycogen. Glycogen, which is also called animal starch, is the storage form used for carbohydrate in animal liver and muscle tissues. Typical glycogen molecules contain from 6000 to 30,000 glucose units.

When starches are partially hydrolyzed, various smaller polysaccharides can be formed, some with single chains of 1,4 linkages and others with some 1,6 linkages causing branching. These materials are essentially low molecular weight starches. They are sometimes called dextrins. Dextrins are used for adhesives such as that used on postage stamps.

Figure 20–10 Amylose starch. (drawn according to the same conventions as Figure 20–9, including the twisting of the bridging oxygen bonds).

Figure 20–11 Amylopectin or glycogen starch. (drawn according to the same conventions as Figures 20–6 and 20–9, including the twisting of the bridging oxygen bonds except for the one 1,6 bond at the branching site).

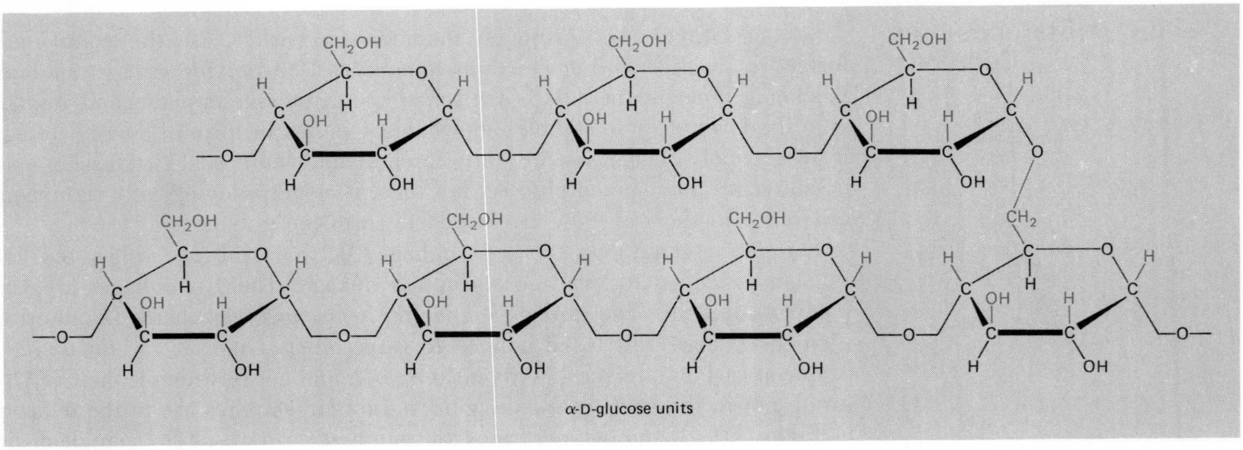

20–1.2 Cellulose.

Cellulose is the polymer that forms much of the structural material in plants. Wood and cotton are examples of cellulose structural materials. Cellulose is a polymer formed by 1,4 linkages between

[Figure: Cellulose structure showing α-D-glucose units]

α-D-glucose units

464   Energy Reservoirs   [20]

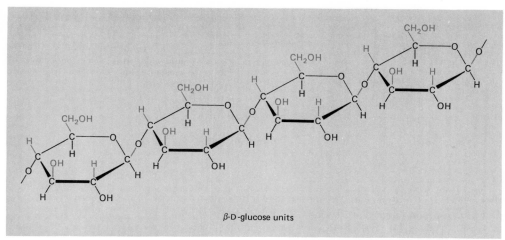

Figure 20-12 Cellulose (drawn according to the same conventions as Figure 20-6. Note that bridging bonds do not have to be twisted to show all units from the same viewpoint when the rings are β instead of α).

Cellulose is a polymer of β glucose rings linked by 1,4 linkages.

glucose units. As such it is identical to amylose starch except for one thing—the type of glucose being linked. Cellulose consists of glucose units in the β form. Figure 20-12 shows a portion of a cellulose molecule with its 1,4 linkages of β-D-glucose units. Typical cellulose molecules contain from 1000 to 6000 glucose units, *all in the β form.*

The β form linkage is significant because the enzymes that hydrolyze the usual α linkages in starch cannot fit the different geometry of the β linkages. As a result, man and most animals and plants cannot break down cellulose to use it as an energy source. When the cellulose polymer is formed, the material in it is lost from the available stored energy pool for biochemical processes. Some species have developed the ability to break the β-1,4 linkages in cellulose, but that ability is rare among life forms on Earth. One notable example is the ability of a protozoan in the digestive system of termites to break the bonds in cellulose.

20-3 Fats

Triglycerides are the principle category of fats.

Triglycerides are esters of three fatty acid molecules with one glycerol molecule.

The other major form for energy storage besides polysaccharides is in the form of fats. Fats are not simply polymers of sugar type units; they are formed from quite different chemical components. We will discuss here the principle category of fats, the **triglycerides.** These compounds are principally hydrocarbon-like, as opposed to the highly oxygenated carbohydrates. In Chapter 22 we will discuss a number of other types of biologically important compounds that tend to be associated with the fats because they are also principally hydrocarbon-like. The group of compounds including the fats and those associated with fats are called by the general name **lipids.**

Triglycerides are esters of organic acids, called fatty acids, and 1,2,3-propantriol, which has the common names glycerin or glycerol. The fatty acids are all single chains (no branching) with even numbers of carbon atoms.

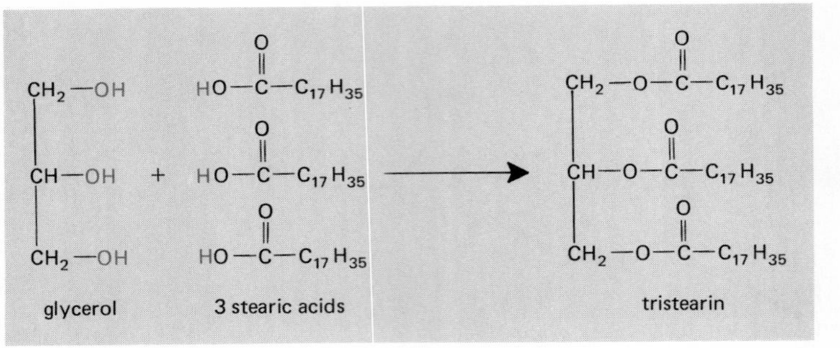

Figure 20-13 Tristearin, a fat.

The condensation of a glycerol molecule with three molecules of stearic acid yields the triglyceride tristearin.

Although all even numbers between 4 and 20 can be found easily in some animal or plant source, the most common lengths are 16 and 18 carbon atoms. Figure 20-13 shows the triglyceride formed from three molecules of the 18-carbon saturated acid, stearic acid, and one molecule of glycerol.

Fats do not have to be made from only one kind of fatty acid. Figure 20-14 shows the triglyceride formed from one molecule of glycerol and one molecule each of stearic acid, palmitic acid (the 16 carbon saturated acid), and oleic acid (an 18 carbon acid with one carbon–carbon double bond).

Unsaturated fatty acids, those containing one or more carbon–carbon double bond, make up an important part of many vegetable fats. The most common unsaturated fatty acid is oleic acid, but several others are also fairly common. Figure 20-15 shows the structures of four important unsaturated fatty acids, each containing 16 or 18 carbon atoms. These structures show a repeating pattern exists in the biochemical synthesis of these fatty acids. Not only are carbon atoms added to the chains in twos, as shown by the even numbers of carbons in every fatty acid, but double bonds tend to be introduced in particular positions. In the group shown in Figure 20-15 every fatty acid has a double bond between the ninth and tenth carbon atoms (with counting starting from the acid end), two have double bonds between the twelfth and thirteenth carbon atoms, and one has a third double bond between the fifteenth

Fatty acids fit regular patterns, all single (unbranched) carbon chains with even numbers of carbon atoms and with relationships in double bond positions of the most common unsaturated fatty acids.

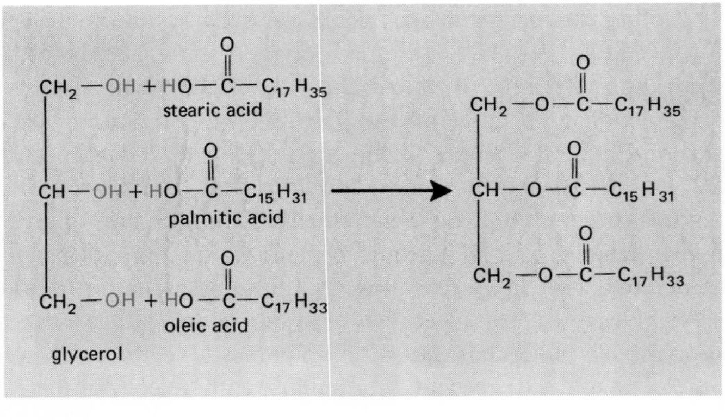

Figure 20-14 Triglyceride with three different fatty acids.

A glycerol molecule must undergo condensation with three fatty acid molecules to form a triglyceride, but the fatty acid molecules do not all have to be the same kind.

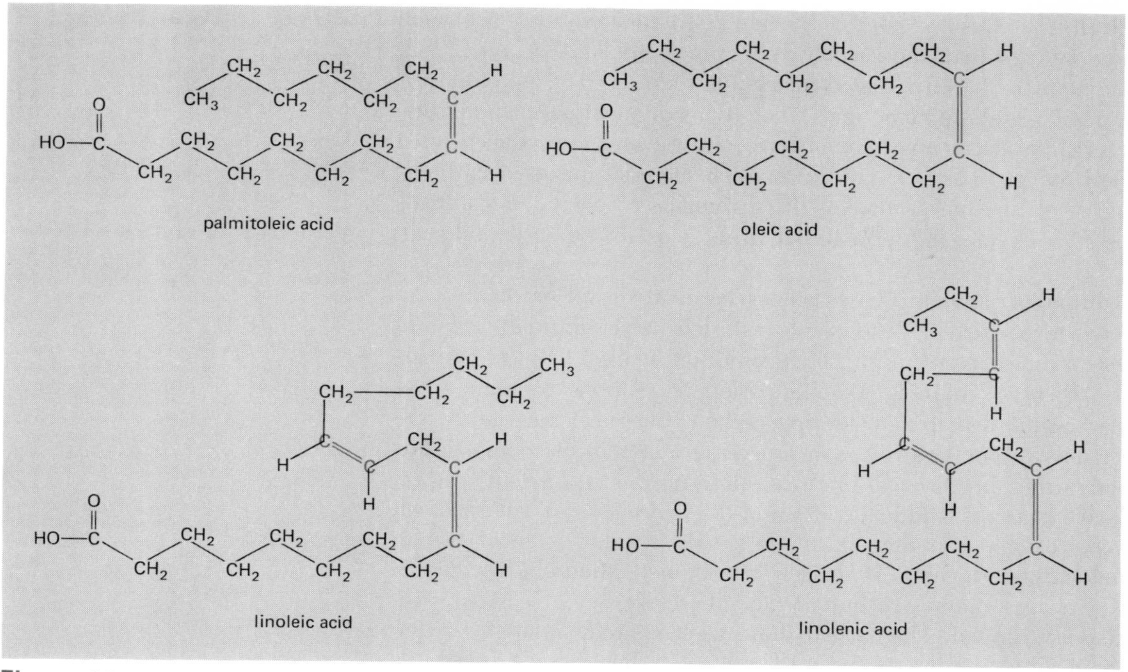

Figure 20-15 Unsaturated fatty acids.

At each of the double bonds, which are shown in color, the carbon chain enters and leaves in a *cis* arrangement. That causes a folding of the carbon chain.

Animal fats are highly saturated but vegetable fats contain mostly unsaturated fatty acids.

and sixteenth carbon atoms. There are other unsaturated fatty acids without double bonds in any of these positions, but it is clear that at least some of the common ones (the four shown) are closely related to each other.

Those fats that contain high fractions of unsaturated fatty acids tend to be liquids (oils), whereas those with mostly saturated fatty acids with long chains (16 or 18 carbons) are solids. It is possible to have liquid fats that are fully saturated, usually when the carbon chains are shorter. Butterfat has a lot of the four- and six-carbon saturated fatty acids.

At one time it was fashionable to hydrogenate vegetable oils so they could be sold as solid shortenings. They brought a higher price as solids. Now research has shown that a high proportion of the unsaturated fatty acids seems to reduce the chance of some heart and circulatory problems, so more of the oils are being used in their highly unsaturated liquid forms. Even the partially hydrogenated vegetable shortenings (which are still produced in large quantities) have fairly high unsaturation. A shortening like Crisco (solid) contains only about 20 to 25% saturated fatty acids whereas 65 to 75% of its fatty acid content is oleic acid and 5 to 10% is linoleic acid. That is much more unsaturated than animal fats such as lard. The unhydrogenated vegetable oils are even more highly unsaturated. Some vegetable oils, notably safflower oil and to a lesser extent corn oil, are much more highly unsaturated than others. It is these unsaturated oils, not the solids, that now command the premium prices.

[20-3] Fats

20-4 Fossil Fuels

20-4.1 Coal.
The fats discussed in Section 20-3 are a more concentrated form of energy storage than the carbohydrates in Sections 20-1 and 20-2. The oxygen content has been reduced in fats, so there is more potential for later reaction with oxidation. Sometimes when living organisms die and decay, part of their organic matter may be converted into even more concentrated forms of stored energy. Large quantities of such material are used as fuels. Since the source of this fuel is the fossilized remains of plants and animals from millions of years ago, the term **fossil fuels** is used to describe all such fuels.

Fossil fuels such as coal and oil are fossilized concentrates (with water removed) of dead plants and animals.

Large quantities of plant life have been accumulated and gradually converted to deposits of carbon. The plant materials were originally rich in carbohydrates, particularly cellulose which could not be used by most animals or the plants as energy sources. Over a period of millions of years these carbohydrates became dehydrated, leaving carbon. Chemical reaction to the form of water plus carbon is a common process in reaction of cellulose. For example, wood burned in a fire will first react to form water (plus some carbon oxides) and leave charcoal behind. After all of the hydrogen content has been removed, excess oxygen can then burn the carbon residue to form carbon monoxide and carbon dioxide. Hydrogen reacts first when the oxygen is limited. When plant remains are buried (and therefore out of contact with air) oxygen is very limited. The carbohydrate itself contains enough oxygen to combine with all of the hydrogen and form water, but there is no excess oxygen left to react with the carbon. Since formation of water is favored, all or very nearly all of the carbon content is eventually left as unreacted carbon. In shallow, relatively recent beds of plant matter the residue is in the form called peat. Longer times and higher pressures convert it to a more completely solid carbon form, which is coal. Different conditions and different starting materials create the various types of coal.

Coal is not pure carbon. Some hydrocarbon materials are also formed (perhaps from fats and proteins in the plant residues) and other elements less reactive than hydrogen may be reduced to the elemental form as the carbon was. Sulfur is a particularly common material in coal. Sulfur is present in all living organisms in the proteins formed using the essential amino acids cysteine and methionine (see Section 19-4). Under the reducing conditions of the plant residues decaying in the absence of oxygen, the sulfur is converted to elemental sulfur. Sulfur content varies in different coals, but sulfur is always a problem in all coal and also in other fossil fuels. When these fuels are burned, the sulfur forms sulfur oxides, which are a serious pollution problem in many areas.

The reducing conditions during fossil fuel formation often also convert sulfur compounds to sulfur.

When carbon is desired for manufacturing (mostly for iron and steel production as noted in Section 16-6) impurities are driven out of coal by heating. The somewhat purified carbon product is called **coke** and the process is called **coking.** Hydrocarbon materials are driven out of the coal by the heating. The hydrocarbon materials, called coal tar, are collected and used as a source of some hydrocarbons and in a variety of other ways. Only certain specific types of coal are suitable for coking. Coking coal must have a very

Carbon for some processes can be obtained by coking coal.

low sulfur content, and it must be pulverized to a fine powder so the coal tars can be efficiently removed.

20-4.2 Oil and Natural Gas.

Some organic residues are converted into hydrocarbons instead of carbon. These hydrocarbons collect in geological formations and are recovered as oil and natural gas. Some of these oil-related materials are in tars or trapped in rock from which they cannot easily be removed.

Coal could form from carbohydrates by loss of water, but the hydrocarbons in oil and gas are more like fats and proteins than carbohydrates.

Although the exact history of oil and gas formation is uncertain, these hydrocarbons may be formed from residues containing more fats than the plant residues that form coal. Fats (and proteins) contain less oxygen than carbohydrates, so the hydrogen content cannot be completely removed as water during oxygen-free decay. Over long periods of time the material may be formed into a large number of quite different organic molecules. These are mostly hydrocarbons as much of the oxygen is removed by formation of water, but crude oils also include some complex oxygen-containing hydrocarbon derivatives. In addition there are compounds that include sulfur, nitrogen, and other elements, including a number of metals. Sulfur is also present in natural gas in the form of H_2S gas. Oil and gas with high sulfur contents (sometimes as much as 1 or 2% sulfur) are called **sour oils** or **sour gas** because of the acid the sulfur can produce in later processing and use, which causes serious problems.

Oil and gas formation leads to a varied mixture of compounds, mostly hydrocarbons.

Sour oils and gas are those with high sulfur content.

20-5 Recovery of Energy Sources

The existence of stored energy in the forms produced by life is not enough to assure the availability of that energy. The energy sources must be recovered in usable form. Some energy sources are easily utilized, whereas others are virtually inaccessible. Also some forms are directly usable by life, whereas other energy sources are only useful via reactions such as combustion.

Sugars are the energy source most easily used by living organisms.

Sugars are the energy sources most directly produced from sunlight energy, and they are the most easily recovered and used by living organisms. Sugars are easily transported in water solutions and are efficiently absorbed in digestion. Metabolism of sugars is described in Section 21-4.

As the energy source is converted into forms other than sugars it becomes less accessible to living organisms.

As sugar is tied up in formation of starches, it becomes less accessible biologically because it must first be converted back to sugars before it can be digested or metabolized. If glucose is made into cellulose instead of starch, it becomes recoverable only to a few organisms. Fats are more biologically recoverable energy sources than cellulose, but fats are not digested as efficiently as sugars and their metabolism requires extra steps.

When the stored energy is converted into fossil fuels, the coal, oil, and gas are not in the chemical forms produced directly by life. These forms are also not directly usable by life, except for some compounds that certain bacteria can use. Therefore coal is simply inert biologically and oil or gas can actually kill off life by interfering with access to the usable food sources and oxygen needed by animal life.

Recovery of energy sources for nonbiological uses is largely a matter of separating the fuel from nonfuel material. Animals separate their food sources through selection and digestion. Similar selection must be practiced with fossil fuels or cellulose. Selecting wood without including rocks or soil is not too difficult, but much of each tree is usually left behind and wasted because of handling and transportation difficulties. Recovery of coal seems more straightforward because the solid coal is fairly easy to handle, but again there is considerable waste of both energy and land in gaining access to the coal and separating it from surrounding material.

The separation problem becomes even more serious with oil and natural gas, which are intimately mixed into other material. Oil and gas are usually recovered through wells instead of mining. Gas will usually flow to a well from which it can be removed, but some rock formations make such flow difficult. Oil is even less easily removed. Sometimes rock formations are shattered to provide more channels for flow to the wells. As oil is removed it is often replaced by ground water. Sometimes water is pumped into the rock formation to help flush the oil out to the producing wells. This technique is called water flooding. Unfortunately, some of the rock remains "wet" with oil and pockets of oil are cut off between flow channels that become filled with water. In many cases two thirds or more of the oil may remain inaccessible to the wells.

When the oil is largely viscous high molecular weight material, recovery becomes even more difficult. Thick, tar-like materials flow to the wells so poorly that very little would be recovered. In some cases the flow has been increased by use of "fire flooding." A fire is started in the underground formation and air is pumped in to keep the fire going. The hot combustion products are forced through to the producing wells and some of the thick oil is carried to the wells with the hot gases. But like water flooding, only the oil near the flow channels is affected. Large pockets of oil are left inaccessible.

Two major sources of oil that will not flow to wells at all have been suggested for future recovery. One is oil shale. Oil trapped in shale rock cannot be easily released, but the entire rock can be mined and then crushed and heated to release the trapped oil. Heating the rock while still in the ground (a procedure similar to fire flooding) has also been suggested. Large deposits of oil shale exist in the western United States. The other major nonflowing oil source is tar sands. Viscous, high molecular weight tars exist in sand formations that could be easily strip mined and processed with little or no crushing, but the tars would be hard to drive out of the sand and hard to handle after they were separated. The tar sands cannot be ignored, however, because there are very large amounts available. The Athabasca tar sands, which cover a large area of Canada north of Edmonton, Alberta, may have more hydrocarbon content in that one area than all the oil fields in the Western Hemisphere. Recovery of tar sands and oil shales is just a matter of time

Separation of fossil fuels from nonfuel is often difficult.

Oil shale and tar sands are major potential energy sources.

and developing technology. The increase in oil prices during the 1973 energy crisis made commercial recovery of nonflowing oil economically attractive.

Nuclear energy is another major energy reservoir. Recovery of nuclear energy by nuclear reactors is described in Section 15–5.

20–6 Problems Associated with Fossil Fuels

20–6.1 Sulfur Oxides.

There are a number of problems associated with our extensive use of fossil fuels, some of which are discussed in Section 21–8. One of the most widespread problems is the sulfur and sulfur compounds associated with fossil fuels. The same oxygen-free conditions that allow the reduced (nonoxidized) forms of carbon (free carbon, hydrocarbons, and hydrocarbon derivatives) to be accumulated also lead to reduced forms of sulfur. Although the percentage of sulfur varies greatly, some sulfur is present in all living organisms from which fossil fuels could form. Therefore at least some sulfur is present in all fossil fuels, unless some process (natural or artificial) has specifically removed the sulfur.

Sulfur burned with fossil fuels forms sulfur oxides which leads to acids harmful to life and property.

Sulfur can cause a number of problems. It can interfere with processes using the fossil fuel, such as oil refining processes or steel production. In Section 20–6.2 we will describe a particular environmental problem (acid mine drainage) caused by sulfur. But the most common problem is the formation of sulfur oxides when fossil fuels are burned. Coal and fuel oils often have fairly high sulfur content, so their use for heating and generating electricity in populated areas produces large amounts of SO_2 (plus some SO_3). SO_2 reacts with water to form sulfurous acid, H_2SO_3, and is eventually oxidized to the strong acid sulfuric acid, H_2SO_4. Therefore the sulfur oxides are serious air pollutants. They both irritate human tissues (eyes, noses, and so on) and cause very acidic rainfall, which dissolves carbonate rocks. The defacing of stone sculptures in cities such as Venice, Italy, has occurred through the effect of acid on the $CaCO_3$ in limestone and marble.

20–6.2 Acid Mine Drainage.

Mines opened to oxygen lead to oxidation of sulfur and acidic mine drainage.

The oxidation of sulfur to form acid can be a problem even when the sulfur is not burned with the fossil fuel. Sulfur in coal beds is in a reducing chemical environment because of the lack of oxygen. When mining opens these beds to the air the combination of sulfur, oxygen from the air, and ground water results in the formation of H_2SO_4 in the ground water flowing through the mine. When this very strongly acidic ground water reaches surface streams (either through mine entrances or via seepage through the ground) it creates acid conditions in the rivers that can kill all stream life and make the water unsuitable for use by downstream communities. Acid mine drainage is a very serious water pollution problem in many coal mining areas.

Although some efforts are made to lessen acid mine drainage by blocking water flow from mine entrances, the only way to prevent the problem is to keep the sulfur in the mines from being exposed to oxygen. If old mines were completely sealed up, the sulfur would remain in the nonoxidized form and no acid would be formed in the ground waters. Sealing of mines is usually difficult, and the amount of oxygen available during mining may already have oxidized enough sulfur to cause problems before the mine can be sealed. One solution to that problem is to keep the mine sealed and free of oxygen even while mining is being done. "Space suit" type outfits are being developed for use by miners in oxygen-free mines. These suits will also protect the miners from coal dust, which appears to cause a debilitating illness called black lung disease. The oxygen-free mines will also be safer because there is no oxygen to support fires or the reactions that cause mine explosions. This attempt to develop oxygen-free mining is an example of man adjusting to his natural environment. The natural environment in underground coal beds is an oxygen-free reducing chemical environment.

Oxygen-free mining could avoid several problems.

20–6.3 Oil Spills.

Recovery of energy resources causes a number of types of environmental disruptions. Sometimes these disruptions are intentional, as in the case of strip mining. In those cases problems can be solved by decision making, leading either to stopping the activity or preplanning actions to reestablish a favorable environment after the disruptive activity has been completed. But some other environmental problems result from accidents whose extent and location cannot be known in advance. Oil spills constitute one of the main categories of accidental environmental crises.

Accidental crises such as oil spills can harm plants and animals.

Large quantities of oil and oil products are transported in ships or barges (which can sink), pipelines (which can leak or break), and railroad cars or trucks (which can have accidents). Large quantities are stored at any given time. Under these conditions, the occurrence of spills is not surprising, but their number, size, and location is uncertain.

Although spills create several kinds of problems, including fire hazards and air pollution, the most persistent problems are due to coating of surfaces. Oil may be adsorbed on plants and animals. That may cut off the organism's access to essential substances (such as oxygen or nutrients) or interfere with important functions (such as the use of feathers for flying or for water birds in floating). Oil may also coat sand, rocks, or soil. Since oil is notoriously hard to separate (perhaps only one third is recovered in oil fields) and is not easily usable biologically, a large spill can permanently change the area in which it occurs.

The best control for oil spills is to contain the oil in the smallest possible area and separate as much of it as possible while it is still fairly concentrated. Devices such as barriers to contain the spread of oil on water and mechanical separators to separate oil and water are being developed, but simple methods such as adsorption of oil onto straw (which can then be collected and burned) are also useful. All methods suffer from the emergency nature of the spills. Leaks, such as a fuel oil tank spill which contaminates a sewage system, are neither anticipated nor easy to trace when discovered. And accidents such as shipwrecks are most likely to occur in remote locations and under bad

Oil spills are hard to clean up and clean-up efforts suffer from the inability to prepare completely and supply help immediately.

weather conditions. Therefore, oil spills will continue to occur and cause damage as long as we continue to transport and use oil. At best, we can reduce the damage by investing thought to anticipate possible problems and hardware to make accidents less likely. We can also stockpile equipment and material needed for possible crises and maintain surveys and citizen alertness to provide early discovery of spills.

Fossil fuels, particularly oil and gas, are limited.

20-6.4 Limited Supplies. The ultimate problem related to use of fossil fuels is simply that we must eventually run out. The present supplies of coal, oil, and gas are the product of millions of years of plant growth. Natural processes cannot even come close to replacing them at the rate of present use. Already there are shortages developing in the most convenient fossil fuel to recover and use—natural gas. Oil reserves will last longer, particularly in the forms that cannot be recovered efficiently by present technology. Many uses of oil and gas will be taken over by coal (particularly coal converted to artificial gas fuel) which may last for several centuries, but eventually man must convert to other energy sources.

Other energy sources must eventually replace fossil fuels, which will be needed as chemical raw materials.

As fossil fuels become scarce, they will be used less as energy sources and more as raw materials for chemicals. Eventually the carbon and hydrocarbon materials will cease to be used as fuels and might more properly be called fossil chemicals instead of fossil fuels. Future generations may have to use precious energy from nuclear power, solar energy, tides, and geothermal sources to synthesize needed hydrocarbons. Electrolysis of water to produce H_2 gas will probably be the first step in the series of reactions needed to produce the reduced chemical compounds such as hydrocarbons. When that becomes necessary, people will look back on our period of history as a time of terrible waste when complex, life-related compounds were consumed for low value uses such as burning for the heat energy released.

Skills Expected at This Point

1. You should be able to define the terms aldose, ketose, triose, pentose, hexose, L-sugar, D-sugar, monosaccharide, disaccharide, starch, cellulose, 1,4 linkage, 1,6 linkage, triglyceride, fossil fuel, and sour oil or gas.
2. You should be able to describe the structure of glucose and other common sugars. You should include the general tendency toward ring structures, a description of α and β forms, and, if asked, show the complete structure of D-glucose.
3. You should be able to describe both the structural difference between starch and cellulose and the biological significance of that difference.
4. You should know and be able to list the components found in natural triglycerides.
5. You should be able to describe the conditions for formation of fossil fuels, the probable source materials for each type of fossil fuel, and the impurity problems and production problems associated with fossil fuels.

6. You should be able to list and explain the problems associated with fossil fuel use that were described.

Exercises

1. List all the ways in which the actual form of D-glucose is different from the structures of the trioses.
2. List all of the forms in which D-glucose appeared in this chapter (singly or as part of a larger unit) and describe how each of them differs from the six-carbon chain aldose structure.
3. Describe (in *your* words, not by copying a figure) the parts and ring linkage arrangements in each of the common disaccharides discussed in this chapter.
4. **(a)** How are unbranched starch chains put together? **(b)** How does chain branching occur in starch?
5. The most common fats are triglycerides. The fact that triesters of glycerol form so extensively is an example of the regularity of nature, repeating the same pattern (use of glycerol with fatty acids). List two other kinds of natural regularity or repetitiveness that show up in the triglycerides.
6. The quantity of coal on Earth is much greater than the quantity of oil and gas. Explain why this should be so. (Careful thought about the Earth and its chemistry will uncover factors beyond the simplest explanation if you want to look beyond the coverage in this chapter.)
7. In each of the following cases list one example of a difficulty in the recovery and use of energy resources which is related to the scientific principle stated.
   (a) Entropy tends to increase.
   (b) Like dissolves in like.
   (c) Some chemical environments are oxidizing environments and others are reducing environments.
8. Fats are known to be higher in calorie count (usable energy in human diet) than carbohydrates. Explain the difference in terms of their chemical compositions.
9. Write the formula for **(a)** a monosaccharide, **(b)** a disaccharide, **(c)** starch, **(d)** cellulose, **(e)** a triglyceride.
10. Draw a sketch showing enough to indicate all the important general features of the structure for each of the compounds in question 9. Label each of those important general features on your sketches.

Test Yourself

1. Explain the meaning of each part of the name α-D-glucose.
2. Draw a structure (the usual way of writing optical isomers is acceptable) of α-D-glucose.
3. Write the formula of the triglyceride involving the 16-carbon saturated acid as the only kind of fatty acid.

4. State briefly the difference between starch and cellulose.
5. Which of the following is α-D-glucose?

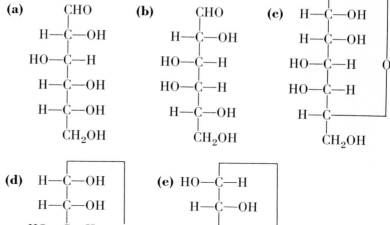

6. Choose the three words that best answer all three questions. What units is cellulose made up of? What ring configuration does it have? Cellulose is linked into a chain by linkages of what type?
   (a) D-galactose; aromatic; α
   (b) D-glucose; β; 1,4
   (c) D-glucose; α; peptide
   (d) D-mannose; α; ester
   (e) D-glucose; α; 1,4
7. Choose the words and phrase that best answer the question and complete the sentence. What must react with fatty acids to form triglycerides? The fatty acids always have
   (a) hydrocarbons, straight chains without branching
   (b) glyceric acid, straight chains without branching
   (c) glycerol, even numbers of carbons
   (d) glucose, even numbers of carbons
   (e) triol, an OH on each carbon
8. Which of the following is a carbohydrate?
   (a) a triglyceride
   (b) coal
   (c) starch
   (d) a polypeptide
   (e) a fatty acid
9. Which of the following is a fossil fuel?
   (a) a triglyceride
   (b) coal
   (c) starch
   (d) a polypeptide
   (e) a fatty acid

10. Which of the following is the common type of fat?
- **(a)** a triglyceride
- **(b)** coal
- **(c)** starch
- **(d)** a polypeptide
- **(e)** a fatty acid

11. Choose the words or phrases that best answer the questions. The formation of what substance from burning fossil fuels has led to defacing of stone sculpture? This substance is formed from burning what impurity in fossil fuels? Why does this impurity tend to be present?
- **(a)** free radicals, aromatic, rings form under the high pressures at which fossil fuels form
- **(b)** solvent, oxygen containing, oxygen is so abundant on earth
- **(c)** base, metal, it is complexed by the hydrocarbons
- **(d)** acid, sulfur, it was reduced to sulfur as the fossil fuels were formed
- **(e)** acid, carbon, fossil fuels are full of soot

21 Using Energy Reservoirs

21–1 Types of High Energy Compounds in Cells

ATP is biochemically important in storing energy and making energy available for reactions.

High energy compounds have a large amount of readily available stored energy per molecule.

The compound adenosine triphosphate, abbreviated ATP, was mentioned in Section 19–10. A reaction sequence for formation of ATP was shown in Figure 19–13, which is repeated here as Figure 21–1. Energy is required to form ATP because each step brings negative charges together. When ATP is hydrolyzed, energy is released as the negative charges are allowed to move apart. ATP is a vitally important substance in biochemistry because energy is stored in the form of ATP and is then available for use in other reactions.

ATP is referred to as a high energy compound in biological systems because the amount of readily available stored energy per ATP is fairly high. Hydrolysis of 1 mole of ATP to ADP (adenosine diphosphate) releases about 8000 cal of energy. That is the step used as an energy source by most biochemical processes. ADP is also a high energy compound. Hydrolysis of 1 mole of ADP to AMP (adenosine monophosphate) releases 6500 cal of energy. In both ATP hydrolysis and ADP hydrolysis a pyrophosphate linkage (phosphate to phosphate) is broken. In fact any pyrophosphate would be a high energy compound. ATP and ADP are merely examples of a general class of high energy compounds, the pyrophosphates. Other general classes of high energy compounds are also possible. ATP is of special interest because it happens to be the high energy compound used most widely in biochemistry on earth.

Phosphate esters are an important class of compounds with some readily available energy.

Another energy source in cells can be illustrated by the hydrolysis of AMP to adenosine; 2200 cal are released when 1 mole of AMP is hydrolyzed. That energy is small enough that AMP is not usually classified as a high energy compound. However, AMP is a representative of an important general class of lower energy compounds in cells, the phosphate esters. Two other phosphate esters, glucose-6-phosphate and 3-phosphoglyceric acid, are shown in Figure 21–2. Each of those compounds releases 3300 cal when 1 mole is hydrolyzed, and both will appear in the metabolism discussion in Section 21–4.2.

You may notice from the above examples that more energy is stored when more negative charges are concentrated. ATP has more concentration of charge than ADP. Neighboring phosphates cause more charge concentration than the alcohol-phosphate combination to form an ester. Other combinations

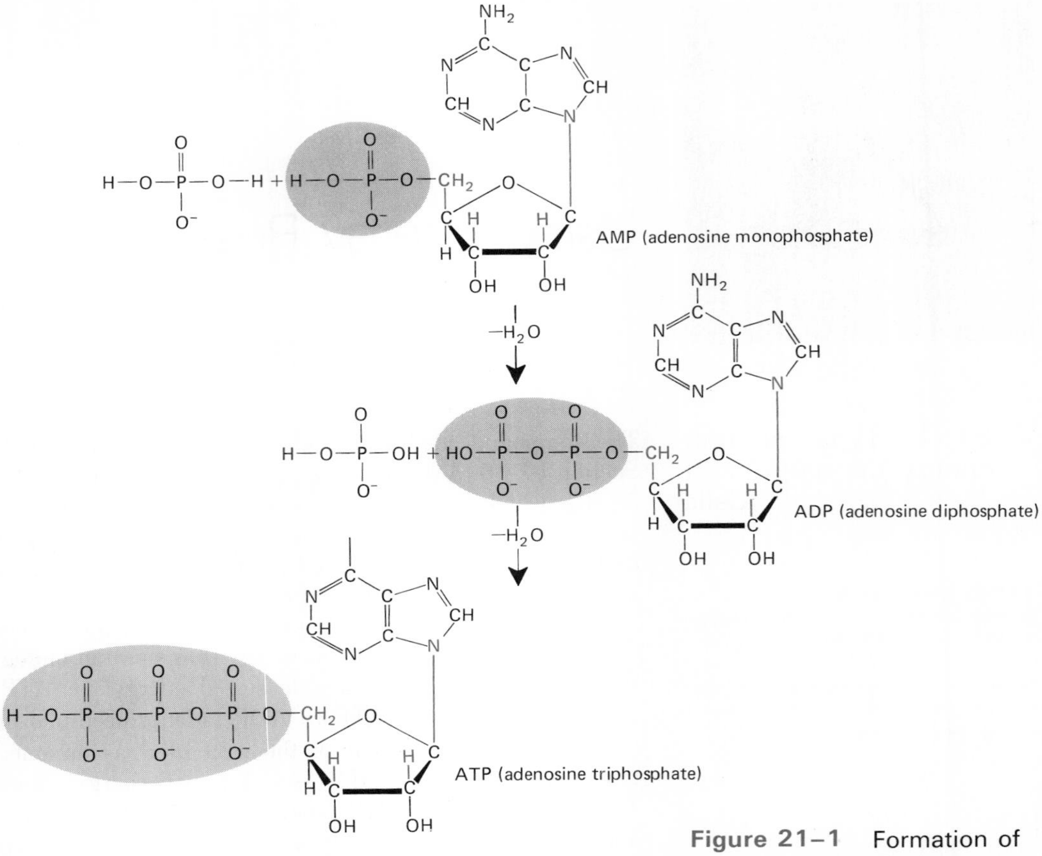

Figure 21–1 Formation of adenosine triphosphate (ATP).

Adenosine monophosphate (AMP) is converted to adenosine diphosphate (ADP) by condensation with a phosphate species. ADP can then be converted to adenosine triphosphate (ATP) by another similar condensation with a phosphate.

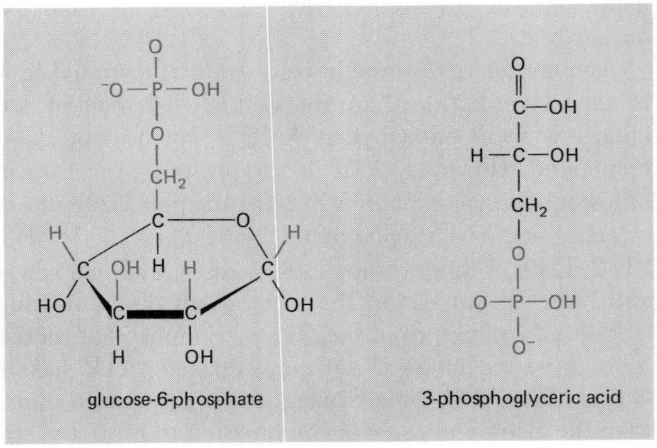

Figure 21–2 Phosphate esters.

Glucose-6-phosphate, shown on the left, is drawn using the same conventions (color for above the ring) as Figure 20–4. In addition, the high energy portions of both glucose-6-phosphate and 3-phosphoglyceric acid (which is shown on the right) are shown in color. The phosphate ester grouping is the high energy portion in each of these molecules.

478   Using Energy Reservoirs   [21]

Bringing together large negative charges leads to high energy compounds of the classes pyrophosphates, organophosphate anhydrides, amine phosphates, and thioesters.

that bring more negatively charged parts close together also qualify as high energy compounds. The failure of phosphate esters to qualify as high energy compounds is due to the relatively small negative charge concentration on alcohol groups.

Organic acids have a larger concentration of negative charge than alcohols, so the anhydrides formed between phosphate and organic acids should store more energy than the phosphate esters. Acetyl phosphate, shown in Figure 21–3, is an example of such an anhydride high energy compound. Hydrolysis of 1 mole of acetyl phosphate releases 10,000 cal. The other example shown in Figure 21–3 represents another class of high energy compounds, an anhydride of phosphate with an acidic enol alcohol position. The enol form of pyruvic acid is shown in Figure 21–4. Phosphoenolpyruvic acid, the second example shown in Figure 21–3, releases 12,000 cal when 1 mole is hydrolyzed. Phosphoenolpyruvic acid is an intermediate in the carbohydrate metabolism discussed in Section 21–4.2.

Another class of high energy compounds is condensation reaction products of phosphate and amines. The example shown in Figure 21–5, creatine phosphate, is a high energy compound found in muscle. Hydrolysis of 1 mole of creatine phosphate releases 10,000 cal. Creatine phosphate will be discussed in Section 21–5.

Another class of high energy compounds does not involve oxygen in the linkage. These are the thioesters where sulfur takes the place of oxygen in an ester linkage. Figure 21–6 shows an example in which a large biologically important unit, called coenzyme A (see Figure 21–14), is linked in a thioester

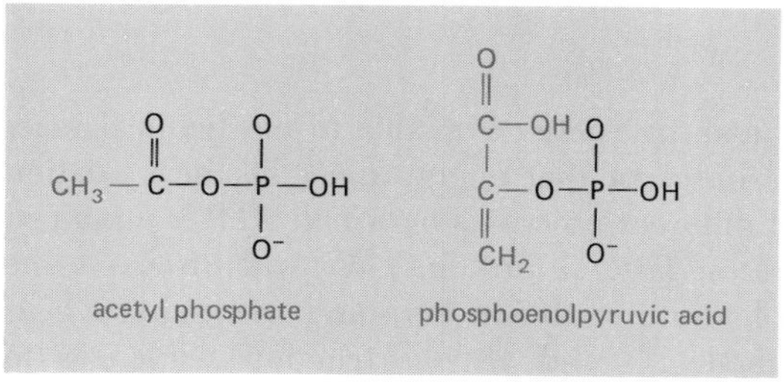

Figure 21–3 Anhydride high energy compounds. (The high energy portion of each of these acid anhydrides is shown in color.)

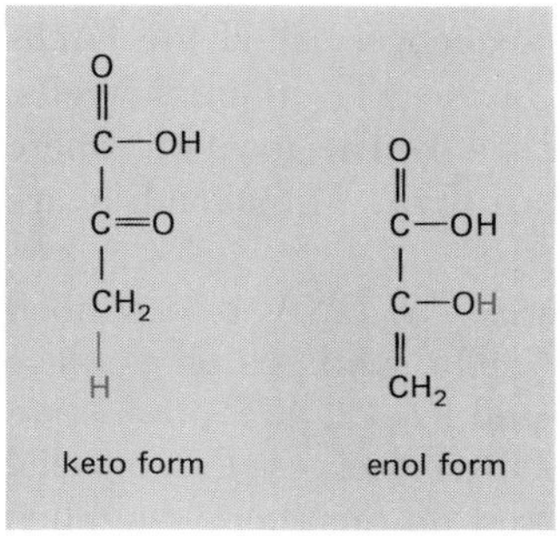

Figure 21–4 Forms of pyruvic acid.

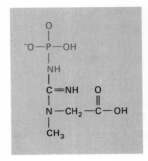

Figure 21-5 Creatine phosphate. (The high energy portion of the molecule is shown in color.)

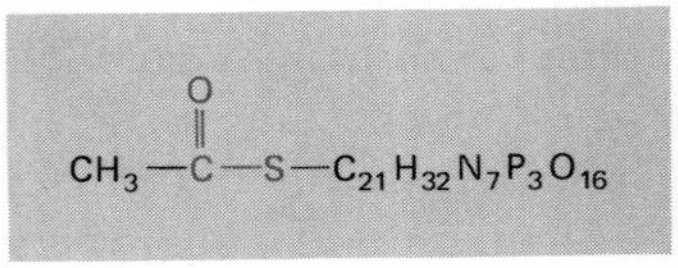

Figure 21-6 Acetyl coenzyme A. (The high energy portion of the molecule is shown in color.)

to an acetic acid unit; 8200 calories are released during hydrolysis of 1 mole of this substance. This compound, acetyl coenzyme A, plays a central role in the metabolism discussed in Section 21-4.4.

This brief discussion should give an indication of the variety of high energy compounds possible in cells. The role of the most important of these high energy compounds, ATP, will be discussed throughout the following sections.

ATP is the most important of the several high energy compounds used in cells.

ATP serves as an intermediate making energy available to biochemical reactions. In animals the ultimate source of that energy must be the oxidation of foods. There are a number of different processes by which ATP is produced during oxidation of carbohydrates, fats, or proteins. We will discuss some of the main pathways in carbohydrate oxidation here and in Section 21-4, but we will leave most of the details and various reaction sequences to coverage by biochemistry courses and texts.

ATP is produced in several different sequences of reactions occurring at sites throughout cells, but there is one reaction sequence, called the Krebs cycle or the citric acid cycle, that produces most of the ATP in animal cells. The Krebs cycle will be described in Section 21-4.4. The location where the Krebs cycle reactions occur is the **mitochondria**. Mitochondria are present in all cells of both plants and animals.

Cells have a number of identifiably different parts. The DNA we described in Section 19-10 is found in the cell nucleus. Protein synthesis takes place on the ribosomes, as mentioned in Sections 19-10 and 19-12. ATP production occurs mainly in the mitochondria. And other parts serve other special functions. Since all the other functions of a cell run on the energy supplied

21-2 Formation of ATP in Oxidation

ATP production in oxidation occurs mainly in the mitochondria.

by ATP, the mitochondria can be thought of as central power plants for the cell.

Plant cells can produce ATP from carbohydrates, fats, or proteins by the same sort of reactions animal cells use. But plants can also harness another energy source, light, through photosynthesis. The process of photosynthesis includes the production of ATP.

21–3 Digestion

Digestion hydrolyzes foods to small absorbable components.

Animals must begin their processes of energy utilization and growth by obtaining food from outside sources. The carbohydrates, fats, and proteins obtained must be broken down into relatively small molecules by digestion before they can be absorbed and used by the animal. Digestion breaks food down by hydrolysis of carbohydrates to monosaccharides, hydrolysis of fats to glycerol and fatty acids, and hydrolysis of proteins to amino acids. A number of enzymes are used to catalyze these hydrolyses. Other components in foods may also be absorbed.

In humans the digestive action of enzymes begins with saliva. Saliva contains an enzyme that hydrolyzes starch to glucose.

Saliva hydrolyzes starch, the stomach provides acid conditions for protein hydrolysis, and the small intestine neutralizes acidity and completes hydrolysis.

In the stomach enzymes are added that react to hydrolyze proteins, and stomach acids create the acidic conditions (pH 1.6 to 1.8) at which the hydrolysis of protein by the enzyme pepsin is most favored. Muscle action mixes the ingredients to improve the digestive process.

When the partially digested food passes from the stomach to the small intestine, the acidity is neutralized by alkaline fluids from the pancreas, gall bladder, and the intestine. These fluids also contain additional enzymes for hydrolysis of proteins, nucleic acids, and carbohydrates plus substances for the digestion of fats. Cellulose, which is not attacked by any of the available enzymes, passes through unaffected. Therefore foods such as carrots, which contain large fractions of cellulose, provide large quantities of "roughage" which passes through undigested.

The presence of fats in the duodenum (the first section of small intestine) stimulates the gall bladder to secrete bile. Some of the bile components (called bile acids and bile salts) have a detergent-like action that emulsifies the fats and makes them more accessible to an enzyme in pancreatic fluid which is needed to hydrolyze the fats. Figure 21–7 summarizes the steps in digestion.

Compounds are often absorbed by the body through active transport across membranes.

The small, water soluble molecules produced by digestion are then absorbed into the body. The absorption processes are often not a simple flow of material through the intestinal walls. Each cell of every plant and animal is surrounded by a membrane. These membranes are much more than a wall to hold the cell together. The cells in the intestine actively incorporate specific substances from one side and force them out the other side. By doing so they act as biological pumps, delivering desired substances in a particular direction. These membranes can build the concentration of sugars, amino acids, fatty acids, glycerol, and other digested foods up to noticeably higher concentrations on the body side than in the intestinal fluid. This pumping action is called **active**

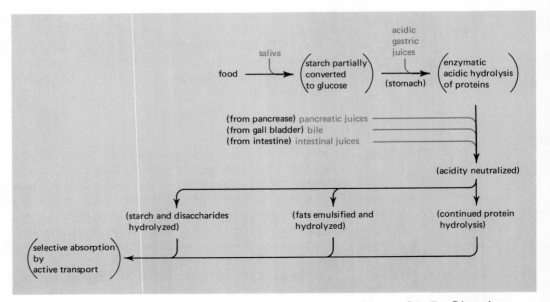

Figure 21–7 Digestion.

transport. Figure 21–8 shows what active transport accomplishes. Active transport also occurs in plant roots, concentrating nutrients from soil, and in various transfers between cells and body fluids. Active transport is not fully understood, even for the simplest and most common substances, but it must require energy. As is true for so many biochemical processes, that energy is generally supplied by ATP.

Because active transport requires interaction of the membranes with particular molecules, the efficiency of the transport varies considerably. For example, among the common hexose sugars, galactose is absorbed most rapidly from the intestines, glucose somewhat slower, and fructose slower yet. Similar

The efficiency of active transport varies among different compounds.

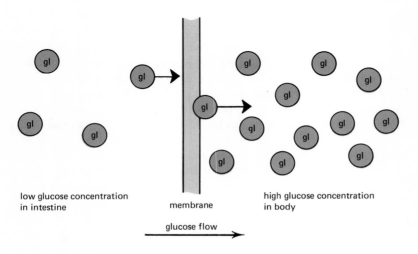

Figure 21–8 Active transport.

Glucose units are shown being actively transported across a membrane from a region of low glucose concentration to a region of high glucose concentration.

variations exist among amino acids. Since human membranes have evolved in an environment where only D-sugars and L-amino acids exist, it would not be surprising if human intestinal membranes were totally incapable of transporting some or all of the L-sugars or D-amino acids. Cell membranes may appear simple compared to other cell components, such as the nucleus with its heredity-carrying DNA, but the membrane function is complex and strongly influences the nature of life on Earth.

21-4 Carbohydrate Metabolism Producing ATP

21-4.1 Limited Scope of This Coverage.

Food metabolism is clearly a complex subject. Different foods may be metabolized in very dissimilar ways. Some food is metabolized for energy production while the same kinds of foods are also metabolized to produce structural material in the organism. The same overall reaction can sometimes be accomplished in different ways. Metabolism is an important subject, and we wish to offer some information about how it occurs. For simplicity, we are limiting the coverage to the main reaction path for a single kind of use, energy production, of a single food, glucose. Because the other common sugars and starches can be converted to glucose by various reactions in cells, this will give us the main energy producing metabolism for all carbohydrates (except cellulose, which is not metabolized by man). It will also allow us to point out a relationship between carbohydrate and fat metabolism.

Metabolism of glucose is important and is also related to fat metabolism.

21-4.2 Embden-Meyerhof Pathway.

Glucose or phosphates of glucose can be obtained from the diet or from hydrolysis of the glycogen starch in which it is stored in animals. Other sugars can be converted to glucose in cells. When glucose is the starting material, the first step is reaction with ATP to form glucose-6-phosphate. When glycogen starch is the starting material, glucose-1-phosphate is obtained by the phosphorolysis of the glycogen and then isomerized to glucose-6-phosphate without going through the stage of free glucose. In either case, glucose-6-phosphate is available for a series of energy releasing reactions.

Energy input from ATP is required at the start of the Embden-Meyerhof pathway.

The principle energy releasing reaction sequence for glucose-6-phosphate is called the Embden-Meyerhof pathway. These reactions can occur under anaerobic conditions—an absence of oxygen. The Embden-Meyerhof pathway produces some ATP but it also produces the starting material for a second reaction sequence that produces more ATP. The second sequence requires oxygen (aerobic conditions).

Figure 21-9 shows the reactions of the Embden-Meyerhof pathway. The first portion of the cycle requires an input of energy. If a glucose molecule is the starting material, one ATP molecule is required to convert it to glucose-6-phosphate. After an enzyme catalyzes the isomerization of the glucose-6-phosphate to fructose-6-phosphate, another ATP molecule is required to convert it to fructose-1,6-diphosphate. Another enzyme then catalyzes the

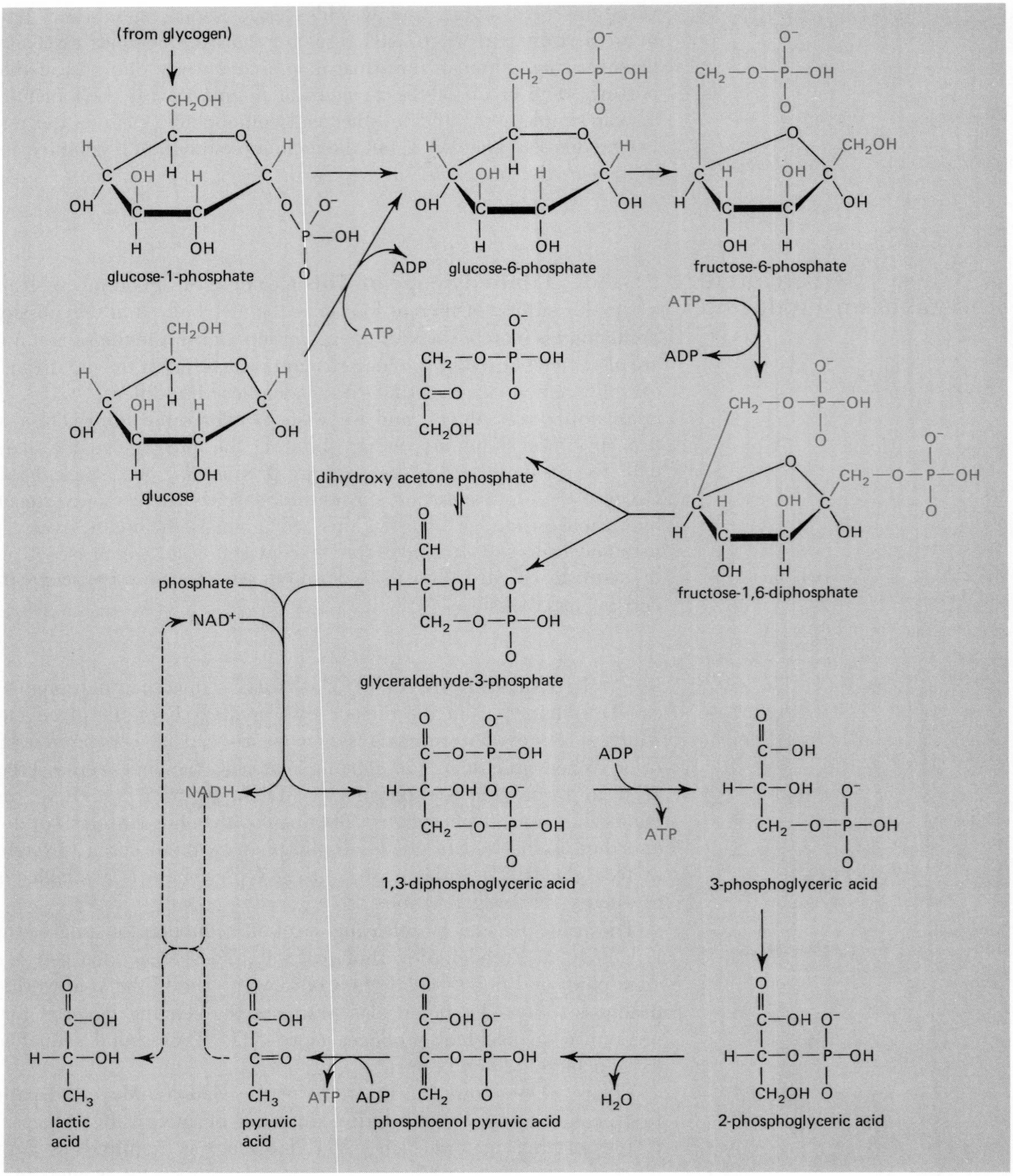
(from glycogen)
glucose-1-phosphate
glucose
glucose-6-phosphate
fructose-6-phosphate
ATP
ADP
dihydroxy acetone phosphate
fructose-1,6-diphosphate
glyceraldehyde-3-phosphate
phosphate
NAD+
NADH
1,3-diphosphoglyceric acid
3-phosphoglyceric acid
H_2O
2-phosphoglyceric acid
phosphoenol pyruvic acid
pyruvic acid
lactic acid

Figure 21-9 Embden-Meyerhof pathway.

The sugars are drawn using the conventions of Figure 20-4 (color for above the ring). The other use of color identifies the high energy forms, ATP and NADH, which are produced or used in the various reactions. When the splitting into two parts is taken into account, the process uses two ATP molecules but produces four ATP molecules plus two NADH molecules (which may or may not be used to form lactic acid).

Niacin is a vitamin needed to form NAD^+ and $NADP^+$.

cleavage of that molecule into two triose phosphates. The triose phosphates can be isomerized, so both can be converted into the glyceraldehyde-3-phosphate needed for the next reaction step. At this point the energy releasing steps begin.

The sequence of reactions leading to energy production requires a reactive substance called NAD^+. NAD^+ is assembled from adenine (see Section 19-9 and Figure 19-9), D-ribose (see Section 20-1 and Figure 20-4), phosphate, and nicotinamide. Nicotinamide can be produced from nicotinic acid. The two compounds, nicotinic acid and nicotinamide, are shown in Figure 21-10 and are an essential nutrient to human beings. (*Note:* The poison in tobacco which is called nicotine is a different compound.) Nicotinic acid is a suitable source for nicotinamide, which is also referred to as niacin. Niacin is a vitamin often associated with the vitamin B group, and niacin deficiency causes the disease pellagra. Figure 21-11 shows how nicotinamide is included in nicotinamide adenine dinucleotide, an important biochemical compound abbreviated NAD^+, and in a closely related substance called $NADP^+$. Both NAD^+ and $NADP^+$ are involved in biochemical oxidation-reduction reactions with their reduced forms, NADH and NADPH, having the nicotinamide portion reduced.

The reactions are perhaps best represented by the equations

$$NAD^+ + 2[H] \longrightarrow NADH + H^+$$
(from some organic compound being oxidized)

$$NADP^+ + 2[H] \longrightarrow NADPH + H^+$$
(from organic compound)

In each case there is originally a $+1$ charge on the nicotinamide ring nitrogen. When a hydrogen is added to a carbon adjacent to that nitrogen, the $+1$ charge can be lost. It appears on an H^+ ion product. In every case *two* hydrogen atoms must be taken from another molecule to reduce NAD^+ to NADH (or $NADP^+$ to NADPH).

Glyceraldehyde-3-phosphate can (under the influence of an enzyme) undergo an oxidation-reduction reaction in which it is oxidized and NAD^+ is

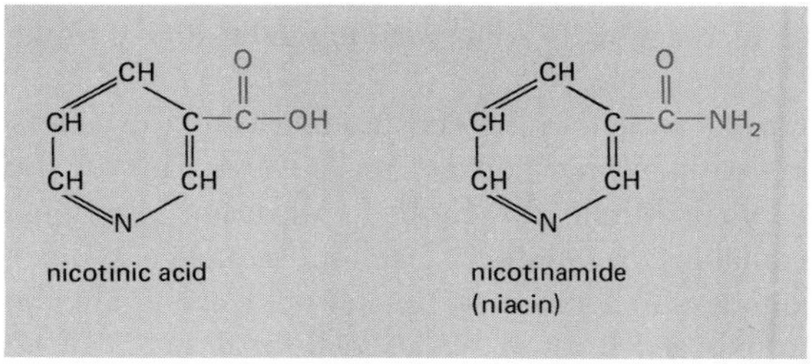

Figure 21-10 Niacin sources.

Nicotinamide is niacin, but nicotinic acid can be easily converted to nicotinamide by humans.

[21-4] Carbohydrate Metabolism Producing ATP

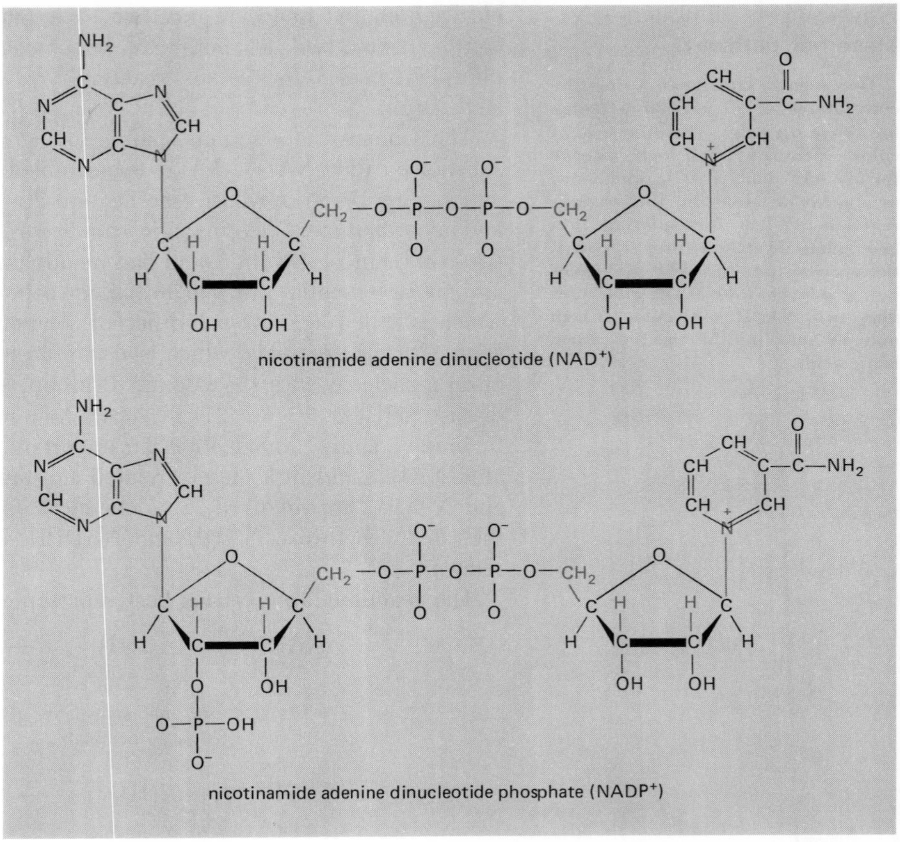

Figure 21–11 NAD⁺ and NADP⁺. (The conventions of Figure 20–4 are used, color for groups above the sugar rings.)

reduced to NADH. Under aerobic conditions this NADH is a major source of usable energy, as will be described below. In the absence of oxygen, the NADH is used up in the last step of the Embden-Meyerhof pathway and is not available to produce ATP. As the glyceraldehyde-3-phosphate is oxidized it is also reacted with phosphate to form a diphosphate product, 1,3-diphosphoglyceric acid. That diphosphate can react with ADP (in the presence of an enzyme) to form ATP and 3-phosphoglyceric acid. Other enzymes then isomerize that product and remove water to form the high energy compound phosphoenolpyruvic acid (see Section 21–1). Another enzyme then causes the phosphoenolpyruvic acid to react with ADP forming ATP and pyruvic acid.

Under aerobic conditions, pyruvic acid is the end product of the Embden-Meyerhof pathway. Reactions with oxygen convert the NADH back to the oxidized NAD⁺ form needed to continue the Embden-Meyerhof reactions, and the pyruvic acid is available for another reaction sequence. Under anaerobic conditions, the Embden-Meyerhof reactions would eventually be stopped by a shortage of the oxidized form of NAD⁺. The reactions can be

Under aerobic conditions the Embden-Meyerhof pathway produces ATP, NADH, and pyruvic acid; under anaerobic conditions it produces ATP and lactic acid.

continued by converting the NADH back to NAD^+ in a reaction not requiring oxygen. This can be accomplished by reducing pyruvic acid to lactic acid. Formation of lactic acid under anaerobic conditions builds up acidity in the cells. Lack of oxygen causing lactic acid buildup was thought to be the cause of muscle soreness after strenuous exercise, but more recent opinion is that enzyme irritation is the real cause of soreness. The low yield of ATP under anaerobic conditions forces a large number of enzyme catalyzed reactions to occur to meet the demand for ATP, and that apparently leads to enzyme irritations as a side effect.

21-4.3 Oxidative Phosphorylation.

The reduction of NAD^+ to NADH in the Embden-Meyerhof pathway is not the only biochemical source of NADH. Other compounds besides glyceraldehyde-3-phosphate can also be oxidized by NAD^+. In the presence of oxygen these reactions are a source of ATP through a series of reactions initiated by the oxidation of NADH back to NAD^+. These reactions are referred to as oxidative phosphorylation because they harness the energy of oxidation to cause addition of phosphate to ADP molecules. The sequence of intermediate molecules and reactions is also called the electron transport system. It transfers electrons from NADH to oxygen via several crucial intermediates. In Section 21-4.5 we will discuss the energetic advantage of breaking reactions such as $NADH + O_2$ down into small steps.

Figure 21-12 shows the reactions of the electron transport system. Enzymes are required throughout the process. One of the intermediate reactions, reduction of FAD to $FADH_2$, can also be carried out directly by reactions

Figure 21-12 The electron transport system.

The reduced forms capable of further reactions to produce ATP eventually are shown in color. The other necessary "fuel," O_2 is also shown in color. The high energy products, ATP, are shown in color on grey backgrounds. (The H^+ formed and used along with NADH has been omitted to keep the diagram to one product or reactant species at each stage.)

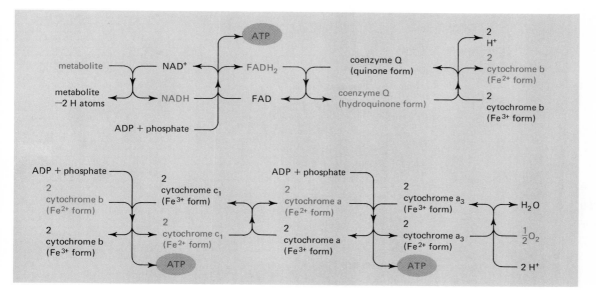

in metabolism of foods. Figure 21-13 shows the structures of FAD, which is flavin adenine dinucleotide and of riboflavin (vitamin B_2) from which it is derived. Coenzyme Q is a reactive hydrocarbon derivative, and the cytochromes are compounds containing protein and a complex of four nitrogen-containing rings in the same large organic molecule with iron called a porphyrin or heme. A related protein-heme compound is the oxygen carrying component of blood, hemoglobin, and will be discussed in Section 21-7. (Chlorophyll, the compound which absorbs light energy in photosynthesis, also has a structure closely related to the cytochromes.)

Aerobic oxidation of NADH (or NADPH or $FADH_2$) by the electron transport system produces ATP.

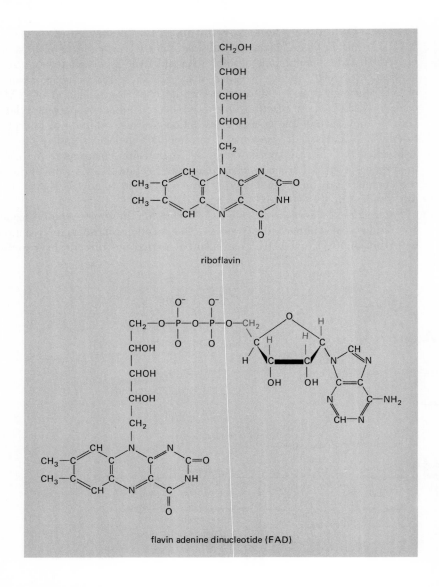

Figure 21-13 Riboflavin and FAD. (Drawn using the conventions of Figure 20-4 for sugar rings—color for groups above the ring.)

488 Using Energy Reservoirs [21]

Each mole of NADH oxidized by the electron transport cycle converts 3 moles of ADP to ATP. One-half mole of O_2 is required for each mole of NADH oxidized. When $FADH_2$ is available separately from metabolic processes, oxidation of $FADH_2$ also requires $\frac{1}{2}$ mole of O_2 per mole of $FADH_2$ oxidized, but only 2 moles of ATP are produced for each mole of $FADH_2$ available. NADPH can also be oxidized with a total of 3 moles ADP converted to ATP for each NADPH reacted.

21-4.4 The Citric Acid Cycle (Krebs Cycle).

When metabolism occurs under aerobic conditions, a large amount of ATP can be produced by oxidation of the pyruvic acid product from the Embden-Meyerhof pathway. (Some lactic acid from the Embden-Meyerhof pathway can be oxidized back to pyruvic acid and used in the citric acid cycle, but most of the lactic acid is converted back to glycogen by the liver and then has to begin its reaction from the start of the Embden-Meyerhof pathway). The oxidation of pyruvic acid is carried out efficiently by the citric acid cycle, or Krebs cycle, of

Figure 21-14 Coenzyme A and acetyl coenzyme A. (The sugar rings are drawn using the convention of Figure 20-4—color for above the ring—and the point where the high energy portion forms is also shown in color.)

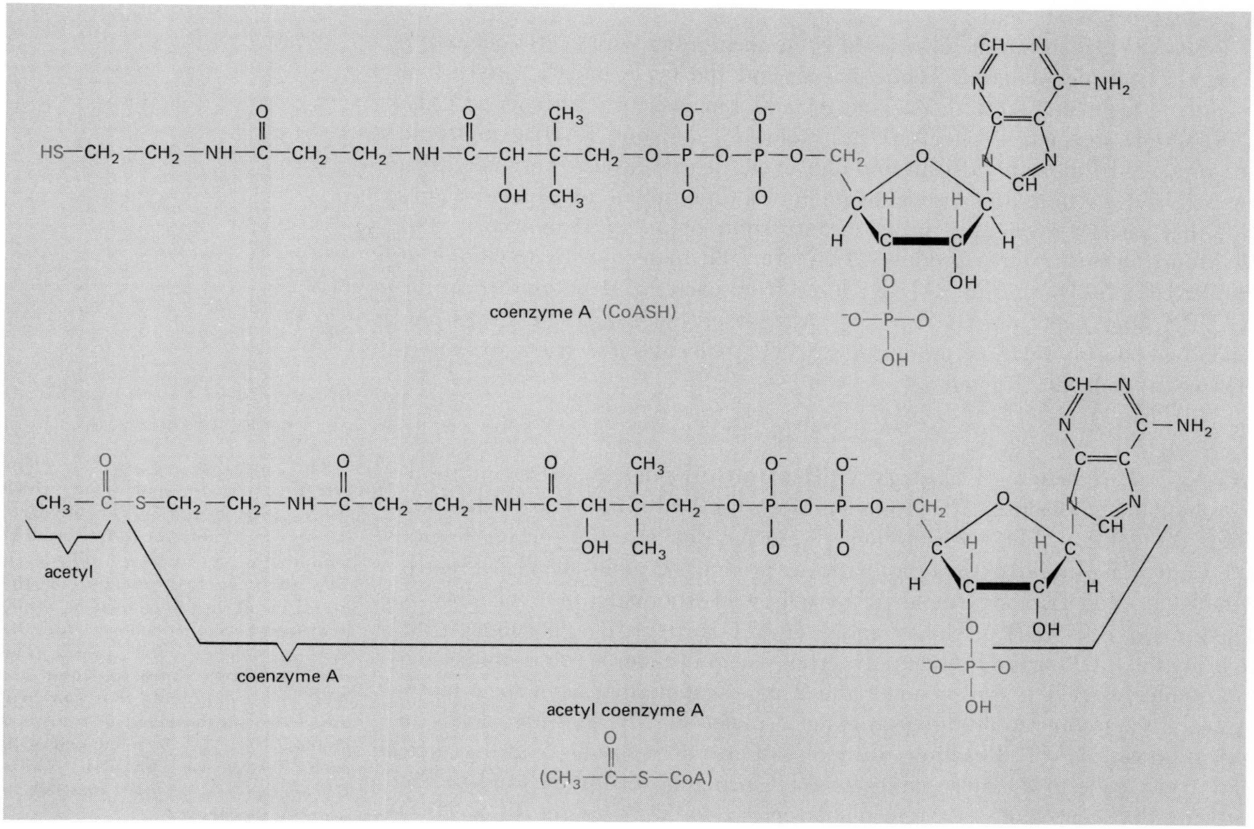

reactions which occur in the mitochondria of cells. It is this effective ATP producing process that gives the mitochondria the role of "powerhouse" of the cell.

The first step in reaction of pyruvic acid by the citric acid cycle is the formation of acetyl coenzyme A, one of the high energy compounds mentioned in Section 21–1. Figure 21–14 shows the structures of coenzyme A and acetyl coenzyme A. We will abbreviate these structures by using CoA to represent all of the coenzyme A except the SH group. The reaction of each pyruvic acid molecule with coenzyme A also reduces one NAD^+ to NADH and produces one CO_2 molecule. Oxidation of the one NADH molecule back to NAD^+ can then produce three ATP molecules by the oxidative phosphorylation described in Section 21–4.3.

Acetyl coenzyme A is really a more general starting material for the citric acid cycle than pyruvic acid because acetyl coenzyme A is also produced in fat metabolism. Acetyl coenzyme A reacts with many compounds to add the two-carbon fragment, acetyl. The fatty acids have only even numbers of carbon atoms because they are all made by such additions of two-carbon units. In metabolism they are also taken apart by two-carbon units in reactions that produce one acetyl coenzyme A molecule for each two-carbon unit removed from the fatty acid chains.

Figure 21–15 shows the citric acid cycle, beginning with acetyl coenzyme A as the starting material. Each turn around the cycle adds a two-carbon fragment (from the acetyl coenzyme A) and removes two carbons as CO_2. Two NADH molecules, one NADPH molecule, and one $FADH_2$ molecule are produced for each turn around the cycle. There is also one step with enough excess energy to cause conversion of a diphosphate to a triphosphate, and that energy eventually leads to formation of an ATP molecule. The electron transport system produces 11 more ATP molecules by oxidation of the NADH, NADPH, and $FADH_2$, three from each NADH, three from the NADPH, and two from the $FADH_2$. Altogether the citric acid cycle plus oxidative phosphorylation produces 12 ATP molecules for each acetyl coenzyme A molecule reacted.

21–4.5 Efficiency of Energy Utilization. Under the best conditions (aerobic), metabolism via the Embden-Meyerhof and citric acid cycles produces 38 moles of ATP for each mole of glucose converted to carbon dioxide and water. The Embden-Meyerhof pathway produces 2 moles of ATP directly, 2 moles of NADH, and 2 moles of pyruvic acid from each mole of glucose. The 2 moles of NADH produce 6 moles of ATP by oxidative phosphorylation to bring the total for the Embden-Meyerhof portion (under aerobic conditions) to 8 moles of ATP. Conversion of the 2 moles of pyruvic acid to 2 moles of acetyl coenzyme A produces another 2 moles of NADH which leads to 6 more moles of ATP. Reaction of 2 moles of acetyl coenzyme A in the citric acid cycle leads to 24 more moles of ATP, bringing the grand total to 38.

Using 8000 cal/mole as the available energy in ATP, we get 38×8000

Aerobic oxidation of pyruvic acid via the Krebs cycle produces a large amount of ATP.

Acetyl coenzyme A can be obtained from reaction of pyruvic acid or by fat metabolism.

Each acetyl coenzyme A can initiate one turn of the Krebs cycle producing 12 ATP molecules.

Figure 21–15 The citric acid cycle (Krebs cycle).

The portions shown in color are the pyruvic acid from which the cycle is begun in carbohydrate metabolism and the high energy compound producing products. The high energy portion of the acetyl coenzyme A (shown in color) provides necessary energy for the step producing citric acid. The other products shown in color are ATP and compounds that can produce ATP via the electron transport system. (One H^+ is produced with each NADH or NADPH. Those H^+ products have been omitted to simplify the figure.)

[21-4] Carbohydrate Metabolism Producing ATP

or 304,000 cal/mole of glucose. That compares to 686,000 cal that are released altogether when glucose is burned with oxygen to form carbon dioxide and water. Therefore, about 44% of the energy has been "saved" in a form usable in later biochemical reactions. The actual percentage kept as usable energy is somewhat uncertain because the energy available from hydrolysis of ATP in cells is not known exactly. Estimates have ranged at least from 7000 to 12,000 cal/mole. If the highest estimate was correct, the efficiency of energy storage could be 66%. In any case, the energy stored in ATP is less than the total energy of oxidation and appears to be near half of the total.

About half of the available energy is converted to the ATP form when glucose is oxidized via the aerobic Embden-Meyerhof and Krebs cycles.

In each reaction there must be a net tendency for the reaction to occur. That tendency for reaction is measured by the free energy change during reaction (see Appendix C–1) and includes both the tendency to increase entropy of the material (see Chapter 11) and the tendency to release energy as heat (which actually increases the entropy of the surroundings). The 686,000 cal/mole of glucose being oxidized by oxygen is the free energy change for the reaction at standard conditions. As the concentrations of reactants and products are varied, different values for the free energy change of the reaction can be obtained. The same variation with concentrations is also true for the hydrolysis reactions of ATP, and that creates the uncertainty about how much energy is available from ATP hydrolysis.

The efficiency at harnessing reaction energy is strongly dependent upon having a good match between the energy available in any given reaction step and the energy being stored. This pairing is illustrated in Figure 21–16 by examples using weights connected by a rope over a pulley. The weights used have values in grams similar to the values in calories associated with the reactions in metabolism. In the first example, a weight of 686,000 g easily lifts an 8000-g weight, but most of the potential energy of the 686,000 g weight is wasted. This example is comparable to using the entire energy from oxidizing a glucose molecule to form a single ATP molecule from ADP and

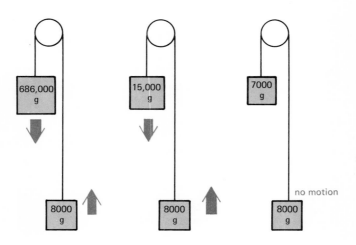

Figure 21–16 Conditions for doing work.

Work is done only in cases such as the first two shown where some excess energy can be wasted in the process.

Some "waste" of energy is necessary in real processes.

phosphate. In the second example, a 15,000-g weight can also lift the 8000-g weight, and less potential energy is wasted. This example is comparable to the energies in some of the actual ATP producing reactions in metabolism. The third example shows an even more closely matched pair of weights, but nothing happens. Because the 7000-g weight in the air is lighter than the 8000-g weight on the floor, it cannot lift it in this simple rope and pulley device. Some "waste" in the form of excess mass for the lifting weight is necessary to assure the process will proceed.

For ideal efficiency, the energy available should be only a little more than needed by each step.

Ideally, ATP should be produced biochemically by a reaction with just a little excess available energy over the amount required, perhaps around 10,000 to 12,000 cal/mole. But the common available energy source, oxidation of glucose, provides 686,000 cal/mole. If ATP production could be directly linked to glucose oxidation, it would certainly succeed in making ATP, as shown by the analogy to it in the first part of Figure 21–16, but the process would be very wasteful. In order to show how the waste is reduced in the real case, we are going to use another analogy in Figure 21–17. In this case we start with a waterfall 686 m high, analogous to the 686,000 cal/mole available from glucose oxidation. We are going to use the water to drive a water wheel doing useful work. But we only have one design available for water wheels, and that water wheel can only use the energy from the water falling 8 m. That is analogous to biochemical reactions that run on ATP energy at only 8000 cal/mole.

Using the entire waterfall in Figure 21–17 to drive our water wheel is wasteful, but we do not have any devices available to use the whole available energy from a single fall. One solution would be to build a new water channel that came down the hillside in a series of cascades, as shown in Figure 21–18. Each of these cascades should be a drop of at least 8 m so we could install

Figure 21–17 Using a waterfall for work.

The 686-m waterfall shown can turn the waterwheel, which requires only an 8-m fall of water, but most of the energy will be wasted.

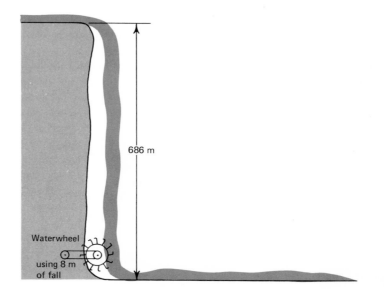

[21–4] Carbohydrate Metabolism Producing ATP 493

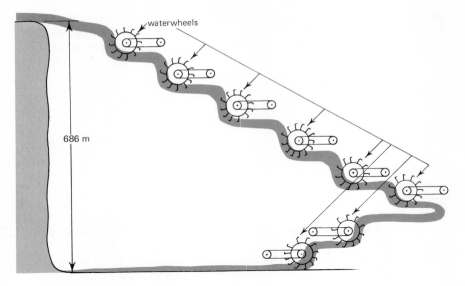

Figure 21–18 Using a series of falls for work.

Diverting the water from one 686-m fall to a series of smaller falls permits use of a larger number of waterwheels and reduces the waste of energy. Each new fall can drive a waterwheel provided the falls are each at least 8 m high (the amount of fall required by each waterwheel).

Figure 21–19 (opposite) The glucose falls water power facility.

The actual conversion of glucose into ATP by aerobic oxidation in the Embden-Meyerhof and Krebs cycles is shown symbolically in this "waterworks." Use of the electron transport system to convert NADH, NADPH, and $FADH_2$ to ATP is symbolized by linking the NAD, NADP, and FAD "units" to the correct number of ATP generators. The first NAD and the first two ATP direct generators in each stream plus the two ATP "pumps" correspond to the Embden-Meyerhof pathway. The next NAD in each branch corresponds to the conversion of pyruvic acid to acetyl coenzyme A. The remainder of each branch corresponds to one turn around the citric acid cycle (Krebs cycle).

The Embden-Meyerhof and Krebs cycles raise efficiency by dividing the glucose oxidation process into a number of usable smaller energy changes.

a "standard model" water wheel. But if any drop was less than 8 m, the water wheel would not turn and the blocked flow would interfere with our system, so we choose to make all drops at least 15 to 20 m to be safe.

However, when we go to the hillside to build our series of cascades, we cannot choose the heights we want. The only available flat spots on the hill for use as the sections are sometimes much further apart. These flat spots are analogous to the available intermediate compounds in metabolism. But the efficiency of our cascade system is saved by technological developments. A new company, Electron Transport Systems, Inc., has come out with some new devices. One, the NAD, can use a large cascade and drive three 8-m water wheels, and a smaller model, the FAD, can use an intermediate sized cascade and drive two 8-m water wheels. Using all available tools, our workers can then assemble the best water wheel system possible on the hillside by our waterfall. The result shown in Figure 21–19 is analogous to the Embden-Meyerhof and citric acid cycles which have been evolved for carbohydrate metabolism by life on Earth.

21–5 Storage of ATP Energy in Creatine Phosphate

The concentration of ATP present in cells at any given time is always small. If ATP is being consumed rapidly, as is the case under strenuous muscular exercise, metabolism of glucose and other foods may not be able to supply ATP as rapidly as it is being demanded. The same strenuous exercise conditions that call for rapid ATP supply also tend to consume the available oxygen more rapidly than it can be supplied. The resulting anaerobic conditions stop the citric acid cycle and sharply reduce the cell's ability to produce ATP quickly. Muscle tissues have a mechanism to reduce this problem by providing ATP from another source.

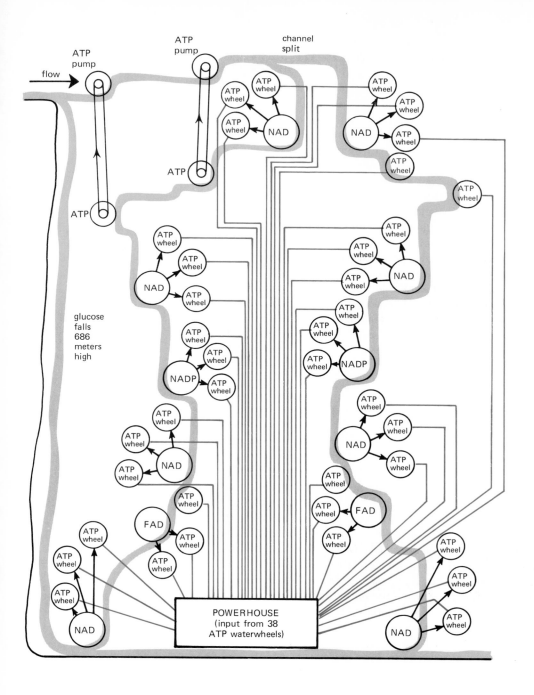

[21-5] Storage of ATP Energy in Creatine Phosphate

When muscles are at rest or low activity, oxidation of glucose, muscle glycogen, and other foods converts much of the ADP to ATP. Under the resulting conditions of relatively high ATP and low ADP, the reaction shown in Figure 21-20 is energetically favored and substantial amounts of creatine are converted to the high energy compound creatine phosphate. The reaction proceeds until an equilibrium is established between the rate of reaction of ATP with creatine and the rate of reaction of the creatine phosphate with the relatively rare ADP. In a cell with a large (by biochemical standards) creatine concentration, the total energy stored in the form of creatine phosphate can become large enough to be quite important.

When exercise causes the ATP concentration to drop, the reaction in Figure 21-20 is thrown out of equilibrium balance. The reaction then occurs in the opposite direction from that shown in Figure 21-20 and serves as an alternate source of ATP. The reaction of creatine phosphate to produce ATP continues until either the ATP level rises from its low concentration or until the creatine phosphate is reduced to a concentration low enough to establish a new equilibrium with the low ATP concentration in the cell. The extra ATP supplied allows a higher level of muscle cell reactions (hence more muscle strength) and a lower level of lactic acid production from the Embden-Meyerhof pathway than would be possible in cells without the creatine phosphate energy storage mechanism.

Formation of creatine phosphate provides a mechanism for storing ATP energy.

21-6 Use of ATP Energy in Muscle Contraction

21-6.1 Components of Muscle.

Hopefully the preceeding sections include enough examples of ATP use or production in chemical reactions to allow you to visualize how all biochemical synthesis and use of foods can be related to energy storage and transfer. But it may be harder to visualize how ATP is used to drive processes where chemical compounds are not the product of interest. Instead of forming chemicals needed by the biological system, some reactions bring about important physical changes. Examples are active transport across membranes (which we have already mentioned), production and use of the electrical impulses associated with brain and nerve activity, and muscular motions. Muscle actions are probably the most obvious of these processes to our external observation, and we choose here to use a brief discussion of muscles as our sample of how biochemical reactions can cause physical changes. We will begin our discussion by describing the most obvious of the components involved in muscle action. (The limited space here makes a complete description of muscle action impossible, so we will omit mention of lesser components and much of the details of the process.)

The two main components in muscle contraction are fibers of actin and fibers of myosin. Actin is a globular protein (see Section 19-13). In the presence of ATP the individual actin molecules can be linked into chains of the bulky globules. These chains occur in pairs which are wound around each other in a double helix, as shown in Figure 21-21. This double helix is much less sharply twisted than the helix structures found in DNA, RNA,

ATP energy can be used to cause physical changes in biological systems.

Figure 21-20 Storage of energy as creatine phosphate.

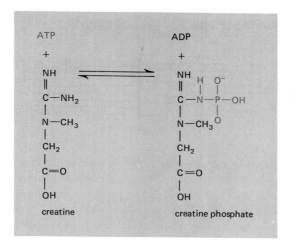

Figure 21-21 Actin double helix.

Actin molecules, each about 55 angstroms (Å) (or, in SI units, about 5.5nm) in diameter, link to form chains. Two such chains are then coiled around each other as shown to form a double helix actin fiber.

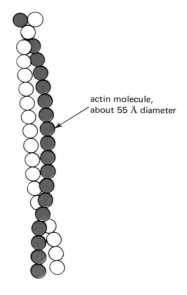

and protein chains. The actin double helix is quite strong and apparently quite rigid. Apparently it does not stretch when stress is applied. Actin fibers (the double helix chains) are anchored on both sides of a connecting plate and large numbers of fibers reach out in parallel alignment into adjoining segments of muscle on each side of the plate. Figure 21-22 shows how the actin fibers are arranged.

The actin fibers intermesh with another set of parallel fibers, the myosin fibers. Myosin is a much larger protein than actin and is divided into three dissimilar regions. Each myosin has a long "tail" which is a fibrous protein (see Section 19-13). These fibrous tails have a large number of ionic sites

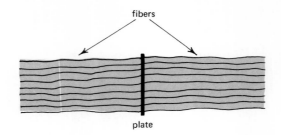

Figure 21-22 Arrangement of actin fibers.

Parallel actin fibers are anchored to a central plate.

that are attracted to oppositely charged sites on other myosin tails. The interaction of the myosin tails hold the myosin fibers together. The other end of each myosin molecule is a "head" of globular protein. This head actually has two separate globular sections. The head of myosin is a catalyst for hydrolysis of ATP and serves as the site where ATP energy is released during muscular contraction. Between the head and tail of myosin is a "neck" region which probably differs significantly from the tail. The neck is a fibrous connection, but (unlike the rather rigidly fibrous tail) the neck must be quite flexible to allow the myosin head to move about as much as it apparently does. The interlocking myosin tails form a fiber (which is larger than the actin fiber). The head portions of the myosin molecules stick out of this fiber at regularly spaced intervals, as shown in Figure 21-23. The myosin heads can interact with surrounding, parallel actin fibers as shown in Figure 21-24.

Muscle contraction occurs by a sliding action of the actin fibers in among the myosin fibers. Muscle relaxation occurs by the actin fibers sliding back out. Figure 21-25 shows a segment of muscle at various stages of contraction.

Actin and myosin, the main components of muscle, are arranged in a particular pattern of fibers.

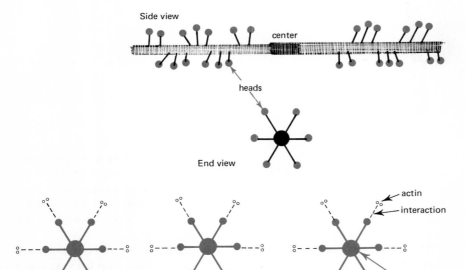

Figure 21-23 Myosin fibers.

Myosin fibers are linked tail to tail in the center. Toward each end the myosin heads, shown in color, appear in a regular pattern. The pattern is shown by a side view and an end view of the same fiber.

Figure 21-24 Actin-myosin arrangement.

The alignment of parallel actin and myosin fibers in muscle are shown by an end view. The myosin fibers, shown in color, are able to link to the actin fibers via reactions between the myosin heads and actin.

Figure 21–25 Muscle positions.

The movement of the myosin (shown in color) and actin fibers relative to each other is shown by these varied muscle conditions.

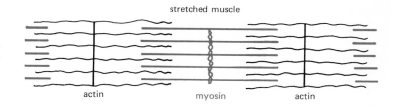

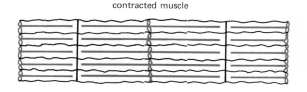

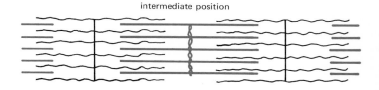

When many myosin heads remain bonded to actin, a muscle is in the rigid condition called rigor.

21–6.2 The Condition of Muscle Rigor. The globular heads of myosin and the globular actin molecules both have reactive sites. When brought into contact, they can bond to each other. Such bonding is possible at any stage of muscle contraction, as shown in Figure 21–26. If bonds form and remain, the myosin and actin fibers are held rigidly to each other. This condition, which allows no movement for further contraction or relaxation, is called rigor.

Rigor is the normal result in all muscle when no ATP is present. When an organism dies, the ATP is eventually consumed with no source of replacement and rigor sets in. That leads to the inflexible muscle condition in corpses called rigor mortis.

Figure 21–26 Muscle rigor.

Two different conditions of muscle rigor are shown. Whether the muscle is stretched or contracted, any time the myosin (shown in color) is bonded to actin at many or all of the myosin heads, which are near actin fibers, the muscle is rigid. That is the condition of muscle rigor.

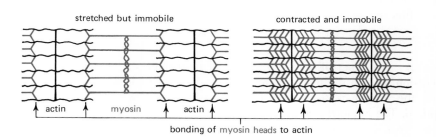

21-6.3 Relaxation in the Presence of Mg^{2+} and ATP.
When ATP is present, the bonds between myosin and actin can be broken and the rigor of the muscle is released. The breaking of the actin-myosin bonds may be due to repulsion between negative charges. ATP can react with sites on the myosin heads. Under the pH conditions in muscle (pH about 6.5), the ATP is present predominently as the -4 ion form. When that negative charge becomes associated with the myosin head, it is repelled from negative charges on the actin fiber (perhaps charges on another protein held within the actin double helix).

However, the effectiveness of ATP in breaking the myosin-actin bonds is dependent on which positive ions are present with the ATP. When Ca^{2+} is present with the ATP, ATP is consumed by a hydrolysis reaction. When the ATP is gone, the myosin bonds to actin and the muscle is again fixed in the condition of rigor. If instead Mg^{2+} is present with the ATP, the $MgATP^{2-}$ ions formed are attached to the myosin heads and cause dissociation from actin, but the ATP does not hydrolyze. $MgATP^{2-}$ ion is hydrolyzed very slowly by myosin. Therefore, $MgATP^{2-}$ allows the muscle to relax by separating the myosin and actin and keeping them from reforming bonds. This condition of Mg^{2+} and ATP, with very little or no Ca^{2+} present, is required to allow a muscle to relax.

Mg^{2+} and ATP^{4-} form $MgATP^{2-}$ which releases actin-myosin bonds and allows muscle relaxation.

21-6.4 Contraction When Ca^{2+} is Added.
When Ca^{2+} ions are added to a relaxed muscle (with Mg^{2+} and ATP present), contraction occurs. ATP is hydrolyzed as the contraction occurs, but it cannot be simply by forming $CaATP^{2-}$ which can react on the myosin. Contraction occurs only if Mg^{2+} is present, and during contraction the myosin and actin become bonded. A possible explanation is that $MgATP^{2-}$ remains on the myosin sites and the Ca^{2+} interacts with the negative charges on the actin fiber. As the Ca^{2+} neutralizes some negative charge, the repulsion between actin and myosin is reduced and actin to myosin bonds form. When the myosin is bonded to actin, the $MgATP^{2-}$ can then be hydrolyzed. Energy from the ATP hydrolysis is used in some way to change the shape of the myosin and pull the actin and myosin fibers toward a greater overlap.

In the presence of excess ATP and Mg^{2+}, the myosin can pick up a new $MgATP^{2-}$ and break the myosin-actin bond again, leaving the chains free to be pulled together by other new bonds that are being formed. Figure 21-27 shows this sequence of reactions during muscle contraction.

Addition of Ca^{2+} to a relaxed muscle causes reactions resulting in muscle contraction.

21-6.5 Release of Ca^{2+} by Nerve Impulse.
From the previous discussion we can see that muscle function depends upon keeping Ca^{2+} out when the muscle is to be relaxed and adding Ca^{2+} when contraction is required. The removal of Ca^{2+} is accomplished by an active transport mechanism across a membrane surrounding the muscle fibers. As is true of many

Ca^{2+} in muscle is controlled by active transport and nerve induced release.

Figure 21-27 Muscle contraction.

(1) The myosin fiber (shown in color) is shown with its heads supplied with MgATP^{2-} which releases bonds to actin. (2) Some Ca^{2+} ion is added and the actin becomes reactive to myosin heads with MgATP^{2-}. (3) A myosin-actin bond forms. (4) Reaction of the ATP from MgATP^{2-} provides energy which pulls the myosin head down and back, thus pulling the entire myosin fiber to the left. (5) Ca^{2+} ion makes another actin site reactive to myosin heads with MgATP^{2-} and additional MgATP^{2-} reacts with the first myosin head, releasing its bond to actin. (6) and (7) Repeat the cycle of (3) and (4).

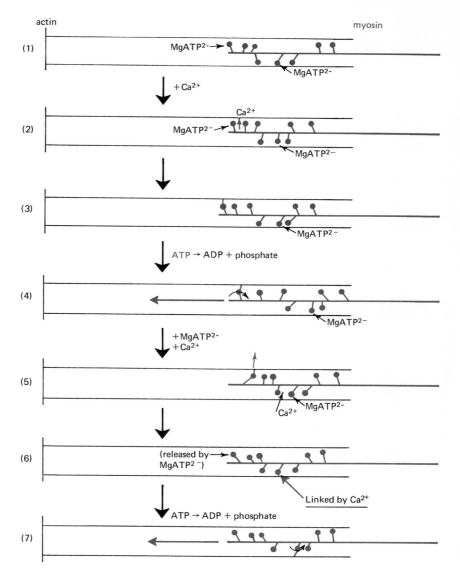

other membrane transport processes, ATP provides the energy for "pumping" Ca^{2+} out of the muscle fibers.

The release of the Ca^{2+} (from the sites where it is stored after being pumped out of the muscle fiber) is set off by arrival of a nerve impulse. Nerve impulses travel by opening up the nerve membranes to diffusion by Na$^+$ ions that had been pumped to one side of the membrane. When the pulse reaches the muscle, reactions occur that cause a similar opening of the membranes to Ca^{2+} diffusion. The Ca^{2+} ions diffusing into the actin-myosin regions then set off the muscle contracting reaction.

Production of ATP in the mitochondria of cells by the citric acid cycle requires a source of oxygen to maintain aerobic conditions. Oxygen has a low solubility in water, as expected for a nonpolar gas (see Section 10–2). Therefore some mechanism is required for carrying oxygen to the cells and keeping it there until needed. The mechanism used is complex ion formation.

In the lungs the limited amount of oxygen that dissolves in the blood is almost completely tied up in the form of a complex of oxygen with hemoglobin in the blood's red cells. The complex is called oxyhemoglobin. That complex keeps the dissolved oxygen concentration below the equilibrium level possible with air and allows more oxygen to be absorbed by the blood (and then complexed by hemoglobin). As the blood passes cells with very low oxygen concentrations, the oxyhemoglobin can release its oxygen and allow it to replenish the supply in the cells.

Hemoglobin is a compound formed from the heme units mentioned in Section 21–6.3 and globular proteins called globin. The hemoglobin unit actually consists of four large protein molecules, each with a globular tertiary structure, held together by a specific quaternary protein structure (see Section 19–6.3). Each of the four protein molecules contains one heme group. Two of the proteins are an identical pair of a type called alpha chains, and the other two are a different identical pair of a type called beta chains. The complete structures of these proteins are now known.

The bulky structure of the hemoglobin keeps most molecules from approaching the iron in the heme units, but one oxygen molecule can reach and be complexed by each of the four hemes. Figure 21–28 shows the structure of the heme portion that actually complexes the oxygen. The heme is bonded to the protein chain at several points.

When oxygen passes into cells, another mechanism is needed to store it so cell activity does not have to depend upon oxygen from passing hemoglobin in the blood. The storage is done by another complex to a heme unit bonded to a protein chain. The storage compound is oxymyoglobin, the complex of

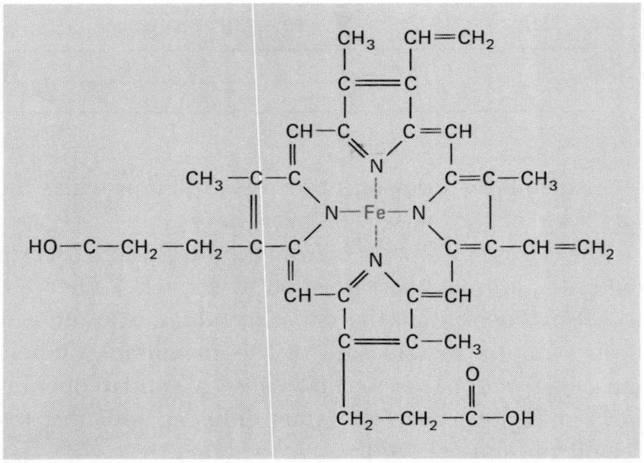

21–7 Oxygen Availability Via Complexes

Oxygen availability is increased by use of the complexes oxyhemoglobin and oxymyoglobin.

Figure 21–28 Heme.

The iron atom, shown in color at the center, can form additional bonds to ligands besides the four nitrogens shown bonded to it in the heme group. The heme groups in hemoglobin can bond oxygen molecules.

oxygen and myoglobin. Myoglobin is very similar to one of the four units of hemoglobin without a quaternary structure bonding it to other units.

21–8 Use of Fossil Fuels

Cellulose can be used as food by some species.

As noted in Sections 20–4 and 20–5, fossil fuels are used extensively as energy sources by our civilization. All of the energy in these fossil fuels comes from biochemical energy storage through photosynthesis, but later reactions have converted the products into forms that cannot be used directly in the usual metabolism by life on Earth. Sometimes the energy in "nonfood" material can be recovered in ways that are biochemically useful. For example, cellulose is in a chemical form inaccessible to metabolism by most plants and animals, but it can be converted back into usable food if the ring linkages are opened and reversed. Protozoans in the digestive tract of termites can accomplish that. Cellulose (in the form of the wood pulp used in paper making) is now being fed to cattle on an experimental basis. Bacteria in the digestive system of the cattle break up the cellulose, and the cattle seem to be doing very well.

Other future efforts to use cellulose and fossil fuels to make foods will no doubt be attempted. But other uses of fossil fuels can also help support life. Food can be conserved by using fossil fuels to provide energy that would have been provided by metabolism. An example is a saving in feed required by hogs if a heater is used to warm their water supply and a mechanical mill is used to grind the corn in their feed so they do not need as much muscle activity (chewing) to eat it.

Fossil fuels are used by society in ways analogous to food use by animals.

The use of fossil fuels by our technological society is somewhat analogous to the metabolism of food by animals. Some of the fossil fuel is used to release energy and some is used to build structural materials. Energy released in some of the reactions is used to drive other processes. Transferable energy forms, such as electrical energy, mechanical energy, gravitational potential energy, or available chemical energy (such as in a flashlight battery), are used to link energy production and consumption. These transferable forms serve a function similar to that of the high energy compounds produced in cells. Electricity is probably the transferable form most analogous to ATP in cells. Electricity can be produced in power houses, taken where it is needed, and used in a very large number of ways.

Conversion of heat energy to electricity is not very efficient.

As we noted in Section 21–4.5 and Figure 21–16, some energy must always be "wasted" in energy conversion processes to assure the reaction will proceed. In carbohydrate metabolism about half of the energy was wasted and an estimated 44% ended up in the desired form, ATP. When fossil fuels are reacted, they are usually burned to produce heat. That heat is then used to produce electricity. But the conversion of heat to electricity is limited by more than just the "equal or larger energy" factor shown in Figure 21–16. Electricity is a low entropy energy form. As noted in Chapter 11, entropy must increase during any spontaneous process. To change some of the heat energy into the form of electrical energy, some other process is required in which entropy rises. The process used is the flow of heat from hot (lower entropy) to cold (higher entropy). But the heat energy flowing away at the cold temperature is "wasted"—it is not converted to the desired form, electricity. (The waste heat also causes thermal pollution problems, see Section 13–5.8). Even under the most favorable conditions achieved by present fossil fuel electrical generating plants, about 65% of the heat is lost. Only 35% or less of the energy is converted into the desired form, electricity.

Notice that a living cell has a higher efficiency (44%) for producing its preferred energy form (ATP) than modern technology has for producing its preferred energy form (electricity). There are other ways of producing electricity more efficiently. The potential energy of water flowing downhill can be harnessed quite completely, and fuel cells in which chemical reactions produce electricity directly (without going through the form of heat) offer hope for a high percentage of the energy as electricity. But our present fossil fuel technology is limited by the heat intermediate and the laws of thermodynamics. Note that it is very wasteful to use the "expensive" energy form of electricity for production of heat, which could be obtained directly from fossil fuels.

Electric heat is wasteful.

Use of fossil fuels also depends upon conditions that lead to reaction. The most abundant fossil fuel, coal, exists in a solid form that is not very accessible to oxygen. If large portions of coal on the inside of pieces remain uncombusted, the efficiency of the fuel use is lowered. If coal was pure carbon, there would be no problem because the outer layers would be continually removed as gaseous products, thus exposing the remaining carbon to oxygen. But real coal leaves an ash, and the ash from burning the outer layers may keep oxygen away from the inner coal. Combustion of coal in large power plants has been made more efficient by pulverizing the coal to a fine powder, but that has created serious problems of ash particles in the smoke (see Section 11–5).

Complete combustion of coal is difficult.

Combustion of liquid fossil fuels is somewhat simpler. Oil is converted into volatile light hydrocarbons (gasoline) which can be vaporized into air and efficiently burned by the oxygen in that air. Because the combustion is so simple, it can be carried out effectively on a much smaller scale than that required for complete combustion of coal. Liquid fuels are therefore suitable for portable power plants, such as automobile engines. The demand for fuel in this convenient form is so great that the less volatile fractions of oil are often processed to break them down into volatile smaller molecules. The cracking reaction described in Section 18–1.3 is used extensively.

Liquid fuels are more convenient in use than solids.

Use of gasoline in internal combustion engines (the type used in automobiles) leads to several problems. First, the combustion is usually incomplete. An internal combustion engine can be controlled by varying the air input and is usually operated with somewhat less available oxygen than needed for complete combustion. As a result, not all of the fuel is burned to carbon dioxide and water. Some carbon monoxide and unburned hydrocarbons are present in the exhaust. These constitute an air pollution problem, particularly from the hydrocarbons. Hydrocarbons are a principal contributor to photo-

Internal combustion engines cause air pollution.

chemical smog in areas such as Los Angeles, and internal combustion engines are the largest source of hydrocarbons in smog areas. Second, the mixture of air and fuel also contains nitrogen, and during combustion some of the nitrogen reacts with oxygen to form nitrogen oxides. Nitrogen oxides are also involved in smog. If the engines are adjusted to provide more oxygen so combustion will be more complete, there is also more oxygen left over to form nitrogen oxides. Third, the efficiency of converting the fuel energy into energy of motion is quite low, perhaps only 15 or 20%. The relatively high temperatures at which waste heat is lost in the exhaust and engine cooling contribute to the low efficiency.

Conversion of coal to gas will become important.

The easiest fossil fuels to burn are the natural gases. Because of their convenience, gas supplies are being consumed faster (as a percentage of total available supply) than the other fossil fuels. But there is a technological solution to satisfy this preferential demand for gas. Chemical reactions with coal to convert it into combustible gases will become a major process in the near future. That should become the main way in which fossil fuels are used before the year 2000, and it should remain a major source of energy for about 200 years. Conversion of coal into convenient liquid fuels will also be important. Production of methanol (CH_3OH), an excellent liquid fuel, is particularly cheap, simple, and likely to become commercially important in the near future.

21–9 Greenhouse and Dust Effects

The very extensive use of fossil fuels is changing the composition of the Earth's atmosphere in a way that could change our climate. As the carbon which has been in the form of coal, oil, or natural gas for millions of years is burned in large quantities, the total amount of carbon dioxide available on Earth is increased. As the concentration of carbon dioxide in the atmosphere rises, some of the extra carbon dioxide is removed by increased growth of plants. There is also an increase in the rate at which carbon dioxide dissolves in sea water and is removed by formation of insoluble calcium carbonate. These mechanisms for carbon dioxide removal lessen the change in atmospheric carbon dioxide concentration each year as burning of fossil fuels continues. Figure 21–29 shows both the seasonal variation in carbon dioxide concentration (showing a drop each year during the plant growing season) and the general increase over several years. The data shown were taken at the Mauna Loa Observatory, Hawaii. It is estimated that atmospheric carbon dioxide has increased from 292 ppm (parts per million parts of air) in 1860 to over 320 ppm now, and that it will increase to 375 ppm by the year 2000.

Carbon dioxide in the atmosphere is increasing and causing a greenhouse effect.

As the atmospheric carbon dioxide increases it causes a **greenhouse effect.** The light reaching Earth from the sun is of shorter wavelengths than the light emitted from Earth into space. As noted in Chapter 11, life on Earth is dependent on the entropy increase in that change to long wavelengths to drive the reactions that support life. The short wavelength portions of sunlight can pass through carbon dioxide without being absorbed—carbon dioxide is

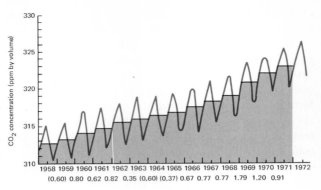

Figure 21-29 Increasing atmospheric carbon dioxide.

Variations in the carbon dioxide content of the atmosphere have been followed at the Mauna Loa Observatory in Hawaii by the Scripps Institute of Oceanography. The sawtooth curve in color indicates the month to month changes in concentrations from 1958 to 1972. The oscillations reflect seasonal variations in the rate of photosynthesis and vary somewhat from year to year. The average values for each year in black show the long term upward trend. [Adapted from "Atmospheric Effects of Pollutants" by P. V. Hobbs, H. Harrison, and E. Robinson, *Science*, 183, 909 (1974), by permission. Data from C. D. Keeling, A. E. Bainbridge, C. A. Ekdahl, P. Guenther, and J. F. S. Chin, *Tellus*, 26, (1974, in press).

transparent to such light. But some of the longer wavelengths of light are absorbed by carbon dioxide. The fraction of the light emitted by Earth that can be absorbed by carbon dioxide is much higher than the fraction of sunlight energy in the same longer wavelengths. As a result, the increasing concentration of carbon dioxide allows sunlight to reach Earth but interferes with Earth's ability to radiate energy into space. The effect is very similar to a greenhouse, where glass is transparent to light entering but prevents radiation out of some long wavelength energy. In both cases the temperature rises until a new, higher rate of energy emission is reached to restore the balance between incoming and outgoing energy. Figure 21-30 shows the operation of the carbon dioxide greenhouse effect.

As the Earth's temperature rises, every natural cycle dependent upon temperature can be affected. One cycle of particular interest is the buildup of ice caps and glaciers and their eventual melting. At higher temperatures, melting is increased, thus reducing the amount of water stored in frozen form. Some such melting is clearly occurring on Earth. Consider this quote about Glacier Bay in Alaska by Park Service Ranger-Naturalist Barbara Minard (from "John Muir's Wild America" by Harvey Arden in *National Geographic*, April, 1973, page 456). "When Vancouver sailed by here in 1794, he saw only a huge wall of ice at the mouth of a small inlet. By the time Muir arrived in 1879, that inlet had become a bay stretching back more than forty miles. In the ninety-odd years since Muir made the first detailed explorations, the glacier named for him has retreated another twenty-five miles." Melting of ice could have serious effects on the level of the oceans. If all the ice caps melted, the oceans would rise on the order of 200 ft* and cover much of the land area most useful to mankind.

Increased temperature could cause disastrous melting of ice caps.

A second effect of rising temperature is increased evaporation from the

---

*J. Spar on page 66 of *Earth, Sea, and Air, a Survey of the Geological Sciences*, (Addison-Wesley, 1962) states that the increased pressure on the oceans would cause a shifting (sea floors dropping, continents rising) that would reduce the rise of the oceans from an original prediction of 186 feet to an actual rise of only about 125 feet.

Figure 21-30 Carbon dioxide greenhouse effect.

A carbon dioxide layer allows the short wavelength light in sunlight to pass. The same carbon dioxide layer absorbs and reemits the longer wavelength light which predominates in the emission of energy from the Earth. Some reemitted light returns to the Earth, thus raising the temperature.

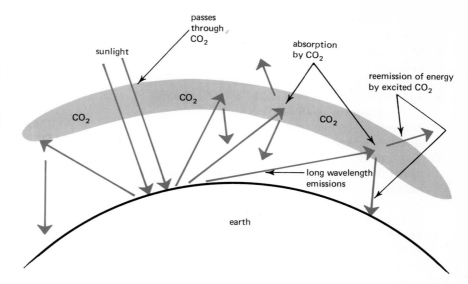

Clouds and dust could lower temperatures by increasing Earth's reflection of sunlight.

oceans. That could lead to increased cloud cover and a greater reflection of sunlight, which could cause the temperature of the Earth to drop.

Between 1880 and 1940 there was an increase of about 0.5°C in the average temperature of the Earth. Since 1940 the average temperature has been dropping (despite the rising CO_2 concentration). The drop reached about 0.3°C by the 1970s.

The recent drop in average temperatures could be influenced by an increase in atmospheric dust. Dust can reflect sunlight away from the Earth and small dust particles can help condense water vapor into droplets in clouds. Dust is created by intensive mechanized farming and by some burning of fossil fuels. It is interesting to note that the turnaround in temperature changes occurred near the time when electrical power generation by combustion of pulverized coal (which produces very fine ash in the stack gases) was being established on a very large scale. Other industrial activities producing small particles in smoke were also expanding in the same period. As we improve controls on such dust pollution in the 1970s, the effects of dust on the climate may be lessened. Unfortunately, we do not really know whether those effects are large or totally insignificant.

Skills Expected at This Point

1. You should be able to explain what is meant by a high energy compound in cells, be able to list some of the several classes of high energy compounds, and be able to describe ATP and the conversion it undergoes when used for energy production.
2. You should be able to describe the phenomenon of active transport.
3. You should be able to list the main segments of the metabolism of carbohydrates (including the starting materials and products), identify

the mitochondria and their special function, and explain the advantage of multiple step processes in efficient use of biochemical energy.
4. You should be able to state the nature of NAD^+, $NADP^+$, FAD, and their reduced forms and the amount of ATP each reduced form can produce in oxidative phosphorylation.
5. You should be able to describe acetyl coenzyme A and its importance to metabolism of both fats and carbohydrates.
6. You should be able to point out the general importance of ATP and describe the operation of creatine phosphate as a reservoir for ATP.
7. You should be able to list the main components in muscle action and their functions; you should be able to state the nature of muscle rigor, relaxation, and contractions and the conditions causing each of them.
8. You should be able to name the compounds involved in oxygen transport and storage and describe their functions.
9. You should be able to compare the use of fossil fuels to produce electricity to the use of foods to produce ATP and describe the sources of inefficiency in each.
10. You should be able to list the problems given in this chapter that are associated with internal combustion engines.
11. You should be able to describe the carbon dioxide greenhouse and dust effects on climate.

Exercises

1. List three types of high energy compounds, showing the bond to be broken in the high energy reaction in each case.
2. Why does ATP provide energy upon hydrolysis to ADP?
3. What reaction sequence produces most of the ATP in animal cells and where does that reaction sequence occur?
4. List the steps (and where they occur) involved in digestion of fats.
5. Briefly describe the phenomenon of active transport.
6. List the reactants and products of the Embden-Meyerhof pathway under aerobic conditions. List them under anaerobic conditions.
7. **(a)** How is acetyl coenzyme A formed in carbohydrate metabolism? **(b)** How is acetyl coenzyme A related to fats?
8. Describe the components and arrangement of NAD^+. (You do not need to write out the complete structure.)
9. List the reactants and products for one turn of the citric acid cycle. How are the original products converted into ATP?
10. Why is the efficiency of ATP production from oxidation of glucose increased by having a many-reaction sequence?
11. Using the facts available in the chapter, list at least two reasons why creatine phosphate is an especially good form for storing extra ATP capacity in cells.
12. List the main components in muscle and describe their arrangement in a relaxed muscle and the conditions that allow the muscle to relax.

13. Describe the sequence of events in a muscle contraction.
14. How does the protein structure of hemoglobin and myoglobin make them more selective for oxygen complexing?
15. List two facts about internal combustion engines that cause them to contribute to smog.
16. Describe the carbon dioxide greenhouse effect.

Test Yourself

1. Write an equation showing the structural changes during energy producing reaction of ATP. Write out the molecular structure, but you may just write the word "adenine" instead of the structure for that portion of the molecule.
2. (a) List the reactants and products of oxidative phosphorylation.
   (b) Name the intermediate that can also serve as a starting material for oxidative phosphorylation and the products formed when it is the starting material.
3. List the conditions required for a practical but efficient conversion of metabolic energy into ATP.
4. What is the condition of muscle rigor and how does it occur?
5. What is hemoglobin and how does it bond O_2?
6. List and compare the approximate efficiencies of energy conversion to the useful form in the processes of glucose oxidation to form ATP in cells and fossil fuel combustion to form electricity in power plants.
7. Which of the following is particularly important as a site for ATP production by cells?
   (a) nucleus
   (b) mitochondria
   (c) membranes
   (d) active transport
   (e) myosin
8. Although several of the reactions of digestion may occur in the stomach and may be favored by the acidic conditions there, one particular reaction is particularly favored in the stomach and particularly dependent upon the acidic conditions for successful completion. Which of the following is the reaction particularly favored by the acidic conditions in the stomach?
   (a) hydrolysis of proteins
   (b) hydrolysis of starch
   (c) hydrolysis of fats
   (d) emulsification of fats
   (e) hydrolysis of disaccharides
9. Active transport is
   (a) use of muscle contraction to create motion.
   (b) transfer of an enzyme from one substrate molecule to another.
   (c) movement of selected molecules in one direction through a membrane.
   (d) movement of food through the stomach and intestines.
   (e) stimulation of bile fluid injection by fats requiring it for digestion.

10. Which of the following is directly involved in oxidative phosphorylation?
    (a) nitroso adenine dimer, called NAD
    (b) creatine phosphate
    (c) greenhouse effect
    (d) flavin adenine dinucleotide, called FAD
    (e) citric acid
11. Which of the following is *not* an advantage obtained by the use of creatine phosphate in muscles?
    (a) Muscle power is increased by more rapid availability of ATP.
    (b) The processes are less dependent upon the rate at which oxygen can be supplied.
    (c) There is less buildup of lactic acid.
    (d) An equilibrium consumes ATP when excess is available and produces ATP when it is needed.
    (e) Oxidative phosphorylation proceeds more rapidly.
12. Muscle contraction is started in real life by
    (a) adding $CaATP^{2-}$ to muscle tissues where ATP had been absent.
    (b) adding $MgATP^{2-}$ to muscle tissues where ATP had been absent.
    (c) adding ATP^{4-} to muscle tissues where ATP had been absent.
    (d) adding Ca^{2+} to muscle tissues containing $MgATP^{2-}$ without Ca^{2+}.
    (e) adding Mg^{2+} to muscle tissues containing $CaATP^{2-}$ without Mg^{2+}.
13. After oxygen has been delivered to cells it is stored as
    (a) oxymyoglobin
    (b) oxyhemoglobin
    (c) creatine phosphate
    (d) ATP
    (e) NADH
14. Gasoline is preferred over coal as a fuel (resulting in a high demand for gasoline) because in comparing the usual ways each is used
    (a) gasoline is easier to burn in simple devices.
    (b) gasoline gives a higher percentage efficiency for useful work.
    (c) gasoline is more abundant and cheaper.
    (d) coal creates air pollution and gasoline does not.
    (e) coal cannot produce temperatures as high as are needed.
15. The greenhouse effect of carbon dioxide occurs because
    (a) the dense CO_2 reduces air turbulence, thus holding in hot air.
    (b) CO_2 reflects sunlight, thus concentrating it on Earth.
    (c) CO_2 prevents dust from settling, thus building up a dust barrier.
    (d) CO_2 allows sunlight in but slows the lower energy radiation of heat away from Earth.
    (e) CO_2 in the atmosphere polymerizes to form a covering over the Earth.

Biologically Active Molecules

22-1 Insulin: A Hormone Balance

Only a few of the many biologically important compounds can be discussed as examples in a brief text.

Because the number of different compounds involved in biochemical processes is so large, we cannot discuss all of them in the brief coverage in this text. We cannot even mention all of the known important substances or examples of all the important types of compounds. Instead in previous chapters we have attempted to select a few of the important general categories and processes and discuss them in a way that brings out mention of a variety of the most important compounds. We will continue that coverage through limited examples in this chapter by identifying and discussing some compounds whose biochemical activity is particularly striking. We choose insulin as our first example because the insulin deficiency disease diabetes and its treatment by insulin are so widespread.

Insulin is a hormone. Hormones serve a regulatory function that maintains a balance between various biochemical processes. Unlike enzymes, which catalyze a specific reaction, hormones may influence a number of related reactions. One possible mechanism for hormone regulation of processes is through effects on the enzymes or on the ability of reactants to reach the enzymes.

Insulin is a protein acting as a hormone regulating blood glucose concentration.

Insulin is a protein that increases the conversion of blood glucose into other forms, the first step toward using glucose in metabolism. The glucose level in blood is controlled by the opposing actions of insulin, which lowers blood glucose, and several other hormones, which raise blood glucose. One of the hormones that causes a rise in the blood glucose level is adrenalin, which is produced in response to crises or excitement. Adrenalin raises the availability of glucose for energy production in the body. A balance between insulin and the other hormones is necessary to keep the blood glucose within the range of variations fitting normal body activity. Figure 22–1 shows how the concentration of glucose in blood is related to body processes.

If insulin is not available in sufficient quantity, glucose is concentrated in the blood. Small excesses in glucose concentration (called hyperglycemia) are normal for a few hours after eating and lead to storage of the excess glucose energy by formation of glycogen and fats. But when there is a lack of insulin, carbohydrate oxidation will not proceed as needed and the condition

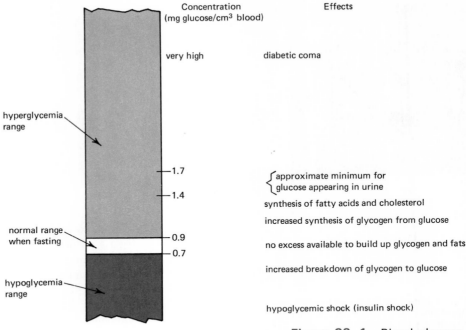

Figure 22–1 Blood glucose levels.

Diabetics have persistent high glucose levels and also produce unusually large concentrations of certain other compounds.

Lack of insulin upsets the complex interrelationships between reactions in carbohydrate metabolism.

of **diabetes mellitus** results. A person who suffers from diabetes has a consistently high blood glucose level. After eating foods that supply glucose, the diabetes sufferer has an even higher blood glucose level which persists for longer than the increased glucose level period for normal people. Diabetics may be identified by the appearance of glucose in the urine, which is caused by blood glucose concentration exceeding the threshold for excretion through urine. Diabetics also produce concentrations of acetoacetic acid, 3-hydroxybutanoic acid, and acetone which lead to their appearance in urine. In severe cases, acetone may also appear in air from the person's lungs.

It is now believed that the unusual concentrations of certain compounds in diabetics are caused by the inability to oxidize acetyl coenzyme A. If the normal reaction of acetyl coenzyme A with oxaloacetic acid fails to occur to start the citric acid cycle, the acetyl coenzyme A can react to form acetoacetic acid CH_3—CO—CH_2—COOH. The acetoacetic acid can then either lose CO_2 to form acetone or be reduced to form 3-hydroxybutanoic acid.

The reason for this metabolism failure is not known with certainty. One suggestion is that insulin promotes the absorption of glucose from the blood by cells. In the absence of insulin the Embden-Meyerhoff pathway has very little glucose in the cells from which it can produce the pyruvic acid. The resulting pyruvic acid can be almost totally consumed by the available coenzyme A to form acetyl coenzyme A. If more pyruvic acid was produced, some could react by a side reaction to add CO_2 and make oxaloacetic acid.

That input of extra oxaloacetic acid may be needed to make up for losses in other side reactions in the citric acid cycle. Without it the cycle slows to a virtual stop for lack of oxaloacetic acid. The above theory may or may not be the ultimate explanation of diabetes, but it does show the sort of complex interrelationships between reactions that must be considered to explain hormone action.

If diabetes is left untreated, byproducts such as acetone may reach such high concentrations that they cause loss of consciousness. The result is a diabetic coma and possible death.

Diabetes is treated by injections of insulin. Because insulin is a protein it cannot be taken by mouth. The normal processes of digestion would simply convert it back to its component amino acids. Direct injection of highly purified insulin avoids the digestion problem. But direct injection of insulin is a very dangerous action. If too much is supplied to the blood at one time, the blood glucose level can be depressed far down into the hypoglycemic region and cause shock. The amount in the blood is kept down by injecting the insulin in muscle tissue, where it can be slowly released to the blood. Nevertheless, insulin overdose remains a serious problem and the cause of numerous deaths. Patients using insulin combat the problem by carrying sweets to serve as a rapid source of blood glucose. When they feel the beginning of shock symptoms such as trembling and weakness, they eat some candy to push their glucose level back to the normal range. If a diabetic loses consciousness from shock, the best emergency treatment is direct injection of glucose into the blood.

Insulin must be injected to be effective, and injection can lead to shock from an overdose.

22-2 Steroids

Steroids are an important class of natural products with a characteristic structure of four interlocked rings.

One of the most important classes of natural products is the steroids. Steroids have a series of four interlocking hydrocarbon rings plus hydrocarbon side chains and other functional groups such as OH. The four rings have the same carbon atoms in all steroids, but variations in amount of double bonding, in ring linkage (similar to the α and β ring variations we described in sugars), and in side chains and functional groups lead to a large number of variations. These variations cause sharp differences in biochemical behavior of the steroids. All steroids are lipids (fat soluble) because of their predominantly hydrocarbon structure.

Cholesterol is the most abundant steroid, but there are many other steroids with different functions.

The most abundant steroid in animals is cholesterol. The structure of cholesterol is shown in Figure 22-2. Cholesterol is thought to be the precursor for many (or all) of the other natural steroids. Cholesterol itself has been shown to be produced from acetyl coenzyme A by a chain forming sequence similar to the sequence producing fatty acids. A chain compound with several double bonds then undergoes ring formations to give the characteristic four rings of steroids.

Steroids are used in the human body as hormones to stimulate antibodies to wall off foreign bodies through inflammation. Another steroid, cortisone (see Section 22-5) reduces inflammation by suppressing antibody formation.

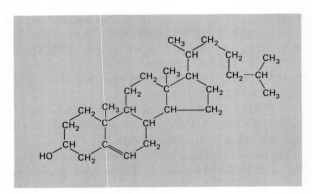

Figure 22-2 Cholesterol, a steroid.

Sex hormones and the artificial sex hormones used in birth control pills are also steroids (see Sections 22-3 and 22-4). The bile acids used to emulsify fats in digestion are steroids, and vitamin D is an example of a nonsteroid that is closely related (it lacks one of the ring closures).

22-3 Sex Hormones

Both male and female sex hormones are steroids.

The characteristic sex hormones of both males and females are steroids. Production of these steroids is stimulated by polypeptide hormones from the pituitary gland, but the hormones produced directly by the sex organs are steroids. Some "male" and "female" steroid hormones are produced in the adrenal gland of both sexes, so each sex has small amounts of the hormones characteristic of the other sex.

The female steroids fall into two main categories associated with different parts of the menstrual cycle. Figure 22-3 shows the structures of estrone, estradiol, and estratriol which are called estrogens. These estrogens are associated with the early part of the menstrual cycle. We have shown three forms to emphasize the point that even in nature there is some freedom to substitute one hormone for another of similar structure. Note the similarities such as the one aromatic ring and the methyl group. The arrangements of ring linkages are also all identical, and the directions of the bonds to hydrogen at the ring junctions are shown (solid line is up, dotted lines down) to indicate the linkage used. This principle of replacements by slightly varied structures will be important in Section 22-4 where we consider artificial substitutes. Estrone and estradiol are both found in urine, particularly during pregnancy. Estradiol is more potent and may be the principal hormone.

Estrogens dominate the early part of the menstrual cycle and progesterone dominates the latter part of the cycle and pregnancy.

Under the influence of estradiol (or the related hormones), a follicle in the ovary ripens and finally produces an egg. After ovulation occurs another hormone, progesterone, is produced in quantity and dominates the cycle. Progesterone and a reduction product, pregnanediol, to which progesterone is converted prior to excretion in urine, are shown in Figure 22-4. The first main function of progesterone is preparation of the uterus for implantation of a fertilized egg on the wall. But progesterone also depresses the estrogen

Figure 22-3 Estrogens.

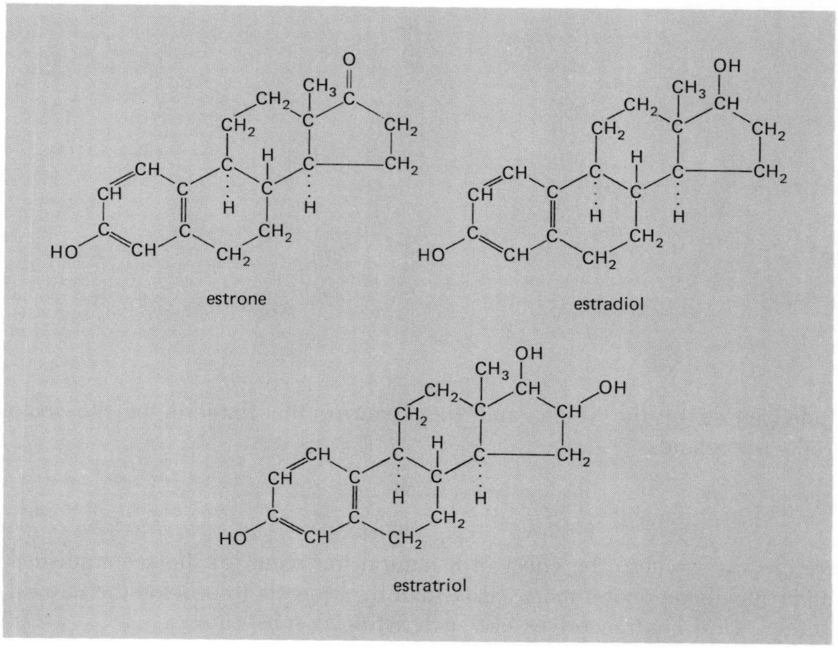

estrone

estradiol

estratriol

Figure 22-4 Progesterone and its reduction product, pregnanediol.

Color is used to show where H atoms have been added when progesterone was reduced to pregnanediol.

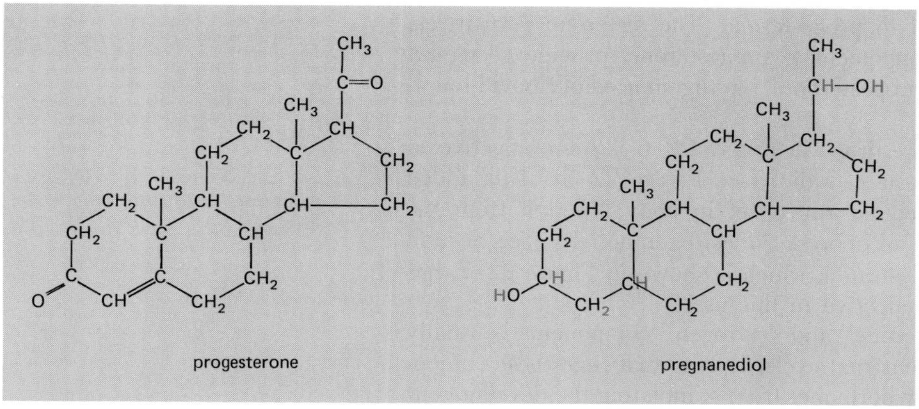

activity and prevents further ovulation. If pregnancy occurs, continued production of progesterone also stimulates the development of the mammary glands in preparation for lactation.

Figure 22-5 shows testosterone and androsterone, two of the hormones associated with males. Testosterone is produced in the testes and may be converted to androsterone and other derivatives in later reactions. These male hormones, called androgens, are similar to the estrogens and progesterone in many features such as the ring arrangements and the side groups in certain positions. The androgens stimulate male characteristics such as beard growth

progesterone

pregnanediol

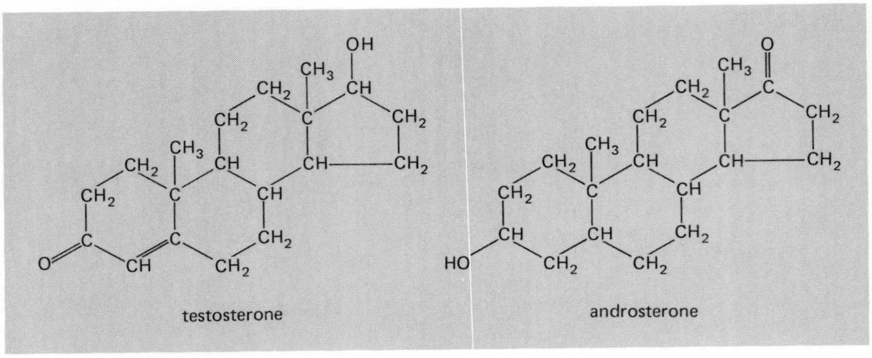

Figure 22-5 Two male hormones.

and deepening of the voice, and they control the function of the male reproductive glands.

Androgens stimulate male characteristics.

22-4 Artificial Substitutes

There are cases where the effect of a natural hormone can be accomplished by an artificial compound more easily than by attempts to provide the natural substance. Oral contraceptives use such artificial substitutes.

Among other actions, the steroid hormone progesterone prevents ovulation. If progesterone could be given to a woman in the early portion of the menstrual cycle, fertility could be prevented. The preferred method for administering the progesterone would be orally, thus avoiding the complications of injections and the need for very high purity material to prevent reactions to impurities in injections. But progesterone is not effective orally. The digestive process must either convert the progesterone to other, less active compounds, or progesterone must fail to be absorbed efficiently.

The antifertility effect can however be achieved by oral administration of other steroids closely related to progesterone in structure. Figure 22-6 shows three such steroids. Two of them, norethindrone and norethynodrel, are related to testosterone (see Figure 22-5) while the third, medroxyprogesterone acetate, is more closely related to progesterone. The testerone derivatives, which share the same ring arrangement as progesterone, provide good oral potency. Norethynodrel is the "progesterone" type agent used in Enovid brand birth control pills.

Ovulation is prevented by progesterone or appropriate substitutes.

The antiovulation effect of the drugs in Figure 22-6 is more effective in the presence of estrogens such as estradiol (see Figure 22-3). This added estrogen is also desirable to reduce any upset in body function from the tendency of the "progesterone" to depress the estrogen activity (see Section 22-3). An artificial estrogen, mestranol, which is shown in Figure 22-7, has proven very potent and orally effective in this use.

An artificial estrogen is added to a progesterone substitute to make effective oral birth control pills.

The combination of "progesterone" and "estrogen" components is usually taken on days 5 to 24 of the menstrual cycle. During that time these components act to prevent the pituitary hormones from stimulating the development

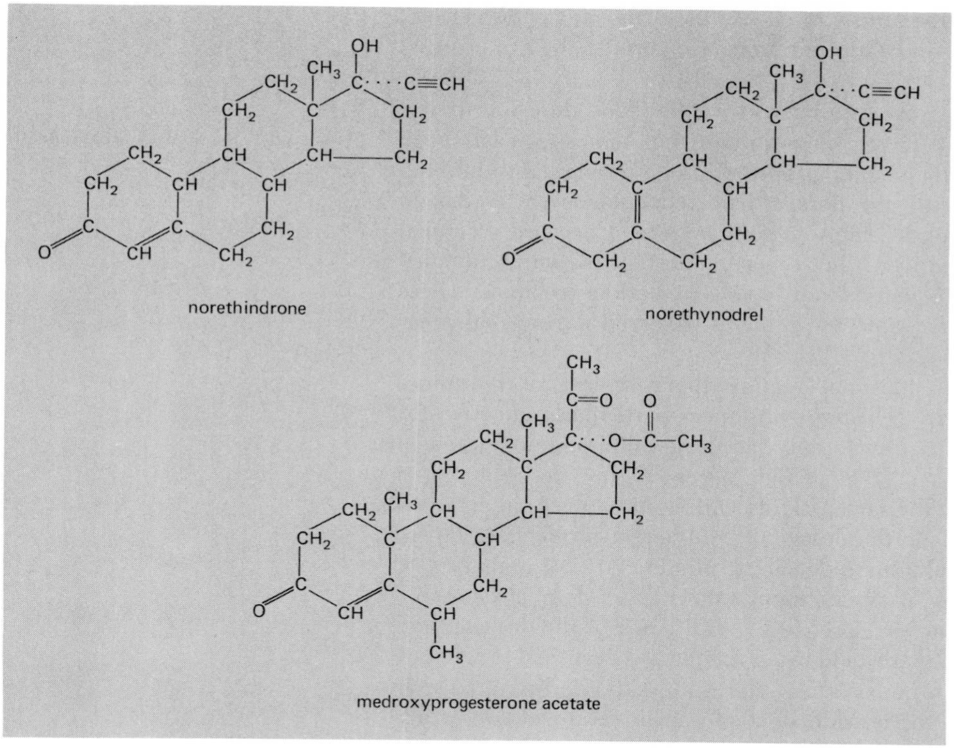

Figure 22-6 Orally active substitutes for progesterone.

Oral birth control pills extend the natural infertile period of the rhythm method.

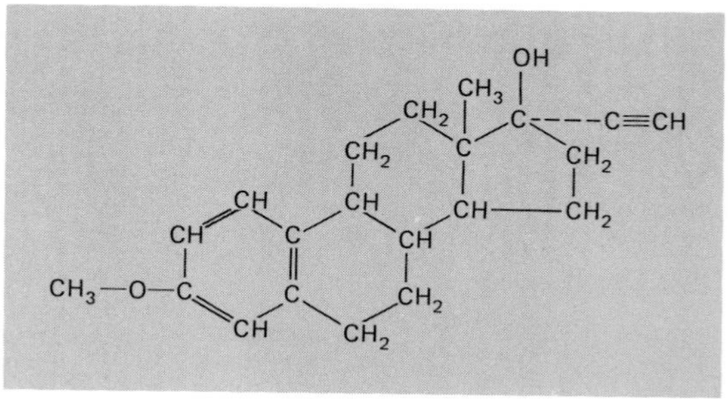

Figure 22-7 Mestranol, an estrogen substitute.

of an ovarian follicle and eventual ovulation. The effect is similar to the blocking action that prevents further ovulation during pregnancy. After day 25 of the cycle the drug can be discontinued and a menstrual period results. The cycle is then too far advanced for development of ovulation during that menstrual cycle.

The cycle amounts to a brief artificial pregnancy—a technological improvement of the rhythm method of birth control (intercourse only during the "infertile period"). The natural type of control (a steroid hormone) is used to provide an infertile period covering the whole cycle. The special

[22-4] Artificial Substitutes 517

advantage of the method is the ability to choose to permit a "fertile period" at any time by simply omitting the drug from use throughout a menstrual cycle.

When oral contraceptives were first introduced, the daily dose was 10 mg of the "progesterone" and 0.15 mg of mestranol, but the dosage has been reduced considerably since then; 1.0 mg of the "progesterone" and 0.05 mg of mestranol is now frequently the dosage. The reduction is partly due to increased confidence—less of a "safety factor" is needed because the reliability of the drug has been proven. But a more important reason is the need to reduce side effects such as nausea and blood circulation problems. These side effects are frequently the same as problems observed during real pregnancy.

Dosage of oral birth control pills has been decreasing, thus reducing the hazard from side effects.

Fortunately birth control does not require the full dose of hormones associated with pregnancy. By selecting substances particularly effective for the purpose at hand (preventing ovulation) and using minimum dosages, side effect hazards can be reduced. It is neither necessary nor desirable for a nonpregnant woman to share the same risk of death by thrombosis in enlarged arteries as pregnant women. As the dosage of contraceptive pills goes down, such secondary effects should be reduced or eliminated. As a matter of philosophy, the drug dosage in situations of known hazards might best be kept so low that there is some (but preferably small) risk of the drug failing. In the case of contraceptives, that would mean risking an occasional pregnancy to eliminate an occasional thrombosis. As that kind of borderline dosage is approached, the thoughtful supervision of a physician aware of any other medical problems becomes particularly vital. Biologically potent substances such as steroids should never be used without competent guidance from a physician supported by an adequate body of research results.

Sometimes the biochemical activity of a compound can be mimicked by substitutes that appear very different in structure. An example is estrogenic activity of diethylstilbestrol, which is abbreviated DES. Figure 22–8 shows DES written in a form to show the maximum structural similarity to the natural estrogens. The structure of estradiol is also shown in Figure 22–8 for comparison. DES has been widely used in feeds and as capsule implants

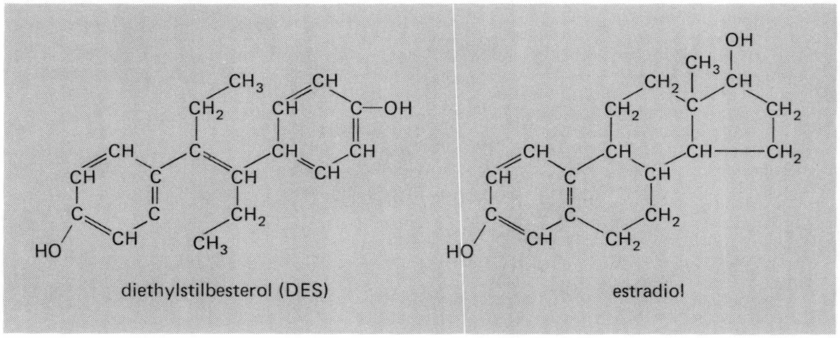

Figure 22–8 Similarity of diethylstilbesterol to estrogens.

in animals because it increases the efficiency with which feed is utilized in meat production. DES was also used as a medication for women having certain problems during pregnancy. (Remember, Section 22–3 and the beginning of 22–4, that progesterone present during pregnancy depresses estrogen activity.)

Diethylstilbestrol has estrogen-like effects but has a harmful side effect.

Unfortunately, we have since discovered that DES can make unborn girl babies more susceptible to cervical cancers developing when they are around 20 years old. Since this discovery, DES has been removed both from drug use in pregnant women and from use in stimulating animal growth. The discontinuation of DES as a growth stimulant is open to critical debate in a world where some people may starve or be retarded by protein deficiency long before they could be threatened by any risk of cancer from DES. Because of such considerations the use of controlled amounts of DES to stimulate growth of cattle was restored in the USA. But the original elimination of DES was a responsible effort to keep foods free of any traces of unnatural hazardous substances, and the incident illustrates the need for constant vigilance to detect the effects of man-made substances on natural systems

22–5 Cortisone and Function of Antibodies and Inflammation

Antibodies are part of natural body defenses.

The human body has several natural defenses against infection. The barrier of the skin is an obvious defense, and the action of white blood cells to surround germs and thus inactivate them is well known. Another type of defense is the formation of antibodies which act against the specific invader. Each human body has a recognition system which helps identify "foreign" substances. Antibodies (a special type of proteins) are then produced that react with the invading foreign particle and deactivate it. A similar effect is sometimes achieved by certain steroid hormones. Invaders such as the tubercule bacillus of tuberculosis may be walled off from the rest of the body by inflammation. Steroid hormones help the body respond to the invaders in an inflammatory way that leads to production of a barrier around the invader.

Cortisone is a steroid hormone lessening inflammation to prevent arthritis and related problems.

But inflammation in body tissues is not always a good thing. When inflammation occurs too easily it can have crippling effects on tissues at points where no walling off of invaders is possible. This is a particularly common problem in connective tissues in joints, muscle, the heart, and skin. The most common of these inflammatory diseases is rheumatoid arthritis, a stiffening of the joints that affects millions of people. The natural defense against these inflammatory diseases is a hormone balance. The inflammation stimulating steroid hormones are balanced against another steroid hormone, cortisone, which lessens inflammation. Cortisone, whose structure is shown in Figure 22–9, is produced in the adrenal gland under stimulation by another hormone (from the pituitary gland) which is called ACTH.

Arthritis and other inflammatory diseases can be relieved somewhat by aspirin, and that is the usual treatment. Severe cases can be treated by injections of cortisone or smaller amounts of ACTH, which then stimulates

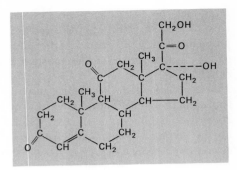

Figure 22-9 Cortisone.

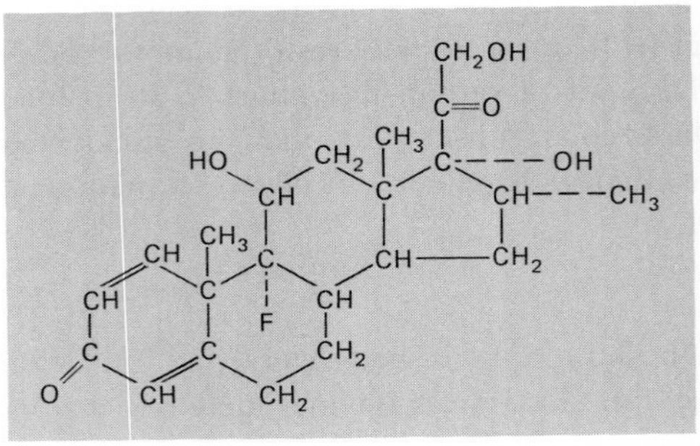

Figure 22-10 9-Fluoro-16-methylprednisolone, a cortisone substitute.

cortisone production. But the cortisone dose levels needed to achieve the desirable effects also led to some problems. For example, in the early days of cortisone therapy some patients developed active tuberculosis. Apparently bacteria that had been contained within inflammatory walls were set free to activate the tuberculosis. Physicians now guard against such side effects.

One of the efforts to reduce complications from cortisone therapy has been the development of substitutes that are effective in much smaller dosage. Figure 22-10 shows the structure of 9-fluoro-16-methyl prednisolone, which is used as a substitute for cortisone. This compound is 100 to 250 times as potent as cortisone and can therefore be used in small doses to treat severe arthritic flareups. Aspirin remains the choice for ordinary continuous treatment.

A cortisone substitute more potent than cortisone can be used to treat arthritis, but aspirin is the choice for continuous treatment.

22-6 Vitamins

22-6.1 Ascorbic Acid, Vitamin C. Vitamins are substances necessary for proper biochemical function which humans need to have supplied from outside sources or special reactions. When vitamins are not available in sufficient quantity, vitamin deficiency diseases result.

Ascorbic acid, which is also called vitamin C, is an oxidation product of hexose sugars. Figure 22-11 shows the structures of ascorbic acid and dehydroascorbic acid, a product that can be readily formed by oxidation of

Figure 22-11 Oxidation of vitamin C.

The two hydrogen atoms shown in color are easily removed by oxidation.

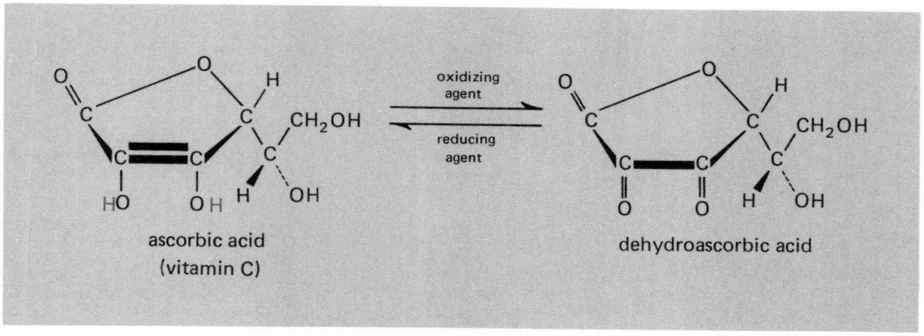

ascorbic acid (vitamin C) ⇌ (oxidizing agent / reducing agent) ⇌ dehydroascorbic acid

ascorbic acid. Vitamin C is probably used by humans in biochemical oxidation-reduction reactions using the simple conversion of ascorbic acid to dehydroascorbic acid and the reverse reaction.

Vitamin C deficiency leads to a disease called scurvy. Symptoms include loss of weight, sore gums (bleeding readily in advanced scurvy), tendency to fatigue, and other similar problems. Scurvy was a serious problem among sailors on sailing ships. Fresh vegetables are an important source of ascorbic acid, and sailors away from land for long periods had a diet deficient in fresh vegetables. The discovery by the British that citrus fruits (specifically limes) prevented scurvy contributed to the establishment of Great Britain as a sea power and led to the name "limey" associated with British sailors. Citrus fruits (principally oranges and grapefruit) are now a common source of ascorbic acid in diet, and ascorbic acid is manufactured and sold cheaply as tablets. As a result, serious cases of scurvy are now very rare.

Ascorbic acid, vitamin C, is easily oxidized and is required to avoid the vitamin C deficiency disease, scurvy.

22-6.2 B Vitamins. At one time it was believed that there was a water soluble substance called vitamin B. We now know that "vitamin B" is really a mixture of a number of essential substances that promote appetite, growth, and general well being. As these substances were separated and identified they were given names such as vitamins B_1, B_2, B_3, and so on. Some of these compounds are no longer identified by their vitamin B names. An example is niacin (nicotinamide) which we described in Section 21-4 and Figure 21-10.

There is a mixture of essential substances in the water soluble material once called vitamin B.

Figure 22-12 shows the structures of four compounds still known as B vitamins. Thiamine (shown as the salt thiamine chloride in Figure 22-12) is vitamin B_1. Thiamine deficiency causes the disease beriberi which involves **nervous system** abnormalities, anxiety, mental confusion, and, if completely denied for long enough, death. Beriberi was common in Asian areas where polished rice made up a large part of the diet. This beriberi (which may also have involved other deficiencies besides thiamine) can be avoided by using the unpolished rice or adding some rice hulls to the diet. Pork, yeast, fresh fruits and vegetables, and whole grains are good sources of thiamine. In this country thiamine is now added to polished rice and to white bread. About 1 mg of thiamine per day is required by humans. There is also some indication

Thiamine (vitamin B_1) deficiency causes nervous system disorders called beriberi.

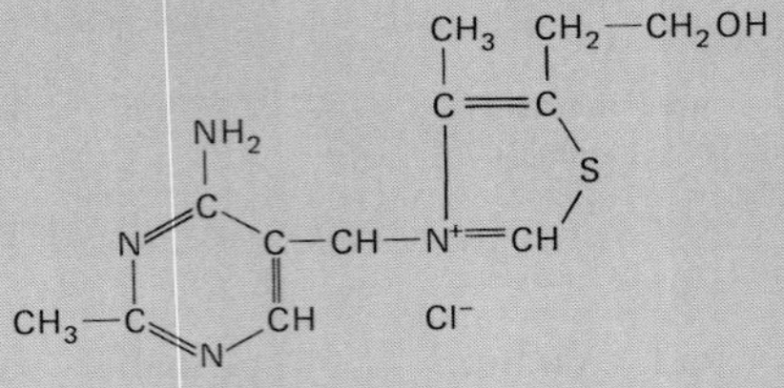

vitamin B₁ as thiamine chloride

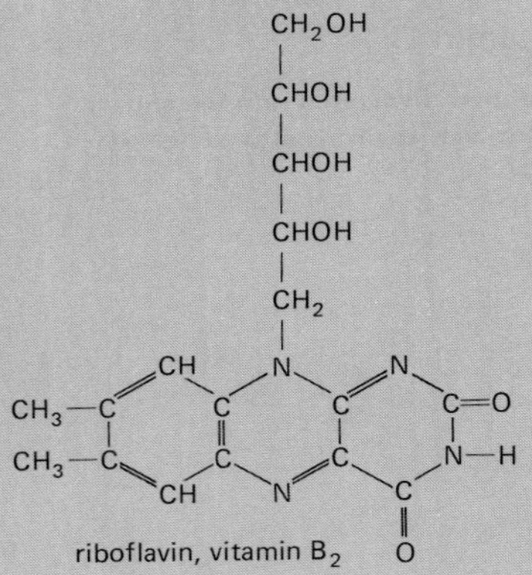

riboflavin, vitamin B₂

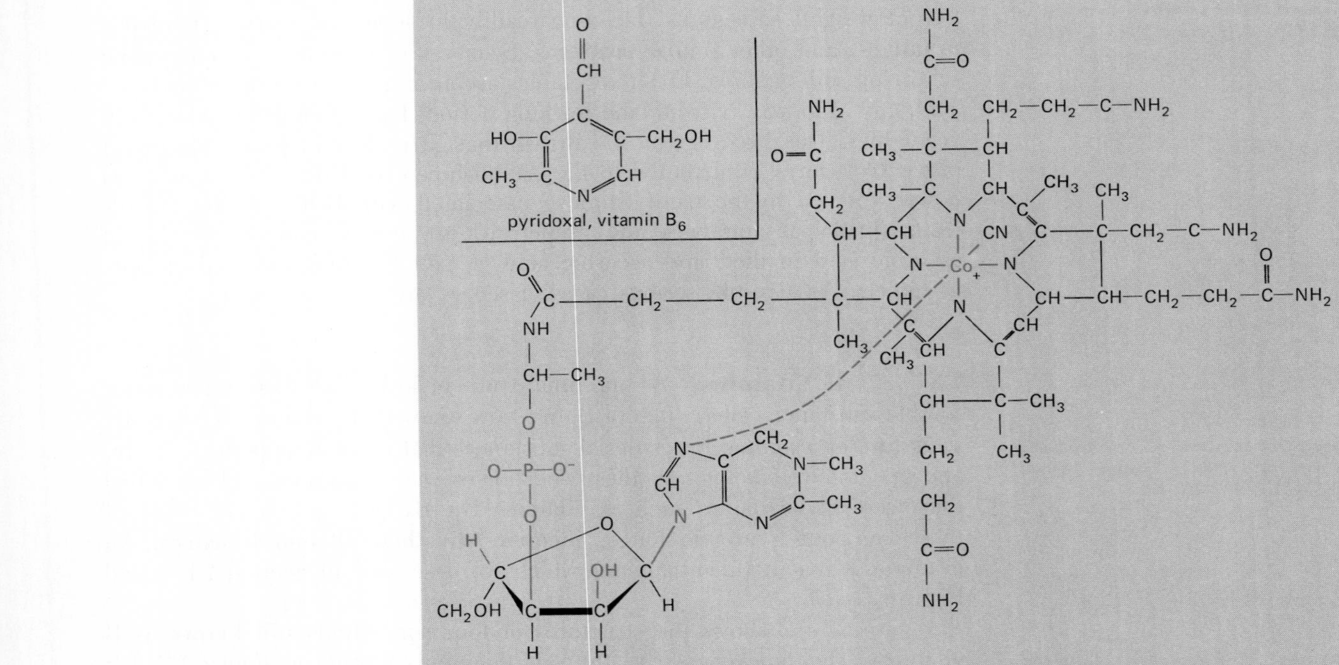

pyridoxal, vitamin B₆

cyanocobalomin, vitamin B₁₂

Figure 22-12 B vitamins.

Thiamine, vitamin B_1, is shown in the form of thiamine chloride. Riboflavin is vitamin B_2, and pyridoxal is vitamin B_6. Cyanocobalamin, vitamin B_{12}, is shown using the conventions of Figure 20-4 for the sugar ring (color for groups above the ring). The cobalt atom and the bonds to the cobalt are also shown in color.

Thiamine functions as a coenzyme for decarboxylation, riboflavin (vitamin B_2) is needed to form FAD, and pyridoxal (vitamin B_6) takes part in a coenzyme for many amino acid reactions.

Cyanocobalamin (vitamin B_{12}) deficiency leads to pernicious anemia, but vitamin B_{12} is always adequately available for those who can absorb it well in digestion.

that thiamine deficiency may contribute to alcoholism. Although psychological factors are no doubt more important, a sound diet with thiamine rich foods such as whole grain cereals and pork might help a reformed alcoholic maintain his resolve.

Thiamine functions as a coenzyme in decarboxylation reactions. Thiamine deficient persons have high blood levels of pyruvic and lactic acid, apparently due to failure to decarboxylate pyruvic acid in the reaction forming acetyl coenzyme A (see Section 21-4.4). Figure 22-13 shows the intermediate believed to be formed by reaction of pyruvic acid with the reactive carbon atom between the sulfur and nitrogen atoms in a diphosphorylated thiamine. This intermediate then leads to decarboxylation and the formation of acetyl coenzyme A.

Riboflavin, vitamin B_2, is required in amounts of about 1.5 mg per day by humans. Riboflavin is used in production of FAD, which is required in metabolism as described in Sections 21-4.3 and 21-4.4 and Figure 21-13. Riboflavin is found in yeast, liver, and wheat germ. Milk and eggs are less concentrated sources of riboflavin but probably are the major riboflavin sources for most Americans.

Pyridoxal, vitamin B_6, may join other former B vitamins in being removed from the list of vitamins. Although pyridoxal is certainly essential, there is evidence that some pyridoxal is synthesized in the human body, either by the body itself or by bacteria in the intestinal tract. Additional pyridoxal from diet may or may not also be necessary. Pyridoxal phosphate is a coenzyme for many amino acid reactions, and other pyridoxal derivatives play essential enzymatic roles.

Cyanocobalamin, vitamin B_{12}, is not usually a problem in nutrition since only small quantities are required (perhaps 10^{-6} g/day) and the body can store some reserve. Vitamin B_{12} is widespread in animal products, but some persons apparently fail to absorb it efficiently in digestion. Deficiency of vitamin B_{12} leads to pernicious anemia. Treatment is given by injections of

Figure 22-13 Intermediate for decarboxylation of pyruvic acid by thiamine.

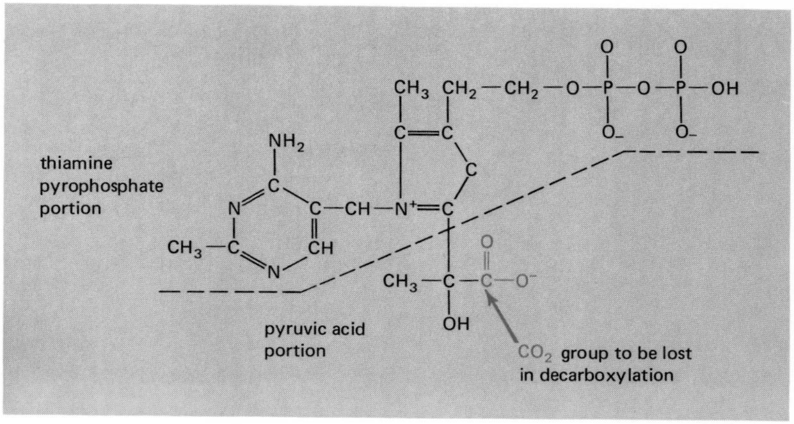

vitamin B_{12}, because deficient persons are not absorbing the vitamin effectively from the digestive tract.

22-6.3 Fat Soluble Vitamins.
Animal fats such as cod liver oil and butter are sources of several fat soluble vitamins, although fats are not the only sources of these fat soluble compounds.

Vitamin A is necessary for growth and maintenance of cells in mucous membranes. Mild vitamin A deficiency leads to night blindness and more severe deprivation causes inflamed eyes and susceptibility to infections in the eyes, respiratory tract, digestive tract, and urinary tract. Vitamin A deficiency also causes sterility by affecting linings of genital organs.

Figure 22-14 shows the structures of vitamin A and the natural pigment β-carotene from which humans can produce vitamin A. Vitamin A is yellow whereas β-carotene is orange, and some yellow or orange vegetables (carrots, squash) are good sources of vitamin A. Vitamin A is also available from many green vegetables. However, heat destroys vitamin A, so cooked vegetables lose their value as sources of the vitamin.

Vitamin A is directly involved in the cycle of reactions causing the most light sensitive kind of vision. When vitamin A is missing, the eye has only the less sensitive kinds of vision used for daytime and color vision, hence the resulting night blindness. Vitamin A can exist in a form with all of the double bonds in its chain portion in the *trans* configuration, or it can be converted by the liver to a form with one *cis* double bond. Figure 22-15 shows how this conversion is used in vision. The form with a *cis* bond is oxidized (by converting NAD^+ to $NADH$) from the alcohol to the corresponding aldehyde, called retinal. That aldehyde (with one *cis* bond) reacts with a protein called opsin to form a light sensitive compound called rhodopsin (visual purple). When rhodopsin is exposed to light, the retinal portion has

Vitamin A is necessary for mucous membrane health and is used in night vision.

Figure 22-14 Vitamin A and a common source.

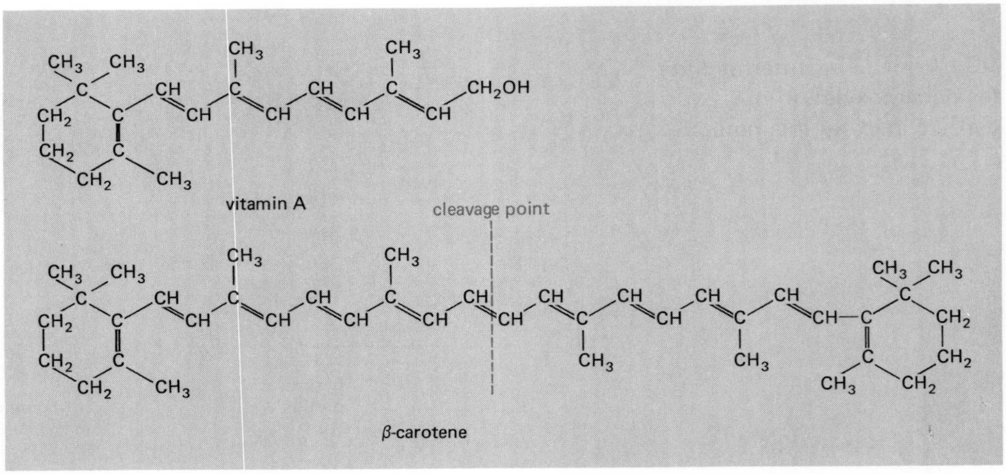

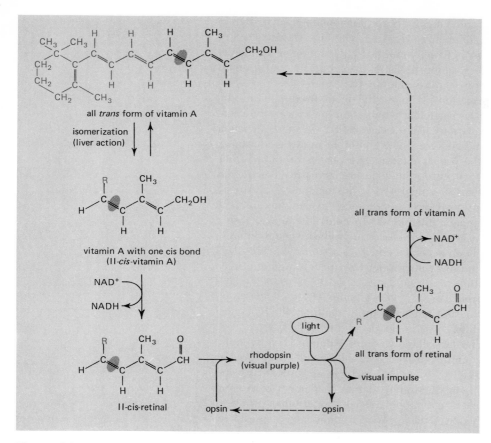

Figure 22-15 Night vision reactions.

The R group (unchanged throughout the process) is shown in color. The double bond where the *cis-trans* changes occur is indicated by the shaded region.

its *cis* bond converted back to the *trans* arrangement. This all-*trans* form is unable to remain bonded to the opsin. The light-induced splitting of rhodopsin into the all-*trans* form of retinal plus opsin initiates the reactions allowing night vision. The all-*trans* form of retinal can then be reacted with NADH to form NAD^+ and the all-*trans* form of vitamin A from which the cycle started.

Vitamin A deficiency is a serious problem in some parts of the world. In India at least one million persons are permanently blind because of eye diseases resulting from severe vitamin A deficiency. But in other areas there may be more excesses of vitamin A than deficiencies. An excess of vitamin A (hypervitaminosis A) produces skin irritations and loss of appetite. It is interesting to note natural diet patterns avoiding such problem. Eskimos, whose diet has adequate vitamin A, generally do not eat the liver from polar bears. Liver is a rich source of Vitamin A, and polar bear's liver contains extremely high levels of vitamin A which would cause hypervitaminosis A. By contrast, if polar bear's liver were available in India, it would probably become a prized delicacy among the vitamin A starved population.

An excess of a vitamin can be harmful.

Vitamin D is another fat soluble vitamin. Several compounds have vitamin D activity. Both of the forms commonly occurring, vitamins D_2 and D_3, are converted from inactive precursors to active forms by the action of ultraviolet light. Figure 22–16 shows the reaction in the skin of humans and animals to irradiate 7-dehydrocholesterol to form vitamin D_3.

Vitamin D controls the formation of bones and teeth from calcium and phosphorus minerals. Vitamin D is particularly important for growing children, who suffer abnormal bone formations in a disease called rickets if deficient in vitamin D. On the other hand an excess of vitamin D can also be harmful. Natural evolution to achieve the proper vitamin D level has produced some interesting effects. Polar bears, with little sun exposure, actually have vitamin D formed externally on their hair and ingest it by licking their fur. Among humans vitamin D adjustment provides the simplest explanation for the development of racial skin pigment differences. In cold, low sun exposure climates a very light skin was necessary for babies to produce enough vitamin D for survival and reasonable bone development. In hot sunny climates dark skin pigments were necessary to lower ultraviolet light penetration and keep the vitamin D levels down within the desired range. In the present day and age when dark skinned children can be given vitamin D supplementation and light skinned people can be protected from sun by clothing, the biological significance of skin pigment differences is clearly obsolete.

Figure 22–17 shows two other fat soluble vitamins, vitamin E and vitamin K. The compound shown for vitamin E is α-tocopherol, one of several similar compounds with vitamin E activity. Vitamin E acts as an antioxidant and may protect some vital cell components from oxidation. Deficiency of vitamin E has been associated with muscular distrophy, sterility, and other problems in animals, but there is no evidence that proves vitamin E is required by humans. At present it appears that if vitamin E is required by humans at all it is adequately supplied. Excess vitamin E does not help humans with any of the problems associated with vitamin E deficiency in animals, so any

Vitamin D, formed by action of ultraviolet light on certain steroids, controls formation of bones and teeth and influences racial skin pigment differences.

Vitamin E is needed by animals but perhaps not by humans.

Figure 22–16 Formation of vitamin D.

The positions involved in the reaction are shown in color.

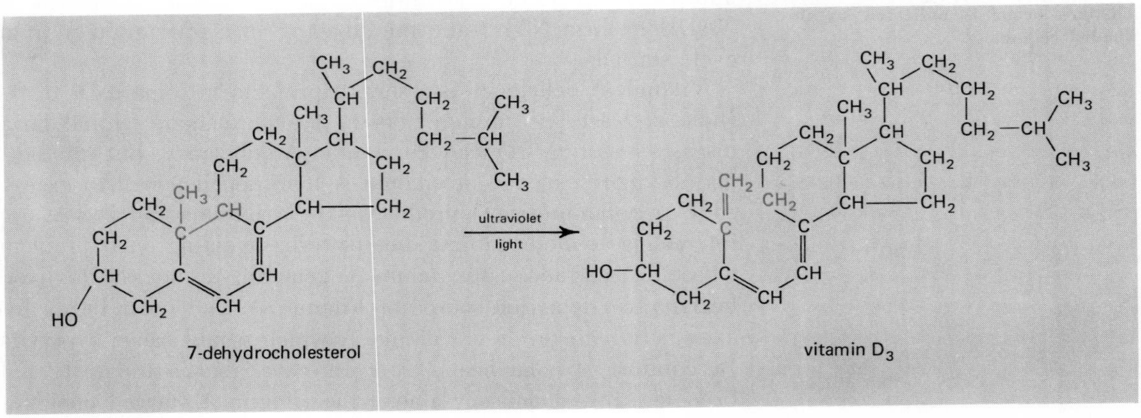

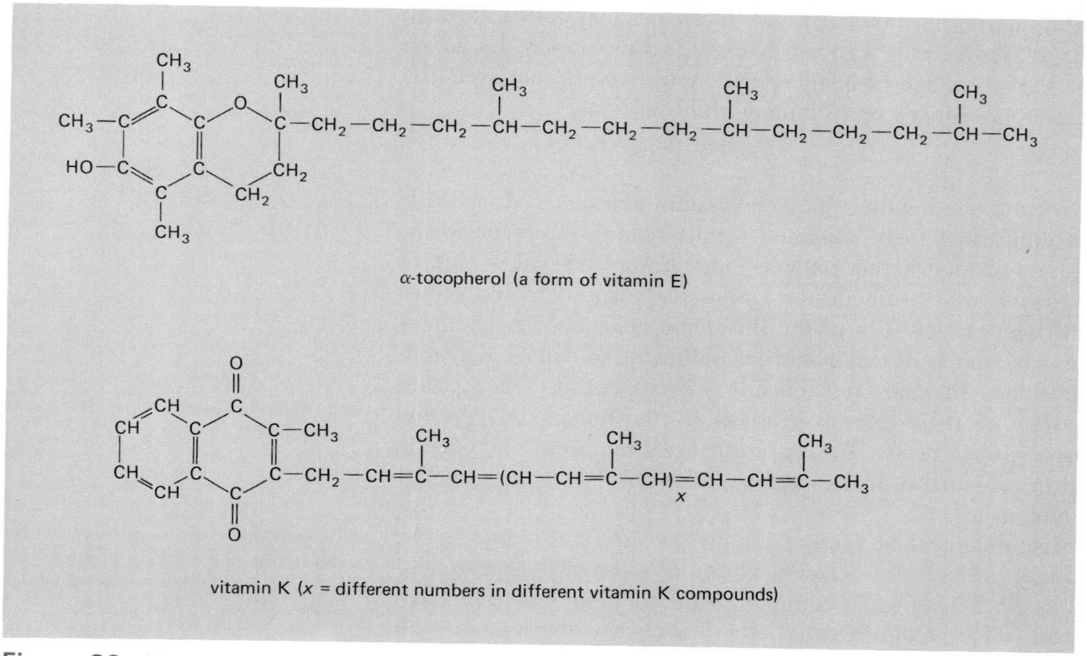

Figure 22-17 Vitamins E and K.

α-tocopherol (a form of vitamin E)

vitamin K (x = different numbers in different vitamin K compounds)

Vitamin K is associated with blood clotting, and similar compounds can prevent clotting by displacing the needed vitamin K.

supplementation of vitamin E for humans is probably more likely to cause hypervitaminosis problems than to provide any benefit.

Vitamin K is essential to blood clotting action. There are several active forms of vitamin K differing only in the length of the side chain as noted in Figure 22-17. Vitamin K provides an interesting example of the use of substitutes to cause *failure* of a process. In some persons excessive clotting causes a serious health hazard or death by blocking key blood vessels. Clotting can be reduced by treatment with drugs bearing enough similarity to vitamin K to be accepted in its place by reactants but not enough similarity to cause completion of clotting. Figure 22-18 shows two such compounds, dicoumarol and warfarin. These compounds have a greater effect on some animals than on others. Warfarin, which merely lessens clotting in humans, causes such

Figure 22-18 Anticlotting agents.

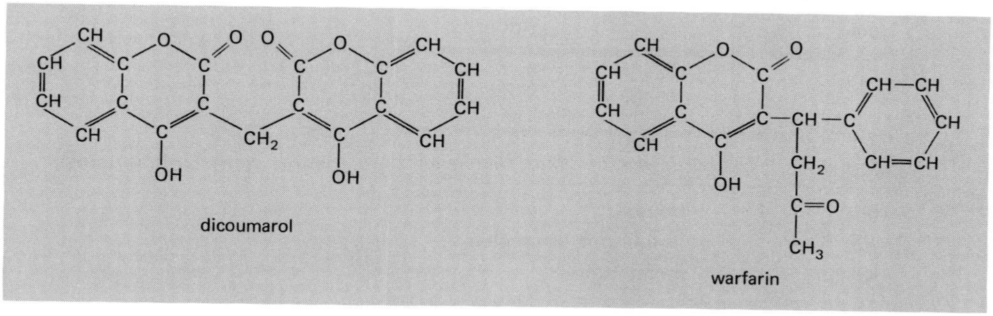

dicoumarol

warfarin

[22-6] Vitamins 527

a great reduction of clotting in rats that it eventually produces death through internal hemorrhages. As a result warfarin is used as a rat poison. Its action is slow but it has the advantage of being species selective and not seriously hazardous to most other animals or to human children.

22-7 Essential Amino Acids

In the preceding section on vitamins, the term vitamin was reserved for those substances which either had to be obtained from outside sources or where there was uncertainty about whether outside sources were needed or not. A similar criterion can be used to divide the amino acids from which proteins are assembled into two classes. Ten of the 20 amino acids used by humans can be synthesized by the body in quantities sufficient to satisfy demand. The other ten must be obtained from outside sources because they either cannot be synthesized or their rate of synthesis by the human body is too low to satisfy normal growth needs. The ten amino acids required from outside sources are called the **essential amino acids.** Table 22–1 lists the essential amino acids for humans.

Humans need adequate amounts of ten essential amino acids from outside sources.

Most animal proteins are good sources for all the essential amino acids. Some important plant proteins are deficient in one or more of these essential amino acids. Table 22–2 lists some common plant food sources that have protein content and which of the essential amino acids are deficient in each source. Among people whose diet is low in animal proteins, a mixture of plant products is needed to avoid amino acid deficiencies. If one (or more) essential amino acid is inadequately supplied, protein synthesis is slowed or stopped at the points where that amino acid is needed. That prevents normal growth and causes serious damage, particularly to children who cannot simply reuse the amino acids from degradation of old cells.

Animal proteins supply all amino acids, but some plant proteins are deficient in certain essential amino acids.

Lysine deficiency is known to cause mental retardation of children (among other effects) and later improvements in diet cannot repair the damage.

Table 22–1
The Essential Amino Acids

arginine	methionine
histidine	phenylalanine
isoleucine	threonine
leucine	tryptophan
lysine	valine

Table 22–2
Amino Acid Deficiencies

Source	Essential Amino Acid Deficiencies
wheat	lysine
corn	lysine and tryptophan
soybeans	methionine

Because of this problem, chemical production of lysine to be added to wheat products was tried for some areas of the world, and production of other amino acids may become desirable in the future to supplement diets with particular deficiencies. These amino acid supplements must be provided cheaply so poor, hungry people can afford to choose foods which let their children grow up with their full natural intelligence and decent health.

22–8 Other Essential Substances

Many other substances are also essential in proper quantities to maintain life.

In this chapter we have only sampled some of the biochemically important substances. Many other substances are also essential to life. Some of these are raw materials needed in substantial quantities for known purposes. Examples would be the calcium and phosphorus minerals needed for bones and teeth. Others are needed only in small quantities for purposes that may or may not be understood. An example is copper, which acts as a poison in large quantities but is essential in small quantities. Others are specific compounds as biochemically potent or even more potent than any we have discussed. An example is the class of compounds called the prostaglandins. These compounds are present in very low concentrations in most organisms and have only recently attracted attention because of their great biochemical effects.

We cannot complete coverage of the really important substances for two reasons. First, we do not have the time or space for all that is known. Second, there obviously must be many important substances that have not yet been discovered or whose functions are not yet known. We can only hope the sampling we have offered is useful in showing the sort of processes involved in life on Earth.

Skills Expected at This Point

1. You should be able to list the effects of high and of low blood glucose levels, describe the role of insulin, and describe diabetes and its treatment.
2. You should be able to describe, in words and structural sketch, the nature of steroids.
3. You should be able to name the principal types of sex hormones and describe their main functions.
4. You should be able to list the types of components in oral contraceptives and describe how the oral contraceptive works.
5. You should be able to list both advantages and hazards associated with use of artificial substitutes for biochemically potent compounds.
6. You should be able to describe the natural balance between inflammation and the antiinflammatory action of cortisone.
7. You should be able to list the compounds described as vitamins in this chapter, list the general function of each, and describe the deficiency diseases or problems associated with at least three of the vitamins.

8. You should be able to describe the hazard of vitamin excesses as well as deficiencies and list two examples.
9. You should be able to explain why meatless diets can result in protein synthesis problems and how these problems can be avoided (without meat).
10. You should be able to relate facts from this chapter with information learned earlier in the course.

Exercises

1. Name and describe the condition caused in humans by insufficient insulin.
2. Why do diabetes sufferers need to carry candy or other sweets?
3. Describe the class of compounds called steroids.
4. List four types of effects caused by different steroids.
5. (a) Name an artificial substitute that is more potent than the compound it is substituted for and list its function and the situation where it is used.
   (b) Name an artificial substitute that fails to produce the activity of the compound it is substituted for and list the functions both of the natural compound and of the substitute under the conditions where it is used.
6. Why has dosage in birth control pills decreased over the past few years?
7. Describe the sources of vitamin A and the function of vitamin A.
8. List one essential amino acid known to be inadequately supplied in some diets and describe the consequences of such a shortage.
9. For each section of this chapter, list one important compound mentioned in that section and describe its function.

Test Yourself

1. List a possible cause of a very low blood glucose level and the possible effects of such a low glucose level.
2. Define the following terms: (a) diabetes, (b) steroid, (c) testosterone, (d) essential amino acid.
3. Describe the operation of an oral contraceptive.
4. Name the compound known as vitamin C and describe its probable function.
5. The vitamin deficiency disease associated with a thiamine deficiency is called
   (a) rickets   (d) beriberi
   (b) scurvy   (e) hypervitaminosis
   (c) diabetes
6. Insulin has the effect of
   (a) curing rickets
   (b) walling off foreign substances
   (c) reducing inflammation

(d) stimulating vitamin C production
   (e) lowering blood glucose levels
7. Cortisone has the effect of
   (a) curing rickets
   (b) walling off foreign substances
   (c) reducing inflammation
   (d) stimulating vitamin C production
   (e) lowering blood glucose levels
8. A shortage of riboflavin could interfere with all but one of the following processes. Which one would not be directly affected?
   (a) oxidative phosphorylation
   (b) production of ATP by the Krebs cycle
   (c) the Embden-Meyerhof pathway
   (d) production of FAD
   (e) formation of $FADH_2$
9. Which of the following is *not* a steroid?
   (a) vitamin A   (d) progesterone
   (b) cortisone   (e) cholesterol
   (c) estradiol
10. Which of the following is a list of essential amino acids that may be lacking in some diets?
   (a) glycine, leucine, and lysine
   (b) lysine, methionine, and tryptophan
   (c) adenine, cytosine, and guanidine
   (d) cortisone, insulin, and prostaglandin
   (e) cholesterol, leucine, and thymine

Changes from Natural Chemistry

23–1 Stimulating Natural Growth

Human civilization has a great impact on the chemistry of the natural environment. Some man-made changes are far from the natural chemistry on Earth, but others are only changes in the amount, location, and timing of processes that are quite normal in the natural order of events. In many cases man simply strives to stimulate natural growth.

Use of fertilizers is a simple and widespread example of growth stimulation. The earliest fertilizers were simply wastes, such as manure, which essentially mimicked nature's cycling of nutrients. With improved knowledge of the limiting nutrients, we can now provide them from other sources. The result is improved efficiency plus less tendency to spread diseases. Chapter 14 described some of the factors involved in using fertilizers wisely and some of the problems, such as oxygen demand in bodies of water. Our present knowledge is providing some help to lessen problems, but it is clear that much more knowledge and a willingness to use it are needed.

The limiting nutrient supplied in the greatest quantity by man to stimulate growth is water. Irrigation seems like such a simple procedure that no scientific knowledge should be necessary. But the basic principles of solutions discussed in Chapter 10 show why irrigation cannot be used indiscriminately without salting out soils. Other more subtle effects on climate may also eventually need to be better understood.

Man stimulates natural growth by providing fertilizers and irrigation.

In addition to supplying nutrients and irrigation, humans also stimulate growth that is usable for mankind by selection of the species allowed to grow. Land is cultivated to remove unproductive plants (weeds) and crops of only a single productive type of plant are sown, protected, and harvested. The maximization of production is also furthered by selective breeding. More productive varieties of wheat, rice, corn, and other plants and animals have been evolved and put into widespread use. Unfortunately, single crop fields and widespread use of a few productive types make the plants more susceptible to epidemics of diseases and pests. The original development of good varieties and growing conditions must be followed by steady redevelopment of new varieties and conditions to prevent a collapse of the newfound productive capacity.

Productive growth is stimulated by selective breeding and cultivation.

The susceptibility of "superior" species to natural enemies can perhaps be illustrated by considering some of the ways survival capacity has been obviously lessened to increase productivity. Wheat production has been sharply increased by breeding programs, and short stalked varieties are particularly successful. The short plants do not waste energy on stalk growth, so in uniform fields they can direct more of their food production to wheat kernels. But in a wild field the less efficient tall wheat would tend to shadow the short varieties so completely that the short varieties would be poor competitors. Similarly, the highest priced present day commercial bulls would not fare well in natural combat with thin, stringy longhorn bulls.

Hormones or hormone substitutes can stimulate growth.

Sometimes natural growth is stimulated by adjustment of the potent biochemical compounds that control growth processes. For example, a hormone has been identified that stimulates a number of processes in human growth. This human growth hormone, which we will call by the abbreviation HGH, may be inadequately supplied in some children whose growth is retarded. If a shortage of HGH is identified as the cause of retarded growth, small amounts of HGH may restore a more normal growth pattern. Of course, such a biochemically potent substance must be used with great caution to avoid harmful excesses.

Hormone adjustments have been used to increase efficiency of meat production from feed. Male hormones stimulate some kinds of development at the cost of disadvantages in other ways. In cattle, the toughness and combativeness in bulls makes the animals harder to control and handle (and lowers quality of the beef after slaughter). Therefore, it is economically advantageous to remove the male hormones. That is done by castrating bull calves to be raised only for slaughter. The resulting steers fatten more quickly because of their lack of the male hormone production.

The idea of avoiding the effects of male hormones can be accentuated by creating a feminine hormone balance. The estrogenic activity of DES (diethylstilbestrol, see Section 22–4) stimulates growth with good feed efficiency. DES was given to steers and other noncastrated animals (notably chickens) until a link with cancer caused abandonment of its use (see Section 22–4). DES was so useful as a growth stimulant that controlled use with steers has been restored.

23–2 Substitutes for Natural Balances

A balance similar to natural conditions can sometimes be achieved with artificial compounds.

Sometimes unnatural intervention is necessary to restore conditions to a state close to that which should have been created by natural balances. Failure of the natural balance can occur because of man-made disturbances or failure of the natural processes to function normally. The treatments of diabetes by insulin and arthritis by cortisone described in Chapter 22 are examples of artificial creation of the usual hormone balances. Intervention with such potent substances can require delicate adjustments to avoid hazards, as noted in Section 22–1. Sometimes it is better to establish a new balance in which a natural compound is replaced by a different one. Section 22–5 describes the use of chemical substitutes for cortisone. The new balance established is essentially the natural one, inflammation versus antiinflammation, but it is achieved by different compounds. The following paragraphs will describe other cases where a balance somewhat similar to the natural balance is reached by artificial means.

Human activities often create conditions quite far from the natural balances. In some cases other human activities can then return the situation to one similar to, or at least not much less desirable than, the natural situation. Collection of sewage into a central sewage system is one human activity that upsets natural balances. Sewage collection is done to prevent the spread of diseases, but concentration of wastes in water and in one area also concentrates the biological oxygen demand, as noted in Section 14–7. If large quantities of untreated wastes are allowed to enter a stream at a single point, the normal excess of oxygen supply over oxygen demand cannot be maintained. Redistribution of the sewage to add small amounts at many points (a more natural condition) so oxygen supply would be more adequate is not practical. But balancing of the one artificial situation of high, localized oxygen demand by high, localized oxygen supply can be done. The secondary sewage treatment plants described in Section 14–7 provide the high oxygen supply needed. The resulting sewage treatment is unnatural both in oxygen demand and oxygen supply, but treating the sewage leaves the water bodies receiving the output much closer to their natural chemistry.

Concentration of BOD in sewage can be balanced by supplying extra oxygen to the sewage treatment process.

Human activities to stimulate growth also often provide an unnaturally attractive situation for diseases and other pests. As a result the pests thrive at much higher levels than they could sustain under natural conditions. Natural conditions often provide pest controls in the form of predators, dispersal of the pest's targets, and the tendency for susceptible species to be replaced by their more resistant neighbors. Some of those controls are lost when single species are raised in high local concentrations as is the case in cultivated fields. Transportation activities by man may also allow pests to become established in areas free of their natural predators.

Pest controls are needed to balance the unnaturally favorable conditions for pests in agriculture.

Sometimes natural enemies can be identified and brought to the area where the pest is a problem. However, if suitable predators are not available, direct action against the pest may be necessary. Section 23–6 discusses ways in which poisons can be used to kill pests. Interfering with reproduction is essentially a duplication of natural control because the limited populations and food supplies under natural conditions lower pest survival and success at reproduction. Many of the pests do not survive to the adult stage where they could reproduce.

Pests can be reduced by natural enemies, by interference with the reproduction of the pests, or by selective breeding for resistance to the pests.

One of the most interesting approaches to pest control allows insects to survive in their environment of abundant food but prevents their reaching adult stage. Development of control methods using that approach is just beginning at the time of this writing. Certain juvenile hormones have been identified in some insects and others are being sought. Those hormones or artificial substitutes for those hormones can be applied widely to the infested areas to prevent sexual maturation of the insects. This approach avoids poisons on the desired crop and also avoids poisoning natural predators. As our understanding of the biochemistry of hormones increases, use of juvenile hormones will probably gain popularity.

Another way to interfere with reproduction is to permit breeding that cannot succeed. Females of many species breed only once. If they breed with sterile males, no offspring are produced. Huge numbers of males of particular pest species have been raised, sterilized, and released in infested areas. Some successful reproduction by the fertile natural males continues, but the reproduction can be made small by using a large excess of sterile males. This approach is particularly useful when the damage is caused by either the juvenile form of the pest (for example, fruit flies) or by the females (for example, mosquitos).

Another way to duplicate natural controls is by breeding for resistance to particular diseases or pests. Under natural conditions the varieties that best resist the existing diseases and pests have a greater survival. The same criterion can be used by the breeder developing new varieties. Instead of selecting only on the basis of productivity, he can consider a combination of productivity and resistance to known and anticipated problems. Responsible plant scientists are particularly concerned about identifying and maintaining the availability of as many different types of resistance as possible.

23-3 Disrupting Natural Problems

23-3.1 Pain Relievers.

Sometimes a change from natural chemistry is desirable to disrupt natural problems. The most common natural problem treated chemically is probably pain. The drugs used to treat headaches and minor aches and pains are called analgesic drugs, and the most widely used one is aspirin.

Figure 23-1 shows the structure of acetylsalicylic acid, commonly known as aspirin. Aspirin is absorbed when taken orally and is effective in relief of some pains, fever, and inflammation. The pains relieved by aspirin are generally those caused by swelling of tissues (which causes pressure on nerves) or minor inflammations and fever. These problems result from colds, flu,

Figure 23-1 Aspirin.

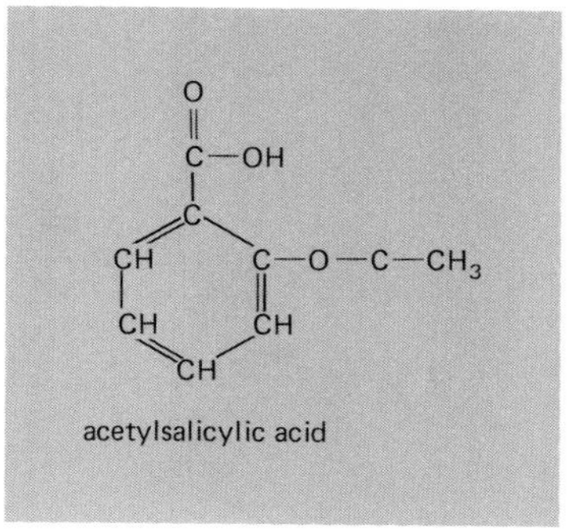

acetylsalicylic acid

arthritis, and other disturbances. By controlling body temperature and reducing swelling and inflammation, aspirin also relieves the pain. Aspirin may also have a chemical effect on the nerves. Aspirin is used in very large quantities because it is effective for most people, cheap, and not hazardous in moderate quantities. In the United States the average person uses about 200 of the standard 5-grain aspirin tablets per year.

Aspirin is sometimes combined with other ingredients. Buffering agents help speed the aspirin's absorption but make no contribution to the eventual analgesic action. The speed of buffered aspirin (Bufferin and other brands) has been widely advertised. But some other additives do help relieve pain. A common combination is the APC tablet combination of aspirin and the two compounds shown in Figure 23-2, phenacetin and caffeine. Phenacetin can also be replaced by a closely related compound, *N*-acetyl-*p*-aminophenol, and other pain relieving compounds can also be used. Phenacetin, caffeine, and some other drugs are **synergistic** with aspirin—they accentuate the effectiveness of the aspirin. Phenacetin produces some of the same effects as aspirin, whereas caffeine probably assists by improving transport of the aspirin in the blood. These combinations are widely advertised as "extra strength pain relievers," but aspirin is the main active ingredient in each case. Phenacetin has in fact been removed from a major "extra strength" pain reliever because of side effects, leaving caffeine as the only "extra" ingredient added to aspirin in that product.

Flavorings are also added to aspirin to make "childrens' aspirin." The almost candylike childrens' aspirins are sold in bottles designed to prevent children from opening them. The cause for concern is the danger of poisoning from an aspirin overdose. The safety of aspirin holds only when dosages are kept moderate. Because aspirin is so widespread (and "adult" aspirin is not sold in hard to open bottles), aspirin overdoses cause a number of deaths.

Although very large doses (perhaps as much as 50 or more tablets) of aspirin are needed before most adults would be seriously poisoned, some people react poorly to small doses. In cases where there is an allergic reaction to aspirin or aspirin simply does not help a particular individual, alternative pain

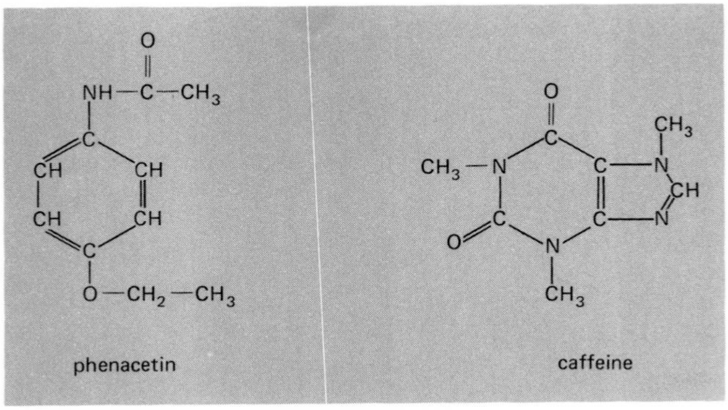

Aspirin provides a cheap, relatively safe means for disrupting the natural problems of pain, fever, and inflammation.

Other drugs are available either to assist or replace aspirin.

Figure 23-2 Phenacetin and caffeine.

relievers are needed. The most common choice is acetaminophen (Tylenol or NoAspirin), a compound similar to phenacetin. It has no advantages for people who respond normally to aspirin.

Some more potent pain relievers exist, but they also have sedative effects. They will be mentioned in Section 23-3.3.

23-3.2 Antihistamines. Allergic responses can cause release of histamine with biochemical consequences that may need to be stopped. Histamine and the closely related amino acid histidine are shown in Figure 23-3. Decarboxylation of histidine would produce histamine. Histamine stimulates small blood vessels to dilate, affects muscles related to breathing and some other functions, stimulates production of large quantities of very acidic gastric juices, stimulates nerve endings, and can cause itching. The effects of histamine can be lessened or eliminated by antihistamine drugs, such as those shown in Figure 23-4. These drugs are all substituted aminoethanes. The drugs are believed to compete with histamine for the active sites on the affected cells.

Antihistamines lessen allergic responses and often cause drowsiness.

Antihistamines tend to have the side effect of producing drowsiness. That is a disadvantage in most uses and can cause a hazard in activities such as driving cars. But it has been turned to advantage in some of the antihistamines. Figure 23-4 shows active ingredients of Dramamine, which is used to prevent motion sickness, and Sominex and Nytol, which are used to treat insomnia.

23-3.3 Sedatives and Tranquilizers. Difficulties with nervous tension and pain can also be treated by drugs that depress the action of the nervous system. These drugs vary from tranquilizers, which may depress only the sympathetic nervous system or some chemicals in the brain, to sedatives, which depress the central nervous system.

Sedatives and tranquilizers lessen pain and nervous tension by depressing the nervous system.

Figure 23-5 shows the structure of reserpine, one of the drugs sometimes used as a tranquilizer. Reserpine has a number of physiological effects, including the ability to lower blood pressure. It is used to treat hypertension (high blood pressure). Hypertension patients using reserpine must also be affected by the depression of some brain chemicals and the central nervous system characteristic of reserpine. This production of unwanted effects is

Figure 23-3 Histamine and histidine.

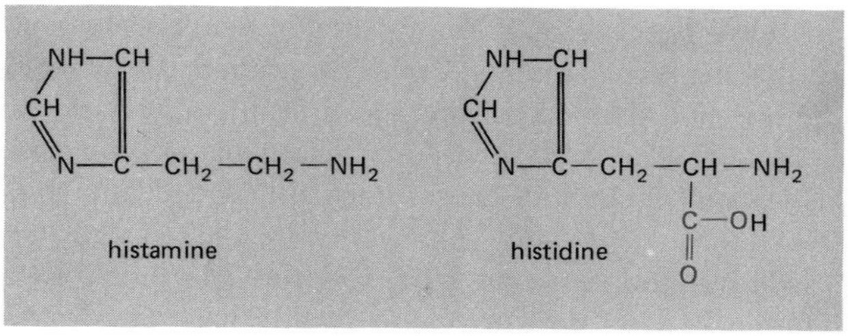

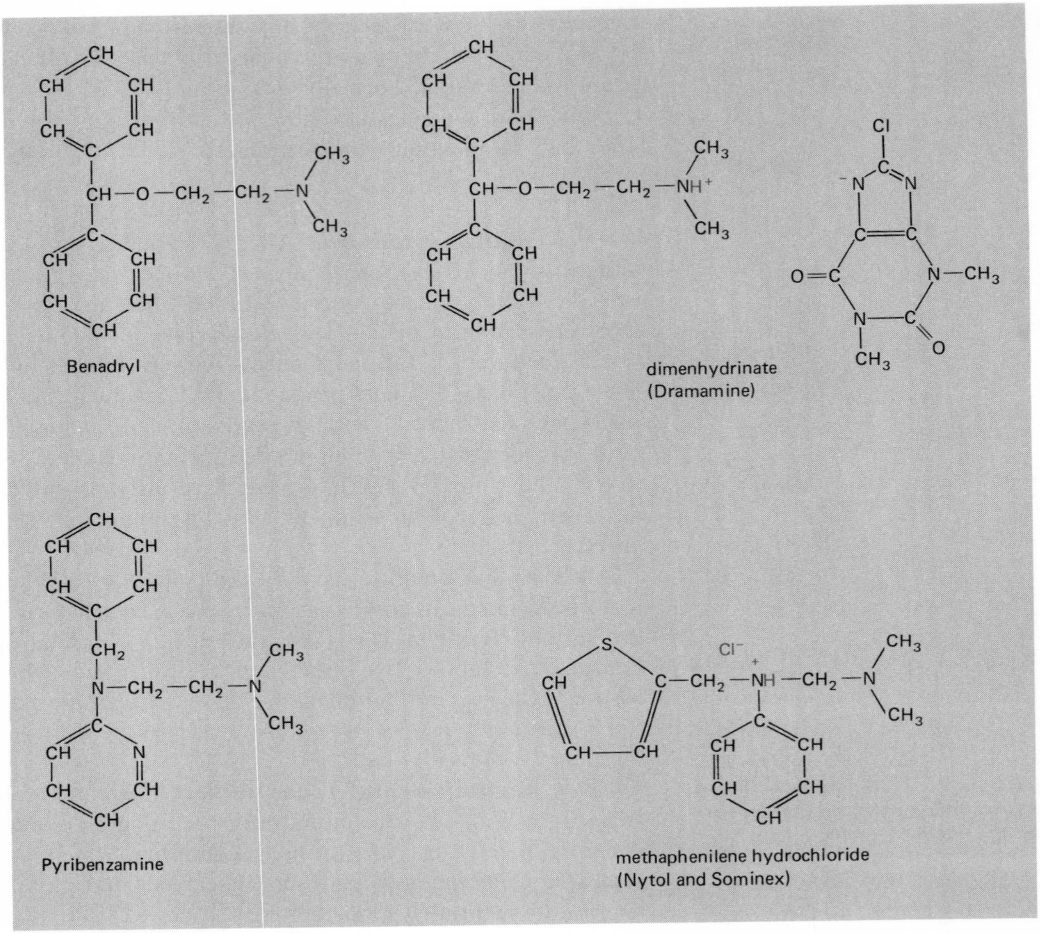

Figure 23-4 Some antihistamine drugs.

indicative of the sort of problems to be expected from use of any nerve depressant drugs.

Figure 23-6 shows the structures of five sedatives. Barbital and Seconal are barbituates. The longer hydrocarbon side chains on Seconal increase its lipid solubility, which affects the speed and duration of its effects. Various other barbituates can be produced, all of which depress the central nervous system. The third compound in Figure 23-6, Meprobamate, is not a member of the barbituate family but has very similar effects. Continued use of barbituates can lead to addiction, and this is a serious social problem. Overdoses of barbituates or the combined synergistic effect of barbituates and alcohol can depress the nervous system enough to cause coma or even death, and such deaths do occur.

Morphine is both a sedative and a potent pain reliever. Morphine is a component of opium. Opium and its derivatives are called narcotic drugs.

Nerve depressants often have undesirable side effects, and drug addiction is a problem with some depressants.

Figure 23-5 Reserpine.

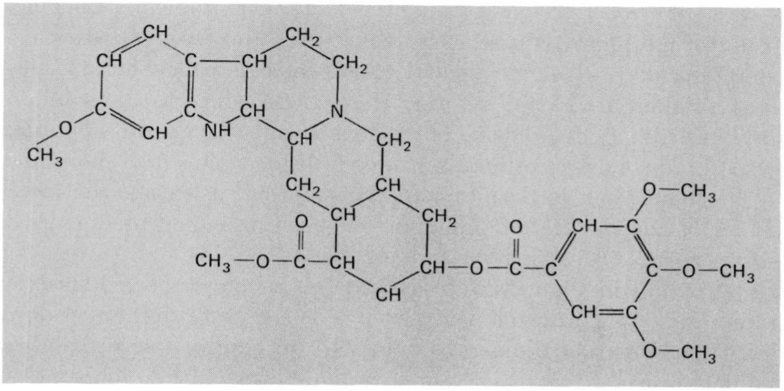

Figure 23-6 Some sedatives.

The portion of morphine shown in color completes another ring which comes closer to the viewer than the other rings shown in black.

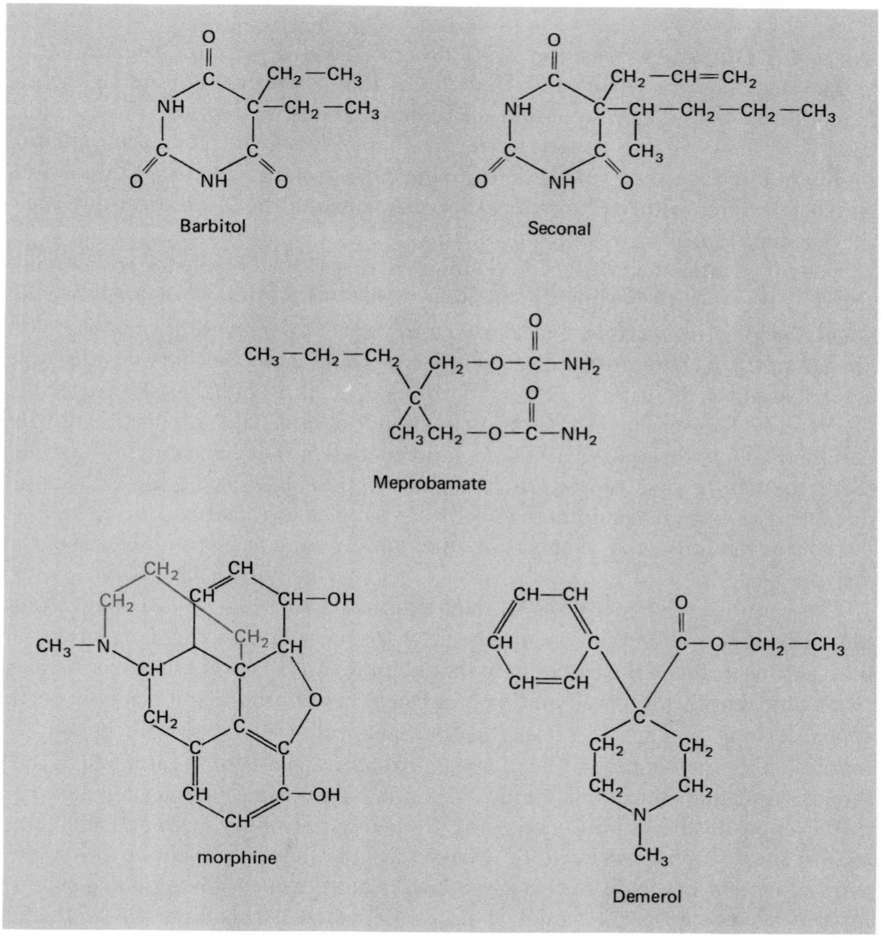

[23-3] Disrupting Natural Problems

Narcotics cause the most serious of all drug addiction problems. Codeine, a methyl ester of morphine, is used as a weaker narcotic. Over 50 years ago another morphine derivative, the diethyl ether which is called heroin, was developed as an alternative pain reliever. Heroin was hailed as the solution to the morphine addiction problem, but it turned out exactly the opposite. The increased lipid solubility of heroin makes it faster acting than morphine (similar to the comparison of Barbitol and Seconal), and it became the more popular choice of addicts. Heroin addiction has been promoted by organized crime and has destroyed many thousands of lives.

The fifth structure in Figure 23–6 is Demerol, a compound synthesized to replace morphine as a pain reliever. Demerol is a useful drug, but it does not eliminate the addiction problem. Doctors try to control the availability of these addictive drugs and to vary prescriptions to avoid development of specific dependences.

23–4 Identifying and Treating Some Hereditary Problems

23–4.1 Diabetes. Section 22–1 described the disease diabetes milletus and its treatment with insulin. Diabetes is one of several hundred genetic-defect diseases that can be passed on to descendants. Hereditary diseases are caused by flaws in the genetic code that either lessen (or totally eliminate) production of a necessary protein or produce the protein with a slight variation which interferes with its function. Diabetes is caused by a failure to produce the hormone insulin in sufficient quantity.

Several general features of hereditary diseases can be illustrated by considering the specific features of diabetes. The seriousness of diabetes varies widely among its victims, but most cases are relatively mild and may not appear until the victim is full grown. That would not be so if diabetes was a total absence of insulin. A person with no insulin could not be expected to survive. Those who closely approach such a total lack of insulin die in childhood. Only those with afflictions mild enough or delayed enough to permit them to mature and reproduce can pass on the defective genes. Because diabetes can be mild or unnoticed in the young, many diabetics have reproduced successfully and diabetes is the most common identifiable genetic disease.

Diabetes is a common hereditary disease because victims usually survive to successful reproduction.

The natural selection process that eliminates those genes causing child mortality affects all hereditary diseases. As we succeed in saving lives of those most seriously affected we eliminate that natural control and permit the disease to become more widespread and more serious. Insulin treatment has now been available long enough to raise a serious possibility that we have begun to increase the incidence of the disease through successful treatment. Even though most diabetes is fairly mild, it is now appropriate to consider moving in the opposite direction on breeding by using a stronger artificial selection against reproduction instead of weakening the natural selection. Genetic counseling is a first step in this direction. A more aggressive approach might involve access to sperm banks of nondiabetics as a substitute for diabetic

Successful treatment of a hereditary disease lessens the natural selection against defective genes.

males desiring children free of their disease. Methods for early detection of the disease and accurate information about the genetic basis for the disease will be helpful to any efforts to lessen the incidence of a hereditary disease.

23-4.2 Sickle Cell Anemia. In discussing diabetes we noted that survival of victims to reproduce was important to continuation of a hereditary disease. But at least one widespread, serious hereditary disease seems to have been associated with other characteristics advantageous to survival. Throughout much of human history in tropical areas resistance to malaria was the most important factor for human survival. Carriers of the gene causing **sickle cell anemia** were part of a racial group that also had a greater malaria resistance, and the gene therefore became fairly common. Present technology now offers other ways to combat malaria so the advantage associated with the sickle cell gene is less useful but the agonizing sickle cell disease remains.

Sickle cell anemia spread in malarial regions because of an association with increased malaria resistance.

The exact genetic nature of sickle cell anemia is known. Victims make hemoglobin molecules that have a substitution of one amino acid (valine) for the usual one (glutamic acid) at the sixth amino acid from the NH_2 end of the two beta chains. The other 145 of the 146 amino acids in the beta chains are identical with normal hemoglobin and the alpha chains (141 amino acids) are completely normal. The substitution of valine for glutamic acid can be caused by a single substitution of uracil for adenine at one of the 438 base positions providing the code sequence on the mRNA used to produce the hemoglobin beta chains. It is believed that sickle cell anemia is caused by that single substitution of uracil for adenine as directed by a single error in a DNA molecule.

Details of the cause and mechanism of sickle cell anemia are known.

Genes for production of hemoglobin are received from both parents, so a person could have two genes for normal hemoglobin (normal), one for normal hemoglobin and one for sickle cell hemoglobin (sickle cell trait), or two genes for sickle cell hemoglobin (sickle cell disease). Presence of sickle cell hemoglobin lowers the solubility of hemoglobin and causes distortion of red blood cells into a sickle shape in victims of sickle cell disease. Those distorted red cells break up and are lost much more quickly than normal red cells. The insoluble crystalline sickle cell hemoglobin also causes a rigidity in the sickle cells and in the parts they break into which leads to plugging of small blood vessels. The result for those with sickle cell disease is anemia from inability to produce new red cells fast enough to replace the rapidly breaking up sickle cells at the rate needed for normal red blood cell concentrations. In addition the victims suffer sickling crises during which blood vessels are blocked causing intense pain and other serious problems. The disease is so serious that children with it do not normally survive to maturity.

People with only sickle cell trait (one sickle cell gene) do not suffer significant impairment. The lowered solubility of their hemoglobin makes their red cells shorter lived than normal, but the shorter lifetime is within the range where production of new red cells can still maintain an adequate supply. The main problem for those with sickle cell trait is the danger that some of their children may receive the sickle cell gene from both parents and therefore have sickle cell disease.

The sickle cell gene became quite common in Africa. The gene is now found mostly in Black people, and its occurrence is high enough to cause a substantial danger of sickle cell disease in Black families. It also occurs in other races but is rare enough that very few children are afflicted with sickle cell disease.

Because we know so much about sickle cell anemia, much of the human suffering it has caused can be prevented in the future. The principal weapon in the fight against sickle cell disease is genetic counseling. Because sickle cell trait causes a significant change in the blood cells, carriers of the gene can be identified by a fairly simple blood test. Such a blood test should now be taken by all Blacks and any others whose family history shows a past problem that could have been sickle cell disease. If both members of a couple carry sickle cell trait, they should not produce children. Their responsible choices are to remain childless, adoption, use of a sickle-cell-free male donor for sperm, or seeking other marriage partners free of sickle cell trait. One fully normal parent is enough to guarantee the children will not suffer sickle cell disease.

Sickle cell disease can be prevented through tests and genetic counseling, and victims can now be given some help.

If a child is born with sickle cell disease, there is no cure. But knowledge about the nature of the problem now makes treatment to provide some relief possible. A chemical alteration of hemoglobin to increase its solubility is needed. Such an alteration can be accomplished by providing cyanate ions, CNO^-, to react with the ends of the protein chains. Treatment with regular doses of sodium cyanate is being attempted. This salt substance can be taken orally so treatment is simple and inexpensive. Over a long period of time the unnatural presence of cyanate may upset other body processes, so it is too early to credit cyanate treatments as a permanent solution. But no side effect of cyanate treatment is likely to be as bad as untreated sickle cell disease. You will note that the substance used for the above treatment is cyanate, CNO^-, and not cyanide, CN^-. Cyanide is a deadly poison that would be fatal in quite small doses.

23-4.3 Galactosemia. Galactosemia is a hereditary deficiency in metabolism, as is diabetes. Whereas diabetes is a deficiency in using glucose, galactosemia is a deficiency in using galactose. The specific problem in galactosemia is known to be a deficiency in the enzyme that converts galactose-1-phosphate to glucose-1-phosphate so it can enter the usual glucose metabolism cycle. Glucose is used in the primary energy storage and utilization processes in human carbohydrate metabolism, so diabetes upsets the entire metabolic system. Galactosemia need not have as great an effect because use of galactose can simply be omitted.

The damage of galactosemia is caused by the digestive system continuing to absorb galactose even when it cannot be used. Galactose is the sugar most rapidly absorbed by active transport across intestinal membranes. That fact

Galactosemia is caused by inability to metabolize galactose and is a serious problem for infants.

has been put to use in evolution of the infant feeding mechanism of man and other mammals. Milk contains lactose, which is hydrolyzed to glucose and galactose. Even the immature digestive system of an infant can absorb the galactose (and to a lesser extent, the glucose) quite efficiently. Galactosemia victims build up high galactose concentrations which interfere with normal body processes. Infants are particularly affected. They tend to vomit when fed milk, fail to gain weight, and may die. Enlarged livers are a side effect. If they survive and continue to receive galactose, victims may be dwarfed, mentally retarded, and suffer eye cataracts.

A diet low in galactose can, if started early enough, prevent damage from galactosemia.

Treatment of galactosemia is accomplished by removing milk and any other galactose sources from the diet. The nutrients in milk can be replaced by a lactose-free artificial substitute. If this treatment is begun when the child is only a few weeks old, the symptoms of galactosemia disappear and no lasting damage is done. As the child grows older, alternate metabolic pathways for utilizing galactose usually develop and a more normal freedom in food selection becomes possible.

The major problem with galactosemia is detection of the disease in the critical first few weeks of life. Because the disease is rare and symptoms resemble those of many more common infant diseases, the blood test necessary for detection of galactosemia may not be done.

There are several known types of hereditary metabolism defects.

Galactosemia is only one of many known possible inborn errors in metabolism. Diabetes is by far the most common and galactosemia may be the most easily treated. Others include several types of glycogen storage diseases and at least one disorder causing an anemia problem which is thought to have helped survival in malarial areas. It seems likely that many more hereditary diseases will be identified as our knowledge increases. Perhaps some of them can be treated as effectively as galactosemia.

Phenylketonuria is an important cause of mental retardation resulting from a defect in phenylalanine metabolism.

23-4.4 Phenylketonuria (PKU). Inborn errors in metabolism are not limited to carbohydrate metabolism. An hereditary abnormality in the metabolism of phenylalanine (one of the essential amino acids) has been identified as a cause of mental retardation. About 1% of all feeble minded persons in institutions are victims of this disorder. The disease is called phenylketonuria (and sometimes abbreviated PKU) because it results in buildup of a phenylketone product (phenylpyruvic acid) to the point where it is excreted in urine.

Phenylalanine can normally be used in the three ways shown in Figure 23-7. Some is essential to build proteins, but the amount used in any given protein molecule is fixed by the number of phenylalanines specified by the genetic code for that protein. When excess phenylalanine is available, some can be converted to tyrosine, another one of the 20 basic amino acids. Tyrosine is used to make proteins, and excess tyrosine is used in several reactions leading to products such as melanin, the dark pigment in human hair and skin. One of the reaction paths of tyrosine leads to reactants for the Krebs cycle and eventual oxidation to carbon dioxide and water. The third phenylalanine reaction forms phenylpyruvic acid. Although some further oxidation

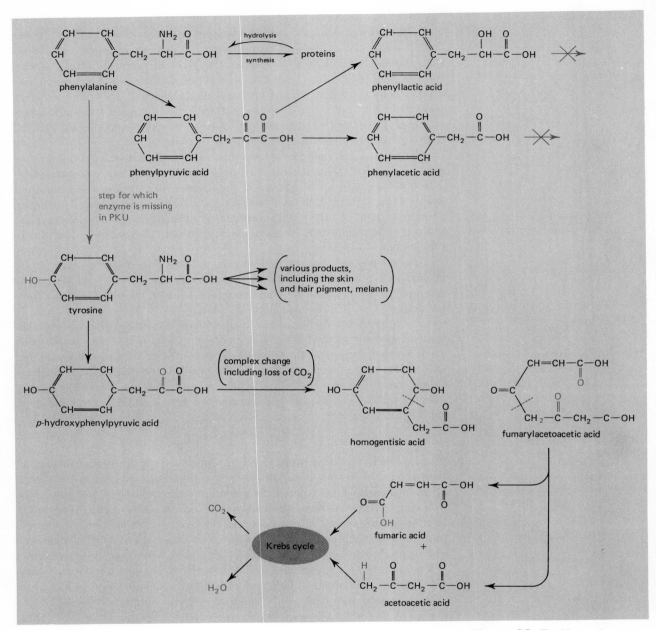

Figure 23-7 Normal metabolism of phenylalanine.

of the phenylpyruvic acid is possible, as shown in Figure 23–7, humans cannot oxidize phenylpyruvic acid to carbon dioxide and water.

When a person fails to make the enzyme for the conversion of phenylalanine to tyrosine, phenylketonuria results. In the absence of the tyrosine formation, all excess phenylalanine must be converted to phenylpyruvic acid and its products. These products cannot be oxidized to carbon dioxide and water, so their concentrations build up until another removal mechanism (rapid excretion in urine) takes over. That contrasts to normal individuals where only a moderate buildup of phenylpyruvic acid would result from that reaction because much of the phenylalanine is used to form tyrosine. A person carrying one normal gene and one defective gene of the PKU type would build up somewhat higher than normal phenylpyruvic acid levels before urine excretion balanced the rate of formation, but the level would not cause mental retardation. Only those receiving defective genes from both parents reach the phenylpyruvic acid levels necessary to cause damage.

The identification and treatment of PKU follow patterns similar to that for galactosemia. Detection is best accomplished by a blood test on young babies. A urine test is also possible for PKU detection, but that requires a delay until the phenylpyruvic acid has reached higher concentrations. Treatment is by limiting phenylalanine in the diet. Phenylalanine cannot be totally removed because it is essential and cannot be synthesized by humans. But by reducing the phenylalanine intake, the level of phenylpyruvic acid can be kept down to the moderate levels found in normal persons who convert part of their phenylalanine to tyrosine.

Phenylalanine is an essential protein, but PKU damage can be prevented by limiting phenylalanine in a victim's diet for the first few years of life.

The mental retardation of PKU victims must occur when they are quite young. The special, low phenylalanine diet has been discontinued after PKU children reach the age of five or six years without causing retardation. The exact age or maturity level at which PKU ceases damaging effects is not known, and physicians are naturally hesitant to risk discontinuing treatment too early.

Because a single PKU defect gene causes some increase in phenylpyruvic acid levels, a general survey of prospective parents may be technically possible. Such a survey could be used for genetic counseling similar to that described for sickle cell trait carriers in Section 23–4.2. At present most counseling is based on identification of families in which PKU cases occur. If a good test for PKU trait is perfected and made generally available, genetic counseling could reduce the frequency of the disease.

23–5 Selective Poisons as Medicines

23–5.1 Medicines. Living organisms create internal chemical environments providing the substances needed for their survival and function. Unfortunately, the environment of energy sources, amino acids, and other substances in cells is also favorable to other living organisms, and various parasite species have developed to take advantage of other organisms. Some of these parasites have limited growth that does not harm their hosts. Examples are some intestinal bacteria which even serve useful functions for humans, such as synthesizing some vitamins. But other parasites severely harm their hosts. Examples are infectious diseases. The harmful parasites vary from viruses, which are less than complete cells, through single celled bacteria to larger organisms such as malarial parasites or parasitic worms. In the presence of competition from such parasites, organisms need more than nutrients to prosper. They also need defenses.

Natural defenses of various types have evolved in higher organisms, and ways of overcoming those defenses have evolved in the parasitic forms. In the natural order a sort of balance is developed between survival of the higher organism in some cases and victory for the infectious diseases in other cases. But modern man considers death or serious disability from infectious disease very undesirable, so technology has been directed toward making conditions less favorable for the diseases. Often that includes changing the chemical environment in cells in a way that assists the natural defenses. The chemical aids used are called medicines.

Infectious diseases are caused by parasites harmful to their host.

Medicines help natural defenses overcome diseases.

23–5.2 Sulfa Drugs. Sulfa drugs were among the first chemical agents developed to help resist infections, and the sulfa drug class is still one of the most important antibacterial medicines. Sulfa drugs are very similar in structure to *p*-aminobenzoic acid, a substance used by bacteria in synthesis of folic acid. Figure 23–8 shows the structures of two of the sulfa drugs and of *p*-aminobenzoic acid and folic acid. The *p*-aminobenzoic acid portion involved in folic acid synthesis is shown in color as is the similar portion of each of the sulfas.

Figure 23–8 Sulfa drugs and related biochemicals.

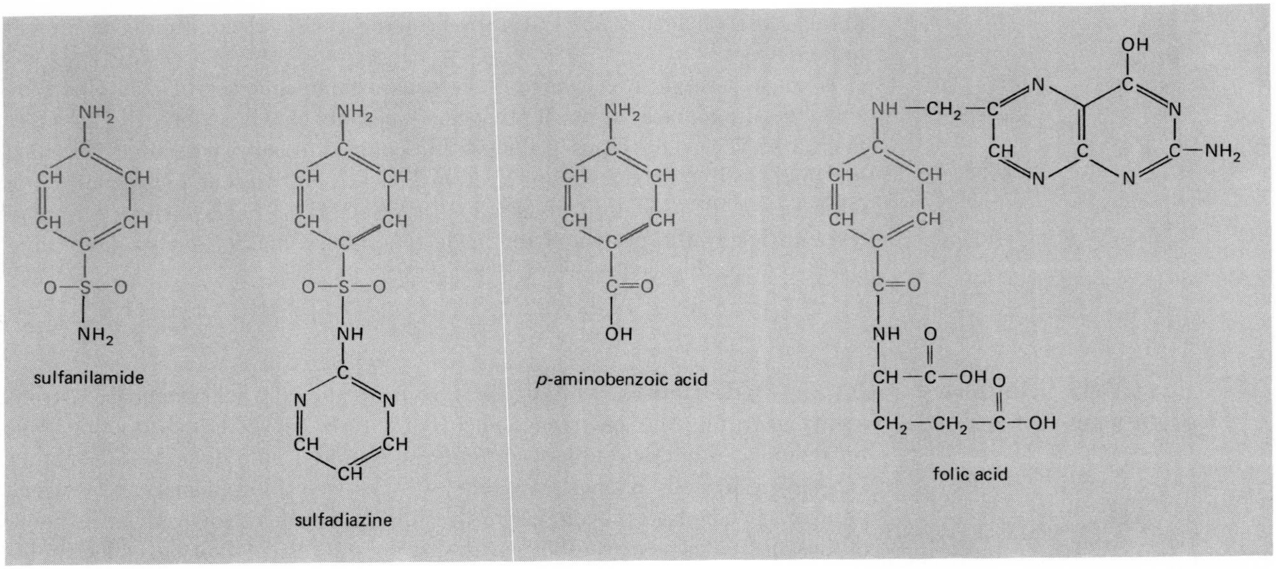

Sulfa drugs interfere with the folic acid synthesis needed by many bacteria.

Sulfa drugs are believed to interfere with folic acid formation by their substitution of a similar, but not identical, segment into the synthesis. (In some cases they may also have other effects.) The resulting product cannot perform the functions of folic acid. Folic acid is necessary to cell functions and is associated with the B vitamins in human nutrition. Humans cannot synthesize folic acid and therefore depend upon foods and the actions of intestinal bacteria to supply the folic acid. Humans must therefore be fairly efficient at absorbing folic acid and either delaying the need for folic acid or storing enough to supply periods of limited outside availability. Humans can survive a day or two without any supply of folic acid at all. Bacteria cannot survive without folic acid. Because bacteria can manufacture their own folic acid, they lack available reserves when sulfa drugs interfere with the folic acid synthesis. Therefore bacterial growth stops in the presence of sulfa drugs, and natural body defenses (such as white blood cells and antibodies) can then eliminate the disease. The sulfa drugs work because they interfere with a process needed by bacteria but not by humans. The sulfa drugs therefore have the effect of being selective poisons.

Substituted sulfanilamides with increased water solubility are more useful as drugs.

Sulfanilamide, one of the sulfa drugs shown in Figure 23–8 shows the basic sulfa structure. However, sulfanilamide leads to problems because of its limited water solubility. It may accumulate in the kidneys during excretion and damage that vital organ. More soluble drugs have been formed by substituting larger organic groups for one of the hydrogens on the sulfonamide nitrogen. These sulfanilamide derivatives are used commercially and (along with sulfanilamide itself) are referred to as sulfa drugs. Sulfadiazine is shown in Figure 23–8 as an example of the sulfanilamide derivatives found to be especially useful as medicines.

Sulfa drugs can kill useful bacteria and endanger the supply of some vitamins.

As is the case with all potent drugs, sulfa drugs should be used only when really needed. If sulfa drugs were used at high dosage for long periods of time, one effect could be to kill off all intestinal bacteria and thus cut off an important source of folic acid and some other vitamins needed by the human body. Intestinal bacteria may be less affected than disease bacteria by the sulfa drugs because they grow more slowly and are not being attacked by natural body defenses. But useful intestinal bacteria must also be affected by the sulfa drug interference with their folic acid production.

23–5.3 Penicillin.

In 1928 Dr. Alexander Fleming made an important discovery while observing bacteria grown on slide plates. One of the plates was accidentally contaminated by a particular kind of blue-green mold, and no bacteria grew in the zone immediately surrounding the mold. Dr. Fleming searched for the substance produced by the mold that prevented bacterial growth. Figure 23–9 shows the structure of penicillin, the substance eventually identified as the cause of the effect. The structure shown is penicillin G, one of several possible variations of the basic penicillin structure. Penicillin was the first of the antibiotics to be discovered. Penicillin is a substance produced by the mold to carry out a natural chemical warfare against certain bacteria.

Penicillin is a natural mold product acting against bacteria.

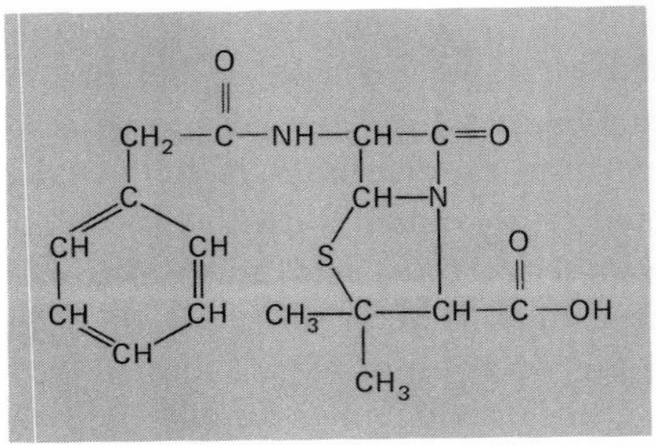

Figure 23–9 Penicillin G.

It is a selective poison, acting by interfering with reproduction by the bacteria. The most affected bacteria include the fast growing streptococci responsible for many serious human diseases.

It is now possible to synthesize penicillin by artificial chemical reactions, but the most economical method of producing penicillin is by growing and then processing the molds. The process has become extremely efficient and large quantities of penicillin are now produced and used for treatment of diseases. In 1945, 100,000 units of penicillin G in a shot cost $20 for those lucky enough to obtain it, whereas 250,000 units of a penicillin G salt in a pill cost only about four cents in the 1970s. As a result, penicillin can now be used with little concern about the cost, and most penicillin is used in more expensive forms, which are somewhat more effective and permit use of much larger dosage.

Penicillin is mass produced cheaply and can be converted to different forms.

Penicillin is both effective against many serious diseases and freer from harmful side effects than most other medicines. But some people have serious reactions to penicillin. Those reactions have become a major concern of doctors using penicillin for treatments. Apparently some people become sensitized to the penicillin molecule and develop antibodies to fight the "invader." The result can be serious allergic reactions in sensitized individuals, sometimes even causing death. Because these reactions are allergic reactions, at least two exposures are needed to cause trouble—one to start the antibody production and a later exposure to which the sensitized person reacts. It should be emphasized that penicillin is an unusually safe drug. Allergic reactions occur in only a small fraction of the people using penicillin. This author and millions of others have had many exposures to penicillin over many years without becoming sensitized. But any substance used as widely as penicillin will create problems for some users and some of the problems with penicillin are very severe.

Allergic reactions to medicines can be serious.

23–5.4 Tetracyclines. The great success of penicillin led to an extensive (and still continuing) search for other natural antibiotic substances. Molds have been a particularly important source. The natural structures have also been varied synthetically to maximize favorable effects. Most of hundreds of antibiotics now known are not useful medicines because they poison humans

A variety of available medicines helps avoid treatment problems.

Figure 23-10 Tetracyclines.

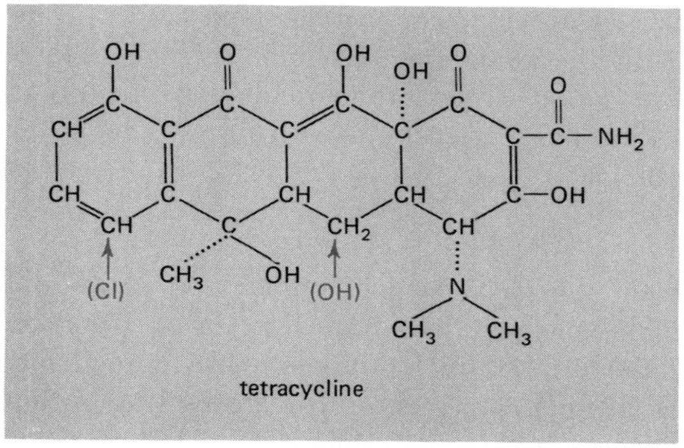

The compound shown in black is tetracycline (Achromycin). If a chlorine is substituted for hydrogen at the position shown by the colored (Cl) and arrow, the compound becomes chlorotetracycline (Aureomycin). If instead the original tetracycline has an OH group substituted for hydrogen at the position indicated by the colored (OH) and arrow, the compound becomes oxytetracycline (Terramycin). The ring arrangement has a specific geometry. The dotted lines in the figure indicate groups bonded in such an arrangement that they are below the rings as the tetracycline is viewed in the way shown. Each of the ring carbon atoms to which one of these "dotted line" bonds is shown is an asymmetric carbon, and only one optical isomer is present in the tetracyclines.

Tetracyclines are a group of antibiotics with a particular four-ring structure.

and other higher animals as well as poisoning bacteria. But several classes of antibiotics have become medically important. The variety of such medicines helps physicians treat cases such as the penicillin-sensitized individuals mentioned in the previous section and illnesses that are resistant to the common medicines. Among these other antibiotics the most widely used general class is the tetracyclines.

The tetracyclines are a family of compounds with various functional groups on the same basic structure of four interlocking rings. Figure 23-10 shows the form known simply as tetracycline and the positions where the structure is changed to form the other two derivative forms most widely used commercially. The marketing practices with these drugs led to a long and complex antitrust suit against major drug companies.

Tetracyclines cause more frequent allergy sensitization than penicillin.

Although tetracyclines are widely used and very important, it is worth noting that they cause allergic sensitization more frequently than penicillin. The combination of potential allergy problems and the possible necessity of tetracycline for treatment of some diseases resistant to other drugs makes physicians cautious about using tetracyclines. They are particularly hesitant to use tetracyclines with young children who may be more susceptible to sensitization than adults.

23-6 Selective Poisons as Pesticides

23-6.1 Pesticides. Selective poisons can be used against higher organisms as well as against bacteria and other parasites. Man is in competition with many other species for the resources of the Earth, particularly the food resources. All animal life is essentially acting as a parasite on the plant life of the Earth, and man is affected by the actions of other species. Those species that compete with man for food (and are not themselves useful as food) or which threaten to destroy useful species are called pests. The success and survival of man are dependent upon limiting the "waste" caused by pests using or destroying food sources. In India where human starvation often occurs, insects are estimated to consume half of all food crops. Common pests are insects, rodents, fungi, and weeds. Poisons that affect pests are called

Nonuseful species that compete for food are pests.

Chemical pest killers are pesticides.

pesticides, meaning pest killers. Pesticides can be classified according to the type of pest they affect with the categories insecticides, rodenticides, fungicides, and herbicides. The action of one rodenticide (warfarin) was described in Section 22–6. In the following sections we will describe a few important insecticides and herbicides.

23–6.2 DDT. It would seem the ideal pesticide should be cheap and fatal to many of the most troublesome pests but have little or no effect on humans. The widely used insecticide DDT (dichlorodiphenyltrichloroethane, the compound shown in Figure 23–11) comes close to that description, but we now know there are ways in which it is not ideal. DDT is not selective enough as a poison. DDT applied as a pesticide may also harm useful species. Useful insects may be killed along with the pests, and DDT (which is quite stable chemically) is concentrated in fats and other lipids and passed up the food chain. The result is a food chain concentration effect similar to that described for mercury in Section 16–12.2. The concentration of the DDT is believed responsible for some developing problems of higher species, such as failure of pelicans to lay eggs with shells strong enough to survive. These problems may also involve other chlorinated hydrocarbons, but the type of problem is the same no matter which particular compounds are responsible. DDT and other chlorinated hydrocarbons are both fairly general poisons and very persistent in nature. As a result, use of DDT is being restricted by law and use of other persistent chlorinated hydrocarbons is the subject of sharp criticism by some concerned citizens.

Total elimination of DDT is not easily achieved. Insecticides play a major role in disease control (by killing the insects which spread diseases such as malaria) as well as protecting food supplies. Substitute insecticides are often less effective, more expensive, and more hazardous to humans. Cost is an especially important factor to low income countries. Use of DDT allowed Ceylon and Madagascar to cut their death rates in half, so such low income tropical countries wish to continue its benefits. The worldwide plight of pelicans and peregrine falcons carries little weight with governments faced by the prospect of millions of cases of malaria if DDT use is discontinued. But continued heavy use of DDT is self-defeating. As DDT becomes a common part of the environment, insects are evolving into DDT-resistant forms. Changes to other insecticides are needed both to protect the environment

DDT poisons many insects, but its persistence and nonselectivity cause problems.

Substitutes for DDT cause other problems and extra expense.

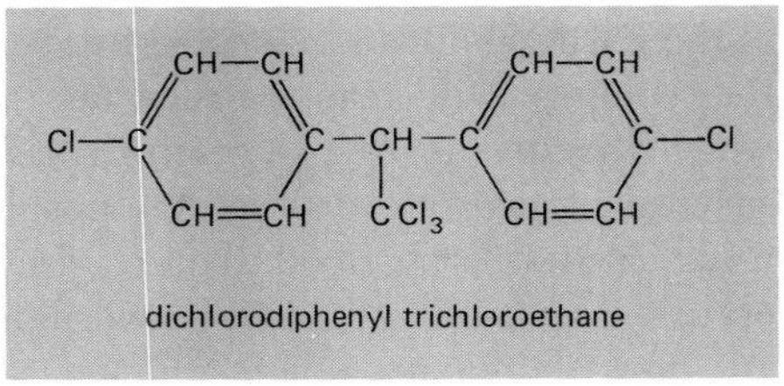

Figure 23–11 DDT.

dichlorodiphenyl trichloroethane

from DDT buildup and to prevent the development of resistant species of insect pests.

23-6.3 Other Insecticides. Many other compounds besides DDT can be used to poison insects. Some are natural plant products, but synthetic poisons can usually be obtained at lower cost and work more effectively. Several of the commercially important insecticides are organochlorine compounds, like DDT. Figure 23-12 shows four of these organochlorine insecticides.

Lindane is one of the early insecticides. It consists of a mixture of isomers formed by chlorinating benzene. One of the arrangements of chlorines relative to the cyclohexane ring is more effective than the other slightly different arrangements, but the manufacturing process also gave the less effective isomers. As a result lindane usage put a large quantity of persistent organochlorine compounds into the environment. Aldrin and dieldrin are also persistent organochlorine compounds that might cause harmful environmental effects. Aldrin and dieldrin are also several times more poisonous to higher animals than DDT. Their principle advantage is that fewer insects have developed resistance to these newer insecticides than to DDT. Methoxychlor is an insecticide closely resembling DDT in structure but much less poisonous to higher animals and perhaps more biodegradable than DDT. However, all organochlorine compounds are somewhat persistent in the environment and possible environmental hazards. At this writing DDT, lindane, aldrin, and dieldrin have all been banned or had their use sharply limited in the United States.

Another class of effective insecticides are certain organophosphorus com-

Organochlorine insecticides are effective but cause environmental hazards.

Figure 23-12
Organochlorine insecticides.

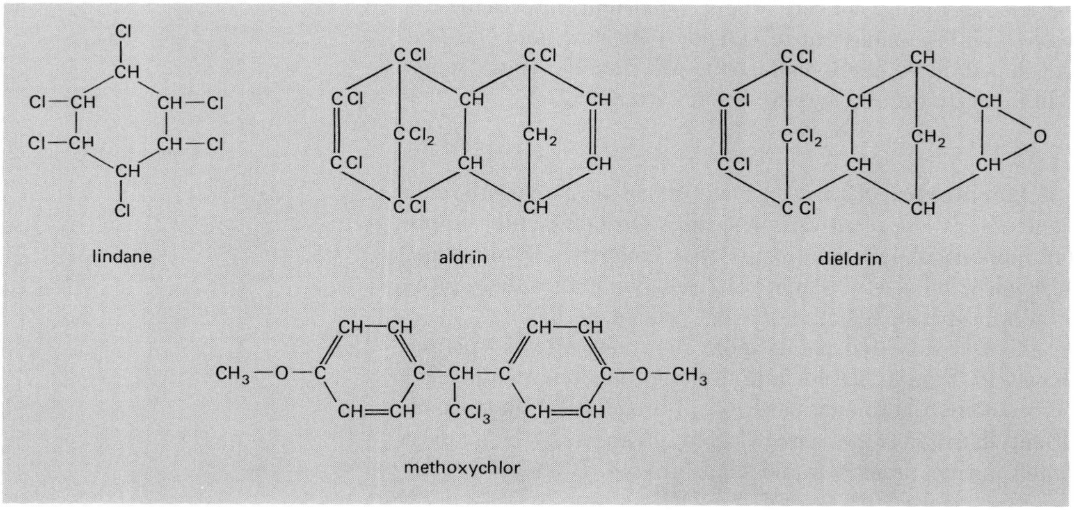

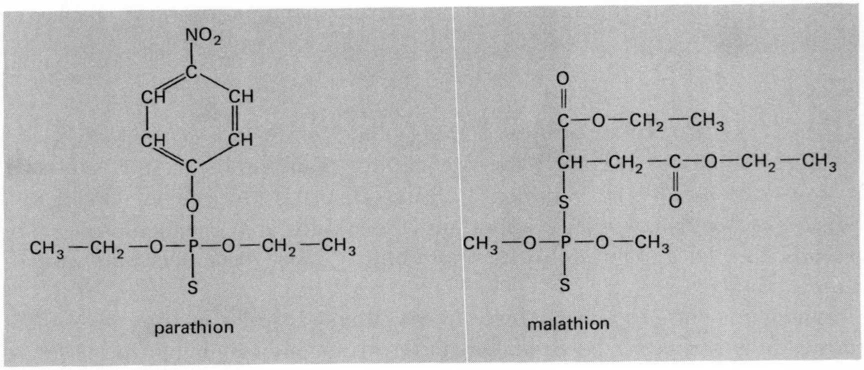

Figure 23-13
Organophosphorus insecticides.

pounds. Figure 23-13 shows two commercially important organophosphorus insecticides, parathion and malathion. These compounds work by inhibiting the enzyme required to regenerate the compound used to transmit nerve impulses from one cell to another. The resulting nerve paralysis leads to death. Since human nerve impulses are transferred by the same mechanism, the organophosphorus insecticides are quite hazardous to humans.

Actually, the hazard to humans varies greatly with different compounds. Some related compounds are very lethal to humans and have been developed as potential chemical warfare weapons (nerve gases). Parathion is also quite dangerous, about 20 times as lethal to higher animals as DDT, but malathion is only about one tenth as lethal to higher animals as DDT. Misuse of parathion has killed fairly large numbers of humans on some occasions. However, if humans can be protected from the parathion for a reasonable period after application, there is no long-term hazard. The organophosphorus insecticides are relatively easily degraded in the environment and are therefore considered nonpersistent insecticides. Rising concern about persistent insecticides is increasing the use of these less persistent insecticides. As new nonpersistent insecticides are developed it is likely that some will involve short range poisoning hazards like parathion and require great care when used.

Organophosphorus nerve poisons are dangerous to humans but not persistent environmentally.

23-6.4 2,4-D, a Herbicide.

Human agriculture involves considerable efforts to establish and assist those plant varieties which most efficiently supply products usable for humans. Any other plants that compete for nutrients, water, and space needed by the useful plants are pests. Sometimes these pests can be controlled by appropriate selective poisons, called herbicides.

Some herbicides affect broadleafed plants more than narrowleafed plants. The compound shown in Figure 23-14 is 2,4-dichlorophenoxyacetic acid (called 2,4-D). It is a common herbicide used to kill broadleafed plants such as dandelions without damaging the narrowleafed grasses. 2,4-D is most effective when applied during times of rapid plant growth. It should be left in contact with the plant leaves for several days. (Rain would wash it off

2,4-D selectively kills broadleafed plants.

Figure 23–14 2,4-D.

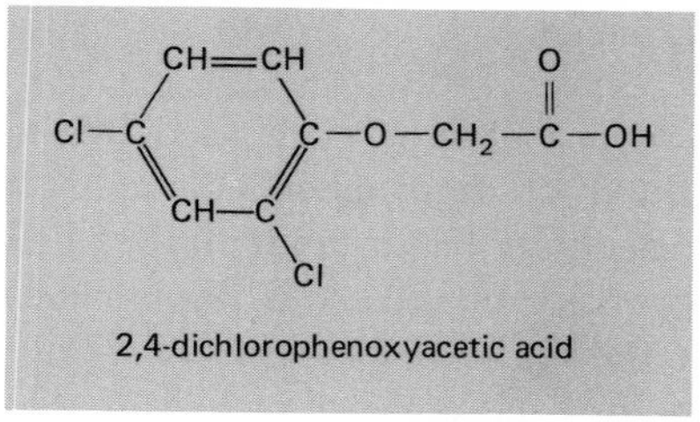

2,4-dichlorophenoxyacetic acid

and lessen the effectiveness.) Fast growing broadleafed plants absorb the 2,4-D more effectively than narrowleafed plants and are eventually poisoned. 2,4-D is often used in the form of alkylamine salts.

23–6.5 Other Herbicides. Just as there are a number of different insecticides working in several different ways, there are several types of commercially important herbicides. These herbicides are useful for different weed problems. Timing of the herbicide application is often critical.

Figure 23–15 shows 2,4,5-trichlorophenoxyacetic acid, called 2,4,5-T. 2,4,5-T is closely related to 2,4-D and is used to control some broadleafed plants such as clover and chickweed, which are resistant to 2,4-D. Application is the same as described for 2,4-D and 2,4,5-T is generally more potent and effective on more broadleafed weeds. Unfortunately, the commercial production of 2,4,5-T has led to a product containing an impurity (dioxin) which has been implicated as a possible cause of birth defects and possibly other problems. As a result 2,4,5-T was removed from use in herbicides sold in the United States.

Other herbicides are available for a variety of specific purposes requiring different selective poisonings.

The compound 4-chloro-2-butynyl-*m*-chlorocarbanilate is an herbicide used to kill a narrowleafed weed. This substance, under the brand name Carbyne, is used to kill wild oats (a grass) in fields of peas or lentils (which are broadleafed). That is exactly opposite to the effect of 2,4-D and shows that differences in plants can lead to various kinds of selectivity in poisoning. Carbyne also illustrates the importance of timing. Farmers are advised to walk through their field to determine the exact time the weeds begin to emerge.

Figure 23–15 2,4,5-T.

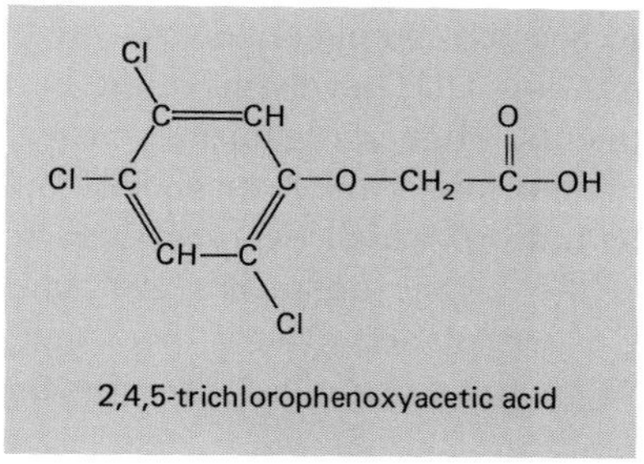

2,4,5-trichlorophenoxyacetic acid

Spraying must then be done at a particular stage of weed growth, normally 4 to 9 days after emergence.

Selectivity of the herbicide is so important that it is even used in the commercial name for one compound. 2-sec-Butyl-4,6-dinitrophenol ammonium salt is sold as Selective Weed Killer. But selectivity has meaning only when the planned use is known. The list of victims of Selective Weed Killer includes several common weeds, such as chickweed and ragweed, but it also includes peas and alfalfa, which might be the crops in need of weeding. All herbicides and other pesticides must be recognized as poisons and used with great care.

23–7 Losses of Selectivity

Medicines and pesticides sometimes lose their effectiveness after long or widespread use. Since these substances are selective, they affect some organisms more than others. If any of the target organisms survive, they must be strongly selected for resistance to the poisoning effects. After several generations of such selection for survival, the target organisms may evolve into strains that are essentially immune to the medicine or pesticide. This selective development of resistance is particularly noticeable in bacteria subjected to medicines and in insects subjected to insecticides. Many generations of bacteria or insects pass during their exposures to poisons, so the selection for resistant strains is strong.

Widespread use of any control leads to the development of natural resistance.

Development of resistance is favored when the dosage of poisons is low so there are some survivors. Therefore it would be better to treat diseases with massive doses of medicines such as penicillin to minimize the chance for survivors to develop resistant strains. Unfortunately, large doses raise problems of possible interference with the "nonsusceptible" host being treated, either by poisoning or by development of allergic reactions. The general conclusion is that dosage is critical. Too much loses selectivity by affecting everything, and too little allows loss of selectivity by permitting development of resistance.

The gradual development of resistance by exposure to low levels of poisons is a particularly serious problem with compounds such as DDT, which are persistent in the environment. As DDT becomes a common substance (at low concentrations) throughout the Earth, we should expect all organisms to evolve to strains resistant to DDT. Then DDT would become completely nonselective. Simple pests (with short lives and exposure to low DDT levels) have the best opportunity to make the necessary adjustments. Higher organisms such as pelicans cannot adjust as quickly because of their longer time per generation. They also are subject to food chain concentration, which exposes them to DDT at higher concentrations. Therefore, the higher organisms are most threatened by any change in the chemistry of our planet, even if the "new" chemical was originally somewhat selective in their favor, as was true for DDT.

Higher organisms are less able to adjust to changing chemistry than are lower organisms.

23-8 Side Effects of Control Measures

One side effect of controls can be loss of natural resistance.

Control measures sometimes lead to unfortunate side effects. The problems with DDT described in Sections 23–6.2 and 23–7 are examples of such side effects. Substances that are poisonous to many organisms can cause serious harm to useful species, and very selective poisons may lose their effectiveness as their targets evolve into resistant strains.

All species (or at least all surviving species) have developed means to cope with their natural surroundings, including natural enemies. One side effect of control measures can be loss of some natural defenses. That can make us very susceptible to diseases and pests and highly dependent upon the success of continued control measures.

The importance of natural resistance can be illustrated by the historic spreading of diseases by colonists from Europe. Natives in the Americas and Africa were susceptible to diseases imported by the colonists. Populations that had been stable or growing for centuries suffered sharp declines. In some places smallpox virtually exterminated native settlements which had no earlier exposure to the disease.

Smallpox was also serious for the colonists, but the continued exposure of Europeans to smallpox over centuries led to natural immunities. Some members of each community of colonists were immune due to earlier exposures, immunities were passed to children via mother's blood and breast feeding and possibly by selective breeding effects of greater survival of those resistant to smallpox. The presence of some immune members in a community also helped protect the nonimmune members by interfering with direct spread of the disease.

By contrast, every member of a native community could catch the disease and pass it on, so the spread of the disease there was explosive. Other European diseases also affected natives more than colonists. This uneven resistance was not one-sided. European colonists had less resistance to malaria than Africans and Rocky Mountain Q fever is an example of a regionally limited disease to which most people have very little resistance. The significant general point is that exposure to a natural problem for which no natural resistance exists can be disastrous.

Current human activities could lead to many serious side effects. We will mention three here as examples of side effects from three different kinds of control measures—medicines, pesticides, and agriculture.

Section 22–1 described treatment of diabetes by insulin, and section 23–4.1 described the genetic nature of the disease and how treatment might increase its incidence. Diabetes is a good example for discussing natural controls because it is so common and because it shows some of the inadequacies of the natural controls. Most diabetics survive to the age of reproduction even without insulin. If that was the only criterion for survival selection, there would be no hereditary preference for nondiabetics. In fact, the selection mechanism is rather weak, which helps make diabetes so common. But there was some selection against diabetics because they were less able to raise their own children to maturity under primitive conditions. Even without insulin our modern society would lessen this natural selection by providing better for surviving children. Availability of insulin can only accentuate this loss of natural selection. Unless some other bias against the genetic defects causing diabetes is introduced, we can expect diabetics to increase gradually to a higher fraction of the population than existed under primitive conditions.

> Availability of insulin could lead to an increase in diabetes.

DDT has been widely used to control mosquitos which transmit malaria. In many areas of the world this has significantly improved human survival and human energy available for useful work. But DDT is affecting some birds and may reduce in numbers or destroy some natural predators of insects. In addition the continued use of DDT is stimulating development of resistant strains of mosquitos, which require higher dosage of DDT for control. Eventually mosquito strains may be unaffected by DDT. If natural controls have been simultaneously weakened, malaria could be spread more effectively than in the days before DDT. An alternative method (or methods) of controlling malaria is needed to replace DDT spraying.

> DDT control of malaria could lower natural resistance and leave man more susceptible to malaria spread by DDT-resistant mosquitos.

Agriculture has increased useful production of crops by eliminating competing plants (by cultivation and herbicides) and by selective breeding to achieve highly efficient varieties. But the same conditions of uniform plantings of productive varieties that favor man's harvest may also favor pests competing for the harvest. Insecticides or other controls are needed to hold down the insects favored by the agricultural methods. Diseases such as fungus growths may also be a problem.

In natural conditions the spread of a particular fungus type is limited by the dispersion of the plants on which it thrives. If a fungus does become widespread, natural selection lets those strains most resistant to the fungus prosper. But commercial crops are strains selected principally for their productivity, not disease resistance, and close purebred plantings allow easy spreading of any disease. As a result, plant disease epidemics are possible and, in fact, likely. Problems are especially likely when a single plant breed is overwhelmingly successful. The widespread use of T cytoplasm hybrid corn and the resulting corn blight epidemic of 1970 described in Chapter 1 of this book is an example.

> Agriculture makes plant disease epidemics more likely.

23-9 Compensating for Artificial Changes

When we know enough to anticipate the effects of artificial changes, we may be able to compensate so as to lessen or eliminate the harmful effects. For example, when plant disease problems are known to exist, selective breeding can be used to choose disease resistance along with high productivity. Wheat is known to be quite susceptible to some fungus diseases, and resistance to these diseases is one of the characteristics sought by wheat breeders.

> Breeding for disease resistance can compensate for disease-favorable conditions in agriculture.

Quarantines are another form of compensation for artificial changes. Man-made travel has broken down natural barriers that prevented spread of some diseases and pests. A quarantine simply replaces the natural distance barrier by an inspection (perhaps with waiting period) or prohibition on movement of substances, such as soil, which might carry pests.

> Quarantines compensate for man-caused movements.

Vaccinations are another mechanism for compensation. In this case we go

Vaccinations compensate for lowered development of natural resistance via disease.

beyond the natural resistance to infection by a few members of a community exposed to a disease. In the absence of the disease there would be no resistance, but by vaccination all (or at least a large fraction) can be "exposed" so they develop immunity to the disease. For serious diseases such as smallpox or measles the advantages of vaccination over having the real disease in the community are tremendous.

Power to cause changes is accompanied by responsibility and a need for knowledge.

There are many other ways in which the potential damage from artificial changes can be lessened or avoided. In each case knowledge about the changes and their effects on the chemical and biochemical environment is necessary to guide technology in the most favorable directions. It is reasonable for us to change our surroundings—much in our natural surroundings is cruel and unsatisfactory. But if we have gained the power to change our world, we have also gained the responsibility to know what we are changing and make sure the changes are for the better.

Skills Expected at This Point

1. You should be able to list examples of changes from natural chemistry and, if asked, you should be able to choose at least one example illustrating each of the following types of change: stimulation of natural growth, substituting for a natural balance, relieving pain, depressing histamine responses, depressing nervous activity, use of a medicine, use of an insecticide, and use of an herbicide.
2. You should be able to describe the cause of each of the examples of changes you use to satisfy skill number 1, how that works to cause the change, and the main results of that change. You should also be able to list some possible side effects of each change.
3. You should be able to explain the nature of the problem and describe the actions possible to counteract the problem for each of the four hereditary diseases described in this chapter.
4. You should be able to describe the general function of medicines and the specific structural pattern and biological effects of sulfa drugs, and list the most important classes of antibiotics.
5. You should be able to list organochlorine and organophosphorus as important classes of insecticides, list the advantages and disadvantages of each, and explain why the mechanism of the organophosphorus compounds makes them hazardous to humans.
6. You should be able to name (short common name) the most widely used insecticide and the common herbicide for broadleafed plants.
7. You should be able to describe the variety of selective actions in herbicides.
8. You should be able to describe the mechanism by which selectivity is lost.

Exercises

1. Give an example of a change made to compensate for some other change and establish a substitute sort of balance. Then describe that example

in moderate detail—cause, how it works, the result, and any significant side effects.
2. Describe the cause and treatment of galactosemia.
3. Define the following terms: **(a)** antihistamine, **(b)** sedative, **(c)** tetracycline, **(d)** medicine, **(e)** insecticide, **(f)** herbicide, **(g)** sickle cell trait.
4. Write the structure of sulfanilamide and use it to explain **(a)** how sulfa drugs work, and **(b)** how other sulfas differ from sulfanilamide and why they are better.
5. What is the cheapest antibiotic (not a sulfa) and what other problems and advantages does that drug offer?
6. List some side effects associated with sedatives or tranquilizers.
7. List some side effects associated with DDT use.
8. List the two types (classes) of insecticides described in this chapter.
9. Selectivity is a word used frequently in this chapter. **(a)** What is meant by selectivity? **(b)** Give some examples of selectivity. **(c)** Describe some failures in selectivity.
10. How do organophosphorus insecticides work?
11. In what ways does a vaccination program resemble natural control of the disease?
12. How did diabetes become so much more common than galactosemia?
13. Why is sickle cell anemia more common among Blacks than among Scandinavians?
14. List the ways agriculture upsets natural balances, side effects of these upsets, and some of the possible compensating changes.

Test Yourself

1. List the changes from natural conditions brought about by aspirin.
2. Write the structure of a sulfa and use it to explain how sulfa drugs work.
3. Why is DDT less effective now than when first used as an insecticide?
4. Why can't PKU be treated by the same method as galactosemia—total exclusion of the problem compound from the diet?
5. Which of the following best explains why tetracyclines provided an important improvement over the earlier situation when penicillin was the only important antibiotic?
   (a) Tetracyclines are less likely to cause allergic reactions.
   (b) Tetracyclines are cheaper.
   (c) A wider choice was needed.
   (d) All of these.
   (e) None of these.
6. Which of the following is one of the main classes of insecticides?
   (a) 2,4-D
   (b) sulfa
   (c) tetracyclines

(d) organochlorine compounds
(e) vitamin K substitutes such as warfarin

7. Sickle cell anemia
   (a) is the most common hereditary disease.
   (b) is caused by an amino acid substitution that makes hemoglobin less soluble.
   (c) can be treated effectively by a carefully controlled diet.
   (d) can occur in pelicans as a side effect of buildup of DDT in the environment.
   (e) fits none of these.

8. Organophosphorus compounds are used
   (a) as antibiotic medicines.
   (b) as herbicides for narrow-leafed plants.
   (c) as insecticides by acting as nerve paralizers.
   (d) as fertilizers to stimulate plant growth.
   (e) as none of these.

9. The most common herbicide for broad-leafed weeds is
   (a) DDT   (d) penicillin
   (b) 2,4-D   (e) insulin
   (c) sulfa

10. The most widely used insecticide has been
    (a) DDT   (d) penicillin
    (b) 2,4-D   (e) insulin
    (c) sulfa

11. Sulfa drugs kill bacteria by
    (a) forming a layer around them that encapsulates them.
    (b) lowering their defenses against antibodies.
    (c) causing a shortage of vitamin C.
    (d) interfering with their reproduction processes.
    (e) interfering with folic acid synthesis.

12. Sedatives
    (a) neutralize the compound causing allergic responses.
    (b) depress the central nervous system.
    (c) are selective poisons killing bacteria.
    (d) are selective poisons killing weeds.
    (e) are medicines used to correct hereditary errors in metabolism.

SI Units

A–1 SI Units

In 1960, the international authority on units, the Conference Generale des Poids et Mesures, C.G.P.M., agreed to adopt the Systeme International d'Unities, or the Internation System of Units, for which the agreed abbreviation is SI in all languages.

Many countries have prepared or are preparing the necessary legislation to make SI the only legally acceptable classification of units. It is apparent that SI is destined to be in worldwide use within a few years. It is also clear that once the International System of Units has been universally accepted it will be a long time before any new system is introduced to replace it.

SI units consist of a set of prefixes used with a small number of basic units (often defined by reference to fundamental physical phenomena) and a larger number of units derived from the basic units. Table A–1 lists prefixes used in SI. Table A-2 lists the basic units. Table A–3 lists derived units and shows how they are related to the basic units.

Table A–1 Prefixes*

Prefix	*Symbol*	*Factor*
tera	T	10^{12}
giga	G	10^{9}
mega	M	10^{6}
kilo	k	10^{3}
(hecto)	(h)	(10^{2})
(deca)	(da)	(10^{1})
(deci)	(d)	(10^{-1})
(centi)	(c)	(10^{-2})
milli	m	10^{-3}
micro	μ	10^{-6}
nano	n	10^{-9}
pico	p	10^{-12}
femto	f	10^{-15}
alto	a	10^{-18}

*The recommended prefixes differ by steps of 10^3. Those prefixes shown in parenthesis are not recommended and should be used only on occasions where the recommended prefixes are inconvenient.

Table A–2 Basic SI Units*

Physical Quantity	Unit	Symbol
length	meter	m
mass	kilogram	kg
time	second	s
electric current	ampere	A
temperature	kelvin	K
luminous intensity	candela	cd
plane angle	radian	rad
solid angle	steradian	sr

*The mole has been recommended as a basic unit.

Table A–3 Derived SI Units

Physical Quantity	Equivalent in Previous Units	Name of Derived Unit	Symbol	Formula (Basic Units) for Derivation
force	$\dfrac{\text{mass} \times \text{length}}{\text{time}^2}$	newton	N	kg m s^{-2}
energy	force × length	joule	J	kg m^2 s^{-2}
power	$\dfrac{\text{energy}}{\text{time}}$	watt	W	kg m^2 s^{-3}
electric charge	elec. current × time	coulomb	C	A s
electrical potential	$\dfrac{\text{power}}{\text{elec. current}}$	volt	V	kg m^2 A^{-1} s^{-3}
electrical resistance	$\dfrac{\text{elec. potential}}{\text{elec. current}}$	ohm	Ω	kg m^2 A^{-2} s^{-3}
electrical capacitance	$\dfrac{\text{elec. charge}}{\text{elec. potential}}$	farad	F	A^2 s^4 kg^{-1} m^{-2}
inductance	$\dfrac{\text{elec. pot.} \times \text{time}}{\text{elec. current}}$	henry	H	kg m^2 A^{-2} s^{-2}
frequency	$\dfrac{\text{number}}{\text{time}}$	Hertz	Hz	s^{-1}
magnetic flux density	$\dfrac{\text{force} \times \text{time}}{\text{elec. charge} \times \text{length}}$	tesla	T	kg A^{-1} s^{-2}
magnetic flux	mag. fl. den. × length^2	weber	Wb	kg m^2 A^{-1} s^{-2}
luminous flux	luminous intensity × solid angle (brightness equiv. by the standard luminosity curve to $\tfrac{1}{685}$ watt of 555 nm wavelength light)	lumen	lm	cd sr (1.46×10^{-3} kg m^2 s^{-3} for 555 nm light)
illumination	$\dfrac{\text{luminous flux}}{\text{length}^2}$	lux	lx	cd sr m^{-2} (1.46×10^{-3} kg s^{-3} for 555 nm light)

A-2 Notes on the Use of SI Units

SI stresses elimination of any unnecessary letters or marks. Therefore:

1. Symbols are sufficient. It should not be necessary to write out the names of universally recognized units.
2. Plurals waste an extra s, so the singular form of each SI unit is the only one used. *Example:* 15 km (not 15 kms).
3. Either a point (like a period) or a comma on the line can be used to indicate the decimal point. *Example:* 1.5 km or 1,5 km.
4. No commas are used (except as decimal points), but long numbers are grouped in threes (starting from the decimal point) for easier reading. *Example:* 9 192 631 770 Hz (not 9,192,631,770 Hz).
5. No degree sign is used with Kelvin temperatures. Example: 273.15 K (not 273.15°K).
6. Use of negative exponents is preferred over use of two lines or a solidus (/). *Example:* 1.5 kg m^{-1} s or 1.5 kg s m^{-1} preferred over 1.5 kg s/m (1.5 kg/m/s is totally unacceptable).
7. Only one prefix should be used for any unit. *Examples:* 555 nm (not 555 mμm); 1 mg (not 1 μkg)
8. The exponent on a unit refers to the whole unit, including the prefix. *Example:* 1 km^2 = 10^6 m^2

A-3 Some Conversion Factors

Some of the other (not official SI units) units we have used can be related to SI as shown in Table A-4.

Table A-4 Conversion Factors

Unit	SI Equivalent
Angstrom (Å)	10^{-10} m
cm	10^{-2} m
cm^2	10^{-4} m^2
liter	10^{-3} m^3
ml	10^{-6} m^3
atm	101 325 N m^{-2}
torr	133.322 N m^{-2}
cal (15°C)	4.1855 J

Additional conversion factors are available in handbooks.

Further Reading

Suggested further reading in connection with SI units is

1. Anderton, P., and P. H. Bigg: *Changing to the Metric System,* 3rd ed. H.M.S.O., London, 1969.

2. McGlashan, M. L.: *Physico-Chemical Quantities and Units*. The Royal Institute of Chemistry, London, 1968.
3. Socrates, G. and L. J. Sapper: *SI and Metrication Conversion Tables*. Newnes-Butterworth, London, 1969.
4. Socrates, G.: SI units. *J. Chem. Educ.*, **46:**710 (1968).

Forms in Which Equilibrium Constants Are Used

APPENDIX B

B–1 Writing and Using K

B–1.1 Writing K. The expression of concentrations equal to K is useful in problems. The law of mass action tells us how to write this equilibrium constant expression, and we can then put in known values for equilibrium concentrations and solve for a missing piece of information, such as a concentration which had not been measured.

The first step in all equilibrium constant problems is writing the equilibrium constant expression. In some cases that is the whole problem, because just looking at the expression can give us a good understanding of what factors favor or disfavor the reaction. A typical problem might read like Example B-1.

Example B–1. What is the equilibrium constant expression for the reaction of N_2 with H_2 to form NH_3?

Our first step must always be to check the equation to be sure it is complete and balanced. If it is not complete and balanced, we must do that before going on. This problem did not give us the equation, so we must write the equation and balance it. N_2 and H_2 are reactants and NH_3 is the product, so we can write:

$$N_2 + H_2 \rightleftharpoons NH_3$$

Sometimes common substances, such as H_2O, are not mentioned in the problem. It is assumed you would know H_2O (and also the H^+ or OH^- ions formed if water should ionize) is available if the reaction is done in water. But in this example N_2, H_2, and NH_3 seem to be the only substances needed. We balance the equation by taking $2\,NH_3$ (to balance the 2 N atoms in N_2) and then taking $3\,H_2$ (to balance the 6 H atoms in $2\,NH_3$).

$$N_2 + 3\,H_2 \rightleftharpoons 2\,NH_3$$

When we have the balanced equation, we write the equilibrium constant expression according to the law of mass action. Product concentrations go on top and reactant concentrations go in the denominator. Each gets its own coefficient as an exponent.

$$K = \frac{[NH_3]^2}{[N_2][H_2]^3}$$

When the balanced equations are available, a large number of equilibrium constant expressions can be completed about as fast as they can be written.

Example B-2. Write the equilibrium constant expressions for the following balanced equations.

Equations	Answers
$PCl_3 + Cl_2 \rightleftharpoons PCl_5$	$K = \dfrac{[PCl_5]}{[PCl_3][Cl_2]}$
$CO_2 + H_2 \rightleftharpoons H_2O + CO$	$K = \dfrac{[H_2O][CO]}{[CO_2][H_2]}$
$2\,NO_2 \rightleftharpoons N_2O_4$	$K = \dfrac{[N_2O_4]}{[NO_2]^2}$
$C_3H_8 + 5\,O_2 \rightleftharpoons 3\,CO_2 + 4\,H_2O$	$K = \dfrac{[CO_2]^3[H_2O]^4}{[C_3H_8][O_2]^5}$

B-1.2 Calculating K from concentrations. When we have become accustomed to writing the expression for K, we can begin to use the expression in problems. To use K in other problems we must first find out what K is. Calculating K from given data about concentrations is therefore our first type of equilibrium constant calculation.

Example B-3. At 426°C the following equilibrium concentrations were found for the $H_2 + I_2 \rightleftharpoons 2\,HI$ reaction: $[H_2] = 1.831 \times 10^{-3}\,M$, $[I_2] = 3.129 \times 10^{-3}\,M$, $[HI] = 1.767 \times 10^{-2}\,M$. Calculate K for the reaction.

First we write the equilibrium constant expression.

$$K = \frac{[HI]^2}{[H_2][I_2]}$$

Then we put in the known concentrations.

$$K = \frac{(1.767 \times 10^{-2}\,M)^2}{(1.831 \times 10^{-3}\,M)(3.129 \times 10^{-3}\,M)}$$

We get the following arithmetic solution.

$$K = \frac{(1.767 \times 10^{-2} \ M)(1.767 \times 10^{-2} \ M)}{(1.831 \times 10^{-3} \ M)(3.129 \times 10^{-3} \ M)}$$

$$= \frac{1.767 \times 1.767 \times 10^{-2} \times 10^{-2} \times M^2}{1.831 \times 3.129 \times 10^{-3} \times 10^{-3} \times M^2}$$

$$= \frac{3.122}{5.729} \times \frac{10^{-4}}{10^{-6}}$$

$$= 0.5450 \times 10^{+2}$$

$$= 54.50$$

In the H_2, I_2, HI case (and all cases where both sides of the equation have the same number of molecules whose concentrations are measured) there are no units for K. Let us see what happens to units in a problem with unequal numbers of molecules on the two sides.

Example B-4. At 25°C, the following concentrations exist at equilibrium for the $N_2 + 3 H_2 \rightleftharpoons 2 NH_3$ reaction: $2.00 \times 10^{-3} \ M \ NH_3$, $1.875 \times 10^{-2} \ M \ N_2$, and $6.66 \times 10^{-2} \ M \ H_2$. Calculate K for the reaction.

$$K = \frac{[NH_3]^2}{[N_2][H_2]^3}$$

$$= \frac{(2.00 \times 10^{-3} \ M)^2}{(1.875 \times 10^{-2} \ M)(6.66 \times 10^{-2} \ M)^3}$$

$$= \frac{(2.00 \times 10^{-3} \ M)^2}{(1.875 \times 10^{-2})(6.66 \times 10^{-2})^3 (M \times M^3)}$$

$$= \frac{4.00 \times 10^{-6} \ M^2}{(1.875 \times 10^{-2})(6.66)^3(10^{-2})^3(M^4)}$$

$$= \frac{4.00 \times 10^{-6}}{1.875 \times (6.66)^3 \times 10^{-2} \times 10^{-6} \ M^2}$$

$$= \frac{4.00 \times 10^{-6}}{1.875 \times 295 \times 10^{-8} \ M^2}$$

$$= \frac{4.00}{554 \ M^2} \times 10^2$$

$$= \frac{7.22 \times 10^{-3} \times 10^2}{M^2}$$

$$= \frac{7.22 \times 10^{-1}}{M^2} = 7.22 \times 10^{-1} \ M^{-2}$$

Each of the two examples for which we have calculated K is a reaction between gaseous molecules. The concentrations of gases usually have fairly small numerical values when molarity is used to measure concentrations. The concentration of a gas can be measured in other units besides molarity. At any given temperature, pressure measures concentration of a gas. Pressure is often a convenient way to measure concentration of a gas. Let us repeat the above two examples using pressures instead of molarities. (Exactly the same concentrations are being used in each case.)

Example B–3a. At 426°C the following equilibrium partial pressures were found for the $H_2 + I_2 \rightleftharpoons 2\,HI$ reaction: $p_{H_2} = 0.105$ atm, $p_{I_2} = 0.1795$ atm, $p_{HI} = 1.012$ atm. Calculate K for the reaction.

$$K = \frac{[HI]^2}{[H_2][I_2]}$$

$$= \frac{(1.013\text{ atm})^2}{(0.1050\text{ atm})(0.1795\text{ atm})}$$

$$= \frac{(1.013)^2\text{ atm}^2}{(0.1050)(0.1795)\text{ atm}^2}$$

$$= \frac{1.027}{(1.885 \times 10^{-2})} = 54.50$$

The numbers from pressures look very different, but that has not changed K at all. K is K. This K is a number without units, and it does not depend upon which units were used to calculate it (so long as they are consistent within the calculation).

B–1.3 K_p and K_c. When the concentration units appear in the value for K, the choice of units will affect the number value for K.

Example B–4a. At 25°C, the following partial pressures exist at equilibrium for the $N_2 + 3\,H_2 \rightleftharpoons 2\,NH_3$ reaction: 0.0489 atm NH_3, 0.458 atm N_2, and 1.63 atm H_2. Calculate K for the reaction.

$$K = \frac{[NH_3]^2}{[N_2][H_2]^3}$$

$$= \frac{(0.0489\text{ atm})^2}{(0.458\text{ atm})(1.63\text{ atm})^3}$$

$$= \frac{(0.0489)^2\text{ atm}^2}{(0.458)(1.63)^3\text{ atm}^4}$$

$$= \frac{0.00239\text{ atm}^2}{(0.458)(4.32)\text{ atm}^4}$$

$$= \frac{2.39 \times 10^{-3}}{1.98\text{ atm}^2}$$

$$= 1.21 \times 10^{-3}\text{ atm}^{-2}$$

This value looks different, but actually 1.21×10^{-3} atm^{-2} and 7.22×10^{-1} M^{-2} are the same at 25°C. Since 1 mole occupies 22.4 liters at 0°C and 1.00 atm, at 25°C and 1.00 atm

$$V = 22.4 \times \frac{298°K}{273°K} = 24.45 \text{ liters}$$

then

$$\frac{1 \text{ mole}}{24.45 \text{ liter}} \text{ at } 1.00 \text{ atm} = 0.0409 \, M \text{ at } 1.00 \text{ atm}$$

or

$$0.0409 \, M/\text{atm} \quad \text{or} \quad 4.09 \times 10^{-2} \, M/\text{atm}$$

If we multiply

$$(7.22 \times 10^{-1} \, M^{-2})\left(4.09 \times 10^{-2} \frac{M}{\text{atm}}\right)^2$$

$$= \left(7.22 \times \frac{10^{-1}}{M^2}\right)(4.09)^2\left(10^{-4} \frac{M^2}{\text{atm}^2}\right)$$

$$= (7.22 \times 10^{-1})(16.7 \times 10^{-4} \text{ atm}^{-2})$$
$$= (121 \times 10^{-1})(10^{-4} \text{ atm}^{-2})$$
$$= 121 \times 10^{-5} \text{ atm}^{-2} = 1.21 \times 10^{-3} \text{ atm}^{-2}$$

Therefore, K is still K, but in this case K has units so its value reflects the units used.

Sometimes the two different common types of units are indicated by subscripts following the letter K. K_c means concentrations (in molarity) are used whereas K_p means pressures (in atmospheres) are used. For the reaction $N_2 + 3 H_2 \rightleftharpoons 2 NH_3$ at 25°C, $K_c = 7.22 \times 10^{-1} \, M^{-2}$ and $K_p = 1.21 \times 10^{-3}$ atm^{-2}. For the reaction $H_2 + I_2 \rightleftharpoons 2 HI$, $K_c = K_p = K = 54.5$, so no subscripts are needed. Because of the choice of standard reference states in thermodynamics, K_p is the conventional way to write K for gases (where measurements are often of pressures) and K_c is the conventional way to write K for reactions in liquid solution (where molar concentrations are used to report the measurements).

Because we are not going to define the standard reference states here, the separate systems for K_p and K_c will not add to our understanding or use of equilibrium constants. Therefore, we choose to use only one form of K, K_c. K_c is the one which can be used for either solutions in liquids or gases. From this point on we will choose to state all of our concentrations in molarity and all of our K values as K_c in those equilibrium problems involving a choice among possible units. The square bracket symbol [ ] will mean concentration in molarity.

B-1.4 Calculating One Unknown Concentration. Once we have determined K, we can use it to find a missing piece of information in another equilibrium of the same reaction at the same temperature. The simplest cases are those where only one concentration is left to be determined. Here is an example using the K value of 54.50 we found in example B–3.

Example B–5. At $426°C$, K for the reaction $H_2 + I_2 \rightleftharpoons 2\,HI$ is 54.50. Measurements on an equilibrium mixture show the concentrations of I_2 and HI are $[I_2] = 7.38 \times 10^{-4}\,M$, $[HI] = 1.354 \times 10^{-2}\,M$. What is the concentration of H_2 in the equilibrium mixture?

First we write the equilibrium constant expression.

$$K = \frac{[HI]^2}{[H_2][I_2]}$$

Then we put in all the known values

$$54.50 = \frac{(1.354 \times 10^{-2}\,M)^2}{[H_2](7.38 \times 10^{-4}\,M)}$$

solving

$$[H_2] = \frac{(1.354 \times 10^{-2}\,M)^2}{(54.50)(7.38 \times 10^{-4}\,M)}$$
$$= \frac{(1.354)^2 \times 10^{-4} \times M^2}{(54.50)(7.38) \times 10^{-4} \times M}$$
$$= \frac{(1.354)^2}{(54.50)(7.38)} M$$
$$= \frac{1.833}{402.2} M = 4.56 \times 10^{-3}\,M$$

B–1.5 Using the Change During Reaction. Sometimes these problems also require us to calculate one or more of the equilibrium concentrations from other information. For example, in the $H_2 + I_2 \rightleftharpoons 2\,HI$ reaction it is much easier to measure $[I_2]$ than $[H_2]$ or $[HI]$. I_2 has a dark violet color and the concentration can be measured by looking at the intensity of that color. (To be more precise, the ability of the gas to absorb certain colors of light is proportional to the I_2 concentration.) But measurement of the I_2 may be enough to tell us what another concentration is, as in this example.

Example B–6. Some pure HI is sealed into a bulb and heated to $426°C$. When equilibrium is established at $426°C$, $[I_2]$ is found to be $4.79 \times 10^{-4}\,M$. Calculate $[H_2]$ and $[HI]$. K is 54.50 at $426°C$.

$$K = \frac{[HI]^2}{[H_2][I_2]}$$

Putting in the known values

$$54.50 = \frac{[HI]^2}{[H_2](4.79 \times 10^{-4} \, M)}$$

We cannot solve one equation for two unknowns. Therefore, we must find one of them ($[H_2]$ or $[HI]$) in some other way. Going back to the problem, we notice that we started with *pure* HI. Therefore $[H_2] = 0$ and $[I_2] = 0$ at the start. During the reaction enough I_2 was formed to make $[I_2] = 0.479 \, M$. The $[I_2]$ *changed* from 0 M to 0.479 M. During the same reaction some H_2 also was formed and some HI was used. The amounts of the H_2, I_2, and HI *changes* are related to each other by the coefficients of the balanced equation. Things on the same side of the equation both go in the same way. Therefore, 1 H_2 is *made* for each 1 I_2 *made*. Things on opposite sides vary in opposite directions. Therefore, 2 HI (from the HI coefficient of 2) are *used* for each 1 I_2 (I_2 coefficient is 1) *made*. These facts can be neatly summarized by making a small table showing *original* concentrations, *changes* in concentration, and *final* concentrations. The *final* concentrations are the ones related to the *equilibrium constant*. The *changes* are the ones related to the *coefficients*. The original concentrations are not related to any special fact, but often the original concentrations are what we know. Here is the table for this example:

	$[H_2]$	$[I_2]$	$[HI]$
original concentration	0	0	?
change	$+1 \times (I_2 \text{ change})$	$+4.79 \times 10^{-4} \, M$	$-2 \times (I_2 \text{ change})$
final concentration		$4.79 \times 10^{-4} \, M$	

Putting in the values for the changes and adding to get the final concentrations:

	$[H_2]$	$[I_2]$	$[HI]$
original	0	0	?
change	$4.79 \times 10^{-4} \, M$	$4.79 \times 10^{-4} \, M$	$-9.58 \times 10^{-4} \, M$
final	$4.79 \times 10^{-4} \, M$	$4.79 \times 10^{-4} \, M$	$? - 9.58 \times 10^{-4}$

We still don't know the $[HI]$, but we can use $[H_2]$ in our earlier equilibrium constant expression.

$$54.50 = \frac{[HI]^2}{(4.79 \times 10^{-4} \, M)(4.79 \times 10^{-4} \, M)}$$

$$[HI]^2 = 54.50(4.79 \times 10^{-4} \, M)^2$$

taking the square roots

$$[HI] = \sqrt{54.50} \times (4.79 \times 10^{-4}\ M)$$
$$= (7.38)(4.79 \times 10^{-4}\ M)$$
$$= 3.54 \times 10^{-3}\ M$$

(We can now also see that the amount of HI originally put in must have been $[HI] = (3.54 \times 10^{-3}\ M) + (9.58 \times 10^{-4}) = 4.49 \times 10^{-3}\ M$.)

In Example B–6 it was not necessary for the original H_2 concentration to be zero. It was only necessary for it to be *known* so we could add the change and get the final concentration. Let us change the problem slightly to illustrate how any original concentration could be used.

Example B–7. Some pure HI is added to a bulb containing $1.00 \times 10^{-3}\ M$ H_2 gas and no I_2. When equilibrium is established at 426°C, $[I_2]$ is found to be $4.79 \times 10^{-4}\ M$. Calculate $[H_2]$ and $[HI]$. K is 54.50 at 426°C.

Starting with the table of concentrations and filling in known values and addition results gives us

	$[H_2]$	$[I_2]$	$[HI]$
original	$1.00 \times 10^{-3}\ M$	0	$x + 9.58 \times 10^{-4}\ M$
change	$+4.79 \times 10^{-4}\ M$	$+4.79 \times 10^{-4}\ M$	$-9.58 \times 10^{-4}\ M$
final	1.479×10^{-3}	$4.79 \times 10^{-4}\ M$	x (the unknown)

Writing K and solving:

$$K = \frac{[HI]}{[H_2][I_2]}$$

$$54.50 = \frac{x^2}{(1.479 \times 10^{-3}\ M)(4.79 \times 10^{-4}\ M)}$$

$$x^2 = 54.50(1.479 \times 10^{-3})(4.79 \times 10^{-4})\ M^2$$
$$= 386 \times 10^{-7}\ M^2$$
$$= 38.6 \times 10^{-6}\ M^2$$

$$[HI] = x = \sqrt{38.6} \times 10^{-3}\ M = 6.21 \times 10^{-3}\ M$$

If we know *all* of the original concentrations, then *any one* of the final concentrations is enough to let us calculate all of the final concentrations. That is one common method for determining the value of K experimentally. Here is an example.

Example B–8. A bulb is sealed containing $4.000 \times 10^{-3}\ M$ H_2, $1.690 \times 10^{-3}\ M$ I_2, and $1.470 \times 10^{-2}\ M$ HI. When the reaction $H_2 + I_2 \rightleftharpoons 2\ HI$ reaches equilibrium at 426°C, the I_2 concentration is $1.25 \times 10^{-3}\ M$. Calculate K for the reaction at 426°C.

The table of given data is

	$[H_2]$	$[I_2]$	$[HI]$
original	$4.000 \times 10^{-3}\ M$	$1.690 \times 10^{-3}\ M$	$1.470 \times 10^{-2}\ M$
change		$-0.440 \times 10^{-3}\ M$	
final		$1.250 \times 10^{-3}\ M$	

Completing the table

	$[H_2]$	$[I_2]$	$[HI]$
original	$4.000 \times 10^{-3}\ M$	$1.690 \times 10^{-3}\ M$	$1.470 \times 10^{-2}\ M$
change	$-0.440 \times 10^{-3}\ M$	$-0.440 \times 10^{-3}\ M$	$+0.880 \times 10^{-3}\ M$
final	$3.560 \times 10^{-3}\ M$	$1.250 \times 10^{-3}\ M$	$1.558 \times 10^{-2}\ M$

$$K = \frac{[HI]^2}{[H_2][I_2]} = \frac{(1.558 \times 10^{-2}\ M)^2}{(3.56 \times 10^{-3}\ M)(1.25 \times 10^{-3}\ M)}$$

$$= \frac{(1.558)^2 \times 10^{-4} \times M^2}{(3.56)(1.25) \times 10^{-6} \times M^2}$$

$$K = \frac{2.427}{4.45} \times 10^2 = 0.5455 \times 10^2 = 54.55$$

Notice that K is not quite the same as in the earlier calculation (Example B–3), but it is quite close. Experimental errors in measuring the concentrations will cause some variation, but K can be determined closely enough to give good results in the many calculations that use it.

B–1.6 Calculating Equilibrium Concentrations from Original Concentrations and K. So far we have carefully limited ourselves to problems where basic arithmetic (addition, subtraction, multiplication, division, and an occasional square root) is able to provide a solution. But the most common type of real equilibrium problem often requires some algebraic manipulations. We will set up one example of such a problem here to show what equilibrium calculations are most useful for and the sort of difficulties that arise. Then in Section C–2 we will come back to this type of problem and discuss shortcuts that can eliminate some of the mathematical difficulties.

After K has been determined at a particular temperature (by the methods in Examples B–3 or B–8), their greatest value is in allowing us to *predict* the results of an equilibrium *before* it is tried. We can then select the conditions that produce the result we wish and plan our future work to be more productive. These predictions can have great economic importance in design of factories to produce fertilizers and other important compounds. The general feature of these problems is that we know all of the original conditions but none of the changes or final concentrations.

Mathematically, a single equation can be solved to find one (and no more than one) unknown. We have several unknown final concentrations. But if we knew any *one* of the changes in concentration, we could find all the changes and all the final concentrations. We can choose to make one of the changes be our algebraic unknown, which we will call x. Then everything can be determined as some combination from the original concentrations and x. That lets us set up an equilibrium constant expression (with the known K) which has only one unknown, x, and which can be solved. Here is a typical example.

Example B-9. A bulb is sealed with $1.50 \times 10^{-3}\ M$ H_2, $3.50 \times 10^{-3}\ M$ I_2, and $5.00 \times 10^{-3}\ M$ HI. If it was heated to $426°C$, what would be the concentrations at equilibrium? At $426°C$ the $H_2 + I_2 \rightleftharpoons 2\ HI$ reaction has a K value of 54.50. Setting up our table of concentrations and selecting the change in $[H_2]$ as x, we get

	$[H_2]$	$[I_2]$	$[HI]$
original	$1.50 \times 10^{-3}\ M$	$3.50 \times 10^{-3}\ M$	$5.00 \times 10^{-3}\ M$
change	$+x\ M$		
final			

Using the equation coefficients and adding to get final concentration, we find:

	$[H_2]$	$[I_2]$	$[HI]$
original	$1.50 \times 10^{-3}\ M$	$3.50 \times 10^{-3}\ M$	$5.00 \times 10^{-3}\ M$
change	$+x\ M$	$+x\ M$	$-2x\ M$
final	$(1.50 \times 10^{-3} + x)M$	$(3.50 \times 10^{-3} + x)M$	$(5.00 \times 10^{-3} - 2x)M$

We can then write

$$K = \frac{[HI]^2}{[H_2][I_2]}$$

and substituting in the final (equilibrium) values find

$$54.50 = \frac{[(5.00 \times 10^{-3} - 2x)M]^2}{(1.50 \times 10^{-3} + x)M(3.50 \times 10^{-3} + x)M}$$

$$= \frac{(5.00 \times 10^{-3} - 2x)^2 M^2}{(1.50 \times 10^{-3} + x)(3.50 \times 10^{-3} + x)M^2}$$

That equation can be solved, but it and others like it are hard enough that we will assume your instructor will not require you to do it at this time. If your instructor discusses the shortcuts in Appendix C, Section C–2, he will probably then expect you to begin to do this kind of problem.

For completeness, the solution to the example is

$$K = 54.50 = \frac{(5.00 \times 10^{-3} - 2x)^2}{(1.50 \times 10^{-3} + x)(3.50 \times 10^{-3} + x)}$$

$$= 54.50 = \frac{25.00 \times 10^{-6} - 20.00 \times 10^{-3} x + 4x^2}{5.25 \times 10^{-6} + 5.00 \times 10^{-3} x + x^2}$$

then

$$2.86 \times 10^{-4} + 2.72 \times 10^{-1} x + 54.5 x^2$$
$$= 2.50 \times 10^{-5} - 2.00 \times 10^{-2} x + 4 x^2$$

and

$$50.5 x^2 + 2.92 \times 10^{-1} x + 2.61 \times 10^{-4} = 0$$

This is a quadratic equation which must be solved by using the formula

$$x = \frac{-b \pm \sqrt{b^2 - 4ac}}{2a}$$

so that

$$x = \frac{-2.92 \times 10^{-1} \pm \sqrt{(2.92 \times 10^{-1})^2 - 4(50.5)(2.61 \times 10^{-4})}}{2(50.4)}$$

$$x = \frac{-2.92 \times 10^{-1} \pm \sqrt{(8.52 \times 10^{-2}) - (5.27 \times 10^{-2})}}{100.8}$$

$$= \frac{-2.92 \times 10^{-1} \pm \sqrt{3.25 \times 10^{-2}}}{100.8}$$

$$= \frac{(-2.92 \times 10^{-1}) \pm (1.80 \times 10^{-1})}{100.8}$$

$$= \frac{-1.12 \times 10^{-1}}{100.8} \quad \text{or} \quad \frac{-4.72 \times 10^{-1}}{100.8}$$

$$= -1.11 \times 10^{-3} \quad \text{or} \quad -4.685 \times 10^{-3}$$

The answer -4.685×10^{-3} would make $[H_2]$ less than zero, which is physically impossible. Therefore $x = -1.11 \times 10^{-3}$ and

$$[H_2] = (1.50 \times 10^{-3} + x)M = [(1.50 \times 10^{-3}) - (1.11 \times 10^{-3})]M$$
$$= 0.39 \times 10^{-3} M = 3.9 \times 10^{-4} M$$

$$[I_2] = (3.50 \times 10^{-3} + x)M = [(3.50 \times 10^{-3}) - (1.11 \times 10^{-3})M]$$
$$= 2.39 \times 10^{-3} M$$

$$[HI] = (5.00 \times 10^{-3} - 2x)M$$
$$= (5.00 \times 10^{-3})(-2)(-1.11 \times 10^{-3})M$$
$$= [(5.00 \times 10^{-3}) + (2.22 \times 10^{-3})]M$$
$$= 7.22 \times 10^{-3} \, M$$

Notice that we did not need to know in advance whether $[H_2]$ went up or down. The sign of the answer takes care of that for us.

Equilibrium calculations can be much more complex than this example. But it is these complex calculations that produce the truly rewarding information in real life. In Section C–2 we will look at ways to handle these important calculations with the least possible work.

B–2 Special Forms of K: K_a, K_b, K_w

Some types of equilibrium calculations which are very common are done using forms of K that do not match the equation for the reaction. These special forms of K are used to reduce the number of concentrations which must be written down. In each case something that does not change has been removed by including it in the K value. Sometimes this also helps eliminate problems in deciding exactly what concentrations should be in a true law of mass action K. Each of these special forms of K is identified by a subscript following the K. The subscripts are chosen to indicate something about the particular class of substances for which that special form is used.

Acids are one class of substances for which a special form of K is used. Acids ionize in water to give the characteristic species for acid, which we call H^+ ion. H^+ never exists alone in water; it associates with one or more water molecules and exists as H_3O^+ ion or some larger grouping such as $H_9O_4^+$ ion (see Section 13–4). By the law of mass action rules K should be for the reaction $H_2O + HA \rightleftharpoons H_3O^+ + A^-$

$$K_{H_3O^+} = \frac{[H_3O^+][A^-]}{[H_2O][HA]}$$

or for the reaction $4\,H_2O + HA \rightleftharpoons H_9O_4^+ + A^-$

$$K_{H_9O_4^+} = \frac{[H_9O_4^+][A^-]}{[H_2O]^4[HA]}$$

We need to know a specific reaction to write K. But we do not need to know whether we form H_3O^+ or $H_9O_4^+$. Whatever is formed is "acid." Therefore, we will just call it H^+ and not worry about how many H_2O units it ties up. We can also remove the $[H_2O]$ from the expressions. In liquid water, $[H_2O]$ is always about the same (55.6 M). Therefore we can rewrite the expressions with the $[H_2O]$ term on the side with the constant.

$$K_{H_3O^+} \times [H_2O] = \frac{[H_3O^+][A^-]}{[HA]} \quad \text{or} \quad K_{H_9O_4} \times [H_2O]^4 = \frac{[H_9O_4^+][A^-]}{[HA]}$$

or rewriting using H^+ to represent the acid species (H_3O^+ or $H_9O_4^+$ or any other)

$$K_{H_3O^+} \times [H_2O] = \frac{[H^+][A^-]}{[HA]} \quad \text{or} \quad K_{H_9O_4^+} \times [H_2O]^4 = \frac{[H^+][A^-]}{[HA]}$$

Because $[H_2O]$ is constant and $K_{H_3O^+}$ is constant, $K_{H_3O^+} \times [H_2O]$ is also constant. The same is true for $K_{H_9O_4^+} \times [H_2O]^4$ or the combination for any other number of water molecules plus H^+. We will simply lump together all of these terms to get the expression for all of the acid forms together. We give this new constant the name K_a.

$$K_a = \frac{[H^+][A^-]}{[HA]}$$

All K_a expressions will have H^+ as one of the products, and no K_a expression will have the solvent concentration $[H_2O]$ among the concentrations needed.

Bases are another class of substances for which a special form of K is used. This special K is called K_b, it always has OH^- as one of the products, and $[H_2O]$ is not among the concentration terms. For the reaction

$$NH_3 + H_2O \rightleftharpoons NH_4^+ + OH^-$$

we write

$$K_b = \frac{[NH_4^+][OH^-]}{[NH_3]}$$

Another special K is called K_w, for the ionization of water. Since $[H_2O]$ is constant and therefore not needed in water or water solution, the reaction we write as $H_2O \rightleftharpoons H^+ + OH^-$ gives the special K:

$$K_w = [H^+][OH^-]$$

Although K_w (and all other K values) changes as temperature is changed, K_w is so common and important in problems that it is worth learning its value at room temperature. At 25°C (close to the temperature for most lab work) K_w has the particularly simple value $K_w = 10^{-14} M^2$. We will use that value for K_w at 25°C and will in fact give all K values as the value at 25°C unless we specifically state that it is the value for a different temperature.

B–3 Hydrolysis

Another special form of K is a hydrolysis constant, K_h. Hydrolysis is the process in which an ion reacts with water to produce either H^+ or OH^-. Hydrolysis constants are all either special forms of K_a or special forms of K_b in which the reactant is an ion that loses some or all of its charge in the reaction. That is just the reverse of the usual situation in a K_a or K_b, where a neutral molecule or ion reacts to take on a larger charge.

Hydrolysis occurs with the salts of weak acids or weak bases. An example would be $NaC_2H_3O_2$, a salt consisting of Na^+ ions and $C_2H_3O_2^-$ (acetate) ions. The acetate ions can be formed by the ionization of the weak acid, acetic acid, $HC_2H_3O_2$. Acetic acid is weak because the $HC_2H_3O_2$ molecules hold the H^+ portion fairly strongly (see Section 13-4). When acetic acid is added to water, only a small fraction of the $HC_2H_3O_2$ molecules lose H^+, so the solution is only weakly acidic. The same strong attraction for H^+ by the $C_2H_3O_2^-$ unit that makes the reaction

$$HC_2H_3O_2 + H_2O \rightleftharpoons H_3O^+ + C_2H_3O_2^-$$

go for only a small fraction can cause the $C_2H_3O_2^-$ ion to pick up H^+. One way to pick up H^+ is to take it away from H_2O in the reaction

$$C_2H_3O_2^- + H_2O \rightleftharpoons HC_2H_3O_2 + OH^-$$

That is a hydrolysis reaction. We could write K for that reaction as

$$K = \frac{[HC_2H_3O_2][OH^-]}{[C_2H_3O_2^-][H_2O]}$$

But we already know we can write this as a special K which includes $[H_2O]$ in the K value.

$$K_b = \frac{[HC_2H_3O_2][OH^-]}{[C_2H_3O_2^-]}$$

Because the charge on the reactant goes down (from -1 on $C_2H_3O_2^-$ to zero as $HC_2H_3O_2$) instead of up, this particular K_b is a K_h.

$$K_h = \frac{[HC_2H_3O_2][OH^-]}{[C_2H_3O_2^-]}$$

You may wonder at this point why we did not just say, "The reaction $C_2H_3O_2^- + H_2O \rightleftharpoons HC_2H_3O_2 + OH^-$ is a hydrolysis with $K_h = [HC_2H_3O_2][OH^-]/[C_2H_3O_2^-]$." We could then quickly introduce other examples of hydrolysis, as in the list given in Table B-1 on the next page. But our development of the $C_2H_3O_2^-$ case showed that the hydrolysis is related to a particular K_a. It is almost the reverse of the $HC_2H_3O_2$ ionization reaction. Each hydrolysis reaction is closely related to a particular acid or base ionization. Table B-2 shows the related pairs for the hydrolysis reactions we have shown so far.

You will notice that each related pair has one K_a and one K_b. When the ionization reaction leads to a K_a, the corresponding K_h is in fact in the form of a K_b. When the ionization reaction leads to a K_b, the related K_h is in the form of a K_a. These pairs are called **conjugates** of each other. Every weak acid forms a product that can act as a base in hydrolysis. $HC_2H_3O_2$ (an acid) forms $C_2H_3O_2^-$ which is called its **conjugate base.**

In comparing the K_a of a weak acid and the K_h of its conjugate base, another feature is apparent. If the $[H^+]$ and $[OH^-]$ are disregarded, the K_a

Table B–1 Some Examples of Hydrolysis Reactions and Constants

Reaction	K_h
$NH_4^+ + H_2O \rightleftharpoons H_3O^+ + NH_3$	$\dfrac{[H^+][NH_3]}{[NH_4^+]}$ (note: this is a K_a)
$H_2PO_4^- + H_2O \rightleftharpoons H_3PO_4 + OH^-$	$\dfrac{[H_3PO_4][OH^-]}{[H_2PO_4^-]}$ (note: this is a K_b)
$HPO_4^{2-} + H_4O \rightleftharpoons H_2PO_4^- + OH^-$	$\dfrac{[H_2PO_4^-][OH^-]}{[HPO_4^-]}$ (note: this is a K_b)
$PO_4^{3-} + H_2O \rightleftharpoons HPO_4^- + OH^-$	$\dfrac{[HPO_4^-][OH^-]}{[PO_4^{3-}]}$ (note: this is a K_b)
$S^{2-} + H_2O \rightleftharpoons HS^- + OH^-$	$\dfrac{[HS^-][OH^-]}{[S_2^-]}$ (note: this is a K_b)
$HS^- + H_2O \rightleftharpoons H_2S + OH^-$	$\dfrac{[H_2S][OH^-]}{[HS^-]}$ (note: this is a K_b)
$SO_4^{2-} + H_2O \rightleftharpoons HSO_4^- + OH^-$	$\dfrac{[HSO_4^-][OH^-]}{[SO_4^{2-}]}$ (note: this is a K_b)
$HSO_4^- + H_2O \rightleftharpoons H_2SO_4 + OH^-$	$\dfrac{[H_2SO_4][OH^-]}{[HSO_4^-]}$ (note: does not work forward. H_2SO_4 is too strong an acid)

Table B2–2

Ionization Reaction	K	Related Hydrolysis	K_h
$HC_2H_3O_2 + H_2O \rightleftharpoons H_3O^+ + C_2H_3O_2^-$	$K_a = \dfrac{[H^+][C_2H_3O_2^-]}{[HC_2H_3O_2]}$	$C_2H_3O_2^- + H_2O \rightleftharpoons HC_2H_3O_2 + OH^-$	$\dfrac{[HC_2H_3O_2][OH^-]}{[C_2H_3O_2^-]}$
$NH_3 + H_2O \rightleftharpoons NH_4^+ + OH^-$	$K_b = \dfrac{[NH_4^+][OH^-]}{[NH_3]}$	$NH_4^+ + H_2O \rightleftharpoons H_3O^+ + NH_3$	$\dfrac{[H^+][NH_3]}{[NH_4^+]}$
$H_3PO_4 + H_2O \rightleftharpoons H_3O^+ + H_2PO_4^-$	$K_a = \dfrac{[H^+][H_2PO_4^-]}{[H_3PO_4]}$	$H_2PO_4^- + H_2O \rightleftharpoons H_3PO_4 + OH^-$	$\dfrac{[H_3PO_4][OH^-]}{[H_2PO_4^-]}$
$H_2PO_4^- + H_2O \rightleftharpoons H_3O^+ + HPO_4^{2-}$	$K_a = \dfrac{[H^+][HPO_4^{2-}]}{[HPO_4^-]}$	$HPO_4^{2-} + H_2O \rightleftharpoons H_2PO_4^- + OH^-$	$\dfrac{[H_2PO_4^-][OH^-]}{[HPO_4^{2-}]}$
$HPO_4^- + H_2O \rightleftharpoons H_3O^+ + PO_4^{3-}$	$K_a = \dfrac{[H^+][PO_4^{3-}]}{[HPO_4^{2-}]}$	$PO_4^{3-} + H_2O \rightleftharpoons HPO_4^- + OH^-$	$\dfrac{[HPO_4^-][OH^-]}{[PO_4^{3-}]}$
$H_2S + H_2O \rightleftharpoons H_3O^+ + HS^-$	$K_a = \dfrac{[H^+][HS^-]}{[H_2S]}$	$HS^- + H_2O \rightleftharpoons H_2S + CH^-$	$\dfrac{[H_2S][OH^-]}{[HS^-]}$
$HS^- + H_2O \rightleftharpoons H_3O^+ + S^{2-}$	$K_a = \dfrac{[H^+][S^{2-}]}{[HS^-]}$	$S^{2-} + H_2O \rightleftharpoons HS^- + OH^-$	$\dfrac{[HS^-][OH^-]}{[S^{2-}]}$
$H_2SO_4 + H_2O \longrightarrow H_3O^+ + HSO_4^-$	goes to completion	$HSO_4^- + H_2O \longleftarrow H_2SO_4 + OH^-$	does not go left to right
$HSO_4^- + H_2O \rightleftharpoons H_3O^+ + SO_4^{2-}$	$K_a = \dfrac{[H^+][SO_4^{2-}]}{[HSO_4^-]}$	$SO_4^{2-} + H_2O \rightleftharpoons HSO_4^- + OH^-$	$\dfrac{[HSO_4^-][OH^-]}{[SO_4^{2-}]}$

and K_h are reciprocals (one fraction equals the other one upside down). In fact, we could write $1/K_a$ and get something quite close to K_h. In the case of $HC_2H_3O_2$ and $C_2H_3O_2^-$, we find

$$\frac{1}{K_a} = \frac{[HC_2H_3O_2]}{[H^+][C_2H_3O_2^-]}$$

and

$$K_h = \frac{[HC_2H_3O_2][OH^-]}{[C_2H_3O_2^-]}$$

We could make the two identical by multiplying $1/K_a$ by $[H^+]$ and $[OH^-]$.

$$1/K_a \times [H^+][OH^-] = \frac{[HC_2H_3O_2]}{[H^+][C_2H_3O_2^-]} \times [H^+][OH^-]$$

$$\frac{[H^+][OH^-]}{K_a} = \frac{[HC_2H_3O_2][OH^-]}{[C_2H_3O_2^-]}$$

$$K_h = \frac{[HC_2H_3O_2][OH^-]}{[C_2H_3O_2^-]}$$

$$= \frac{[H^+][OH^-]}{K_a}$$

But $[H^+][OH^-]$ is something we already know: the special form of K called K_w! We even know its value at 25°C, $10^{-14} M^2$. Therefore we can say

$$K_h = \frac{[H^+][OH^-]}{K_a} = \frac{K_w}{K_a}$$

and at 25°C

$$K_h = \frac{10^{-14} M^2}{K_a}$$

Because of that simple relationship, we do not need tables of K_h values. We can always find K_h quickly from K_a or K_b and the known value of K_w. In every case, K_a of a conjugate base is K_w/K_a of its conjugate acid. For the conjugate acid of a weak base, K_h is always K_w/K_b of its conjugate base. You can prove this to yourself with the examples in Table B-2.

Table B-3

Acid	K_a	Conjugate Base	$K_h = \frac{K_w}{K_a}$
$HC_2H_3O_2$	$1.8 \times 10^{-5} M$	$C_2H_3O_2^-$	$5.6 \times 10^{-10} M$
HCN	$7 \times 10^{-10} M$	CN^-	$1.4 \times 10^{-5} M$
H_2CO_3	$4.3 \times 10^{-7} M$	HCO_3^-	$2.3 \times 10^{-8} M$
HCO_3^-	$5.6 \times 10^{-11} M$	CO_3^{2-}	$1.8 \times 10^{-4} M$

The concept of hydrolysis and the simple method for getting K_h from K_a or K_b are important in many chemical systems, including some important in biological processes. Table B–3 lists some examples of the relationship between K_a of weak acids and K_h of their conjugate bases.

B–4 Solubility Products

Another special form of K occurs with solubility problems. When one of the substances involved in an equilibrium is a solid, the concentration of the solid is a constant factor that can be included in the constant. Impurities in the solid (a solid in solid solution) could cause some variation, but the nonvarying cases involving pure solids are presumed to be the normal situations. When a pure solid is in contact with a solution, the rates of forward and reverse reactions also depend upon the surface area of the solid, but that factor cancels out in K because it affects both reactions equally. The resulting K for dissolving a solid is not a fraction; the only reactant (the solid) is a constant factor included in K. Therefore, only the product concentrations remain. Since that series of numbers multiplied together (a mathematical multiplication product) represents the solubility of the substance at that temperature, it is named the **solubility product** and given the symbol K_{sp}.

Table B–4 lists some examples of solubility products.

Table B–4 Solubility Products

Reaction	K_{sp}
$C_{12}H_{22}O_{11}(s) \xrightleftharpoons{H_2O} C_{12}H_{22}O_{11}$ (in solution)	$K_{sp} = [C_{12}H_{22}O_{11}]$
$NaCl \xrightleftharpoons{H_2O} Na^+ + Cl^-$	$K_{sp} = [Na^+][Cl^-]$
$CaCl_2 \xrightleftharpoons{H_2O} Ca^{2+} + 2\,Cl^-$	$K_{sp} = [Ca^{2+}][Cl^-]^2$
$Al_2(SO_4)_3 \xrightleftharpoons{H_2O} 2\,Al^{3+} + 3\,SO_4^{2-}$	$K_{sp} = [Al^{3+}]^2[SO_4^{2-}]^3$

Skills Expected at This Point

1. You should be able to use the equilibrium constant expression to solve for any one value (of K or a concentration) when all the other values are known.
2. You should be able to construct a table of original concentrations, changes, and equilibrium concentrations and use it to solve for K, or a missing concentration, or the extent of reaction if you are given information as adequate as the data for the examples in Appendix B and the problem does not require a quadratic (or more difficult than quadratic) solution.
3. You should recognize and be able to use K_a, K_b, K_w, K_h, and K_{sp}, and you should be able to state why each of these differ from a simple law of mass action K.

Exercises

Based on "light reading"

1. (a) List the special forms of K mentioned in this appendix. (b) What makes them "special"?
2. What is hydrolysis?

Based on mastery

3. Write the following equilibrium constant expressions:
   (a) K_a for HCN
   (b) K_b for NH_3
   (c) K_h for $NaC_2H_3O_2$
   (d) K_{sp} for $Al_2(SO_4)_3$
4. Formic acid, $HCHO_2$, is at equilibrium with $[HCHO_2] = 0.100\ M$, $CHO_2^- = 0.0100\ M$, and $H^+ = 0.0200\ M$. (*Note:* The solution was slightly acidic before the formic acid was added.) Calculate K_a for formic acid.
5. Acetic acid, $HC_2H_3O_2$, has a K_a of $1.85 \times 10^{-5}\ M$. Calculate the $[C_2H_3O_2^-]$ in a pH = 4 solution containing $10^{-2}\ M\ HC_2H_3O_2$.
6. K_b for NH_3 is $1.80 \times 10^{-5}\ M$. Calculate K_h for NH_4^+ ions.
7. 10^{-2} moles of H_2O, 10^{-3} moles of Cl_2, 10^{-4} moles of O_2, and zero HCl are placed in a 10.0 liter tank. When equilibrium is reached for the reaction $\quad 2\ H_2O(g) + 2\ Cl_2(g) \rightleftharpoons 4\ HCl(g) + O_2(g),\quad [HCl] = 10^{-5}\ M$
   (a) Set up a table showing all the concentrations (original and final) and changes in concentrations.
   (b) Calculate K for the reaction.
8. At 100°C, 33.4 g of $PbCl_2$ will dissolve in 1 liter of pure water. Calculate K_{sp} for $PbCl_2$.

Test Yourself

1. Write the equilibrium constant expressions for the following gas phase reactions (the equations have not been balanced).
   (a) $N_2 + H_2 \rightleftharpoons NH_3$
   (b) $CO + O_2 \rightleftharpoons CO_2$
   (c) $NH_3 + O_2 \rightleftharpoons NO + H_2O$
   (d) $HCl + O_2 \rightleftharpoons H_2O + Cl_2$
   (e) $C_6H_{14} + O_2 \rightleftharpoons CO_2 + H_2O$
2. Calculate the concentration of undissociated $HC_2H_3O_2$ in a solution containing $10^{-3}\ M\ C_2H_3O_2^-$ ions and $10^{-3}\ M\ H^+$ ions. K_a for $HC_2H_3O_2$ is $1.85 \times 10^{-5}\ M$.
3. A bulb is filled with $1.00 \times 10^{-3}\ M$ HI gas and $1.00 \times 10^{-4}\ M\ H_2$ gas (no I_2). After heating to reach an equilibrium for the $H_2 + I_2 \rightleftharpoons 2\ HI$ reaction, the bulb contains $1.05 \times 10^{-4}\ M\ I_2$. Calculate K for that equilibrium.
4. If at equilibrium in the reaction of H_2O and Cl_2 to give HCl and O_2, HCl is $10^{-2}\ M$, O_2 is $10^{-3}\ M$, H_2O is $10^{-3}\ M$, and Cl_2 is $10^{-2}\ M$, then K is (all answers assumed to be in units consistent with all concentrations in molarity)
   (a) 10^{-4}
   (b) 10^{-3}
   (c) 10^{-1}
   (d) 10^{+1}
   (e) none of these

5. For equilibrium calculations about the reaction
$$C_2H_3O_2^- + H_2O \rightleftharpoons HC_2H_3O_2 + OH^-$$
we would use

(a) $K = \dfrac{[HC_2H_3O_2][OH^-]}{[C_2H_3O_2^-][H_2O]}$

(b) $K = \dfrac{[C_2H_3O_2^-][H_2O]}{[HC_2H_3O_2][OH^-]}$

(c) $K = \dfrac{[H_2O][HC_2H_3O_2]}{[C_2H_3O_2][OH]}$

(d) $K_a = \dfrac{[H^+][C_2H_3O_2^-]}{[HC_2H_3O_2]}$

(e) $\dfrac{K_w}{K_a} = \dfrac{[HC_2H_3O_2][OH^-]}{[C_2H_3O_2^-]}$

6. K_a, K_b, K_h, K_w, and K_{sp} are all forms of K where
   (a) something constant has been put in the K instead of among the concentration terms.
   (b) acids or bases are involved.
   (c) the law of mass action would be incorrect.
   (d) gas pressures are used instead of molarity for the concentrations.
   (e) none of these.

APPENDIX C

Quantitative Basis and Applications of Equilibrium

C-1 Free Energy and Equilibrium

C-1.1 Equilibrium Dependence on Relative States.

In Section 12-5 we pointed out how differences in the activation energies of the forward and the back reactions contribute to establishing the conditions necessary for equilibrium to be achieved. Small concentrations of the materials reacting via a low activation energy will be enough to balance the reaction rate against the higher activation energy reaction even though quite large concentrations are available of the reactants for the higher activation energy reaction.

At any fixed temperature, the equilibrium remains the same even when the activation energies are changed. We can show experimentally that addition of a catalyst speeds the reactions but does not affect the equilibrium at all. We know (Section 12-3.3) that catalysts lower the activation energy. We also know (Section 12-4) that the activation energy of the back reaction must be lowered by the same amount. The experimental fact that both forward and back reactions can be sharply changed in rate (as they must be changed by the drop in activation energies) without affecting the equilibrium suggests that the equilibrium is determined by some other, more fundamental property than the reaction rates. Since the equilibrium remains constant when catalyst is added, this fundamental property must be one that also remains constant during addition of catalyst.

The most obvious fundamental property of the equilibrium reaction that remains constant during addition of catalyst is the *difference* between the two activation energies. We know that this difference influences the amounts of reactants and products required at equilibrium. That difference in activation energies is in fact the difference in the total energies of reactants and products. We can therefore conclude that the equilibrium might be determined by the difference in energies of the reactants and products.

The relationship to energies in the above paragraph is an important part of the basis for equilibrium, but it is not complete. Activation energies contribute to the equilibrium condition, but there are also other factors. These factors must also be considered in the fundamental property upon which equilibrium depends. However, the general pattern of the conclusion reached from the energy factor remains true. Equilibrium is determined by the conditions for the reactants and the products alone; the particular way they are reacted (catalyzed or uncatalyzed) has no influence at all. What we need is a description of the condition of the reactants, including energy and any other significant terms, and a similar description of the condition of the products. These conditions are referred to as the thermodynamic **states.** Equilibrium is determined by the **change of state** in the process. When the change in state causes equal tendencies for reaction in each direction, equilibrium is established.

Thermodynamics is concerned with a number of properties called **state functions** which describe the thermodynamic state of any particular sample of matter. Some of these state functions are familiar to us. Temperature, pressure, and volume are examples. The energy content of the matter is another state function. And the entropy factor we described in Chapter 11 is also a state function. We will see that energy and entropy are the factors that cause reactions to occur and establish the conditions of equilibrium.

However, changes in the other state functions can also affect the equilibrium. It is possible, through thermodynamics and use of calculus, to determine the exact relationships between these state functions at both equilibrium and nonequilibrium conditions. For our purposes here, we will limit ourselves to the relationships that determine the equilibrium under the single most common condition for chemical reactions on Earth. The conditions will be constant temperature (the temperature of the surroundings) and constant pressure (the atmospheric pressure at the time of reaction). The surroundings will take away any extra heat or supply any needed heat to maintain the temperature, and the volume of the reacting material will expand or contract as needed to maintain the constant pressure (atmospheric pressure).

Let us now consider what changes of state contribute to **tendencies for reaction.** We can then determine when the reaction tendencies will balance each other and establish equilibrium.

C-1.2 The Tendency to Release Energy as Heat: Enthalpy.

In Section 11–1 we described several natural processes for converting one form of energy into another form of energy. The total energy does not change in these processes, but some of the energy conversions tend to occur spontaneously. That shows that energy has a greater tendency toward occurring in some forms than in others. In particular, we find energy tends to be converted from fixed forms, such as the potential energy of a particular object, to dissipated forms, such as heat spread over a large amount of material. There are numerous examples all around us showing this natural tendency to dissipate energy. Let us consider a few such examples.

Figure C–1 shows three examples of spontaneous energy transformations. Heavy objects tend to fall, converting gravitational potential energy (a fixed form) to motion of objects struck in the fall or to heat (dissipated forms); a coiled spring unwinds, converting its energy of coiling (a fixed form) to mechanical motion (a more dissipated form); wood burns in air, con-

Figure C–1 Spontaneous energy transformations.

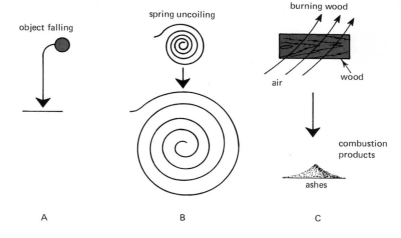

verting its chemical potential energy (a fixed form) to heat and light (dissipated forms).

When heavy objects fall, they are held more strongly to the Earth and their total energy content must be lowered by an amount equal to the energy dissipated as motion and heat. When a spring uncoils, its total energy content must be lowered by an amount equal to the mechanical energy lost. (The metal in the spring is held together more strongly in the uncoiled position where the metal atoms have returned to their original, most energetically favorable positions, which are the positions where they are held with the greatest strength.) When wood burns, the product molecules are held together by greater bonding forces than the reactants, and the total energy content has been lowered by that increased bonding. The energy content lowering just equals the energy dissipated to other forms.

Figure C–2 shows the changes in energy content for these three trans-

Energy content changes in spontaneous transformations.

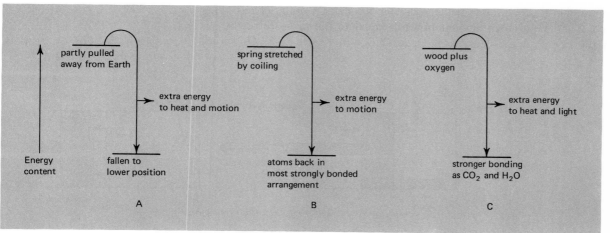

[C–1] Free Energy and Equilibrium

formations. Notice that "low energy" always means "strongly bonded." High energy (in the fixed, potential energy form) would mean that more of the energy needed to escape the attractive forces completely was already there. Therefore, low energy means the arrangement is harder to break up. A rocket escaping from Earth requires more fuel to lift it from sea level (as at Cape Kennedy) than it would from a higher energy position such as the high mountain plateaus of Wyoming's Great Divide Basin. It has been seriously suggested that a high elevation launch site be considered to help lower the cost of repeated space trips. Similarly, if the components in wood and air were to be broken down to unbonded atoms, it would be easier (require less outside energy) before the wood was burned.

The examples in the previous paragraph lead to the following conclusion: *When more than one energy content is available, matter tends to go to the lowest energy state and release the remaining energy to dissipated forms.* That is a completely general principle and, unless it is overcome by the other general principle causing change (Section C–1.3), it will determine the direction in which changes take place.

If reactions occurred in conditions of constant volume, the dissipation of energy could be measured by the lowering of energy content. In the burning of wood that lowering would be equal to the increased chemical bonding energies. Such measurements are interesting, particularly if one wanted to compare various bond strengths, and any good book on thermodynamics will discuss that situation in some detail. But in the real world we rarely do things at constant volume. We usually work at constant pressure (the atmospheric pressure). Certainly most burning of wood is done in open air at atmospheric pressure.

Under conditions of constant pressure, part of the energy from the stronger chemical bonding is not dissipated as heat. Instead it is used to do work. The products of burning wood, carbon dioxide and water vapor, occupy more volume as a gas than the oxygen consumed in the reaction occupied. As the wood burns, the products formed must push back the surrounding gas (at the atmospheric pressure) to make room for the extra volume of gas being formed. That requires an amount of work equal to the pressure of the gas times the extra volume being formed, as shown in Figure C–3. This work

Figure C–3 Energy needed to do work at constant pressure.

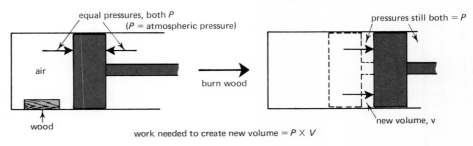

Figure C-4 Energy release by volume contraction.

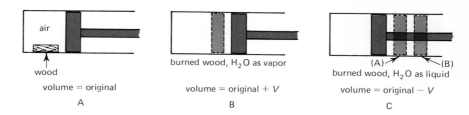

is stored as potential energy, just as the work used to lift a heavy object is stored as potential energy of that object. If the volume is reduced at a later time, that stored potential energy can be released.

Figure C-4 shows what would happen if the water from burning wood was first formed as a vapor and later condensed to liquid water. As shown in Figure C-4, when water condenses the pressure of the surrounding gas provides energy as work to compress the gases from the reaction into their new smaller volume. Table C-1 shows how the energy involved in work affects the heat being dissipated by the processes.

Because the spontaneous energy transformations are those that dissipate energy, the conversion of energy from one fixed form (energy content of the molecules) to another fixed form (PV potential energy) does not contribute to the tendency for reaction. The energy converted to heat is equal to the loss in the total of the fixed energies. Therefore the *total* of the fixed energies, molecular energy content plus PV energy, becomes the important and useful quantity for reactions at constant pressure.

The various thermodynamic state functions are given names and symbols.

Table C-1 Effect of *PV* Work on Heat Dissipated

	Wood and Air	Burned Wood, Water as Vapor	Burned Wood, Water as Liquid
volume	original	original $+ V$	original $- V$
total work done on surrounding air	none	$P \times V$	$P \times (-V)$
energy content (molecules)	original	(original $-$ extra chemical bonding)	(original $-$ extra chemical bonding $-$ bonding of liquid)
energy content (PV potential energy)	original ($P \times$ total original volume)	original $+ PV$	original $- PV$
energy dissipated as heat	none	extra chemical bonding $- PV$	extra chemical bonding $+$ bonding of liquid $+ PV$

[C-1] Free Energy and Equilibrium

Table C-2
Effects of Energy and Volume Changes on Enthalpy

	Wood + Air	Burned Wood Water as Vapor	Burned Wood, Water as Liquid
volume	V_0	$V_0 + V$	$V_0 - V$
energy	E_0	$E_1 = E_0 - E_{chem}$ (extra chemical bonding)	$E_2 = E_1 - E_{liq}$ (liquid bonding)
PV	PV_0	$P(V_0 + V)$	$P(V_0 - V)$
H	$E_0 + PV_0$	$E_1 + P(V_0 + V)$ $= E_0 + PV_0 - E_{chem} + PV$	$E_2 + P(V_0 - V)$ $= E_0 + PV_0 - E_{chem} - E_{liq} - PV$
ΔH	0	$-E_{chem} + PV$	$-E_{chem} - E_{liq} - PV$
heat released $= -\Delta H$	0	$E_{chem} - PV$	$E_{chem} + E_{liq} + PV$

The internal energy (which we described above by the crude term "molecular energy content") is called energy and given the symbol E. Pressure and volume are given the symbols P and V. The sum of E plus PV is given the name enthalpy and the symbol H. When one of these quantities changes during a process, the symbol delta (Δ) is used to represent "the change in." Therefore, for reactions at constant pressure and temperature we can state that ΔH is the factor contributing to the tendency for reaction. Since heat is released as H goes down, the heat released must equal $-\Delta H$. The energy dissipation factor contributes to the tendency for reaction when ΔH is negative. Since both E and V can vary, but P is constant for our condition of constant temperature and pressure, $\Delta H = \Delta E + P \Delta V$. These various facts are used to retabulate the information of Table C-1 into Table C-2, showing how the enthalpy term, H, holds the key relationship.

We have now determined how to find whether energy factors favor a process—we simply look for a negative ΔH. The more negative ΔH is, the more strongly energy favors the process. The more positive ΔH is, the more strongly energy disfavors the process. We can therefore now go on to look for nonenergy factors that may work with or against the ΔH factor.

C-1.3 The Tendency to Mix. In Section 11-2 we described the universal tendency toward mixing and the thermodynamic property entropy, represented by the symbol S, which increases during spontaneous processes. You may wish to review Section 11-2 to strengthen your understanding of entropy and its tendency to increase. The tendency to increase entropy is the other driving force causing reactions to occur.

The overall tendency for reaction at constant pressure and temperature is determined by the sum of the effects of the two parts, the tendency to lower the enthalpy and the tendency to raise the entropy. We need a quantitative relationship between the two to make a summation in consistent units possible. The relationship should relate the change in entropy, ΔS, to the change in enthalpy, ΔH, during the reaction. Both ΔS and ΔH are related to heat flow, and that leads to the necessary relationship to units. $-\Delta H$ is equal to the heat flow from the material to the surroundings, so ΔH must have the same units as heat, calories or some other suitable unit of energy. We will simply state here that ΔS is equal to the heat which would flow if the reaction was completely reversible divided by the absolute temperature

$$\Delta S = \frac{\text{reversible heat}}{T}$$

Entropy has been defined so that it would have that relationship. We will not attempt a detailed justification of that definition, but we can show that it is a good one for at least one case of increasing entropy, heat flow from hot to cold. ΔS must be positive for a spontaneous process involving no energy change, and heat flow is such a spontaneous process. Since the heat lost from the hot object is equal to the heat gained by the cold object, the temperature difference makes the ΔS *gained* at the cold object larger than the ΔS *lost* at the hot object. The overall ΔS is

$$\Delta S_{\text{total}} = \Delta S_{\text{cold}} + \Delta S_{\text{hot}} = \frac{+ \text{ heat transferred}}{T_{\text{cold}}} + \frac{- \text{ heat transferred}}{T_{\text{hot}}}$$

Since heat transferred/T_{cold} is a larger fraction than heat transferred/T_{hot}, the total ΔS is positive, as required for the spontaneous change.

C–1.4 Free Energy: The Overall Tendency to React.

Since the tendency for reaction (under constant pressure and temperature) depends both upon the tendency to release heat (a lowering of the enthalpy, H) and the tendency to mix (a raising of the entropy, S), an overall term should be possible that involves both H and S. This overall term is called the **free energy** and is given the symbol G. Free energy has units of energy (as H does), so it is necessary to multiply S by the absolute temperature T to make it consistent in units. When that is done and the directions of the changes are set to fit observed tendencies, we get the following equations.

$$G = H - TS$$

and for changes at constant temperature and pressure

$$\Delta G = \Delta H - T\Delta S$$

Notice that both of the natural tendencies (ΔH negative and ΔS positive) contribute to *negative* values of ΔG. Therefore, values of ΔG that are negative mean the process tends to occur. The more negative the value for ΔG, the stronger the tendency for that process.

If we write a reaction (or other change) and find that ΔG would be positive, we have written it in the wrong direction. By turning it around we would get a negative ΔG and therefore have a reaction with a tendency to occur. As a result, positive values of ΔG always mean the reaction (or other process) occurs in the opposite direction (when pressure and temperature are constant).

ΔG, the change in free energy, is therefore a measure of what will happen at constant temperature and pressure. A negative ΔG means the process occurs as written, and a positive ΔG means it occurs in the reverse direction.

C-1.5 The Free Energy Condition for Equilibrium.

We have seen that the sign of ΔG (positive or negative) tells the direction of change, but there is one other possibility. ΔG can also be zero. If ΔG is zero, there is no net tendency for reaction in either direction. The reactions in the forward and back directions must therefore be balancing each other out, and an equilibrium exists. In fact, the condition, $\Delta G = 0$, is the necessary condition for equilibrium at constant temperature and pressure. Any nonzero value of ΔG, positive or negative, shows some net tendency in one direction that prevents equilibrium.

The equilibrium condition, $\Delta G = 0$, can be used to determine some interesting relationships. For instance, since $\Delta G = \Delta H - T \Delta S$ we can say that at equilibrium $0 = \Delta H - T \Delta S$. That equation can be rearranged to the form $\Delta H = T \Delta S$ (at equilibrium). If we know ΔH for an equilibrium process, we can solve for ΔS. ΔH is an easily measured heat flow in most cases whereas ΔS is harder to measure. At $100\,°C$, ΔH for the evaporation of water is 540 cal/g or 9720 cal/mole. Therefore, we can calculate ΔS as follows.

$$\Delta H = T \Delta S \text{ (at equilibrium)}$$
$$9720 \text{ cal/mole} = 373\text{K} \times \Delta S$$
$$\Delta S = \frac{9720}{373} \text{ cal/deg mole} = 26 \text{ cal/deg mole}$$

Calories per degree mole are sometimes also called simply **entropy units,** since they are a common unit for the changes in entropy.

We are principally concerned here with your being able to use ΔG to identify the condition for a process at constant temperature and pressure (the most common conditions on Earth). If ΔG is negative, the process can occur; if ΔG is zero, the process is at equilibrium, if ΔG is positive, the process goes in the other direction.

Our secondary concern is that you recognize the potential for precise calculations about the conditions by using the thermodynamic terms. Let us take one specific example as an illustration. In our example above where we calculated ΔS for evaporation of water, we found an increase in entropy of 26 entropy units was needed to balance out the requirement for heat flow in the unfavored direction (in) to provide the heat of vaporization. When a gas is formed, the increase in entropy depends upon the volume the gas occupies. The larger the volume, the greater the disorder and therefore the larger the positive value for ΔS. The volume per mole depends (at a fixed temperature) on the pressure. The lower the pressure the higher the volume and the larger the ΔS. If the entropy for liquid water is known and the entropy of the gas is calculated from the known equilibrium vapor pressure, we must get the same 26 entropy units for ΔS of the evaporation. The equilibrium vapor pressure must be whatever value is needed to make that so, because that is necessary to make $\Delta G = 0$ and achieve equilibrium.

But one of the terms in the above balance can be changed. The entropy of liquid water does not have to be the entropy of *pure* liquid water. We could have a solution. A solution is more disordered than a pure substance, so a solution has a higher entropy. But the heat of vaporization from the solution is still essentially the same. (As long as the solution is fairly dilute, the intermolecular forces to be broken during evaporation must remain about the same as in the pure liquid.) And if ΔH is still 9720 cal/mole, ΔS at equilibrium must still be 26 entropy units. To keep ΔS at 26 entropy units, the entropy of the gas formed must go up by as much as the entropy of the solution was higher than the pure liquid. That can be done by *lowering the vapor pressure* at equilibrium and requiring each mole of gas to occupy more volume. And the amount of extra volume needed is determined by the amount of entropy increase in the solution compared to the pure liquid.

You may recall that in Section 10–5.1 we used a model of molecules getting in the way to explain the lowering of vapor pressure in solutions. But the model had flaws when examined closely. Why should tiny Na^+ ions interfere as much as large sugar molecules? Now we can offer a better explanation based on thermodynamics. A tiny Na^+ ion can contribute as much to the entropy (by the uncertainty about where it is in the solution) as a large sugar molecule, and our illustration in the previous paragraph shows that entropy is the determining factor.

Because the thermodynamic terms can be handled in precise mathematical equations, many specific numerical relationships can be calculated from them. For instance, we could go on with the illustration above and derive Raoult's law, which was shown by a graph in Section 10–5.1. Or we could use the relationship of entropy to concentrations (in gas or liquid solutions) and $\Delta G = 0$ to derive the law of mass action shown in Section 12–6. And these calculations, tied in to the basic tendencies of nature (energy degradation to heat lost and mixing), often give us much better agreement with natural facts than we can get from qualitative models such as our "solute gets in the way" vapor pressure lowering model.

C–2 Using K in More Complicated Problems: Simplifying Approximations

C–2.1 Simplifying Approximations. Calculations involving equilibrium constants can be mathematically simple in some cases and very complex in other cases. When the problems are complex much of the difficulty is often in parts of the calculation that have only small effects on the answer. In many of these problems it is possible to simplify the calculations by making **simplifying approximations.** Simplifying approximations involve substituting a false idea, which happens to allow easy calculation, for the true facts, which are hard to use in the calculation. A good simplifying approximation is one that saves a lot of work without causing much error in the result.

There are at least three reasons for the very frequent use of simplifying approximations in equilibrium problems. First, we need to save time and effort. This is not just a matter of laziness. Equilibrium calculations are involved in many real and practical problems, and we need solutions promptly so we can use the answers and go on to other real problems. A single mathematically perfect answer would not be enough if we needed to have finished solving a dozen important calculations in the same working time.

Second, the answers found by "perfect" mathematical calculations are not correct anyhow. All of our information comes from measurements that include some errors. It is useless (and we would be deceiving ourselves) to calculate an answer to ten significant figures from a K value that was not accurately known to more than two or three significant figures. Therefore, we accept errors from our calculations that are as large as our original measurement errors and sometimes we even accept somewhat larger calculation errors.

Third, we need to be aware of the limited accuracy of our answers. When we know we have introduced some errors in our calculations, it is easier for us to keep alert to all the sources of error. We also tend to be more aware of the need to worry about how large the errors are. If the errors become too large, we can then try to find new ways to obtain more accurate answers.

In this section we are going to work examples of a number of common problem types using simplifying approximations. In each category we will discuss first what usually helps and second the conditions when the simplification fails. In describing the ways to simplify problems we will have to identify the changes that have a significant effect on the mathematics and we will have to notice which approximations we are actually using in each case. When we look at the failures we will be forced to consider the sources of error and the accuracy needed in our answers. These basic questions (essentially "What are you doing?" and "What do you have when you finish?") are probably much more important to a science student than the mere ability to solve a particular problem. Therefore, our effort to push you to think about these questions in the following sections is not an accidental sidelight to the problem solving; concern about what the approximations are and when they work is the central issue you need to study here.

A conscious effort to "be lazy" and look for the easiest methods will bring up these important questions and teach us a lot about both equilibrium and accuracy in science.

C–2.2 Acid Ionizations.
Weak acids are only partially ionized. The extent of their ionization is often the point of interest in problems. In these problems we usually know the equilibrium constant and the amount of the weak acid available. We do not know the final (equilibrium) concentrations. All of the final concentrations can be set up in terms of one variable, so the problem is mathematically possible but rather time consuming. Here is an example of the setup for the "usual" acid ionization problem:

Example C–1. K_a for acetic acid ($HC_2H_3O_2$) is 1.85×10^{-5} M. Calculate [H^+] for a solution of 0.100 mole of $HC_2H_3O_2$ in 1.00 liter of solution.

Setting x as the [H^+] formed, we get the following *changes* in concentration as some of the molecules ionize.

	[$HC_2H_3O_2$]	[H^+]	[$C_2H_3O_2^-$]
original concentrations	0.100 M	—	—
change	$-x$	$+x$	$+x$
final concentrations	0.100 M $- x$	x	x

The K_a expression is

$$K_a = 1.85 \times 10^{-5} \, M = \frac{[H^+][C_2H_3O_2^-]}{[HC_2H_3O_2]} = \frac{(x)(x)}{(0.100 \, M - x)}$$

The x^2 term is not too bad, but the (0.100 M $- x$) term is what makes the problem a quadratic equation requiring a rather tedious solution in this form (all units are M)

$$1.85 \times 10^{-5} = \frac{x^2}{0.100 - x}$$

$$(1.85 \times 10^{-6}) - (1.85 \times 10^{-5} \, x) = x^2$$

$$x^2 + (1.85 \times 10^{-5} \, x) - (1.85 \times 10^{-6}) = 0$$

to be solved by quadratic formula

$$x = \frac{-b \pm \sqrt{b^2 - 4ac}}{2a}$$

The above problem could be simplified if we could *forget about subtracting x from 0.100 M*. Therefore, the **simplifying approximation** we want to try is to *ignore the effect of ionization* on the acid concentration. With that approximation we get

$$K_a = 1.85 \times 10^{-5} \, M = \frac{[H^+][C_2H_3O_2^-]}{[HC_2H_3O_2]} = \frac{(x)(x)}{0.100 \, M}$$

$$1.85 \times 10^{-5} \, M = \frac{x^2}{0.100 \, M}$$

$$x^2 = 1.85 \times 10^{-6} \, M^2$$

$$x = \sqrt{1.85 \times 10^{-6} \, M^2} = \sqrt{1.85} \times 10^{-3} \, M = 1.36 \times 10^{-3} \, M$$

The quadratic formula is not needed to solve the above problem with the simplifying assumption. Similar problems are usually solved by using the assumption that the extent of ionization can be ignored when writing the concentration of the acid.

There is also another simplifying assumption made in the acid ionization problem above. We assumed there was no original $[H^+]$ and that the final $[H^+]$ was equal to the extent of ionization of the acid, x. Actually, there is also $[H^+]$ from the ionization of water. Neutral water has 10^{-7} M H^+. After ionization of the acid there is somewhat less than 10^{-7} M H^+ from water. (The added H^+ causes a LeChatelier's principle shift of the water ionization equilibrium toward less ionization.) The actual $[H^+]$, coming from both the acid and water, can only be found by considering both equilibrium processes in a simultaneous equilibrium calculation such as those described in Section C–2.4. But in most acid ionizations the amount of H^+ from water is so small compared to the amount from the acid that the water contribution can be ignored. That simplifying approximation, *that the ionization of water can be ignored*, lets us avoid the complexities of simultaneous equilibrium calculations.

When the Approximations Fail. Each of the simplifying approximations in acid ionization calculations causes some error. If these errors are small compared to the accuracy we need, use of the approximations to save work is wise. If one (or both) of the errors is large enough to prevent the necessary accuracy, we need to rework the calculation more accurately. We can estimate the size of the errors by using our answer to determine how far from the truth each approximation was.

In the problem about ionization of 0.100 M $HC_2H_3O_2$, ignoring the ionization of water is a very good approximation. The $[H^+]$ of 1.36×10^{-3} M from the acid makes the contribution of H^+ from water quite insignificant. We know that the H^+ from water is less than 10^{-7} M. Actually, the H^+ from water equals the $[OH^-]$. In the presence of 1.36×10^{-3} M H^+ from the acid

$$K_w = [H^+][OH^-] = 10^{-14} \, M^2$$
$$1.36 \times 10^{-3} \, M \, y = 10^{-14} \, M^2$$

$$y = \frac{10^{-14} \, M^2}{1.36 \times 10^{-3} \, M} = 7.35 \times 10^{-12} \, M = \text{extent of ionization of water}$$

Therefore, the fractions of H^+ from water was only $(7.35 \times 10^{-12})/1.36 \times 10^{-3}$ or about 5.4×10^{-9}. Ignoring that contribution, 5.4 out of each billion, caused an error much smaller than the uncertainty in our original information. For instance, the 0.100 M $HC_2H_3O_2$ is only known within one part in 100.

Although the accuracy needed in answers varies depending on the use to be made of the result, we can choose to use 1% (one part in 100) as the largest error acceptable in problems where we have no stated accuracy requirement. The ionization of water will contribute less than 1% of the H^+

in acid ionizations (or OH^- in base ionizations) if the resulting pH is less than 6 or more than 8. Only when the pH is near 7 (between 6 and 8) will it be necessary to consider the ionization of water. We will work an example of the problem arising when pH is between 6 and 8 as part of Section C–2.4, Simultaneous Equilibrium.

The other simplifying assumption, ignoring the fact that some $HC_2H_3O_2$ was used up by the ionization, is not so successful. The extent of ionization (1.36×10^{-3} M) is more than 1% of the original acid (0.100 M).

$$\frac{1.36 \times 10^{-3} \, M \text{ ionized}}{0.100 \, M \text{ total}} = 1.36\% \text{ ionized}$$

In this case the error is so close to 1% we would probably choose to ignore it and accept the inaccuracy. But for learning purposes, let us proceed with reworking the problem as an example of what can be done if the error is too large to be allowed.

We can, of course, do the problem by the quadratic equation method. (At least the success of the other simplifying assumption on water ionization will save us from the mathematical horror of a quadratic as only one part of a simultaneous equilibrium problem.) But we are trying to "be lazy" so we can learn about possible shortcuts. There is, in fact, another way to avoid the quadratic solution. We can use **successive approximations.** In this method we use the answer from one attempt to make the approximation for a second try. Instead of assuming no reduction of the 0.100 M H $C_2H_3O_2$ (as we did in the **first approximation**), we will assume that the 0.100 M $HC_2H_3O_2$ is reduced by 1.36×10^{-3} M (the answer for extent of ionization found in the first approximation). The calculation then becomes the following **second approximation.**

$$K_a = \frac{[H^+][C_2H_3O_2^-]}{[HC_2H_3O_2]}$$

$$1.85 \times 10^{-5} \, M = \frac{(x)(x)}{0.100 \, M - (1.36 \times 10^{-3} \, M)}$$

$$1.85 \times 10^{-5} \, M = \frac{x^2}{0.09864 \, M}$$

$$x^2 = (1.85 \times 10^{-5})(0.09864 \, M^2)$$
$$= 1.825 \times 10^{-6} \, M^2$$
$$x = 1.825 \times 10^{-3} \, M$$
$$= 1.35 \times 10^{-3} \, M$$

We can check the error in this second approximation by comparing the new answer's effect on the approximation to the approximation we used.

$$\frac{\text{error}}{\text{total}} = \frac{(0.100 \, M - 1.35 \times 10^{-3} \, M) - (0.100 \, M - 1.36 \times 10^{-3} \, M)}{(0.100 \, M - 1.36 \times 10^{-3} \, M)} =$$

$$\frac{1 \times 10^{-5} \, M}{0.09864 \, M} = \frac{1}{9864} = \text{less than } 1\%$$

If the error was still greater than 1%, we could go on to a third approximation, and even more than three approximations can be made if needed. If the first approximation gives a fairly small error, a second approximation will usually give an acceptable answer and save work. If the first error is large, it may be easier to do one "exact" calculation than the many approximations that might be needed to get acceptable accuracy.

C-2.3   Buffer Problems.   Acid ionization calculations are simplified when the concentration of the acid remains essentially constant—unaffected by the ionization. The calculation becomes even simpler if *two* concentrations stay essentially constant. That situation can be accomplished if two of the concentrations have large amounts from sources other than the ionization and the extent of ionization is small.

In the case of a weak acid such as $HC_2H_3O_2$, the amount of nonionized acid put in is usually "large" compared to the "small" extent of ionization. The same situation can be accomplished for the negative ion, $C_2H_3O_2^-$ ion in the $HC_2H_3O_2$ case, by putting it in as a salt. A salt, such as $NaC_2H_3O_2$, will ionize completely. If that ionization provides a "large" amount of the negative ion compared to the "small" amount of ionization, the effect of the ionization on the negative ion concentration can be ignored as a simplifying approximation. We will also assume the contribution to H^+ from water is small enough to be ignored in all buffer problems. Even at pH values near 7, the ability of a buffer to "control" H^+ concentration makes ignoring the contribution of water a good simplifying approximation, as we will show below.

In cases where a weak acid and its salt (or a weak base and its salt) are present in large concentrations, the acidity (and basicity) is controlled at an almost constant level. These solutions are called **buffers** and their ability to control acidity makes them very useful.

Example C-2.   Calculate the conditions for a buffer solution made by adding 0.100 mole $HC_2H_3O_2$ and 0.100 mole $NaC_2H_3O_2$ to water making a liter of solution. The salt ionizes completely to give 0.100 M Na^+ and 0.100 M $C_2H_3O_2^-$ ion. We therefore have

	$HC_2H_3O_2$	H^+	$C_2H_3O_2^-$
original concentrations	0.100 M	—	0.100 M
change upon ionization	$-x$	$+x$	$+x$
final concentrations	0.100 $M - x$	x	0.100 $M + x$

The equilibrium expression is therefore

$$K_a = 1.85 \times 10^{-5} \, M = \frac{[H^+][C_2H_3O_2^-]}{[HC_2H_3O_2]} = \frac{x(0.100 \, M + x)}{(0.100 \, M - x)}$$

In that form, the problem would require a quadratic formula solution. But we can simplify by using the simplifying approximations that x can be ignored compared to 0.100 M. That gives us the following approximated final concentrations:

$$HC_2H_3O_2 = 0.100 \ M \quad H^+ = x \quad C_2H_3O_2^- = 0.100 \ M$$

And the equilibrium expression becomes

$$K_a = 1.85 \times 10^{-5} \ M = \frac{[H^+][C_2H_3O_2^-]}{[HC_2H_3O_2]} = \frac{(x)(0.100 \ M)}{0.100 \ M}$$

That is very easily solved as follows:

$$1.85 \times 10^{-5} \ M = \frac{(x)(0.100 \ M)}{0.100 \ M}$$

$$[H^+] = x = 1.85 \times 10^{-5} \ M$$

Notice that *when the weak acid and salt are equal in concentration, $[H^+]$ is exactly equal to K_a*. We will see below that even when the acid and salt are unequal, a buffer always gives a value (for H^+ in acid-salt buffers, for OH^- in base-salt buffers) *near to the value of K*.

Example C–3. Consider the effect of adding 10.0 ml of 0.020 M NaOH to 10.0 ml of the buffer above (0.100 M $HC_2H_3O_2$ plus 0.100 M $NaC_2H_3O_2$). The NaOH would react with H^+ ions. Since H^+ concentration is almost held constant by the buffer, $HC_2H_3O_2$ will ionize to form H^+ in the amount needed. That will reduce the $HC_2H_3O_2$ concentration and form additional $C_2H_3O_2^-$ ions. The process would be

	$HC_2H_3O_2$	H^+	$C_2H_3O_2^-$
original concentration	0.100 M	$1.85 \times 10^{-5} \ M$	0.100 M
quantity	$0.100 \times 10 \ M$ ml	$1.85 \times 10^{-5} \times 10 \ M$ ml	$0.100 \times 10 \ M$ ml
change from reaction with 10.0 ml 0.020 m NaOH	$-0.020 \times 10 \ M$ ml	?	$+0.020 \times 10 \ M$ ml
final amount in 20 ml	$(0.100 \times 10 \ M$ ml $-0.020 \times 10 \ M$ ml $= 0.080 \times 10 \ M$ ml	?	$(0.100 \times 10 \ M$ ml $+ 0.020 \times 10 \ M$ ml) $= 0.120 \ M$ ml
final concentration	$\dfrac{0.080 \times 10 \ M \ \text{ml}}{20 \ \text{ml}}$ $= 0.040 \ M$	x	$\dfrac{0.120 \times 10 \ M \ \text{ml}}{20 \ \text{ml}}$ $= 0.060 \ M$

$$K = \frac{[H^+][C_2H_3O_2^-]}{[HC_2H_3O_2]} = \frac{(x)(0.060 \ M)}{0.040 \ M} = x\tfrac{3}{2}$$

$$x = K\tfrac{2}{3} = (1.85 \times 10^{-5} \ M)\tfrac{2}{3} = 1.23 \times 10^{-5} \ M$$

[C–2] Using *K* in More Complicated Problems: Simplifying Approximations

Therefore, addition of 10.0 ml of 0.020 M NaOH to 10.0 ml of 0.100 M $H_2C_2H_3O_2$ plus 0.100 M $NaC_2H_3O_2$ buffer changes H^+ concentration only from 1.85×10^{-5} M to 1.23×10^{-5} M. The success of the buffer in preventing large changes in acidity or basicity is shown by comparing that small change to the effect of the same NaOH solution added to pure water.

Example C–4. Consider the effect of adding 10.0 ml of 0.020 M NaOH to 10.0 ml of H_2O.

	H^+	OH^-
original concentrations	10^{-7} M	10^{-7} M
effect of adding 10 ml of 0.020 M NaOH		
— added amount	—	0.020×10 M ml
final volume	20 ml	20 ml
final concentration	x	$\dfrac{0.020 \times 10\ M\ \text{ml}}{20\ \text{ml}}$ $= 0.010\ M = 10^{-2}\ M$

$$K_w = [H^+][OH^-] = (x)(10^{-2}\ M)$$
$$= 10^{-14}\ M^2 = (x)(10^{-2}\ M)$$

$$x = \frac{10^{-14}\ M^2}{10^{-2}\ M} = 10^{-12}\ M$$

$$[H^+] = 10^{-12}\ M$$

The H^+ concentration in water is reduced to $1/10^5$ as much (one one hundred thousandth) by the same NaOH which only reduces H^+ concentration in the buffer to two thirds as much as the original.

When we listed the simplifying approximations at the beginning of this section, we stated that the contribution of water could always be safely ignored. We can now prove that statement by calculating the effect of water on a buffer at pH = 7 (the condition where the ionization of water is at its maximum). If we had a buffer of a weak acid (which we will call HA) with $K_a = 10^{-7}$ M, even a very dilute buffer solution would probably have at least 10^{-2} M concentrations of HA and of a salt such as NaA.

Example C–5. Consider the effects of approximations on the H^+ concentration of 10^{-2} M HA and 10^{-2} M NaA given $K_a = 10^{-7}$ M. Using our assumptions we get

$$K_a = 10^{-7}\ M = \frac{[H^+][A^-]}{[HA]} = \frac{[H^+](10^{-2}\ M)}{10^{-2}\ M}$$

$$[H^+] = 10^{-7}\ M$$

But when we do not make any simplifications, we get

$$K_a = 10^{-7} M = \frac{[H^+][A^-]}{[HA]} = \frac{(x+y)(10^{-2} M + x)}{(10^{-2} M - x)}$$

where $x =$ H$^+$ from HA and $y =$ H$^+$ from H$_2$O.

If x is still negligible compared to 10^{-2} M, that gives

$$10^{-7} M = \frac{(x+y)\, 10^{-2} M}{10^{-2} M}$$

$$(x+y) = [H^+] = 10^{-7} M$$

No matter what the amount of H$^+$ from H$_2$O, the amount from HA simply shifts enough to keep the [H$^+$] at 10^{-7} M. In this particular case, x becomes zero. But even when x is not zero, y (which can never be more than 10^{-7} M) cannot make x become large enough to change [HA] and [A$^-$] enough to change the answer significantly. Every other case where the buffer is working as a buffer gives the same result: the H$^+$ contribution from water does not significantly affect the result.

When the Approximations Fail. Buffer problems do have two simplifying approximations that can fail. First, we assume that the extent of reaction, x, does not change the concentration of the acid (or base) that has not ionized. Second we assume that the extent of reaction, x, does not change the concentration of the ion needed to complete the buffer.

The most obvious way for these approximations to fail is to have too little of the acid or the salt used as a source of the ions.

If there is a large amount of a weak acid but only a small amount (compared to the ionization) of the salt, we have the same sort of acid ionization problem discussed in Section C–2.3.

Example C–6. Calculate H$^+$ for a solution which is 1.00 M HC$_2$H$_3$O$_2$ ($K_a = 1.85 \times 10^{-5}$ M) and 10^{-6} M NaC$_2$H$_3$O$_2$. We would have (with every quantity in molarity units)

	H^+	$C_2H_3O_2^-$	$HC_2H_3O_2$
original	—	10^{-6}	1.00
change	$+x$	$+x$	$-x$
final	x	$(10^{-6} + x)$	$(1.00 - x)$

Here x is clearly *not* going to be small compared to 10^{-6} M, so we cannot make the "buffer" assumptions. We can ignore x compared to 1.00 M and get the following acid ionization problem requiring a quadratic solution.

$$K_a = 1.85 \times 10^{-5} M = \frac{(x)(10^{-6} + x)}{1.00} M$$

If an approximation is made with the quantity $(10^{-6} + x)$, the number to be ignored must be the 10^{-6} because that is smaller than x in this case. We could test for that possibility by trying the solution ignoring one part of the $(10^{-6} + x)$ at a time. Ignoring x compared to 10^{-6} gives

$$K_a = 1.85 \times 10^{-5}\,M = \frac{(x)(10^{-6}\,M)}{1.00\,M}$$

$$x = \frac{1.85 \times 10^{-5}\,M}{10^{-6}} = 18.5\,M \quad \text{a ridiculous answer}$$

Because $18.5\,M$ is *not* negligible compared to $10^{-6}\,M$, that approach must be discarded. Ignoring 10^{-6} compared to x gives

$$K_a = 1.85 \times 10^{-5}\,M = \frac{(x)(x)}{1.00\,M}$$

$$x^2 = 1.85 \times 10^{-5}\,M^2$$
$$x = 4.3 \times 10^{-3}\,M$$

Checking our assumption, we find that $10^{-6}\,M$ is small compared to $4.3 \times 10^{-3}\,M$, so the answer is acceptable and we do not have to solve for the answer by the quadratic equation. That assumption works only because the $10^{-6}\,M$ $NaC_2H_3O_2$ is so completely ineffective in causing any buffering.

A low concentration of the weak acid (or weak base in a base-ion pairing) can also lead to loss of buffering. In that case we would have a hydrolysis problem (see Section B-3 for a description of hydrolysis).

Example C-7. Consider a solution containing $10^{-2}\,M$ $NaC_2H_3O_2$ and only $10^{-6}\,M$ $HC_2H_3O_2$. If the approximations used for buffers held, we would have

$$K_a = 1.85 \times 10^{-5}\,M = \frac{[H^+][C_2H_3O_2^-]}{[HC_2H_3O_2]} = \frac{[H^+]\,10^{-2}\,M}{10^{-6}\,M}$$

$$[H^+] = 1.85 \times 10^{-9}\,M$$

That would be a basic solution, showing the dominance of hydrolysis by $C_2H_3O_2^-$ over ionization by $HC_2H_3O_2$. The net amount of hydrolysis is shown by the OH^- concentration formed, which we can calculate from K_w and the H^+ concentration as follows:

$$K_w = 10^{-14}\,M^2 = [H^+][OH^-] = (1.85 \times 10^{-9}\,M)[OH^-]$$

$$[OH^-] = \frac{10^{-14}\,M^2}{1.85 \times 10^{-9}\,M} = \frac{1.0 \times 10^{-15}}{1.85 \times 10^{-9}}M = 5.4 \times 10^{-6}\,M$$

The contribution from water to OH^- must be equal to the H^+ remaining from water ($1.85 \times 10^{-9}\,M$), so the $5.4 \times 10^{-6}\,M$ must be almost entirely from the hydrolysis of $C_2H_3O_2^-$. But our "buffer" setup assumed that the $HC_2H_3O_2$ concentration did not change significantly from $10^{-6}\,M$. An answer requiring $5.6 \times 10^{-6}\,M$ of $HC_2H_3O_2$ to be formed in the hydrolysis is inconsistent with that assumption. Therefore, treating the solution as a buffer is shown to be invalid.

If instead we set the same problem up as a hydrolysis problem, we have (ignoring ionization of water)

$$K_h = \frac{K_w}{K_a} = \frac{[HC_2H_3O_2][OH^-]}{[C_2H_3O_2^-]}$$

(All values in molarity)	$C_2H_3O_2^-$	$HC_2H_3O_2$	OH^-
original concentration	10^{-2}	10^{-6}	—
change	$-x$	$+x$	$+x$
final concentration	$10^{-2} - x$	$10^{-6} + x$	x

That gives (with all numbers consistent with use of molarity units)

$$K_h = \frac{10^{-14}}{1.85 \times 10^{-5}} = \frac{(10^{-6} + x)(x)}{(10^{-2} - x)}$$

Because we already know (from our attempt using buffer assumptions) that x is somewhere near 10^{-6} M, we can see that x can be ignored compared to 10^{-2} M, but not compared to 10^{-6} M. That leaves us the following form requiring quadratic solution. (This problem is worked as an example of quadratic solutions.)

$$\frac{10^{-14}}{1.85 \times 10^{-5}} = \frac{(10^{-6} + x)(x)}{10^{-2}}$$

$$5.4 \times 10^{-10} = \frac{(10^{-6} + x)(x)}{10^{-2}}$$

$$5.4 \times 10^{-12} = (10^{-6} + x)(x) = 10^{-6}x + x^2$$

$$x^2 + 10^{-6}x - 5.4 \times 10^{-12} = 0$$

substituting in the quadratic formula

$$x = \frac{-b \pm \sqrt{b^2 - 4ac}}{2a}$$

$$x = \frac{-10^{-6} \pm \sqrt{(10^{-6})^2 + 4 \times 1 \times 5.4 \times 10^{-12}}}{2.1}$$

$$= \frac{-10^{-6} \pm \sqrt{10^{-12} + 21.6 \times 10^{-12}}}{2}$$

$$= \frac{-10^{-6} \pm \sqrt{22.6 \times 10^{-12}}}{2}$$

$$= \frac{-10^{-6} \pm \sqrt{22.6} \times 10^{-6}}{2}$$

$$= \frac{-10^{-6} \pm 4.75 \times 10^{-6}}{2}$$

Of the positive and negative choice, one is the answer and the other has to be impossible. If the sign is minus

$$x = \frac{-10^{-6} - 4.75 \times 10^{-6}}{2} = -2.87 \times 10^{-6}$$

That is impossible because it would make $HC_2H_3O_2$ less than nothing. If the sign is plus

$$x = \frac{-10^{-6} + 4.75 \times 10^{-6}}{2} = \frac{+3.75 \times 10^{-6}}{2} = 1.87 \times 10^{-6}$$

so $[OH^-] = 1.87 \times 10^{-6} M$.

The above examples showed how buffering could fail if one of the two necessary "large" concentrations was not large enough compared to the extent of reaction. Although we used a weak acid and its salt, similar cases could be created with a weak base and its salt. In those cases (weak acid and salt or weak base and salt), the solutions *could* be buffers if the concentrations were large enough. But some other combinations of an acid and its salt (or base and its salt) cannot ever be buffers because the acid (or base) is not weak. For example, a solution of the strong acid HCl and NaCl is never a buffer. The HCl is always completely ionized (or at least so close to completely ionized that we cannot tell the difference). The extent of ionization is never negligible compared to the HCl not ionized. The combination of the strong base NaOH and NaCl would be similar: the NaOH would always ionize completely. The H^+ and OH^- concentrations of these solutions are easy to calculate because no equilibrium needs to be considered for the strong acid or strong base. But the same absence of an equilibrium needing consideration makes these combinations unable to serve as buffers.

C-2.4 Simultaneous Equilibria. One of our common simplifying approximations has been to ignore the contribution of water to H^+ or OH^- concentrations. If that approximation fails, we are left with a problem involving two related equilibria simultaneously. In other cases, even when water's contribution can be ignored, we may have simultaneous equilibria of two weak acids or two weak bases. Solubility problems (see Section C-2.5) may also involve simultaneous equilibria.

When two equilibria are interacting with each other, there are two equilibrium constant expressions. Those two expressions give us two equations, and the combination of two equations can be solved for two unknowns. The two unknowns are usually the extent of the first equilibrium reaction and the extent of the second equilibrium reaction. Solution usually involves two parts. First, we must solve for a relationship between the two unknowns. Then we must use that relationship to replace one of the unknowns in an equilibrium constant expression (either one of the two) and solve for the other unknown. Some examples follow.

First let us consider a case where water ionization is not negligible but there is only one equilibrium to be considered. In this case we will not need simultaneous solution of two equilibrium constant expressions.

Example C-8. Calculate the H^+ concentration of a $10^{-7}\ M$ HCl solution. HCl is a strong acid, so it will ionize completely to give us $10^{-7}\ M\ H^+$. No equilibrium constant needs to be considered to reach that result. But the $10^{-7}\ M\ H^+$ concentration would give us a pH of 7. That is in the range (pH = 6 to pH = 8) where the ionization of water will account for more than 1% of the actual H^+ concentration. Therefore the equilibrium constant for water ionization must be considered. The addition of $10^{-7}\ M\ H^+$ from the HCl will shift the $H_2O \rightleftharpoons H^+ + OH^-$ reaction toward less H^+ and OH^- from water ionization than the $10^{-7}\ M$ values in pure water. To find out how much less, we must solve the equation

$$K_w = 10^{-14}\ M^2 = [H^+][OH^-]$$

If we call the extent of water ionization x in this case, OH^- concentration equals x and H^+ concentration equals the sum of x (from water) plus $10^{-7}\ M$ (from HCl). Therefore

$$10^{-14}\ M^2 = (x + 10^{-7}\ M)(x)$$
$$x^2 + 10^{-7}\ Mx - 10^{-14}\ M^2 = 0$$

$$x = \frac{-b \pm \sqrt{b^2 - 4ac}}{2a} = \frac{-10^{-7}\ M \pm \sqrt{10^{-14}\ M^2 - 4(1)(-10^{-14}\ M^2)}}{2 \times 1}$$

$$x = \frac{-10^{-7}\ M \pm \sqrt{5 \times 10^{-14}\ M^2}}{2}$$

since $x = [OH^-]$, which cannot be less than zero

$$x = \frac{-10^{-7}\ M - \sqrt{5 \times 10^{-14}\ M^2}}{2}$$

is impossible so

$$x = \frac{-10^{-7}\ M + \sqrt{5 \times 10^{-14}\ M^2}}{2}$$
$$= \frac{-10^{-7}\ M + \sqrt{5} \times 10^{-7}\ M}{2} = \frac{(\sqrt{5} - 1) \times 10^{-7}\ M}{2}$$
$$= \frac{\sqrt{5} - 1}{2} \times 10^{-7}\ M = \frac{2.235 - 1}{2} \times 10^{-7}\ M = \frac{1.235}{2} \times 10^{-7}\ M$$
$$= 6.18 \times 10^{-8}\ M \quad \text{or} \quad 0.618 \times 10^{-7}\ M$$

Therefore $[H^+] = 10^{-7} + x = 10^{-7} + 0.618 \times 10^{-7}\ M$

$$[H^+] = 1.618 \times 10^{-7}\ M$$

Next, let us consider a case where the ionization of water is important and the other source of H$^+$ (or OH$^-$) is also an equilibrium that must be considered. In that case the ionization of water will be affected by the other reaction (as it was above by the H$^+$ from HCl) and the other equilibrium will be similarly affected by the H$^+$ (or OH$^-$) from water. This is a true simultaneous equilibrium problem. Let us consider a case involving two sources of H$^+$ ions, the weak acid HCN and H$_2$O.

Example C-9. Calculate the H$^+$ concentration of a 10^{-4} M HCN solution in water. K_a for HCN is 7×10^{-10} M.

A quick check of the approximations usually used for K_a problems shows that H$_2$O ionization cannot be ignored.

$$K_a = 7 \times 10^{-10} \, M = \frac{[\text{H}^+][\text{HCN}]}{[\text{HCN}]} = \frac{(x)(x)}{10^{-4} \, M}$$

$$x^2 = 7 \times 10^{-14} \, M^2$$
$$x = 7 \times 10^{-7}$$

This is not acceptable because it is too close to 10^{-7} M to ignore H$_2$O ionization.

Therefore, we must proceed to define *two* unknowns and solve for one in terms of the other. Let us call the extent of HCN ionization x and the extent of water ionization y. We then get the following

$$[\text{H}^+] = x \text{ (from HCN)} + y \text{ (from H}_2\text{O)}$$
$$[\text{OH}^-] = y \text{ (from H}_2\text{O, none from HCN)}$$
$$[\text{CN}^-] = x \text{ (from HCN, none from H}_2\text{O)}$$
$$[\text{HCN}] = 10^{-4} \, M - x$$

Because x is near 10^{-7} M, we can ignore x compared to 10^{-4} M without causing as much as 1% error, so we will use

$$[\text{HCN}] = 10^{-4} \, M$$

The two equilibrium constant expressions are therefore

$$K_a = 7 \times 10^{-10} \, M = \frac{[\text{H}^+][\text{CN}^-]}{[\text{HCN}]} = \frac{(x+y)(x)}{10^{-4} \, M}$$

$$K_w = 10^{-14} \, M^2 = [\text{H}^+][\text{OH}^-] = (x+y)(y)$$

Multiplying each side of the K_a expression by 10^{-4} M gives

$$(7 \times 10^{-10} \, M)(10^{-4} \, M) = \frac{(x+y)(x)}{10^{-4} \, M} \, 10^{-4} \, M$$

$$7 \times 10^{-14} \, M^2 = (x+y)x$$

In all simultaneous equilibrium problems there is a factor common to both equilibria. In this case it is the H$^+$ concentration (equal to $x + y$). If that common term can be eliminated, the problem will be greatly simplified to x terms and y terms with no terms containing both x and y. That simplification is done by *dividing* one of the equations by the other as follows

$$\frac{7 \times 10^{-14} \, M^2 = (x+y)(x)}{1 \times 10^{-14} \, M^2 = (x+y)(y)}$$

$$7 = \frac{x}{y} \quad \text{or} \quad x = 7y$$

All simultaneous equilibrium problems can be simplified by such a division. The result of that division can then be used to solve one of the equations.

$$1 \times 10^{-14} \, M^2 = (x+y)(y)$$
$$1 \times 10^{-14} \, M^2 = (7y+y)(y) = (8y)(y) = 8y^2$$
$$y^2 = \frac{1 \times 10^{-14} \, M^2}{8} = 0.125 \times 10^{-14} \, M^2$$

or
$$1.25 \times 10^{-15} \, M^2 \quad \text{or} \quad 12.5 \times 10^{-16} \, M^2$$
$$y = \sqrt{12.5 \times 10^{-16} \, M^2} = \sqrt{12.5} \times 10^{-8} \, M$$
$$= 3.53 \times 10^{-8} \, M$$
$$x = 7y = 24.71 \times 10^{-8} \, M = 2.47 \times 10^{-7} \, M$$
$$x + y = [\text{H}^+] = 8y = 2.82 \times 10^{-7} \, M$$

In order to show that the procedures for simultaneous equilibria are general and not dependent on water ionization being one of the equilibria, let us do one more example involving two weak acids simultaneously.

Example C–10. Calculate the H$^+$ concentration of a solution (in water) which is $1.00 \, M$ in HCN and $0.100 \, M$ in NH$_4$Cl. K_a for HCN is $7 \times 10^{-10} \, M$ and K_b for NH$_3$ is $1.8 \times 10^{-5} \, M$.

In this problem NH$_4^+$ ions will be acidic by hydrolysis, so there is a simultaneous production of H$^+$ from two weakly acidic sources, NH$_4^+$ and HCN. Let us call the extent of HCN ionization x and the extent of NH$_4^+$ hydrolysis y. We then get the following

$$[\text{H}^+] = x + y + \text{H}_2\text{O contribution} = \text{approx. } x + y$$
$$[\text{CN}^-] = x$$
$$[\text{HCN}] = 1.00 \, M - x = \text{approx. } 1.00 \, M$$
$$[\text{NH}_3] = y$$
$$[\text{NH}_4^+] = 0.100 \, M - y = \text{approx. } 0.100 \, M$$

$$K_a = 7 \times 10^{-10} \, M = \frac{[\text{H}^+][\text{CN}^-]}{[\text{HCN}]} = \frac{(x+y)x}{1.00 \, M}$$

$$7 \times 10^{-10} \, M^2 = (x+y)x \tag{A}$$

$$K_h = \frac{K_w}{K_b} = \frac{[\text{H}^+][\text{NH}_3]}{[\text{NH}_4^+]}$$

$$\frac{10^{-14} \, M^2}{1.8 \times 10^{-5} \, M} = \frac{(x+y)(y)}{0.100 \, M}$$

$$\frac{10^{-15} \, M^2}{1.8 \times 10^{-5} \, M} = (x+y)y$$

$$0.556 \times 10^{-10} M = (x+y)y \qquad (B)$$

dividing equation A by equation B

$$\frac{7 \times 10^{-10} M}{5.56 \times 10^{-11} M} = \frac{(x+y)x}{(x+y)y}$$

$$\frac{x}{y} = \frac{7}{5.56 \times 10^{-1}} = 1.26 \times 10$$

$$x = 12.6\, y$$

substituting in equation B

$$5.56 \times 10^{-11}\, M^2 = (x+y)y$$
$$5.56 \times 10^{-11}\, M^2 = (12.6\, y + y)y = 13.6\, y^2$$
$$y^2 = \frac{5.56}{13.6} \times 10^{-11}\, M^2 = 0.409 \times 10^{-11}\, M^2 = 4.09 \times 10^{-12}$$
$$y = 2.02 \times 10^{-6}\, M$$
$$[H^+] = x + y = 13.6y = (13.6)(2.02 \times 10^{-6}\, M) = 2.75 \times 10^{-5}\, M$$

C-2.5 Solubility Calculations. A common question about solubility is how much of a given substance can dissolve in a given amount of solution. For ionic salts there is more than one concentration involved, and the solubility can be affected by the presence of another source of one or more of the ions. The solubility product constant, K_{sp}, allows calculation of the resulting solubilities.

If a salt is added to pure water, the amounts of each ion present at equilibrium (saturated solution) are in a fixed ratio to each other according to the coefficients of the balanced equation. That fact can be quickly verified by setting up a table of the original concentrations, change, and final concentrations such as this one for the reaction $CaCl_2(solid) \rightleftharpoons Ca^{2+} + 2\, Cl^-$.

	$[Ca^{2+}]$	$[Cl^-]$
original concentration (pure water)	0	0
change	$+x$	$+2x$
final concentration	x	$2x$

When those concentrations are substituted into the K_{sp} expression, the same coefficients appear again as exponents on the concentrations. For the $CaCl_2$ case the result is

$$K_{sp} = [Ca^{2+}][Cl^-]^2 = x(2x)^2 = x(2x)^2 = (x)(4x^2) = 4x^3$$

Some people find the double use of the coefficients confusing, but it occurs because there are two distinctly different ways the coefficients are used. One (the use as exponents) is always present because of dependence of equilibrium constants (even special forms like K_{sp}) on the law of mass action rules (Section 12–6). The other use of the coefficients comes from the dependence of the change during reaction on the coefficients. If, as in the example above, the change is the only source of the final concentrations, the final concentrations show the ratios of the coefficients.

When a salt is added to a solution already containing one of the ions of the salt, the total concentrations of each ion (original plus the amount from dissolving salt) must be used.

Example C–11. If solid $CaCl_2$ was added to 0.100 M NaCl solution, find the solubility limit.

Call the concentration of $CaCl_2$ that will dissolve (in molarity) x

$$[Ca^{2+}] = x$$
$$[Na^+] = 0.100\ M$$
$$[Cl^-] = 0.100\ M + 2x$$
$$K_{sp}(\text{of } CaCl_2) = [Ca^{2+}][Cl^-]^2 = x(0.100\ M + 2x)^2$$
$$K_{sp} = x(0.0100\ M^2 + 0.400\ Mx + 4x^2)$$
$$K_{sp} = 0.0100\ M^2 x + 0.400\ Mx^2 + 4x^3$$

That problem can be solved mathematically, but only with difficulty. It would be much easier if we could use the simplifying approximation that $2x$ is negligible compared to 0.100 M or that 0.100 M is negligible compared to $2x$. We are not going to give you the value for K_{sp} and work out an answer to this example. But in problems like this one it would probably be well worth the time to check each of the possible simplifying approximations and, if necessary, try second or even third approximations on the better of the simplifying approximations to get an acceptable answer without having to solve the exact equation.

Solubility problems can also involve simultaneous equilibria.

Example C–12. Calculate the solubility of AgCl and AgBr when both solids are added to the same sample of water. Using x as the solubility of AgCl and y as the solubility of AgBr, we would conclude

$$K_{sp}\ AgCl = [Ag^+][Cl^-] = (x + y)x$$
$$K_{sp}\ AgBr = [Ag^+][Br^-] = (x + y)y$$

Dividing one K_{sp} equation by the other would begin a solution very similar to the simultaneous equilibria problems in Section C–2.4.

Since K_{sp} problems do not introduce any new types of approximations, we have elected not to work out any complete numerical solutions here. We feel they would be repetitive of the principles in the earlier sections. If your instructor wants you to repeat these principles to gain skill at solving ionic equilibrium problems, he may give you further examples, including a number of K_{sp} problems. If so, he will no doubt give you additional help and explanations, and may assign one of the chemistry problem manuals covering this area as a supplemental text.

Skills Expected at This Point

1. You should be able to state the thermodynamic basis for equilibrium at constant pressure and temperature, identify each contributing thermodynamic term, and describe the nature of each of the tendencies toward reaction.
2. You should be able to describe the characteristics of a good simplifying approximation and list the simplifying approximations commonly used in any of the types of problems described in Section C–2.
3. You should be able to solve acid or base ionizations, buffer problems, simultaneous equilibria, and solubility problems using the common simplifying approximations where applicable, and you should be able to check whether the approximations were reasonable.
4. You should be able to describe the successive approximation method of problem solving and show how you would put the numbers in for a second approximation calculation.

Exercises

Based on "light reading"
1. **(a)** What are the two factors causing tendency to react at constant pressure and temperature? **(b)** What is free energy? **(c)** What is the free energy condition for equilibrium?
2. List any one simplifying approximation and describe when and how it would be useful.

Based on mastery
3. Write equations for each of the following: **(a)** definition of enthalpy; **(b)** how enthalpy and entropy changes each are related to heat flow; **(c)** the condition for equilibrium at constant pressure and temperature.
4. The heat of fusion of H_2O is 80 cal/g at 0°C. Calculate ΔS for melting 1 mole of H_2O at 0°C.
5. K_a for HCN is 7.0×10^{-10} M. Calculate the pH of a 0.143 M HCN solution.
6. Calculate the pH if the same HCN solution in problem 5 had enough NaCN solid dissolved in it to make it 1.00×10^{-3} M in NaCN.
7. Calculate the OH^- concentration in a solution which is 0.500 M in NH_3 and 0.100 M in NaCN. K_b for NH_3 is 1.8×10^{-5}. K_a for HCN is 7.0×10^{-10}. (*Hint:* Start by converting K_a for HCN into the constant you need to work with CN^- instead of HCN.)

Note: The relatively brief coverage in this book and limited number of examples are aimed primarily at people just surveying the subject. If you really must master equilibrium calculations, you must work *many* more sample problems. Chemistry problem books have hundreds of these problems.

Test Yourself

1. Define each of the following terms: **(a)** free energy; **(b)** the units needed for entropy; **(c)** buffer.

2. From data tables we determine that a certain reaction A + B $\rightleftharpoons$ C + D would release 2,550 cal/mole of A reacted at 25°C and would cause a decrease of 9.25 cal/mole deg in entropy per mole of A reacted at 25°C if all concentrations are equal. Starting with equal concentrations, does the reaction:
   (a) go (A + B $\longrightarrow$ C + D)
   (b) go backwards (C + D $\longrightarrow$ A + B)
   (c) go nowhere (A, B, C, and D at equilibrium)
3. (a) Calculate the extent of ionization of a 0.5 M NH_3 solution. K_b for NH_3 is 1.8×10^{-5} M.
   (b) Calculate the H^+ concentration in the solution.
4. Calculate how much $NaC_2H_3O_2$ must be added to 1 liter of 10^{-3} M $HC_2H_3O_2$ to make a neutral solution (pH = 7). K_a for $HC_2H_3O_2$ is 1.85×10^{-5} M.
5. K_{sp} for AgCl is approximately 10^{-13} M^2. From that we can calculate that the solubility of AgCl when 0.100 mole of AgCl is placed in 1.00 liter of 0.100 M NaCl solution is
   (a) 10^{-12} M (c) $\sqrt{10^{-13}}$ M (e) none of these
   (b) 10^{-7} M (d) 10^{-13} g/ml
6. In problem 5, there is one simplifying assumption that *makes the problem easier* to solve and which causes very little error. What is that simplifying assumption?
   (a) The contribution from water is small and can be ignored.
   (b) The ionization of AgCl reduces the AgCl concentration but by so little that the reduction can be ignored.
   (c) NaCl has no effect on the AgCl solubility.
   (d) The contribution to Cl^- from AgCl is so small compared to the Cl^- from NaCl that the AgCl contribution can be ignored.
   (e) None of these.

Nomenclature Variations: Acids and Compounds Containing Only Nonmetals

APPENDIX D

D–1 Nomenclature in This Book

The last part, pages 25–28, of Section 2–6 describes the basic system of inorganic nomenclature and Chapter 17 describes organic nomenclature. The naming variations given here cover the other areas of inorganic nomenclature which are used in the text. Further variations would be needed to name the complexes discussed in section 16–3 but have been omitted as nonessential to this book.

D–2 Acid Nomenclature

As noted in section 13–4, water is necessary for compounds such as HCl to act as acids in the usual way. Therefore, we assign each acid compound its conventional name when present without water and a new acid name when in water. The acid names are related to the compound names as follows:

1. No mention is made of the positive part of the compound, which is hydrogen in every case.
2. The negative part is changed by keeping the root portion of the name but changing the ending.
   (a) All *-ate* endings are changed to *-ic acid*.
   (b) All *-ite* endings are changed to *-ous acid*.
   (c) Names ending in *-ide* have a prefix, *hydro-* inserted in front of the root name and also have the ending changed from *-ide* to *-ic acid*. Table D–1 lists some examples of how the system works.

Table D–1 Acid Names

Compound	Name Outside of Water	Name in Water (Acid Name)
H_2SiO_3	hydrogen silicate	silicic acid
NHO_2	hydrogen nitrite	nitrous acid
HCl	hydrogen chloride	hydrochloric acid

The naming of acids is quite regular in the way the endings are changed, but there is some inconsistency in the root names used. Two of the elements which form particularly important acids have longer root names used in acid naming than in ordinary compound naming. In both elements (sulfur and phosphorus) two extra letters from the element name are left in when acids are named. For sulfur, that makes the root name the entire element name instead of *sulf-*; for phosphorus, the acid root name is *phosphor-* instead of *phosph-*. Therefore, hydrogen sulfate in water is sulfuric acid, hydrogen sulfite in water is sulfurous acid and hydrogen sulfide in water is hydrosulfuric acid. The most common phosphorus acid, phosphoric acid, corresponds to the compound name hydrogen phosphate.

D–3 Nomenclature of Nonmetal Compounds

The nomenclature system of Chapter 2 is poor for compounds between two nonmetals because neither is very positive. Often several different compounds can form via covalent bonding, so a method to tell them apart is needed. The systematic approach via listing the different positive oxidation states fails partly because the "positive" part is not at all like a clear cut positive unit and partly because covalent bonding may lead to more than one compound with the same elements and the same apparent oxidation states. The compounds NO_2 and N_2O_4 (very different substances) illustrate the difficulty. Both would be named nitrogen (IV) oxide and neither has anything that is even close to a $+4$ ion.

The dilemma of nonmetal compound names is solved by simply rewriting the formulas in words. The number of atoms is indicated by a prefix on each part. The prefixes are based on greek numbers and are also used elsewhere when counting is useful in names. These prefixes, which are listed in Table D–2, are found in many organic chemical names in Chapter 17. If we gave the system for naming complexes, you would find these same prefixes are part of that system. Table D–3 shows the system used to assign the systematic names to the oxides of nitrogen.

Table D–2
Common Prefixes Used for Counting

Number	Prefix	Number	Prefix
one	mono-*	six	hexa-
two	di-	seven	hepta-
three	tri-	eight	octa-
four	tetra-	ten	deca-
five	penta-	twelve	dodeca-

Note: If the thing being counted starts with an a or o, the prefixes ending in a or o drop that final vowel.
*Used if the number one is "unusual", as in carbon monoxide. In other cases the absence of a prefix means one.

Table D–3
The Oxides of Nitrogen

Formula	Name
N_2O	dinitrogen oxide
NO	nitrogen oxide
N_2O_3	dinitrogen trioxide
NO_2	nitrogen dioxide
N_2O_4	dinitrogen tetroxide
N_2O_5	dinitrogen pentoxide

Answers to Exercises

Chapter 2

1. (a) water evaporating, ice melting; **(b)** wood burning, iron rusting
2. Any single substance produced by the chemical linkage of two or more elements is a compound **3.** C, carbon; Si, silicon; Na, sodium **4.** $KClO_3$, potassium chlorate; Cu_2O, copper(I) oxide **5.** $Pb(C_2H_3O_2)_2$; Fe_2O_3
6. $NCl_3 + 3\ H_2O \longrightarrow NH_3 + 3\ HOCl$ **7. (a)** $2\ CO + O_2 \longrightarrow 2\ CO_2$;
(d) $2\ C_2H_6 + 7\ O_2 \longrightarrow 4\ CO_2 + 6\ H_2O$.

Chapter 3

1. meter, liter, gram; liter = 10^{-3} m^3 and 10^{-3} liter water (at max density) = 1 g **2.** milli = 10^{-3}; centi = 10^{-2}; kilo = 10^3 **3.** meter, m; liter, l; gram, g; millimeter, mm; centimeter, cm; kilometer, km; (similar for ml, cl, kl, mg, cg, kg) **4.** Centiliters, cl is not the cube of an exact decimal fraction of a meter **5. (a)** 3.55 m × 100 cm/1 m = 355 cm; **(c)** 452 kg(1000 g/1 kg)(1000 mg/1 g) = 4.52×10^8 mg; (answer double circled) **(f)** (2.5×10^{-2} liter)(1000 ml/1 liter)(1 cm^3/1 ml) = 25 cm^3
6. (a) 1 ml = 1 cm^3; **(b)** 1 ml H_2O = 1 g **7. (a)** yd $\longrightarrow$ ft $\longrightarrow$ in $\longrightarrow$ cm $\longrightarrow$ m; **(d)** lb $\longrightarrow$ g $\longrightarrow$ ml **8. (a)** 91.4 m; **(b)** 398 km
9. (a) 3; **(b)** 2 **10. (a)** $°C = \frac{5}{9}(°F - 32)$; **(b)** $°F = \frac{9}{5}°C + 32$; **(c)** $°K = °C + 273.15$ **11. (a)** 20°C; **(b)** 95°F; **(c)** 37°C; **(d)** 300°K
12. 36.8 ml **13.** 37,500 cal

Chapter 4

1. ^{12}C as 12 **2. (a)** 34.97 amu; **(b)** 34.97 g; **(c)** 6.022×10^{23}; **(d)** 1.722×10^{22} **3. (a)** the average atomic weight of the naturally occurring mixture of Cl atoms; **(b)** 35.46 **4. (a)** 44; **(c)** 218; **(e)** 310 **5. (a)** 70%; **(c)** 2.00%; **(d)** 5.4% K_2O **6. (a)** 8.16 g **7.** Mg_3N_2 **10.** 142 g **12.** 69%
13. (a) 28.2 liters **15.** negative, ΔH is defined as the heat going in, which is a negative number in this case **16.** -41.7 kcal/mole **17.** 87.8 g

Chapter 5   **2. (a)** The tube is a cathode ray tube, with vacuum and high voltage applied. The path of the rays is blocked by a barrier with a slit in it. A thin beam is formed by rays passing through the slit. The beam strikes a curved fluorescent backing, causing a glow which shows the path of the beam. The beam goes in a straight line from negative to positive electrodes when there is no magnetic field. A magnetic field deflects the beam away from the original straight line.   **3. (b)** The ratio of charge to mass for electrons was known from Thomson's experiment, but an independent measurement of one of them (charge or mass) was needed so the charge and mass could be known. The oil drop experiment provided an opportunity to determine the electronic charge.   **4. (b)** Rays flow from negative to positive; the rays travel in a straight line, so the target blocking some rays casts a sharp shadow; rays striking the walls give the same glow as seen in all other cathode ray tubes. **(d)** A single element may produce several different charge to mass ratios for its ions, indicating isotopes of different masses; compared to ^{12}C as 12, all isotopes have masses close to whole numbers.   **5. (a)** 7 p, 7 n, 7 e^-; **(b)** 17 p, 20 n, 17 e^-; **(c)** 92 p, 146 n, 92 e^-; **(d)** 1 p, 2 n, 0 e^-; **(e)** 16 p, 16 n, 18 e^-; **(f)** 4 p, 5 n, 2 e^- **6. (a)** near zero mass, -1 charge; **(b)** 1 mass unit, $+1$ charge   **7. (b)** the negatively charged electrode; **(d)** the tiny central region containing the positive charges and most of the mass of an atom; **(f)** the number of positive charges on the nucleus (equal to the number of protons)   **8.** see figures in the chapter.

Chapter 6   **1.** order was by behavior even when that did not match atomic weight order; gaps were left for missing elements   **2. (a)** outermost electrons in the same shell; **(b)** same configuration of electrons beyond the last closed shell   **3.** n, related to distance from nucleus, integer values of 1, 2, 3, etc.; l, related to shape of spacial distribution, values of 0, 1, 2, etc., up to maximum of $n - 1$ **4.** each electron must have a unique set of the four quantum numbers (Pauli exclusion principle) and there are six possible sets when n is 2 and l is 1 (2 p electrons)   **5. (d)** $1s^2 2s^2 2p^6 3s^2 3p^6 4s^2 3d^{10} 4p^6 5s^2 4d^{10} 5p^6 6s^2 4f^{14} 5d^{10} 6p^3$ **6.** n is 3 and l is 2   **7. (a)** spherically symmetrical—equal probability in all directions; **(b)** perpendicular; **(c)** two   **10.** identical shape, but with all ionization potentials higher and with the breaks (shell, subshell, half-filled subshell) one element sooner in each case   **11.** bringing in an electron adds to crowding and is therefore more affected than taking an electron away, which relieves crowding   **12. (a)** Li, nuclear charge effect; **(b)** K, shell effect; **(c)** Na, charge and shell effects; **(d)** F^- electron repulsion (crowding); **(e)** Na, shell and repulsion effects; **(f)** F^-, nuclear charge effect between isoelectronic species.

Chapter 7   **1.** conductors under these conditions: **(a)** always; **(b)** when fluid (molten or in solution in a liquid); **(c)** never   **2. (a)** conduct electricity; conduct heat; can be reshaped without breaking **3. (a)** electron affinity of Cl makes some energy available, attraction of Ca^{2+} ions to Cl^- ions in the $CaCl_2$ crystals makes energy available; **(b)** ionization potential of Ca costs energy (there is also energy cost to convert Ca and Cl_2 into the form of free atoms) **4.** conduction of current by ionic solutions; movement of separate "parts" (the ions) in opposite directions as current flows **5.** see the Section 7–6, "Chemical Evidence for Ions" **6.** acids are sources of H^+, bases are sources of OH^- **7. (a)** $+2$; **(b)** $+1$; **(c)** -2 **8.** $+2$, there are only two electrons in the $n = 4$ shell, the shell which needs to be lost to form a small positive ion **9. (a)** $CaSO_4$ precipitates because Ca^{2+} and SO_4^{2-} ions are present in those conducting solution and will react in the same way they did when $CaCl_2$ and Na_2SO_4 solutions are mixed

10. (a) Na· + :C̈l: ⟶ Na⁺ + :C̈l:⁻; **(c)** ·C̈· + 2 :C̈l:C̈l: ⟶ :C̈l:C:C̈l:
$$$$
(with :C̈l: above and below the central C)

11. (a) covalent (electronegativity difference is $3.5 - 3.0 = 0.5$); **(b)** polar covalent (difference 1.0); **(c)** ionic (difference 3.0) **12. (b), (d),** and **(e)** Work this one by writing down the dot diagrams

13. (a) H:C̈:H (with H above and below), tetrahedral; **(c)** O₂N:NO₂ structure with two N atoms bonded, each N bonded to two O atoms, trigonal around each N; **(i)** :C̈l:—P—:C̈l: with four additional :C̈l: groups (octahedral), octahedral

---

Chapter 8

1. large scale operation with elaborate equipment for efficient use of cold product to cool incoming air **2.** paramagnetism of O_2 **3. (a)** $CH_4 + 2\,O_2 \longrightarrow CO_2 + 2\,H_2O$; **(b)** $2\,Na + O_2 \longrightarrow Na_2O_2$; **(c)** $FeO + H_2 \longrightarrow Fe + H_2O$ **4. (a)** $+2$; **(b)** $-3, +5, +3$ (less certain); **(d)** -1 and positive states, probably $+1, +3, +5, +7$; **(f)** $+2, +3, +4, +5$ **5.** All halogens form -1 ions **6.** ionization energies drop as one moves down the alkali metal column **7.** the outer shell of Cu has only one electron, which can be taken away from the effective $+1$ charge of the kernel fairly easily and made available for sharing in a metallic crystal; the six electrons in the outer shell of S are held much more strongly by the $+6$ charge of the kernel (and also because they are in a lower shell) and cannot be shared well enough to give metallic behavior **8.** Bi is more able to give up electrons and be metallic; shell effect **9.** diagonal relationship between the first two rows of eight **10.** d electron level in third shell (P) but not in second shell (N) **11.** no d level available for Ne **12.** increasing difficulty in reaching higher oxidation states as one moves to the right across the series **13.** need to lose outer shell

Chapter 9

1. (a) 1.45×15.2 ft^3 **2.** $740 \times 323/298$ torr **3.** $2.50 \times 263/303$ liters **5.** $10.0(1.5 \times 760)/720$ g **7.** $3.50(150/300)(750/150)$ liters **8.** molecules occupy space and exert forces. **9.** attractions in liquids are large enough to prevent the escape of molecules from the fixed volume of liquid, even though there is some freedom to move about **10.** crystals have a fixed pattern, allowing no movement except vibrations around those positions; amorphous solids lack the pattern but are viscous enough to resist the changes in positions seen in liquids **12. (a)** ice melts to absorb heat; **(b)** water evaporates to restore the vapor pressure, but with a continuing leak the evaporation continues until the water is gone or until the engine cools enough to permit a new equilibrium.

13.

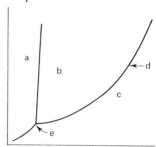

Chapter 10

2. CH$_3$OH and H$_2$O, NaCl and H$_2$O, I$_2$ and C$_6$H$_{14}$, CH$_3$OH and C$_6$H$_{14}$, I$_2$ and CH$_3$OH

3.

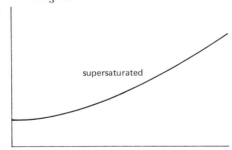

4. (a) high gas pressure and low temperature

5.

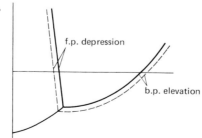

6. (a) $3.0\ M$ Na$_2$SO$_4$

7. (a) 3.75 m Na$_2$SO$_4$ 8. (a) 5.5, 4.0, and -1; (b) 12.5; (c) 7.0. 9. (a) 10^{-9} M; (b) 10^{-5} M 10. (a) $-2.79°C$; (b) $-4.1°C$ 12. interferes with water intake by osmosis 15. Land plants must have inward water flow to make up for evaporation whereas water plants need only reach osmotic balance 16. $55.5 \times \frac{1}{2} \times 0.750/0.500$ ml.

Chapter 11

4. (a) lower; (b) no; (c) yes 5. (a) 6. iceberg, longer wavelength quanta are possible because of the lower rate at which energy must be lost.

Chapter 12

1. contact, activation energy, steric arrangement 2. (a) speed rate; (b) double rate; (c) triple rate; (d) 16 times rate; (e) no effect

5. (a) $= \dfrac{[HCl]^2}{[H_2][Cl_2]}$ (b) $K = \dfrac{[H_2O]^2}{[H_2]^2[O_2]}$ (c) $K = \dfrac{[H_2O]^2[SO_2]^2}{[H_2S]^2[O_2]^3}$

Chapter 13

5. a hydrogen nucleus shared between pairs of electrons on two small, highly electronegative atoms 6. empty spaces are required to permit maximum hydrogen bonding 9. (a) an acid that ionizes completely; (b) an acid that is only slightly ionized to form H$^+$; (c) electron pair acceptor; (d) in water, pH over 7; (e) reaction with water to produce excess H$^+$ or OH$^-$ 10. (a) reaction of the Lewis base NH$_3$ with the Lewis acid BF$_3$ 13. CO$_2$ + H$_2$O $\rightleftharpoons$ H$_2$CO$_3$; H$_2$CO$_3$ $\rightleftharpoons$ H$^+$ + HCO$_3^-$; H$^+$ + CaCO$_3$ $\rightleftharpoons$ Ca^{2+} + HCO$_3^-$ 15. a suspension of particles (or drops) small enough to remain suspended through Brownian motion (but not small enought to be units in solution) 17. high heat capacity and heat of vaporization

Chapter 14

1. (a) water; (b) N, P, or K; (c) O$_2$ 2. N$_2$, very strong triple bond, not very reactive; P$_4$, strained single bonds, very reactive, or giant red P molecule, single bonds, more reactive than N$_2$ 3. both the highest oxidation state, favored with excess oxygen available; NO$_3^-$ because small sized atoms set up good π bonding, which allows resonance stabilization and permits a good bonding pattern which keeps the net charge on the ion low; PO$_4^{3-}$ because the larger P cannot π bond as well, so an additional single bond is favored even though it causes a higher ionic charge 7. plant and animal remains and wastes, dissolved minerals in soil water; nitrogen compounds, acidified phosphate rock, potassium salts

Chapter 15 1. (a) $^{214}_{84}\text{Po} \longrightarrow {}^{210}_{82}\text{Pb} + {}^{4}_{2}\text{He}$; (b) $^{210}_{83}\text{Bi} \longrightarrow {}^{210}_{84}\text{Po} + {}^{0}_{-1}e$; (c) $^{40}_{19}\text{K} + {}^{0}_{-1}e \longrightarrow {}^{40}_{18}\text{Ar}$; (d) $^{12}_{6}\text{C}$; (e) $^{24}_{11}\text{Na}$ 2. (a) $\frac{1}{16}$ g 3. α, β^-, β^+, γ, ec, fission, n emission 5. 2, 8, 20, (28), 50, 82, 126 6. (a) preference for equal $p + n$ (at low numbers); (b) preference for even numbers (and coulomb repulsions) 8. no coulomb barrier 9. lower their energy, thus increasing their wavelength 11. dependence of reaction efficiency on moderation, and formation of some delayed neutrons by n decay radioactivity 14. (a) $^{8}_{4}\text{Be}$ is skipped in the main fusion sequence; (b) $^{4}_{2}\text{He}$ units in the main fusion sequence leads to $^{32}_{16}\text{S}$, whereas $^{34}_{16}\text{S}$ must be made by a side reaction

Chapter 16 1. forms substances that are (a) easily condensed and (b) of low density 3. (a) something reacting with itself to form two different oxidation states; (b) when special stability of one (or both) product(s) makes the combination of products more stable than the original state 5. an electron pair donor used in formation of complexes 6. linear, sp; tetrahedral, sp^3; octahedral, d^2sp^3; square planar, dsp^2 7. (a) Cr^{2+} has too many electrons to allow octahedral complexing without forcing electrons to pair up; (b) V^{3+} has a smaller nuclear charge to attract the ligand electron pairs 8. (b) leaching, flotation, magnetic separation, chemical processing 9. (a) Al_2O_3; (b) Fe_2O_3, Fe_2S_3, FeS; (c) Cu_2S, CuS, Cu_2O, CuO 10. silicon oxide is more acidic, thus soluble in weaker base 11. ore, transportation of ore, purification of ore (equipment and chemicals), electrolysis equipment, electricity, and carbon anodes (material and labor) 14. (b) removal of impurities from the product 16. the amounts of carbon and other substances in blast furnace iron are uncontrolled and variable, whereas definite, known amounts are required 18. high carbon steel should be harder and more brittle 19. a solution of two (or more) metals 21. acidity makes the basic lead oxide dissolve 22. mercury is not metabolized for energy by organisms in the food chain and is kept and concentrated in the successive organisms

Chapter 17 1. single bonds to noncarbon atoms, single bonds to carbon atoms, double or triple bonds 2. (a) sp^3; (b) sp^2; (c) sp 3. (a) a pair of electrons shared between the parallel p orbitals of two atoms; (b) several pairs of electrons shared in resonance among the parallel p orbitals of several atoms (6 atoms in the simplest aromatics) 4. compounds with a general feature in common and differing only in the length of the saturated hydrocarbon group attached 5. (a) butane; (b) methylpropene; (c) 3-methylpentane; (d) 2,4,4-trimethylhexane; (e) benzene (or 1,3,5-cyclohexatriene) 7. 2-methylhexane; 3-methylhexane; 2,2-dimethylpentane; 2,3-dimethylpentane; 2,4-dimethylpentane; 2,2,3-trimethylbutane; 3-ethylpentane; 7 total branching isomers in addition to the unbranched heptane 9. (a) see answer to No. 7;

(b) H₂C=CH(CH₃)—CH₃ (cis) and H(CH₃)C=C(CH₃)H (trans);

(c) Cl—CH(H)—Br and Br—CH(H)—Cl

11. (a) chloromethane; (b) 2-methylpropan-1-ol; (c) propanal; (d) 2-pentanone; (e) methanoic acid; (f) methyl propanoate; (g) ethylpropylether; (h) 2-aminobutane

Chapter 18

1. need to break an atom away before a new bond can form **2.** polar or polarizable character, with ability to interact with a negative charge **3.** the π electrons polarize the Br_2, whereas HBr is already polar **5.** polarize hydrogen and alkene and bring them together **7.** $CH_3\text{—}CH_2Cl \longrightarrow CH_2\text{=}CH_2 + H^+ + Cl^-$ **8. (b)** $CH_3\text{—}COOH + CH_3\text{—}CH_2OH \longrightarrow CH_3\text{—}CO\text{—}O\text{—}CH_3$ **9.** ethene + HCl $\longrightarrow$ chloroethane; chloroethane + Mg $\longrightarrow$ ethyl Grignard; ethyl Grignard + 3-pentanone $\longrightarrow$ (then add H_2O) $\longrightarrow$ 3-ethyl-3-pentanol **11.** Double bonds left after reaction (they are all *cis* double bonds in rubber).

Chapter 19

1. proteins, nucleic acids, and polysaccharides **3. (a)** valine, leucine, isoleucine, and phenylalanine-hydrophobic interactions; **(b)** serine, aspartic acid, glutamic acid, glutamine-hydrogen bonding, ionic interactions, dipole-dipole interactions **4.** —C—C(=O)—NH—C— **6.** tertiary is between parts of the same polypeptide; quaternary is between two or more separate polypeptides **7.** ring compounds with more than one element among the atoms forming a ring **8.** cytosine (C), guanine (G), adenine (A), thymine (T), and uracil (U); C, G, A, and T in DNA, C, G, A, and U in RNA; C≡G, A=T, and A=U **9.** *bases* (C, G, A, or T) linked to *deoxyribose* which is linked to two *phosphate* units, each of which is linked to another deoxyribose with the same sort of linkages as the first deoxyribose **10.** by hydrogen bonding of the base pairs (as in No. 8) **12.** the shape of the globular protein makes the enzyme specific (only certain substances fit in) and catalytic (polar sites are exposed in specific positions).

Chapter 20   1. ring structure; additional carbons, each with alcohol group and a hydrogen; optical isomers possible at each additional carbon (glucose being one specific set of choices) and at the ring closure   3. sucrose is α-D-glucose and β-D-fructose linked by an ether linkage from the 1 position on glucose to the 2 position on fructose (replacing the OH groups in the positions which determine alpha or beta form); lactose is α-D-glucose and β-D-galactose with a 1,4 linkage, 1 position on galactose to 4 position on glucose; maltose is a 1,4 linkage between two α-D-glucose units   4. (a) 1,4 linkages of α-D-glucose units; (b) 1,6 linkages to the 6 position of units already in a chain of 1,4 linkages   5. even numbers of carbons on the fatty acids; appearance of a double bond at the same position in several of the most abundant unsaturated fatty acids   6. the carbohydrate precursors likely for coal are produced (by plants) in greater abundance than fats or proteins which might lead to oil; also, the cellulose form used for the plant fibers is less usable by most organisms than fats or proteins   7. (c) acid mine drainage after air (oxidizing) is introduced into coal mines (reducing before air availability)   8. carbohydrates are already partially oxidized whereas fats are mostly oxidizable carbon and hydrogen with much smaller proportions of oxygen   9. (a) $C_6H_{12}O_6$; (b) $C_{12}H_{22}O_{11}$; (c) $(C_6H_{10}O_5)_x$; (d) $(C_6H_{10}O_5)_x$; (e) $(C_{16}H_{31}-COO)_3-C_3H_5$

Chapter 21   2. the concentration of negative charge is lessened   3. citric acid cycle, in mitochondria   5. material is taken from the side where its concentration is low, interacts with the membrane, and (in any energy consuming process) selectively released from the membrane on the side where its concentration is higher   6. aerobic conditions—glucose, ADP, phosphate, and O_2 to form pyruvic acid and ATP (8 ATP per glucose used); anaerobic conditions—glucose, ADP, and phosphate to form lactic acid and ATP (2 ATP per glucose used)   8. niacin, adenine, ribose, and phosphate; adenine on ribose with a diphosphate linkage to another ribose on which a nicotinamide is bonded   9. (pyruvic acid, coenzyme A, 3 ADP and 3 phosphates to acetyl coenzyme A plus CO_2 plus 3 ATP, then) acetyl coenzyme A, 12 ADP, and 12 phosphates to form coenzyme A, 12 ATP, and 2 CO_2. Oxidative phosphorylation.   11. low molecular weight; good high energy bond   12. actin and myosin; separate parallel strands when relaxed; ATP available and absence of Ca^{2+} allows relaxation   14. the shape allows O_2 to reach the heme site for complexing but excludes most other possible ligands   15. hydrocarbon combustion is incomplete; some nitrogen oxides are formed

Chapter 22   1. diabetes; excess blood glucose accumulates   2. glucose may be needed quickly to overcome an excessive depression of blood glucose after insulin is taken   3. steroids are hydrocarbons (or derivatives) with a particular set of four interlocking rings of carbon atoms which is produced and used by living organisms **5. (a)** mestranol, an estrogen used to maintain female hormone balance in patients using oral contraceptives **6.** the limits needed to be effective are now better known **7.** β-carotene from carrots, squash, and other plants is converted to vitamin A in the body; vitamin A reacts with another substance to form a light-sensitive compound used in vision **8.** lysine; inadequate production of some proteins causing damage such as mental retardation

Chapter 23

1. Sewage plants substitute for the natural oxidation of small, dispersed quantities of wastes by pumping air (or oxygen) through liquid to which bacteria have been added. Side effects are sludge produced and an effluent with concentrated nutrients instead of the natural dispersion. **2.** inability to convert galactose to glucose; diet avoiding sources of galactose **3. (a)** a substance that reduces the allergic type effects of histamine; **(e)** a poison to insects, or to certain insects **5.** penicillin; allergy reactions and inability to affect some diseases; less allergic reactions than some other drugs **6.** drowsiness (even when not desired), addiction **7.** development of resistance by insects; buildup in the environment because of slow degradation **9. (a)** affects certain species but not others; **(c)** herbicides for crabgrass may interfere with some types of grasses **10.** blocking nerve impulses **11.** antibodies are developed **12.** diabetes victims usually survive to the age for reproduction **13.** the increased malaria resistance of a racial group which included carriers of sickle cell trait was more important in tropical areas where malaria occurs than the disadvantages of the sickle cell trait.

Index

Numbers and letters in brackets, e.g., [560t], and [334f] refer to tables and figures in the text.

A

a (prefix), [560t]
A (adenine), 445, 449
A (electricity), [561t]
A=T (bonds), 445, 447, 448
A=U (bonds), 447, 451
A, 562t
Abbreviations for amino acid names, 431, [432t–434t]
Ability to function, 452
Ability to support growth, 293
Abnormal bone formation, 526
Absence of oxygen, 468, 483, 486
Absolute temperature, 40, 166, 244, 589
Absolute zero, 40
Absorb light, 411
Absorbed in digestion, 469, 481, 512
Absorption of energy, 315
Abstraction reactions, 410
Abundance of elements, 287, 310, 323–26, 331–37, [334f], 351
Accept electrons, 269, 270, 408, 417
Accessible reactive sites, 453
Accidental environmental crisis, 472
Accuracy requirements, 592, 594, 596
Acetaminophen, 537
Acetate, [27t], 577
Acetate fabrics, 393
Acetic acid, 268, 394, 577, 593
Acetic anhydride, 395
Acetoacetic acid, 512
Acetone, 393, 512, 513
Acetyl coenzyme A, 480, 480f, 489f, 490, 491, 512, 513, 523
Acetylene, 385
Acetylphosphate, 479, [479f]
Acetylsalicylic acid, 535, [535f]
Achromycin, [599f]
Acid(s), 114–15, 200–202, 267–70, 271, 431, 469, 575, 610
    acetic, 268
    Arrhenius, 115, 158, 268
    Brönsted-Lowry, 268
    conjugate, 579
    destroys tissues, 395
    dissociation constraints of, 575–76, [579t]
    electrolytic properties of, 114–16
    groups, 428
    irritating humans, 361
    Lewis, 269
    naming of, 610, [610t]
    neutralization of, 114–15
    nomenclature, 610
    reactions in water hardness, 271–72
    from smelters, 361
    strong 112, 268
    structure of, and strength, 270
    weak, 116, 268, 269, 271
Acid
    anhydride(s), 395, 415, [415f]
    catalyzed, 415f
    chloride, 421, 427, 428
    in rivers, 471
    damage to plants, 361
    ionization problems, 593–96, 599
    mine drainage, 362, 471–72
    root names, 610
Acidic
    behavior, 406, 414, 444
    conditions, 481
    ground water, 471
    oxides, 270, 348, 353, 355, 361
    open hearth process, 355
    rainfall, 291, 471
    solutions, 200–201, 268, 270
Acidity, 481, 487
ACTH, 519
Actin, 496, 497, [497f], 500
Actin double helix, 496, [497f], 500
Actin fiber, 498, 500
Actin-myosin arrangement, [498f]
Actin-myosin bonds form, 500
Activated carboxylic acid, 428, 434
Activated sludge process, 298
Activating group, 434
Activation, 420
Activation energy, 241–42, 244, 246–50, 583
Active molecules, fraction, 242–47
Active reactants, 242
"Active sites," 452
Active transport, 212, 481–83, [482f], 496–500, 542
Actively incorporate substances, 481
Adding
    Ca^{2+}, 500
    Cl_2, 404
    H^+, 406
    HCl, 404, 406
    H_2O, 405, 406, 407, 418
Addiction, 538
Addition reaction 404, 406, 413, 416, 417, 418, 419, 421–23
1,4-addition, 422
1,4-addition polymerization, 422
Addition chain polymerization, 421–23, [423f]
Addition to double bond, 404
Addition of H_2 408, 413
    to carbonyl, 416
Addition of hydrogen atom, 409
Addition of phosphate, 487
Addition of polymers, 421–23
Addition of two-carbon units, 490
Adenine, [441f], 442 [443f], 444, 445, [447t], 485
Adenine-thymine bonds, 445, [455f]
Adenine-uracil bonds, 447
Adenosine, [443f], 477
Adenosinediphosphate, 477, [478f]
Adenosine(-5'-) monophosphate, [443f], 477, [478f]
Adenosinetriphosphate, 291, 444, [444f], 477 [478f]
Adhesives, 464
Adjacent amine and carboxylic groups, 431
Adoption, 542
ADP, [444f], 477, [478f], 486, [487f], 489, [491f], 496
Adrenal gland, 514, 519
Adrenalin, 511
Adsorbed alkene, 408
Adsorbed, H_2, 408
Adsorption, 248, 278
Absorption of oil onto straw, 472
Advantage of even-even isotope, 322
Advantage(s)
    of low areas, 270
    of higher concentrations, 240
Aerobic bacteria, 298
Aerobic conditions, 483, 486, 490, 502

Aerosol, 275
Africa and Africans, 555
Age of Earth, 307
Aggregates, 276
Agriculture, 552, 555, 556
Air, 192
Air bubbles, trapped, 213
Air mixed into water, 299
Air pollution, 231, 232, 361, 471, 472, 504
Air pressure, 164
Air turbulence, 229, 237
Airborne ash, 232
Airplanes, 229
Al^{3+}, 349
Ala, [432t]
Alanine, 430, [431f], [432t]
Alaska, 506
Alcohol(s), 391–92, 402, 404, 413, 416, 417, 442, 479, 524, 538
Alcohol hydrogens, 414
Alcohol-phosphate combination, 477
Alcoholic beverages, 392
Alcoholism, 523
Aldehyde(s), 392–93, 456, 524
Aldehyde end of sugar, 461
Aldehyde form, 460
Aldehyde sugar (Aldose), 457, 461
Aldin, 551, [551f]
Alfalfa, 554
Algae, 217, 295, 296
Algebraic manipulations, 572
Algebraic unknown, 573
Alicyclic compounds, 389
Align potential reactants, 452
Aliphatic compounds, 378
  with rings, 389
Alk, 380
Alkali metals, 150, 337
Alkaline earth metals, 337
Alkaline fluids, 481
Alkane(s) 378–79, 380–83, 403, 408
  reactions, 401–403
Alkane-bromine mixture, 410
Alkene(s), 379, 383–85, 403, 404, 406, 407, 408
Alkyl chain fragments, 402
Alkyl compounds, 380
Alkyl groups, 382
Alkyl hydrogens, 414
Alkyl radical, 402, 409, 410
Alkylamine salts, 553
Alkyne(s), 380, 385–86
Allergy problems, 536, 537, 548, 549, 554
Alloys, 359
Al_2O_3, 347, 349, 351
Alpha-amino acids, 428
Alpha arrangements, 460, 465
Alpha chains (hemoglobin), 541
$alpha$-D-glucose, [459f], 460, [460f], 461, 462, [462f], 464
$alpha$-D-sugars, [459f]
alpha decay, 305, 313
alpha emissions, 304, 305
alpha form (sugars), [459f], 460, 461, 463
alpha helix, 436, 445, 451
alpha-ketoglutaric acid, [491f]
alpha particles, 79, 304, 315
alpha side of ring, 462
alpha-tocopherol, 526, [527f]

Alternate arrangements, 381
Alternate metabolic pathways, 543
Alternate source of ATP, 496
Alternative pain relievers, 536–37
Alto-, [560t]
Aluminum, 347–51, 366
  form metal, 350
  oxide, 347
  silicates, 347
  source, 347
Alvarez, 308
Americans, 523
Americas, 555
Amide linkages, 419, 434
Amine(s) (group), 395–96, 427, 428, 434, 442
Amino-, 395
Amino acid(s), 395, 428–34 [432t–434t], 438, 439, 440, 449, 481, 483, 513, 528–29, 537, 541, 543
  deficiencies, [528t]
  reactions, 523
  sequence, 435, 436
  site for bonding of, 446
  supplements, 529
2-Aminobutane, 395
Aminomethane, 395
Ammonia, 121, 286
Ammonium ion, bonding in, 121
Amorphous solid, 182, 183
Amorphous sulfur, 182
Amount of carbon dioxide, 505
Amount, surface, 244
AMP, [443t], [444f], 477, [478f]
Ampere, [561t]
Amphoterism, 158, 347–48
Amylopectin (starch), 464, [464f]
Amylose (starch), 464, [464f], 465
Anaerobic conditions, 483, 486, 487, 494
Anaerobic bacteria, 298
Analgesic drugs, 535, 536
Analogies
  ants, 1
  car on hill, 222
  cavemen, 6–9
  chimpanzees, 2
  dating, 238–41
  football, 119
  "glucose falls water power facility," [495f]
  man-woman interreactions, 238–49
  matchmakers, 248
  party, 242
  reaction rate, 236
  rocket launching, 174, [175f]
  social, 238
  waterfall, 493–95
Analysis of soil, 292
Andosterone, 515, [516f]
Androgens, 515
-ane, 380, 385
Anemia, 541, 543
Angles of chemical bonds, 440
Anhydride(s), 479
  goes through skin, 395
  high energy compounds, [479f]
  not acidic, 395
  phophate with enol, 479
Animal fats, 394, 419, 467, 524

Animal protein(s), 526
Animal starch, 464
Anode, 72, 111, 350
Anode sludge, 360
Answers to exercises, 612–20
Anthracene, [388f]
Antibacterial medicines, 546
Antibiotics, 547, 548
Antibodies, 451, 452, 519, 547, 548
Antibonding orbitals, 141
Anticodon(s), 446
Antieutrophication efforts, 296
Antifertility effect, 516
Antifreeze, 392, 405
Antihistamines, 537, [538f]
Anti-inflammation, 534
Antiknocking property, 383
Antiovulation effect, 516
Antioxidant, 526
Antitrust suit, 549
Anxiety, 521
APC tablets, 536
Apparent oxidation states, 611
Appetite, 521
Applications of equilibrium, 236, 592–607
Approximations, 592
  in models, 202
Approximation(s) fail(s), 594, 599, 602
Arden, H., 506
Area, surface, 244
Areas of charge concentration, 267
Arg, [432t]
Arginine, [432t], [528t]
Argon, 192
Arguments of Science, 12
Aromatic
  compounds, 378, 389
  hydrocarbons, 387–89
  nomenclature, 389
  ring(s), 387, [388f], 388, 514
Arrangement
  of actin fibres, [498f]
  of bases, 447
  of carbon and iron, 357
  change in, 242
  of materials, 230
  disordered, 226
  mixing of 225, 226
  position, 241
Arrhenius, S., 111
  acid-base theory, 115, 158
  theory of ionization, 111
Arrow representation, [411f], 412, 413
Arthritis, 519, 520, 533, 536
Ascorbic acid, 461, 520–21, [521f]
Ash, 504
Asia, 293, 521
Asp, [432t]
Asparagine, [432t]
Aspartic acid, [432t]
Asp-NH$_2$, [432t]
Aspirin, 519, 520, 535, [535f], 536, 537
Aspirin poisoning, 536
Assembly line workers, 449
Assisting natural defenses, 546
Associated ions, 267
Assumption, simplifying, 203
Assymetric carbon, 457
Astatine, 150

-ate, 610
Athabaska tar sands, 470
Atm, 60, 164, 562
Atmosphere(s), 292
 composition of, 192
 pollutants in, 231, 232
 pressure (atm), 60, 164, 568
 ozone in, 143
 water in, 259, 273
Atmospheric carbon dioxide, 505
Atmospheric dust, 507
Atmospheric pressure, 586
Atomic energy levels, 227
Atomic masses, 47
Atomic mass units, 47
Atomic model
 according to wave mechanics, 92–95
 of Bohr, 87–89
 of Dalton, 20–22
 of Rutherford, 79
 of Thomson, 78
Atomic nucleus, 79
Atomic number, 81
 relationship of, to periodic table, 86
Atomic orbitals, 92–95
 hybrid, 129
 overlap of, 125–26
Atomic radius, [152f]
Atomic size, 100, 152
Atomic structure, 304
Atomic theory, 20–22
Atomic weight(s), 48
 correction of, by Mendeleev, 86
Atoms
 early ideas concerning, 20–21
 electrons in, 78–82
 properties of, 71–82
 structure of, 78–79
ATP, 444, [444f], 477, [478f], 480–81, 482, 483, 486, 487, [487f], 489, 490, [491f], 493, 496, 499, 500, 501, 504
 number from each glucose, 490
 production, 480, 483–94
ATP energy, 493, 498
ATP producing process, 490, 493
Attached amino acid, 451
Attack on carbonyl carbon, 421
Attack by electrophilic reagent, 406, 417
Attack by nucleophilic reagent, 420
Attacking double bonds, 405
Attractions
 between ions, 237
 between protons and neutrons, 321
 to water, 277
Attractive forces, 172, 192, 295
Automobiles, See car(s)
Availability of energy, 469
Availability of glucose, 511
Availability of insulin, 556
Available chemical energy, 493, 503
Available electron levels, 339
Available energy in ATP, 490
Available entropy increase, 231
Available intermediate compounds in metabolism, 494
Available oxygen, 494
Available phosphate, 292
Available supply (fuels), 505
Average kinetic energy, 171, 174

Average rainfall and snowfall, 216
Average temperature of the Earth, 507
Avogadro's hypothesis, 58–59
Avoiding barriers, 237
Auger procuss, 308, [309f]
Aureomycin, [549f]
Automatic shut down, 320
Automobile engines, 504
Automobile seat covers, 422
Azide, 26

B

β, 403
β⁻, 309, 310, 311, 313
β⁺, 309, 310, 311, 313, 315, 316, [316f], 324
 for mapping, 316, [317f]
β-carotene, 524
β decay, 306, 308
β form (sugars) [459f], 460, 461, 462, [462f], 465
β-D-glucose, [459f], 460, [460f], 465
β linkages, 465
B vitamins, 521–24, [522f], 547
Babies, 545
Back reaction, 248–49, 583
Backbone (polypeptide), 435, 436
Backwaters, 270
Bacteria, 271, 295, 297, 363, 469, 503, 520, 523, 546, 547, 548, 549, 554
Bacterial degradation, 298
Bacterial proteins and RNA, 446
Bad moderators, 319
Balance
 between hormones, 511
 between processes, 511
 forager to scavenger fish, 295
 of forward and reverse rates, 175, 250
 in Great Salt Lake, 215–217
Balanced equations, 254, 564, 565
Balancing equations, 28–30
Balancing of energy levels, 311, 312
Balmer series, 88f
Ban on NTA, 364
Band, 109
BaO₂, 146
Barbitol, 538, [539f], 540
Barbituates, 538
Barges, 472
Barium peroxide, 146
Bark, 299
Barometer, 163, 164f
Barriers, 472, 519
 avoiding, 237
 to water, 271
Base(s), 200–201, 267–70, 287, 291, 413, 431, 576
 Arrhenius, 115
 Brönsted-Lowry, 268
 in code positions, 446
 conjugate, 577, 579
 dissociation constants of, 575–76, [579t]
 electrolytic properties of, 114–16
 Lewis, 269
 neutralization of, 114–15
 organic, 439–42, [441f]

 strong, 116, 268
 structure of, and strength, 270
 weak, 116, 268, 269
Base pairs, [445f], 446, 447, 448
Base-sugar units, 443, [443f]
Base-sugar-phosphate units, 443, [443f], 444
Basic groups, 420, 428
Basic oxides, 270, 348, 353, 355
Basic oxygen process, 355–56, [355f]
Basic solutions, 200–201, 268
Basis
 for interaction, 109–11, 119–21, 238
 thermodynamic, for equilibrium, 236
Bauxite, 347, 349
Bayer process, 346, 347–49, [347f], 351
$^{8}_{4}$Be, 322, 324, 331, 333
Bear River, 216
Beard growth, 515
Becquerel, H., 304
Beef, 533
Benadryl, [538f]
Bend toward excess neutrons, 312
"Bent bond" structures, 462, [463f], [464f]
Bent structure, 130, [132f]
Benzene, 128, [128f], 261, 378, [387f], [388f], [389f], 440
Beriberi, 521
Beryllium abundance, 325
Bessemer process, 354, [354f]
Beta particles, 304
BF₃, 316
Bicarbonate ion in water hardness, 271–73
Bile, Bile acids, and Bile salts, 481
Binding energy, 249, 322
Biochemical
 activity, 511, 518
 effects, 529
 energy storage, 503
 environment, 557
 oxidation-reduction reactions, 485, 521
 reactions, 242, 387, 456, 477, 493
  cause physical changes, 496
 synthesis, 496
Biochemically potent, 529, 533
Biochemistry, 419, 427
 of hormones, 535
Biodegradable, 551
Biological
 building blocks, 427–53
 deserts, 294
 effects, alcohol, 391–92
 environment, 230
 heterocyclic compounds, 440–42, [441f]
 interactions, 396
 oxygen demand, 297, 298, 299, 534
 pumps, 481
 significance of skin pigment, 526
Biologically active molecules, 511–29
Biologically important amino acids, 428, [432t–434t]
Biologically important compounds, 465
Biologically important linkages, 427
Biologically important optical isomers, 387

Biologically important oxygenated hydrocarbons, 391
Biologically potent substances, 512
Biologically recoverable energy sources, 469
Biologically significant differences, 463
Biologically significant molecules, 428
Biphenyl, [388f]
Biopolymers, 419, 427, 434–39, 442, 444
Bird droppings, 294
Birth control, 514, 518
Birth defects, 553
Black lung disease, 472
Black people, 542
Blast furnace, 352–53 [352f]
Blast furnace iron, 353–56
Blindness, 392, 525
Blister copper, 360
Blocking mine entrances, 471
Block(s) of energy, 227, 315
Blood, 439
    buffering action in, 237
    circulation problems, 518
    clotting action, 527
Blood glucose level, 511–13, [512f]
Blood test, 542, 545
"Blow," 354
Blue flame, 354
Blueprint, 449
Boats, 231
BOD, 297, 300
Bohr, N., 88
Bohr atom, 89
Boiling, 178–79
Boiling chips, 179
Boiling point, 184, 386
Boiling point elevation, 207–210
Bombarding nuclei, 316
Bond(s), 106–128, [238f], 500
    axis, 376, 377
    between carbons, 374
    coordinate covalent, 121
    covalent, 119–21, 238, 239, 262, 270
    directions, 289
    double, 120, 285
    forces between protein segments, 438–39, [438f]
    forms between amino acids, 451
    hydrogen, 193, 194, 262–67
    ionic, 109–11
    metallic, 107–108
    multiple, 120
    P_2 molecule, 289
    pi, 126, [126f], 133
    polar, 121
    sigma, 126, [126f], 263
    strength, 285, 289, 586
    triple, 120
Bond angle, 138–39
Bond energy, 142
1,4-bonding (sugars), 462
Bonding, 237
    between sugars, 461
    covalent, 119–35, 238, 239
    of dissimilar atoms, 238
    of d^2sp^3, 341
    hydrogen, 286
    models, 339
    to sp^3 hybrid, [376f]
    structure shorthand, 377
    with carbon, 373–77
Bonding orbitals, 141
Bones, 291, 529
Borderline
    dosage, 518
    ionic-covalent, 122–23
    metal-nonmetal, [108f], 151–52
Boron, 293
Bottom decay, 293
Bottom lands, 270
Boundary, mineral, 337
Boyle, R., 164
Boyle's law, 165, 167
Br_2, 402, 404, 411
Brackish estuaries, 213
Brain and nerve activity, 496
Branched starch, 464
Branching isomers, 381, 384
Branching in polymer chains, 463, 464
Branching position, 381
Brass, 359
Breakapart, 411
Break β-1,4-linkages, 465
Breaking actin-myosin bonds, 500
Breaking bonds, 285, 401, 409
Breaking of cells, 213
Breathing, 236
Breeding (programs), 533, 535, 540
Brick lining, 355
Bring negative charges together, 477
British sailors, 521
Brittle cementite, 356
Broadleafed plants, 552, 553
Broken natural barriers, 556
Bromide, [27t]
Bromine, 149–50, 404
Bromine-alkane mixture, 411
Bromine atoms, 410, 411
Bromo-, 390
Brönsted-Lowry theory, 268
Bronze, 359
Brownian motion, 277
Bubble formation, 178–79, 213
Buffer(s), 237, 596, 597, 598, 599, 602
Buffer assumptions, 599
Buffer problems, 596–602
Buffering fails, 602
Building blocks of nylon, 396
Building blocks of proteins, 395
Bulky CH_3 group, 422
Bulldozers, 231
Bulls, 533
Buried plant remains, 468
Burn off impurities, 353–56
Burning coal, 233
Burning fossil fuel, 471, 507
Burning of wood, 586, 587
But-, 381
1.3-butadiene polymerized, 423
Butadiene units, 422
Butane(s) 381t, 381, 382, 384
Butanone, 393
1-butanthiol, 396
Butene, 384
1-butene, 384
2-butene, 384
2-butene arrangement, 423
Butylethanoate, 395

Butyl mercaptan, 396
1-butyne, 385
2-butyne, 385
Butter or Butterfat, 467, 524

C

c, (speed of light), 227
c (SI Prefix), [560t]
C (SI unit), [560t]
C (chem. symbol), [23t]
C (genetic code), 445, 449
Ca^{2+}, 500
$CaCO_3$, 353, 471
    and water hardness, 271–73
Caffeine, 536 [536f]
C≡G (bonds), 445, 446, 448, 451
$Ca(H_2PO_4)_2$, 291
Cal, 562
Calcium, 526, 529
Calculating equilibrium concentrations, 572–75
Calculating K, 565–67
Calculating an unknown concentration, 568–69
Calculation errors, 592
Calculation(s) involving equilibrium constants, 592
Calculation of solubilities, 606
Calculus, 202, 584
California, 296
Calm backwaters, 270, 271
Calm lakes, 299
Calorie(s), 41, 589
Calories per degree mole, 590
Campus socialites, 240
Canada, 297, 364, 470
Canal rays, [75f]
Cancer, 288
Candela, [561t]
Candy, 513
CaO, 353
Cape Kennedy, 586
Capillary fall, 172–73, [173f]
Capillary rise, 172–73, [173f]
$Ca_3(PO_4)_2$, 291
Capsule implants (DES), 518
Captured electron, 308
Car(s), 229, 232, 248, 287
Car on hill analogy, 222
Car pools, 232
Car radiator boiling, 179
Carbohydrate(s), 456, 468, 469, 481, 483
Cabohydrate metabolism, 479, 483–94, 503, 542, 543
Carbohydrate oxidation, 480
Carbon, 286, 353, 505
    anodes, 350, 351
    chains and rings, 286
    deposits of, 274, 468
Carbon-12, as atomic weight standard, 47–48
Carbon-carbon bonds, 374, 377, 402, 403
Carbon chains, 373, 377, 380
    combined, 414
Carbon char, 406, 416
Carbon compounds, 373
Carbon dioxide, 231, 344, 353, 505

hardness in water, and, 271–73
Carbon dioxide greenhouse effect, 506, [507f]
Carbon dispersed in iron, 356
Carbon for manufacturing, 468
Carbon-halogen bonds, 416
Carbon monoxide, 339, 353, 504, [611t]
Carbon-oxygen bonds, 416, 417
Carbon with a negative charge, 416
Carbon residue, 468
Carbon rings, 377, 378, 388
16-carbon saturated acid, 466
18-carbon saturated acid, 466
18-carbon unsaturated acid, 466
Carbonate hardness, 271–73
Carbonate(s), [27t], 313
Carbonium ion, 406, 407–408
   forms, 407
   reaction, 407
   structures, 409
Carbonyl(s), 339, 392, 394, 416, 417, 430, 442, 457, 459
Carboxyl (group), 393, 428, 431, 434
Carboxylic acid(s), 393–94, 395, 414, 420, 421, 427, 428, 434
   activated, 428, 434
Carbyne, 553
Carp, 295
Carrots, 481
Carry code, 446
Carrying salts away, 273
Case hardening, 358, [358f]
Casein, 276, [277f]
Castrating bull calves, 533
Catalysis, 116
   chemical equilibrium and, [249f]
   mechanism of, 247
Catalyst(s), 144, 247–48, 249, 252, 285, 287, 408, 413, 414, 416, 420, 421, 423, 498, 583
   enzymes as, 264–65
   negative, 248
Catalyze, 511
Catalyze isomerization, 483
Cathode, 72, 111, 350
Cathode ray tubes, 72–76
Cattle, 288, 503, 533
Causeway, gravel, 216, 217
Causing nuclear reactions, 317
Caves, formation, [274f]
Cell(s), breakup, 212
Cellulose structural materials, 464
Celsius temperature scale, 38–40
Cementite, 356
Center of positive charge, 416
Center of star, helium, 324
centi, 33, [560t]
Centigrade temperature scale, see Celsius
Centimeter, 33
Central atom, 130, 133
Central nervous system, 537
Central sewage system, 534
Cervical cancer, 519
Ceylon, 550
C.G.P.M., 560
CH_4, 378, [378f], 381t, 381
C_2H_6, 378, [378f], 381t, 381, 384
C_3H_8, 378, [378f], 381t, 381, 384
C_4H_{10} [381t]

C_5H_{12}, [381t]
C_6H_{14}, [381t]
C_7H_{16}, [381t]
C_8H_{18}, [381t]
C_9H_{20}, [381t]
$C_{10}H_{22}$, [381t]
C_2H_4, 379, [379f], 384
C_3H_6, 379, [379f], 384
C_4H_8, 379, [379f], 384
C_2H_2, 380, [380f]
C_3H_4, 380, [380f]
CH_3-, 382
C_2H_5-, 382
Chains (protein), 435–39, 496
Chains attach to main chain, 382
Chain polymerization of propene, 421, [422f]
Chain reaction, 319, 409, 410, [410f]
Chain transfer reactions, 410
Chance for reaction, 242
   neutrons, 319
Chancourtais, B., 85
Change(s)
   in arrangement, 242
   in concentrations, 570, 573, 593
   climate, 505
   in deep water, 295
   during reaction, 607
   in enthalpy, 589
   in entropy, 589, 590
   in free energy, 590
   in environment, 230
   irreversible, 223, [223f]
   from natural chemistry, 532–37
   nonspontaneous, [223f], 228
   spontaneous, 222–26, 228, 231, [223f]
   in ring forms, 461
   in state, 584
   in structure, 452
   surroundings, 557
   types of fish, 295
   reversible, 223, [223f]
Changing conditions, 10, 271
Changing Earth's atmosphere, 505
Changing lake level, 216, 217
Characteristic four rings of steroids, 513
Characteristics of polymers, 423
Charcoal, 468
Charge of an electron, 76, 78
Charge concentration areas, 267
Charge-to-mass ratio, 76, 80–81
Charges allowed to move apart, 477
Charges concentrated, 477
Charles' law, 165–66, [165f]
CH_3-CH_3 type shorthand, 377
C_3-CH(NH_2)-C_2H_5, 395
CH_3COOH, 394
C_2H_5COOH, 394
C_4H_9COOH, 394
CH_3-COO-CH_3, 395
CH_3-COO-C_4H_9, 395
C_3H_7-COO-C_2H_5, 395
Checking assumptions, 600
Chemical change, 16
Chemical abstracts, 382
Chemical activation, 428
Chemical alteration of hemoglobin, 542
Chemical corrosion, 359
Chemical energy, 228–33, 585, 586

Chemical environment, 546, 557
Chemical equations, 28–30, 56, 564
   balanced, 28, 254
   balancing of, 28–30
   complete, 28
Chemical equilibrium, 236, 250–54
Chemical equivalents, 56
Chemical evidence for ions, 114–17
Chemical family 102, 149–55
Chemical forms produced by life, 469
Chemical kinetics, 237–48
Chemical nature of water, 259, 271, 295
Chemical production of lysine, 529
Chemical reactions, 237
   energy dependence of, 242
Chemical removal of nutrients, 296
Chemical separations, 346, 347
Chemical warfare weapons, 552
Chemistry, 13, 16
Chemistry of nitrogen and phosphorus, 284
Chewing, 503
CH_3 groups, 421
C_4H_{10} isomers, 381
Chickens, 533
Chickweed, 553, 554
Child mortality, 540
Children, 528, 529, 533, 541, 542, 549
Children's aspirin, 536
"Chinese syndrome," 320
Chlorates, [26t], 144
Chloride, [26t]
Chlorinated benzene, 557
Chlorinated hydrocarbons, 404
Chlorinated hydrocarbon pesticides, 366
Chlorination reactions, 404
Chlorine, 149–50, 217
Chlorite, [26t]
Chloro-, 340
Chloroalkane, 416
4-chloro-2-butynl-m-chlorocarbanilate, 553
Chloroethane, 422
Chloroform, 391
Chlorohydrin process, [404f], 405
Chloromethane, 391
Chlorophyll, 488
Chlorotetracycline, [549f]
CH_3NH_2, 395
C_5H_5N, 460
$C_2H_3O_2^-$, 577, 596
CH_3-O-CH_3, 394
CH_3-O-C_2H_5, 394
C_2H_5-O-C_2H_5, 394
CH_3OH, 391, 505
C_2H_5OH, 391
C_3H_7OH, 391
Choice of addicts, 540
Cholesterol, 513, [514f]
CHR, 435
Chromium, 157, 338, 359
C_4H_9SH, 396
cis-, 385
cis-aconitic acid, 491f
cis arrangement, 423, 524, 525
cis isomer, 385, 423
cis pattern like natural rubber, 423
cis-polybutadiene, 423, [423f]
cis-trans isomers, 385, [385f], 386
Cities, 297

Citric acid, [491*f*]
Citric acid cycle, 480, 489–90, [491*f*], 494, 502, 512, 513
Citrus fruits, 521
Civilization, 332
Class of compound, 380
Classes of derivatives, 396
Class(es) of high energy compounds, 477, 479
Classes of polymers, 423
Clay, 347
Clean reduction step, 351
Clean surfaces, 365
Clear water, 294
Cleveland, 299
Climate, 259, 278, 280
Close planting, 292
Closed shell, 102, 119–121, 123, 154, 285 340
Closed shell configuration, halfway between, 373
Cloud of electron concentration, 407
Clouds, 507
Cloud seeding, 183
Clover, 553
Cloverleaf structure, 446
cm, 562
cm^2, 562
CN$^-$, 542
CNO$^-$, 542
CO, 353, 354
CO$_2$, 184, 284, 292, 293, 326, 344, 348, 353, 505, 506
and water hardness, 271–73
CoA, 490
Coagulation of colloids, 276
Coal, 352, 361–62, 377, 468–69, 470, 471, 473, 504, 505
as a fuel, 230, 231
burning techniques, 232
deposits, 361
dust, 472
mining, 361
mining areas, 471
pulverized, 232
supplies, 473
tars, 468, 469
CoASH, 491
Coating of surfaces, 472
Cobalt, 157
Cod liver oil, 524
Code, 439, 446, 447, 461, 541
Code carrying site, 446, 451
Codeine, 540
Codon(s), 446, 449
Coefficient, 570
Coefficient as exponent, 254, 564, 606
Coenzyme, 523
Coenzyme A, 479, [489*f*], 490, 512
Coenzyme Q, [487*f*], 488
CoF$_6^{3-}$, 343
Coiled helical chains, 445
Coiled sections of t-RNA, 446
Coiled spring, 584
Coke, 352, 362, 468
Coking, 468
Colds, 535
Collapse
of solid ice structure, 254

of puckered hexagons, 265
Collected wastes, 298
Collection of wastes, 297
Colligative property, 202–211
Collision frequency, 239
Colloidal dispersions, 193, 275–78
of butterfat, 276
gases in liquids, 275
liquids in solids, 276
hydrophilic, 275–76
hydrophobic, 276
stabilizing, 276
in water, 275–78
Colloidal particles, size, 275
Color of ions, 114
Color television set phosphors, 159
Columns of the periodic table, 85
Coma, 538
Combination of chains by H$_2$O removal, 414
Combination of iron and carbon, 356
Combination of orbitals, 340
Combination of "progesterone" and "estrogen" 516
Combinations of active sites, 452
Combinations, nonreactive, 248
Combustion, 146, 430, 469
of coal, 502
of liquid fuels, 504
of pulverized coal, 507
Commercial addition reactions, 404
Commercial crop, 556
Commercial granulated sugar, 461
Commercial nickel separation, 339
Commercial phosphate fertilizer, 291
Commercial polymers, 413, 419–23, 427
Commercial production
of ether, 414, 415*f*
of aluminum, 347
of nylon, 415, 419–26
of 2,4,5-T, 553
Commercial recovery of nonflowing oil, 471
Commercial synthetic fibers, 421
Commercial sugar production, 461
Commercially important free radical chain reactions, 411
Commercially important herbicides, 553
Commercially important insecticides, 551, 552
Common names, 381, 382, 385, 388, 465
Completely folded proteins, 439
Complex calculations, 575
Complex formers, best, 340
Complex ions, 124–25, 149, 338–44, 364, 428, 502
Complex reactions, rate of, 244
Complex organic bases, 434–42
Complex organic molecules, 377
Complex of oxygen with hemoglobin, 502
Complexing agent, 363–64
Complexing of Cr^{3+}, [341*f*]
Complex ring ligands, 339
Complexing metals, 339
Complications of injections, 516
Complimentary bases, 449
Components of muscle, 496–98
Compounds between nonmetals, 611

Compounds, 19
associated with life, 373, 377
composition of, 19–20
covalent, 119–21
ionic, 109–11, 237, 239
naming of, 25–27
related to sugars, 461
of Xe and Kr, 155
Compressing gas, 165, 179
effect on vapor pressure, 176
Concentrate, 197, 231
salts by evaporation, 273
Concentrated fissionable material, 319
Concentrated forms, 231
of energy storage, 468
Concentrating fissionable material, 319
Concentration, 64, 196–211, 240, 249
of aluminum in Earth's crust, 346
of animals, 293
of ATP, 494
below equilibrium, 502
of charge, 477
of dead plants, 297
effective, 241, 242
effect on rates, 242
of electrons, 405
away from bond axes, 377
of free radicals, 410
of gas(es), 566, 567
of glucose in blood, 571
increasing, 240
lowering, of reactants, 248
multiplying, 240
of negative charges, 444
of people, 297
rate dependence on, 239–41
of salts, 212
terms, 250
of useful substances, 232
of wastes, 534
Condensation
preparation of colloids by, 275
of steam, 279
of vapor, 171, 175, 507
Condensation polymer, 419–21, 423, 427
Condensation reaction, 413, 414–15, [415*f*], 418, 419, 421 427, 434, **442**, 443, 461
amine and carboxylic acids, 415, 419, 420
of phosphate and amine, 479
Condensation reaction reagent, 444
Condensed phase, 171
Condition of products, 584
Condition of reactants, 584
Conditions
for doing work, [492*f*]
of equilibrium, 584
favorable, 242
for moderating, 319
N$_2$ reaction, 285
necessary, 237
Conductor, electric, 106–7, 111, 359
Conduction bands, 108–9
Conference General des Poids et Measures, 560
Configuration effect, 150
Co(NH$_3$)$_6^{3+}$, 342
Conjugate, 577

Conjugate acid, 579
Conjugate base, 577, 579
Connecting layer, 276
Connecting plate, 497
Connective tissue, 519
Consecutive groups of 3 bases, 451
Conserving food, 502
Conservation of energy, law of, 17, [18f]
Conservation of momentum, 316
Consistent quality, 419
Constant concentration, 596
Constant mass number, 306
Constant mass of Universe, 314
Constant motion, 167
Constant pressure, 584, 586, 587
Constant temperature, 584
Constant total charge, 306
Constant volume, 586
Consume oxygen, 295
Contact(s), 239, 247
  effective, 242
  frequency of, 239, 243
  number of, 240
Contraceptive pill(s), 518
Contracted muscle, [499f]
Contractile proteins, 452
Contraction, 453, 500
Contribution of water, 602
Control H^+ concentration, 596
Control measures, 555
Control methods, 534
Control rods, 319
Control systems, nuclear reactor, 321
Controlling body temperature, 536
Controlling nuclear reactors, 319
Conversions
  of coal, 505
  between α and β sugars, [460f]
  of blood glucose, 511
  heat to electricity, 279, 488, 503
  liquid to solid, 261–62
Conversion factors, [562t], 562
Conversion relationships, in problem solving, 35–37
Convert back to sugars, 469
Convertibility of energy, 227, [238f]
Converting hydrogen to helium, 324
Converting NADH to NAD^+, 487
—COO^-, 429
—COOH, 393, 427, 428, 429, 430
Cooked vegetables, 524
Cool deep waters, 278, 279, 280, 295
Coolant choice, 279
Cooling
  by evaporation, 174
  towers, 280
Coordinate covalent bond, 121, 149–50, 289, 428
Copper, 293, 335, 338, 344, 354–62, 538
  deficiency, 362
  impure, 360
  low abundance, 359
  mining, 344
  oxidation states of, 158
  oxides, 344, 359
  reduced, 360
  sulfide, 359
Copper-tin alloys, 359
Copper zinc alloys, 359

Corn, 10, 292, 528t, 532
  oil, 467
Corn belt, 292
Corpses, 499
Cortisone, 519–20, [520f], 533, 534
  therapy, 520
Cosmic rays, 267
Cost
  in entropy, 232
  indispersal of energy and material, 231
"Cottage industries," 563
Cotton, 464
Coulomb, 561t
Coulomb barrier, 317
Coulomb repulsion, 312–13, 316
Covalent bonding, 119–35, 238, 239, 262, 267, 270, 285, 409, 438, 448, 611
Cr^{2+}, 338, 341
Cr^{3+}, 338, 341
Cr^{6+}, 338
Cracking, 402–403, 504
$Cr(H_2O)_6^{3+}$, 340
Cream, 276
Creatine, 496
Creatine phosphate, 479, [480f], 494–98, [497f]
Creation of order, 230
Crick, F., 445
Crippling effects, of arthritis, 519
Crisco, 467
Critical point, 179, 184
Critical pressure, 179
Critical temperature, 179
Crops, 532, 535, 549, 554
Crops, irrigating, 273
Crowding effect, 100
Crude oil, 469
Crushing, 345
Crust of Earth, oxygen concentration, 325
Cryolyte, 350
Crystal(s), 182
Crystal pattern, 183, 356
Crystal structure(s), 182, 421, 449
  of Al_2O_3, 340
  ice, [266f]
  metallic, 207
  relationship to supersaturating, 196
Crystalline sulfur, 183
Crystallized sugar, 461
Cu^{2+}, interaction with water, 338
Cubic centimeter, 34 [36t], 37
$Cu(H_2O)_2^{2+}$, 340
$Cu(H_2O)_6^{2+}$, 343
Cultivated fields, 534
Cultivation, 297, 556
Cultural environment, 230, 232
Cumulative poisons, 364–66
CuO, 359, 360
Cu_2O, 359, 360
Current, electric, 72, 106, 111
Curve of stability, 312–13, 421
CuS, 359
Cu_2S, 359
Cuyahoga River, 299
Cyanate ion, 542
Cyanate treatment, 542
Cyanide, 27t, 366, 542
Cyanocobalamin, [522f], 523

Cyclic hydrocarbons, 390
Cycling of nutrients, 532
Cyclo, 340
Cyclohexane, 390
  ring, 557
Cyclohexene, 390
Cyclopentane, 390
Cytochromes a, a_3, b, and c, [487f]
Cytosine, [441f], 442, 444, 445, [447t]
Cytosine-guanine (bonds), 445 [445f]

D

D, 430
2,4-D, 552–53, [553f]
Δ, 588
d electrons, 90, 161–164
d electron levels, 337, 339, 340
d electron level availability, 153, 155
d orbitals, 94
d-isomers, 386
d subshell, 98
Dacron, 421
Daily dose, 518
Dairy plant, 296
D-Alanine, 430 [431f]
Dalton, J., 21, 22
Dalton's atomic theory, 21–22
Damage from artificial changes, 557
Damage of galactosemia, 542
D-amino acid(s), [429f], [431f], 430, 431, 483
Dams, 9, 10, 213, 278, 299
Dandelions, 552
Dangerous foods, 366
Dangerous vapor pressure, mercury, 364
Dark and light skin, 526
Darkrooms, 248
Data about concentrations, 565
DDT, [365f], 366, 550–51, [550f], 554, 555, 556
DDT-resistant species, 550
D-deoxyribose, [459f], 460, 461
Deactivated, 242
Dead fish, 296, 366
Death, 296, 392, 513, 521, 527, 536, 538, 546, 548, 552
Death rates, 550
Debris, mining, 231
deca-, [560t], 611t
Decane, 381t
Decarboxylation of histadine, 537
Decarboxylation reactions, 523
Decay, 293
Decay chains, 306, 313
Decay energy, 309
Decay rate, 307
Decay of $^{234}_{90}$TH, 306
Decay of $^{235}_{92}$U, 307
deci-, [560t]
Deep ocean, 294
Deep water, 278, 279
  decay in, 298
Deep wells, 271
Deepening voice, 516
Defective genes, 540, 545

Defenses, 545
Deficiency in using galactose, 542
Deficiency in using glucose, 542
Deficiency of essential amino acids, 528
  of vitamin A, 524, 525
  of vitamin B_{12}, 523
  of vitamin D, 526
Deficient diet, 521
Deficiency of essential amino acids, 528
Definite composition, law of, 19–20, [20f], 21, 43, 51
Degradation, natural, 271
Degradation of sewage, 296
Degrading bacteria, 299
Degrading energy, 229, 230
Dehydration, 212
Dehydroascorbic acid, 520, [521f], 521
Dehydrocholesterol, 526
Dehydrogenation, 414
Delayed afflictions, 540
Delayed neutrons, 319
Delivery of oxygen, 230
Delta(s), 278, 588
Demerol, [539f], 540
Democritus, 21
Denaturation, 452
Dense material, concentration in Earth's core, 325, 335
Density, 41, 172, 179, 181, 385,
  ice less than water, 265, 278
  properties of water, 265, 278, 280, 294
Deoxyribonucleic acids, 444–51
Deoxyribose, 442 [442 f], 444 [447t], 449, [459f]
Dependence
  on concentration, 239–41
  of K_{eq}, 251
  of rates, 239–41
  on light energy, 411
  temperature of energy distribution, [245f]
Dependence on water supply, 270
Depleted oxygen, deep water, 295
Depleted soil, 292
Deposit of ore, 366
Deposition of new soil, 270
  stopped, 271
Depression of nervous system, 537, 538
Derivatives of pyrimidine and purine, 440–42, [441f]
Derived units (SI), 560, [561t]
DES, 518, [518f], 519, 533
Descendants, 449
Deserts, 259
Desirable locations, flood plains, 270
Destruction of plant life, 361
Destructive mining, 352
Detergents, 11, 276, 296, 363, 452
Detergent-like action, 481
Deuterons, 316
Develop antibodies, 548
Develop immunities, 557
Dextrins, 464
dextro-, 430
D-fructose, [459f], 460, 461, [462f]
D-galactose, [459f], 460, 461, [462f]
D-glucose, [459f], 460, [460f], 461, [462f]
  crystallized, 460
D-glyceraldehyde, 457, [458f]

di-, 383, 390, [611t]
Diabetes, 511–13, 533, 540, 541, 543, 555, 556, 580
Diabetic coma, 513
Diacid chloride, 421
Diacids, 420, 421
Diagonal relationships, 153–54
Diamagnetic, 99, 141
Diamines, 420
Diamond, 183
Diatomic molecule, 119
1,2-dibromopropane, 391
Dichlorodiphenyltrichloroethane, 550, [550f]
4,4-dichloro-3-iodo-butene, 391
2,4-dichlorophenoxyacetic acid, 552, [553f]
Dichromate ion, [135f]
Dicoumarol, 527, [527f]
Dieldrin, 551, [551f]
Diene, 384, 423
1,3-diene, 422
Diet, 525, 528
Diethyl ether, 394, 540
  of morphine, 540
Diethylstilbestrol, 518, [518f], 533
Differences
  in activation energy, 583
  in attractive forces, 299
  in binding force, 322
  in biochemical behavior, 513
  due to coulomb repulsion, 313
  emphasizing, 238
  in energy, 249, 583
  in polymer chain, 422
  in reactivity, 408
  in tempering, 358
Diffraction, x-ray, 86
Diffuse heat, 232
Diffusion,
  of gases, 178
  in liquid, 174, 215, 217
Digestion, 211–13, 469, 470, 481–83, [482f], 516, 523
Digestive system, 543
Dihydroxyacetone, 457, 457f
Dihydroxyacetone phosphate, 484f
2,3-dihydroxypropanal, 457, [457f]
1,3-dihydroxypropanone, 457, [457f]
Dilemma of nonmetal compound names, 611
Dilute solutions, 208
Dilution, 197, 198, 231, 232
  of Great Salt Lake, 216
Dimenhydrinate, [538f]
Dimer, 287–88
Dimethyl-, 384
Dimethylbenzenes, 389, [389f]
Dimethyl ether, 394
Dinitrogen oxide, 20f, [611t]
Dinitrogen pentoxide, [20f], [611t]
Dinitrogen tetroxide, [20f,] [611t]
Dinitrogen trioxide, [20f], [611t]
Diol, 392, [404f], 405
1,4-dioxane, 440
Dioxin, 553
Dipole, 121
Dipole-dipole interactions, 438
Diphosphate product, 486

Diphosphoglyceric acid, [484f], 486
Diphosphorylated thiamine, 523
Direct injection of insulin, 513
Direct oxidation by air, 405
Direction of helix, 436
Disaccharide, 461–63
Discharge wastes, 293
Discovery of alpha helix, 445
Discovery of DNA structure, 445
Disease(s), 10, 293, 532, 534, 548
  resistant to drugs, 549
Disease-causing bacteria, 297, 298
Disease control, 550
Disease(s) imported by colonists, 555
Disease resistance, 556
Disorder, and entropy, 224, 226
Disordered arrangement, 226
Dispersed
  colloid, 276
  forms, 231, 232
  heat, 231
  particles, 275
Dispersion
  of liquid in liquid, 276
  on a molecular scale, 199
  of molecules or ions, 192
  preparations of colloids by, 275–76
  spontaneous, 276
Displacement reaction, 402, 412, [412f], 416, 418
Dispersion of plants, 556
Disposal problems, 337
Disposal into water, 271, 297
Disproportionation, 338
Disruptive activity, 472
Dissimilar atoms, bonding, 238
Dissimilar groups, 385
Dissimilar myosin regions, 497
Dissimilar reaction sites, 434
Dissipation of energy, 584, 586, 587
Dissolving, 259
  of carbonate rocks, 471
  of ionic substances, 267
Dissolved
  carbon, 357
  CO_2, 291
  iron, 295
  oxygen, 295, 299
  salts, 293
Dissolving acidic oxides, 348
Dissolving in water, 344
Dissociation from actin, 500
Distribution of elements, 304
Disulfide bridge, 438
Division of equations, 605
dl, 386
D-lactic acid, 430, [430f]
dl-3-methylhexane, 386
D-mannose, [459f], 460
DNA, 442–51, 460
  replication, 448
  structure, 445
Dobereiner, 85
Donation of H^+, 406, 428
Donor of electron pair, 269
Dosage, 554
Dosage, reduced, 518
D-sugars, 457, [458f], 458, 483
Double arrow, 412

Double bond, 210, 276, 285, 384, 403, 405, 421
  addition to, 404
  between carbons, 379
  carbon-oxygen, 392
  formation, 402
  rigidity, 376
Double bond attacks halogen, 407
Double helix, 446, 496, 497, [497f]
Double use of coefficients, 605
Downstream communities, 471
Drainage basin, 216
Dramamine, 537, [538f]
Drawing arrows, 411
D-ribose, [459f], 460, 485
Drinking water, 278
Drowsiness, 537
Drugs, 535
Drugs (anticlotting), 527
Drugs, birth control, 516, 517
Drug dosage, 518
Drug use, by pregnant women, 519
Dry areas, irrigation of, 273
"Dry ice," 184
dsp^2, 340, 343
d^2sp^3, 340, 341, 343
Duodenum, 481
Duplicate natural controls, 535
Duplication of DNA, 445, 448, [448f]
Duplication of identical proteins, 436
Dust, 275, 505, 507

E

e, 76
e/m, 76–78, 80–81
E, 588
$E = hc/\lambda$, 227
E_{act}, 242
E plus PV, 588
Earth, 227, 228, 230, 259, 333, 506
  ability to radiate energy, 506
  magnetic field of, 226
Easily degraded, 552
Easily polarized, 406
Eastern United States, 364
Ecological system change, 271
Economic factors, 350, 351
Ecuador, 294
Edmonton, Alberta, 470
Effect of β decay, 306
Effect of catalyst, 249
Effect of changes on enthalpy, 588t
Effect of PV work, [587t]
Effect on rates of concentration, 241, 242
Effect of wastes on water, 293
Effective concentrations, 242
Effective contact, 242
Effective electrophilic reagents, 406
Effective molality, 210
Effective reaction, 242
Effective sharing, 238, 239
Effective solvent, 260
Egg(s), 514, 523
Einstein's theory of relativity, 314
Ejecting electrons, 308
Electric charge, 77–78
Electric conductance, 106

Electric current, 72, 106, 111, 350
  relation to chemical reactions, 111–17
Electric field, 76, 304
Electric heat is wasteful, 504
Electrical charges, 439
Electrical conductivity, 106–117
Electrical contacts, 364
Electrical energy, 233, 503
Electrical evidence for ions, 113–14
Electrical forces, 407
Electrical imbalance, 263
Electrical impulses, 496
Electrical wiring, 359
Electricity, 72, 233, 350, 503, 504
  generation of, 213, 230, 232, 233, 504, 507
Electrode, 72, 106, 111, 350
Electrode reactions, 107
Electrolysis, 72, 111–14, 145, 349–50, 364, 365
  of Al_2O_3, 350
  Faraday's Laws of, 72, 111
  run in series, 360
  of water, 473
Electrolytic production of hydrogen and oxygen, 146
Electrolytic reduction, 349–51, 356
Electromagnetic waves, 87, 304
Electron(s), 76–78, 81–82, 304
  charge of, 76, 78
  charge to mass ratio of, 76
  energy levels, 308
  mass of, 76, 316
Electron
  affinity, 99–100
  capture, 308, [309f], 313
  concentration around oxygen, 417
  concentration in π bond, 405
  groupings, 130
  -loving reagents, 405
  pair(s), 409
    acceptor, 269, 270, 339
    donor, 269, 270, 339
  probability distribution, 92–94
  repulsion model, 129–35
  sharing, 119
  shifts, 308, 411–12
  spin, 90
  transport system, 487, [487f], 489, 490
Electron volt, 95–96
Electronegative halogen, 416
Electronegative oxygen, 417
Electronegativity, 121–23, 285, 289
  simplified approximations, 122–23, [123f]
Electronic configurations, 90–92, 156, 159, 286
  periodic table and, 101–102
  rules for assigning, 90–91
  table of, 91–92
  transition metals, 339
Electronic rearrangement, 171
Electrons moving, 411, 412
Electrophilic reagent(s), 405, 407, 408, 409, 417, 428
  attacking double bonds, 406
Electroplating, chromium on steel, 366
Electrorefining, 360, [360f]
Elements, 19

  beyond uranium, 318
  found on earth, 321
  general trends, 140–60
  inner transition, 159–60
  noble gas, 154–55
  periodic classification of, 85–87, 101–102
  prediction of new, 86
  representative, 149–54
  transition, 155–58
Elemental sulfur, 361, 468
Eliminating free radicals, 411
Eliminating mathematical difficulties, 572
Eliminating natural control, 540
Elimination of DDT, 550
Elimination (reaction), 412–14, 418
Embalming, 393
Embden-Meyerhof pathway, 483–87, [484f], 489, 490, 494, 496, 512
Emergency diabetic treatment, 513
Emergency nature of spills, 472
Emissions, rate of, 228, 230
Emitted radiation, 304
Empirical formula, 24–25
Empty holes in water, 264
Empty orbitals, 339, 441
Emulsifying agent, 276, [277f]
Emulsifying fats, 481
Emulsion, 276
-ene, 380, 383
Energy, 16, 41, 222, 314, 588
  activation 241–42, 244, 246, 247, 249
  available, 309
  average, 171, 174, 244
  biological, 291
  blocks of, 227
  to break bonds, 409, 410
  chemical, 228–33
  concentration of, 226
  conservation of, law of, 17, [18f], 222
  conserving attitude, 232
  content, 584, 585
  conversions, [238f], 584
  converted to heat, 587
  convertibility, 227
  crisis, 471
  degradation of, 229, 230
  difference, 249, 583
  dissipated, 585
  distribution of, 227, 245, [245f]
  electrical, 233
  emitted by stars, 324
  "escape of," 223
  factor(s), 583, 588
  flow, 227, 230
  of helium fusion, 324
  from hydrolysis of ATP, 492, 500
  from sunlight, 456
  gravitational, 222, 230
  human use of, 230
  hydroelectric, 279
  ionization, 95
  kinetic, 17, 41, 170, 222
  loss of, 245
  mass, 314–15
  minimum, 244
  of molecules, 244
  of motion, 237, [238f]
  zero, 40, 244, 259

629

needed to make a positron, 309
of neutrons, 319
in "nonfood" material, 503
of oxidation, 487
potential, 17, 41, 222
produced in stars, 322
production, 483, 485
relations, 61–64
released, 263, 285, 308
   by volume contraction, [587f]
releasing reaction sequence, 483
reservoirs, 427, 456–73
source(s), 204, 297, 452, 468, 469, 473, 477, 480, 503
   recovery of, 469–71
storage, 291, 456, 468–69
   in plants and animals, 463
stored, 477, 496, [497f]
   in ATP, 492
storing reactions, 463
transfer, 496
usable in biochemical reactions, 492
use of, 230, 481
"wasted," 229, 503
of water, 259
Energy dependence of chemical reactions, 242
Energy level approach, 314
Energy level shifts, 155–58
Energy levels
   atomic, 90, [91f], 93–95
   close, 313
   factors affecting, 97–100
   higher for neutrons, 313
   higher for protons, 312–13
   in hydrogen, 88–89, [89f]
   in metals, 108
   in oxygen, 143, [142f]
   protons and neutrons, 311, 312
   in transition metals, 155–58
Energy–mass relationship, 314–15
Energy storage molecules, 456, 465
Energy transfer agent, 419
Energetic advantage of small steps, 487
Energetically favored products, 338
Energetically favored shape, 439
Engagement, 248
Engineering project, 217
English system of measurement, 35–37
Enlarged liver, 543
Enol form, [441f], 442
Enol form of pyruvic acid, 479, [479f]
Enriched water, 296
Enrichment, nutrients, 294
Enrichment, surface waters, 295
Enthalpy, 584–88, 589, 591
Enthalpy, change in, 62
Entropy, 107, 127, 202, 224–33, 287, 288, 456, 588, 589
   cost, 232
   decreases, 226–28
   decreasing reactions, 366–67, 456
   defined, 589
   factor, 584
   high, 226, 228, 230
   increase(s), 224–28, 230–31, 504, 505
      conservation of, 232
      in solution, 591
   increasing reactions, 456

   low, 227, 228
   principles, 232
   waste of increase, 232
   zero, 224
Entropy units, 590
Environment, 230
   mercury in, 365
Environmental consequences, 217
Environmental crisis due to accident, 472
Environmental disruptions, 472
Environmental hazards, 551
Environmental problem, 471
Enzyme(s) (catalysts), 428, 448, 451, 452, 465, 481, 483, 485, 486, 487, 511, 523, 542, 545, 552
Enzyme catalyzed reactions, 487
Enzyme irritation, 487
"Equal or larger energy" factor, 503
Equal protons and neutrons, 311
Equations, chemical. *See* Chemical equations
Equation balancing by inspection, 28–30
Equilibrium 175, 202, 217, 250–54, 288, 357, 486, 583, 584, 590
   applied, 236
   calculations, 572
   concentrations, 564, 569
   condition, 250, 590
   at constant T and P, 590
   dependence on relative states, 583–84
   mathematical applications, 236
   mixture of α and β rings, 460
   principles of, 236
   problems, 592
   shifts, 176, 250
   of volative materials in water, 273
Equilibrium constant(s), 251–54
   calculation, 565
   K^c and K^p, 567–68
   problems, 564–80
   thermodynamic basis, 236
Equilibrium constant expression(s), 564, 565, 569
Equivalent, 198
Erosion, 273–75, 278
Error in a DNA, 541
Errors in base sequence, 449
Errors from calculations, 592
"Escape" of energy, 223
Escape from fission region, 319
Eskimo(s), 260, 525
Essential amino acids, 468, 528–29, [528t], 543
Essential nutrient, 485
Essential substances, 529
Establish equilibrium, 584
Ester(s), 394–95, 414, 465, 479
   linkage broken, 418
Estimate size of errors, 594
Estradiol, 514, [515f], 516, 518
Estratriol, 514, [515f]
Estrogen(s), 514, [515f], 516, 518, 519
Estrogenic activity of DES, 518, 533
Estrogen substitute, 516, [517f]
Estrone, 514, [515f]
Estuaries, 213
Ethanal, 393
1,2-ethandiol, 392, 405
Ethane, [381t], 384

Ethanoic acid, 394
Ethanoic anhydride, 395
Ethanol, 116, 391
Ethene, 384, 404, 406, 409
Ethers, 394, 414
Ethyl, 382, 384
Ethyl alcohol, *See* Ethanol
Ethylbenzene, [389f]
Ethylbutanoate, 395
Ethylene glycol, 392, [404f], 405
Ethylene oxide, 405
2-ethyl-1-pentene, 384
Ethyl radical, 40
Ethyne, 385
Europeans, diseases of, 555
Europium, 159
Eutrophication, 294, 296, 297
Evaporation, 174–75, 205, 212, 215–17, 591
Evaporative salt concentration, 273
Even-even isotopes, 311
Even n isotopes, 311
Even numbered elements, 310, 331
Even numbers of carbon atoms, 465, 490
Even numbers of neutrons, 310
Even proton-odd neutron isotopes, 311
Evolution, 217, 543
Exchange, bottom and surface, 294
Exchange forces, 321
Excited state, 306, 315
Exclusion principle, 89
Excretion in urine, 545
Exercise, 496
Experiment, 6, 7, 11
Explanation of diabetes, 513
Exponential numbers, 37–38, 200, 243–44
Exponents, 562, 564
Exposed land, 217
Exposure to acids and bases, 452
Exposure to oxygen, 299
Extent of ionization, 116, 593, 595, 596, 602
   ignored, 593–94
Extent of reaction, 599
Eye diseases, 525, 543

F

f, [560t]
F, [561t]
$^{18}_{9}$F, 316

F_2, 402
f electron levels, 159
f electrons, 90, 159
f orbitals, 94, 159
f subshell, 98
Factor common to both equilibria, 604
Factor in reaction rates, 285
FAD, 487, [487f], 488, [488f]
$FADH_2$, 487, [487f], 489, 490
Fahrenheit temperature scale, 38–40
Failure to decarboxylate pyruvic acid, 523
Failure to dimerize, 288
Failure of natural process, 533
Fall, spontaneous, 237
False alarm shutdown, 320–21
Families, organic compounds, 378
Family of elements, 102, 149–54

Family history, 542
Family relationships, 150–55
Farad, [561t]
Faraday, 111
Faraday, M., 72, 77, 111
Faraday's law of electrolysis, 72, 111
Farm land, irrigated, 273
Farming, 275
Fast-moving streams, 270
Fat soluble, 513
    vitamins, 524–28
Fatalities, 392
Fate of lakes, 278
Fats, 419, 465–67, 468, 469, 481
    metabolism 466, 490
Fatty acids, 465–68, 481, 490
Favorable binding energy, 249
Favorable conditions, 242
    for ether formation, 414
Favorable hydrogen bonding, 451
Favored arrangements, 338
Favored numbers, 313
Fe^{2+} and Fe^{3+}, 295
Fe_3C, 356–58
Feeble-minded persons, 543
Feed efficiency, 519, 533
Feeds, 518
Female hormones (steroids), 514, [515f]
Feminine hormone balance, 533
Femto, [560t]
Fermi, E., 317–18
Ferric, [26t]
Ferrous, [26t]
Fertile period, 518
Fertility prevented, 517
Fertilizers, 217, 284, 287, 292, 294, 296, 403, 532, 573
Fertilizing effect, 294
Fertilizing of lakes, 293
Fever, 535
Fibers, 419, 496, 497, 498
Fibrous connection, 498
Fibrous materials, 451
Fibrous proteins, 451, 452, 453, 497
    "cables," 451, [451f]
    rearrangement of, 453
Field, electric or magnetic, 74, 76
Fields, 534
Figures, significant, 37–38
Filling in of lakes, 278
Final concentrations, 570, 593
Fingernails, composition of, 419
Fire, 468
Fire flooding, 470
Fire hazards, 472
Fire, damaging to steel, 358
First approximation, 595, 596
First-order reactions, 244
First row of transition metals, 337
First transition series, 155–58
Fish, 294, 295
    breeding, 271
    desirable or game types, 295
    forager and scavenger, 295, 296
    kills, 213, 295, 299
Fishing area, 294
Fission, 308, 309, 313, 318–21
Fission rate changes, 319
Five atom rings, 440, 458

Five-carbon sugars, 456
Fixed energies, 587
Fixed forms, 584
Fixed volume, 172
Flashlight batteries, 503
Flat bottom lands, 270
Flavin adenine dinucleotide, 488, [488f]
Flavor, drinking water, 295
Flaws in genetic code, 540
Fleming, A., 547
Flexible fibrous connection, 498
Flexible tubing, 359
Flotation, 345, [345f]
Floating soap, 275
Floods, 270, 271, 278
Flow
    of energy, 227, 230
    of heat, 224, 226, 237, 504
    of liquid, 173, 265
    osmotic, 210–11, 212
Flood Plains, 270, 271, 275
Fluoride, [27t]
Fluorine, 149–50
Fluoro-, 390
9-fluoro-16-metylprednisolone, 520, [520f]
Foam(s), 275, 345
Fog, 280
Fold into secondary and tertiary structure, 451
Folded chains (protein), 436
Folic acid, 546, [546f], 547
Follicle, 514
Food, 232, 295, 470, 503, 543, 549
    metabolism, 483
    preservatives, 248, 288
    processing, 299
Food chain concentration, 365–66, [365f], 550, 554
Football analogy, 119
Forces
    of attraction, 171
    intermolecular, 170, 194
    London. See Van der Waals
    Van der Waals, 170–71
    within nuclei, 314
Forest fires, 361
Formaldehyde, 393
Formic acid, 394
Forms in which equilibrium constants are used, 564–80
Formula(s)
    chemical, 23, 49–58
    empirical, 24–25, 49–50, 55
    in words, 611
    molecular, 24, 50
    from percent composition, 55
    from weights, 54–55
Formula unit, 49–50
Formula weight, 50
Forward reaction(s), 248–49, 583
Fossil chemicals, 473
Fossil fuels, 230, 468–69, 470, 471, 473, 503, 505
Fossil fuel electrical generating plant, 504
Fossil fuel technology, 504
Fossilized remains, 468
Four bonding pairs on carbon, 373
Four-carbon saturated fatty acids, 467
Four different groups attached, 386, 429

Four ligand complexes, [344f]
Fraction of molecules, 203–207
    active, 242–47
Fractional distillation of air, 144
Franklin's experiments with electricity, 72
Free copper metal, 359
Free energy, 389–91, 583
Free energy change, 492
Free energy condition for equilibrium, 590–91
Free flowing streams, 299
Free glucose, 483
Free metal, 336
Free protons, 267
Free radical, 402, 408–11, 412
Free radical chain reaction, 409, 410, 421
Free radical concentration, equilibrium, 410
Free radical formation, 409, 410, 411
Free radical mechanism, 409–11, 418
Free reactants, 248
Freezing, 180, 182, 263, 278
Freezing point(s), 180
    depression of, 207–10
Frequency of reactant contacts, 239, 243
Fresh fruits and vegetables, 521
Fresh water lakes, 278
Friction, 222, 229, 262, 275
Fructose, [469f], 482
Fructose-1,6-diphosphate, 483, [484f]
Fructose-6-phosphate, 483, [484f]
Fruit, 521
Fruit flies, 535
Fuel cells, 504
Fuel oil, 471
Fuels, 229, 230, 377, 468, 473, 504
Fumaric acid, [491f], [544f]
Fumarylacetoacetic acid, [544f]
Function of antibodies, 519–20
Functions of proteins, 451–53
Fungicides, 550
Fungus diseases, 549, 556
Fundamental particles, 81–82
Furnace designs, 232
Fusion, heat of, 180, 183
Fusion nuclear, 321–25

G

G (genetic code), 445, 449
G (thermodynamics), 589
G (SI prefix), [560t]
ΔG, 589, 590
Gamma rays, 304, 306, 309, 315
Gain electrons, 148, 359
Gain of energy, 245
Galactose, [459f], 462, 482, 542, 543
Galactosemia, 542–43
Galactosemia victims, 543
Galactose-1-phosphate, 542
Gall bladder, 481
Gamma rays, 304, 306
Gas(es), 58–61, 163–71, 259
    concentration, 175
    mixing, 237
    mixtures, 192, 225
    natural (fuel), 469, 470, 473

631

pressure, 163–64, 167
quantity, 163
soluble, 194
solutions of, 192–94, 224
stripped by Sun, 325
temperature, 163, 167
volume, 163
Gas constant, 166
Gas law calculations, 168–70
Gas laws, 164–66
Gas problems, 168–70
Gas relationships, 166, 167
Gaseous HCl, 267
Gaseous oxides, 335
Gaseous products, 504
Gaseous wastes, 231
Gasoline, 203, 383, 403, 504
Gastric juices, 362
Gel, 276
Gelatin, colloidal properties of, 275
Gene carriers, 542
General principle, energy, 586
General survey of prospective parents, 545
Generating of electricity, 471
Generators, water in, 279
Genes, 540, 541, 545
Genetic basis for disease, 541
Genetic code, 461
Genetic counseling, 540, 542, 545
Genetic defect diseases, 540–45
Genetic materials, 291
Genital organs, 524
Geography, 259
Geological formations, 469
Geology, 259
Geometrical pattern, 263
Germs, 519
Giant molecules, SiO_2, 50, 289
giga-, [560t]
Girl babies, 519
Glaciers, 506
Glacier Bay (Alaska), 506
Glass, 506
Glassblowers, mercury danger to, 364
Globular proteins, 452, 453, 496, 498, 502
Glu, [432t]
Glucose, [459f], 464, 469, 482, 483, [484f], 492, 511, 542, 543
"Glucose Falls water power facility," [495f]
Glucose in urine, 512
Glucose-1-phosphate, 483, [484f], 542
Glucose-6-phosphate, 477, [478f], 483, [484f]
Glue, 275
Glu-NH_2, [433t]
Glutamic acid, [432t], 541
Glutamine, [433t]
Gly, 432t
Glyceraldehyde, 457, [457f], 460
D-, 457, [458f]
L-, 457, [458f]
Glyceraldehyde-3-phosphate, [484f], 485, 486, 487
Glycerin, 465
Glycerol, 392, 465, [466f], 481
Glycine, 421, [432t]
Glycols, 392

Glycogen, 464, [464f], 483, 489
Glycogen storage diseases, 543
Gold, 336, 360
Gold colloid, [277f]
Gold panning, 345
Graham's law, 167
Grain alcohol. See Ethanol
Gram, 2
Gram atomic weight, 49
Granulated sugar, 461
Grapefruit, 521
Graphite, 183
Grasses, 552
Gravel fill in Great Salt Lake, 214–17
Gravitational contraction, 323–24
Gravitational energy, 222, 230, 503, 584
Gravitational stripping, 325
Gravity concentration, 323–26
Great Britain, 521
Great Divide Basin, 586
Great Salt Lake, 214–17, [214f]
Greek numbers, 611
Green wood sticks, 360
Greenhouse effect, 505–507
Grignard reagents, 416, 417, [417f]
Grinding corn, 503
Ground state, 306, 315
Ground state electronic configuration, 90–92, 156
Ground water, 271–75, 471, 472
Groups,
of electrons, 130
of the periodic table, 85, 149
of three bases, 451
Group I, 150
Group VII, 149–50
Group VIII, 154–55
Growing children, 526
Growing polymer chain, 447
Growing season, 293
Growth, 521, 528
explosion, 294
increase, 292
limit, 292
limiting factors, 284
of new cells, 445
stimulation, 519, 532, 533
Growth, plant, 231, 284
Guanine, [441f], 442, 444, 445, [447t]
Guide technology, 557
Gutta-percha, 423

H

h, 227
H (thermodynamics), 62, 588, 589
H (unit), [561t]
H^+, 115, 148, 158, 267, 405, 406, 407, 417, 428, 431, 564, 575, 576, 593, 594, 596, 597, 598, 602, 605
in acid-salt buffers, 597
acceptor, 268
donor, 268, 428
as electrophilic reagent, 405
equal to K_a, 597
ion product, 485
H^-, 148
$_1^2H^+$, 316

-H, 428, 429
H_2, 287, 473, 565–75
addition, 408, 413, 414
elimination, 413, 414
from cracking, 403

$_1^1H$ burnup in fusion, 324

$_1^1H$ converted to $_2^4He$, 324

ΔH, 62, 588, 589, 591
ΔH_f, 62
Haber process, 287, 403
Hair, 419
Half-filled orbitals, 341
Half-filled subshell effect, 98
Half-life, 307 [308f]
Hall Process, 349–51, [351f]
Halogen(s), 149–50, 401, 404, 406, 407, 408, 413
Halogen substituents, 390
Halogenated hydrocarbon derivatives, 390–91
Hard coating, 358
Hard water, 271–73, 296
HCl, solutions, 194, 267
Hardness, 357
of cementite, 356
Hardness ions, 363–64
Harm environment, 551
Harming useful species, 550, 555
Harmful parasites, 546
Hazards from mercury, 364–66
$HC_2H_3O_2$, 593, 596
HCl, 407, 421, 603, 610, [610t]
addition, 418
addition to ethene, 406
elimination, 413, 414, 418
HCN, 604, 605
HCO_3 in water hardness, 271–73
HCOOH, 394

$_2^4He$, 305

$_2^4He^{2+}$, 304, 316

Headaches, 35
Health, 529
Heart, 519
Heart attacks, 364
Heat, 222, 229, 230, 237, [238f], 252, 584, 585
diffuse or dispersed, 231, 232
disposal problem, 280, 362
flow, 224, 226, 237, 589
released, 233, 263
waste, 230
wasted, 504
Heat absorbing brick work, 355
Heat capacity, 279
Heat of formation, 61–62
Heat of fusion, 180, 183
Heat of neutralization, 114–15
Heat of sublimation, 181
Heat of vaporization, 174–76, 179, 591
Heating, 452, 471, 503
Heavy metal poisoning problems, 297
Hecto, [560t]
Helium, 148
Helix, 436, 445

Hell's canyon, 299
Heme, 488, 502
Hemoglobin, 439, 452, 502, 541
   structure, 502
Henry, [561t]
Henry's law, 195, 203, 213
Hepta-, [611t]
Heptane, [381t], 383, 384
Hereditary abnormalities, 543
Hereditary deficiency in metabolism, 542
Hereditary diseases, 540–45
Hereditary patterns, 448
Hereditary preference for nondiabetics, 555
Hereditary transfer of genes, 449
Heredity, 439
Heredity, molecular basis, 448
Heroin, 540
Hertz, [561t]
Hess's law, 62
Heterocyclic compounds, 439–42
Heterocyclic ring form (sugars), 442
Heterogeneous mixture, 18
hexa-, 383, [611t]
Hexane, [381t], 383, 384
Hexagonal rings, puckered, in water, 264, [266f]
Hexose sugars, 520
Hexoses, 457, 458, 460
HGH, 533
HI, 565–75
High blood pressure, 537
High charged positive ions, 308
High energy, 366, 585
   light, 456
High energy bombardment, 267
High energy compounds, 477–80, 486, 496, 503
High energy molecules, 244–45
High energy neutrons, 312, 319
High energy quanta, 228
High entropy, 226, 456
High galactose concentrations, 543
High heats for water, 260, 279
   capacity, 263, 279
   fusion, 263
   vaporization, 263
High temperature(s), 401, 403, 409, 420, 428, 505
High temperature, high pressure equipment, 421
High unsaturation, 467
Higher concentrations, advantages of, 240
Higher energy state, 250
Higher entropy, 226
Higher fraction of diabetics, 556
Higher organisms, 365–66
Higher rest mass, 315
Higher than normal phenylpyruvic acid levels, 545
Highly unsaturated oils, 451
His, [434t]
Histamine, 537, [537f]
Histidine, [434t], 440, [528t], 537, [537f]
History of oil and gas formation, 469
HNO_2, [610t]
H_2O, 284, 564, 604
H_2O constant, 576
H_2O molecules, 338

$H_9O_4^+$, 575
H_3O^+, 267, 405, 575
Hogs, 503
Holes, empty, in water, 264, 265
Homogeneous mixture, 19
Homogenized milk, 276
Homogentisic acid, [544f]
Homologous series, 378, 384, 402
Hormones, 451, 452, 511, 513, 516, 517, 519, 533, 540
Hormone adjustments, 533
Hormone balance, 511–13, 519
Hospital formula accident, 212
Host(s), 545
H_3PO_4, 291
H_2S, 298, 469
H_2SiO_3, [610t]
H_2SO_3, and H_2SO_4, 361, 471
Huckleberry bushes, 361
Human activities, 534
Human cells, 446
Human diseases, 271, 548
Human energy, 556
Human growth hormone, 533
Human interference, 296
Human intestinal membranes, 483
Human nerve impulses, 552
Human proteins, 446
Human settlements, 271
Human starvation, 549
Human suffering, 542
Human survival, 541, 556
Human technology, 231
Human use of energy, 230
Human wastes, 293, 296
Hund's rule, 98
Hybrid orbital symbolism, 340
Hybrid orbitals, 126, 129, 374–76
Hydrates, 267
Hydration of hydrogen ion, 267
Hydrides, 148, 340
Hydrocarbon derivatives, 390–96, 401, 469
   involving nitrogen, 395
Hydrocarbon materials, 468
Hydrocarbon-like, 465
Hydrocarbons, 377–90, 403, 469, 473
   naming, 380–90
Hydrochloric acid, [610t]
Hydroelectric, 229
Hydrogen, 148–49, 335, 610
   attack on double bond, 408
   chloride, [610t]
   content, 468, 469
   covalent bond, 149
   free radical, 409
   fusion, 323–24
   halide, 413
   heated by gravity, 323
   made positive, 408
   molecular orbitals, 141
   nitrate, [610t]
   nucleus, sharing, 263
   phosphate, 610
   position in periodic table, 148
   preparation of, 145
   silicate, [610t]
   sulfide, 610
   sulfate, 610
Hydrogen abundance, 324, 325

Hydrogen atom(s), 410, 428
   Bohr theory of, 88–89
   energy levels, 88
Hydrogen bond, 263, 265
   in solubility, 193, 194
Hydrogen bonding, 266, 274, 279, 286, 287, 296, 426–38, [437f], [438f], 445, 447, 448, 449, 450, 451
   between DNA chains, [445f]
Hydrogen chloride, solubility, 194
Hydrogen–hydrogen bond, 285
Hydrogen ion, 115, 148
Hydrogen–nitrogen bond, 285
Hydrogen–oxygen bond, 285
Hydrogen peroxide, 145
Hydrogen phosphate units, 444
Hydrogenate vegetable oils, 467
Hydrogenation, 408, 414, 416
Hydro-ic acids, 610
Hydrolysis, 268, 418, 576–80, 600, 605
   ADP and AMP, 477
   ATP (to ADP), 477, 492, 498, 500
   carbohydrates, fats, and proteins, 481
   creative phosphate, 479
   esters, 418, [418f]
   ethers, 418
   glycogen, 483
   $MgATP^{2-}$, 500
   of NCl_3 and PCl_3, 153–54
   nucleic acids, 481
   starch, 462
Hydrolysis constant, 576, [578t]
Hydrolysis products of DNA and RNA, 447
Hydrolysis reaction(s), 577, [578t]
Hydrolyzing, 465
Hydronium ion, 267
Hydrophilic, 439
Hydrophilic colloids, 275–76, [276f]
Hydrophobic, 439
Hydrophobic colloids, 276, [276f]
Hydrophobic interactions, 438–39
Hydrosulfuric acid, 610
Hydroxide, [27t], 115
3-hydroxybutanoic acid, 512
p-hydroxyphenylpyruvic acid, [544f]
Hyperglycemia, 511, [512f]
Hypervitaminosis A, 525
Hypervitaminosis problems, 527
hypo-, 26
Hypochlorite, 26
Hypoglycemia, [512f], 513
Hypothesis, 12

I

I_2, 565–75
i-butane, 381
-ic acid endings, 394, 610
Ice
   crystals, 183
   density, 260
   floating, 260
   heat of fusion, 263
   not slippery, 260
   skating, 260
   structure of, 262–66
   traffic hazards, 280

weight breaking wires, 280
Idaho, 299
-ide, 26
Ideal gas, 166
Ideal gas law, 166, 170, 177
  deviations from, 166–67
Ideal gas thermometer, 167
Ideal pesticide, 550
Identification of PKU, 545
Idle acid plant, 362
Ileu, 432*t*
Illegal alcoholic beverages, 392
Immiscible liquids, 244
Immune members in community, 555
Immunity to medicine or pesticide, 544
Immunities passed to children, 555
Implantation of egg, 514
Impurity, 553
Impurity minerals, 351, 353
Inability to oxidize acetylcoenzyme A, 512
Inactive reactants, 242
Inadequacy of natural controls, 555
Inborn error in metabolism, 543
Incomplete combustion, 504
Incomplete reaction, 269
India, 525
Induced nuclear reactions, 315, 316
Industrial activities producing smoke, 507
Industrial boon, 217
Industrial detergents, 296
Industrial effluents, 299
Industrial salt plants, 217
Inertness, biological, 469
Inert gas. *See* Noble gas
Infant feeding, 543
Infections, 452
Infectious diseases, 546
Infections, 452
Infertile period, 517
Inflamed eyes, 524
Inflammation, 513, 514–20, 534, 535
Inflexible muscle condition, 499
Inhibitions in behavior, 392
Inhibitors, 248
Initiation reactions, 410, 421
Injecting acid, 395
Injection
  of cortisone, 519
  of glucose, 513
  of insulin, 513
Inner transition elements, 159–60
Inner sphere complex, 342
Inorganic colloids, 276
Inorganic nomenclature, 25–27, 610–11
Input of energy, 483
Insecticides, 550–52, 554, 556
Insect pests, 551
Insects, 549, 554
Insertion reaction, 402, 416
Inside protein, 439
Insoluble, 291, 295, 331
  sickle cell hemoglobin, 541
Instability of $^{8}_{4}$Be, 322
Insulation, 237
Insulin, 511–13, 533, 555
Insulin deficiency disease, 511

Insulin overdose, 513
Insulin treatment, 540
Intensity of color, 569
Intentional power excursion, 320
Interactions
  along a polypeptide, 438
  with double bond, 405
  of myosin tails, 498
  potential for, 237
  with salt, 295
Intercourse, sexual, 517
Interest in common, 238–39
Interfere with reproduction, 534, 535, 548
Interlocked rings, 388
Interlocking hydrocarbon rings, 513
Interlocking myosin tails, 498
Intermediate for decarboxylation, [523*f*]
Intermediate chain structure, 460
Intermediate stabilized by emitting energy, 315
Intermolecular forces, 170, 194, 591
Internal combustion engines, 504, 505
Internal energy, 588
Internal polarization, 417
Internally ionic, 431
International System of Units, 560
International Union of Pure and Applied Chemistry, 381
Interruptable power, 350
Intestinal bacteria, 546, 547
Intestinal membranes, 483, 542
Intestinal walls, 481
Intestines, 212, 481
Inverse proportionality, 172
Iodide, [27*t*]
Iodo-, 390
Ion(s), 25, 111–18, 207
  associated with water, 267
  chemical evidence for, 114–17
  complex. *See* Complex ions
  electrical evidence for, 113–14
  formation of, 96, 109–11, 117–18
  in mass spectrometer, 80
  naming, 25
  in oceans, 273
  sizes of, 101, 117–18, 210
Ionic, 107
Ionic attractions, 237
Ionic bond, 109–11, 121, 438
Ionic character, 106
Ionic compounds, 109–11, 237, 239
Ionic-covalent borderline, 122–23
Ionic mechanism, 413, 414, 418
Ionic radius, 101
Ionic reactions, 238
Ionic route, 409
Ionic salt of aluminum, 350
Ionic sites, 497
Ionization, 111–17, 158
  effect on solutions, 209–10
  partial, 210
  of water, 405, 576, 594, 598, 599, 600
  ignored, 596
  of weak acid, 577
Ionization energy. *See* Ionization potential
Ionization potential, 95–98
  first, 96, [97*f*]
Iron, 356, 502

oxidation states of, 157–58
production of, 231, 239, 351–56
recovered in scrap, 367
Iron–carbon mixtures, 356–59, [357*f*]
Iron ores, 351
Iron oxides, 351
Iron and steel production, 351–59, 362, 468
Irreversible change, 223, 223*f*
Irreversible process, 222, 249
Irrigation, 212, 273, 278, 292, 532
Irritation of human tissues, 471
Irritating effect, 239
Iso-, 381
Isocitric acid, [491*f*]
Isoelectronic, 101
Isoleucine, [432*t*], [528*t*]
Isomerize, 483, 486
Isomers, 380, 381
Isooctane, 203, 383
Isoprene, 422
Isotopes, 48, 81, 310
  on curve of stability, [312*f*], 321
  formed by fusion, 325
-ite, 26, 610
IUPAC, 381
IUPAC names, 382

J

J, [561*t*]
Joints, 519
"John Muir's Wild America," 506
Joule, 41 [561*t*]
Juvenile form, 535
Juvenile hormones, 534, 535

K

k (SI prefix), [560*t*]
k (rate constant), 243
K (temperature), 40, [561*t*]
$^{40}_{19}$K, 307, 308
K (equilibrium), 251–54, 564, 569, 572, 592
K_a, 291, 575–76, 577, 593
$1/K_a$, 579
K_b, 575–76, 577
K_c, 567–68
K_{fp}, 208
K_h, 576, 577, 579
  of conjugate base, 577
K_p, 567–68
K_{sp}, 580, 606, 607
K_w, 575–76
Kelvin temperature scale, 40, 166, [561*t*], 562
Kentucky, 274
Kernel, 123
Keto form, [441*f*], 442
Ketone(s), 391–92, 442, 457
Ketone sugar, 457
Ketose, 457, 461
kg, [561*t*]
Kidneys, 547
Kilo-, 33, [560*t*]

Kilocalories, 61
Kilogram, 33, [561t]
Kilometer, 34
Kinds of secondary structure, 436
Kinetic energy, 17, 41, 170, 222
   average, 170
Kinetic molecular theory of gases, 167–68
Kinetic theory, 167
Kinetics, chemical, 236
Known hazards, 518
Known important substances, 511
Known universe, 436
Krebs cycle, 480, 489–90, [491f], 543

L

λ, 227
L, 430
l (quantum number), 88, 92
l quantum number effect, 98
Lab safety, 365
Labeling standards, [53f]
Laboratory equipment, 421
Lactation, 515
Lactic acid, [484f], 487, 523
   buildup, 487
   formation, 487, 496
Lactose, 461–62, [462f], 543
Lactose free artificial substitute, 543
Lake(s), 259, 278–79, 293, 294, 295, 296
   aging, 295
   level changes, 216, 217, 279
   turning over, 278
Lake Mendota, 296
Lake Tahoe, 296, 298
Lakes and streams, 293, 298
L-alanine, 430, [431f]
L-amino acids, [429f], 430, 431, [431f], 483
Lance, 355
Land plants, 217
Lanthanide contraction, 166
Lard, 467
Laundry detergents, 363
Law
   of conservation of energy, 17 [18f], 222
   of definite composition, 19–20
   of gases, 164–66
   of mass action, 236, 251–54, 575, 591, 607
   of multiple proportions, 20
   of random chance, 244
   of solutions, 214
   of statistics, 244
   of thermodynamics, 504
Layers of liquid water, 261
Leaching, 344, [344f]
Lead, 362–64
Lead-glazed pottery, 363
Leaks, oil, 472
Leaded gasoline, 363
Least productive lakes, 294
Leaving group, 412, 421
Le Chatelier's principle, 176–77, 180–81, 185, 201, 261
Le Chatelier's principle shift, 348, 357, 594

Left, right, above, and below convention, 430
Length, 32
Lentils, 553
Less accessible biologically, 469
Less favored complexes, 342
Less reactive metals, 359
Leu, [432t]
Leucine, [432t], [528t]
Leucippus, 21
Levees, 271, [272f]
Level changes in lake, 216, 217
Levels of complexity, 427
levo-, 430
Lewis acid, 269, 339
Lewis base, 269, 339
Lewis dot diagrams, 123–24, 125, 262
Lewis, G. N., 124
Lewis theory of acids and bases, 269
l-glyceraldehyde, 457, [458f]
l-glucose, 387
Life, 226, 227, 228, 230, 232, 269, 285, 429, 494, 529
Life related compounds, 473
Ligand(s), 339–41, 408, 428
Light, 227, 284, 292, 293, 481, 585
   absorbed, 294
   emitted from Earth, 505
   energy, 401, 411
   quantum nature of, 227
   skin color and, 526
   speed of, 227
   wavelengths, 505
Light compounds, 333
Light elements, 336
Light metals, 333, 347
Light switch controlled reaction, 411
Like alkanes, 404
Like dissolves in like, 193, 295, 296, 438
Limestone, 231, 271, 273, 353, 471
"Limey," 521
Limit access, 452
Limit growth, 292, 293
Limiting factor, 284
Limiting nutrients, 292, 293, 294, 532
Limiting phenylalanine in diet, 545
Lindane, 551, [551f]
Line spectrum of hydrogen, 87, [88f]
Linear complex, 340
Linear molecules, 130, [138f], 264
Lining up p orbitals, 376
Link with cancer, 533
Link energy production and consumption, 503
Link molecules, 419
1,4-linkage, 462, 463, 464, 465
1,6-linkage, 463, 464
Linkage, breaking of, 418
Linking phosphates, 290
Linoleic acid, 467, [467f]
Linolenic acid, [467f]
Lipid solubility, 538, 540
Lipids, 465, 513
Liquid(s), 171–74, 259, 279
   behavior of, on mixing, 192–93
   boiling of, 178–79
   capillary fall in, 172–73, [173f]
   capillary rise in, 172–73 [173f]
   diffusion in, 174

   freezing of, 180
   immiscible, 244
   solutions of, in liquids, 192–93
   solutions of, in solids, 194
   supercooled, 185
   surface tension of, 172, [172f]
   vapor pressure of, 174–77
   viscosity, 173–74, [174f]
Liquid fats, 467
Liquid oxygen, 140
Liquid phase, 171
Liquid sulfur, 182
Liquid wastes, 231
Liter, 32
Lithium, 150, [153f]
Lithium, abundance of, 325
Liver, 464, 523, 525
Living cells, 446
Living organisms, 456, 468, 471
l-lactic acid, 430, [430f]
lm, [561t]
Locate tumor, 316, [317f]
Log (Logarithms), 201
   negative, of molarity, 201
London forces. *See* Van der Waals forces.
Lone pair(s) of electrons, 130, 149, 153, 262, 263, 267, 286, 287, 289, 338, 339, 341, 407, 420, 428
Long chain polymers, 434
Long term planning, 232
Long wavelength radiations, 228, 456
Longer peptide chains, 434
Longer wavelength light, 506
Longest single carbon chains, 382
Longhorn bulls, 533
Los Angeles, 232, 505
Lose a hydrogen, 402
Loss
   of buffering, 600
   of consciousness, 513
   of electrons, 147
   of energy, 245
   of H^+, 348, [348f]
   of natural defenses, 555
   of natural selection, 555
Losses of selectivity, 554
Lost nutrients, 294
L-sugar, 457, [458f]
Lumen, [561t]
Lungs, 232, 502, 512
Lux, 561t
lx, 561t
Lyman series, [88f]
Lys, [432t]
Lysine, [432t], [528t]
Lysine deficiency, [528t]

M

μ [560t]
M. *See* Molarity.
M (prefix), [560t]
m (electron mass), 76
m (quantum number), 89, 99
m (solutions). *See* Molality.
m (prefix), [560t]
m (unit), [561t]

Madagascar, 550
Madison, Wisconsin, 296
"Magic numbers," 313–14, 323, 325
  similarity to closed shells, 313
Magnesium, 118, [153*f*], 311, 416
Magnesium-halide group, 416
Magnetic field, 76, 345
  Earth's, 226
Magnetic iron ore, 345, 352
Magnetic properties, 74, 99
  of oxygen, 141, 143
Magnetic sorting, 361
Magnetism,
  dia-, 99, 141
  para-, 99, 141
Magnets, 225, 226
Main energy producing metabolism, 483
Main fusion sequence, 324, 325
  abundance pattern from, 325
Malaria, 546, 550, 555, 556
Malathion, 552, [552*f*]
Male donor, 542
Male hormones, 514, 515, [516*f*], 533
Male reproductive glands, 516
Malic acid, [491*f*]
Maltose, 462, [463*f*]
Mammals, 543
Mammary gland, 515
Mammoth Cave, 274
Manganese, 157, 359
Manganese dioxide, 24–25, 144, 145
Man-made changes, [272*f*], 532, 533
Mannose, [459*f*]
Manufacture of soap, 419
Manure, 293, 532
Mapping by β^+, 316, 317*f*
Marble, 471
Marketing practices, tetracyclines, 549
Markonikov's rule, 407
Marriage, 240
Marshes, salt, 213
Mass, 16, 32, 40, 314
  conservation of, law of, 16, 41
  converted to energy, 314
  of an electron, 76
  energy, 314–15
  real, 314
  rest, 314
  total, 314–15
Mass action, law of, 236, 251–54
Mass number, 81, 305, 306
Mass spectrometer, 80, 267
Materials, arrangement of, 230
Matchmakers, 248
Mathematical applications of equilibrium, 236
Mathematical equations, 591
Matter, 16, 40
  states of, 163
  wave nature of, 89
Maximation of production, 532
Maximum binding per nucleon, 322, 324, 331
Maximum density, 260, 266, 278
Maximum disorder, 225, 226
Maximum hydrogen bonding, 436
Maximum number of bonds, 401
Maximum number of bonding electrons, 374

Maximum separation of negative charges, 262, 264
Measles, 557
Measure mass, 314
Measurement errors, 592
Measurement numbers, 37
Measuring pressure, 164
Measurements, 592
Meat production, 519
Mechanical crushing, 345, 503
Mechanical motion, 584
Mechanical separations, 345, [346*f*], 352, 367
Mechanical separators, 472
Mechanism for carbon dioxide removal, 505
Mechanism for carrying oxygen, 502
Mechanisms, 404, 405–12, [415*f*]
Mechanized farming, 507
Medication, 519
Medicine, 545–48, 554, 555
Medroxyprogesterone acetate, 516, [517*f*]
Mega-, [560*t*]
Melanin, 543
Melt nuclear reactor, 320
Melting
  heat for water, 260
  of ice caps, 506
Melting point, 180, 182, 185, 207, 260, 386
Melting point curve, 185
Membranes, 210, 481, 483, 496, 500, 501
Mendeleev, D., 85–86
Menstrual cycle, 514, 516, 517, 518
Mental confusion, 521
Mental retardation, 363, 528, 543, 544
Meprobamate, 538, [539*f*]
Mercaptans, 396
Mercury, 164, 362, 364–66
  poisoning, 364–66
Mercury barometer, 163–64
Mercury-carbon bonds, 365
Mesabi region, 352
Mesons, 321
Message on mRNA, 446
Messenger RNA, 446
Mestranol, 516, [517*f*], 518
Met, 433*t*
Meta-, 389
Metabolism, 461, 477, 480, 483–95, 503
  aerobic conditions, 489
  failure, 512
  of foods, 488
  of glucose, 483–95
  of phenylalanine, 543, [544*f*]
Metal disposal, 366–67
Metal poisoning problems, 362
Metal smelter, 361
Metallic bond, 107
Metallic crystals, 107–108, 150, 331
Metallic properties, 106–109
Metalloids, 152
Metals, 106–109, 151–52, 270, 331–67, 469
  alkali, 156
  chemical reactions, 336
  conduction bands, 108–109
  in Earth's crust, 336
  oxides, 270
  softness, 109
  strength, 109

  transition, 155–58
Metaphosphate, 290
Metaphilene hydrochloride, 538
Meter, 32, 33
Methane, [381*t*]
Methanol, 391, 392, 505
Methanoic acid, 394
Methionine, [433*t*], 468, 528*t*
Methoxychlor, 551, [551*f*]
Methyl, 382
Methyl ester of morphine, 540
Methylacetate, 395
Methylacetylene, 385
Methylbenzene, 388, [389*f*]
2-methyl-1, 3-butadiene, 422
Methylbutanal, 393
Methylbutanoic acid, 394
Methylcyclopentane, 390
Methylethanoate, 395
Methylethyl ether, 394
3-methylhexane, 386
2-methylpentane, 383
3-methylpentane, 383
Methylpentanes, 382
Methylpropane, 382
2-methylpropene, 384, 385
2-methylpropyl, 382
Metric system, 32, 33
Meyer, L., 85
Mg^{2+}, 416, 508
$MgATP^{2-}$, 500
Micro, [560*t*]
Microscope, 277
Migrations, fish, 213
Migrations, salt, 217
Mild reducing agents, 360
Milk, 276, [277*f*], 461, 523, 543
milli-, 33, 560*t*
Millikan, 77
Millikan's oil drop experiment, 77–78
Milliliter, 33
Millimicrons, 275
Minard, B., 506
Mineral buildup, 337
Mine explosions, 472
Minimum dosage, 518
Minimum energy, 244, 246, 316
Mining, 361, 470, 471, 472
Mining debris, 231
Minnesota, 352
Mirror images, 386–87
Miscibility, 193
Missouri, 279
Mists, 280
Mitochondria, 480, 490, 502
Mix, tendency to, 192, 224
Mixed gas, 225
Mixing, 224, 226, 278, 279, 294
  of arrangements, 225, 226
  of gases, 237
  spontaneous, 224
Mixtures, 18
  of essential substances, 521
  heterogeneous, 18
  homogeneous, 19
  of gases, 192
ml, 562
mm Hg, 164
Model(s), 13, 202–203, 412, 591

oil drop (proteins), 439
Moderators, 319, 320
Modern technology, 231
Molal boiling point elevation, 208
Molality, 206, 210, 211
Molar, 197
Molar volume of gas, 58–61
Molarity, 64, 196–202, 566, 568
 in exponential form, 200–201
Molds, 547–548
Mole, 48, 50, 56, 58, 64, 111, 163
Mole fraction, 203
Mole weight. See Molecular weight(s).
Molecular aggregates, 265
Molecular basis for heredity, 448
Molecular energies, 244, 588
Molecular formula, 24, 50
Molecular geometry, 128–35
Molecular orbital theory, 129, 141–43, 285, 288
Molecular shapes, 128–35
Molecular weight(s), 49
 of proteins, 435
Molecule(s), 22, 24
 covalent, 119–35, 193
 fraction active, 242–47
 giant, 289
 high energy, 244
 low energy, 244–45, 250
 nonpolar, 133, 193
 P_4, 289
 polar, 133, 144, 171, 193
 used to store energy, 456
Molten cryolyte, 350
Molten iron, 353–56
Momentum conservation, 316
mono-, 611
Monoclinic sulfur, 183
Monophosphate, 444
Monosaccharides, 456–61, 463, 481
More neutrons than protons, 312
More reactive form, 427
Morphine, 538, [539f], 540
Moseley, 87
Mosquitos, 535, 556
Motion, 222
 energy of, 237, [237f]
Motion sickness, 537
Mountainous areas, 361
mRNA, 446, 449, 450, 451, 541
Muddy water, 271
Multiple bonding, 289, 373, 376
Muir, J., 506
Multiple covalent bonds, 120, 421
Multiple proportions, law of, 20, [20f], 21, 47
Multiplying-concentrations, 240
Muscle, 479, 496, 519
 action, 481, 496, 503
 contraction, 453, 496–501, [501f]
 fibers, 500, 501
 function, 500
 movements, 291, 496
 positions, [499f]
 proteins, 452
 rigor, 499, [499f]
 soreness, 487
 strength, 496
 tissues, 464, 494, 513

work, 452, 453
Muscle contracting reaction, 501
Muscular dystrophy, 526
Music, soft, 247, 249
Myoglobin, 452, 503
Myosin, 496, 497, 498, [498f], 500
Myosin-actin bonds, 500
Myosin and actin fibers held rigidly, 499

N

N_2, 284, 285, 564, 566–68
n (prefix), [560t]
N (unit), [561t]
n (moles of gas), 163
n (quantum number), 89
n quantum number effect, 98
Na^+, 501, 577
Na_3AlF_6, 350
N-acetyl-p-aminophenol, 536
$NaC_2H_3O_2$, 577, 598
NAD+, [484f], 485, [486f], 487, [487f], 490, 524, 525
NADH, [484f], 485, 486, 487, [487f]
NADP+, 485, [486f], 491
NADPH, 485, 489, 490, 491
NaI, 183
Names. See Naming; Nomenclature
Naming
 acid anhydrides, 395
 acids, 610
 alkanes, 380, [381t]
 alkenes, 380
 alkynes, 380
 carboxylic acids, 393
 complexes, 611
 esters, 394–95
 ethers, 394
 hydrocarbons, 380–90
 optical isomers, 429–31
 organic compounds, 380–96
  basis of, 378
 system, organic, 380
 variations, 610
Nano-, [560t]
Nanometers, 275
Na_2O_2, 145, 146
NaOH, 348, 598
Naphthalene, 388, [388f]
Narcotics, 538, 540
Narrow leafed plants, 552, 553
National Geographic, 506
Natural antibiotics, 548
Natural balance(s), 533, 534
 upsetting, 213, 230
Natural gas, 230, 361, 377, 469, 470, 473, 505
Natural rubber, 385, 422, 423, [423f]
Natural sciences, 11
Natural selection, 540, 556
Naturally occurring elements, 309
Nature's preferences, 310–14
Nausea, 518
n-butane, 381
NCl_3 hydrolysis, 154
Necessary condition for equilibrium, 590
Negative catalysts, 248
Negative charge concentration, 479

Negative charges, 291
Negative halogen, 407
Negative ion, 288, 289
Neighboring phosphate, 477
Neon, 155
Nerve activity, 496
Nerve depressant drugs, 538
Nerve impulse, 500–501, 552
Nerve paralysis, 552
Nervous system, 537
Nervous system abnormalities, 521
Nervous tension, 537
Net charge, electrical, 263, 276
Net flow, 210
Net salt increase, 273
Net tendency for reaction, 492
Neutral H_2O molecules, 407
Neutral solutions, 200–201, 268
Neutralization of acids and bases, 114–15, 348
Neutron(s), 81–82, 305, 317
 becomes proton, 311
 emissions, 309, 312
 escape, 319, 320
 induced reactions, 317
  fission, 318, 319
 from nuclear reactions, 317
 produced by decay, 319
Neutron number, 81–82
Nevada, 296
New conditions, 295
New radioisotopes, 315, 317, 318
New York, 232
Newlands, J., 85
Newton, [561t]
—NH—, 395, 435, 440
—NH_2, 395, 427, 428, 429, 431, 436
NH_3, 286, 287, 296, 298, 564, 566–68, 605
—NH_3^+, 429
NH_4^+, 605
NH_4NO_3, 292
$(NH_4)_2SO_4$, 292
Niacine, 485, [485f], 521
Niacine deficiency, 485
Nickel, 157, 359, 408
Nickel family, 414
Nickel surface, 408
Nickel tetracarbonyl, 339
$Ni(CN)_4^{2-}$, 340
$Ni(CO)_4$, 339–40
Nicotinamide, 485, [485f], 521
Nicotinamide-adenine dinucleotide, 485, [486f]
Nicotinamide-adenine dinucleotide phosphate, [486f]
Nicotinamide portion reduced, 485
Nicotine, 485
Nicotinic acid, 485, [485f]
Night blindness, 524
Night vision, 525
Niobium, 159
Nitrate(s), [27t], 288, 290
"Nitric oxide", [20f]
Nitriloacetic acid, 297, 363–64
Nitrogen, 284–88, 292, 293, 294, 469
 halides, 288
 usable, 287, 292
Nitrogen dioxide, [20f], [611t]

637

Nitrogen lone pair, 421
Nitrogen oxide, [611t]
Nitrogen (IV) oxide, 611
Nitrogen oxides, [20f], 287, 505, [611t]
Nitrous acid, [610t]
nm, 275
NO, 287, 288, 409, [611t]
NO_2, 287, 288, 611, [611t]
N_2O, [611t]
N_2O_4, 288, 611, [611t]
N_2O_3, [611t]
N_2O_5, [611t]
NO^-_2, 288
NO^-_3, 288
Noble gas compounds, 155
Noble gases, 154–55, 333
Nomenclature, 25–27, 377–96, 610–11
  acids, 610
  negative parts, 25–26
  nonmetal compounds, 610–11
  organic, 377–96
  positive parts, 25
  prefixes (mono-, di-, tri-, etc.), 610
  See also Naming
Nonane, 381t
Nonaromatic ring compounds, 389–90
Nonbiological uses (energy), 470
Noncarbonyl end, 548
Nonconductor, 106, 111
Nonenergy factors, 588
Nonflowing oil source, 470
Nonhydrocarbon substitutes, 390
Nonidentical groups added, 404
Nonmetals, 151–52, 198, 270
  oxides, 270–71
Nonpersistent insecticides, 552
Nonpolar gas, 502
Nonpolar molecule, 133, 193
Nonpolar R group, 438
Nonpolarizable, 408
Nonreactive combinations, 248, 289
Nonreactivity of N_2, 285
Nonselective, 554
Nonspontaneous change, 223f, 228, 357
Nonvolatile solutes, 204, 273
Norethindrone, 516, [517f]
Norethynodrel, 516, [517f]
"Normal", 381
Normal boiling point, 260
Normal growth pattern, 553
Normal hemoglobin, 541
Normal seasonal variations, water transport cycle, 270, 273
Northwest arm, Great Salt Lake, 215–17
NTA, 297, 363–64
Nuclear atom, 79
Nuclear charge effect, 97, 98, 99, 100, 155
Nuclear energy, 318–21, 471
Nuclear equation, 305, 306
Nuclear fission, 318–21
Nuclear fuels, 160
Nuclear reactions, 315, 317, 318, 319
Nuclear reactor, 230, 309, 319–21, 471
Nuclear wave functions, 321
Nucleic acids, 439, 442–51
Nucleon, binding per, 322
Nucleophilic reagents, 412, 416, 417, 420, 428
Nucleoproteins, 452

Nucleotides, 443, [443f]
Nucleus, atomic, 79, 316
Nucleus of living cells, 445
Number of contacts, 240
Number of isotopes, 310 [310t]
Number 1 carbon (sugar), 461
Number by prefix, 611
Numbering carbons, 384, 385, 388
Nutrient-rich bottom waters, 294
Nutrients, 212, 270, 284–300, 532, 543
  for animal life, 295
  stirred from bottom, 295
Nutrition, 523, 547
Nylon, 415, 419–21, [420f], 428
Nytol, 537, [538f]

O

Ω, 561t
o-, 389
—O— group, 394
O_2, 140–47, 225, 285
Obsolescence, 10
Obstruction, 202
Ocean, 259, 279, 293, 294
Octa-, [611t]
Octahedral, 130, [131f], 340
Octahedral complex, 341, 342, [347f]
Octane rating, 383
Oct-5-ene-2-yne, 386
Odd molecules, 287
Odd n isotopes, 311
Odd numbered elements, 310, 344
Odd-odd isotopes, 311
Odd proton, even neutron isotopes, 311
Offensive wastes, 231
Official system, organic names, 381
Offshore islands, 294
Offspring, 448
OH^-, 115, 158, 268, 270, 405, 564, 576, 602
OH unit, 391
Ohm, [561t]
-oic acid, 393
Oil(s), 230, 231, 377, 467, 469, 470, 473, 505
  adsorbed on plants and animals, 472
  refining, 403, 471
  reserves, 473
  sources, 476
  spills, 472–73
  transported, 472
Oil drop experiment, 77–78
Oil drop theory, proteins, 439
Oil related materials, 469
Oil shale, 470
"Oily" R groups, 439
-ol, 391
Oleic acid, 466, [466f], [467f]
-one, 393
Open chain aldehyde structure, 460
Open hearth process, 354–55, [355f]
Opening membrane to diffusion, 501
Opening sugar ring, 460
Opium, 538
Opposing actions, 511
Opposite of addition, 412

Opposite character, 238
Opposite reaction, 413, 418
Opsin, 524, [525f], 525
Optical isomers, 386–87, 457
  of amino acids, 429, [429f], [431f]
Optical isomerism, 429–31, 442
  lactic acid, 430, [430f]
Oral administration, steriods, 516
Oral contraceptives, 516, 518
Orally active substitutes, 516, [517f]
Oranges and orange juice, 363, 521
Orbitals, 92–95, 119, 339
Orbits, 88
Order
  of amino acids, 435, 436
  of arrangements, 428
  of bases, 449
  of elements by abundance, 332
Order, of a chemical reaction, 244
Orderliness, 224, 225, 226, 230
Ore, 231
  enrichment, 337, 352
Organic acids, 393, 465, 479
Organic chemical names, 393
Organic chemistry, 269, 373–423
Organic reactions, 401–423, 428
Organic solvents, 452
Organisms, decay of, 468
Organisms, incorporating lead, 363
Organochlorine insecticides, 551, [551f]
Organophosphate, 290
Organophosphorus insecticides, 551–52, [552f]
Origin of elements, 304
Original concentration, 570
Orlon, 421
Ortho-, 389
Orthophosphate, 291
-ose, 457
Osmium, 158
Osmosis, 210–11, 217, 297
Osmotic balance, 212
Osmotic flow, 210–11, 212
Osmotic pressure, 211, 212
Ostwald, W., 117
Outer shell, 101, 107, 118
Outer sphere complex, 342, 343, [343f]
Outside sources, 528
Ovary, 514
Overenrichment of waters, 363
Ovulation, 514
Oxaloacetic acid, [491f], 512
Oxidation, 147, 349
  of alkenes, [404f]
  of ascorbic acid, 520–21
  of $FADH_2$, 489
  of foods, 480, 496
  of glucose, 496
  of muscle glycogen, 496
  of NADH, 487, 490
  of pyruvic acid, 489
  of sulfur, 471
  of wastes, 297
Oxidation number. *See* Oxidation state.
Oxidation-reduction, 147–48, 485
Oxidation state(s), 25, 117–18, 149, 157–59, 270, 285, 288, 611
  of inner transition elements, 159
  transition metals, 337, 338

Oxidative phosphorylation, 487–89, [487f], 490
Oxides, [27t], 270–71
  acid-base behavior of, 270
Oxidizable materials, 299
Oxidized, 147, 149
  by NAD+, 487
Oxy acids, naming of, 610
Oxyhemoglobin, 502
Oxygen
  abundance, 309, 324–26
  availability via complexes, 502–503
  balance, 299
  bridge(s), 290, 291, 458, 459, 461
  concentration in Earth's crust, 325
  containing hydrocarbon derivatives, 391–95
  content, 468
  delivery of, 230
  depletion, 298, 299
  dominance, 326
  extraction of, from air, 140, 144
  insertion, 402
  laboratory preparations, 144–45
  magnetic properties of, 140–41
  molecular orbitals, 142–43
  molecular structure of, 141–44
  occurrence of, 140
  reactions of, 146–47
  requirements, 298
  supply, 299
  transport and store, 339
  use of, 231
Oxygen-free mines, 472
Oxygen–oxygen bond, 285
Oxygen-rich planet, 297
Oxymyoglobin, 502
Oxytetracycline, [549f]
Ozone, 143–44, 146
  protective layer, 144

P

P, 588
p-, 389
p (prefix), [560t]
π bond, 126, 141, 143, 289, 290, 376, 377, 387, 402, 405, 408, 409, 411, 413, 417, 422
  electron concentration, 405, 407
  openness to attack, 405
π bonding resonance, 388
π cloud rings, 388
p electrons, 90, 92
P_4 molecules, 289
p orbitals, 93, [95f], 126, 289, 290, 374, [375f], [376f]
p notation (solutions), 201
p subshell, 98
$^{234}_{91}$Pa, 306

Pain and pain relievers, 535–37
Paint, 248, 363
Paired protein chains, 437
Pairing of bases, 445, [455f], 447, 448
Pairs of opposite reactions, 418

Palladium, 414
Palmitic acid, 466, [466f]
Palmitoleic acid, [467f]
p-aminobenzoic acid, 546, [546f]
Pancreas and pancreatic fluid, 481
Paper production, 299, 503
Para-, 389
Parallel alignment, 496
  of actin and myosin fibers, 497, [498f]
Parallel protein chains, 437
Paralysis, 397
Paramagnetism, 99, 140, 156
Parasite(s), 545, 546, 547
Parathion, 552, [552f]
Partial ionic character, 116
Partial ionization, 210
Partial miscibility, 193
Partial oxidation, 406
Partial pressure, 567
Partially hydrogenated vegetable oils, 467
Partially oxidized hexose, 461
Particle accelerators, 316
Particular set of ring forms, 461
Party analogy, 242
Passengers, single, in vehicles, 232
Passing electric current, 349
Path for energy production, 483
Pattern of amino acids, 435
Pattern of four quantum numbers, 313
Pattern, need for in forming crystals, 182, 184
Patterns for mRNA's, 449
Pauli exclusion principle, 89
Pauling, L., 445
$^{206}_{82}$Pb, $^{207}_{82}$Pb, and
$^{208}_{82}$Pb, 306

PCl_3 hydrolysis, 153
Peat, 468
Pelicans, 550, 554
Pellagra, 485
Penicillin, 547–48, 554
Penicillin G, 547, 548, [548f]
penta-, 383, [611t]
Pentane, [381t]
Pentene, 384
Pentoses, 457, 458, 460
Pepsin, 481
Peptide linkage and chains, 434
Peptization, 275
per-, 26
Percent reacted, 57
Percentage composition, 51–54
Perchlorates, [26t]
Peregrine falcons, 550
Performance of gasolines, 363
Perfumes, 394
Periodic table, 85–87, 150, 284, 289
  electronic configurations and, 101–102
  long form, 87, [86f]
  size trends, 100
  See also inside front cover
Periods of high neutron production, 320
Permanent antifreeze, 392, 405
Permanently changed areas, 472
Permanganate ion, 158, [159f]
Pernicious anemia, 523
Peroxides, 143, 145, 146, 158, [159f]

Persistent insecticides, 552
Peru, 294
Pest controls, 534
Pesticides, 404, 549–54, 555
Pests, 532, 534, 535, 549, 550, 552, 556
pH, 200–202, 500, 595, 598
pH_3, 289
Phase, 171, 179, 180, 184
Phase diagram(s), 184–86
  of benzene, [261f]
  of carbon dioxide, [184f]
  of sulfur, [185f]
  of water, [262f]
Phase transitions, 171, 179, 180–82, 184
Phe, 433t
Phenacetin, 536, [536f], 537
Phenanthracene, [388f]
Phenylacetic acid, [544f]
Phenylalanine, [433t], [528t], 543, [544f], 545
Phenylketone product, 543
Phenylketonuria, 543–45
Phenyllactic acid, [544f]
Phenylpyruvic acid, 543, [544f], 545
Phosph-, 610
Phosphate, 11, [27t], 290, 291, 363, 442, 444, 479, 485
  available, 292
  in detergents, 11, 296, 297
  removal, 296, 298
Phosphate esters, 458, 477, [478f], 479
Phosphates of glucose, 483
Phosphite, 290
Phosphoenolpyruvic acid, 479, [479f], [484f], 486
2-phosphoglyceric acid, [484f]
3-phosphoglyceric acid, 477, [478f], [484f], 486
Phosphor-, 610
Phosphoric acid, 291, [447t], 610
Phosphorolysis of glycogen, 483
Phosphorus, 284, 288, 289, 292, 293, 294, 353, 526, 529, 610
Photochemical energy, 505
Photon, 227
Photonuclear reactions, 315
Photosynthesis, 481, 488, 503
Physical change, 16, 496
Physical contact, 239
Physical sciences, 16
Physics, 13, 16, 227
Pi bond, 126, 141, 143, 289, 290, 326
Pico-, [560t]
Picric acid, [116t]
Pigment, 543
"Pigs" and pig iron, 353
Pipelines, 472
Pituitary gland, 514, 519
Pituitary hormones, 516
PKU, 543–45, [544f]
PKU defect gene, 545
PKU victims, 545
Planar molecules. See Trigonal
Planck, M., 227
Planck's constant, 227
Planning, long term, 232
Plant(s)
  disease problems, 556
  growth, 231, 284, 293, 295

639

killing, by salt, 212
land, 217, 270
proteins, 528
residues, 293, 468, 469
roots, 482
salt water, 216–17
Platinum, 414
Platonic relationships, 239
Pleated sheet, 436, 437–38, [438f]
Plugging of small blood vessels, 541
"Plum pudding" atomic model, [78f]

PO_3^{3-}, 290

PO_4^{3-}, 290

Pockets of oil, 470
pOH, 201, 202
Poison(s), 297, 316, 392, 529, 534, 536, 542, 548, 549, 550, 554, 555
Poison insects, 551
Poisoning hazards, 552
Poisoning of natural predators, avoiding, 535
Poisonous metallic elements, 362–66
Polar bears, 525–26
Polar covalent bond, 121, 133
Polar materials, 416, 431
Polar molecule, 133, 144, 171, 193, 194, 263, 413
Polar R group, 439
Polar water, 438
Polarizable, 170–71, 407, 408, 413
Polarized, 406
light, 386
Polarized roles, 238
Polished rice, 521
Pollutants, 232, 233
Pollution, 337
air, 231, 232
problems, 298, 299, 362, 468
water, 11, 146, 231
Polyethylene, 421
Polymer chains lined up, 420
Polymer formation, 419
Polymers, 419–23, 427
$(H_2O)_x$, 263
of glucose, 463, 464
Polyphosphates, 290, 291
Polypeptide chain, 435, 436, 438
Polypeptides, 434–35, [435f]
backbone, 435, 436
Polypropylene, 421
Polysaccharide(s), 461, 463, 464, 465
Polyvinylchloride, 413, 422, [422f]
Population explosion, 296
Popular drugs, 392
Pork, 521, 523
Porous membranes, 210, 291
Porphyrin, 488
Portable power plants, 504
Position arrangements, 241
Position isomers, 384
Position numbers, 382
Positive carbonyl, 417, 420
Positive charge, 263
Positive electrons, 309
Positive emissions, 304, 312
Positive halogen atom, 407
Positive ions, 287, 406, 407

Positive oxidation state, 289, 349
Positive parts of water, 295, 405
Positive rays, 75
Positron, 309, 316, [316f]
Possible codons, 449
Possible variation of life, 431
Postage stamps, 464
Potassium, 102, 284, 292, 346
Potassium pentanoate, 394
Potassium sulfate, 217
Potato processing plants, 299
Potato starch and peelings, 299
Potential energy, 17, 41, 222, 504, 587
wasted, 492, 493
Potential for interactions, 237
Potential for precise calculations, 590
Power of cells, 481, 490
Power excursions, 320
Power level drops, 320
Power plants, 271, 279
Practical problems, 592
Precipitate, 155
Precipitation, 115, 214, 216, 217
of salts, 115, 215, 217
Predators, 534
Predictions, economic importance of, 573
Preference for Cr^{3+}, 338
Preference for equal numbers, 311
Preference for even numbers, 316, 331
Preference for 5 or 6 membered rings, 460
Preference for magic numbers, 313
Preferential demand for gas, 505
Preferentially adsorbed coating, 344
Prefixes, 383, 390, 611
SI, 560, [560t]
used for counting, [611t]
Pregnancy, 514, 515, 518, 519
Pregnanediol, 514
Preplanning actions, 472
Prepurification, 351
Prescriptions, 540
Preservatives, food, 248
Pressure, 163–64, 167, 213, 567, 584, 586, 588, 591
critical, 179
dependence of gas solubility on, 194–95, 196
effect, ice skating, 262
from environmental groups, 297
measurement, 164
osmotic, 211
partial, 177–78
solidified by, 261
vapor. See Vapor pressure
water, 213
Primary structure (protein), 435–36
Primary treatment, 298, 299
Principal sugars in life, 458
Principle of maximum multiplicity, 98
Principle of replacements, 514
Prized delicacy, 525
Pro, [433t]
Procedures for simultaneous equilibria, 605
Process(es)
commercial, 419
irreversible, 222
natural, 231
reversible, 222, 226, 248–49

spontaneous, 237
Process of energy utilization, 481
Process needed by bacteria, 547
Production
of aluminum, 346–51
of ATP, 480, 483–94, 502
of copper, 359–61
of crops, 556
of electric power, 230, 232
of goods, 230
of FAD, 523
of iron, 351
of neutrons, 319
of nylon, 416, 427
of progesterone, 515
of wastes, 231
Productive varieties, 532
Progesterone, 514, 516, 518, 519
Prohibition on movement, 556
Proline, [434t], 440
Propanal, 393
1,2-propandiol, 392, 405
Propane, 203, [381t], 384
Propanoic acid, 394
1-propanol, 391
2-propanol, 391
Propanone, 393
1,2,3-propantriol, 392, 465
Propene, 384, 404, 405
Propylene glycol, 392, [404f], 405
Propylene oxide, 405
Propyne, 385
Prostaglandins, 529
Protecting food supplies, 550
Protecting fruit plants, 279
Protein "factories," 446, 449
Protein polymer, 431
Protein-heme compounds, 488
Proteins, 419, 428, 429, 434–39, 444, 445, 449–53, 468, 469, 481, 497, 511, 513, 519, 524, 528
alpha helixes, 436, 445, 451
deficiency, 431
formation, 434
functions, 451
in milk, 276
natural, 435
number possible, 435–36
shape, 435
structure, 435–39
synthesis, 528
uniqueness, 427
Proteins, grouping together of, 439
Proton(s), 81–82, 305, 316
acceptor, 268
donor, 268
shifts, 267
Proton becomes neutron, 311, 312
Proton number. See Atomic number
Proton-neutron attractions, 321
Proton-neutron balance, 312
Proton-neutron ratio, 321
Protozoan(s), 465, 503
P-T relationship, 167
Public transportation, 232
Puckered hexagons (in water), 264
Pulp mill, 299
Pulsed reactors, 320
Pulverized coal method, 232, 504, 507

Pumice rock, 275
Pumping of Ca^{2+}, 501
Punched tape, 449
Pure copper, 360
Pure liquid water, 591
Pure metal product, 351
Pure waters, 293
Purebred plantings, 556
Purification step, 359
Purines, 439–42, [441f]
Purposely "wasted" water, 273
Putting nuclei together, 321
PV energy, 587
P-V relationship, 165
Pyramidal, 130, [132f]
Pyribenzamine, [538f]
Pyridine, 440, 460
Pyridoxal, [522f], 523
Pyrimidine and purine bases, 440–42, [441f]
Pyrimidines, 439–42, [441f]
Pyrophosphate(s), 290, 477
Pyrophosphate linkage, 477
Pyruvic acid, 479f, [484f], 486, 490, [491f], 512, 523
   decarboxylation, [523f]

Q

Quadratic equation, 574, 593, 597, 601
Quanta, 88, 227, 228
Quantity (of gas), 163
Quantitative basis for equilibrium, 583–91
Quantum nature of light, 227
Quantum number(s), 89–90
Quarantines, 556
Quaternary structure, 438–39, 452, 502
Quenching of steel, 358

R

R-, 402, 428, 429, 438
R, 166
Racial group, 541
Racial skin pigment differences, 526
Rad, [561t]
Radian, [561t]
Radiating, 230
Radiations, 230, 231, 304
Radioactive
   decay, 305, 307
   isotope, 304, 313, 315
   substances, 304
Radioactivity, 79, 304
Radioisotopes, 304, 307
Ragweed, 554
Railroad cars, 472
Railroad, transcontinental, 214
Rain water, acidic, 273, 293, 344
Rainfall, 216, 217, 271, 279, 284, 291, 292, 296, 552
Rainstorms, 275
Raising blood glucose, 511
Random chance, 244, 245
Range of activation energy, 246
Raoult's law, 202–203, [204f], 216, 217, 591

Rapids, 299
Rare element, 304, 336
Rare-earth elements. *See* Inner transition elements
Rat poison, 528
Rate constant, 243, 251
Rate dependence, 253
   on concentration, 239–41
Rate determining step, 241, 244
Rate of emissions, 228, 230
Rate of increase, fission, 319
Rate of radiations, 230
Rate of reaction, 236, 239, 241, 242, 244, 248
Raw material, 377, 473, 529
React with a double bond, 408
React with phosphate, 486
Reactant(s), 28, 237–39, 247
   active, 242
   contracts, frequency, 239, 243
   inactive, 242
   lowering concentration of, 248
   nucleus, 315
   suitable, 237
Reactions
   of alkanes, 401–403
   of alkenes, 403–405
   of ATP with creatine, 496
   back, 248–49
   balance between, 250
   biochemical, 242
   between gaseous molecules, 566
   at carbonyls, 416–17
   in cells, 483
   of cellulose, 468
   chemical, 237
   with CO, 353
   at constant P and T, 588
   of creatine phosphate with ADP, 496
   effective, 242
   first order, 244
   forward, 248–49
   of H_2 with double bonds, 408
   of halogens with alkenes, 406
   of HCl with propene, 407
   of hydrocarbon derivatives, 412–21
   ionic, 238
   increasing rates of, 247
   involving proteins, 427
   of life, 456
   with oxygen, 486
   N_2, 285
   neutralization, 114–15
   order of, 244
   oxidation-reduction, 147–48
   path, energy production, 483
   to penicillin, 548
   with phosphates, 443
   probability for, 243
   at proteins, 456
   range of activation energy, 246
   rates of, 236, 239, 241, 242, 244, 248
   reversible, 249
   at room temperature, 411
   second order, 244
   sequence, 421
   significant in life, 419
   site, 452
   spontaneous, 237

   at standard conditions, 492
   stoichiometry of, 56–58
   tendencies balance, 584
   in water, 564
Reactive alkenes, 403
Reactive amino acids, 428
Reactive free radicals, 409
Reactive groups, 427, 431
Reactive intermediates, formation of, 420
Reactive metals, 333
Reactive molecule, 287, 288
Reactive sites, 452, 499
Reactivity, suitable, 241, 247
Reactivity
   of aluminum, 349
   of amines, 427–28
   of double bonds, 405
   of organic acids, 427–28
Reactor control, 319–20
Real costs, 280
Real gases, 166
   deviation from ideal, 166–67
Recognition system, 519
Recombination of colloids, 226
Records, 248
Recoverable ores, 361
Recovery of energy sources, 469–71
Recreation, 278, 296
Recycled gases, 298
Recycled sludge, 298
Recycling, 366
Red blood cells, 502, 541
Reducing pyruvic acid, 487
Reducing agent, 149, 452
   CO, 353
Reduction, 147–48, 149, 349–53, 468, 471, 473
   of clotting, 528
   of FAD, 487
   of NAD^+, 487, 490
   product, 514
Reduction step, 349
   clean, 349
   for aluminum, 349–51
   for copper, 360
   for iron, 352–53
Reference points, 39
Reflection of sunlight, 507
Reformed alcoholic, 523
Region of β^+ emitter, 316, [317f]
Region of electrical charge, 407
Region of maximum binding, 322, 324
Region of stable isotopes, 312
Regionally limited disease, 555
Regions of population density, 364
Regular patterns, 436
Regulatory functions, 511
Regulatory proteins, 452
Related pairs for hydrolysis, 577
Related to testosterone, 516
Relationship between unknowns, 602
Relationship, carbohydrate and fat metabolism, 483
Relationship, energy and wavelength, 317
Relative abundance(s), 310, 324–26, 332, [334f]
Relative number, solute and solvent, 209
Relaxation (muscle), 500
Release of energy by fission, 322

641

Release of energy by fusion, 322, 323
Relevance, 10
Reliability of drug, 518
Relief pain, 536
Removal of carbon dioxide, 505
Removal of heat, 279–280
Removal of H_2O, 414, 468
Removal of H_2S, 361
Removal mechanism, 545
Repeating pattern, 466
Replacement, 401
Replication, 447–49
Representative elements, 149–54
Repressed inhibitions, 392
Reproduce, 540
Repulsion between negative charges, 500
Repulsion between protons, 321, 322, 323
Repulsion overcome, 323
Repulsions of He nuclei, 324
Repulsive force, 276
Residue from supernova, 324
Residues containing fats, 469
Resistance to
   corrosion, 366
   diseases, 535
   insecticides, 551
   known problems, 535
   malaria, 541
   poisoning, 554
   smallpox, 555
Resistant strains, 534, 554, 555
Resonance, 125–28, 287, 288, 289, 290, 378, 387, [387f], 388, 389, 422, 440
   in benzene, [128f]
   in nitrate, [128f]
   in ozone, 143
   in sulfur dioxide, [127f]
Resonance energy, 127–28
Resources of the Earth, 549
Rest mass, 314–16
Retarded growth, 533
Retarded by protein deficiency, 519
Retinal, 524, 525, [525f]
   11-cis, [525f]
Return of heated water, 280
Reversal, left for right, 380
Reverse osmosis, 212
Reversible change, 223, [223f]
Reversible process, 222, 226, 248–89
   coagulation, 276
Reversible reactions, 249
Rhenium, 310
Rheumatoid arthritis, 519
Rhodopsin, 524, 525, [525f]
Rhombic sulfur, 182
Rhythm method of birth control, 517
Riboflavin, 488, [488f], [522f], 523
Ribonucleic acids, 444–51
Ribose, 442, [442f], [443f], 444, [447t], 449, [459f]
Ribosomal RNA, 446
Ribosomes, 446, 449, 450, 451, 452, 480
Rice, 532
Rice hulls, 521
Rickets, 526
Right hand–left hand pairs, 387
Rigid arrangements, 442, 449
Rigidity of double bond, 376, 385
Ring forms (sugars), 442, 460, 461

Ring formations, 513
Ring isomers, 460
Ring junctions, 514
Ring linkage, 513, 514, 515
Ring linkages reversed, 503
Ring opening (sugars), 460
Ring structure, 442, 458, 459, [459f], 460, 461
Rings turned over, 462
Rivers, 231, 259, 273
   salty, 273
RNA, 442–51, 460, 461
Road mapping of problems, 36
Roasting, 359, 361
Rock formations, 259, 273
Rocks, sedimentary, 248
Rocky Mountain Q fever, 555
Rodents, 549
Rodenticides, 550
Role of surfaces, 244, 248
Roller bearings, 358
Room temperature, 401
Root names, 26, 380, 610
Rope and pulley device, 493
Roughage, 481
Rows of the periodic table, 85
rRNA, 446
Rubber, 183
   structure of, 423
Rubidium, 104
Ruby Falls Cave, 274
Rules, alkane names, 382
Rules, systematic nomenclature, 383
Running, 232
Rust, formation of, 146
Rust inhibitors, 248
Rutherford, E., 78
Rutherford's nuclear atom experiment, 79

S

s, 561t
S. See Entropy
σ bond, 126, 141–43, 376
s electrons, 90, 92–93
s orbital, 93f
   dart board analogy to, [94f]
s subshell, 98
Safety factor, 518
Safflower oil, 467
Sailors, 521
Saliva, 481
Salmon, 213
Salt(s), 25, 106, 114
   buildup, 273
   killing of plants, 212, 273
   naming, 25–27
   of organic acids, 394
   nonvolatile, 273
   of weak acids, 577
   of weak bases, 577
Salt marshes, 213
Salt producers, 217
Salting effect (on soils), 273, 532
Salty land, 216
Sample analysis control, 356
Saturated bonding, 401, 404
Saturated fatty acids, 466, 467

Saturated hydrocarbons, 378, 379, 412
Saturated solution, 195, 214, 217
Scandium, 157
Science, 5, 11–13
Science fiction, 326, 431
Scientific method, 6–9, 22
Scientists, 233
Scurvy, 521
Sea birds, 294
Sea of electrons, 107
Sea level, 584
Sealing of mines, 472
Seasonal changes in gasoline, 203
Seasonal temperature variations, 278
Seasonal variations in carbon dioxide, 505
Seawater, concentration of salt in, 273
2-sec-Butyl-4,6-dinitrophenol ammonium salt, 554
Secluded spots, as catalysts, 248
Seconal, 538, [539f], 540
Second, [539f], 540, [561t]
Second approximation, 595, 596
Second order reactions, 244
Secondary effects (drugs), 518
Secondary effects of environment, 280
Secondary reactions in supernovas, 325
Secondary structure (proteins), 436–38, 452
Secondary treatment (sewage), 298, 534
Sedative(s), 537–40, [539f]
Sedimentary rocks, 236
Seepage, in mines, 471
Segments of muscle, 496
Selective breeding, 532, 555
Selective development of resistance, 554
Selective poisons, 547, 548, 550, 552, 555
   as medicines, 545–49
   as pesticides, 549–54
Selective Weed Killer, 554
Selectivity of hydrogen bonding, 449
Selectivity in poisoning, 553
Semiconductor, 152
Sensitive vision, 524
Sensitized, 548
Separation problem, 470
Septic tank, 298
Sequences of bonded atoms, 435
Sequence of reactions, 451
   during muscle contraction, 500
Ser, [432t]
Series of cascades, 493–94, [494f]
Series of fissions, 319
Serine, [432t]
Serious disability, 546
Set of three bases, 446, 451
Settling pond, 299
Sewage, 298, 534
   treatment, 296, 298, 363–64, 534
Sex hormones, 514–16
Shape(s)
   of bonding arrangements, 375
   molecular, 128–35
   orbital, 92–95
   of proteins, 435
   requirements, 242
Sharing of electrons, 119, 152, 158, 238, 239, 339
Sheet of protein, 438
Shells, 98

SH group, 396, 490
Shift in equilibrium, 176, 181, 250
Shipping costs, 352
Shipwrecks, 472
Shock symptoms, 513
Short peptide chains, 434
Short varieties, wheat, 533
Short wavelength portions of sunlight, 505
Shortage of NAD+, 486
Shortages developing, 473
Shortcuts, 572
Shortening, 467
Shorter hydrocarbon chains, 403
Shoving electrons around, 412
Shutdown nuclear reactor, 320
SI units, 33–35, 40, 41, 48, 275, 560–62
Sickle cell anemia, 541–42
 hemoglobin, 544
Sickling crisis, 541
Side chains, 382
Side effect hazards, 518
Side effects, 536, 548, 555–56
 of control measures, 555–56
 of NTA, 297
Side reactions, 513
Sigma bond, 126, 141–43, 376
Significant figures, 38, 592
Silent switches, 364, 365
Silicic acid, [610t]
Silicon, 333, 347–48, 353
Silicon oxide, 348
Silt, 273, 275, 295
Silver, 359, 360
Silver catalyst, 405
Similar structure, 514
 DES and estrogens, 518, [518f]
Simple models, 203
"Simple" solutions (complex problems), 297
Simple sugars, 456
Simplifying approximations, 592, 593, 594, 597, 599, 602, 607
Simplifying assumptions, 594
Simultaneous detection, 316
Single bond, 119, 290, 376
Single bonded O, 417
Single chains, 465
Single crop fields, 532
Single sugars, 461
SiO_2, 50, 353
Site for amino acid, 446
Site carrying 3 base code, 446
Sites involved in chemical reactions, 452
Six atom rings, 440, 458
Six-carbon diacid, 419
Six-carbon diacid chloride, 421
Six-carbon diamine, 419, 421
Six-carbon saturated fatty acids, 467
Six-carbon sugars, 456
Six-membered carbon ring, 387
Sizes
 of atoms, 100, 151–52, [152f], 286
 effect of transition elements, 152
 of ions, 101, 117–18, 151–53, [152f]
 of transition elements, 155, [156f]
Skin, 419, 519, 526

Skipping of $^8_4$Be, 331

Skunk oil, 396

Slag, 353
Sludge, 298
Small error, 596
Small intestine, 481
Small ions, need for, 118
Smallpox, 555, 557
Smell, 396
Smelters, 361
Smog, photochemical, 143, 287, 505
Smoke, 275, 362, 504, 507
Smoking meats, 288
Smokestack, 232
Snake River, 299
Snow, 216, 259, 273, 279
SO_2, 471
SO_3, 471
Soap, 276, 419
Social analogy, 238–43
Social problem, 538
Sodium, 104, 346
Sodium acetate, 394
Sodium chlorate, 51, 55, 56, 57–58, 60
Sodium chloride, 25, 56, 57–58, 64, 214
Sodium cyanate, 542
Sodium ethanoate, 394
Sodium hexafluoroaluminate, 350
Sodium hydroxide, neutralization of, 115
Sodium ion, 150
Sodium peroxide, 145, 146
Sodium sulfate, 64, 214
Soft music, 247, 249
Softer iron, 356
Softness in metals, 109
Soil, 259, 275, 278, 292
 bacteria in, 287
Solar system, 324
Solid(s), 180–86, 259, 263, 580
 amorphous, 182, 183
 catalysts, 248
 crystalline, 182
 melting of, 180
 solutions of, in liquids, 194, 195–213
 solutions of, in solids, 194
 sublimation of, 181
 transitions involving, 180–82
Solid angle, 228
Solid geometry, 264
Solid shortenings, 467
Solid wastes, 232
Solids filtered off, 348
Solubility, 193, 195, 214, 287, 348
 of Al_2O in cryolyte, 350
 of hemoglobin, 541
Solubility curve, 195
Solubility equilibrium shifts, 348
Solubility problems, 580, 602, 606–607
Solubility product constant, 580
Soluble compounds, 337
Soluble gases, 194
Soluble nitrates and phosphates, 298
Soluble orthophosphates, 271
Soluble salts, 288, 293
Solute, 193
 nonvolatile, 204
Solution to SO_2 emission problem, 361
Solution(s), 19, 64, 192–217, 224, 591
 acidic, 200–201
 basic, 200–201
 boiling points of, 208–209

 concentration of, 64–66
 containing cyanide, 366
 electrolytic properties of, 106–107
 formation of, 192
 freezing points of, 208
 of gases in gases, 192, 193, 224
 of gases in liquids, 194–95
 of gases in solids, 194
 of liquids in liquids, 192–93
 of liquids in solids, 194
 neutral, 200–201
 saturated, 195–214
 of solid in liquids, 194, 195–213
 solid (of solids in solids), 194
 stoichiometry of, 64–66
 supersaturated, 196
 unsaturated, 195
 vapor pressure of, 202–205, [202f], [204f]
Solvent(s), 193, 393
 transfer, 205
 vapor pressure, 204
Sominex, 537, [538f]
Sour gas and oils, 469
South America, 294
Soybeans, [528t]
sp, 340, 375, [375f], [376f]
sp^2, 374, [375f], 376, 387
sp^3, 340, 343, 374, [374f], 375, [376f], 378
sp^3d^2, 129
Special forms of K, 574–76, 580

Special preference for $^4_2$He, 322

Species-selective poisons, 528
Specific directions for proteins, 446
Specific Grignard reagent, 417
Specific pairing of bases, 446
Specific position bonding, 407
Specific quaternary structure, 439
Specific reactivity, 452
Spectrometer, mass, 80, 267
Spectrum, 87–88, 88f
Speed addition, 406
Speed of light, 227
Speedup or slowdown of fission rate, 319
Sperm banks, 540
Spills, oil, 472–73
Spillways, 213
Spin, electron, 90
Spontaneous change, 222–26, [223f], 228, 231, 589
Spontaneous energy transformations, 584, [585f], 587
Spontaneous fall, 237
Spontaneous fission, 318
Spontaneous ignition, 289
Spontaneous mixing, 224
Spontaneous processes, 237, 588
Square bracket symbol, 568
Square planar structure, 130, [132f], 340, 343
sr, [561t]
SST's, 144, 232
Stable isotopes, 306, 310, [311t], 311, 312, [312f], 313
Stable oxidation states, 338
Stable ring structures, 458
Stage of weed growth, 554
Stainless steels, 359, 366

643

Stalactite, 274
Stalagmite, 274
Stamps, 464
Standard enthalpy of formation, 62
Standard reference states, 568
Standard temperature and pressure, 59–60, 169
Stannic, [26t]
Starch, 275, 392, 409, 461, 463–64, 465, 469
   hydrolysis, 462
   simplest form 463, [464f]
Stars, 313, 321, 323–24
Start code, 450
Start protein formation, 449
State function, 584, 587
State of maximum disorder, 225
Statistical laws, 244
"Stealing" an H^+, 267, 407
Stearic acid, 466, [466f]
Steel, 231, 351, 353, 356–59
   commercial, 359
   continuous process units, 279
   mills, 353–56, 362, 471
   variety, 356, 359
Steelhead trout, 213
Steers, 533
"Stem" of cloverleaf, 446
Steradian, [561t]
Steric factors, 241, 242–43
Sterile males, 535
Sterility, 524, 526
Steroids, 513–20
Stimulating male characteristics, 515
Stimulating animal growth (cattle), 519
Stimulating natural growth, 523–33
Stirring of nutrients, 293
Stoichiometry
   of mixtures and reactions, 52–54, 56–58
   of pure substances, 51–52, 54–55
Stomach, 362, 481
   acids, 481
Stop codon, 449
Stop protein formation, 446, 449
Storage
   of ATP energy, 494–96
   compounds, 502
   of energy, 444
   form for carbohydrates, 464
   of glucose, 511
Store oxygen, 452
Stored energy, 291, 465, 469, 477
Stored water, release, 279
Storms, mixing by, 214–15
STP, 59–60, 169
Strained bond relations, 289
Strands, protein, 451
Stratified lake, 295
Stratify, 294
Stream rocks, 275
Strenuously reactive conditions, 401
Streptococci, 548
Strip mine, 470, 472
Stripping ores, 351
Strong acid, 116, 268, 291, 406, 471, 602
Strong attractions, 275, 339
   for electrons, 362, 405
   for hydrogen, 414
Strong base, 116, 268
Structural materials, 419, 451, 464

Structural pattern change, 452
Structural proteins, 451
Structure(s)
   around nitrogen, 289
   of biopolymers, 427
   collapse, 358
   of DNA, 445
   of glyceraldehyde, 457, [457f]
   of hemoglobin, 502
   H_2O molecules, 263
   man's, 270
   of penicillin, 547
   of pentoses and hexoses, 457
   of polymers, 423
   of sucrose, 461, [462f]
Structures showing bent bonds, 462, [463f], [464f]
Study of hydrocarbons, 377
Sublimation, 181
Subshell, 98
Subshell effect, 98
Substances essential to life, 529
Substituent groups, 382, 383, 390
Substitute for cortisone, 520, [520f], 534
Substitute hormone, 514
Substituted aminoethanes, 537
Substitutes
   for natural balances, 533–35
   for progesterone, [517f]
Substitution of amino acids, 541
Substitution of uracil for adenine, 541
1,2-,1,3-, or 1,4-substitution, 389, [389f]
Substitution reactions, 401, 402, 410, 412, 416, 418
   with bromine, 409
Successive approximations, 595
Succinic acid, [491f]
Succinyl coenzyme A, [491f]
Sucrose, 461, [462f]
D-sugar, 457, [458f], 458
L-sugar, 457, [458f]
Sugar beets and sugar cane, 461
Sugar(s), 392, 442, 456–63, 469, 481, 542
   converted to glucose, 483
   derivatives, 461
   five-carbon, 456
   geometry, 442
   ring form, 442
   ring structures, [459f]
   six-carbon, 456
   three-carbon, 456
Sugar-base units, 443, [443f]
Sugar-base-phosphate units, 443, [443f], 444
Suitable reactants, 237
Suitable reactivity, 241, 247
Sulf-, 610
Sulfa drugs, 546–47, [546f]
Sulfadiazine, [546f], 547
Sulfanilamide, [546f], 547
Sulfate ion, 26, [27t]
   structure of, 125, [134t]
Sulfide ion, [27t]
Sulfides, 361
Sulfite ion, [27t], [134t]
Sulfur, 353, 361–62, 396, 468, 469, 471, 472, 479, 610
   amorphous, 182
   in coal deposits, 362, 471

   content, 468, 469
   exposed to air, 362
   liquid, 182
   monoclinic, 183
   oxides, 468, 471
   phase diagram for, [185f]
   plastic. See amorphous
   problems, 361–62
   required for growth, 292
   rhombic, 183
Sulfur dioxide, 126–27, 269
   electronic structure of, 126–27, [127t]
   polar nature, [133f]
   resonance in, 126–27, [217f]
Sulfur hexafluoride, [135t]
Sulfur trioxide, structure of, [134t]
Sulfuric acid, 361, 471, 610
   electrolysis, 145
Sulfurous acid, 361, 471, 610
Sum of protons plus neutrons, 305
Sun, 227, 321, 324
Sun exposure, 526
Sunlight, 231, 456, 469
Supercooled liquid, 185
Superheating, 179
Supernova explosion, 324, 325
   secondary reactions, 325
Superoxides, 146
Superphosphate of lime, 291
Supersaturated solution, 196, 213
Supersonic transport aircraft, 144, 232
Supplement diets, 529
Supplementation of vitamin E, 527
Supplies
   coal, gas, and oil, 473
Supply of alkenes, 403
Supplying nutrients, 532
Support life, 259, 503
Suppress antibody formation, 513
Surface adsorption, 248
Surface area, 244, 580
Surface effects, 244, 248
Surface erosion, 274–75
Surface regions, 293
Surface tension, 172, [172f], 179, 196
Surface water, 294
Surface water, pollution, 271
Survival, fish and plants, 280
Surviving children, 556
Susceptibility
   to diseases and pests, 532, 555
   to infections, 524
   of "superior" species, 533
Suspended particles, 275
   settling of, 278
   soil, 270, 271
Suspension, colloidal, 275
Sustaining life, 227, 230, 232
Symbols, chemical, 22–23, [22t], [23t], inside covers
Sympathetic nervous system, 537
Synergism, 536, 538
Synthesis
   of an alcohol, 417, [417f]
   fatty acids, 466
   of poisons, 551
   of proteins, 449–51, [450f]
Synthesizing vitamins, 546
Synthetic fibers, 419, 421

Synthetic rubber, 385
Syrups, 461
System for naming optical isomers, 430
System International d' Unities, 560
Systematic names, 381, 382, 384, 611
Systematic pattern of names, 381, 382
   rules, 383
Swamps, 294
Sweden, 297, 364
Swelling of tissues, 535

T

T (genetics), 445, 449
T (SI prefix), [560t]
T (SI unit), [561t]
T (temperature), 584
T cytoplasm, 10, 556
Table of concentrations, 571
Table Rock Dam, 279
Tall wheat, 533
Tantalum, 159
Tar sands, 470
Target organisms, 554
Tars, 403, 469, 470
Technical development, 232
Technological society, 503
Technological solution, 505, 517
Technology, 3–5, 473, 546
Teeth, 529
Television, color, 159
Temperature, 32, 38–40, 163, 167, 174, 176, 226, 230, 243, 244, 263, 285, 576, 584
   absolute, 40, 166
   Celsius, 38–40
   centigrade. *See* Celsius
   critical, 179
   of the Earth, 507
   equilibrium constant and, 252
   Fahrenheit, 38–40
   Kelvin, 40, 166
   patterns in lakes, 279
   reaction rate and, 244–47
   rises, 506
   scales, 38–40, 166
   solubility and, 195
   vapor pressure and, 176
   volume of gas and, 165–66
Temperature dependence of energy distribution, [245f]
Temperature effects, 244–47, 252
   on fraction active, 246
Tempering, 358
Temporary hardness, 271–73
Ten-carbon diacid, 421
Tendencies of nature, 591
Tera-, [560t]
Termination reactions, 410, 411
Termites, 465, 503
Terramycin, [549f]
Tertiary treatment, 298
Tesla, 561t
Test for unsaturated hydrocarbons, 404
Test NTA effects, 297
Testes, 515
Testing on a large scale, 297
Testosterone, 515, 516, [516f]
tetra-, 363, [611t]
Tetracycline(s), 548–49, [549f]

Tetrahedral, 130, [131f], 262, 289, 340, 342, 343, 374, [376f], 378
   around oxygens, 264
   arrangement, 263
   distorting, 263
   with one lone pair, 130, [132f]
   with two lone pairs, 130, [132f]
$^{232}_{90}$Th, 306, 318
   half life, 307
$^{234}_{90}$Th, 305, 306
Thaw, 278
Theory, 12
Theory of acid and bases,
   Arrhenius, 115, 158, 268
   Brønstad-Lowry, 268
   Lewis, 269
Theory of ionization, 111–17
Thermal pollution, 279, 504
Thermalization, 317
Thermocline, 294, 295
Thermodynamic basis for equilibrium, 236
Tying up reactants, 248
Time, 32
Time lag, 319
Timing of herbicide application, 553
Titanium, oxidation states of, 157
Titration, 198–99
Tobacco, 485
Tolerance of salt, 213
Toluene, 388, [389f]
Tools, 9
   models as, 202
Torr, 562
Total abundance, 310
Total charge, 306
Total energy content, 585
Total energy of oxidation, 492
Toughness, 356, 359
Trace elements, 293
Trade names, 423
Trains, 229, 231
Tranquilizers, 537–40
Trans, 385
Trans arrangement, 423, 524, 525
Trans-2-butene, 385
Trans isomers, 385, 423
Transcontinental railroad, 214
Thermodynamic states, 584
Thermodynamics, 237, 584, 586, 591
   implications of, to environmental problems, 231–33
   living state and, 226, [229f]
   principles, 236–37
Thiamine, 521
Thiamine chloride, [522f]
Thiamine deficiency, 521, 523
Thick oils, 403, 470
Thioesters, 479
Thiols, 398
Thiosulfate, 26
Third approximation, 596
Thomson, J. J., 76
   atomic model of, [78f]
   e/m experiment of, 76
"Those that have get more," 408
Thr, [432t]
Three-carbon sugars, 456, 457

Three consecutive bases, 449
Three monosaccharides, 461
Threonine, [432t], [528t]
Threshold for excretion (glucose), 512
Thrombosis, 518
Thymine, 440, [441f], 442, 444, 445, 447, [448t], 449
Transfer electrons, 487
Transfer of food and nutrients, 211–13
Transfer RNA, 446
Transfer of salt, 216
Transfer of solvent, 205
Transferable energy, 503
Transport oxygen, 452
Transportation, 337, 351, 534
Transition elements, 155–58, 333, 337, 339, 340
Transport, active, 212
Transport cycle, water, 259, 270–71
Transportation, public, 232
   systems, location, 270
Trapped air bubbles, 213
Trapped oil, 470
Travel, 556
Treatment areas, 299
Treatment of disease, 548
Treatment to enrich ore, 337
Treatment by insulin, 511
Treatment of PKU, 545
Trestle in Great Salt Lake, 214
tri-, 611
Triads, 85
Trichloromethane, 391
2,4,5-trichlorophenoxyacetic acid, 553
Triglycerides, 465–66, [466f]
Trigonal, 130, [131f]
   bipyramid, 130, [131f]
   bipyramid with two lone pairs, 130, [132f]
   with one lone pair, 130, [132f]
2,2,4-trimethylpentane, 383
Triol, 390
Triose(s), 457, [457f]
   phosphate esters, 458, 485
Triphosphate, 290, 419
   unit on adenine, 444, [444f]
Triple bond, 120, 284, 376
Triple point, 185
Triple superphosphate of lime, 291
Trisaccharide, 461
Tristearin, [466f]
tRNA, 446, [447f], 449, 450
   for phenylalanine, 449
Tropical countries, 541, 550
Trout, 279
Truces, 231, 295
Trucks, 472
True bonding arrangement, 126
Try, [434t]
Tryptophane, [434t], 440, [528t]
Tuberculosis, 519, 520
Tuf-syn, 423
Turbulence, 229, 237, 238f, 270, 271, 275
Turnaround in temperature change, 507
Turning nuclear reactor off, 320
Two carbon units removed, 490
Two-dimensional sketches, 430
Two equations solved for two unknowns, 602

645

Two sugars, 461
Tyr, [433t]
Tyrosine, [433t], 543, [544f], 545

U

U (genetics), 449
$^{233}_{92}$U, $^{235}_{92}$U, and $^{238}_{92}$U, 305, 306, 307, 318, 319
Ultraviolet light, 526
Ultraviolet radiation, atmospheric ozone and, 143–44
UMP, [443f]
Unbalanced isotopes, 311
  more neutrons, than protons, 312
Unburned hydrocarbons, 504
Uncertainty, 224, 227
Underground streams, 274
Undesirable materials, in ore, 337
Undesirable silicon, 347
Undissolved aggregates, 275
Undissolved chunks, erosion of, 274
Unequal sharing, 121
Unfilled low lying energy levels, 339
Unhydrogenated vegetable oils, 467
Uniform plantings, 556
Unit factors, 35–37
United States, 551, 553
Units, 32, 560–62, 566, 589
  conversions, 34–37
Universe, 323
University, 240, 436
Unmarried young people, 240
Unnatural bulk of material, 300
Unnatural hazardous substances, 519
Unpaired electron, 287, 409
Unpolished rice, 521
Unreacted carbon, 468
Unreacted metal, 335
Unsaturated fatty acid(s), 466, [467f], 467
Unsaturated hydrocarbons, 379, 389, 401
Unsaturated ring, 440
Unsaturated solution, 195, 216, 217
Unshared electron, 287, 402, 409
Untreated wastes, 534
Unusual bonding, 286
Unusual concentrations, 512
Unwanted effects, 294, 537
Upgrading blast furnace iron, 353–56
Upper atmosphere ozone, 143–44
Upsetting biological balances, 288, 362
Upsetting chemical balance, 300
Upsetting natural conditions, 230
Uranium, 304
Uranium-238, 305
Uracil, [441f], 442, [443f], 444, [447t], 449
  substitution of, 447
Uridine, [443f]
Uridine-3'-monophosphate, [443f]
Urine, 514, 543, 545
Utah, 214–17
Uterus, 514

V

V (voltage), [561t]

V (volume), 588
Vaccination(s), 556, 557
Val, [532t]
Valence bond hybrid, 340
Valence bond model, 129
Valence electron(s), 123
Valence shell, 123
Valence shell electron pair repulsion theory, 129–35
Valine, [432t], [528t]
Valley, 361
Valuable metals, 360
Van der Waals forces, 170–71
Vanadium, 157, 338, 359
Vancouver, 506
Vapor pressure, 591
  equilibrium, 164, 174–78, [175f], [204f]
  of solids, 181–82
  of solutions, 202–205, [208f], [209f], 211, [211f]
Vapor pressure curve, 176, 182, 184, 202, [208f], [209f], [211f]
Vaporization, 174–75
  heat of, 174–76, 179
Variations with concentration, 492
Variations, random chance, 245
Vegetable oils, 467
Vegetables, 521
Vehicles, single passengers in, 232
Veins of ore, 337
Very poor electrophilic reagent, 408
Vibrations, 237, [238f]
Vinyl chloride, 413, 422
Viscosity, 173, [174f], 183, 260, 275
  of water, 267, 274
Viscous high molecular weight material, 470
Vision, 392
Vitamin deficiency disease(s), 520
Vitamins, 520–28
  A, 524–25, [524f]
  B, 521
  B_1, 521, [522f]
  B_2, 488, [522f], 523
  B_6, [522f], 523
  B_{12}, [522f], 523–24
  B group, 485
  C, 461, 520–21, [521f]
  D, 526, [526f]
  E, 526, [527f]
  K, 526, 527, [527f]
Volatility, 394
Volatile short chain hydrocarbons, 403, 504
Volcanic activity, 333
Volt, voltage, [561t]
Volume, 32, 163, 584, 588
  as a gas, 586
  molar, of a gas, 60, 591
Volumetric flask, [65f]
Volunteer clean up group, 367
VWF, 170–71

W

W, [561t]
Walking, 232
Walling off foreign bodies, antibodies, 513

Warfarin, 527, [527f], 528
Waste(s), 532
  animal, 296
  of energy degradation, 229
  of entropy increase, 232
  forms of, 232
  gaseous, 231
  heat, 230, 353
  heat disposal, 294
  human, 293, 296
  liquid, 231
  offensive, 231
  production of, 231
Waste disposal, 11
  in water, 271
Wasted fossil fuels, 470, 473
Wasted potential energy, 492
Wasteful use of electricity, 504
Wasteful process, 493
Water, 259–80, 299, 532
  atmospheric vapor, 280
  bonding in, 262–67
  condensed, 587
  conditioner, 297
  as coolant, 279–80
  density of, 259–66
  electrolysis of, 145
  evaporating, 260, 272, 292
  flooding (oil wells), 470
  hard, 271–73
  heat capacity, 263, 279
  heat of fusion of, 263
  heat of vaporization, 263
  hydrogen bonding in, 262–67
  inlets from lake, 295
  ionization, 594, 603
  layer formed on ice, 261–62
  loving, 439
  maximum density of, 260, 266, 278
  melting, 260
  natural sources of, 259
  normal boiling point of, 260
  percentage composition of, 20
  pH of, 201
  phase diagram of, [262f]
  physical characteristics, 259
  polar molecules, 194, 263, 438
  released from dams, 279
  sea, 273
  self ionization of, 268
  solubility in, 260
  soluble, 337, 481, 521
  solutions, 469
  storage, 278
  stored in frozen form, 506
  supply, 295, 296, 364, 366, 503
  transport cycle, 259, 270, 271, 337
    seasonal variations, 270, 273
  turbulence, 277
  use by plants and animals, 271
  viscosity, 267, 274
  warming, 260, 279–80
  waste, 273, 279–80
Water plants, 217
Water pollution, 11, 146, 231–32, 271, 361, 362, 471
Water table, 271
Water wheel doing work, 493
Waterfall analogy, 493–95

Watson, J., 445
Watt, [561t]
Wave function, 89
Wave nature of light, 227
Wave nature of matter, 317
Wavelength, 88, 227
Wb, [561t]
Weak acid, 116, 268, 269, 271, 577, 593, 596
Weak base, 116, 268, 269, 579
Weaker narcotic, 540
Weakly acidic, 344, 348, 577, 605
Weathering action, 248
Weber, [561t]
Weeds, 549, 553
Weight
   atomic. See Atomic weight(s)
   as distinguished from mass, [35f]
   formula, 50
   mole. See Molecular weight(s)
   molecular. See Molecular weight(s)
Weight relations, 19, 47, 51–58
Weights connected over a pulley, 492, [492f]
Wells, 271, 470
Western Canada, 361
Western Hemisphere, 470

Western United States, 470
Wetting agent, 276
Wheat, [528t], 529, 532, 533, 556
Wheat breeders, 556
Wheat germ, 523
White blood cells, 519, 547
White bread, 521
White River, 279
"White water" rivers, 297
Whole grain cereals, 523
Whole grains, 521
Willard Bay, 216
Wind circulation patterns, 361
Wire, metals for, 359
Women and men, as reactants, 239
Wood, 299, 419, 464, 468, 584, 585
   alcohol, 392
   pulp, 503
   pulp mill, 299
Work, 452, 586, 587
   at constant pressure, [586f]
Working material, 461
Worms, 365
Worn out electroplating solution, 366
Writing K, 564–65
Writing optical isomers, 429–31, [429f], [430f], [431f]

Wyoming, 586

X

Xenon, 155
X-rays, 87, 446, 452
Xylenes, 389, [389f]

Y

Yeast, 521, 523
Yield of a reaction, percent, 57
-yl, 382, 384
-yne, 380, 385
Young men and women, 240

Z

Zero, absolute, 166
Zero entropy, 224
Zinc, 158
   in relation to transition elements, 166
Zn^{2+} complexes, [343f]
ZnO, 359
$Zn(NH_3)_4^{2+}$, 340
Zwitterion, 429, 434

Names, Symbols, and Atomic Weights
(weights based on carbon-12 as 12)
(alphabetical listing)

Name	Symbol	Atomic number	Atomic weight	Name	Symbol	Atomic number	Atomic weight	Name	Symbol	Atomic number	Atomic weight
Actinium	Ac	89	(227)	Krypton	Kr	36	83.80	Tantalum	Ta	73	180.9479*
Aluminum	Al	13	26.9815	Lanthanum	La	57	138.9055*	Technetium	Tc	43	(99)
Americium	Am	95	(243)	Lawrencium	Lr	103	(257)	Tellurium	Te	52	127.60*
Antimony	Sb	51	121.75*	Lead	Pb	82	207.2	Terbium	Tb	65	158.9254
Argon	Ar	18	39.948*	Lithium	Li	3	6.941*	Thallium	Tl	81	204.37*
Arsenic	As	33	74.9216	Lutetium	Lu	71	174.97	Thorium	Th	90	232.0381
Astatine	At	85	(210)	Magnesium	Mg	12	24.305	Thulium	Tm	69	168.9342
Barium	Ba	56	137.34*	Manganese	Mn	25	54.9380	Tin	Sn	50	118.69*
Berkelium	Bk	97	(247)	Mendelevium	Md	101	(256)	Titanium	Ti	22	47.90*
Beryllium	Be	4	9.01218	Mercury	Hg	80	200.59*	Tungsten	W	74	183.85*
Bismuth	Bi	83	208.9806	Molybdenum	Mo	42	95.94*	Uranium	U	92	238.029
Boron	B	5	10.81	Neodymium	Nd	60	144.24*	Vanadium	V	23	50.9414*
Bromine	Br	35	79.904	Neon	Ne	10	20.179*	Xenon	Xe	54	131.30
Cadmium	Cd	48	112.40	Neptunium	Np	93	237.0482	Ytterbium	Yb	70	173.04*
Calcium	Ca	20	40.08	Nickel	Ni	28	58.71*	Yttrium	Y	39	88.9059
Californium	Cf	98	(249)	Niobium	Nb	41	92.9064	Zinc	Zn	30	65.37*
Carbon	C	6	12.011	Nitrogen	N	7	14.0067	Zirconium	Zr	40	91.22
Cerium	Ce	58	140.12	Nobelium	No	102	(254)				
Cesium	Cs	55	132.9055	Osmium	Os	76	190.2				
Chlorine	Cl	17	35.453	Oxygen	O	8	15.9994*				
Chromium	Cr	24	51.996	Palladium	Pd	46	106.4				
Cobalt	Co	27	58.9332	Phosphorus	P	15	30.9738				
Copper	Cu	29	63.546*	Platinum	Pt	78	195.09*				
Curium	Cm	96	(245)	Plutonium	Pu	94	(244)				
Dysprosium	Dy	66	162.50*	Polonium	Po	84	(210)				
Einsteinium	Es	99	(249)	Potassium	K	19	39.102*				
Erbium	Er	68	167.26*	Praseodymium	Pr	59	140.9077				
Europium	Eu	63	151.96	Promethium	Pm	61	(147)				
Fermium	Fm	100	(255)	Protactinium	Pa	91	231.0359				
Fluorine	F	9	18.9984	Radium	Ra	88	226.0254				
Francium	Fr	87	(223)	Radon	Rn	86	(222)				
Gadolinium	Gd	64	157.25*	Rhenium	Re	75	186.2				
Gallium	Ga	31	69.72	Rhodium	Rh	45	102.9055				
Germanium	Ge	32	72.59*	Rubidium	Rb	37	85.4678*				
Gold	Au	79	196.9665	Ruthenium	Ru	44	101.07*				
Hafnium	Hf	72	178.49*	Rutherfordium†	Rf†	104	(261)				
Hahnium†	Ha†	105	(260)	Samarium	Sm	62	150.4				
Helium	He	2	4.00260	Scandium	Sc	21	44.9559				
Holmium	Ho	67	164.9303	Selenium	Se	34	78.96*				
Hydrogen	H	1	1.0080*	Silicon	Si	14	28.086*				
Indium	In	49	114.82	Silver	Ag	47	107.868				
Iodine	I	53	126.9045	Sodium	Na	11	22.9898				
Iridium	Ir	77	192.22*	Strontium	Sr	38	87.62				
Iron	Fe	26	55.847*	Sulfur	S	16	32.06				

Numbers in parentheses are mass numbers of the most stable isotopes of radioactive elements.

All atomic weights are reliable to ±1 in the last digit except those marked *

*Atomic weights where the reliability is ±3 in the last digit.

SOME FUNDAMENTAL CONSTANTS

Gravitational constant	g	9.8067 m/sec^2
Speed of light	c	2.9979 × 10^{10} cm/sec
Electronic charge	e	1.6021 × 10^{-19} coulomb
Electronic rest mass	m	9.1091 × 10^{-28} g
Planck's constant	h	6.6256 × 10^{-27} erg sec
Faraday constant	$\mathcal{F}$	9.6487 × 10^4 coulomb/mole e
Gas constant	R	0.082056 liter atm/mole deg
		8.3143 joule/mole deg
		1.9872 cal/mole deg
Avogadro's number	N	6.0225 × 10^{23} molecules/mole

†The names and symbols for elements 104 and 105 are in dispute. Kurchatovium, Ku, for 104 and Niels Bohrium for 105 have been proposed by Russian workers, but their evidence of discovery was vague and contained inaccuracies. The names given here are the ones assigned by a group which provided more convincing evidence.